Adhesion between polymers and concrete

bonding · protection · repair

Adhésion entre polymères et béton

revêtement · protection · réparation

Adhesion between polymers and concrete

bonding · protection · repair

Adhésion entre polymères et béton

revêtement · protection · réparation

Proceedings of an international symposium organized by RILEM Technical Committee 52 – Resin Adherence to Concrete and Laboratoire Central des Ponts et Chaussées, Paris.

Palais des Congrès
Aix-en-Provence
France
September 16 – 19, 1986

Editor

H.R. Sasse

SPRINGER-SCIENCE+BUSINESS MEDIA, B.V.

Originally published by Chapman and Hall Ltd in 1986
Softcover reprint of the hardcover 1st edition 1986

ISBN 978-0-412-29050-3

British Library Cataloguing in Publication Data

Adhesion between polymers and concrete :
bonding, protection, repair.
1. Polymer-impregnated concrete
I. RILEM II. Laboratoire central des
ponts et chaussées
620.1'36 TA443.P58

ISBN 978-0-412-29050-3 ISBN 978-1-4899-3454-3 (eBook)
DOI 10.1007/978-1-4899-3454-3

Scientific committee
Comité scientifique

G. Stehno
Building and Industrial Construction
University Innsbruck
Technikerstrasse 13
A-6020 Innsbruck

P.D. Steijaert
Institute TNO for Building Materials and Building Structures
P.O. 49
NL-2600 AA Delft

V. Ukraincik
Gradjevinski Institute for Concrete and Masonry Structures
Janka Rakuse 1
YU-41000 Zagreb

P. Ursella
Chemical Laboratories
Centrale Italcementi
Via Camozzi 124
I-24100 Bergamo

K. Zavliaris
Etan S.A.
4 Thetidos Street
P.O. Box 94
GR-612 Athens

Organization/Organisation:
N. Tchang
Laboratoire Central des Ponts et Chaussées
58 Boulevard Lefebvre
F-75732 Paris Cedex 15

Preface
Préface

Adhesion is a phenomenon architects and civil engineers are not very familiar with. In other disciplines knowledge about surface properties and the background of bonding energies is also far from satisfactory; nevertheless there are many important applications in concrete engineering, where adhesion is necessary for success and durability. These include:

- coating and painting
- repair of concrete surfaces
- bonding of fresh to old concrete
- crack injection
- glueing of precast elements
- glueing of steel to concrete, etc.

In 1981 RILEM established the technical committee 52-RAC 'Resin Adherence to Concrete'. The main aims of the committee's work were

- to collect research results and practical experiences
- to initiate and coordinate research programs
- to develop, on a scientific base, test methods for field and for laboratory purposes.

One of the results of the committee's work is a state-of-the-art report, which will be presented orally as a General Report at the International Symposium ISAP '86, and will be printed either in the RILEM journal *Materials and Structures* or separately. Several test recommendations have been elaborated and will be prepared as drafts for the participants of ISAP '86. These are:

- direct tensile test
- pull-off test
- direct shear test
- slant shear test
- four-point bending test
- dynamic loading test
- thermal compatibility test (two versions)
- injectibility test.

The purpose of the Symposium is to exchange information on a scientific, technical and practical level between international experts and to make known the work of the RILEM Technical Committee 52-RAC. The members of the

Scientific Committee of the Symposium hope that ISAP '86 will be the start of a constructive cooperation between experts from all over the world in order to get more information about adhesion phenomena and their practical applications.

Adhésion est un phénomène avec lequel les architectes et les ingénieurs des travaux publics ne sont pas très familiarisés. De même dans d'autres disciplines la connaissance des propriétés de surface et des raisons fondamentales des énergies de liaison est loin d'être satisfaisante. Néanmoins il y a beaucoup de domaines importants d'application en matière de construction en béton où l'adhésion est nécessaire pour le succès et la durabilité:

- revêtement et peinture
- réparation des surfaces en béton
- adhérence du béton frais au béton ancien
- injection de fissures
- collage d'éléments préfabriqués
- collage de l'acier au béton, etc.

En 1981 la RILEM a initié le comité technique 52-RAC 'Adhérence des Résines au Béton'. Les objets principaux du comité étaient:

- le recueil de résultats scientifiques et d'expérience pratique
- l'initiation et la coordination de programmes des recherches scientifiques
- l'élaboration scientifique de méthodes d'essai pour l'utilisation au chantier et au laboratoire.

Un des résultats du comité est le Bilan Technique qui sera présenté comme rapport général au Symposium International ISAP '86 et qui sera publié dans le journal de la RILEM *Matériaux et Structures* ou séparément. Quelques recommandations d'essai ont été élaborées et seront proposées aux participants d'ISAP '86:

- l'essai de traction directe
- l'essai d'adhérence
- l'essai de compression cisaillement directe
- l'essai de compression cisaillement
- l'essai de flexion centrée
- l'essai de flexion dynamique
- l'essai de variations de température (deux versions)
- l'essai d'injectabilité.

L'objet du Symposium est la communication entre experts internationaux au niveau scientifique, technique et pratique ainsi que la présentation du travail du Comité Technique 52-RAC. Les membres du Comité Scientifique du Symposium espèrent que ISAP '86 sera le point de départ d'une coopération constructive entre experts du monde entier pour plus d'information sur le phénomène d'adhésion et sur ses conséquences pratiques.

H.R. Sasse

Aachen, Germany June 1986

Contents
Table des matières

Section 1

Science of adhesion

Science de l'adhésion

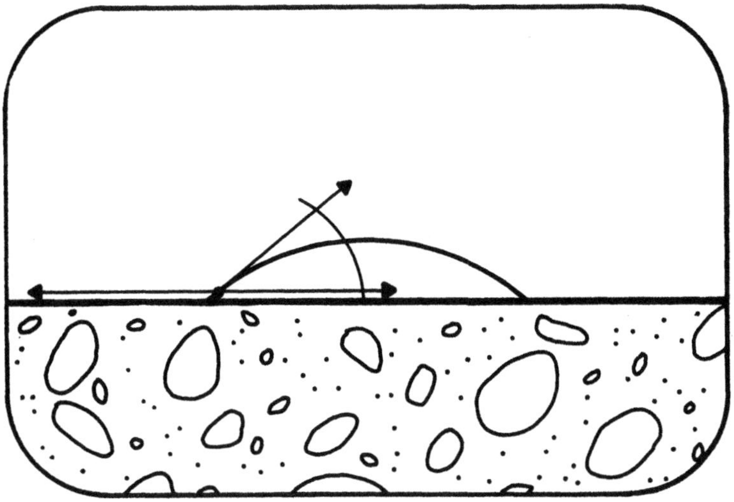

DURABILITY ANALYSIS OF ADHESIVE JOINTS

A.H. CARDON and C.C. HIEL
Faculty of Engineering - Department of Applied Continuum Mechanics
Free University Brussels (V.U.B.)

Abstract
The analysis of the quality of the adhesion of an epoxy material to concrete, or other materials, may be defined by the properties and thickness of a transition zone between the bulk adhesive and the pure adherend.
Direct measurements of those properties and dimensions are very difficult without a strong interaction between the measurement technique and the obtained results. Mixed numerical experimental methods seems appropriate to obtain valuable information on the transition zone.
A defined quality of adhesion on a test specimen can be compared with real adhesion by dynamic test methods.
The influence of environmental factors on the quality of the adhesion are also measured on a test specimen and are the basis of a durability analysis of an adhesive joint taking into account classical dimensional analysis.
Key words : Viscoelasticity, Adhesion, Environmental Factors, Durability.

1. Introduction

The adhesion problem needs a system formulation taking into account the nature of the adherend the nature of the adhesive and the interactive forces between adherend and adhesive. Some of those interactive forces are on atomic and molecular level, other interactions of a purely mechanical nature are on the macroscopic level. Every structural application of an adhesive is on the macroscopic level and the test methods must be on the macroscopic level taking into account all the necessary information on the microscopic level. A possible macroscopic characterisation of the quality, or degree of adhesion is obtained by the introduction of a transition zone along the geometrical interface between the adhesive and the adherend. The thickness of the transition zone is the sum of the lengths in the adherend and the adhesive where interactive forces of any nature change the mechanical nature of the original continuum. Only indirect measures of the depth of this transition zone and its thermomechanical properties seem actually possible without a significant perturbation of the characteristics of this interface layer.

2. Formulation of the basic adhesive problem

The system formulation of the adhesion problem needs the combination of the thermomechanical properties of the adherend, those of the adhesive and the results of the surface preparation of the adherend. This combination results in the existence of an interactive region along the interface adhesive-adherend.
The thermomechanical properties of this transition layer and its thickness are the result and in fact the macroscopic effect of this interaction. The properties of this transition zone are characteristic of the quality of the realised adhesion.

If we consider :

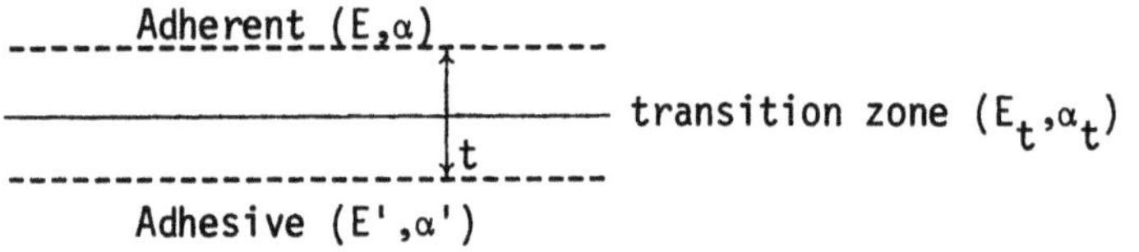

it is necessary to have a test method in order to obtain values for (t, E_t, α_t) for the definition of the quality of the adhesion. When $t = 0$ there is no practical adhesion. If $t \to \infty$, combined with a continuous variation of E_t and α_t from (E,α) to (E',α') we can consider a complete, or total, adhesion.

It is also useful to consider the part of t in the adhesive layer, t_A. If we have and adhesive layer between two adherents such that $t > t_A^{(1)} + t_A^{(2)}$ we have a thick adhesive,and if $t \leqslant t_A^{(1)} + t_A^{(2)}$ the adhesive is thin.

For a thin adhesive layer the properties of the transition zone and the influence of the environmental factors on those properties, controls the quality of the realised adhesion.

For a thick adhesive layer the properties of the bulk adhesive are important for the quality of the obtained adhesion.

3. Measurement of the properties of the transition zone

The characteristics of the transition zone are a function of the properties of the adherend, the properties of the adhesive and the surface preparation. The environmental factors, such as the temperature, the moisture and the radiations, play an important role on the properties of the interface region.

Even if for different adherends and adhesives we have some indications on the thickness of the interface regions, it is actually impossible to have a direct measure of the properties of the transition zone without an interaction with those properties.

It seems only possible actually to obtain the properties of the interface homogenised region starting from the comparison between the numerical results of a test case started from arbitrary values of $(t, E_t$ and $\alpha_t)$ with the experimental results of a multilayered medium adherend-transition zone-adhesive under a specific loading condition.

This comparison is possible on the basis of the results of a finite element program compared with the experimental results of a pure bending or a pure shear test, or by dynamical methods based on the modal analysis of the free vibration of a multilayered plate.

Recently a method was developed by W. De Wilde and H. Sol for the measurement of the mechanical characteristics of a composite laminate that can be applied for the measurement of the characteristics of the interface layer between adherend and adhesive.

The influence of the surface preparation on those characteristics is only possible on a semi-empirical basis by comparison of the characteristics of the interface region on a same test specimen and under the same loading conditions with different surface preparations.

The effects of the environmental factors must also be introduced by the comparison of the characteristics of the interfacial region on a given test specimen and under the same loading conditions with different values of the environmental loading.

For the adhesion between polymers and concrete the direct mechanical locking seems the most important factor and the measurement of the properties of the interface region is possible from the overall behaviour of the multilayer adherend-transition zone-adhesive or by the testing of slices perpendicular and parallel to the interface plane.

4. Durability analysis of the adhesion

An essential problem for adhesion between two adherents is the life time and more generally the durability analysis of the realised adhesion under general mechanical and environmental loading.

The durability analysis needs a possible extrapolation from short time tests, under different loading conditions, to long term behaviour.

The adherend may be elastic or viscoelastic. The adhesive is generally viscoelastic. This viscoelastic nature of the adhesive implies a viscoelastic nature of the realised adhesion. The viscoelasticity of the transition zone may be the basis for the analysis of the long term behaviour of the realised adhesion under the influence of mechanical and environmental loading.

If we consider the one-dimensional linear viscoelastic constitutive equation :

$$\varepsilon(t) = D_0 \, \sigma(t) + \int_0^t \Delta D(t-u) \, \frac{d\sigma(u)}{du} \, du \quad ,$$

when the material is considered as non-ageing, the influence of temperature and moisture can be introduced by the way where the instantaneous compliance, D_0, and the transition zone $\Delta D(t-u)$ depend on the variations of those environmental factors.

The application of the Time-Temperature-Superposition-Principle, generalised to include the stress level as accelerating factor, as suggested by H. Brinson and coworkers, can give an interesting basis for a durability analysis of the realised adhesion.

5. Influence of the geometrical shape

Every specific test of an adhesive joint is performed under special shape conditions that implies given stress distributions in the joint. It is necessary to perform a stress analysis of the real adhesion, where we consider only situations where the adhesive layer must ensure stress transfer between the two adherents, in order to define the significant test specimen for a durability analysis.

Especially for the study of the influence of the environmental factors a dimensional analysis is needed between the test specimen configuration and loading and the real situation in order to obtain results that can be extrapolated to real long term behaviour.

Stress concentrations on the test specimen must be avoided and properties of free boundaries to the volume of the adhesive must be in agreement with the similitude rules if we want to obtain significant results for a long term behaviour.

6. Conclusions

For a given complete adhesion system composed of the adherend, the adhesive, the surface preparation and the processing, the properties of the transition zone must be measured by mixed numerical-experimental methods on a test specimen designed in order to respect the similitude rules.

In the case of a thin adhesive layer the properties of the transition zone are the fundamental control elements of the realised adhesion.

For a thick adhesive layer the properties of the bulk adhesive are the control elements of the realised adhesion.
In both cases the influence of the environmental factors on the controlling element must be carefully analysed.

A method of long term behaviour analysis based on short time tests is possible starting from a generalised Time-Temperature-Stress Superposition Principle where the moisture can also be introduced as accelerating factor.

The influence of the surface preparation and the processing must be measured separately under the same similitude conditions.

Acknowledgements

This research has received financial support from the Belgian National Science Foundation (F.K.F.O.) and the Research Council (OZR) of the Free University of Brussels (V.U.B.).

References

Cardon, A. and Hiel C. (1982) Détermination du comportement des matériaux composites sous conditions dynamiques. Cahiers du Groupe Français de Rhéologie, Tome V, 6, Février 1982, 283-291.

Boulpaep, F., Cardon, A. and Hiel, C. (1982) Dynamic methods for the determination of the viscoelastic properties of epoxy materials used for refections of civil engineering structures. Rheol. Acta 21, 420-422.

Cardon, A. and Boulpaep, F. (1983) Adhesion entre les matériaux traditionnels et les résines époxy. IABSE reports, volume 46, 211-216.

De Wilde, W. and Sol, H. Anisotropic material identification using free vibration analysis on plates. Proceedings of the E-MRS Conference, Strasbourg, Novembre 1985 (to be published).

Brinson, H. (1985) Durability predictions of adhesively bonded composite structures using accelerated characterisation methods". Composite Structures 3, (Ed. I. Marshall), Elsevier Applied Science Publishers, 1-18.

Cardon, A. Mechanical behaviour and durability predictions of adhesively bonded joints. 9th Annual Meeting of the Adhesion Society, Feb. 1986, Hilton Head (SC).

STRESSES DUE TO PHYSICAL AND CHEMICAL ACTIONS IN POLYMER COATINGS ON A CONCRETE SUBSTRATE

M. GÜNTER, H. K. HILSDORF
Institut für Massivbau und Baustofftechnologie, Universität Karlsruhe, Germany

Abstract
Polymer coatings on concrete surfaces repeatedly exhibit blisters, cracks or delaminations. Various processes had been proposed to explain the causes of such damage. However, the magnitude of stresses acting on the interface between coating and substrate is relatively unknown. This is particularly true for stresses caused by chemical and physical actions. In a theoretical and experimental research program an attempt is made to quantify such stresses. Two processes causing stresses are described, osmotic pressure and capillary pressure. Two test set-ups were developed to determine the magnitude of such stresses. The results so far show that osmotic processes and capillary suction can result in pressures up to 45 and 2 bar respectively acting on the interface between concrete and coating. But also smaller pressures may cause blisters or delaminations due to stress concentrations at local defects.

Key words: Concrete substrate, Resin coatings, Capillary pressure, Osmotic pressure, Stress concentrations

1. Problem statement

Resin coatings are applied to concrete surfaces e.g. to ascertain a high resistance to chemical attack, to obtain leak proof structures or to enhance the durability of concrete surfaces of low quality. Such coatings may have a thickness ranging from approximately 100 μm up to several mm. Surface coatings can fullfil their functions satisfactorily over an extended period of time only if there is good bond between the concrete substrate and the coating. Thus, for proper planning and execution of such measures sufficient knowledge on bond properties as well as practical experience are required. Sasse and Fiebrich (1983) gave an extensive literature survey on these aspects.

In previous studies it had been found that soon after the application of resin coatings on a concrete surface stresses may develop in this composite system which are caused by various physical and chemical properties of the coating and of the substrate. A review of this subject was given by Klopfer (1976). Of particular significance in this context is the pore system of the concrete as well as the pore solution in the hydrated cement paste. In Table 1 various mechanisms leading to stresses in the composite system are summarized together with the type of damage they may cause and the conditions under which they are relevant.

Table 1. Stresses acting on coating-concrete substrate interfaces.

	cause of stresses	conditions under which stresses may occur	types of damage
1	chemical attack and saponification	- compounds which can be saponified - pore solution in hydrated cement paste	- delamination - dissolution
2	capillary suction and compression of air in concrete pores	- pores, partially filled with air - water adsorption	- delamination - blisters
3	water and solvent vapor pressure	- pores, partially filled - temperature increase	- delamination - blisters
4	osmotic processes	- water soluble compounds in the coating - semipermeable membranes - pore solution in hydrated cement paste	- folding - delamination - blisters
5	internal stresses and restraint	- temperature change - differences in thermal, swelling or shrinkage properties of coating and of substrate	- folding - cracks - delamination

However, little knowledge exists on the magnitude of stresses which may be caused by the various mechanisms given in Table 1. Therefore, quite frequently the cause of damage occuring in structures with surface coatings cannot be sufficiently explained.

In a research program the various mechanisms which may lead to damage of coated systems are studied both theoretically and experimentally. The magnitude of stresses caused by such mechanisms is estimated and compared to the bond strength between coating and concrete substrate. The project is not completed yet, however, some theoretical considerations as well as some preliminary results will be presented in this paper. The ultimate objective of this work is the formulation of recommendations regarding composition and properties of resin coatings for concrete surfaces, allowable defects in such coatings as well as requirements regarding properties and preparation of concrete substrates.

2. Osmotic pressure

2.1 Theoretical background

Osmotic processes occur when two solutions of different concentration are separated by a semipermeable membrane i.e. only the solvent can penetrate the membrane. Due to the concentration gradient the solvent migrates from the location of higher concentration of the solvent through the semipermeable membrane to the location of lower concentration. This osmotic flow continues until the chemical potential of the solvent is the same on both sides of the membrane. In order to reach equilibrium pressure has to be developed in the solution with the lower concentration of the solvent. This pressure compensates the reduction of the chemical potential of the solvent due to its dilution in the solution. The pressure difference acting on the membrane is refered to as osmotic pressure (Moore and Hummel, 1976).

The equilibrium condition can be described as follows:

$$\int_0^{\pi} V_L^m dP = -RT \ln(P_{L,\beta}/P_{L,\alpha}) \qquad \text{(eq. 1)}$$

where V_L^m = partial mol volume of the solvent in the solution β

$P_{L,\beta}$ = vapor pressure of the solvent in the solution β

$P_{L,\alpha}$ = vapor pressure of the solvent in the solution α

R = gas constant

T = temperature

α = solution with a higher concentration of the solvent

β = solution with a lower concentration of the solvent

π = osmotic pressure

Under the assumption that V_L^m is independent of pressure eq. 1 may be solved resulting in the following expression for the osmotic pressure:

$$\pi = \frac{1}{V_L^m} [RT \ln(P_{L,\alpha}/P_{L,\beta})] \qquad \text{(eq. 2)}$$

2.2 Applications

The processes described in the previous section can also occur in coated concrete surfaces (Klopfer, 1976), (Ettel, 1979). The conditions under which this is possible are described in Table 2.

Table 2: Conditions for the development of osmotic pressure in concrete surface-coating systems.

solvent	semipermeable membrane	solute
-pore solution of the hydrated cement paste -external moisture	-concrete surface with primer -resin coating	-soluble organic substances -saponification products -hardeners -inorganic salts e.g. from dirt, impurities

Osmotic processes may cause swelling or softening of the coating as well as a reduction of bond strength between coating and concrete substrate. This may cause the formation of folds in the coating due to swelling and in particular the formation of osmotic blisters. The magnitude of the osmotic pressure which may cause the formation of such blisters depends on the impermeability and flexural rigidity of the coating as well as on the bond strength between the coating and the concrete substrate. Danger of damage due to osmotic pressure is increased if external water acts on the system.

Osmotic processes may be minimized by proper formulation of the coating material. The risk of damage is greatly enhanced in case of poor workmanship and insufficient hardening of the coating. Incomplete mixing and hardening of the components of the coating, impurities and saponification products may result in the formation of nuclei with water soluble substances in the coating or in the interface between concrete substrate and coating, thus enhancing the possibility of the development of osmotic pressures.

Osmotic processes take place only if semipermeable membranes are present. In the case of a coated concrete surface region such membranes may be formed by the coating itself or by the concrete surface-primer system. However, the semipermeable action of a coating or of the concrete-primer system depends on such a large number of parameters that the actual magnitude of osmotic pressure can not be directly calculated on the basis of eq. 2. Therefore, some experiments on the magnitude of osmotic pressure are inevitable. They are described in section 4.1 and 5.1.

3. Pressure in concrete due to capillary suction

3.1 Theoretical background

When the internal surfaces of a porous body are wetted the forces acting on the surfaces are altered depending on the surface energy of the solid and of the sorbate. In the following a simple system is used to demonstrate in which way these forces lead to stresses in a coating-substrate interface (Figs. 1a and 1b). In the vicinity of the solid forces of adsorption act between the solid and the sorbate resulting in the curvature of the free surface of the sorbate (Fig. 1a). Due to this curvature the surface tension in an element at the surface of the sorbate has a resultant p_c normal to the surface (Kaufmann, 1963):

$$p_c = \sigma \left(\frac{1}{r_1} + \frac{1}{r_2}\right) \qquad \text{(Laplace-equation)} \qquad \text{(eq. 3)}$$

where r_1, r_2 radii of curvature acc. to Fig. 1a
σ = surface tension of the sorbate in air

The resultant p_c acts towards the center of curvature of the sorbate meniscus. Therefore, for wetting sorbates surface tension acts at the point of contact between the meniscus and the solid, and the sorbate is drawn into the surface pores of the solid (Fig. 1 b).

For capillaries with a circular cross-section the following relation is valid:

$$p_c = \frac{1}{R}(2\sigma\cos\delta) \qquad \text{(eq. 4)}$$

where R = radius of the capillary
δ = contact angle

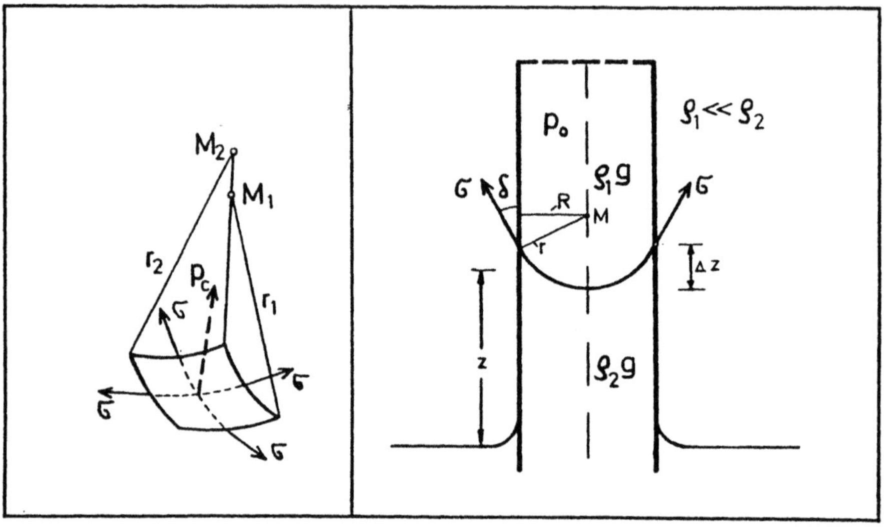

Fig. 1a: Curvature of a surface element

Fig. 1b: Wetting sorbate in a capillary

For an open, circular capillary the meniscus of the sorbate is exposed to the atmospheric pressure p_o. Therefore, the sorbate rises in the capillary until equilibrium between p_c, p_o and gravity forces is reached. From equilibrium conditions and under the assumption that $\rho_1 << \rho_2$ and $\Delta z << z$ the capillary rise of the sorbate, z, can be calculated:

$$z = 2\sigma\cos\delta/(R\rho g) = p_c/(\rho g) \qquad \text{(eq. 5)}$$

where ρ = density of the sorbate
g = acceleration of gravity

If a sorbate is drawn into a capillary which is closed at the other end pressure is developed in the space above the meniscus due to compression of the enclosed air (Sell and Renzow, 1975). Then the following equilibrium equation holds true:

$$\sigma\cos\delta \cdot 2\pi R = \pi R^2\rho g z + \Delta p\pi R^2 \qquad \text{(eq. 6)}$$

and $$\Delta p = \frac{1}{R}[2\sigma\cos\delta] - \rho g z = p_c - \rho g z \qquad \text{(eq. 7)}$$

where Δp = additional pressure above the meniscus of the sorbate.

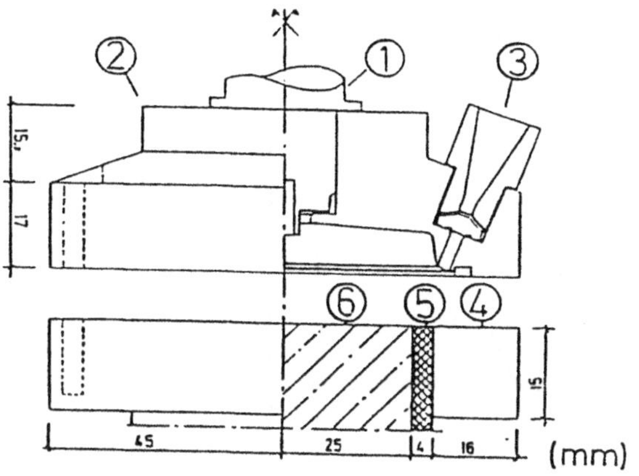

1 pressure transducer
2 pressure cell
3 breather screw
4 sample containment (steel ring)
5 epoxy resin
6 mortar sample

Fig. 3: Test set-up capillary pressure

5. Experimental results

5.1 Osmotic pressure

Preliminary tests showed that a concrete surface treated with an epoxy base coating or with some primers normally used in practice will have semipermeable properties. Furthermore, it was found that the development of osmotic pressure with time depends to a large extent on the flexural rigidity of the coating. In addition to inorganic and organic solutes, the saponification products of some plasticizers contained in the coating and some common epoxy hardeners proved to have osmotic properties with regard to the pore solution of hydrated cement paste. Figs. 4.a and 4.b show some experimental results, both for rigid and for flexible systems. Saponification products of a commercial plasticizer (Fig. 4.a) and an epoxy hardener (Fig. 4.b) were separated from the pore solution of a hydrated cement paste by means of a mortar disk, thickness 2.5 mm. One side of the mortar was treated with a commercial primer. According to Figs. 4.a and 4.b osmotic pressures up to 21 and 46 bar, respectively have been observed for the rigid system. The maximum pressures after 45 days in the flexible systems were approximately 0.03 bar. The differences in the results between rigid and flexible systems are of practical significance in so far as pressures acting on stiff coatings with good bond will be much higher than the pressures to be expected in blisters or in flexible coatings with low bond strength or local defects (see also section 6).

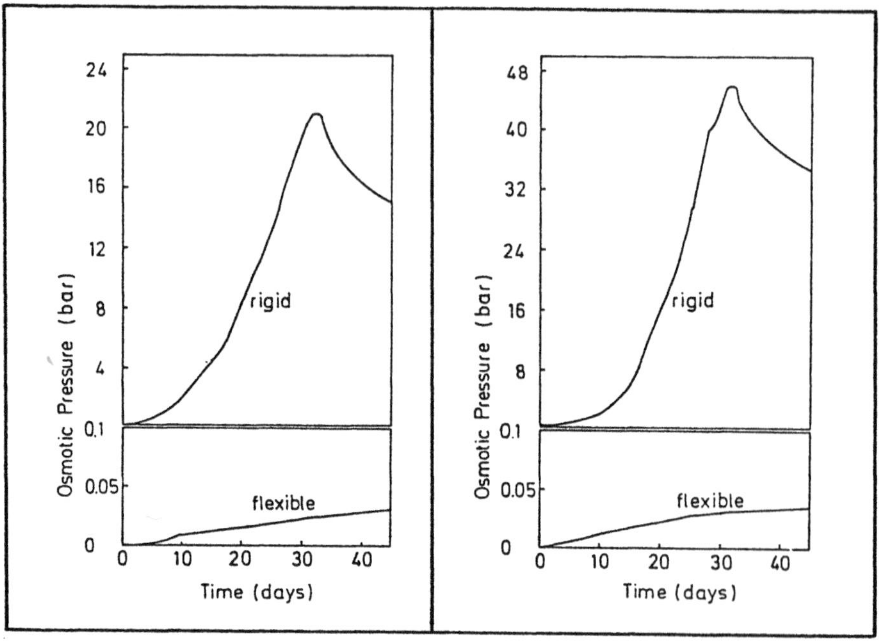

Fig. 4a: Osmotic pressure for a saponification product

Fig. 4b: Osmotic pressure for an epoxy hardener

5.2 Capillary pressure

In Fig. 5 the capillary pressure acting on the interface between concrete and coating is given as a function of time for two different mortar samples A and B. The bottom sides of the samples were exposed to water. Both samples were taken from a larger specimen which had a w/c-ratio of 0.65 and which was cured for a period of several months. However, mortar A was taken from the bottom side of the specimen, whereas mortar B was taken from the top side as cast. Consequently, both samples have different pore structures. Prior to the exposure to water both samples were dried at a relative humidity of 65 percent over a period of 2 months.

According to Fig. 5 mortar A initially shows a slower rate of pressure development. However, it reaches a final value which is almost twice as large than that of mortar B. This can be explained by the difference in pore structure of the two samples: The pore size distribution as determined by mercury porosimetry is given for both mortars in Fig. 6. From this it follows that mortar B has a coarser pore structure than mortar A. As a consequence of this difference capillary pressure in mortar A is higher than that in mortar B. This is in agreement with the theoretical considerations given in section 3.1. Using eq. 7 an "effective average pore radius" of 730 nm can be calculated for mortar A and a corresponding value of 1400 nm for mortar B.

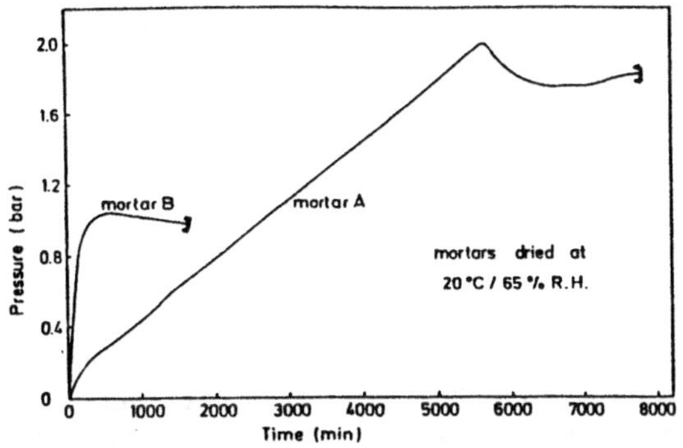

Fig. 5: Capillary pressure caused by capillary suction of mortar samples A and B

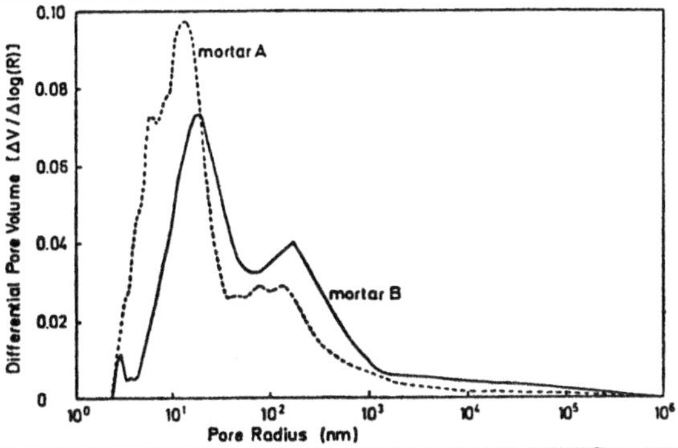

Fig. 6: Pore size distribution in mortar A and B

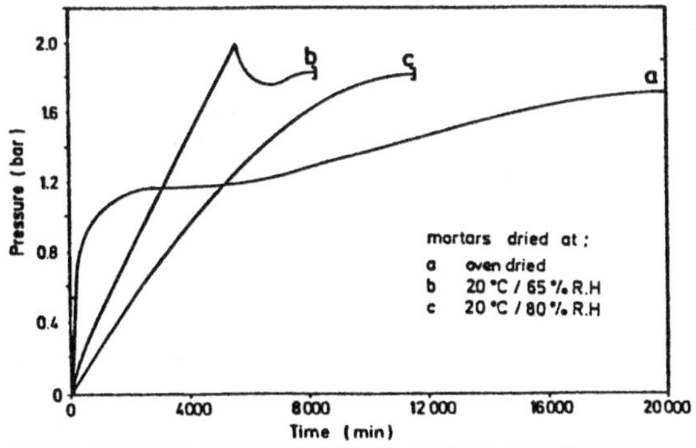

Fig. 7: Capillary pressure caused by capillary suction of mortar samples A; various moisture states

The effect of moisture content prior to exposure to water can be seen from the results given in Fig. 7. There, the development of capillary pressure is given for mortar samples A, which were either oven dried (a), dried at 65 percent r.h. (b) or dried at 80 percent r.h. (c). The initial rate of capillary pressure development depends on the moisture state of the mortars prior to exposure to water. However, after prolonged periods of time the pressure appears to approach values which are independent of the moisture state prior to the exposure to water. This is not surprising because after some time the moisture state in the mortar samples is the same irrespective of their initial moisture states.

In Fig. 8 the development of pressure is shown when the bottom side of a mortar sample B is exposed to water pressure of 1.5 bar. The pressure observed corresponds roughly to the sum of the external water pressure of 1.5 bar and the capillary pressure as given in Fig. 5.

The experiments are continued in order to study in more detail the effect of pore structure, temperature and thickness of the mortar specimens.

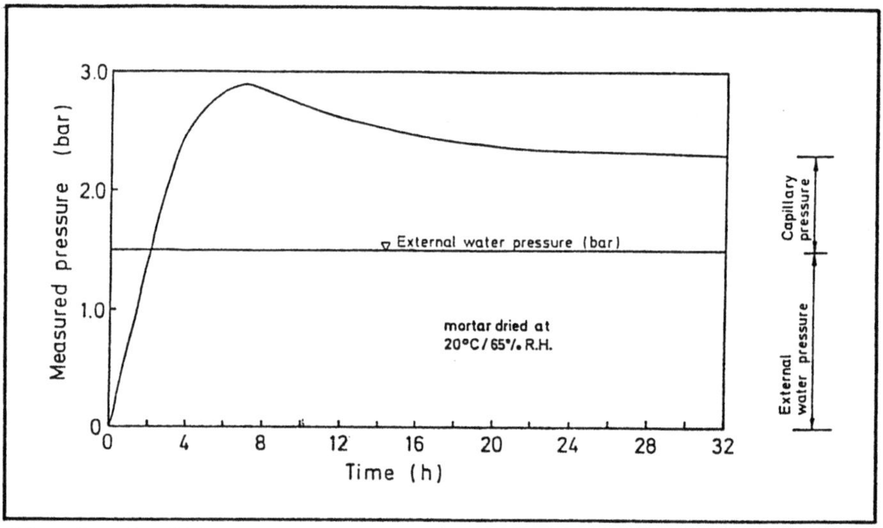

Fig. 8: Capillary pressure caused by capillary suction and external water pressure, mortar B

6. Local defects

The experimental results presented in the previous sections indicated that in the case of a rigid coating osmotic pressures acting on an organic coating may approach the bond strength between coating and the concrete substrate. On the other hand the osmotic pressure developed for flexible coatings as well as the capillary pressure are at least one order of magnitude below the pressure which may exceed the concrete-coating bond strength. Nevertheless, even such mechanisms may cause damage of coated concrete surfaces:

Due to dirt, loose particles, oil or insufficient hardening of the coating, local defects may occur in the interface between concrete and surface coating. As a consequence of such defects osmotic or capillary pressures as described in the preceding sections may lead to stress concentrations which exceed the bond strength between surface coating and concrete substrate even if the actual pressure is well below this value. To estimate the actual magnitude of such stress concentrations concepts of fracture mechanics have to be applied.

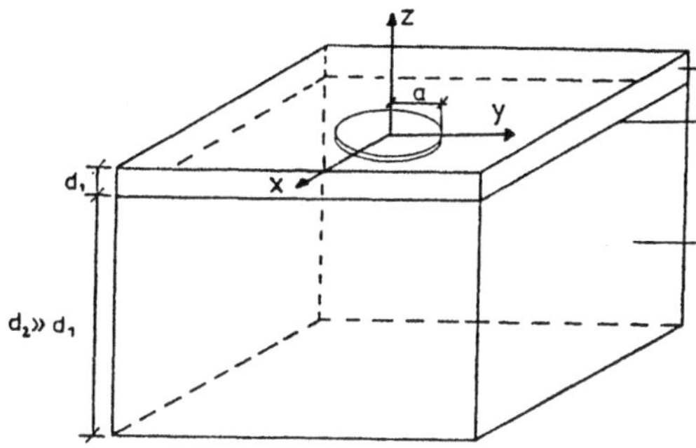

Fig. 9.1: Local defect in the interface substrate-coating

Fig. 9.1 shows the actual geometry of a substrate-coating system with a local defect of circular shape. At this time, no solution for the stress concentrations in the vicinity of such a defect is available, since in the past fracture mechanics concepts had been applied primarily to homogeneous systems of materials with elastic properties where only localized yielding occurs. Fig. 9.2 shows a penny-shaped crack in a homogeneous system representing a local defect. The distribution of stresses σ_z in the vicinity of the defect is given in Fig. 9.3, when the defect itself is subjected to a unit pressure p = 1.

Since the stress concentration at the crack tip approaches infinity, local yielding has to occur. Nevertheless, from Fig. 9.3 it follows that in a homogeneous material local stresses may be up to one order of magnitude higher than the pressure acting on the crack itself. Presently, more detailed calculations with more realistic geometries and materials properties are carried out using the finite element method. It is hoped that these results together with further experiments will enable us to estimate the combination of pressure and size of defects which can cause delamination of coatings on concrete surfaces.

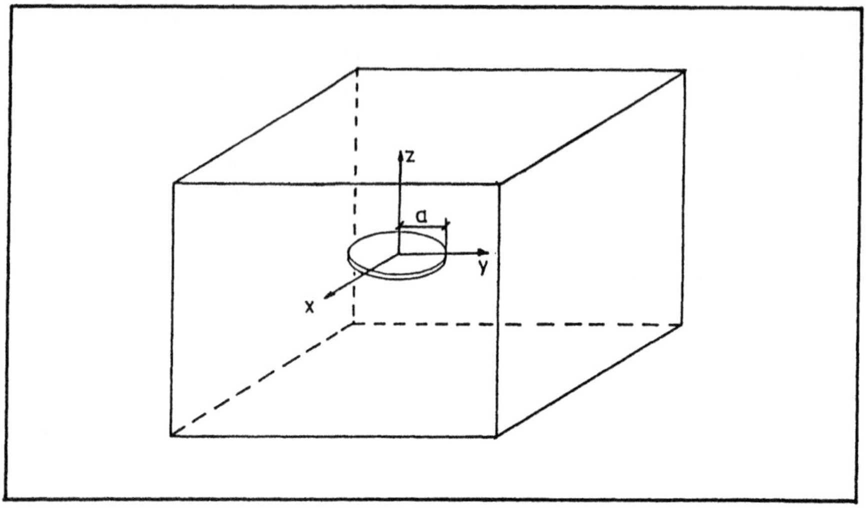

Fig. 9.2: Penny-shaped crack in a homogeneous system

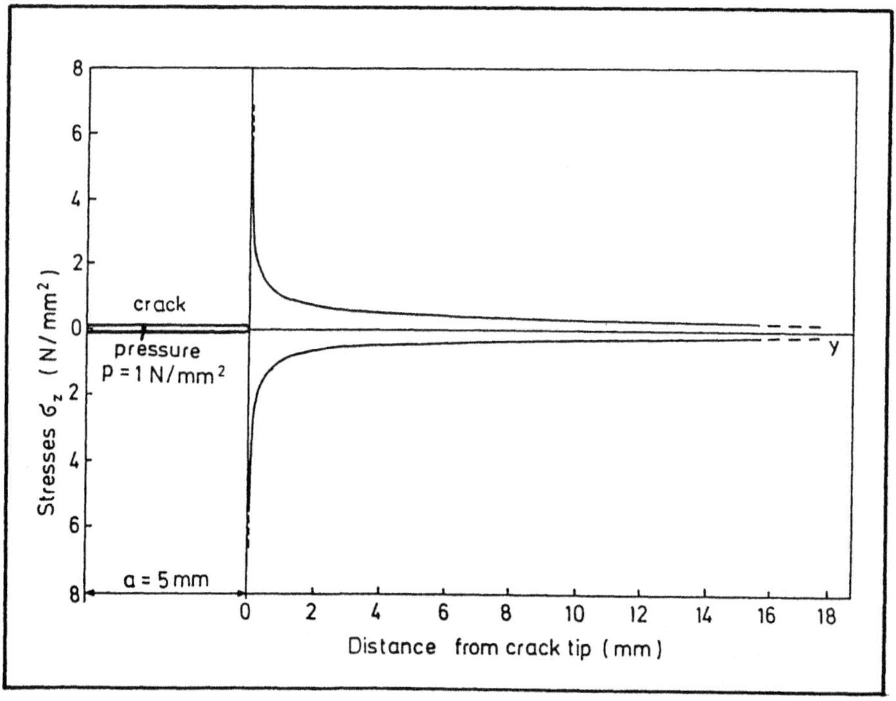

Fig. 9.3: Stresses in the vicinity of a penny-shaped crack due to internal pressure p = 1

7. Summary

- In this paper the mechanisms leading to damage of organic coatings on concrete substrates are summarized.
- Some theoretical background regarding osmotic pressure and capillary pressure is given.
- In experiments it is shown that osmotic pressure in such systems may reach values up to 45 bar thus approaching the bond strength between a concrete substrate and organic coatings. Osmotic pressure is much less if the coating is flexible e.g. after the formation of local blisters.
- Capillary pressures up to 2 bar have been observed due to capillary suction. If an external water pressure acts on the surface of the concrete opposite to the coated surface, capillary pressure and external pressure are superimposed. The magnitude of capillary pressure can be related to the pore structure of the hydrated cement paste.
- Local defects in the coating or in the coating-substrate interface may lead to local stress concentrations resulting in damage or delamination at average stresses well below the bond strength between coating and substrate.

References

Ettel, W.P. and Ahlner, M. (1979) Blasenbildung bei Epoxidharzbeschichtungen und ihre Ursachen. Baustoffindustrie, 6, 15-17

Kaufmann, W. (1963) Technische Hydro- und Aeromechanik. Springer-Verlag, Berlin

Klopfer, H. (1976) Anstrichschäden. Bauverlag GmbH, Wiesbaden und Berlin

Moore and Hummel (1976) Physikalische Chemie. Verlag Walter de Gruyter, Berlin - New York

Sasse, H.R. and Fiebrich, M. (1983) Bonding of Polymer Materials to Concrete. Materials and Structures, 94, 293-301

Sell, P.J. and Renzow, D. (1975) Bestimmung des Benetzungsverhaltens von Pigmenten. Progress in Organic Coatings, 3, 323-348

EFFECTS OF CELLULOSE ETHER ON BOND BETWEEN MATRIX AND AGGREGATE IN CONCRETE

K. HAYAKAWA Organic Chemical Division,
Shin-Etsu Chemical Co., LTD
T. SOSHIRODA Depertment of Architectural Engineering,
Shibaura Institute of Technology

Abstract
The influence of cellulose ether on the strength of mortar and on the anisotropy of concrete are discussed by obtaining relationships between cement-void ratio and compressive strength and by comparing two tensile strengths in the casting direction and direction perpendicular to the casting. - It may be stated that the strength of mortar is little influenced by both cellulose ether and/ or defoamer and the addition of cellulose ether improves bond strength between cement matrix and aggregate by decreasing defect under aggregate due to bleeding.
Key words: Cellulose ether, Anisotropy, Defoamer, Bleeding, Splitting test, Aggregate matrix bond, Cement-void, Mortar strength.

1. Introduction

Recently cellulose ether, which has excellent controlling effects of bleeding and high viscosity, has become to be widely used as a segregation controlling admixture for under water concrete and asbleeding controlling admixture for inverted concrete. This paper describes the results of experiments to investigate the effects of cellulose ether on bond between cement matrix and aggregate in concrete from the view point of segregation controlling.

2. Outline of experiments

The experiments are divided into three series as follows.

In series I , the influence of cellulose ether on the strength of mortar was examined. The relationships between cement-void ratio and compressive strength were obtained changing curing conditions, test conditions and curing time.

In series II , the influences of cellulose ether on the bond strength between mortar matrix and model aggregate (made of mortar), and on the mode of failure were examined.

In series III , the influence of cellulose ether on the anisotropy relative to the casting direction were examined.

Series I
Table 1 shows materials to be used and Table 2 shows mix proportions keeping S/C (absolute volume ratio) = 2, and W/C = 45 % (by weight) constant. The tests were undertaken in four curing-test conditions as follows: water curing → wet test (W → W), water curing → dry test (air curing at 20°C, 60 ~ 65 % RH during three days, W → D), air curing → dry test (D → D), and air curing → wet test (water curing during three days, D → W). Size of test specimen was 4 × 4 × 16 cm.

Table 1. Materials used (Series I)

Cement	Model aggregate	White cement
	Mortar matrix	Ordinary Portland cement
Fine aggregate	Place of origin	Kisarazu
	Specific gravity	2.60
	Fineness Modulus	2.73
Cellulose ether	Methyl cellulose (MC type) 〈S. Company〉	
Defoamer	Nonionic surface active agent 〈S. Company〉	

Table 2. Mix proportions of mortar and properties of fresh mortar (Series I)

	Content of Cellulose ether (C × %)	Content of Defoamer (C × %)	Proportions of absolute volume (ℓ/m³)			Properties of fresh mortar		
			W	C	S	Flow (mm)	Air content (%)	Bleeding (cc/cm²)
N	0	0	315	221	443	204	2.4	0.17
M2	0.2	0	270	192	380	180	16.5	0
M4	0.4	0	253	178	355	152	22.0	0
M2d	0.2	0.05	312	219	438	184	3.9	0
M4d	0.4	0.1	306	215	430	162	5.2	0

Series II
Table 3 shows the mix proportions, molding methods, curing conditions and strengths, of model aggregates and mortar matrix. Fig. 1 shows the size of test specimens, casting directions and loading positions. To examine the degree of defects under aggregate caused by bleeding, model aggregates were set in two ways, that is vertical and horizontal

positions as shown in the figures.

Table 3. Proportions, strength and curing conditions of model aggregate and mortar matrix

Model aggregate (Series II)

Aggregate	W/C (%)	S/C	Strength (kgf/cm²)		Methods of molding and curing
			Tensile strength	Compressive strength	Molding with Vibrator Air-curing during 24 hr. at 20°C Water-curing during 24 hr. at 90°C Water-curing during 14 d. at 20°C
A	35	1.1	37.8	595	
B	45	3	25.9	359	

Mortar matrix (Series II)

	Content of Cellulose ether (C × %)	Content of Defoamer (C × %)	W/C (%)	S/C	Unit weight (kg/ℓ)	Strength (kgf/cm²)	
						Tensile strength	Compressive strength
N-43	0	0	43	2	2.23	32.3	497
M2d-43	0.2	0.05	43	2	2.15	30.5	441
M2d-39	0.2	0.05	39	2	2.16	23.1	487
M4d-39	0.4	0.1	39	2	2.16	40.7	543

Methods of molding and curing
Molding with tamping rod, Air curing during 24 hr. at 20°C
Water-curing during 27 d. at 20°C

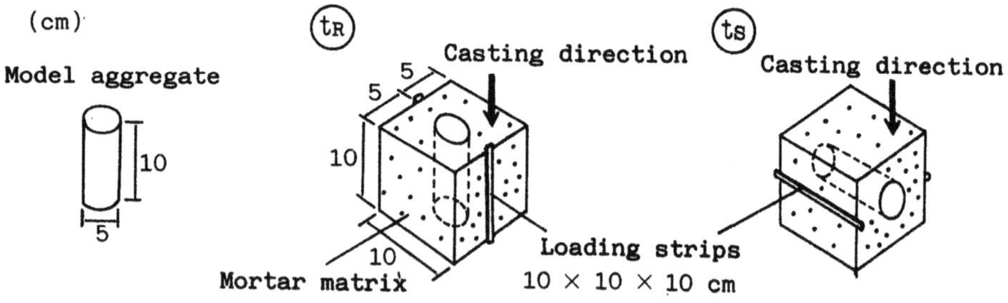

Fig. 1. Loading positions, casting directions, notations of splitting tensile strength, and size of test specimen on splitting test (Series II)

Series III

Table 4 and Table 5 show materials and mix proportions respectively. The extent of defects under coarse aggregate caused by bleeding was examined by comparing the tensile strength of the two directions, i.e. casting direction and the direction perpendicular to this, from test result of 15 cm cube specimens as shown in Fig. 2.

Table 4. Materials used (Series III)

Cement	Ordinary Portland cement		
Fine aggregate	Place of origin	Kisarazu	
	Specific gravity	2.60	2.43
	Absorption	1.58	—
	Fineness Modulus	2.73	2.84
Coarse aggregate	Place of origin	Ōi river	Atsugi
	G max (mm)	25	40
	Specific gravity	2.65	2.65
	Absorption	1.2	1.3
	Fineness Modulus	6.81	7.36
Superplasticizer	Naphthalene Sulfonate (P. Company)		
Cellulose ether	Methyl cellulose type (MC type) 〈S. Company〉		
Defoamer	Nonionic surface active agent 〈S. Company〉		

Table 5. Mix proportions (Series III)

Series	G max (mm)	S/a (%)	W/C (%)	Mix proportions of Base concrete (kgf/ cm²)			
				W	C	S	G
KP- I	40	40	50	165	330	720	1145
KP- II	40	29.8	46	168	365	548	1290
KP- III	25	35.8	50	176	352	627	1195
KP- IV	25	32	45	179	396	548	1235
KP- V	40	28	45	185	410	454	1275

Table 5. Content of additive, Properties of fresh mortar

Series	Base concrete B		Superplasticized concrete F			Cellulose ether concrete M		
	Air content (%)	Slump (cm)	Content (C × %)	Air content (%)	Slump (cm)	Content (C × %)	Air content (%)	Slump (cm)
KP- I	2.8	2.0	2.5	1.7	15.5	0.2	7.4	5.0
KP- II	0.8	4.0	2.5	0.9	13.8	0.2	7.5	9.0
KP- III	1.9	6.0	0.5	1.9	16.5	0.2	7.4	9.5
KP- IV	1.4	10.0	1.0	0.8	20.5	0.2	6.8	10.0
KP- V	0.8	12.5				0.2 (Defoamer 0.05)	1.6	13.5

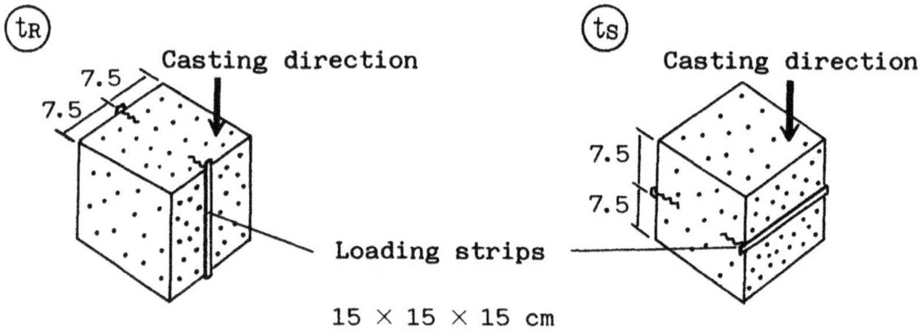

Fig. 2. Loading positions, casting directions, notations of splitting tensile strength, and size of test specimen on splitting test (Series III)

3. Results and Discussions

(1) Series I

Fig. 3 shows the relationships between cement-void ratios and compressive strengths at 28 days and 162 days. The cement-void ratios are obtained from a formula of C/(W+A) (C: absolute volume of cement, ℓ/m^3, W: absolute volume of water, ℓ/m^3, A: absolute volume of air, ℓ/m^3). The strengths of mortar adding no defoamer were low but the reduction of strength can be improved by using defoamer. In every conditions of curing, testing and/or curing time, a good relationships between the cement-void ratios and compressive strengths are obtained as shown in Fig. 3. This indicates that strengths of mortar may be little influenced by both cellulose ether and the nonionic defoamer.

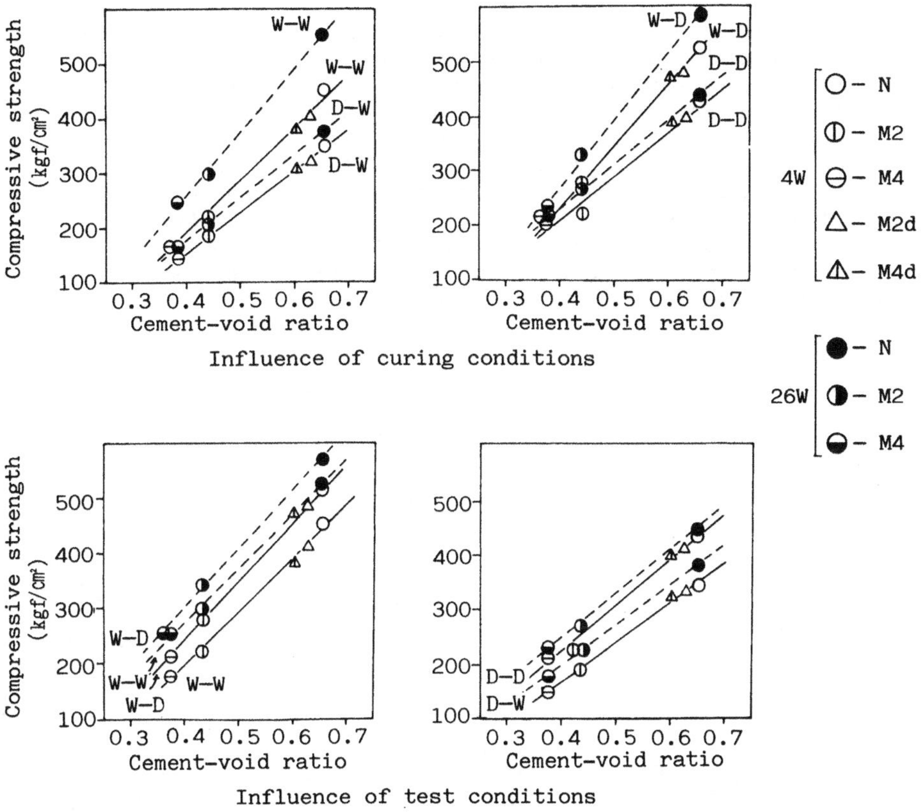

Fig. 3. Relationships between cement-void ratio and compressive strengths

(2) Series II

Fig. 4 shows various modes of failure after splitting tests of model concrete. In the case of model aggregate A having higher strength, there were no failure were observed. Concerning anisotropy t_R was not more than t_S (Fig. 5) in the case of adding cellulose ether, while t_R was more than in the case of plain model concrete. Bond failures were found on even the upper parts of aggregate in the case of M4d with comparatively higher content of cellulose ether. This indicates that bond strength might decrease due to defect under aggregate caused by bleeding in the case of plain concrete, while because of little bleeding in the case of the addition of cellulose ether, bond strength might not decrease. In the case of model aggregate B having lower strength, there are many failures of aggregate. Both the number of the failures of aggregate and the splitting strength in the case of the addition of cellulose ether are higher than those in the case of plain concrete. This indicates that cellulose ether tends to improve the bond strength.

(3) Series III

Fig. 7 shows $t_S - t_R$ relationships. A formula $t_S = 0.937\ t_R$ shown in the figure is based on about a hundred data of conventional concretes. The relationships " $t_R > t_S$ " are shown in the cases of base concrete and superplasticized concrete, while the relationship " $t_R \leqq t_S$ " in the case of concrete containing cellulose ether. This indicates that bleeding decreases remarkably by the addition of cellulose ether and subseguently the defect under aggregate caused by bleeding decreases. In the case of the addition of cellulose ether, it is found that considerable areas of the surfaces of coarse aggregates are covered with cement paste at the failure surfaces after splitting tests. This indicates that cellulose ether improves the bond strength.

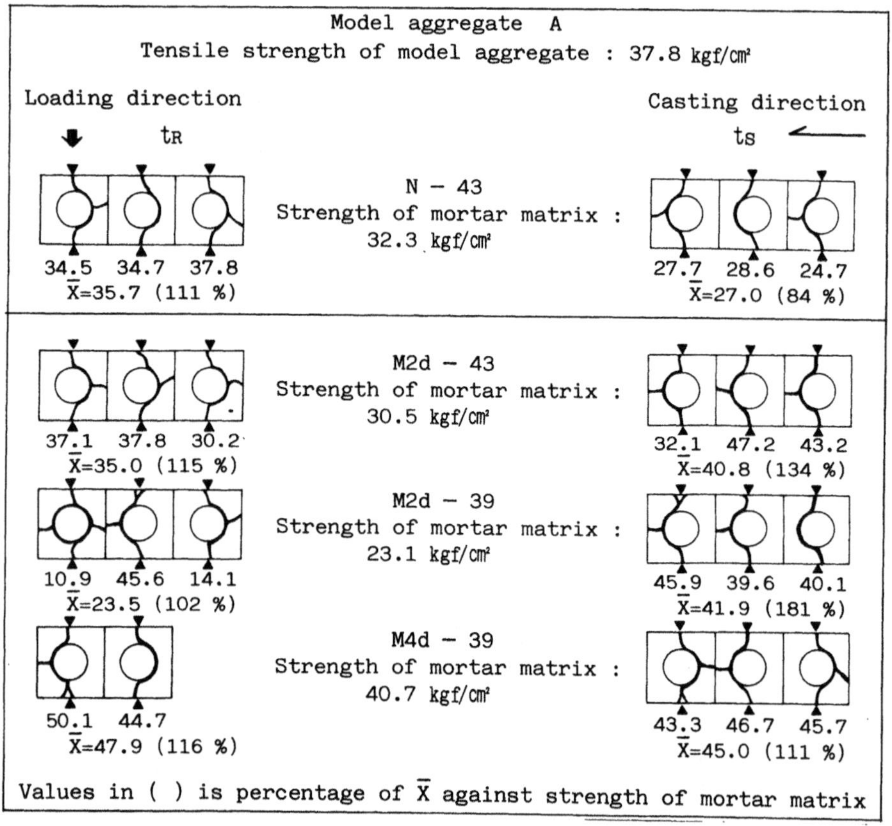

Fig. 4 Modes of failure after splitting tests of model concrete

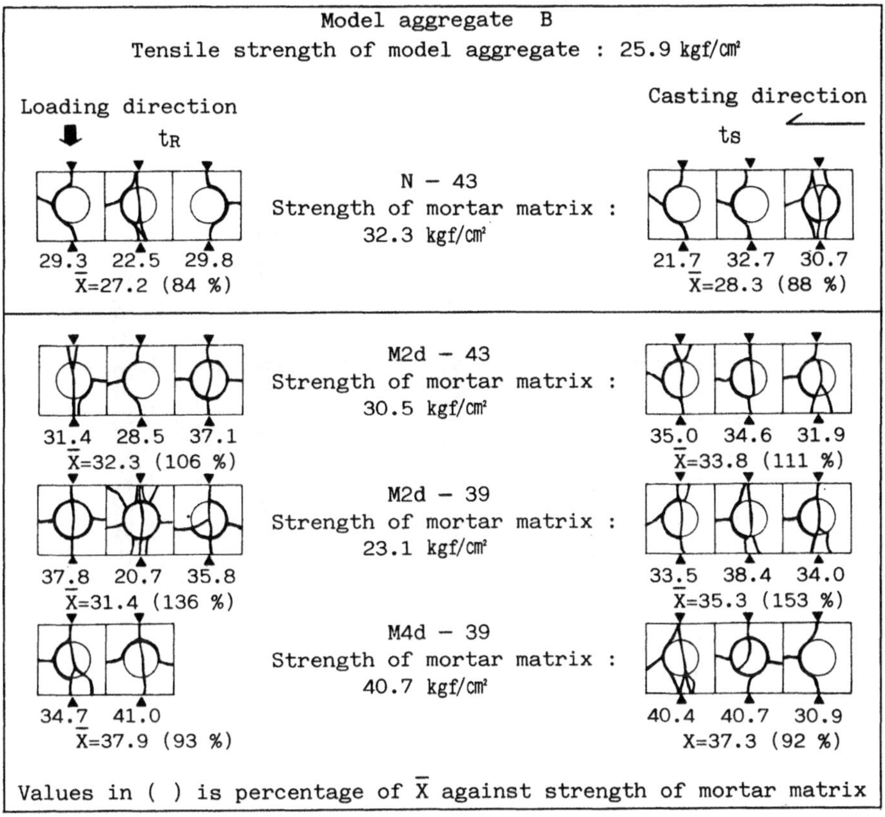

Fig. 4. Modes of failure after splitting tests of model concrete

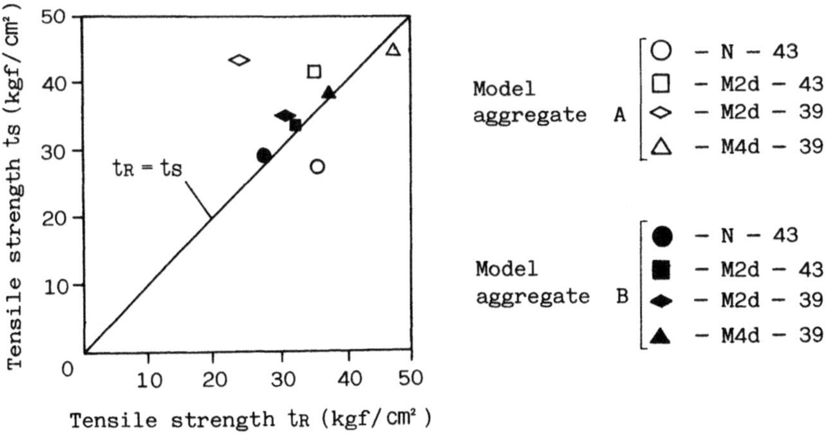

Fig. 5. Relationship between ts and tR of model concrete

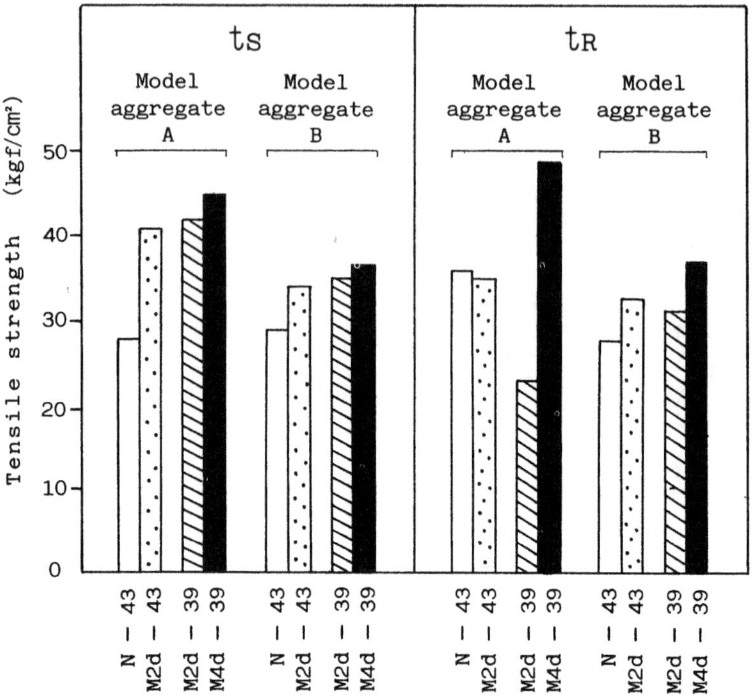

Fig. 6. Tensile strength of model concrete

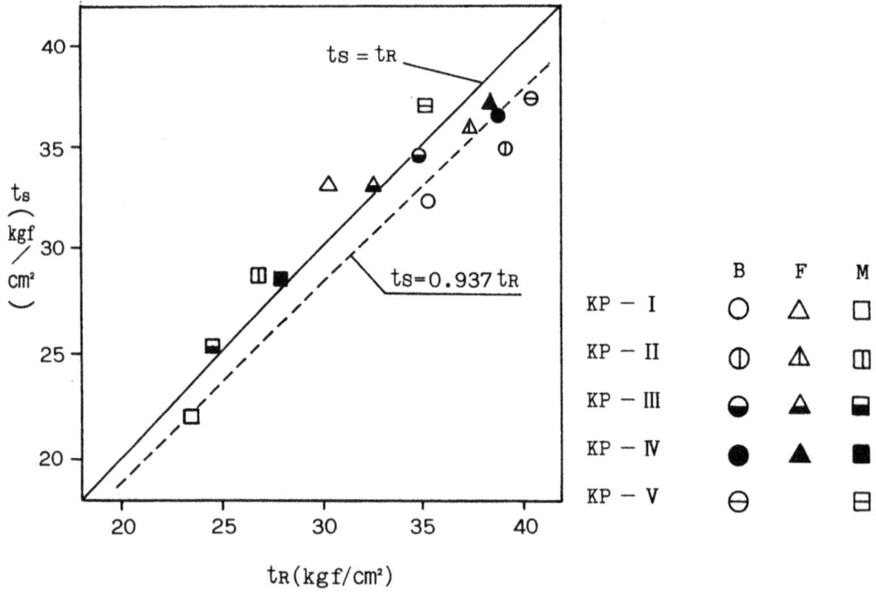

Fig. 7. Relationship between ts and tR

4. Conclusions

Summing up the main results it may be stated that:
(1) In every conditions of curing, testing and curing time, the strength of mortar is little influenced by both cellulose ether and/or defoamer.
(2) The addition of cellulose ether to concrete improves reduction of bond under aggregates in concrete by controlling bleeding.
(3) The addition of cellulose ether to concrete increases adhesion strength of matrix and aggregate.

References

Haga, T., Sogo, S. and Miura, N. (1984) Trans. Japan Concrete Institute, 6, 45-48
Nakazato, Y., Motohashi, K. and Ohno, T. (1984) Trans. Japan Concrete Institute, 6, 41-44
Soshiroda, T. (1974) Cement & Concrete, 330, 2-8
Hayakawa, K. and Soshiroda, T. (1985) Trans. Japan Concrete Institute, 7, 17-24

INTERFACIALE FRACTURE ENTRE LES MATÉRIAUX POLYMÈRES ET LE BÉTON ARMÉ

M.JARONIEK, T.NIEZGODZINSKI
Institut de Mécanique Appliquée, Université Technique de Łódź, Pologne

Résumé

L'ouvrage a été realisé pour caractériser la propagation d'une fissure dans la zone de contact entre le béton armé de fibres métalliques (le bétonac) et le béton polymère. Les essais ont été faits sur des éprouvettes selon la specification ASTM E 813-81 et des éprouvettes DCB (une poutre double console) avec des fissures artificielles dans la zone de contact des materiaux. Le calcul a été effectué en utilisant la méthode des éléments fini. La technique expérimentale consiste à solliciter l'éprouvette et à effectuer les mesures des déformations en utilisant l'enregistreur x-y et les mesures photoélastiques.

1. Introduction

Quand on étudie l'adhésion entre les matériaux polymères et le béton il est utile de concentrer son attention sur la mécanique de la rupture et la propagation d'une fissure dans la zone de contact de ces materiaux.
La mécanique linéaire de la rupture et l'état de contrainte autour l'origine de la fissure (comprise entre deux milieux $M_1(E_1,\nu_1)$ et $M_2(E_2,\nu_2)$ - fig.1) est connue par les travaux: M.L.Williams, J.R.Rice, G.C.Sih, G.R. Irwin, D.L.Clements, J.R.Willis, C.Atkinson, E.Viola, A.Piva, A.Tommaso, ...
Se basant sur ces travaux l'état de contrainte au voisinage du front de fissure au point de coordonnées (r,Θ) s'expriment par les relations:

$$\sigma_{yy}^{(1)} = \frac{K_I^{(1)}}{\sqrt{2\pi r}} \cos\frac{\Theta}{2}\left[1+\sin\frac{\Theta}{2}\sin\frac{3\Theta}{2}\right] + \frac{K_{II}^{(1)}}{\sqrt{2\pi r}} \sin\frac{\Theta}{2}\cos\frac{\Theta}{2}\cos\frac{3\Theta}{2}$$

$$\sigma_{xx}^{(1)} = \frac{K_I^{(1)}}{\sqrt{2\pi r}} \cos\frac{\theta}{2}\left[1 - \sin\frac{\theta}{2}\sin\frac{3\theta}{2}\right] - \frac{K_{II}^{(1)}}{\sqrt{2\pi r}}\sin\frac{\theta}{2}\left[2 + \cos\frac{\theta}{2}\cos\frac{3\theta}{2}\right] + (k-1)T$$

$$\sigma_{xx}^{(2)} = \frac{K_I^{(2)}}{\sqrt{2\pi r}} \cos\frac{\theta}{2}\left|1 - \sin\frac{\theta}{2}\sin\frac{3\theta}{2}\right| - \frac{K_{II}^{(2)}}{\sqrt{2\pi r}}\sin\frac{\theta}{2}\left[2 + \cos\frac{\theta}{2}\cos\frac{3\theta}{2}\right] + \frac{\mu_2}{\mu_1}(k-1)T \qquad (1)$$

$$\tau_{xy}^{(1)} = \frac{K^{(1)}}{\sqrt{2\pi r}}\sin\frac{\theta}{2}\cos\frac{\theta}{2}\cos\frac{3\theta}{2} + \frac{K_{II}^{(1)}}{\sqrt{2\pi r}}\cos\frac{\theta}{2}\left[1 - \sin\frac{\theta}{2}\sin\frac{3\theta}{2}\right]$$

$$\mu_i = \frac{E_i}{2(1+\nu_i)}$$

Quand la fracture se produit les facteurs d'intensité de contrainte sont égales:

$$K_{Ic}^{(1)} = (2\pi r_o)^{-\frac{1}{2}}\, \sigma_{\theta\theta}^{(1)}(\theta_o^{(1)}, r_o)$$
$$K_{Ic}^{(2)} = (2\pi r_o)^{-\frac{1}{2}}\, \sigma_{\theta\theta}^{(2)}(\theta_o^{2}, r_o) \qquad (2)$$
$$K_b = (2\pi r_o)^{\frac{1}{2}}\left[\sigma_{\theta\theta}^2(0, r_o) + \sigma_{r\theta}^2(0, r_o)\right]^{\frac{1}{2}}$$

$$0 \le \theta_o^{(1)} \le \pi \qquad -\pi \le \theta_o^{(2)} \le 0$$

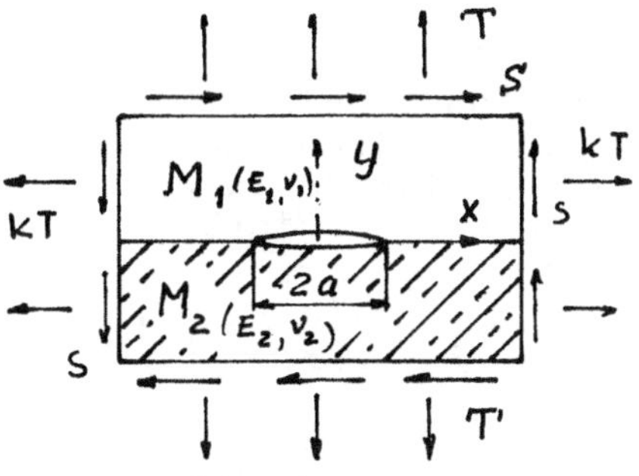

Fig. 1

où:

$K_{Ic}^{(i)}$ est le facteur d'intensité de contrainte critique correspondant à la rupture de milieu i

K_b est le facteur de liaison (the bonding factor)

Irwin a montré (pour un milieu homogène) que dans les points sur les lignes isochromatiques où $\frac{\partial \tau_m}{\partial \theta_m} = 0$ on peut evaluer le facteur d'intensité de contrainte K_I.

A partir de la théorie de Irwin, (pour deux milieux) les facteurs d'intensité de contrainte sont égales respectivement:

$$K_{Ic}^{(1)} = 2\tau_m^{(1)}\sqrt{2\pi r_1}\; f_1(\theta_1)$$
$$K_{Ic}^{(2)} = 2\tau_m^{(2)}\sqrt{2\pi r_2}\; f_1(\theta_2) \qquad (3)$$

où: τ_m est donné par la relation de la photoélsticité:

$$\tau_m = \frac{1}{2}(\sigma_1 - \sigma_2) = \frac{1}{2} f \cdot n$$

f - la valeur de frange

n - l'ordre de ligne isochromatique

$$f_1(\theta_i) = \frac{1}{\sin\theta_i}\left[1 + \left(\frac{2}{3\,tg\,\theta_i}\right)^2\right]^{-\frac{1}{2}} \cdot \left(1 + \frac{2\,tg\frac{3\theta_i}{2}}{3\,tg\,\theta_i}\right)$$

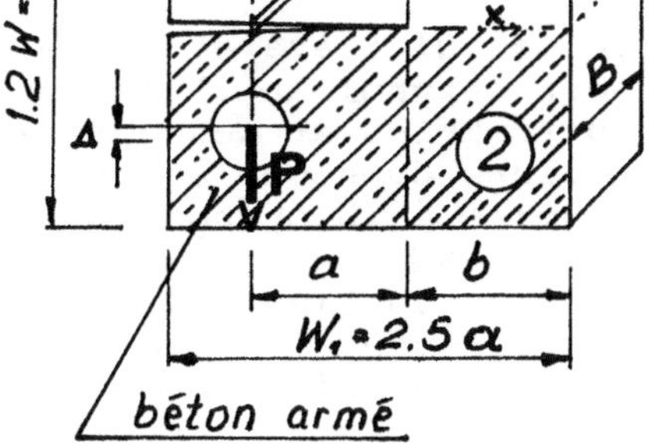

Fig.2

L'énergie de rupture - l'intégrale de Rice J on peut déterminer par la formule:

$$J = \frac{2A}{b \cdot B} \qquad (4)$$

où:

A - la surface comprise entre l'axe des Δ et la courbe P(Δ)

B - la largeur de l'éprouvette

Δ - les déplacements des points d'application des forces P.

L'intégrale de Rice on peut de même évaluer le long du contour autour du centre de la fissure:

$$J = \int_S \left(\tfrac{1}{2} \sigma_{ij} \varepsilon_{ij} \, dx_2 - T_i^{\,n} \frac{\partial u_i}{\partial x_i} \, ds \right) \qquad (5)$$

L'état de tensions (et des déformations) on peut déterminer en utilisant la méthode des éléments fini. Dans ce cas à partir de la méthode de calcul numerique l'énergie de rupture - l'intégrale de Rice s'exprime par la relation:

$$J = \sum_{i=1}^{k} \Big\{ \frac{1}{2} \Big[\frac{1}{E_i} \left(\sigma_{yy_i}^2 - \sigma_{xx_i}^2 \right) - \frac{\tau_{xy_i}^2}{G_i} \Big] n_{1i} - \Big[\frac{\tau_{xy_i}}{E_i} \left(\sigma_{xx_i} - \nu \sigma_{yy_i} \right) n_{2i} + \left(\tau_{xy_i} n_{1i} + \sigma_{yy_i} n_{2i} \right) \frac{\Delta v_i}{\Delta x_i} \Big] \Big\} \Delta s_i \qquad (6)$$

2. Résultats de calcul

Le calcul des tensions et des déformations a été effectué en utilisant la méthode des éléments fini.

La figure 3 montre les dimmensions et les éléments d'éprouvette de system "compact" composé de deux milieux: M_1 (E_1, ν_1) et M_2 (E_2, ν_2) et soumise à la traction de deux forces P. Les propriétés mécaniques des materiaux sont égales respectivement:

- module d'élasticité
  E_1 = 12400 MPa
  E_2 = 24800 MPa
- coefficient de Poisson
  ν_1 = 0,3 ν_2 = 0,15

Le calcul a été effectué pour déterminer l'état de contraintes et des déformations dans une éprouvette soumise à la charge unité (P= 1 daN).

La distribution des contraintes agissants sur la face entre deux milieux M_1 et M_2 représent la figure 4.

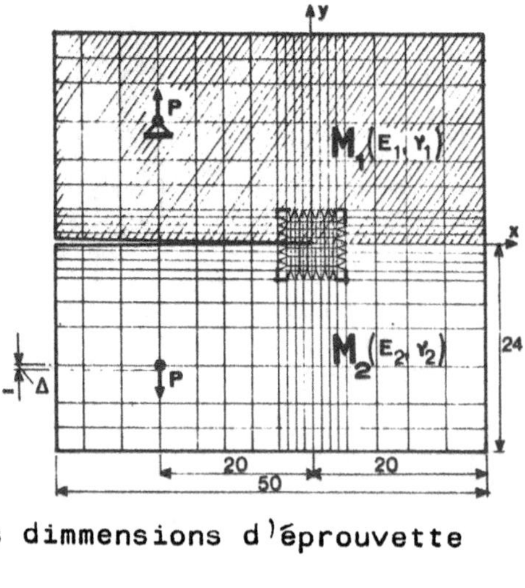

Fig.3 Les dimmensions d'éprouvette

D'après les résultats de calcul on peut de déterminer la différence des tensions principales en chaque point du corps:

$$\sigma_1 - \sigma_2 = 2\tau_m$$

et de même la distribution des lignes isochromatiques. La figure 5 montre la distribution des lignes isochromatiques dans l'éprouvette "compact" soumise à l'action de la charge unite P= 1 daN et la figure 6 montre les lignes isochromatigues au voisinage de l'extrémité d'une fissure.

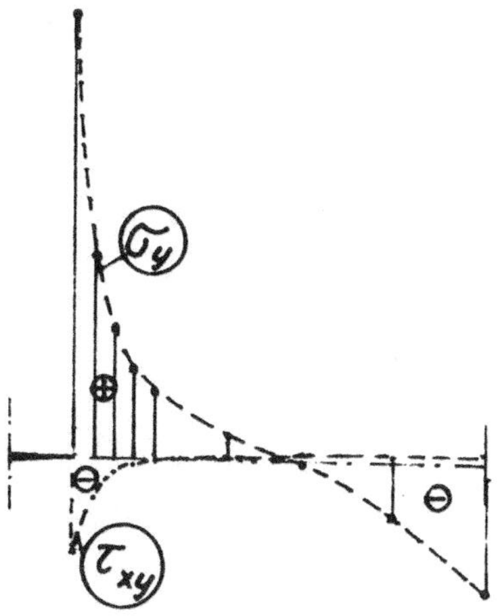

Fig.4 Les contraintes normales σ_y et tangentielles τ_{xy} sur la face entre M_1 et M_2

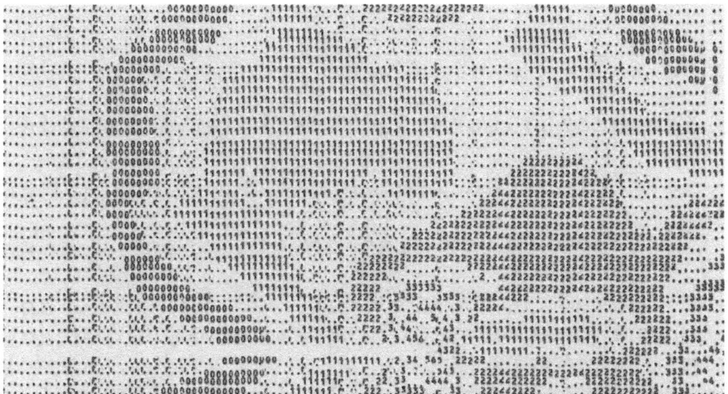

Fig.5 Les lignes isochromatiques d'après de calcul numeriques

Les résultats de calcul permetent de déterminer:
$K_I^{(i)}$, $K_{II}^{(i)}$ - les facteurs d'intensité de contrainte en utilisant les équations (1),(2) et (3)

Fig.6 Les isochromatiques au voisinage d'une fissure

K_b - le facteur de liaison ("the bonding factor") d'après (2)

J - l'intégrale de Rice en utilisant les relations (4) et (6).

En suposant l'élasticité linéaire du materiau on peut evaluer l'ordre de grandeur des facteurs d'intensité de contrainte critique $K_{Ic}^{(i)}$, $K_{IIc}^{(i)}$ et K_{bc} et l'intégrale de Rice correspondant à la rupture de milieu (i) de l'éprouvette, ou à la rupture de liaison à partir des relations:

$$K_{jc}^{(i)} = \beta \cdot K_j^{(i)}$$

$$J_c = \beta^2 \cdot J$$

$$\beta = \frac{P_c}{P}$$

où:

P_c - la force critique

P - la force unité.

Les résultats du calcul sont présentés dans le tableau 1.

Tableau 1. Résultats de calcul

	M_i	K_I	K_{II}	K_D	J	K_{IC}	K_{IIC}	J
		$(MN/m^{3/2})10^{-2}$			N/m	$MN/m^{3/2}$		N/m
CT	M_1	0,227	0,185	0,215	3,191	1,47	1,21	126
	M_2	0,249	0,250		3,155	1,61	1,62	
DCB	M_1	0,719	0,558	0,681	31,9	1,59	1,29	131
	M_2	0,785	0,790		31,55	1,74	1,75	

3. Technique expérimentale

La technique expérimentale consiste essentiellement à solliciter l'éprouvette par deux forces P et à effectuer les mesures suivantes:

1° mesures des forces P en fonction des déplacements Δ (la courbe P(Δ))en utilisant l'engistreur X-Y

2° mesures photoélastiques en utilisant les couches de vernis photoélastiques

D'après les résultats des essais on peut déterminer:

P_c - la charge critique correspondant à la rupture de l'éprouvette

$K^{(1)}_{IC}$, $K^{(2)}_{IC}$ - les facteurs d'intensité de contrainte critique à partir de relation (3)

J - l'intégrale de Rice d'après la relation (4) et P(Δ)

La différence des tensions principales sur la sourface d'éprouvette on peut déterminer à partir de la relation:

$$\sigma_1 - \sigma_2 = n \cdot f^{(i)} \qquad (7)$$

ou: $f^{(i)} = \frac{E_i}{E} \cdot \frac{1+\nu}{1+\nu_i} \cdot f$ - la valeur de frange sur la surface de materiau i

E_i, E, ν_i, ν - les modules d'élasticité et les coefficients de Poisson, respectivement du matériau i et du revêtement photo-élstique

f - la valeur de frange du rêvetement

n - l'ordre de ligne isochromatique

On a remarqué que l'influence de renforcement d'éprouvette par le revêtement photoélastique on peut eliminer en utilisant le couches de vernis photoélastiques composé de deux parties (fig.7).

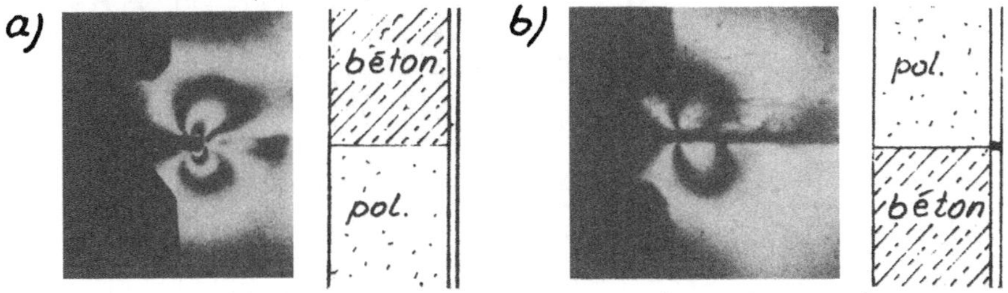

Fig.7. Les lignes isochromatiques dans le revêtement photo-élastique: a) avec l'entaille b) composé

ainsi les conditions dans la zone de contact dans le revêtement sont presque les mêmes comme entre le béton polymère et le béton armé.

4. Résultats expérimentaux

Les essais ont été faits sur des éprouvettes composées de deux materiaux: le béton armé de fibres metalliques (les fibres ont ete reparties d'une facon uniforme dans le volume du béton) et le béton polymère. Les essais ont été faits sur des éprouvettes avec des fissures artificielles dans la zone de contact des matériaux. Ces fissures 0,1 mm d'épaisseur ont été faites pour situer le traject de la fracture dans la zone de contact des matériaux.
Les éprouvettes et la technologie de collage du béton polymère au béton armé ont été realisées par prof.dr L.Czarnecki dans l'Université Technique de Varsovie.
Les propriétés mécaniques des éprouvettes CT("compact tension")et DCB sont presentés dans le tableau 2.
Tableau 2. Propriétés mécaniques des matériaux

Éprouvette	Béton polymère		Béton armé		Revêtement photo-élastique		
	E_1 MPa	ν_1	E_2 MPa	ν_2	E MPa	ν	f MPa/o.is.
CT	12400	0.3	24800	0.15	3450	0.34	3.08
DCB	14200	0.3	2860	0.15			

La figure 8 montre une variation de l'effort P en fonction de déplacement - obtenue en utilisant l'enregistreur X-Y. La figure 9 représent les lignes isochromatiques enregistrée par le revêtement photo-élastique correspondant à la propagation de fissure.Les résultats de trois essais sur

Tableau 3.
Résultats expérimentaux

Éprouvette	$K_{IC}^{(1)}$ MN/m$^{3/2}$	$K_{IC}^{(2)}$ MN/m$^{3/2}$	J_c N/m
CT	1.405	2.17	135
	1.09	1.48	123
	1.67	2.62	231
DCB	1.52	2.47	183
	1.05	2.14	156
	0,92	1,96	96

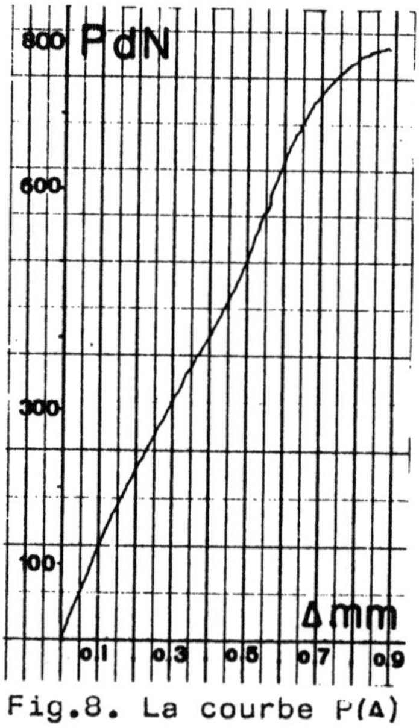

Fig.8. La courbe P(Δ)

des éprouvettes CT et DCB sont indiquées dans le tableau 3

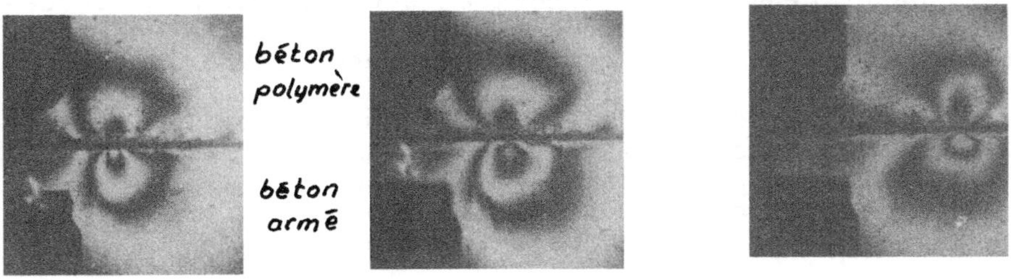

Fig.9. Propagation de fissure

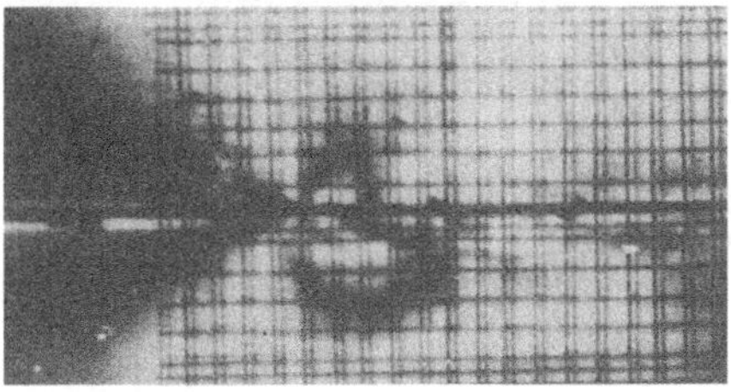

Fig.10. Debut d'apparition de fissure

La méthode photo-élastique permet de detecter tres tôt la fissure et de suivre leur developpement. Par example la fig.10 représent les isochromatiques dans le revêtement collé sur l'éprouvette DCB quand la fissure commence à se propager.

5. Conclusions

D'après les résultats des essais et de calcul il est possible de conclure que ces méthodes on peut appliquer pour caractériser l'interfaciale fracture entre les materiaux polymères et le béton armé.
Le valeur du facteur d'intensité de contrainte du béton atteint le valeur critique et la résistance à la fissuration d'éprouvette est caracterisée par la rupture du béton armé.

ACKNOWLEDGMENT

This research was supported by MRI.23 Problem in cooperation with Institute of Fundamental Technological Research Polish Academy of Sciences.
Polymer concrete composite material has been done in the Institute of Civil Engineering Warsaw Technical University under the supervision of Prof.L.Czarnecki.
The authors are grateful to Prof.A.M.Brandt for his scientific support.

ABSTRACT

This paper rewiews the numerical and experimental procedure used to determine the state of stresses along the interface in the concrete and polymerconcrete in the neighbourhood of a propagating crack. Using the photoelastic measurement results and from a load-displacement courve it is posible to determine the stress intensity factor K_{IC} and the Rice integral J.
The J integral can be determined from numerical calculations using the finite element method.

Bibliographie

Williams,M.L 1959. The stresses around a fault or crack in dissimilar media.Bull.Seismol.Soc.Amer.49,199-204
Rice,J.R and Sih,G.C 1965. Plane problem of cracks in dissimilar media.J.Appl.Mech.E32,418-423
Clements D.L 1971. A crack between dissimilar anisotropic media. Int.J.Eng.Sci.9,257-265
Willis J.R. 1971. Fracture mechanics of interfacial cracks.J.Mech.Phys.Solids 19,353-368
Piva A. and Viola E. 1979 Biaxial load effects on a crack between dissmilar media.Nota Tecnica n.35 I.S.C.B.
Czarnecki L. 1985. Structure and fracture in polymerconcretes some phenomenological approach.Euromech Colloquium 204.

ADHÉSION, COLLAGE ET MÉCANIQUE DE LA RUPTURE

D. MAUGIS et M. BARQUINS
E.R. Mécanique des Surfaces du C.N.R.S., L.C.P.C., Paris

Résumé
L'adhérence des solides est un sujet particulièrement interdisciplinaire, où la physique des surfaces, la chimie, la mécanique et la rhéologie sont liées d'une manière qui peut sembler inextricable. Le but de cet article est de donner une vue générale du sujet et d'éclaircir les principaux concepts. On rappelle quelques données de base sur les forces de liaisons, l'énergie superficielle et l'énergie d'adhésion des solides. Il est montré que la force d'adhérence pour séparer deux solides élastiques en contact se calcule par la mécanique de la rupture, et qu'elle dépend non seulement des forces d'adhésion mais aussi de la géométrie, des constantes d'élasticité et de la méthode de mesure. On examine le cas des solides viscoélastiques, pour lesquels la force d'adhérence dépend considérablement de la vitesse et de la température, de sorte qu'un joint collé peut avoir parfois une résistance plus élevée que celle du substrat.
Mots-clés : Adhésion, Adhérence, Collage, Mécanique de la rupture, solides viscoélastiques.

1. Adhésion et adhérence

La résistance d'un joint adhésif ou collé est en général éprouvée par un contrôle destructif. Rompre un tel joint, c'est créer, à partir d'une entaille ou d'un défaut préexistant, une fissure et provoquer sa propagation. L'initiation ainsi que le déplacement de cette fissure nécessitent l'application d'une contrainte ou d'une déformation, de sorte que le système emmagasine de l'énergie élastique. La propagation de la fissure implique la rupture des liaisons assurant la cohésion des matériaux. Ainsi, lorsque la fissure se propage, une partie de l'énergie élastique stockée est restituée tandis que de l'énergie de surface est consommée.

La cohésion des solides est assurée par un certain nombre de forces d'attraction bien connues (ioniques, covalentes, métalliques, hydrogène, Van der Waals). Le travail pour rompre ces liaisons le long d'un plan imaginaire et séparer de manière réversible et isotherme les deux parties d'un même solide, est égal à deux fois l'énergie superficielle du solide pour chaque unité d'aire rompue. Cette énergie superficielle γ, qui est caractéristique de la nature des liaisons coupées, est de l'ordre de 1000 à 3000 mJ/m^2 pour les solides ioniques et est inférieure à 100 mJ/m^2 pour les édifices moléculaires.

De même l'adhésion entre deux solides peut être due à des forces d'attraction de même nature. Toutefois, l'énergie nécessaire pour séparer réversiblement deux solides 1 et 2 sur une unité d'aire, prend en compte non seulement les énergies γ_1 et γ_2 dépensées pour créer les unités d'aire de 1 et de 2, mais aussi l'énergie interfaciale γ_{12} qui est récupérée. La quantité $w = \gamma_1 + \gamma_2 - \gamma_{12}$ est l'énergie d'adhésion de Dupré, qui caractérise l'adhésivité des surfaces. Il ne faut pourtant pas en déduire que la connaissance de w permet de prévoir la force nécessaire pour séparer deux solides en contact. En effet, cette force, dite d'adhérence par opposition aux forces d'adhésion qui assurent le contact, dépend de l'intervention d'un grand nombre de paramètres directement liée au fait que la séparation ne s'effectue pas en bloc mais progressivement par propagation d'une fissure. Durant la propagation, la rupture des liaisons interfaciales consomme de l'énergie tandis que de l'énergie élastique est libérée et un travail irréversible est dissipé en tête de fissure. Ainsi, la force d'adhérence dépend non seulement de la nature des liaisons coupées mais également de la géométrie du système, des propriétés mécaniques et rhéologiques des matériaux, de l'appareil de mesure et de sa raideur, de la méthode d'essai (déplacement ou charge ou vitesse de déplacement imposé) et de l'environnement (température, hygrométrie, milieu réactif) ; c'est pourquoi l'énergie d'adhésion de Dupré ne représente qu'une valeur seuil de l'énergie qu'il faut fournir au système pour provoquer la séparation et à ce titre, elle peut être considérée comme une constante des matériaux en présence, à un instant donné.

2. Adhérence des solides élastiques

Le contact adhésif de deux solides élastiques peut être étudié selon deux points de vue : la thermodynamique et la théorie de l'élasticité (Maugis et Barquins 1980). Considérons le contact de deux solides sur l'aire A (Fig. 1) provoqué par l'action d'un ressort de raideur k_m

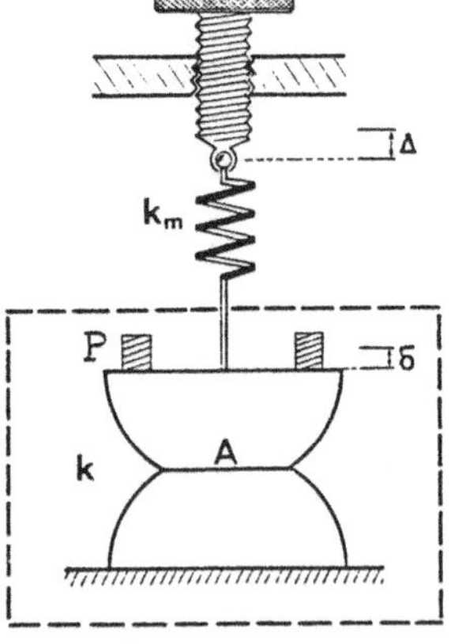

Fig. 1 - Contact d'équilibre entre deux solides élastiques

ou d'un poids mort P ($k_m = 0$). L'aire de contact est susceptible de varier à charge imposée, ou à déplacement Δ imposé, le rapprochement δ des deux solides étant égal à Δ si la raiseur k_m est infinie. Le bord du contact peut être considéré comme une tête de fissure, se propageant en mode I, qui recule ou avance quand l'aire de contact augmente ou diminue.

Maugis et Barquins (1980) ont montré que l'équilibre du système à charge imposée ou à déplacement imposé correspond respectivement aux extréma des énergies libres de Gibbs et de Helmoltz et qu'il est donné par la relation :

$$G = w \qquad (1)$$

où G représente le taux de restitution de l'énergie mécanique défini à partir des énergies élastiques U_E et potentielle U_P par :

$$G = \left(\frac{\partial U_E}{\partial A}\right)_\Delta = \left(\frac{\partial U_E}{\partial A}\right)_P + \left(\frac{\partial U_P}{\partial A}\right)_P \qquad (2)$$

L'équation 1 est le critère d'équilibre de Griffith (1920) qui lie deux des trois variables , P et A. Cet équilibre est stable sous une contrainte donnée (par exemple $P = C^{te}$ ou $\Delta = C^{te}$ si le potentiel thermodynamique correspondant est minimum, ce qui donne la condition générale de stabilité :

$$\left(\frac{\partial G}{\partial A}\right)_\Delta > 0 \ , \ d\Delta = 0$$

Ainsi la stabilité dépend de la raideur k_m de l'appareil de mesure : le ressort emmagasine de l'énergie qu'il peut restituer pour faire avancer la fissure. Pour une raideur finie, il existe une force P_C correspondant à la limite de stabilité et il a été montré (Maugis et Barquins 1980) que cette force décroit de façon monotone lorsqu'on passe de la charge imposée ($k_m = 0$) au déplacement imposé(k_m ∞). Pour $P = P_C$ l'équilibre devient instable et la fissure accélère jusqu'à la réparation complète, à charge constante P_C. La force P_C est la force d'adhérence et correspond à

$$G = w \quad , \quad \left(\frac{\partial G}{\partial A}\right)_{\Delta \text{ ou } P} \leqslant 0 \qquad (3)$$

L'équilibre défini par l'équation 1 peut être rompu par une modification de charge P ou de déplacement Δ . Quand G est différent de w, l'aire de contact évolue spontanément de façon à décroître le potentiel thermodynamique. Si $G < w$, A croît et la fissure recule et inversement, si $G > w$, A décroît et la fissure avance. Quand la fissure s'étend de dA, l'énergie mécanique libérée est égale à GdA, tandis que la rupture des liaisons interfaciale requiert l'énergie wdA, de sorte que l'excès d'énergie (G-w)dA est converti en énergie cinétique s'il n'y a pas de processus dissipatif, et la fissure accélère jusqu'à la séparation complète.

L'évaluation de l'énergie élastique, et par là-même celle de G, nécessite un soin particulier pour tous les systèmes avec énergie superficielle puisque les forces d'attraction moléculaire provoquent des déformations élastiques tout comme les forces extérieures, c'est pourquoi la variation d'énergie élastique n'est plus égale à $Pd\delta$ mais vaut :

$$dU_E = Pd\delta + GdA$$

Johnson et coll. (1971) ont montré que l'on pouvait calculer l'énergie élastique dans le cas d'un poinçon sphérique appliqué contre un massif élastique, en ajoutant un déplacement de corps rigide à la solution classique de l'élasticité (Hertz 1881). Barquins et Maugis (1982) ont établi que cette méthode est tout à fait générale pour des poinçons axisymétriques de forme quelconque et qu'elle peut être prouvée par la théorie de l'élasticité : le calcul des contraintes et des discontinuités de déplacement au bord du contact conduit à des formules identiques à celles de la mécanique de la rupture. Ainsi la force d'adhérence élastique P_c peut être calculée pour un grand nombre de géométries, soit avec la méthode de G par un bilan d'énergie, soit par une analyse des contraintes en tête de fissure qui permet de déterminer le facteur d'intensité des contraintes K_I , ce terme étant relié à G par

$$G = kK_I^2 \qquad (4)$$

où k = $(1-\nu^2)/E$ en déformation plane et k : 1/E en contrainte plane, E et ν représentant le module d'Young et le coefficient de Poisson du matériau.

L'adhérence de poinçons sphériques et plans a été étudiée par Johnson et coll. (1971), Kendall (1971) et Maugis et Barquins (1978, 1983). Il a été établi que pour un poinçon axisymétrique en contact avec un demi-espace élastique sous la charge P et sur une aire de rayon a, le taux de restitution de l'énergie mécanique vaut

$$G = \frac{1-\nu^2}{E} \frac{(P_1-P)^2}{8\pi a^3} \qquad (5)$$

où P_1 représente la charge qu'il faudrait appliquer pour observer le même rayon a pour un contact non adhésif. Il doit être noté que la force d'adhérence n'est pas nécessairement proportionnelle à w, en particulier, pour les poinçons plans, sphériques et coniques, les forces d'adhérence sont proportionnelles respectivement à $\sqrt{w}$, w et w^2.

Depuis que l'on a compris que l'adhérence des solides est un problème de mécanique de la rupture (Ripling, Mostovoy et Patrick, 1964), un certain nombre d'essais destructifs ont été mis au point pour éprouver la résistance des joints adhésifs ou collés, tels que le pelage, le test de double torsion ou l'essai DCB (double cantilever beam). Pour le pelage d'un ruban élastique de largeur b et d'épaisseur h, sous l'action d'une force P qui fait un angle θ avec l'interface, un bilan d'énergie donne aisément (Kendall 1971) :

$$G = \frac{P}{b}(1-\cos\theta) + \left(\frac{P}{b}\right)^2 \frac{1}{2Eh} \qquad (6)$$

L'indépendance du taux G avec la longueur du ruban conduit à un équilibre indifférent sous la force d'adhérence P_c. Il en est de même pour un essai de double torsion (Evans 1972) pour lequel :

$$G = \frac{3P^2\ell^2(1+\nu)}{Eb^2h^3} \qquad (7)$$

où b et h représentent la largeur et l'épaisseur de l'éprouvette et ℓ, la distance entre les points d'application des efforts P/2 qui provoquent la torsion de chacune des deux moitiés du corps d'épreuve. Pour le test DCB à charge imposée, le taux de restitution de l'énergie vaut (Fichter 1983) :

$$G = \frac{12P^2L^2}{Eb^2h^3} \left(1 + \frac{2h}{3L}\right)^2 \qquad (8)$$

où b et h sont la largeur et l'épaisseur de la poutre et L, la distance entre le point d'application de la force P et le fond d'entaille ou la tête de fissure. Il faut noter que lorsque le test DCB est réalisé à déplacement imposé, on a $(\partial G/\partial A)_\delta > 0$, de sorte qu'aucune force d'adhérence ne peut être définie.

Ainsi, du point de vue des concepts, le problème de l'adhérence des solides élastiques est complétement résolu. Il ne fait intervenir que la géométrie du système, des propriétés élastiques, l'énergie d'adhésion et éventuellement la raideur de la machine de mesure. Quant à la méthode du bilan d'énergie, elle présente l'avantage de permettre une étude aisée de l'influence des précontraintes et des contraintes résiduelles dont l'effet majeur est de réduire la force d'adhérence (Kendall 1973). En particulier, dans le contact des surfaces rugueuses, pour lesquelles les aspérités les plus importantes sont comprimées de sorte que l'énergie élastique stockée est plus élevée que pour une surface lisse, l'adhérence peut même disparaître (Fuller et Tabor 1975).

3. Adhérence des solides viscoélastiques

L'adhérence des solides purement élastiques est un cas d'école, car ces matériaux n'existent pas. Des solides comme les élastomères, qui peuvent supporter jusqu'à 800 % d'élongation, sont viscoélastiques et dissipent de l'énergie lorsqu'ils sont soumis à des cycles de déformation, et en particulier en tête de fissure où les contraintes et les vitesses de déformation sont élevées. C'est pourquoi, sous l'action de la force G-w ou de l'excès d'énergie (G-w)dA, la fissure n'accélère pas continuellement comme dans un solide élastique, mais adopte une vitesse limite qui dépend de la température.

Si l'on suppose que les pertes dans le volume du matériau sont négligeables par rapport à celles en tête de fissure, de sorte que les déplacements restent élastiques et que G est calculable par la théorie de l'élasticité ; si l'on suppose également que le freinage visqueux dû aux pertes est proportionnel à l'énergie de Dupré w, ce qui signifie que ces pertes ne peuvent apparaître que si l'interface lui-même est capable de transmettre des contraintes (Gent et Schultz 1972, Andrews et Kinloch 1973), on peut écrire (Maugis et Barquins 1978) :

$$G - w = w\phi(a_T v)$$

où le premier membre représente le moteur de la fissure, et le second, le freinage visqueux. Le facteur sans dimension $\phi(a_T v)$ est donc une propriété du matériau viscoélastique pour un mode donné de propagation (mode I d'ouverture, par exemple) qui ne dépend que de la vitesse v de propagation, et de la température T à travers le facteur a_T de translation de la transformation WLF (Williams, Landel et Ferry 1955) donné par :

$$\log_{10} a_T = \frac{-8,86(T-T_o)}{101,5+T-T_o} \qquad (10)$$

où T_o est une température de référence prise 50 K au dessus de la température de transition vitreuse. Lorsqu'on étudie l'adhérence en fonction du paramètre $a_T v$, on obtient ainsi une courbe maitresse. Pour l'instant, la fonction $\phi(a_T v)$ doit être déterminée expérimentalement, mais elle a été corrélée à la variation avec la fréquence de la partie imaginaire du module d'Young, $E''(\omega)$. De plus un modèle de Zéner parabolique donnant $E''(\omega) \sim (\omega\tau)^n$ à basse fréquence conduit à $\phi(a_T v) \sim v^n$ à basse vitesse. Une confirmation théorique partielle a été récemment présentée par Coussy (1986).

L'intérêt principal de l'équation 9 est que les propriétés de surface qui apparaissent dans w sont clairement découplées des propriétés élastiques, de la géométrie du système et des conditions de chargement qui interviennent dans le calcul de G, et des propriétés viscoélastiques qui sont décrites par $\phi(a_T v)$. Une fois que la fonction $\phi(a_T v)$ est connue (par des expériences de pelage, par exemple) l'équation 9 permet de prédire n'importe quel effet, tel que la cinétique de détachement à charge imposée, à déplacement imposé ou encore à vitesse de déplacement imposée, pour n'importe quelle géométrie où G est calculable. Bien évidemment, des expériences à l'équilibre ($v = 0$, c.a.d. $\phi(a_T v) = 0$) donnent l'énergie d'adhésion de Dupré.

Prenons quelques exemples avec l'adhérence d'un poinçon sphérique en verre sur du polyuréthanne. L'énergie d'adhésion de Dupré ($w = 65$ mJ/m^2) a été déterminée en mesurant les rayons a de contact d'équilibre sous différentes charges P après un temps d'application de 10mn, et en utilisant l'équation 5 avec $P_1 = 4a^3E/3(1-\nu^2)R$ (Johnson, Roberts et Kendall 1971). La figure 2 (Maugis et Barquins 1978) montre la

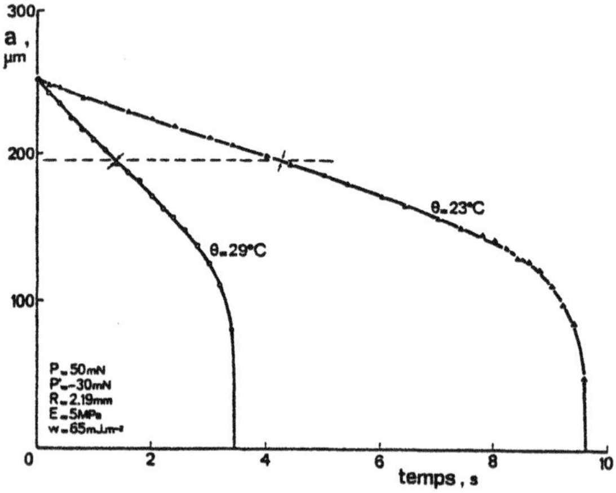

Fig. 2 - Courbes de déchargement de P = 50mN à P' = -30mN d'une bille de verre sur du polyuréthanne pour deux températures différentes.

variation du rayon de contact pendant une expérience de décollement à charge imposée pour deux températures : une augmentation de température, qui a pour effet de diminuer les pertes viscoélastiques, produit un accroissement de la vitesse de la fissure. A chaque point expérimental, G et v = da/dt peuvent être déterminés et par suite la fonction $\phi(a_T v)$ représentée sur la figure 3 (Maugis et Barquins 1978). La même courbe maîtresse a été obtenue avec d'autres types d'essais, chargements et déchargements cycliques (Barquins et Wehbi 1986), déplacements imposés pour des appareils de mesure de raideurs différentes (Barquins 1983) et vitesses de déplacement imposées (Barquins et Maugis 1981) et divers types de géométries : poinçons plans (Maugis et Barquins 1978), poinçons sphériques avec méplats (Maugis et Barquins 1983) et pelage avec précontraintes (Barquins 1984) ou sans (Maugis et Barquins 1978).

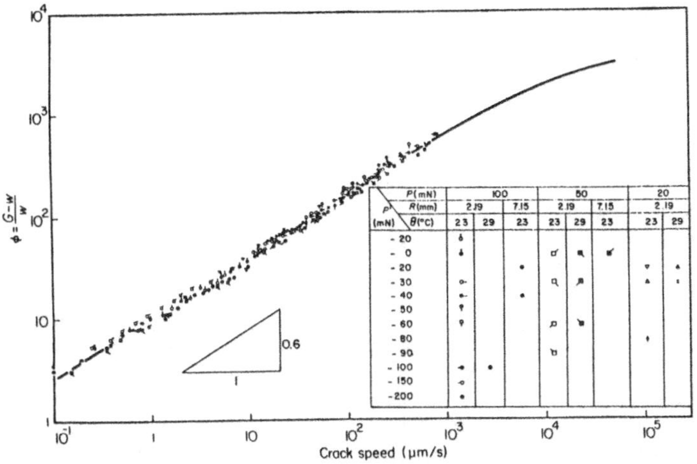

Fig. 3 - Courbe maîtresse pour la cinétique d'adhérence du verre sur le polyuréthanne (w = 65 mJ/m^2, θ = 23°C).

Dans les exemples précédents, le temps de contact initial était maintenu constant et choisi long par rapport à la durée nécessaire au décollement complet. En effet, la force d'adhérence croît généralement avec le temps de contact, et Barquins (1982) a mis en évidence une augmentation apparente de l'énergie d'adhésion w avec le temps de contact entre le verre et le polyuréthanne (figure 4). Entre deux élastomères ou deux polymères, cet effet est interprété par une interdiffusion de chaines par reptation (auto-adhésion), et la théorie et l'expérience donne une variation de l'énergie de rupture variant comme la racine carrée du temps. Un tel mécanisme est peu vraisemblable pour le contact verre-polyuréthanne, d'autant que l'énergie w semble croître comme la puissance 1/10 du temps. En fait, la décroissance du moteur G-w de la fissure avec le temps de contact ne résulte pas d'une augmentation de w, mais d'une diminution de G par relaxation viscoélastique des contraintes dans les rugosités confinées.

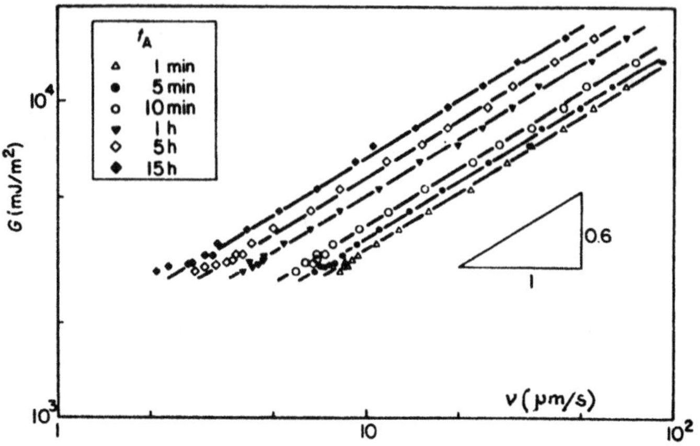

Fig. 4 - Relation entre taux de restitution de l'énergie et vitesse de fissure pour différents temps de contact d'une bille de verre sur le polyuréthanne.

L'énergie élastique emmagasinée à l'interface décroît donc avec le temps et l'adhérence augmente (Barquins 1982).

Dans la figure 3, la fonction $\phi(a_T v)$ varie en fonction de la puissance de $a_T v$ pour un large domaine de vitesses de fissure, un résultat souvent obtenu pour le pelage des matériaux caoutchoutiques (Gent et Petrich 1969), Kendall 1973, Roberts 1979). Toutefois il est bien connu que dans un solide viscoélastique les pertes décroissent à fréquences élevées, de sorte que le comportement redevient élastique. On doit donc s'attendre à ce que la fonction $\phi(a_T v)$ croisse d'abord puis diminue quand la vitesse de la fissure augmente. Cependant la vitesse des fissures est limitée à la vitesse des ondes de Rayleigh, et d'autres mécanismes de freinage, comme l'émission d'ondes acoustiques doivent intervenir à ces vitesses élevées. La courbe $\phi(a_T v)$ doit donc prendre une forme sigmoïde avec un maximum et un minimum. Dans ce modèle proposé par Maugis (1985), la branche entre les deux extrema a une pente négative ($dG/dv < 0$) et correspond à un processus instable (Fig. 5). Si une valeur constante de G est appliquée, la fissure

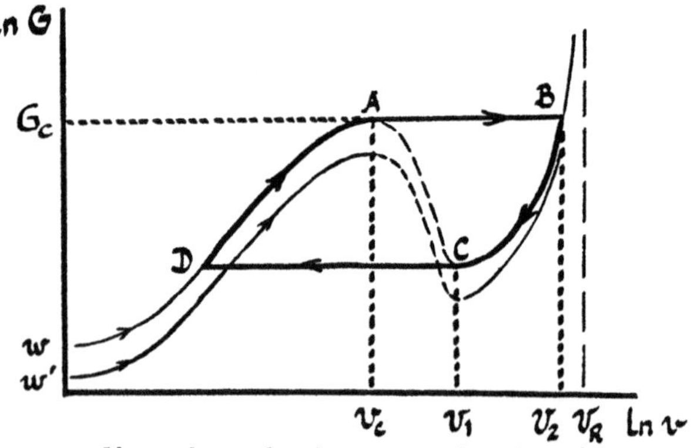

Fig. 5 Conséquence d'une branche à pente négative (AC) dans la courbe G(v) : stick-slip et fragilisation par adsorption.

prend une vitesse constante v(G) tant que $G < G_c$. Quand $G = G_c$ (point A) la vitesse saute brusquement de la valeur v_c à $v_2 >> v_c$ sur la seconde branche stable à plusieurs ordres de grandeur plus élevée de sorte que le joint apparaît fragile. Lorsqu'une vitesse moyenne v comprise entre v_c et v_1 est imposée, le modèle permet de comprendre qu'un cycle d'hystérésis ABCDA s'installe en donnant un mouvement de stick-slip, le freinage et l'accélération brutale provoquant de l'émission acoustique. Pour le pelage, le calcul conduit à des longueurs d'onde entre chaque phase de stick-slip qui augmentent avec la longueur de la bande pelée comme l'a observé Aubrey (1979). Khandani, Barquins et Maugis (1986) ont récemment montré que l'augmentation de la vitesse imposée de v_c à v_1 fait passer le régime de stick-slip d'un mode d'oscillations de relaxation à un mode d'oscillations harmoniques. Ce modèle permet aussi d'intéressantes comparaisons avec la fragilisation par adsorption ou par ségrégation intergranulaire (Maugis 1985). Comme les pertes viscoélastiques sont proportionnelles à w, une réduction de l'énergie de w à w' (Fig.5), par adsorption ou ségrégation, abaisse l'ensemble de la courbe G(v) et un chargement qui donnerait une propagation subcritique pour des valeurs élevées de w, peut donner une propagation catastrophique à des valeurs de w plus faibles.

4. Rôle des liaisons chimiques

L'examen de la littérature fait apparaître que la valeur G_o du taux de restitution de l'énergie en dessous de laquelle aucune propagation lente n'est possible, est généralement très supérieures aux valeurs thermodynamiques 2γ ou w pour des ruptures cohésives et adhésives respectivement. Ce ne sont que pour des fissures interfaciales et en absence de liaisons chimiques fortes que l'on peut obtenir des valeurs de G_o inférieures à $2 J/m^2$. En présence de liaisons interfaciales solides, avec agents de couplage, on atteint aisément des valeurs de $100 J/m^2$.

Abragon et Gent (1975) ont montré, qu'en augmentant la densité de liaisons covalentes entre du verre et un élastomère en faisant varier la proportion de vinylsilane dans l'éthylsilane, le taux de restitution de l'énergie est considérablement plus augmenté aux faibles vitesses qu'aux vitesses élevées et que les courbes G(v) ne sont pas parallèles (Fig.6), G_o étant proportionnel à la concentration de groupes vinyl à la surface du verre traité. D'autres expériences de Chang et Gent (1981) indiquent que G_o est aussi fonction du degré de réticulation, G_o étant d'autant plus élevé que l'élastomère est moins réticulé. Ces résultats ne peuvent pas être expliqués par la théorie de Lake et Thomas (1967) qui suppose que lorsque la fissure se propage, elle doit couper des liaisons C-C après avoir préalablement étiré toutes les liaisons de la chaîne de masse M_c comprise entre deux points de réticulation. Cette théorie qui conduit à la proportionnalité de G_o avec $\sqrt{M_c}$ (Carré et Schultz 1984) :

$$G = w\sqrt{M_c}\left[1 + \phi(a_T v)\right] \qquad (11)$$

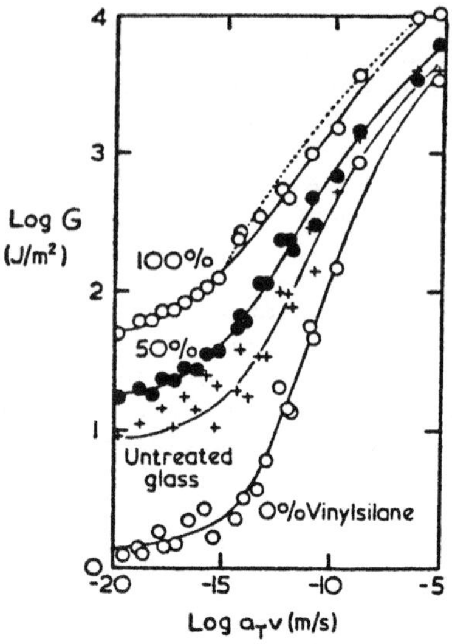

Fig. 6 - Cinétique d'adhérence du polybutadiène contre le verre pour différents traitements de surfaces.

entraîne une simple translation des courbes G(v) sans modification de la pente. En conséquence ni l'équation 9, ni l'équation 11 ne rendent compte des phénomènes observés. Pour varier la pente des courbes G(v) il faut probablement ajouter à l'équation 9 un terme qui tienne compte à la fois du nombre de liaisons primaires et de la longueur des chaînes au voisinage de l'interface. Une voie pour comprendre ces problèmes passe par l'étude de la cicatrisation des interfaces.

Jud, Kausch et Williams (1981) ont montré que la valeur du taux de restitution de l'énergie critique G_c pour refracturer brutalement une fissure partiellement cicatrisée croissait comme la racine carrée du temps de contact pour atteindre la valeur du polymère massif après une certaine longueur d'interdiffusion. Ce sont ces chaînes interdiffusées qu'il faut arracher au moment de la refracture. La même variation est donnée par la théorie de la reptation de De Gennes (1980). Si l'on admet que la dissipation irréversible qui accompagne la propagation de la fissure se fait aussi bien par pertes viscoélastiques en tête de fissure que par arrachage de chaînes entre lèvres de la fissure, on est amené à réécrire l'équation 9 sous la forme :

$$G - w = w\phi(a_T v) + NfL \qquad (12)$$

où N est le nombre de liaisons primaires fixant solidement les chaînes au substrat et permettant de les arracher du polymère, f la force pour extraire les chaînes de leur tube et L leur longueur. L'équation 12, dont le terme additif dans le second membre représente G_o-w, est en parfait accord avec les résultats de Ahagon et Gent (1975) et Chang et Gent (1981) sur la proportionnalité de G_o à la concentration des groupes vinyl et sur l'influence de la réticulation. Toutefois l'examen de la figure 6 montre que la force f doit probablement varier avec la vitesse de la fissure, éventuellement par un mécanisme de marche au hasard biaisé des chaînes. Si le modèle décrit par l'équation 12 était vérifié, il permettrait de réconcilier différentes théories de l'adhérence comme la théorie de l'interdiffusion, la théorie de la liaison chimique, la théorie des couches de faible cohésion et celle du facteur rhéologique.

5. Collage.

Il est bien connu que l'adhérence de deux surfaces sales et non moléculairement lisses est généralement négligeable, mais aussi que l'insertion entre elles d'un fluide visqueux conduit à une adhérence élevée comme dans le cas des cales Johanson. En effet, si l'on examine l'adhérence entre deux disques plans et parallèles de rayon R, séparés par un liquide de viscosité η et d'épaisseur h, le temps nécessaire pour observer la séparation est liée à la force appliquée par (Hoekstra et Fritzuis 1951) :

$$F = \frac{2\pi R^2 \gamma \cos\theta}{h} + \frac{3\pi\eta R^4}{4h^2}\frac{1}{t} \qquad (13)$$

où γ est l'énergie superficielle du liquide et θ l'angle de mouillage. Le premier terme est la force quasi statique d'adhérence (Loi de Laplace) et le second, le terme rhéologique, déduit de la loi de Stefan de l'écoulement visqueux, qui est prédominant quand $\eta/t > h\gamma\cos\theta/R^2$. C'est ce terme rhéologique qui explique le mécanisme d'action des adhésifs sensibles à la pression. Toutefois, le premier terme d'énergie superficielle est fondamental en ce sens que s'il n'y a pas mouillage, les surfaces ont tendance à se repousser dès que la pression est relachée.

On conçoit ainsi parfaitement qu'un bon adhésif doit mouiller et nettoyer le substrat, avoir une énergie d'adhésion élevée et une grande viscosité, et ne pas avoir de contrainte résiduelle. Pour mouiller le substrat, l'adhésif est appliqué sous forme liquide, puis sa viscosité est augmentée par refroidissement, évaporation, oxydation ou polymérisation. Le mouillage peut être amélioré en augmentant la rugosité, puisque les angles de mouillage θ' et θ d'une surface rugueuse ou lisse sont différents, selon la loi de Wenzel :

$$\cos\theta' = r\cos\theta \qquad (14)$$

où r est la rugosité. Cette rugosité favorise également l'accrochage mécanique entre l'adhésif et le substrat. Venables (1984) a montré, par exemple, que certaines dissolutions anodiques produisent à la surface des métaux des films d'oxyde dont la porosité et de la rugosité microscopique favorisent l'engrènement des polymères au substrat, en formant des liaisons beaucoup plus rigides que si les surfaces étaient lisses. La présence de pores et de crevasses impose que l'adhésif liquide que l'on applique doit avoir une faible viscosité et une énergie de surface élevée afin d'augmenter le flux capillaire dans les cavités et d'éviter la formation de poches de gaz. Cette énergie de surface γ élevée doit être compatible avec un faible angle de contact θ, c'est-à-dire que l'énergie d'adhésion de Dupré donnée par l'équation de Young-Dupré :

$$w = \gamma(1+\cos\theta) \qquad (15)$$

doit être élevée. Il faut donc maximiser cette énergie d'adhésion par un choix judicieux des traitements de surface des substrats et des groupes terminaux réactifs pour obtenir des liaisons chimiques solides. Le comportement du film adhésif, une fois durci, est généralement celui d'un solide viscoélastique, dans lequel les contraintes résiduelles très souvent développées lors de la solidification peuvent être évitées en choisissant des couples colle-substrat avec des coefficients de dilatation thermique très voisins.

Dans l'industrie, on souhaite pouvoir réaliser des collages avec des joints épais de façon à élargir les tolérances d'usinage et diminuer les pressions d'accostage, et par suite réduire les prix de revient. Malheureusement l'expérience montre qu'en général la résistance du joint est accrue par diminution de l'épaisseur. Il a été montré (Anderson, Bennett, Devrier 1977) que c'est le développement de contraintes d'arrachement aux extrémités de joints testés en cisaillement, croissant avec l'épaisseur, qui entraine une diminution corrélative de la résistance. De plus l'examen des faciès des zones rompues a clairement montré que la rupture devient adhésive quand l'épaisseur augmente. Dans le cas d'un joint mince, au contraire, la rupture est le plus souvent cohésive. Quand la couche d'adhésif est faible, l'énergie élastique qui peut y être emmagasinée est négligeable et la force d'adhérence quasistatique peut être calculée par la mécanique de la rupture, par exemple, par l'équation 8 pour la poutre double cantilever. Mais les pertes viscoélastiques en tête de fissure ne feront intervenir que l'adhésif, et l'énergie d'adhésion sera le double de l'énergie superficielle puisque la fissure se propage dans l'adhésif. Enfin, si le taux critique G_c de restitution de l'énergie mécanique correspond bien au pic des pertes viscoélastiques, on comprend aisément qu'un joint collé puisse avoir une résistance plus élevée que celle du substrat.

6. Conclusion

Dans l'adhérence des solides viscoélastiques, la mécanique de la rupture, les phénomènes de surface et les effets rhéologiques sont intimement liés. La force nécessaire pour séparer deux solides dépend non seulement de la nature des liaisons mais également de leurs propriétés élastiques, de leurs géométries et de la méthode de mesure. Si l'adhérence des solides élastiques est maintenant tout à fait claire, celle des matériaux viscoélastiques est en net progrès. En ce qui concerne le problème du collage, il apparaît du point des concepts très voisin de celui de l'adhérence de deux solides.

Références

Ahagon, A. et Gent, A.N. (1975). Effect of interfacial bonding on the strength on adhesion, J. Polymer Sc. : Polym. Phys. Ed., 13,1285

Anderson, G.P., Bennett, S.J. et De Vries, K.L.(1977) Analysis and testing of adhesive bonds Academic Press,Londres, pp198-250.

Andrews, E.H. et Kinloch, A.J. (1973) Mechanics of adhesive failure, Proc.R.Soc.Lond. A332, 385.
Aubrey, D.W. (1979) Viscoelastic basis of peel adhesion, dans Adhesion 3 (ed. K.W. Allen), Applied Science Publishers, Londres, pp 191-205.
Barquins, M. et Maugis, D. (1981) Tackiness of elastomers, J. Adhesion, 13,53.
Barquins, M. et Maugis, D. (1982) Adhesive contact of axisymmetric punches on an elastic half-space : the modified Hertz-Huber's stress tensor for contacting spheres, J.Méc.Théo.Appl., 1,331.
Barquins, M. (1982) Influence of dwell time on the adherence of elastomers. J. Adhesion 14,63.
Barquins, M. (1983) Influence of the stiffness of testing machines on the adherence of elastomers , J. Appl. Polymer Sci.28,2647.
Barquins, M. (1984) Kinetics of the spontaneous peeling of elastomers J. Appl. Polymer Sci. 29,3269.
Barquins, M. et Wehbi, D. (1986) Study of adherence of elastomers by cyclic unloading experiments, J. Adhesion, sous presse.
Carré, A. et Schultz, J. (1984) Polymer-Aluminium adhesion:II Role of the adhesive and cohesive properties of the polymer, J. Adhesion, 17,135.
Chang, R.J. et Gent, A.N. (1981) Effect of interfacial bonding of the strength of adhesion of elastomers, I. Self-Adhesion, J.Polymer Sc. : Polym.Phys. Ed., 19,1619.
Coussy, O.(1986) Un modèle de viscoélasticité confinée en mécanique de la rupture. Propagation subcritique, C.R. Acad.Sc.Paris 302,53
De Gennes, P.G. (1980) Sur la soudure des polymères amorphes C.R. Acad. Sc. Paris, 291,219.
Evans, A.G. (1972) A method for evaluating the time-dependent failure characteristics of brittle materials and its application to polycrystalline alumina, J. Mat. Sc., 7, 1137
Fichter, W.B. (1983) The stress intensity factor for the double cantilever beam, Int. J. Fracture, 22,133.
Fuller, K.N.G. et Tabor, D. (1975) The effect of surface roughness on the adhesion of elastic solids, Proc. R. Soc. Lond., A345,327
Gent, A.N. et Petrich, R.P. (1969) Adhesion of viscoelastic materials to rigid substrates, Proc. R. Soc. Lond., A 310,433.
Gent, A.N. et Schultz, J. (1972) Effect of wetting liquids on the strength of adhesion of viscoelastic materials, J.Adhesion, 3,281.
Hoekstra, J. et Fritzius, C.P. (1951) Rheology of adhesives, dans Adhesion and adhesives (ed. N.A. de Bruyne et R. Honwink), Elsevier Publishing Company, Londres, pp 33-90.
Johnson, K.L., Kendall, K. et Roberts, A.D. (1971) Surface energy and the contact of elastic Solids, Proc. R. Soc. Lond. A 324,301
Jud, K., Kausch, H H et Williams, J.G. (1981) Fracture mechanics studies of crack healing and welding of polymers, J. Mat. Sci., 16,204
Kendall, K. (1971) The adhesion and surface energy of elastic solids, J. Phys.D : Appl. Phys. 4, 1186
Kendall, K. (1973) Shrinkage and peel strength of adhesive joints, J. Phys. D : Appl. Phys. 6, 1782.

Khandani, B., Barquins, M. et Maugis, D. (1986) Propagation instable de fissure dans les solides viscoélastiques, C.R. Acad. Sc. Paris à paraître.
Lake, G.J. et Thomas, A.G. The strength of highly elastic materials, Proc. R. Soc. Lond., A300,108.
Maugis, D. et Barquins, M. (1978) Fracture mechanics and the adherence of viscoelastic bodies, J. Phys. D : Appl. Phys., 11,1989
Maugis, D. et Barquins, M. (1980) Fracture mechanics and adherence of viscoelastic solids, dans Adhesion and adsorption of polymers (ed. L.H. Lee). Plenum Publishing Corporation, New York, pp 203-277.
Maugis, D. et Barquins, M. (1983) Adhesive contact of sectionally smooth-ended punches on elastic half-spaces : theory and experiment, J. Phys. D : Appl. Phys., 16,1843.
Maugis, D. (1985) Subcritical crack growth, surface energy, fracture toughness, stick-slip and embrittlement, J. Mat. Sci., 20,3041
Ripling, E.J., Mostovoy, S. et Patrick, R.L.(1964) Measuring fracture toughness of adhesive joints, Materials Research and Standards, 3,129.
Roberts, A.D. (1979) Looking at rubber adhesion, Rubber Chem. Techn., 32, 23
Venables, J.D. (1984) Adhesion and durability of metal-polymer bonds, J. Mat. Sci., 19,2431.
Wang, S.S., Mandell, J.F. et Mc Garry, F.J. (1978) An Analysis of the crack tip stress field in DCB adhesive fracture specimens, Int. J. Fracture, 14,39.
Williams, M.L., Landel, R.F. et Ferry, J.D. (1955) The temperature dependence of relaxation mechanismes in amorphons polymers and other glass-forming liquids, J. Amer.Chem.Soc., 77,3701.

ADHESION FORCES OF POLYMER MODIFIED CONCRETE AND PLAIN CONCRETE TO STEEL IN MOULDS AND IN REINFORCED CONCRETES

J. MŁODECKI Division of Concrete, Building Research Institute, Warsaw, Poland

Abstract
The paper discusses types and conditions of adhesion forces of polymer modified concrete and plain concrete to steel occurring in moulds and in reinforced concretes. Particular attention was paid to hydroxyls -OH, present in the concrete and evoking forces of specific adhesion. The author reports the test results on adhesion and abhesion to steel of the cement and vinyl-maleic modified mortar.
Key words: Polymer modified concrete, Adhesion to steel, Admixtures to concrete, Mortars.

1. Introduction

A knowledge of adhesion forces of concrete to steel is very important when establishing the conditions in which high concrete adherence to steel has to be obtained as well as when determining ways of adhesion prevention in steel moulds.

The relationship between mechanical properties of solids and molecular forces were recognized especially in the first half of the present century /1/. The purpose of the paper is to refer these relations to the bond force of cement concrete and polymer-modified concrete to steel. Thus, it is necessary to define and discuss the adhesion forces of concrete to steel on the basis of tests and analysis of structure and concrete, polymer and steel properties and also phenomena connected with their adherence.

2. Adherence of polymer modified concrete to steel

Adherence of concrete to steel in moulds and in reinforced concrete is a case of adherence between solids, one of them was formed as a result of setting and hardening of liquid or semiliquid substance / concrete mix /, poured on the surface of a second substance being in solid state / steel /. The following factors influence forming and degree of adhesion of concrete to steel: properties of steel and its surface, properties of concrete mix and concrete, especially wettability of steel by concrete mix, adsorption, adhesion, accuracy of adhering of concrete mix and concrete to steel. All these phenomena should be taken into account when looking into the question of the concrete -

steel bond.

There are some kinds of forces occurring in the adhesion phenomenon of concrete to steel. These are first of all: specific adhesion, mechanical adhesion and capillary forces. Specific adhesion is caused by intermolecular, varied interactions that are of electromagnetic nature. Intermolecular binding forces depend on the structure and properties of the molecules adhering to each other, on adherence and external conditions.

Adhesion of polymer modified concrete to steel depends on many factors and particularly on a polymer type and on a quantitative proportion of a polymer to cement. It also depends on preparation technology of a cement-polymer mix, adhesion performance and on external conditions, particularly humidity. In particular, adhesion requires high particle approximation of the materials to be bonded. Two binders are present in the polymer-cement concrete. A cement binder is influenced by a liquid polymer which is next solidified and becomes a solid body. However, polymer molecules separate the cement molecules and a complicated system is formed. Of particular significance will be the structure and properties of each polymer with special attention towards functional groups and contents of emulsifiers in a dispersion. Dispersion forces / London's / come up to high values in big-molecule compounds and are increased when molecular weight is increased. Their share in the bond of polymer modified concrete-steel will be considerable.

3. Specific features of concrete and steel as adhesives and substrates.

When analyzing the structure and properties of concrete mix and concrete it can be stated that among the products of cement hydration there are many hydrated minerals having assymetric structure. What is more, many of them contain hydroxyls -OH such as e.g.: portlandite $Ca/OH/_2$, hilebrandite Ca_2 $[SiO_3$ /OH/] OH, which transforms into the compound called afvilite Ca_3 $[Si_2O_6OH]$ OH when $Ca/OH/_2$ is further being released. Afivilite is an acid-basic calcium pyrosilicate.

In the process of further transition of silicates under the influence of water and oxygen, aliphatic silicates appear in a hardened cement paste, i. e. monoclinic kronotite $Ca_6Si_6O_{17}$ $/OH/_2$ with fibrous cross-section, fibrous tobermorite Ca_4 $[Si_3O_8$ /OH/] OH_2 and also hydrated ribbon silicate - girolite $Ca_4Si_6O_{15}$ $/OH/_2$. Some amount of unreacted cement is always left in the concrete. In general, there are about 20 minerals and chemical compounds in the concrete cement paste. The vast majority of molecules of these compounds have asymmetric structure.

Iron atoms in steel form a regular side-centred space lattice in which metallic bonds occur. On an electron configuration on the orbits of the iron atom K_2 L_8 M_{14} N_2 it is seen that a lone pair of electrons on most external 4th orbit is not strongly attracted by the atom nucleus Fe /3/ in view of a large distance. Iron is an electronegative element / its electronegativity is 1,8 /. A large amount of micropores and microfissures appear on the steel surface. Air filled micropores are the cause of inaccurate wettability of the steel surface with concrete mix or a release agent; therefore the adsorptivity surface is diminished.

4. Types of forces occuring in concrete adhesion to steel
   Intermolecular van der Waals' interactions

All asymmetric molecules form stable dipoles. Interaction dipole-dipole will first of all be occuring in the concrete-steel bond. In view of rare occurrence of non-polar molecules / with symmetric structure / in the concrete, it is estimated that percentage of intermolecular interactions, of dipole-dipole induced type in the concrete-steel bond, will be small.

Capillary forces

Capillary forces are the result of intermolecular interactions. They occur, among others, between two parallel smooth surfaces being very close to each other when water or another liquid is present.

Hydrogen bond

Presence of a large number of hydroxyls -OH in concrete creates the conditions of a hydrogen bond when the concrete contacts the steel. The bond exists between oxygen atoms of hydroxyls and iron atoms in the steel. A necessary condition for the hydrogen bond is considerable electronegativity of atoms of adherent bodies and presence of a lone pair of electrons in one of these atoms. A lone pair of electrons on the most outer 4th orbit of the iron atom is not strongly attracted by the Fe atom nucleus in view of a large distance. As a result of high electronegativity of the oxygen atom, the -OH bond is polarized and a proton around which electron charge density is small, is strongly attracted by a lone pair of electrons of the Fe atom. The hydrogen bond takes place mainly as a result of electrostatic interactions of a proton with electrons of a lone pair of electrons. There are created the conditions for occurrence of the hydrogen bond between oxygen of hydroxyls -OH and Fe atom in accordance with the scheme Fe...H-O

Mechanical adhesion - in the case discussed, it results from the roughness of the concrete interlocking with the roughness of the steel surface.

5. Interaction of bonding forces in adhesion phenomenon of concrete to steel

A given type of bonding force occurs to a small or larger extent and depends on a given case of a concrete bond with steel. Anyway, the forces of specific adhesion are of great importance in the adherence phenomenon.

In steel non-absorptive moulds or reinforcements the mechanical adhesion forces, through cogs, are small. However capillary forces occur because water from concrete capillaries is not absorbed by an undercoat and is present between the concrete and mould surface.

After all these discussions we come to the conclusion that these are, before all, the hydroxyls -OH which generate specific adhesion of concrete to steel. The hydroxyls are numerous in the concrete. However, so far it has not been determined if the hydroxyls act when hydrogen bonds with the Fe atoms in steel or when the intermolecular interaction dipole-dipole takes place. Adhesion of concrete to steel is high both when steel surface is smooth and rough. On the basis of the above mentioned discussion the author is of the opinion that the hydrogen bond is of major importance in the specific adhesion phenomenon of concrete to steel. However, van der Waals' forces have their share too, especially dipole-dipole type as well as capillary forces. Each given case has its own range of forces.

6 Adhesion conditions and protection of steel moulds against concrete adherence

Looking into the adhesion problem of plain and polymer modified concrete to steel it is advisable to pay attention to the adhesion / antiadhesion / phenomenon occurring between these materials. This phenomenon is very important when steel moulds are to be protected against concrete adherence.

Adhesion depends on a number of conditions. Surface tension of a liquid / water in a concrete mix/ should be higher than the value of a corresponding surface layer of a solid body /steel/. As a result, a wetting angle will be large. Hydrophobicity will be the result of the combination of these factors.

In order to attain adhesion, the condition $\gamma_s < \gamma_c$ must be satisfied where: γ_s - surface energy of a solid body, γ_c - surface tension of a liquid. In case of steel and concrete it is quite the contrary. That is why it is necessary to put a layer of hydrophobic substance, having low surface tension, on a steel surface, having high surface energy. A release agent having low surface tension is advantageous because of two reasons: 1/ has good wetting properties thanks to which exact adhesion to steel surface is attained, 2/ the adhesion condition is satisfied: surface tension of an outer layer on a solid body surface / layer of a release agent/ is lower than surface tension of liquid: $\gamma_s < \gamma_c$.

In order to achieve weakest adhesion work W_a between a liquid and solid body the following condition must be satisfied: γ_s and γ_c must be low and interfacial tension γ_{sc} should be high /1/.

$W_a = \gamma_{cg} + \gamma_{sg} - \gamma_{cs}$ where:

γ_{cg} - surface tension of liquid with gas border

γ_{sg} - surface energy of solid body with gas border

γ_{cs} - interfacial tension with liquid and solid body border

Moreover, it can be ascribed that chemically lowered concrete adherence to the walls of a steel mould is achieved by bonding /saturation/ of active hydroxyls, present in a concrete mix, with the aid of active carboxyls, occuring in a release agent. Neutral salts are produced but they have no adhesive properties, /3/.

7. Author's tests

7.1 Adherence of cement and polymer modified mortars to steel

7.1.1 Test procedure

Adherence of mortars to steel was determined by the pull-out test of the rods embeded in concrete. The fig.1 shows the device according to T.M. Kelly and D.E. Bryant. The rods diameter was 9 mm and they were embeded at 100 mm depth. The rods were pulled out after 1st and 7th day of hardening. The portland cement was used - 35 Chełm, Vistula sand - graining up to 2 mm and dispersion of vinyl-maleic polymer DPM - 50 from industrial production. The dispersion contained ca. 50 % of water, its viscosity was 2000 - 4000 mPas, reaction: pH = 3,5 - 5. The chemical composition of the dispersion was adjusted to cement mixes. It does not coagulate in their medium and is alkaliresistant.

The control cement mortar of 1:3 was made /1 part of cement and 3 parts of sand by weight /. Its water-cement ratio $\frac{w}{c}$ = 0,55. The dispersion DPM-50 in quantities of 2, 4, 6 and 10 % of the polymer in relation to the cement mass was added to the mortar with the compositon mentioned above. The dispersion contained ca 50 % of polymer. In order to diminish air contents in polymer modified mortars

Fig. 1 Apparatus for testing concrete adherence to steel by means of the pull-out method of steel rods.

the sand and cement mix were wetted initially with a large part of mixing water /4,5/. The polymer dispersion was being added together with the remaining small quantity of the mixing water used in a quantity needed for gaining the established constant mix consistency. Simultaneously, cube samples were being made with the edge of 10 cm and used with the purpose of determining their compressive strength. The mortars were made in a drum mixer. The control mortar has been cured in humid conditions and polymer-modified mortars in air dry conditions. The test results are shown in Table 1.

7.1.2 Analysis of test results

Three vinyl-maleic polymer modified mortars are characterized by higher adhesion to steel than adhesion of the control cement mortar without an admixture both after 1st and 7th day of performance. Together with an increase of the polymer amount up to 6 %, adherence of polymer modified mortar is higher in both terms. The Z-5 mortar containing 10 % of polymer has achieved after 1 day lower, and after 7 days higher adherence to steel than the control cement mortar. With increasing of a polymer quantity in the polymer modified mortar, the rate of hardening is slowlier. An adherence growth is not big but at the same time an amount of air in the polymer modified mortar is increased.

Table 1. Results of tests on cement vinyl-maleic mortar and control mortar

Mortar type	quantity of polymer % of cement mass	water cement index $\frac{w}{c}$	diminished amount of mixing water %	consistency drop of immersed cone cm	air quantity in fresh mortar %	Boundary stress at pull-out of rods 1st day MPa	7 days MPa	Compressive strength 1st day MPa	7 days MPa
Z-1	0	0,55	-	6,4	7,0	0,68	1,68	3,9	13,4
Z-2	2	0,45	18,2	6,0	11,7	0,85	1,75	4,5	15,2
Z-3	4	0,44	20	6,5	12,0	0,90	1,78	4,7	14,2
Z-4	6	0,43	21	5,0	12,6	0,93	1,84	5,8	15,1
Z-5	10	0,415	21,9	6,3	15,7	0,52	1,97	2,1	13,2

Better adherence of mortars to steel might be expected if the mortars are mixed in a whirling mixer /5/, and/or using anti-foaming agents. Moreover, the dispersion reaction of polymer should not be acid /§§ 3 and 6/. After 28 days of curing excess adhesion forces appeared and it was impossible to determine them on the device used.

The cement vinyl-maleic mortars have in general shown a slightly higher compressive strength than the control mortar without an admixture.

7.2 Concrete adhesion to steel

On the basis of the tests it was found that particularly effective protection of surface of steel moulds against concrete adhesion can be obtained when a mineral oil is used /e. g. machine oil 10/ with an addition of a small amount of weak organic acid e. g. oleic acid and surface active agent /3/.

Oleic acid reacts with hydroxyls of calcium hydroxide and cement minerals /§ 2.2/ with creation of neutral products which do not show adhesive action. This reaction contributes to facilitation of concrete separation from steel. The surface active agent diminishes surface tension of a mineral oil /§ 4/.

7.3 Conclusions

1. By adding a dispersion of vinyl-maleic polymer to a cement mortar we may, in certain determined conditions, achieve better adherence of the mortar to steel.
2. The cement-polymer mortars made were characterized by excessive air quantity. Lowering of air quantity in these mortars would cause further increase of their adherence to steel.
3. By diminishing surface tension of a mineral oil and by chemical bonding of hydroxyls -OH, being present in the concrete, effective protection of steel moulds against concrete adherence was obtained.
4. Knowledge of bonding forces of concrete to steel makes it possible to determine the conditions of obtaining both high adhesion to steel and effective adhesion from steel of the cement concrete and polymer modified concrete, what was confirmed in the experiments performed.

References

Houwink, R. Solomon G., Adhesion and adhesives Vol. 1, Adhesives 1965, Vol.2 Applications 1967, Elsevier Publishing Company, Amsterdam, London, New York.

Grzymek J., Teoria wiazania i twardnienia spoiw gipsowych i anhydrytowych według opracowań polskich W: Zagadnienia chemii budowlanej i materiałów budowlanych, Zeszyt 2, PMN, Warszawa 1964

Młodecki J., Nowe środki antyadhezyjne do form i deskowań stosowanych w prefabrykacji i budownictwie betonowym, Wyd.ITB, Warszawa 1985

Młodecki J., Chemical and technological methods of decrease of air content in the polymer-modified concrete mix, referat na IV Miedzynarodowy Kongres "Polymers in concrete" Darmstadt, 19-21 września 1984, s. 289-292

Młodecki, J., Zastosowanie dyspersji polioctanu winylu Vinacet do wytwarzania polimerobetonów. Przeglad Budowlany 1981, nr. 5, s. 255-260 .

Section 2

Surface properties of substrate
Propriétés des surfaces subjectils

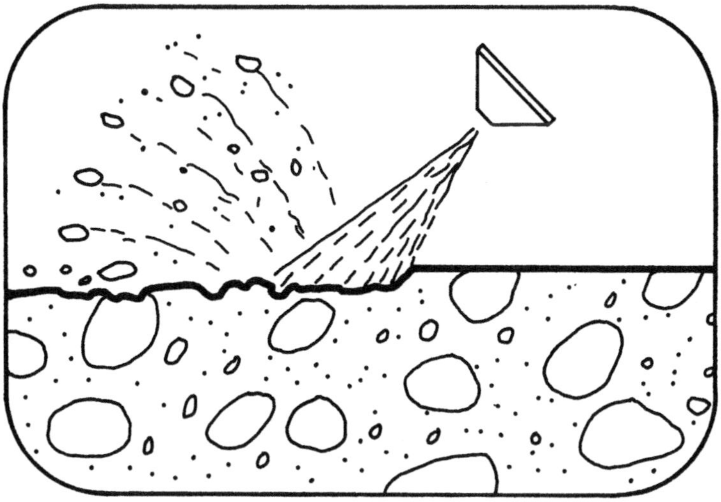

THE ADHESION BETWEEN CONCRETE AND FORMWORK

A. MAZKEWITSCH and A. JAWORSKI
Institute of Civil Engineering, Gorki, UdSSR
Baustoffinstitut, Technical University Munich

Abstract
The bonding of concrete to formwork influences the quality of the concrete surface. The theory of bonding is investigated by scientists and is described. The bond strength of the hardened concrete to the surface of the formwork depends on many parameters, from which the most important are the material of the formwork and the parting agents, the roughness of the surface, the type and quality of the concrete, the way and time of compaction and the curing conditions. The bond strength was evaluated - besides many other parameters - for customary formwork materials and new plastic materials.
Key words: adhesion, bond strength, concrete, formwork, plastics.

The adhesion and the friction between concrete and formwork can lead to a decrease in the quality of the concrete. To prevent this it is important to know the physical-chemical processes and interactions, which are occuring within the contact area.

In similarity with the classical glued connections the concrete and formwork can be thought of as a "contact couple", where the concrete acts as the glue. In this case the bond strength can be seen as the result of mechanical bond, pure adhesion, cohesion and of the contraction of the concrete. The first three factors increase the bond strength and the contraction decreases it.

The amount of the contraction depends on many technological factors (e.g. consistency of the fresh concrete, the conditions of the hardening etc.) as well as on the physical-mineralogical properties of the cement.

The mechanical anchorage of the concrete is connected with the roughness and the porosity of the formwork. It is also related to the properties of the material of the formwork (hydrophilic or hydrophobic) /1/.

It is our opinion that, when estimating the effect of a material, not only its roughness, but also the size and form of protrusions and depressions must be taken into account. In the case of extended but gentle unevenness an increase of the bond strength only comes with the increase of the actual contacting area. If the protrusions and depressions are like hooks an anchorage occurs simultaneously. The technological factors of the concrete (e.g. consistency, method of compaction etc.) have a considerable influence on the mechanical anchorage, the adhesion and the bond strength (see Fig. 1).

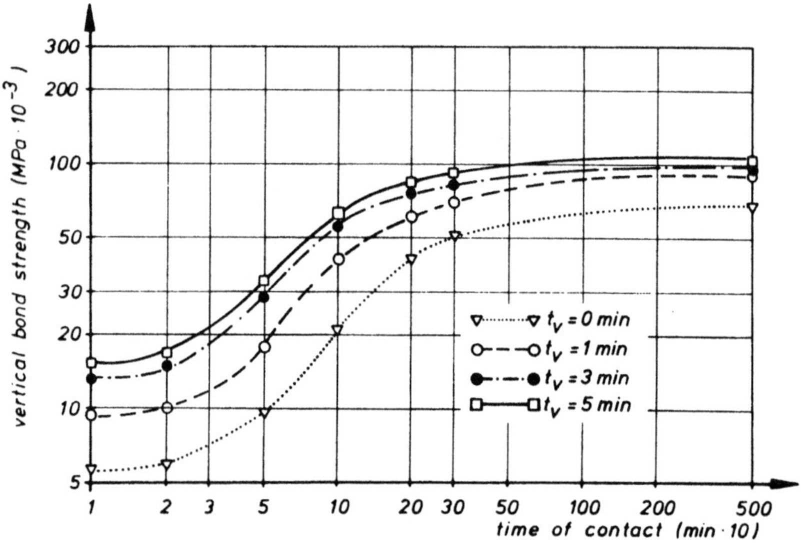

Fig. 1. Influence of the time of vibration on the bond strength between cement mortar and steel.

During vibration the viscosity of the fresh concrete is reduced in the contact-zone. Simultaneously the pressure on the formwork increases. This leads to a better closing of all unevennesses of the formwork and increases the bond strength.

For the contact couple "concrete-formwork" the mechanical anchorage is in practice only realised in connection with other factors. The anchorage is based on the physical-chemical interaction of the microstructure of the cement paste with the of the surface of the formwork. And so the mechanical component of the bond strength can be traced back to the adhesion within the microscopic points of contact.

The amount of the bond strength between concrete and formwork depends also to a great extent on the cohesion of the cement-mortar, which itself is determined by the strength capacity of the cement, its mineralogical components and by the conditions during the hardening of the concrete (see Fig. 2).

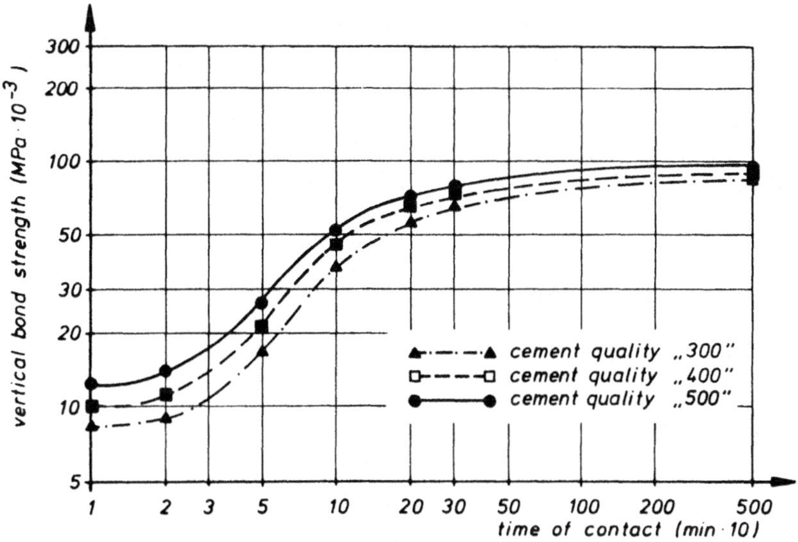

Fig. 2. Influence of the cement quality on the bond strength between cement mortar and steel; time of vibration: 1 min.

The adhesion and cohesion are two interconnected parts of the process of the forming of the contact-zone. It can be assumed that the most important component of the bond strength is the adhesion. The fundamental of the adsorption-adhesion are given in the papers by Mark-Loren, Dbroin, Stawermann et al /2/.

According to the diffusion-theory of adhesion by S.S. Wojuzky /3/ the forming of the contact-zone between two different materials is the result of the reciprocal intrusion of molecules of one material into the surface of the other material. The cause therefor is seen in the Brown's movement. The occuring of the full size diffusion-component is not very likely for the contact-couple "concrete-formwork".

B.W. Derjagin and his coworkers worked on the electrical theory of the adhesion /5, 6/. This theory explains the influence of contacting bodies on each others as the result of the forming of electrical double-layers on the surfaces. Under practical conditions such a double-layer can have a regular or mosaic-like structure (Fig. 3a and b). From the statements of B.W. Derjagin it can be derived that the possibility of the forming of the doble-layer increases by intensive friction between the contacting surfaces. Within the contacting zone between concrete and metallic formwork the occuring of a double-layer is very

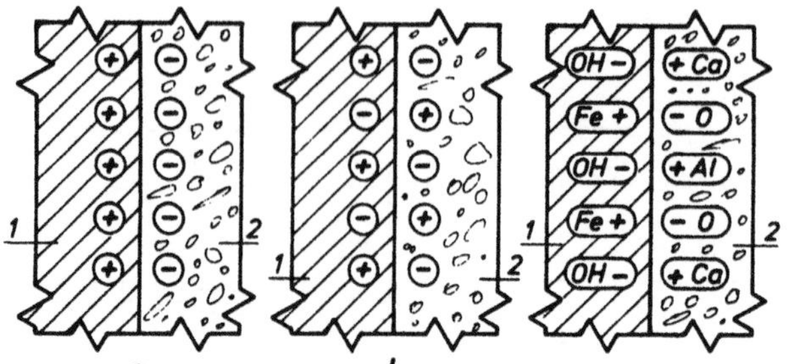

Fig. 3. The scheme of the forming of the electrical double-layer
a - for classical glue with regular structure
b - similar to a mosaic structure
c - for contact between concrete and steel.

likely, as there is a dielectricum due to the intense friction) and a metal. In the contact-zone concrete has OH^- groups and ions as Ca^{2+}, Al^{3+} Si^{4+}, O^{2-} and others. The steel surface - without corrosion products - contains also OH^- groups, free electrons (electron-gas) and ions of Fe^{2+}. Theoretically the formation of a double mosaic-layer is possible (see Fig. 3c).

According to the opinion of B.W. Derjagin /5/ any reciprocal influence at the level of molecules (electrical, adsorptive etc.) has to be treated like a chemical connection. Therefore the forming of new products is possible as to result of the chemical reaction of the concrete on the material of the formwork. The experiments of D.M. Tschernenko /6/ proove this for the contact between cement-paste and metal.

For the contact-couple "concrete-formwork" all conditions for the occuring of adsorptional, electrical, diffusional and chemical processes are practically given. The difference within the technical conditions during the time of contact does not allow direct or single answer to the basis of the adhesion.

Tests were conducted using the systems
- cement paste - cement-stone
- fresh and hardened cement mortar
- fresh and hardened concrete
- fresh and hardened lightweight-concrete

to examine the influence of more than 20 factors on the bond to formwork. The results are partly contained in Figures 1 and 2 and in Table 1.

Based on the obtained results the predominant factors for a reduction of the bond-strength between concrete and formwork can be given as follows:
- the application of effective parting-agents to cover the surface of the formwork

- the use of concrete which does not require vibration e.g. by adding super-plastiziser to the concrete.

Table 1. The vertical bond strength between formwork and concrete (age 1 and 3 days).

Material of Formwork	Slump (cm)	Conditions of Hardening: Temperature (°C)	Conditions of Hardening: Relative Humidity (%)	Vertical Bond Strength $Mpa \cdot 10^{-3}$: Age 1 day	Vertical Bond Strength $Mpa \cdot 10^{-3}$: Age 3 days
Steel without Parting Agent	4-6	16-18	80-90	185	489
Steel with Parting Agent	4-6	16-18	80-90	47	142
Glass-fibre Reinforced Plastic	4-6	16-18	80-90	21	26
"Textolit"	4-6	16-18	80-90	9	10
Multilayered Compressed Paper	4-6	16-18	80-90	18	18
Fluorplastic	4-6	16-18	80-90	9	10
Veneer with a Layer of "ZNIIOMTF"	4-6	16-18	80-90	31	43

References

/1/ Simon, A.D.: The Adhesion of a fluid and the moistening. Moskow, Chimija, 1974

/2/ Debroin, N.; Guwink, R.: The adhesion of resin cement and solder, Moskow, Nauka, 1963

/3/ Wojuzkij S.S.: The diffusions-theory of the adhesion. Resins and technology of glueing. Moskow, Nauka, 1960.

/4/ Derjagin, B.W.; Krotowa, N.A.: Adhesion. Moskow, Leningrad, Nauka, 1949

/5/ Derjagin, B.W.; Krotowa, N.A.; Smilga, W.P.: The adhesion of solid bodies. Moskow, Nauka, 1974.

/6/ Tschernenko, D.M.: The investigation of some questions on the bond between concrete and metal. Leningrad, LPJ; M.J. Kalinin, 1957.

Section 3

Properties of bonding materials

Propriétés d'adhérence des matériaux

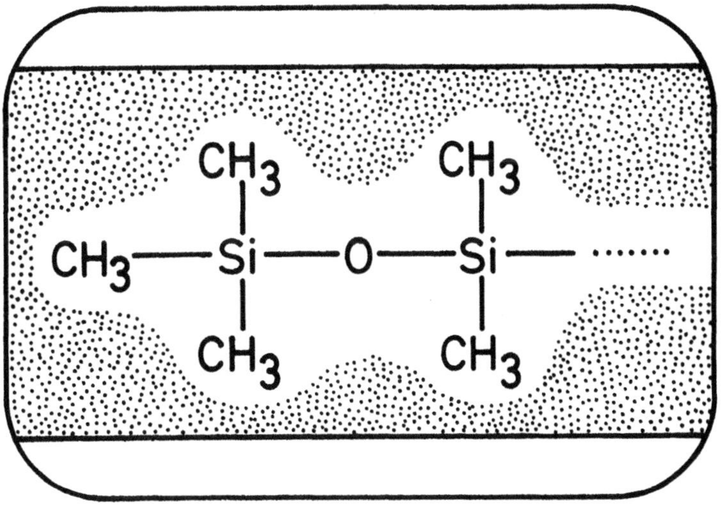

EFFECT OF TEMPERATURE ON THE CREEP BEHAVIOUR OF POLYMER MORTARS

PROF.DR. R. SUBRAHMONIA AYYAR
Department of Civil Engg., Indian Institute of Technology,Bombay, India.

Mr. SURESH N. JOSHI
Structural Engineering Department,
Victoria Jubilee Technical Institute, Bombay, India.

Abstract
Polymer concrete composites which are fast replacing conventional materials of construction possess several versatile properties like high strength, early setting time and high resistance to corrosive and erosive actions. However, polymers are temperature sensitive materials and consequently their mechanical properties are largely affected by variations in temperature. At the same time their creep strains also increase several times with the increase in temperature. Therefore, detailed investigations have to be conducted to study the quantum and pattern of variation of not only the important mechanical properties but also of creep of these materials in terms of temperature variation. The data obtained from these investigations will be of immense use for defining the permissible range of temperature to which the materials can be exposed under the operational conditions of the structure in which they are used. It will also help to theoratically predict the extent of creep likely to develop in the material at a given temperature in terms of the corresponding creep at room temperature.
Key words :Creep, Epoxy, Polyester, Polymer, Stress-strength Ratio, Temperature, Ultimate Compressive Strength.

1. Introduction

Polymer concrete composites which encompass the entire spectrum of polymer modified concrete, polymer impregnated concrete and polymer concrete are fast replacing conventional concrete as construction materials particularly in structutres which are subjected to severe deleterious effect of corrosive and erosive forces. This is mainly due to the several advantages these composites possess such as high strength, excellent resistance to chemical action and abrasion, reduced porosity etc.

However, despite these advantages, the mechanical properties of polymer composites are highly susceptible to temperature variations. Besides temperature variation, other parameters which affect these mechanical properties are the nature and amount of polymer content and also the mode and degree of hardening. These polymer composites also undergo at high temperatures considerable degree of creep which is a complex function of the ambient temperature, stress level and the duration of loading.

A study of this creep behaviour under elevated temperatures and long duration of loading is therefore of very great significance from the practical point of view. This paper, however, concentrates only on the study of the effect of temperature on the important mechanical properties of polymer composites prepared from the two polymers which are extensively available in India on a commercial level viz. Epoxy and Polyester resins. The results of the investigations with special reference to variations in ultimate compressive strength and the creep strains at different temperatures are highlighted and suitable conclusions drawn therefrom.

2. Choice of materials

Polymer mortar, rather than polymer concrete, was preferred for the investigations particularly in view of the better homogeneity of the mix and the consequent improvement in the reliability of the test results. The choice of the polymer binders was also restricted to those which are manufactured and marketed under strict conditions of quality control and consequently possess well established material properties. Indigenous polymers commercially available in the Indian market were preferred with a view to make the results of the investigations directly relevant and applicable to the Indian construction industry. As a result of extensive search, two polymers were finally chosen viz. Epoxy resin (manufactured by Dr. Beck and Co.) and Polyester resin (manufactured by Bakelite - Hylam). Silica (quartz) flour was chosen as the filler material because it is not only chemically inert but has also got excellent binding properties with the resin. Trial tests were conducted on polymer mortar specimens with several grading patterns with a view to select that grading which will give maximum ultimate compressive strength at room temperature (of 30°C).

3. Experimental Investigation

This phase of the investigation consisted of the following sequence of operations :

i) Selection of the mix proportions and the mode of curing for getting the maximum ultimate compressive strength at room temperature (of 30°C).

ii) Choice of the specimens for the tests.

iii) Casting and curing of the specimens.

iv) Test for determination of the ultimate compressive strength at different temperatures.

v) Fixing of the stress-strength ratio for the creep test.

vi) Creep test involving measurement of creep strains under sustained compressive load (corresponding to a preselected stress-strength ratio) and controlled temperature conditions for a sufficiently long period by which time the secondary creep gets practically stabilised.

Pilot tests conducted on the cured specimens of 4cm x 4cm x 4cm size prepared with several binder-filler ratios, indicated that the optimum proportion for maximum ultimate compressive strength in case of both the resins was 1:3.8. Hence the same binder-filler ratio was chosen for the subsequent phases of detailed investigation.

A prism of 4cm x 4cm x 16cm as specified in DIN-1164 was chosen for the creep test specimen. All the specimens required for one test series were cast from the same mix at one time to give uniformity in casting and homogeneity in composition. Consistent with the test requirements, the specimen sizes were also kept as small as possible with a view to ensure overall economy in the cost of experimentation. The casting of the specimens was done at room temperature. The mortar was compacted in the metal moulds through hand compaction by giving 25 evenly distributed blows with a standard tamping rod. The mortar was allowed to set in the moulds for 24 hours at room temperature.After that the specimens were cured as follows :

a) Epoxy mortar specimens were cured for 24 hours in an oven at a constant temperature of 80°C.

b) Polyester mortar specimens were cured for 4 hours in an oven at a constant temperature of 100°C.

These two distinct modes of curing for the two polymers were chosen based on the directions given by the manufacturers and confirmed through trial tests under various curing conditions for getting maximum ultimate compressive strength of the test specimens at room temperature.

Since the creep strain observations are to be taken for different temperature conditions under a preselected stress-strength ratio, it becomes imperative to determine the ultimate compressive strength for each of the temperature conditions under which the creep studies are to be done. Consequently the ultimate compressive strengths were determined over a temperature range of 30°C to 90°C at intervals of 10°C. In order to ensure that isothermal conditions were obtained throughout the body of the specimen before the compressive load was applied, the specimens were exposed to the test temperature for a period of 24 hours preceding the commencement of the test. The values of the ultimate compressive strengths for the two polymer motars determined in the temperature range of 30°C to 90°C were used in deciding the level of stress to be applied to the creep test specimens at a predetermined stress-strength ratio.

The creep tests on the specimens necessitated prior in depth consideration of several important factors like the loading set-up to be used, the type of heating system for the specimen, the instrumentation both for the control and measurement of the temperature of the specimen as well as for the measurement of creep strains. The loading set-up which was designed and fabricated by the authors consisted of a very rigid cross-head at the top connected to an anchor slab below by means of 4 vertical mild steel rods of 5cm diameter each. The cross-head was fabricated from rolled steel I-sections of size 300 x 150 adequately stiffened by mild steel plates. The specimen was kept vertically between the cross-head and the anchor slab and the load applied axially to it through a screw jack fitted below it. A high capacity proving ring inserted between the specimen and the screw jack, continuously monitored the load applied.

The oven for heating the specimen was fabricated out of good quality teak wood planks 3cm in thickness. The size of the oven (36cm x 9cm x 25cm) was kept as small as possible with a view to have uniform temperature inside the oven. The heating was done by means of ordinary electric bulbs of filament type. The voltage of these bulbs was calibrated against the required temperature. The wattage and the arrangement of the bulbs were adjusted to ensure uniformity of temperature over the entire body of the specimen (to an accuracy of $\pm$ 1°C as recorded by thermocouples located at top, middle and bottom of the specimen). Sliding glass doors were provided on the front and back of the oven to facilitate measurement of creep strains. The oven was of a revolving type which makes it possible to measure the creep strains on all the four faces of the specimen. The temperature of the specimen was measured by copper-constantan thermocouples. Two such thermocouples were provided on mutually perpendicular adjoining faces at the middle of the gauge length. One thermocouple was connected to an electronic temperature indicator-cum-controller which was having an accuracy of $\pm$ 2°C while the other was connected to a differential potentiometer which indicates the temperature in terms of millivolts. The copper-constantan thermocouples were calibrated in terms of millivolts for the required temperature range taking 0°C as the datum.

The creep strains were measured by a demountable extensometer of gauge length 5cm. This extensometer can record an axial strain to an accuracy of 1.98×10^{-5}. Brass studs with accurately drilled axial holes were glued on the four faces of the specimen at a chosen gauge distance of 5cm with M-seal compound, a thermosetting epoxy-based material which has a good bond with the specimen even at 100°C. The creep strains were measured on all the four faces of the specimen and the average of all these 4 sets of readings was taken for the subsequent analysis of the results. The creep strain measuring set-up and the instrumentation are shown in figure No. 5 and figure No.6.

From the available literature, it was noted that the creep behaviour of the material significantly depends upon three major parameters viz. the stress level to which the specimen is subjected in terms of the ultimate strength (stress-strength ratio), the duration of loading and the temperature of the specimen. Preliminary tests were conducted on the two polymer mortar specimens to find out for each temperature the maximum stress-strength ratio at which the specimen collapses through extensive creep strains, within a short period of about 6 hours. The results of these tests gave a rough thumb rule as follows:

The product of the temperature in degrees centigrade and the maximum stress-strength ratio at which the creep failure takes place at that temperature is approximately equal to 16 for epoxy mortar and 25 for polyester mortar specimens. Thus for example, for creep studies on epoxy mortar specimens, at a temperature of 50°C, the maximum stress -strength ratio permissible is 16/50 = 0.32 while for polyester mortar at 70°, the maximum permissible stress-strength ratio is 25/70 = 0.36. Therefore a common value of 0.3 was chosen for the stress-strength ratio in subsequent creep studies on both the mortars.

The strain measurements were taken at intervals of every hour for the first 6 hours and then at intervals of 24 hours. These observations were continued upto 21 days by which time it was noticed that the rate of increase in creep strains was neglegibly small. Hence further readings were discontinued.

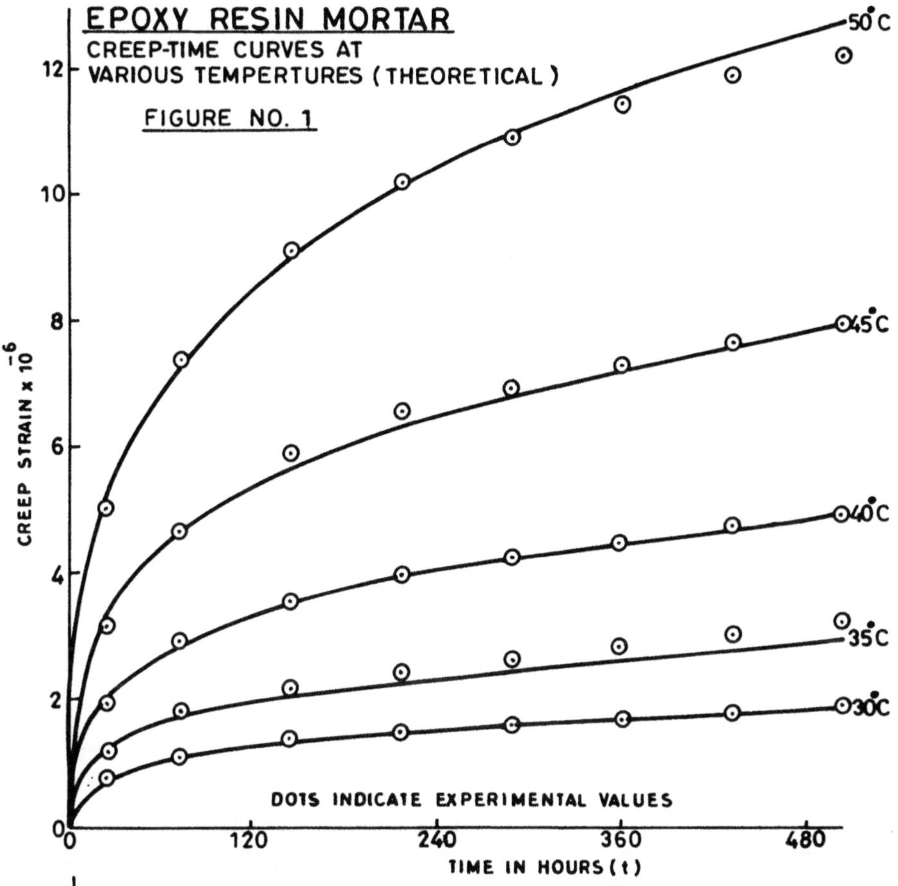

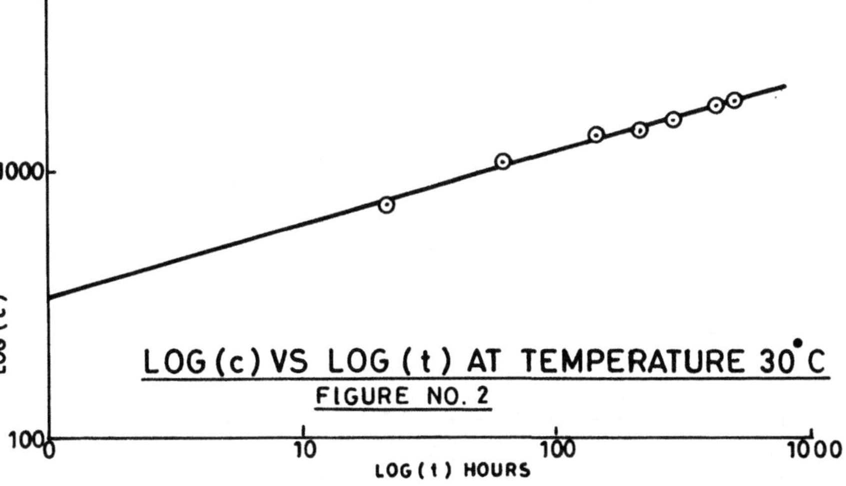

LOG (c) VS LOG (t) AT TEMPERATURE 30°C

FIGURE NO. 2

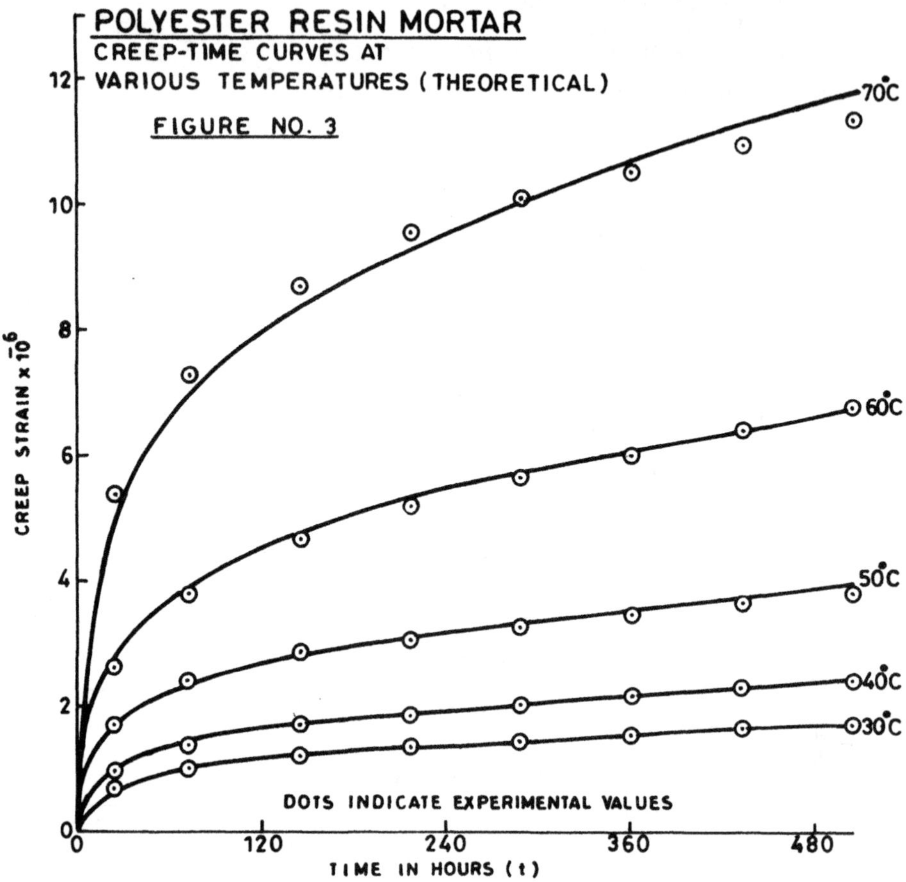

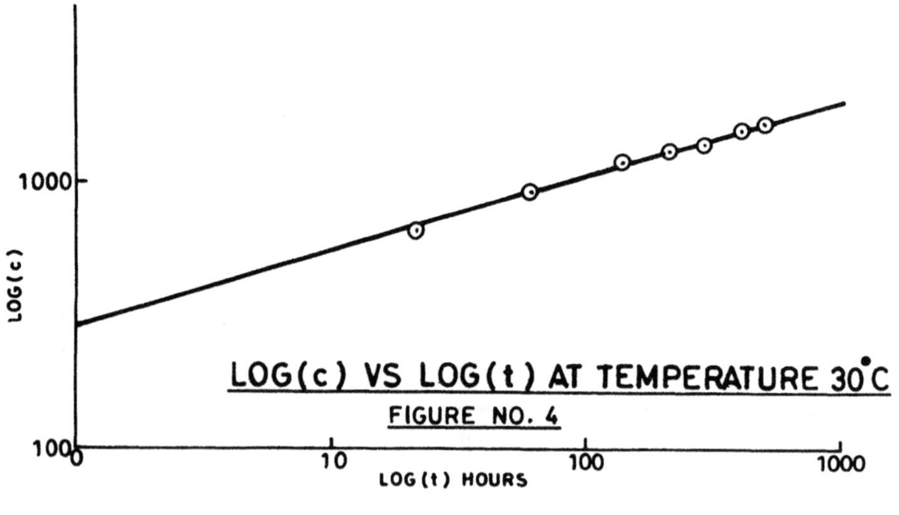

LOG(c) VS LOG(t) AT TEMPERATURE 30°C

FIGURE NO. 4

Figure No. 5 : Loading set-up

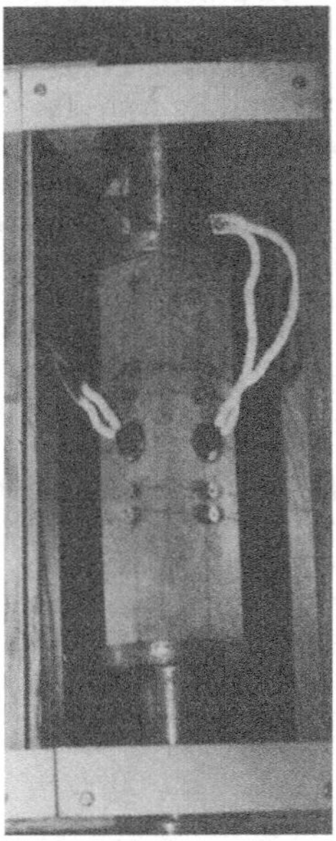

Figure No. 7 : Close-up view of the specimen

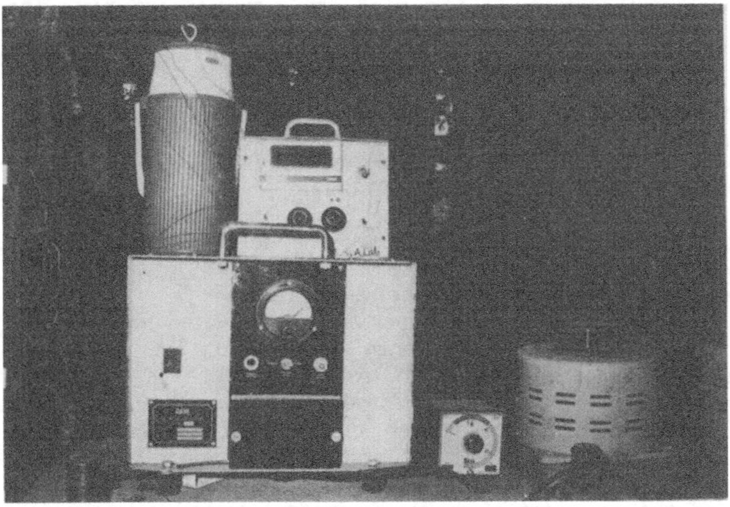

Figure No 6 : Instrumentation set-up

4. Development of a generalised expression for creep

i) Expression for creep as a function of duration of loading :
It is a well known fact that the two parameters which play a significant roll in deciding the pattern and quantum of creep strains in any polymer composite are the duration of loading and the ambient temperature. Since the present investigation concentrates on the effect of temperature only, it was decided to keep the duration of creep measurement constant as 21 days for all the tests. This in trun necessitates the development of a theoretical expression for the determination of creep at room temperature of 30°C when the specimen is loaded over a period of 21 days at the preselected stress level (decided by a stress-strength ratio of 0.3).

With the above objective in mind, the creep measurements in micro-strains recorded at room temperature over a period of 21 days, were plotted against time of loading in hours, seperately for each of the two polymer mortars. These curves are shown in figure No. 1 and figure No. 3 respectively. An exhaustive analysis of creep-time data at room temperature showed that the creep strain can be closely represented by a power expression of the type

$$C = K\,(t)^n \quad \text{..............} \quad (1)$$

where, C = Creep strain developed over a period of time 't' in hours while K and n are constants depending upon the properties of the polymer used. A plot of log C versus log t was prepared for the creep-time curves obtained from the experimental data. These plots are shown in figure No. 2 and figure No. 4 respectively for epoxy and polyester mortars. Since these plots have a straight line variation the value of K could be computed or scaled off as the intercept of the plot with the creep axis while 'n' could be obtained as the slope of the plot. The values of 'K' and 'n' for the two polymer mortars obtained by this procedure are : For epoxy mortar, K = 330, n = 0.28 and for polyester mortar K = 290, n = 0.28.

The theoretical values of creep strains as computed by using the expression given in equation (1) for the values of K and n given above, showed very good agreement with the experimentally recorded strains, the maximum error being of the order of 10% only.

It was therefore decided to use this theoretical expression for creep at room temperature as the basis for the development of the expression for creep in terms of the temperature variation.

ii) Creep as a function of temperature variation :
The pattern and quantum of creep strains at elevated temperatures depend not only on the temperature range but also on the corresponding loss of ultimate compressive strength over that range as compared to that obtained at room temperature. Consequently any theoretical formulation of an expression for creep strain in terms of temperature should also take into account the reduction in ultimate compressive strength.
Keeping this aspect in mind a plot of temperature (T) versus creep strains (recorded over a period 21 days) was made for both the polymer mortars. Since these plots were of curvilinear nature, several power expressions of T were tried (starting with T^2) to get the best curve fit. As a result of exhaustive trials the final expression for creep at any temperature

Table No. 1

1(a) Variation of ultimate compressive strength with temperature

Temperature in °C (T)	Ultimate Compressive Strength in (Kg/cm^2)		Fractional reduction in ultimate Comp. strength with respect to that at 30° C (p)	
	Epoxy	Polyester	Epoxy	Polyester
30	1000.0	1050.0	0	0
35	937.5	-	0.0625	-
40	875	995	0.125	0.0525
45	817.5	-	0.1875	-
50	755	934.5	0.245	0.110
60	650	871.5	0.350	0.170
70	540	807.5	0.460	0.231
80	425	756	0.575	0.280
90	320	693	0.680	0.340

1(b) Comparison of the theoretical and experimental creep strains at various temperatures

Temperature in °C (T)	[Creep Strains at 21 days]x10^{-6}				Percentage error	
	Epoxy		Polyester		Epoxy	Polyester
	Exprimental	Theoretical	Exprimental	Theoretical		
30	1900	-	1750	-	-	-
35	3250	2963	-	-	-8.80	-
40	5000	4778	2450	2577	-4.44	+5.21
45	8000	7920	-	-	-1.00	-
50	12250	12920	3825	4138	+5.47	+8.20
60	-	-	6800	7040	-	+3.50
70	-	-	11325	12459	-	+8.90

was obtained as [creep at room temperature] x $(T^2)^p$ i.e.

$$C_T = C_{T_o} \times T^{2p} \quad \text{..............} (2)$$

where C_T = Creep at any temperature, T

C_{T_o} = Creep at room temperature, To = 30°C

p = Fractional loss of ultimate compressive strength at temperature, T with respect to that at room temperature, T_o = 30°C

C_{T_o} is given by equation (1) i.e. $C_{T_o} = K\,(t)^n$

The theoretical values of creep in micro-strains for specimens subjected to a load (corresponding to a stress-strength ratio of 0.3), over a period of 21 days, computed using equation (2) are shown in Table No. 1. This table shows the variation of ultimate compressive strength with temperature [table 1 (a)], the experimentally recorded creep strains at various temperatures, the theoretically computed creep strains at those temperatures using equation (2) and the percentage deviation of the theoretical strains compared to the experimentally recorded ones [table 1(b)]. The tabulated results also show that the reduction in ultimate compressive strength has a linear variation in terms of temperature for both the polymers.

5. Conclusions

The results of the investigations show that it is possible to predict the value of creep strain caused by temperature variation for any given polymer binder to a reasonable degree of accuracy with the help of a generalised expression which is function of the temperature and the fractional loss of ultimate compressive strength over that temperature range.

The utility of such an expression lies in the fact that knowing the prescribed limit of creep strains for a given structural element over the expected ambient temperature conditions which it is likely to be subjected to, the margin of safety can be estimated with reasonable degree of accuracy without conducting cumbersome long term creep experiments.

Selected references :

Ayyar, R.S. and Deshpande, S. N. : 'Creep Studies on Polymer Mortars', Proceedings of the Fourth International Congress on Polymers in Concrete, Darmstadt, W. Germany September 19-21, 1984, pp. 487-493.

J. Hristova : 'Behaviour of Polyester Polymer Concrete at Temperature upto 70°C', Proceedings of the Third International Congress on Polymers in Concrete, May 13-18, 1981, Vol. II, pp. 1151-1160, Published by College of Engineering, Nihon University, Koriyama, Japan (1982).

Okada, K., Koyanagi, W., and Yonezawa, T, : 'Thermodependant Properties of Polyester Resin Concrete', Proceedings of the First International Congress on Polymers in Concrete, London, U.K., 1975, pp. 210-215.

THE EFFECT OF CHEMICALS ON THE ADHESION OF EPOXY RESIN TO AGGREGATES ON THE BASIS OF QUARTZ

W.P. ETTEL
Technical University Leipzig, Section Technology of Building Production

Abstract
The characteristics of epoxy resin mortars are essential influenced by the adhesion of epoxy resin to aggregates on the basis of quartz. From investigations can be concluded that epoxy resin mortars which are penetrated by chemicals (HCl; NaOH) show a low durability of the adhesion between epoxy resin and quartz and therefore they change their properties very rapidly.
Key words: Epoxy resin, Adhesion, Quartz, Mortar.

Results

Due to its C-O and C-N bonds and hydroxyl groups, epoxy resin has a strongly polar character. Owing to their SiO-bond, the fillers, generally quartz, have an ionicity of nearly 51 per cent. These characteristics result in a high adhesion force between epoxy resin and quartz and in the susceptibility especially to polar media, which form solvation sheaths around both the polar groups of the epoxy resins and the unsaturated bonds on the quartz surface. This can be proved as follows:

- In partly coating big quartz grains and in storing them in acids and bases at 60 °C; the resin mortar separates without any mechanical stress.

- In testing the strength of mortars which are not loaded by chemicals, the fracture goes through the quartz grain, but this is not the case with mortars which are chemically loaded (Fig. 1).

The interface between resin and quartz is a preferred migration path for polar media. There are also small schliere-shaped pores (< 1 µm) within the resin-filler interface region which support the permeation.

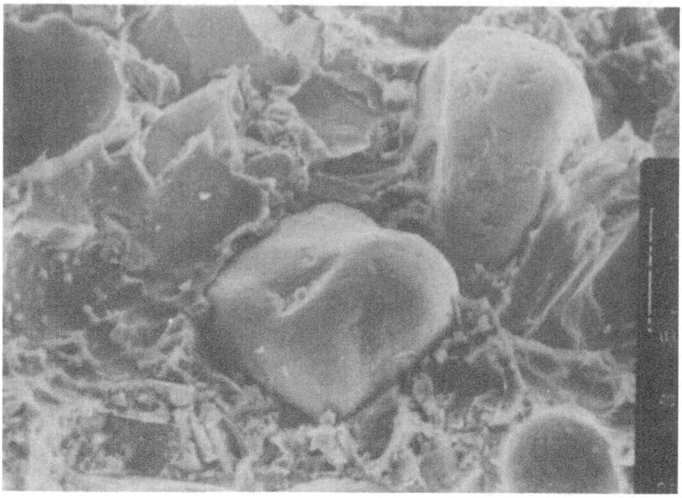

Fig. 1 - SEM-photograph of an epoxy resin mortar attacked by an acid (enlarged 200 times)

While the resistance to chemicals at the mixing proportion: binding agent-filler of 1:1 was distinctly better than that at mixing proportions from 1:3 to 1:5, a significant difference in the resistance to chemicals between the mixing proportions from 1:3 to 1:5 could not be found. In comparison with heptane and benzene, which are non-polar and nearly non-polar organic media, the fillers in the mortar reduce the permeability; the permeation is lower than that in pure resin.

From investigations of the mechano-thermal behaviour of chemically loaded epoxy resin mortars subject to torsional vibration measurements it can be derived that the adhesion between resin and quartz is reduced very rapidly.

In measuring the shear modulus after 7 days, the shear modulus G (and thus the modulus of elasticity, too) was already of very low orders of magnitude (Fig. 2).

The values of shear modulus after 28 days and after 6 months, resp., of treatment with HCl do not differ essentially. Similar results were obtained by storing in 1 per cent HCl and 10 per cent NaOH, resp. Owing to the test specimen thickness of only 2 mm, the real properties of the epoxy resin mortars penetrated by chemicals and thus the low durability of the adhesion between epoxy resin and quartz under the effect of polar media are characterized by this investigation method.

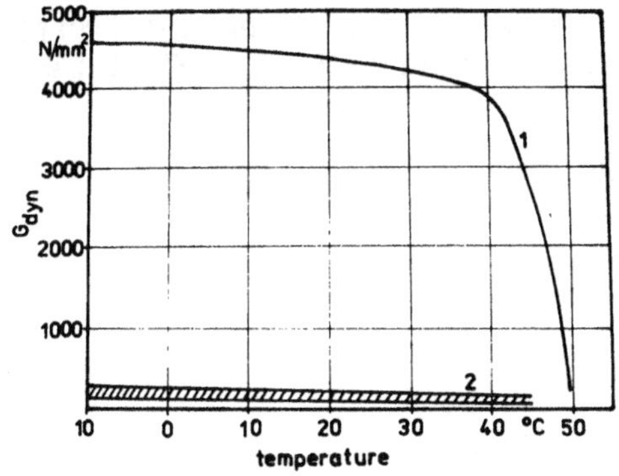

Fig. 2 - Shear modulus function of epoxy resin mortars (Epilox T 20-20; DPTA; max. particle size: 1 mm; mixing proportion: 1:5) after storing in 10 per cent HCl

ESSAI DE COLLES A BETON A BASE DE RESINES EPOXYDES COMPARAISON ENTRE EPROUVETTES FABRIQUEES AU LABORATOIRE ET SUR CHANTIER

F. HUGENSCHMIDT
CIBA-GEIGY Bâle

Résumé
La température exerce une influence marquée sur le comportement des plastiques thermodurcissables et notamment des résines époxydes. Cette influence se répercute sur le choix des méthodes d'essai ainsi que sur la fabrication et le stockage des éprouvettes. Trois méthodes d'essais sont appliquées à des éprouvettes fabriquées respectivement en laboratoire et sur chantier. Les résultats sont comparés statistiquement. Il en ressort que, si les éprouvettes sont fabriquées avec soin, elles donnent des résultats comparables. On propose donc des mesures en vue de réduire la dispersion des résultats. La détermination de la résistance initiale des adhésifs sur le chantier est non seulement essentielle pour la qualité de l'exécution des travaux, mais aussi d'un grand intérêt didactique pour les exécutants.
Mots-clés: résines époxydes, colles béton/acier, méthodes d'essai, laboratoire, chantier, statistique (traitement).

1. Introduction

Les adhésifs et mortiers à base de résines époxydes présentent un profil de propriétés particulièrement adapté à l'assemblage solidaire entre:

- éléments préfabriqués en béton
- béton frais et béton ancien
- acier/béton et acier/acier.

De multiples réalisations, dont les plus anciennes remontent déjà à une vingtaine d'années, témoignent de la fiabilité de ces assemblages. De nombreux ponts dont la superstructure est constituée de voussoirs préfabriqués collés, des ouvrages renforcés par des armatures plaquées en sont des exemples.

Les propriétés des résines réactives et, parmi elles, des résines époydes diffèrent sensiblement de celles des matériaux à assembler. Les variations de température extérieure influent peu sur le comportement de béton en cours d'élaboration et pratiquement pas sur le béton durci (Une chute du module E du béton n'apparaît guère que vers 350°C).

En revanche, les systèmes époxydes, comme tous les plastiques thermodurcissables, subissent fortement l'influence de la température ambiante tant au cours de leur mise en oeuvre et de leur durcissement qu'à l'état durci.

La température agit sur:

- la viscosité du liant, donc l'ouvrabilité du mortier
- la réactivité, donc la vitesse de durcissement; mais de la réactivité dépend aussi la fourchette des températures de mise en oeuvre admissibles, ainsi que les durées d'utilisation correspondantes
- la taux de réticulation d'un mortier
- les résistances mécaniques et les modules
- le comportement à température élevée.

Le durcissement est un processus exothermique. L'élévation de température qu'il peut engendrer dépend de la quantité de matière (épaisseur du joint), de la température des pièces à coller et de leur pouvoir de dissipation de la chaleur.

Les caractéristiques du liant doivent aussi être prises en considération pour la fabrication des éprouvettes. Les méthodes d'essai elle-mêmes doivent parfois être adaptées. En tout état de cause, on n'obtiendra de résultats comparables et reproductibles que si l'on a pris en compte l'influence de la température.

Le contrôle régulier de l'adhésif et la comparaison des résultats expérimentaux avec les valeurs théoriques sont essentiels pour bien connaître les particularités d'un matériau. C'est aussi la seule façon d'atteindre la qualité recherchée dans l'exécution et d'inciter ceux qui en sont chargés à y apporter le plus grand soin.

2. Fabrication des éprouvettes au laboratoire

Il n'est question ici que de méthodes d'essai intéressant les armatures plaquées et le contrôle de la qualité sur le chantier. Il faut par ailleurs distinguer les essais sur l'adhésif proprement dit et des essais sur des éprouvettes composites. [1/2]

On mesure la résistance à la compression de l'adhésif sur des cubes de 40 x 40 x 40 mm, la résistance à la flexion et le module d'élasticité sur des prismes de 40 x 40 x 160 mm. Ces dimensions sont relativement importantes eu égard à l'épaisseur usuelle des joints de colle ($\leq$ 2 mm). Des éprouvettes plus petites souffriraient cependant d'un autre défaut. Les colles à béton étant généralement thixotropes, le remplissage complet des moules est malaisé et les inclusions d'air sont pratiquement inévitables. Sur des éprouvettes de 10 x 10 x 30 mm, la présence de pores réduirait la section de façon indéterminée, entraînant une forte dispersion des résultats.

Pour éliminer les effets préjudiciables de l'exothermie, les moules en acier sont refroidis à l'eau pendant le durcissement (fig. 1). A défaut, il est recommandé de n'emplir d'adhésif qu'une empreinte sur deux du moule. Le durcissement peut alors s'effectuer à la température souhaitée sous conditions constantes (chambre climatisée / bahut refrigerant).

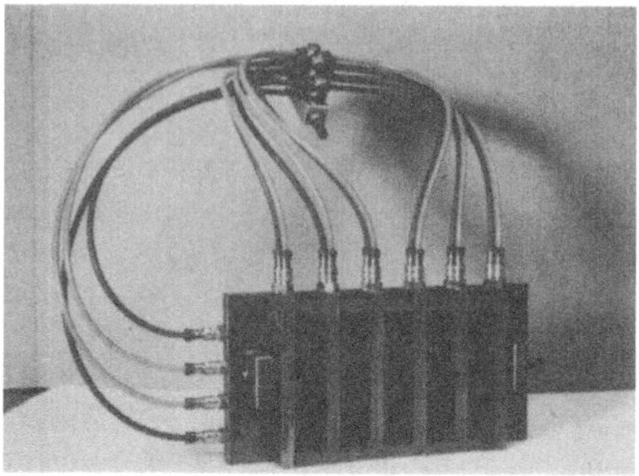

Fig. 1.

Les essais ont lieu à la température ambiante. Comme ils sont de courte durée, la température de l'éprouvette ne varie pas au point de modifier les résultats. Pour la résistance à la flexion, on applique la charge en 4 points (fig. 2). (Aux laboratoires d'essais on applique la charge normalement en 3 points). En règle générale, on détermine la résistance à 7 jours.

Fig. 2.

Pour la pratique, il importe avant tout de connaître l'évolution de la résistance en fonction de temps et de la température. Des éprouvettes soumises à de fortes déformations au début du durcissement ne conviennent pas à cet effet. La résistance à la compression et la résistance à la flexion ne peuvent donc donner une mesure de la croissance de la résistance.

Un essai simple et rationnel consiste en revanche à déterminer la résistance à la traction de pastilles en acier de 25 mm de diamètre, collées sur une barrette (fig. 3). Une rainure à leur périphérie leur donne une assise verticale. Le joint de colle a une épaisseur de 1.5 mm.

Fig. 3.

L'adhésif en excès est expulsé par le trou central. Les surfaces de collage ont été décapées par grenaillage. Le dispositif d'essai est représenté (fig. 4). L'adhésif durcit sous conditions reproductibles et l'influence de l'exothermie est exclue. La rupture se produit dans l'adhésif et à l'interface adhésif/acier (défaillance de l'adhésion ou de la cohésion).

Fig. 4.

La fig. 5 montre l'évolution de la résistance d'un adhésif pour placage d'armatures, aux températures respectives de 15°C, 23°C et 30°C.

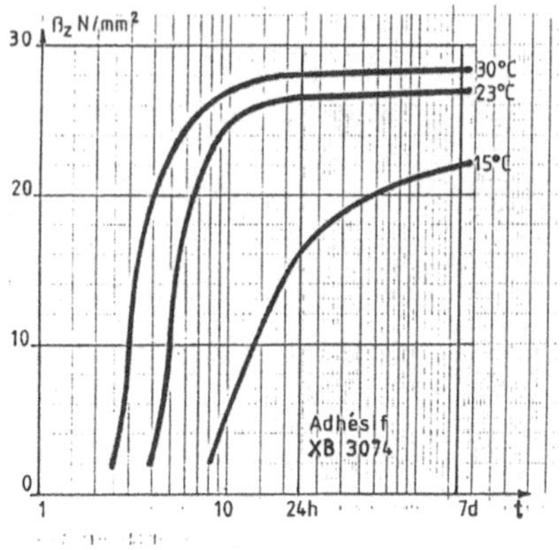

Fig. 5.

3. Fabrication des éprouvettes sur chantier et contrôle de la résistance initiale

On admet que les adhésifs sont livrés en emballages prédosés, dispensant de la pesée des composants sur le chantier.

Pour l'entrepreneur comme pour le bureau d'études, la réponse aux trois questions suivantes revêt un intérêt primoridal:

a) L'adhésif est-il bien mélangé?
b) En combien de temps durcit-il sous les conditions données (température/humidité de l'air)?
   Dans quelle mesure l'armature rapportée commence-t-elle à travailler solidairement avec l'élément à renforcer?
   Quand peut-on enlever les étais provisoires?
c) L'adhésif atteint-il les caractéristiques annoncées par le fabricant?

Ces questions posent le problème de la qualité du travail, pendant et après son exécution. Voici ce que nous pouvons y répondre.

a) Les deux composants de l'adhésif doivent être de couleur différente, blanche et noire en général. Lorsque l'adhésif présente une teinte grise uniforme, sans la moindre traînée, le mélange est achevé. On doit alors procéder sans délai à l'application.
b) Il s'agit ici de connaître les caractéristiques mécaniques initiales et l'augmentation de résistance de l'adhésif. L'essai d'arrachement de pastilles en acier collées sur le béton, schématisé fig. 6, s'est révélé particulièrement approprié. On peut renoncer à fraiser dans le béton une rainure circonscrivant la pastille. Dans ce cas, l'étendue de la surface de rupture n'est pas déterminée à l'avance, mais le procédé se trouve sensiblement simplifié.

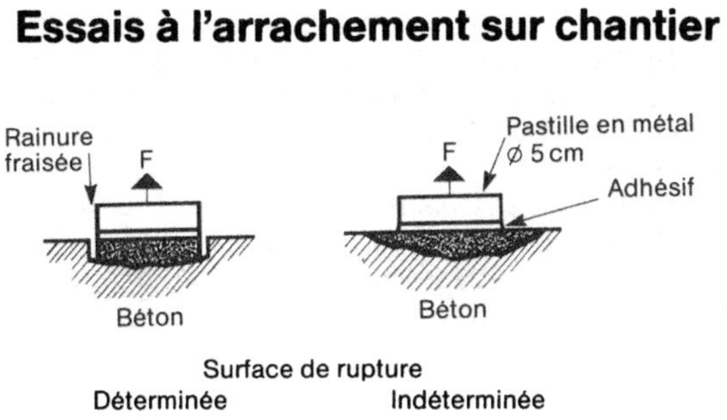

Fig. 6.

La réalisation de l'essai est simple. A la fin de chaque étape de travail, - en principe chaque jour -, et à proximité immédiate de la dernière tôle d'acier posée, on colle au moins 5 pastilles en acier sur la surface de béton prétraitée. L'essai d'arrachement a lieu en fonction de la température moyenne effective, dans la mesure où l'on attend une rupture entièrement localisée dans le béton (fig. 7). Pour cela, il faut connaître d'une part, la résistance à la traction du béton à renforcer et, d'autre part, la croissance de la résistance de l'adhésif en fonction de la température et du temps (cf. § 2, fig. 5).

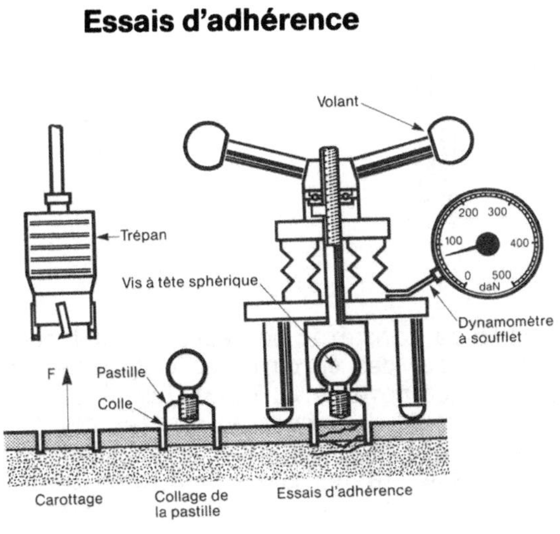

Fig. 7.

Le plus souvent, l'essai d'arrachement peut être effectué au bout de quelques heures. Si la rupture se situe entièrement dans le béton, les travaux reprennent dès le lendemain.
Si la rupture n'a pas lieu dans le béton, il faut avant tout en rechercher les causes.

c) Pour déterminer la résistante finale de l'adhésif, on fabrique des éprouvettes à intervalles réguliers (fig. 8). Le format 40 x 40 x 160 mm est également commode à préparer sur chantier. Il faut fabriquer chaque foix au moins 3 éprouvettes et veiller à ce que leur température ne s'emballe pas sous l'effet de l'exothermie. On y parvient en ne remplissant qu'une cavité sur deux du moule en acier (fig. 1). On mesure la résistance à la flexion sur le prisme entier et la résistance à la compression sur les deux morceaux.

Eprouvettes

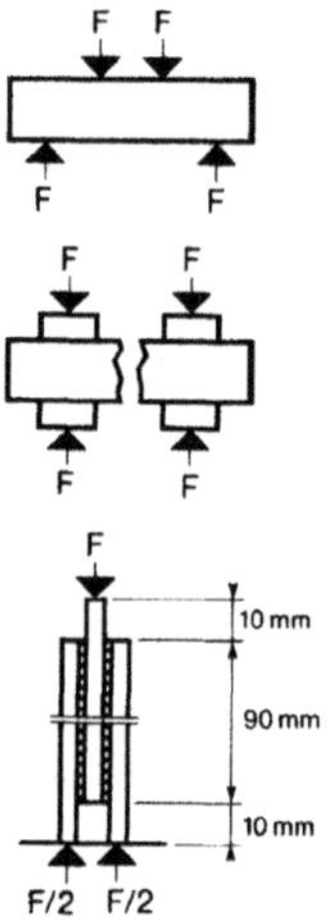

Prisme en adhésif 4 x 4 x 16 cm³
Résistance à la flexion

«Cubes» en adhésif
Résistance à la compression

3 plaques d'acier 5 x 100 x 100 mm³
Collées par l'adhésif
Résistance au cisaillement

$$\tau_m = \frac{F}{2 \times 90 \times 100} \quad \frac{[N]}{[mm^2]}$$

Fig. 8.

Il est souhaitable de compléter les essais sur l'adhésif proprement dite par des essais sur des éprouvettes composites, par exemple en mesurant la résistance moyenne au cisaillement par compression d'éprouvettes formées de 3 plaquettes d'acier collées (fig. 8), d'une épaisseur optimale de 5 mm. Plus minces, elle ont tendance au flambage; plus épaisses, elles augmentent trop le bras de levier interne. On prépare les surfaces à coller comme les tôles d'armature, par sablage et application éventuelle d'une couche primaire anti-rouille. De même, on ponce et on dégraisse les surfaces avant l'encollage.

On étale l'adhésif aussi uniformément que possible et on presse les trois plaquettes de manière à expulser l'excès d'adhésif et à assurer leur parallélisme. Les deux joints doivent avoir la même épaisseur.

Avant de monter les éprouvettes dans la presse, il convient de vérifier que les surfaces d'appui sont bien parallèles et perpendiculaires à l'axe longitudinal. Les rectifier par ponçage, le cas échéant. La plaquette intermédiaire ne doit pas être décalée de plus de 10 mm. Là encore, 3 éprouvettes au moins sont nécessaires pour chaque essai.

Sur le chantier toutes les éprouvettes doivent être entreposées à l'abri, à environ 20°C. Il faut éviter les températures inférieures à la température minimale de durcissement, de même que les températures trop élevées, qui engendrent une post-réticulation trop importante.

En règle générale, les essais on lieu 7 jours après la fabrication des éprouvettes, dans un laboratoire d'essais agrée.

4. Comparaison des résultats collectés à l'EMPA (EMPA = Laboratoire fédéral d'essai des matériaux et Institut de recherche)

Le laboratoire fédéral poursuit depuis 1973 un programme de recherche sur le renforcement de structures en béton par des armatures plaquées [2]. Il a par ailleurs participé à des essais à grande échelle menés dans le cadre de réalisations pratiques et à d'autres essais visant à l'obtention des agréments nécessaires en RFA.

Qu'il s'agisse du programme de recherche EMPA ou d'essais à la demande, l'adhésif a été dans tous les cas l'objet de contrôles réguliers: résistance à la flexion, à la compression et au cisaillement. Les éprouvettes sont représentées fig. 8. Seule différence, les tôles des éprouvettes de cisaillement avaient à l'origine une épaisseur de 3 mm au lieu de 5 mm. Ces éprouvettes étaient fabriquées dans les laboratoires EMPA.

Dans le cas des applications pratiques à grande échelle où EMPA avait un rôle de consultant, les éprouvettes ont été fabriquées sur chantier (environ 850 au total).

Les quelque 1250 valeurs obtenues ont été traitées statistiquement en vue de répondre aux questions suivantes:
- Y a-til une différence significative entre les éprouvettes obtenues en laboratoire et celles qui sont fabriquées sur chantier?
- Entre quelles limites le coefficient de variation évolue-t-il?
- Quelles valeurs minimales peut-on raisonnablement fixer?
- Quelles sont les mesures à adopter pour réduire la dispersion des résultats?

L'ensemble des résultats, traités statistiquement, est présenté dans le tableau A, en fonction du type d'essai, des conditions de fabrication et de l'âge des éprouvettes.

Tableau A		colles epoxydes		
Origine des éprouvettes		laboratoire	chantier	laboratoire et chantier
Age des éprouvettes		7 d	7 d	ind.de l'âge
Résistance à la traction par flexion				
Nombre d'éprouvettes	n	71	173	404
Moyenne	$\bar{x}$ (N/mm^2)	32	37	36
Ecart-type	s (N/mm^2)	7	10	12
Coef. de variation	v (%)	23	26	32
Fractile de 5%	(N/mm^2)	18	19	13
Résistance à la compression				
Nombre d'éprouvettes	n	71	173	404
Moyenne	$\bar{x}$ (N/mm^2)	81	95	90
Ecart-type	s (N/mm^2)	8	13	17
Coef. de variation	v (%)	10	13	18
Fractile de 5%	(N/mm^2)	65	70	57
Résistance au cisaillement				
Nombre d'éprouvettes	n	60	234	439
Moyenne	$\bar{x}$ (N/mm^2)	7.4	8.6	8.1
Ecart-tpye	s (N/mm^2)	3.2	3.8	3.5
Médiane	x (N/mm^2)	5.9	7.4	7.9
Valeur dominante	D (N/mm^2)	5.0	5.9	5.8

Les figures 9, 10 et 11 donnent une interprétation graphique des valeurs de **la résistance à la flexion** et précisent le nombre d' éprouvettes, la moyenne, les fractiles de 5% et la distribution normale (courbe de Gauss). Aux éprouvettes fabriquées au laboratoire correspondent la valeur moyenne la plus basse mais aussi le coefficient de variaiton le plus faible. Un très faible pourcentage d'éprouvettes présente une résistance $\beta_t \leq$ 22 N/mm^2 (respectivement 6%, 3% et 8%).

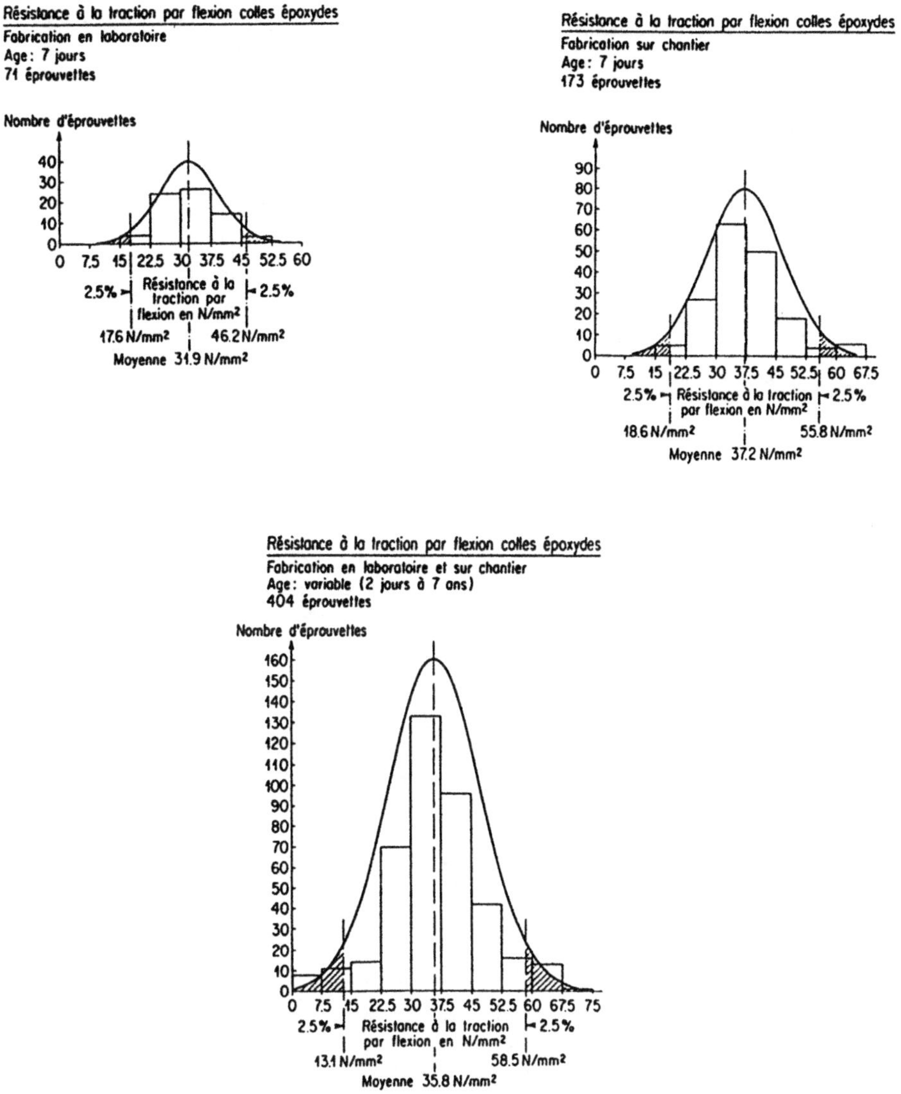

Fig. 9. 10. 11.

Exprimés dans les figures 12, 13 et 14, les résultats de **r é s i s t a n c e à l a c o m p r e s s i o n** montrent une distribution analogue. Par rapport à l'essai précédent, une plus grande fraction d'éprouvettes présente une résistance $\beta_c < 75$ N/mm^2 (28%, 8% et 18%).

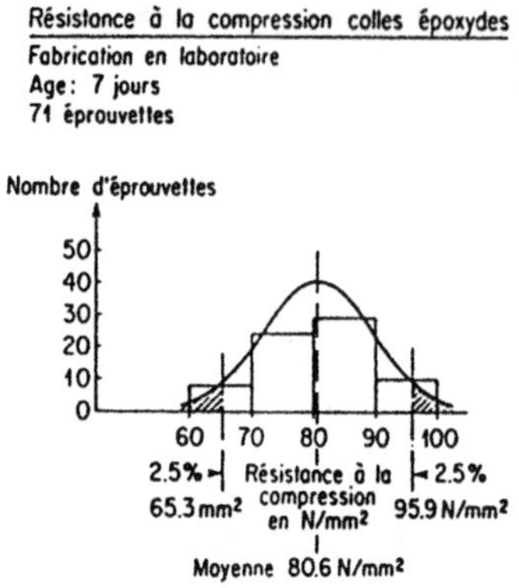

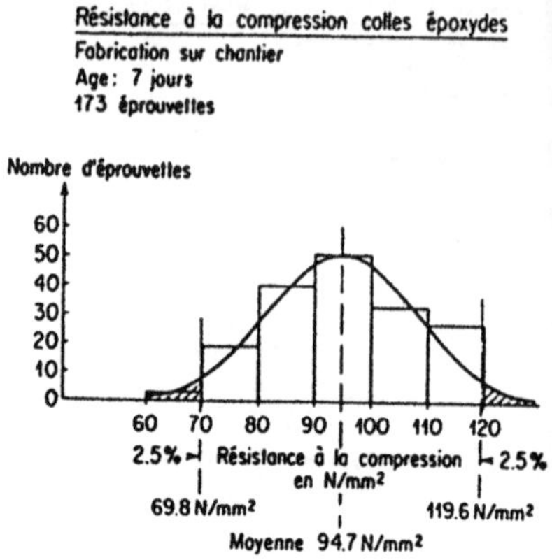

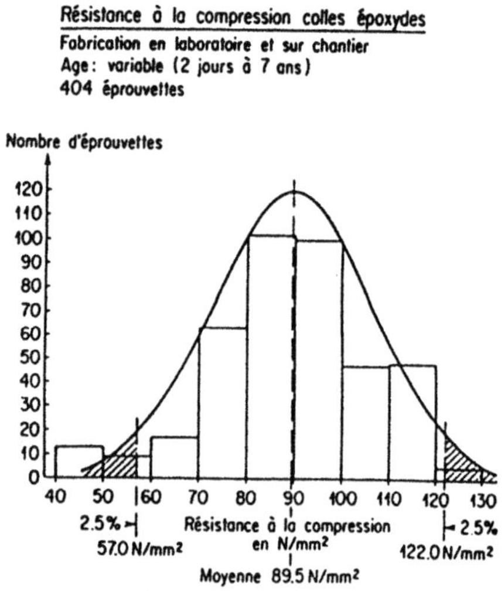

Fig. 12. 13. 14.

Les fig. 15, 16 et 17 sont consacrées à **la résistance au cisaillement**, dont les valeurs ne se répartissent plus selon un distribution normale. Seule quelques points sont inférieurs à une résistance au cisaillement $\tau \leq 4$ N/mm^2 (9% et 6%). L'effectif maximal se situe entre 4 et 6 N/mm^2, exprimé par la valeur dominante (mode) D. La position de la médiane montre que 50% de tous les résultats sont supérieurs à 7 N/mm^2. Toutes les tôles étaient sablées et imprimées.

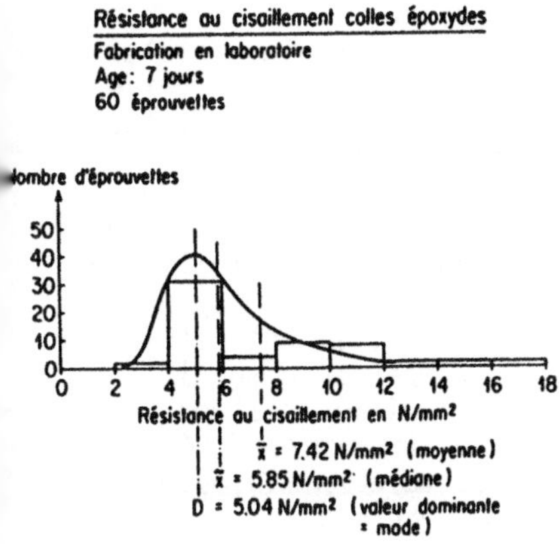

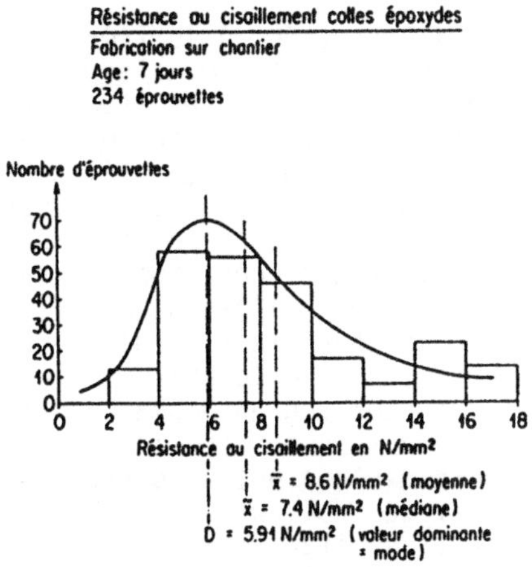

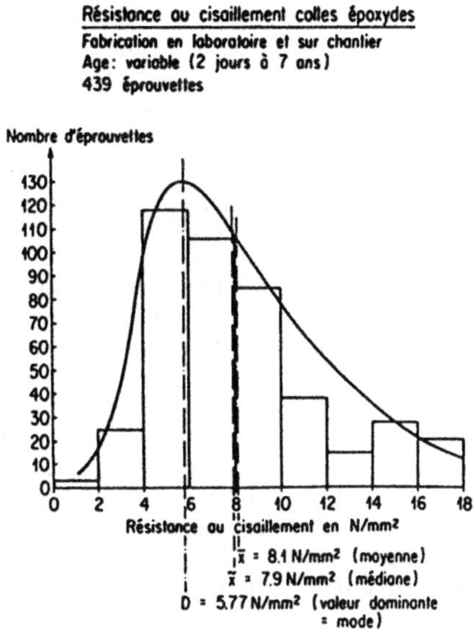

Fig. 15. 16. 17.

5. Conclusions

Six colles différentes ont été utilisées. La seule dont la composition complète connue est celle qui intervient dans le programme de recherche EMPA. Il s'agit d'une colle de réactivité moyenne, utilisable entre 15°C et 30°C. Quant aux autres, on sait seulement que ce sont généralement des systèmes plus réactifs, donc plus étroitement réticulés.

On ne connaît pas plus précisément les conditions de durcissement et de stockage des éprouvettes fabriquées sur chantier. Un assez grand nombre d'entre elles ont été testées plus de 7 jours après le durcissement de l'adhésif.

5.1 Résistance à la flexion et à la compression
Ce sont les éprouvettes de laboratoire qui ont les plus faibles coefficients de variation. La résistance à la compression varie moins que la résistance à la flexion. Les deux essais ainsi que les types d'éprouvettes utilisés semblent bien adaptés.

On peut cependant atteindre une meilleurs concentration des résultats, donc une dispersion moindre, par une spécification plus affinée des conditions de fabrication des éprouvettes: meilleure maîtrise de l'exothermie, séjour des éprouvettes en atmosphère contrôlée, fixation de la durée de durcissement. Il est absolument nécessaire de connaître l'évolution de la température sur le chantier (en un emplacement judicieusement choisi), soit par enregistrement continu, soit par des relevès très rapprochés. Les éprouvettes qui s'écartent sensiblement des valeurs minimales exigées doivent subir un nouvel examen: recherche de retassures ou autres défauts dans les surfaces de rupture, et être rejetées, le cas échéant.

5.2 Résistance au cisaillement
Toutes les tôles, rappelon-le, ont reçu une couche primaire au chromate de zinc, à base d'époxyde ou d'époxyde/isocyanate. En règle générale, on obtient des caractéristiques plus élevées avec des tôles sablées mais non imprimées. Le fait que plus de 50% des résultats (fig. 17) se situent au-delà de 6 N/mm^2 n'en est que plus remarquable. La rupture intervient presque constamment dans la couche primaire. La grande dispersion des résultats indique toutefois que le soin apporté à la préparation des éprouvettes a une influence considérable sur les résultats. On a également constaté que la résistance augmentait en raison inverse de l'épaisseur du joint. Il nous paraît cependant peu souhaitable de chercher à réduire à l'extrême l'épaisseur des joints alors que des joints de 1-2 mm sont de règle dans la pratique. Il semble plus judicieux de spécifier l'épaisseur du joint avec une plus grande précision. On devrait y parvenir en perçant les trois carrés de tôle d'un trou central et en insérant entre chacun d'eux une rondelle de l'épaisseur définie pour le joint. Après l'encollage, on serre les tôles au moyen d'une vis centrale. Leur rigidité est suffisante pour que l'excès d'adhésif soit expulsé et qu'il subsiste un joint de colle bien fourni. Des essais sont éventuellement indiquées. Il reste à déterminer, à quel niveau fixer la résistance au cisaillement minimale.

L'auteur exprime sa vive reconnaissance à MM. Dr. Marc Ladner et Christoph Weder, de l'EMPA, pour le traitement et la communication des résultats.

Références

Directives pour l'essais des adhésifs et des mortier à base de résines époxydes Araldite pour le bâtiment. Ciba-Geigy Basel.
Ladner, M. et Weder, Ch. (1981) Geklebte Bewehrung im Stahlbetonbau
Richtlinie für Zulassungsprüfungen an Reaktionsharzklebern und Primern für angeklebte Laschenverstärkungen von Stahlbetontraggliedern. Entwurf. Technische Universität Braunschweig, Institut für Baustoffe, Massivbau und Brandschutz.

INCIDENCE DU CALAGE DES TURBOGROUPES ET COMPRESSEURS SUR LA DISPONIBILITE DES CENTRALES DE PRODUCTION D'ENERGIE

M. KAEMPFFER
Ceilcote

Dans l'exposé qui suit nous nous intéresserons à l'optimalisation des caractéristiques d'appui des turbogroupes et compresseurs sur leur support génie-civil. Ces appuis participent à la stabilité de la marche de la ligne d'arbre tournant, stabilité qui accroit notablement la sécurité et la disponibilité des centrales de production d'énergie.

Les fondations des grosses machines tournantes se distinguent des autres constructions en béton armé par les efforts dynamiques auxquels elles sont essentiellement soumises.

La conception et la construction de ces fondations devront conduire à éviter tout phénomène de résonance avec la machine. Pour cela la fréquence propre des fondations devra rester nettement en dessous de celle de l'excitateur.

La notion de fréquence propre est une notion fondamentale qui intervient dans tout problème de vibration. En effet, lorsqu'une structure est excitée par des forces périodiques (machine tournante), il se produit une amplification des déplacements et des contraintes dans le cas où l'une des fréquences propres de la structure est proche de la fréquence d'excitation. Par exemple, la majeure partie des vibrations aux paliers et dans la ligne d'arbre d'un turbogroupe en régime établi est constituée par des ondes stationnaires à fréquence de rotation dues au balourd mécanique et thermique inévitable. Le reste se compose de vibrations dont les fréquences sont en majorité des multiples entiers de la fréquence de rotation.

Le phénomène de résonance se traduit par un mauvais fonctionnement, une fatigue accrue du matériel - en même temps qu'une gêne pour le personnel - voire une défaillance causée par des avaries ou déclenchements intempestifs des systèmes de sécurité, ce qui entraîne des frais de réparation et des pertes dues à l'indisponibilité entrant pour une large part dans l'équation des coûts établis pour la définition de la fiabilité optimale du système technique.

En cas d'apparition de fissures ouvertes et continues dans les fondations, dues aux contraintes appliquées (phénomène pouvant être aggravé par la pénétration d'eau, d'huile machine ou autre produit éventuel) il est à craindre leur propagation sous l'action des charges dynamiques par opposition à un bon comportement aux charges statiques, ce qui causera à plus ou moins brève échéance des dommages plus importants au niveau du génie civil et de la machine elle-même.

Par exemple, une seule fissure verticale traversant la structure en modifiera son inertie, d'où risque d'entrer en résonance.

L'utilisation d'une masse de fondation suffisamment importante et d'une masse de calage anti-vibratoire permet d'éviter ces dégâts.

CALAGE AVEC MORTIER A BASE DE RESINES SYNTHETIQUES

Dans la pratique il y a une lacune, au sens propre et figuré, entre la conception du constructeur de machines d'une part et de l'ingénieur génie civil d'autre part, pour la mise en place et le calage de machines. L'emploi d'un groupe de matériaux de calage souvent méconnu, bien que d'une grande importance, permet de combler cette lacune. Il s'agit du groupe de matériaux dénommé GROUT.

Le Grout est un composant performant, faisant partie intégrante du système de fixation de la machine. La disponibilité de celle-ci dépend de son bon positionnement, ainsi que de la transmission des charges statiques et dynamiques aux fondations par l'intermédiaire du Grout . L'effet anti-vibratoire ne peut être obtenu qu'avec un Grout à base de résines synthétiques. On emploie habituellement comme matériau de calage des mortiers à base de résines époxy, qui offrent l'avantage d'être résistantsà l'eau, à l'huile, aux solvants, aux produits chimiques, ainsi qu'aux températures élevées, et ont une excellente adhérence à la fonte, à l'acier et au béton. Ils sont faciles à manipuler, d'une bonne aptitude à la mise en place et durcissent rapidement (courte période d'immobilisation).

Caractéristiques physiques :

module d'élasticité en compression	14 à 20 x 10^3	M.Pa
résistance à la compression	96	M.Pa
résistance à la traction directe	18	M.Pa
résistance à la traction en flexion	30	M.Pa
adhérence à l'acier (en traction)	15	M.Pa
adhérence au béton	meilleure que béton/béton	

Ces Grouts époxy peuvent, en fonction de leur composition, présenter des propriétés physiques et des caractéristiques de mise en oeuvre différentes. Cette possibilité constitue un avantage, car elle permet d'adapter à chaque cas d'utilistion la structure idéale, c'est-à-dire le meilleur compromis entre des caractéristiques physiques appropriées et une capacité de mise en oeuvre satisfaisante.

Pour faire le bon choix il convient de prendre en considération tous les élements depuis la conception du système de supportage de la machine jusqu'au type de service, en passant par les conditions de fonctionnement. Une structure de Grout non adaptée au cas d'utilisation concernée pourra entraîner une dégradation prématurée des fondations et/ou de la machine.

Outre un module d'élasticité élevé, le Grout époxy ideal pour les installations à terre devrait offrir une résistance en compression et à la flexion élevée, compte tenu des températures ambiantes et d'utilisation. Il devrait avoir une bonne tenue au fluage et à la corrosion. En outre, il devrait présenter une bonne adhérence sur les fondations et être facile à manipuler. Ce Grout devrait également avoir une faible réaction exothermique, car un matériau ayant un pic exothermique élevé aura tendance à se fissurer.

Il existe trois méthodes principales de pose pour les grosses machines tournantes, qui sont :

* calage intégral par Grout époxy
* calage par plaque d'embase en Grout époxy et plaque d'entretoise métallique
* calage par plaque d'embase et plaque d'entretoise en Grout époxy

Ces trois méthodes sont les plus couramment employées pour l'installation de machines à terre. Jusqu'à présent le calage par plaques d'embase et plaques d'entretoise en Grout époxy était surtout utilisé dans des installations maritimes et très rarement à terre.

Avec le calage intégral le socle de la machine repose entièrement sur le Grout. Cette méthode est la plus simple, la plus éprouvée et s'emploie pour les installations les plus variées, parce qu'elle consolide l'assise de la machine et qu'il y a une meilleure répartition des contraintes sur les fondations. De plus, l'adhérence du Grout époxy sur le front du socle machine améliore la transmission des charges dynamiques sur les fondations. Cette méthode de calage empêche toute infiltration entre la machine et ses fondations évitant ainsi la corrosion.

Le calage intégral présente toutefois l'inconvénient que la machine ne peut être recalée que moyennant des frais élevés, mais cela n'a pas d'importance dans la mesure où le calage est fait correctement, donc définitivement. Un autre inconvénient est l'absence de refroidissement par circulation d'air entre la machine et les fondations.

Le calage par plaques d'embase en Grout époxy et plaques d'entretoise métalliques présente l'avantage de pouvoir réaligner, si nécessaire, la machine sans refaire les cales époxy. De plus, il y a circulation d'air entre la machine et ses fondations, ce qui contribue à diminuer les contraintes thermiques dans le béton.

L'utilisation d'un calage par plaques d'embase et plaques d'entretoise métalliques augmente considérablement les coûts d'installation. Ceci peut être évité par l'utilisation de ces deux éléments fabriqués en Grout époxy sur le site même. Les plaques d'embase et les plaques d'entretoise en Grout époxy assurent un très bon contact avec le socle de la machine et éliminent toute possibilité de dégradation susceptible de se produire suite aux vibrations (ondes transversales).

Les plaques d'embase avec les plaques d'entretoise époxy à haute élasticité, habituellement employées avec les moteurs Diesel marins, n'ont encore jamais été utilisées pour les montages à terre. On emploie depuis plus de 15 ans déjà des plaques d'embase avec des plaques d'entretoise en Grout époxy fortement chargées à faible élasticité pour les installations à terre. Ces éléments en Grout époxy doivent être, en comparaison, plus grands que ceux en acier, afin de maintenir le niveau de contrainte dans les limites acceptables (environ 3 à 4 M.Pa). Elles peuvent également avoir une épaisseur plus importante que celles en acier (env. 50 mm) pour maintenir une bonne circulation d'air entre la machine et ses fondations.

Le critère de réussite de chaque type de calage au Grout, indépendamment de la méthode d'installation, réside dans le choix du matériau. Aucune formulation ne peut convenir à toutes les situations, et aucun critère de choix n'a une importance telle que les autres puissent être négligés.

Les coûts du matériau de calage sont négligeables par rapport au prix d'une machine importante. Si l'on considère l'incidence du calage par le Grout époxy sur la disponibilité de la machine, le prix en demeure un facture secondaire.

PROPERTIES OF POROUS CONCRETE INJECTED WITH POLYMER AND CEMENTITIOUS MATERIALS

E. NÄGELE, U. SCHNEIDER, A. BRANDAU
Department of Civil Engineering, University Gesamthochschule Kassel
Mönchebergstr. 7, D - 3500 Kassel, FRG

Abstract
Porous concrete cylinders 15/15 cm were made mainly with coarse aggregate, which resulted in a porosity of the specimens of about 40 Vol. %.
The specimens were cured according to German standards. After 28 days one half of the specimens was injected in the mould with a epoxide material and the rest with a cementeous material. Reference specimens were not injected.
The compressive strength of the reference specimens was determined prior to injection and the compressive strengths of the injected specimens were determined half an hour, two hours and 28 days after injection. Porosities were determined at the same time using both density measurements and Hg-Porosimetry.

Key words: Concrete Injection, Repair, Cementitious Materials, Polymer Materials

1. Introduction

During the post-war period in Germany the reconstruction of many buildings was performed by using low quality concrete. For those concretes e.g. crushed bricks were used as aggregate and the cement content was comparatively low. Nowadays many of these concrete structures mostly 3 to 5 storey dwelling buildings indicate cracks and other types of damages. There is an urgent need for repair or reconstruction. Fig. 1 shows such a concrete sample which is drilled from the wall of a 5 storey building. Typically such concrete is very porous and the strength gradients along a simple wall can be very high. Table 1 shows some typical results of strength tests of such walls. For the repair of such porous concrete structures the injection of hydraulic or polymeric binders is an appropriate method. Many different materials have been proposed and used for this purpose.

Fig. 1 Sample of post-war low quality concrete

Table 1. Results of stength tests of post-war low quality concrete

Nr.	1	2	3	4
Compressive strength N/mm²	17,8 15,5 20,5 27,1 13,9 11,3	7,9 9,3 7,3 5,3 4,6 13,2	12,9 19,4 16,1 30,9 23,1 31,7	19,9 13,0 28,5 24,5 27,4 24,5
Mean N/mm²	17,7	7,9	22,4	23,0
Std. Dev. N/mm²	5,6	3,1	7,7	5,7
Var.Koeff.	0,32	0,39	0,34	0,25

However, until now, there has been no attempt to compare the properties of various classes of impregnating materials and to evaluate systematically which class of material provides an optimum technical solution for concrete injection. Therefore, different types of impregnation materials suited for concrete injection have been compared under controlled and reproducible conditions.

2. Materials used and Tests performed

2.1 Porous Concrete

As a basic material porous concrete cylinders of 15 x 20 cm were used. They were made from Ø 8 mm single-size crushed clay brick aggregates and a German standard portland cement PZ 35 F. The cement content was 196 kg/m³ and the water/cement-ratio was 0.53 throughout the tests. The density of the porous material was 1,34 kg/dm³. The porosity of all cylinders was in average 38 Vol.% and all values laid between 36 Vol.% and 40 Vol.%.

The specimens were cured for 28 days according to German standards i.e. 7 day moist, thereafter storage at 20°C and 65 % relative humidity up to 28 days. Typical specimens ready for injection are shown in fig. 2.

Fig. 2 Specimens of porous concrete ready for impregnation

2.2 Impregnation Materials

The specimens were impregnated using the materials specified in table 2.

2.3 Test Procedure

28 days after casting the compressive strength of 12 specimens was tested and the other specimens were impregnated with the materials according to table 2. 12 comparison specimens have been used for the

Table 2. Materials used for the impregnation

Nr.	Material	Mix Design
1	Cement Paste	W/C = 0,7, Silicate admixture 1
2	"	W/C = 0,8, Silicate admixture 1
3	"	W/C = 0,8, Silicate admixture 2
4	"	W/C = 0,6, no admixtures
5	Polyurethane	foaming
6	"	non-foaming
7	Epoxide	

materials 1 to 4 and 3 comparison specimens for the materials 5 to 7. For all materials the compressive strength was tested 28 days after impregnation. For the materials Nr. 1, 2 and 3 additional tests 2 hours after impregnation were performed. The aim of these tests was to detect a possible strength reduction due to the interaction between the dry concrete and the wet cement paste. Fig. 3 shows the device being used for impregnation.

Fig. 3 Impregnation Device

Fig's 4 and 5 show cement paste impregnated and epoxide impregnated specimens respectively.

Fig. 4 Cement-paste impregnated specimen of porous concrete

Fig. 5 Epoxide impregnated specimen of porous concrete

The impregnation was performed in the following way:

The specimen was equipped with a packer and inserted in the apparatus shown in fig. 3 and mounted to the pump. The impregnation material was prepared in a storage tank and the tank was connected to the pump. Then the impregnation was started and continued until the specimen was filled up completely. The maximum pressure was 5 bar. Then the specimen was taken out of the device and covered with a plastic plate and stored under 20°C/65 r.H. until the strength tests. Prior to the strength tests the upper part of the specimen (5 cm), containing the packer, was cut off. Thus, for the strength tests standard 15/15 cm cylinders were used.

3. Results and Discussion

The results of the strength tests are summarized in table 3. The material No."0" comprises the reference data of non-impregnated specimens after 28 days of storage.

Table 3. Results of strength-tests

Material Nr.	Strength of N/mm^2 after the impregnation			Impregnation material
	0	2 hours	28 days	
0	4.5	-	-	none
1	-	2.8	12.1	cement paste
2	-	3.0	10.1	cement paste
3	-	2.8	11.2	cement paste
4	-	-	30.3	cement paste
5	-	-	4.9	polyurethane
6	-	-	6.11	polyurethane
7	-	-	41.3	epoxide

From table 3 the following conclusions can be drawn. If highly porous concrete is impregnated with cement paste, the compressive strength is reduced in the first few hours, due to the fact that the concrete is tested under wet conditions in this case. Wet concrete always has a lower strength than dry concrete. The ultimate strength that can be obtained with an injection of cementeous material is predominantly determined by the water cement ratio of the impregnation mix. The lower the W/C-ratio, the higher the strength of the impregnated specimens.

If the water-cement-ratio exceeds about 0,6 a sedimentation of the fine grains occurs and stabilizing admixtures have to be added to obtain homogenous and workable mixes. However, the stabilisation seems to cease during the impregnation process due to effects unknown in nature up to now. No significant effect of the admixtures on the strength of the impregnated specimens has been observed.

Impregnation mixes made of foaming polyurethanes have no significant effect on the strength of the porous concrete. This result has been expected as the foamed polyurethane has a comparatively low strength. Non foaming polyurethane, still a comparatively soft material increases the strength to about 50% compared to the reference specimens. Highest strength values have been obtained using epoxide as an impregnation material. However, this material is also the most expensive one. Therefore its application in damaged structures is very limited due to economical reasons.

4. Summary

Cement and polymers have been used for impregnating porous concrete specimens. The highest compressive strengths were obtained with epoxide materials, whereas foaming polyurethanes had no effect on the strength values.

If cementitious materials are used in the first few hours, i.e. before the set of the cement paste, a strength reduction occurs, which is due to the re-wetting of the dry concrete matrix. The water-cement-ratio of the paste has been found to be the dominant factor affecting the ultimate strength of the repaired structure. Stabilizing admixtures showed no significant effect on the strength development.

For special repair pruposes epoxides should be used, whereas low W/C-pastes is the preferred material for the common use in concrete repair. With little effort it is possible to reach the strength of normal quality concrete by simple impregnation techniques. Thus, impregnation is an appropriate method for concrete repair if the porosity of the damaged structure is comparatively high.

References

K.H. Conrad et al. "Lebenserwartung von Stahlbeton", Bericht IBF, Erftstadt, 1978.

Int.Symp. "Determination and Protection of stone monuments Experimental methods", Unesco, RILEM, Paris, 1978.

B. Dartsch "Konservieren - Sanieren - Restaurieren -, Beton-Verlag, Düsseldorf, 1978.

Kunststoffe für die Sanierung von Natursteinen, 108.Beilage zum Schweizer Baublatt, Nr. 35 (1981).

K.H. Wesche "Baustoffe für tragende Bauteile", Bd.II, Beton, Bauverlag Wiesbaden, 1981, 2. Aufl.

H.R. Sasse "Baustoff-Handbuch der Altbausanierung", Elsner Verlag, 1975.

DURABILITE DES SCELLEMENTS CHIMIQUES
PROBLEMES LIES A LA SAPONIFICATION DES RESINES POLYESTER

Yves RIZOULIERES
Francis MIGLIORI
Ingénieurs au Laboratoire du C.E.T.E. MEDITERRANEE

RESUME

La présente communication met en évidence au moyen d'essais de fluage réalisés en laboratoire et d'analyses chimiques, les problèmes de durabilité des scellements chimiques à base de résines polyesters lorsqu'ils sont réalisés dans le béton en présence d'une humidité permanente.

Les essais de fluage à +20°C montrent qu'il se produit un fluage important et évolutif qui conduit à terme à une diminution importante de la résistance à la traction des scellements réalisés dans le béton et conservés en milieu humide. Les essais d'extraction conduisent alors à une rupture par glissement des ancrages.

Une étude comparative des variations de la fonction ester sur différents échantillons analysés montre qu'à partir des variations de densité optique et par application des lois de BEER et LAMBERT, on a pu mettre en évidence une forte diminution de la fonction ester sur les scellements conservés en milieu humide, ce qui traduit un phénomène de saponification des résines polyesters.

MOTS CLES : scellements chimiques, résines polyester, fluage, densité optique, saponification

1 - INTRODUCTION

Le principe du scellement chimique consiste à mettre en oeuvre une résine prédosée conditionnée dans une cartouche de verre prête à l'emploi.

La cartouche est introduite dans un trou foré dans le béton puis brisée par enfoncement d'une tige d'ancrage filetée, laquelle, animée d'un mouvement de rotation permet de réaliser le mélange des composants de la résine de scellement.

Les avantages du procédé sont indéniables :

- pas de dosage sur le chantier des composants de la résine de scellement,
- facilité de réalisation des scellements (en plafond notamment),
- rapidité de pose,
- possibilité de mise en service de l'ancrage quelques heures après l'exécution du scellement

Dans la plupart des cas, la résine commercialisée pour ce type de scellement est une résine polyester.

La crainte que l'on peut avoir, lorsque l'ancrage reste en contact prolongé avec l'eau, est qu'il se produise en milieu basique une saponification de la résine polyester susceptible d'entraîner à terme la ruine du scellement.

Cette crainte était évoquée dans une note d'information publiée en 1979 par le SETRA qui conseillait aux Maîtres d'Oeuvre de limiter l'emploi des scellements avec résines polyester aux travaux provisoires, et d'utiliser des résines époxydes pour les ancrages définitifs.

Pour mettre en évidence ce problème de durabilité des scellements chimiques à base de résines polyester lorsqu'ils sont réalisés dans le béton et soumis à une humidité permanente, une étude spécifique a été entreprise au laboratoire du C.E.T.E. MEDITERRANEE.

Cette étude effectuée à partir d'un produit de scellement polyester commercialisé a consisté à réaliser :

- des essais de fluage en traction à + 20°C sur des ancrages réalisés en milieux secs et humides,
- l'extraction des scellements après suivi de leur comportement,
- l'étude chimique des variations de la fonction ester sur des échantillons des produits de scellement extraits

2 - PREPARATION DES EPROUVETTES SUPPORTS ET REALISATION DES SCELLEMENTS

Les blocs d'essais en béton de dimensions 63 x 40 x 20 cm^3 ont été confectionnés au Laboratoire du C.E.T.E. MEDITERRANEE à partir de la formule de composition suivante :

Formule type "COPLA"

Ciment CPA 45 R PM, usine de l'Estaque......... 400 kg/m³
Sable 0,1/4 mm concassé calcaire............... 910 kg/m³
Gravillon 7/16 mm concassé calcaire............ 300 kg/m³
Gravier 16/22 mm concassé calcaire............. 650 kg/m³
Eau de gâchage................................. 205 l/m³

La résistance moyenne à la compression du béton déterminée à 28 jours d'âge sur éprouvettes cylindriques 16 x 32 cm était égale à 48 MPa et la densité du béton était égale à 2,40.

Après décoffrage, les blocs ont été conservés dans l'eau à +20°C jusqu'à 14 jours d'âge du béton.

A partir de cette échéance, les blocs destinés aux essais en milieu sec ont été conservés dans une salle régulée à +20°C et 50 % d'humidité relative.

Ceux destinés aux essais en milieu humide ont été conservés en semi-immersion dans l'eau à +20°C.

La foration des blocs a été réalisée par roto-percussion à 28 jours d'âge du béton dans les conditions définies par le fabricant du produit de scellement aux emplacements définis sur la figure 1.

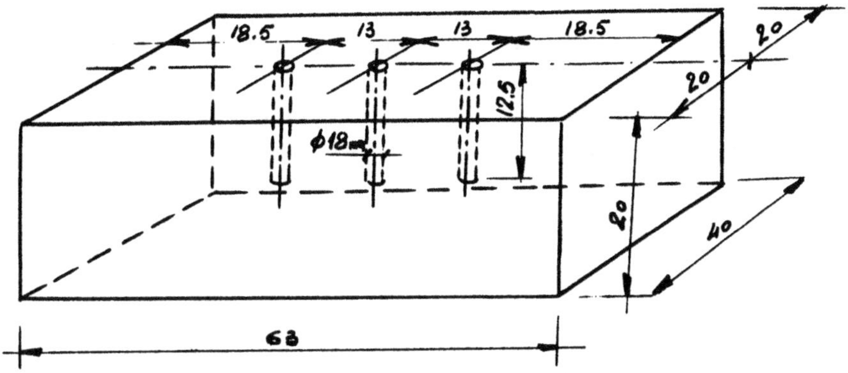

Fig. 1 : Emplacement des scellements

Le nettoyage des évidements a été réalisé par soufflage à l'air comprimé sec et déshuilé, dans les forages destinés aux essais en milieu sec.

Les forages destinés aux essais en milieu humide ont été remplis d'eau puis nettoyés juste avant la réalisation des scellements par soufflage à l'air comprimé.

La réalisation des scellements a été effectuée par introduction d'une cartouche de verre dans chaque forage, et mise en place par roto-percussion des tiges d'ancrage filetées pour briser l'enveloppe de verre et mélanger les composants du contenu (résine polyester + sable quartzeux enrobé de durcisseur).

3 - EXTRACTION DES SCELLEMENTS TEMOINS

Avant d'appliquer la charge, on a respecté le temps de prise et de durcissement du produit de scellement : attente minimum de 30 minutes à +20°C après réalisation du scellement.

Les essais d'extraction ont été effectués sur 3 scellements réalisés en milieu sec et sur 3 scellements réalisés en milieu humide.

Ces 6 scellements ont été sollicités en traction jusqu'à rupture par application, à l'aide d'un vérin annulaire, d'une charge croissant à la vitesse de 1,60 KN/s.

Tous les essais dont les résultats sont donnés par le tableau de la figure 2, ont conduit à la rupture des tiges d'ancrage.

BLOC N° 1 - MILIEU SEC			BLOC N° 2 - MILIEU HUMIDE		
N° scellement	Charge de rupture (KN)	Mode de rupture	N° scellement	Charge de rupture (KN)	Mode de rupture
1 - 1 1 - 2 1 - 3	100 98 100	Rupture des barres d'ancrage	2 - 1 2 - 2 2 - 3	100 100 102	Rupture des barres d'ancrages
Moyenne	99 KN	-	Moyenne	101 KN	-

Figure 2 - Résultats des essais d'extraction des scellements témoins

4 - METHODOLOGIE DES ESSAIS DE FLUAGE

Pour le suivi du fluage en traction des scellements chimiques dans le temps nous avons utilisé un matériel d'essai mis au point et fourni par le Laboratoire de la Société HILTI au LICHTENSTEIN.

Il s'agit de "pots de charge" munis chacun :

- d'un ressort compensateur étalonné permettant d'exercer et de maintenir sur chaque tige d'ancrage une force de traction équivalente au couple de serrage prescrit par le fabricant du produit de scellement,
- d'un comparateur gradué en 1/100e mm pour mesurer les déplacements dus au fluage

Le schéma de la figure 3 donne le principe de l'appareillage.

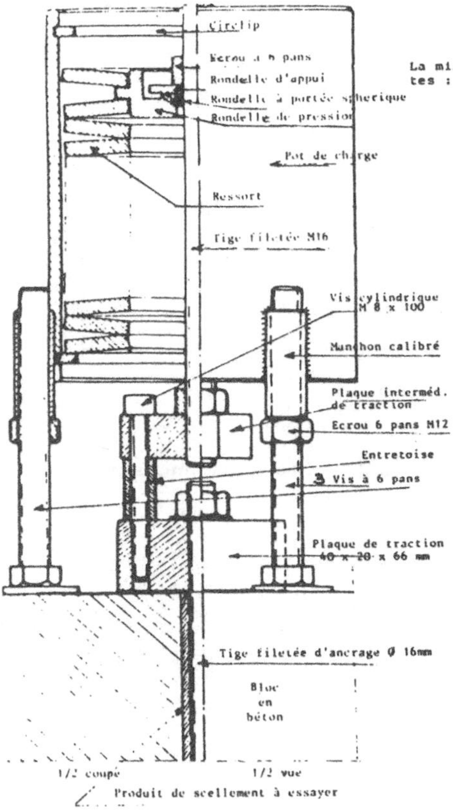

La mise en charge est effectuée selon les modalités suivantes :

- Mise en place de rondelles de calage en acier sous les pieds du pot de charge.
- Positionnement du pot de charge et assemblage de la tige de traction de celui-ci avec la tige d'ancrage du scellement.
- Réglage des pieds du pot de charge de manière qu'aucun effort latéral ne soit transmis par la tige de traction au pot de charge.
- Visšage de la tige de rappel sur la tige de traction.
- Mise en place du manchon d'appui et de la douille sur l'écrou six pans permettant la mise sous-tension du pot de charge.
- Mise en place du vérin hydraulique.
- Mise sous tension de la tige de traction et de la tige de rappel à la valeur de 43 KN par l'intermédiaire d'une pompe hydraulique équipée d'un manomètre.
- Vissage de l'écrou six pans à l'aide d'une tige de serrage et de la douille jusqu'à ce que le serrage de l'écrou compense la force de traction du vérin (le manchon d'appui présente une lumière sur 130° permettant l'engagement de la tige de serrage dans la douille à six pans).
- Mise en place du comparateur pour le suivi des déplacements dans le temps. Le point de lecture est situé sous la face inférieure plane de la tige de traction.
- Immersion complète des blocs dans l'eau à +23°C ou +50°C de manière à assurer un contact direct entre l'eau et la surface libre du scellement.

Fig 3 - Schéma de principe de l'appareillage de traction constante

La mise en charge de chaque scellement a été effectuée au vérin annulaire de manière à obtenir un enfoncement de plateau mobile de chaque "pot de charge" de 18 mm, ce qui correspond à une force de traction sensiblement égale à 40 KN.

En fonction des caractéristiques d'étalonnage des ressorts, les forces de traction développées pour la sollicitation de chaque ancrage sont récapitulées dans le tableau de la figure 4.

BLOCS N° 3 et 4 - MILIEU SEC		BLOCS N° 5 et 6 - MILIEU HUMIDE	
N° scellement	Force de sollicitation en traction (KN)	N° scellement	Force desollicitation en traction (KN)
3-1	40,5	5-1	40,7
3-2	40,3	5-2	41,0
3-3	40,5	5-3	40,7
4-1	40,5	6-1	40,0
4-2	40,5	6-2	41,0
4-3	41,3	6-3	41,7
Moyenne	40,6	Moyenne	40,8

Figure 4 - Forces de sollicitation en traction des ancrages

5 - RESULTATS DES ESSAIS DE FLUAGE

5.1. - ESSAIS EN MILIEU SEC

Le comportement des six scellements réalisés et conservés en milieu sec (20°C et 50 % d'humidité relative) est satisfaisant.

Le graphique de la figure 5 montre que le fluage croît rapidement jusqu'à 22 à 25/100e de mm pendant les 4 jours qui suivent la mise en charge.

Il croît ensuite plus lentement jusqu'à 33 à 35/100e mm pendant les 4 mois qui suivent la mise en charge puis se stabilise entre 36 et 40/100e mm au cours du deuxième semestre suivant la mise en charge.

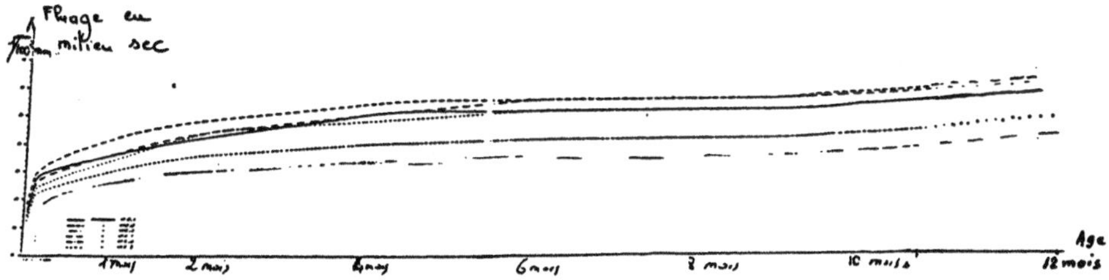

Figure 5 - Essais de fluage en milieu sec

A l'échéance d'un an, les six scellements ont été soumis à des essais d'arrachement.

Chaque essai a conduit à une rupture brutale de la tige d'ancrage hors de la zone de scellement avec quelques éclatements de surface du béton des blocs supports.

La force de rupture mesurée dans chaque cas était sensiblement égale à 100 KN.

5.2. - ESSAIS EN MILIEU HUMIDE

Dans la première partie de l'expérimentation, la conservation des éprouvettes de fluage en milieu humide a été effectuée par immersion jusqu'à mi-hauteur des blocs de béton support dans l'eau à +20°C.

Dans la deuxième partie de l'expérimentation, les blocs ont été conservés en immersion complète.

Le graphique de la figure 6 montre qu'à l'échéance d'un mois, le fluage mesuré est plus faible en milieu humide qu'en milieu sec (19/100e mm en milieu humide et 27/100e mm en milieu sec).

Mais les différences de comportement apparaissent surtout à partir de 4 mois de conservation puisque pour les éprouvettes conservées en milieu sec, la vitesse de fluage se stabilise autour de 0,055 mm/an alors que pour les éprouvettes conservées en semi-immersion la vitesse de fluage mesurée entre 4 et 12 mois est sensiblement 3,6 fois plus élevée (0,196 mm/an).

A l'échéance de 12 mois, pour mieux mettre en évidence l'influence de l'eau sur le comportement des scellements nous avons immergé complètement les éprouvettes.

Le graphique de la figure 6 montre qu'il se produit alors une augmentation importante du fluage. Entre 12 et 18 mois, celui-ci double pratiquement sur toutes les éprouvettes. Il passe en moyenne de 45/100e mm à 91/100e mm.

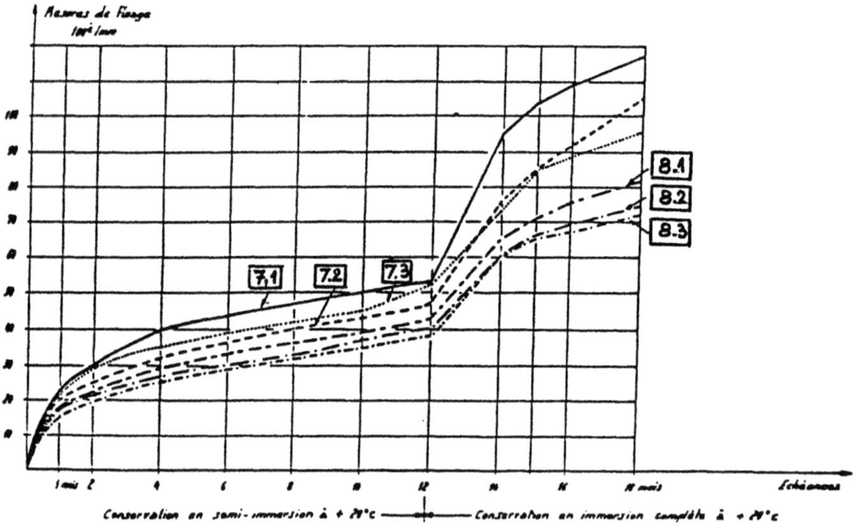

Fig 6 : Essais de fluage en milieu humide

A l'échéance de 18 mois, nous avons réalisé l'extraction des scellements. Le tableau de la figure 7 montre qu'il se produit dans tous les cas un glissement des scellements dans le béton avec une diminution moyenne de 20 % des charges de rupture par rapport aux témoins.

Numéro de scellement	Force de sollicitation (en KN)	Mode de rupture
7-1	81,5	Glissement des scellements
7-2	69,9	dans le béton
7-3 (*)		-
8-1	79,2	Glissement des scellements
8-2	84,0	dans le béton
8-3	87,0	
Moyenne	80,3	-

(*) Scellement non extrait (conservé pour carottage)

Fig 7 : Extraction à 18 mois des scellements soumis aux essais de fluage en milieu humide

6 - ETUDE DE SAPONIFICATION

Nous avons effectué l'étude du degré de saponification des produits de scellement en suivant par spectroscopie infra rouge la variation d'intensité de la fonction ester sur la vibration de valence (C = 0) à 1 730 cm^{-1}.
Nous avons appliqué les lois de BEER et de LAMBERT.

Rappelons brièvement que la loi de BEER exprime une relation entre l'intensité d'absorption et la concentration de l'échantillon et que la loi de LAMBERT exprime une relation entre l'intensité d'absorption et l'épaisseur traversée.

En combinant ces deux lois, on obtient une loi générale qui relie l'absorption d'une radiation incidente mono-chromatique à la concentration et à l'épaisseur de la substance absorbante. On traduit cette loi par les équations :

$$I = Io\ 10^{-kcd}$$

$$kcd = \log \frac{I}{Io} \quad \text{(densité optique)}$$

I et Io sont les intensités de la lumière transmise et de la lumière
incidente,
K est le coefficient d'extinction
C est la concentration de l'échantillon
d est l'épaisseur traversée

Dans cette étude, nous nous sommes bornés à effectuer une comparaison des densités optiques à savoir la mesure expérimentale de $\log \frac{I}{Io}$ pour des échantillons analysés en phase solide dans des concentrations rigoureusement identiques (5 mg de produit pour 200 mg de bromure de potassium).

Les échantillons analysés ont été prélevés en surface des scellements extraits après avoir éliminé les traces de carbonates provenant du béton.

Le tableau de la figure 8 et les graphiques de la figure 9 illustrent les variations des densités optiques mesurées en fonction des échantillons analysés.

Echantillons		Densité optique $\log \frac{I}{Io}$
Cartouche de résine		0,49
Cartouche broyée		0,22
Scellement témoin	Haut	0,22
	Milieu	0,22
	Bas	0,20
Scellement ayant subi les essais de fluage en milieu sec	Haut	0,22
	Milieu	0,20
	Bas	0,32

Echantillons — Nature	N°		Densité optique $\log \frac{I}{Io}$
Scellements ayant subi les essais de fluage en milieu humide	7 - 1	Haut	0,12
		Milieu	0,12
		Bas	0,12
	7 - 2	Haut	0,09
		Milieu	0,07
		Bas	0,15
	7 - 3	Haut	0,10
		Milieu	0,12
		Bas	0,11
	8 - 1	Haut	0,08
		Milieu	0,10
		Bas	0,12

Fig. 8 : Mesures des densités optiques sur les scellements extraits

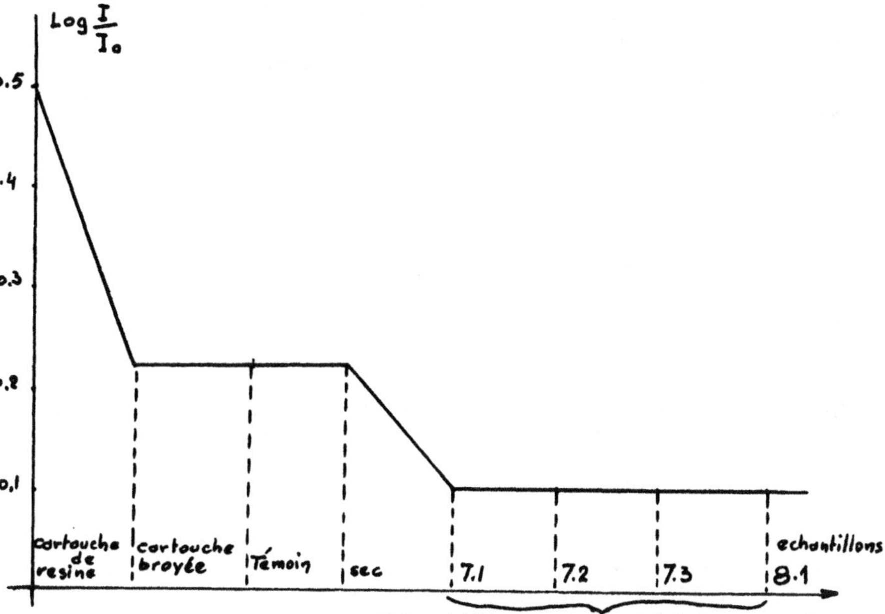

Echantillons ayant subi les essais de fluage en milieu humide

Fig. 9 : Graphes des variations de densité optique en fonction des échantillons analysés

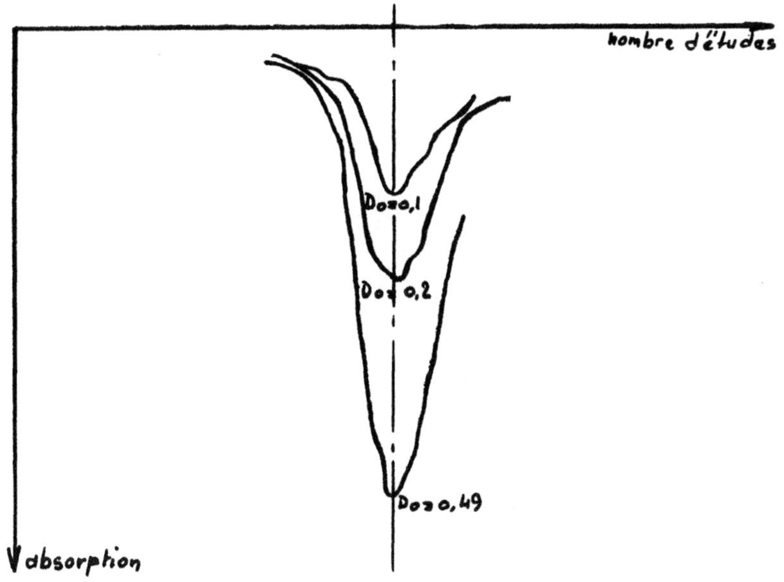

Fig. 10 : Variation de la fonction ester (C = 0)

Ces variations importantes des densités optiques permettent de mettre en évidence une forte diminution de la fonction ester sur les échantillons ayant subi les essais de fluage en milieu humide ce qui traduit le phénomène de saponification de la résine polyester.

7 - CONCLUSIONS

Les essais de fluage à + 20°C montrent qu'il se produit un fluage important et évolutif des scellements chimiques à base de résines polyester lorsqu'ils sont réalisés dans le béton et conservés en présence d'une humidité permanente.

Les essais d'extraction conduisent alors à une rupture par glissement des scellements dans le béton avec une diminution importante des charges de rupture par rapport aux témoins.

L'étude comparative des variations de la fonction ester sur les scellements extraits montre qu'à partir des variations de densité optique et par application des lois de BERR et de LAMBERT on a pu mettre en évidence une forte diminution de la fonction ester sur les échantillons conservés en milieu humide ce qui traduit le phénomène de saponification de la résine polyester et confirme les craintes que l'on avait sur la durabilité de ce type de scellement.

EFFECTS OF CELLULOSE ETHER ON HOMOGENEITY OF CONCRETE IN STRUCTURES -- RELATING QUALITY VARIATIONS AND CONSTRUCTION JOINTS

T. SOSHIRODA Shibaura Institute of Technology
K. HAYAKAWA Shin-Etsu Chemical Co., Ltd.
K. YODA Shibaura Institute of Technology
M. TANAKA Shibaura Institute of Technology

Abstract
Quality variations of concrete containing cellulose ether in full-size models of structures relative to the depth and width were examined by strength, absorption and density of cores drilled from the models and pulse velocity and rebound number of each part of the models comparing with conventional concretes. Quality variations of concrete containing cellulose ether in the model structure were smaller than those of conventional concretes. It may be caused by the effect of cellulose ether on controlling segregation of concrete.

Otherwise strengths of construction joints in concrete specimens were examined by splitting tests relating addition of cellulose ether. Strengths of construction joints were increased when the old concrete was containing cellulose ether. It may be caused by little laitance because of controlling bleeding.

As the results of the experimental investigations it may be stated that the addition of cellulose ether improves heterogeneity characteristics of concrete in structures.
Key words: Cellulose ether, Concrete, Heterogeneity, Quality variation, Full-size model of structure, Construction joint, Segregation, Bleeding.

1. Introduction

Concrete in structures has heterogeneity characteristics because of lack of uniformity in manufacturing, instability during fresh, compaction by self-weight, several kinds of defects due to bad workmanship, construction joints macroscopically, and so on.

Therefore it may be stated that the controlling segregation of fresh concrete is an effective approach to keep homogeneity of concrete in structures.

Hayakawa and Soshiroda (1985) reported the results of laboratory tests that the addition of cellulose ether improves heterogeneity of concrete along the depth and defects below the aggregate particles by controlling segregation and bleeding.

In this paper further investigations on the effects of cellulose ether on homogeneity of concrete using full-size wall-type models of structures and models of construction joints are described.

2. Effect of cellulose ether on variations in quality of concrete in structures

2.1 Experimental details

Three panels of model walls were cast vertically with single reinforcement in plywood forms as shown in Fig. 1. One panel was made of one kind of mixture, then three kinds of concrete as shown in Table 2 were supplied by a ready-mixed concrete plant. The walls were cured under moist condition, demolded at 12-day, tested by two non-destructive methods at 25-day and drilled to core specimens at 27-day. The cores were taken out from different positions in a wall. The size and position of cores and test items are given in Fig. 1 and Table 1. Splitting tensile strengths were obtained on two ϕ 15 x 15 cm cut from a core by two directions as shown in Fig. 2: one is the casting direction and the other is the direction perpendicular to it.

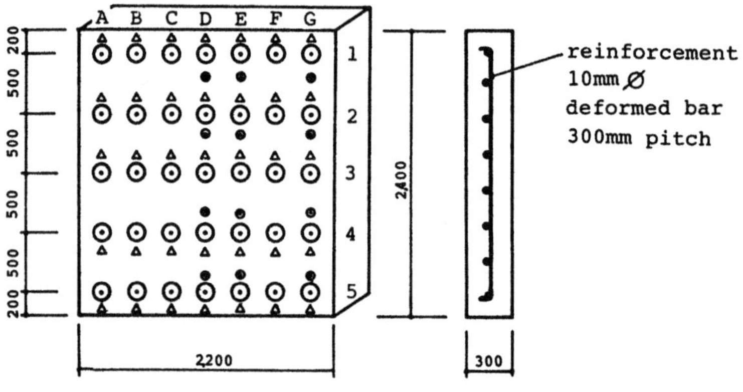

Fig.1. Model wall

Table 1. Test items and positions.

Mark of position	Test item	Size of core	
O A,D,F-raw	Compressive test	Ø15x30cm	
O B,C,E,G-raw	Tensile test	Ø15x30cm	15x15cm x2
o	Water absorption	Ø 7x30cm	
▲	Rebound number by Schmidt hammer		
•	Ultrasonic pulse velocity by Pundit		

Table 2. Mix data of concrete.

Mix	W/C (%wt.)	s/a (%vol.)	Water content (l/m^3)	Additive AE&WR (g/m^3)	CE (g/m^3)	DF (g/m^3)
H	37.5	34.6	173	1,155	—	—
N	57.0	44.8	189	830	—	—
C	57.0	44.8	189	830	664	332

Materials: Cement — Normal Portland cement
Fine aggregate — River sand
Coarse aggregate — Crushed stone maximum size 40mm
Additives: AE&WR — Air entraining and water reducing agent
CE — Cellulose ether (Methyl cellulose type)
DF — Deformer (Nonionic surface active agent)

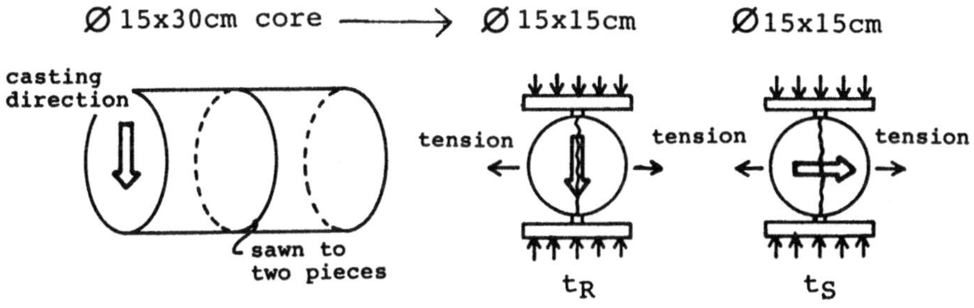

Fig.2. Splitting tensile test method and notation.

2.2 Test results

The test results of fresh concrete are given in Table 3.

The bleeding capacity of the mix C is the smallest among three mixes, showing 1 / 8 of that of the mix N. It is found for the addition of cellulose ether to control bleeding remarkably. The bleeding capacity of the mix H shows 1 / 4 of that of the mix N because of less water and more cement content.

Table 3. Test reslts of fresh concrete.

Mix	Slump (cm)	Air content (%)	Bleeding capacity (%)
H	7.5	4.2	0.11
N	19.0	3.9	0.44
C	16.0	4.1	0.06

A statistical analysis of the test results was made to study the effect of the addition of cellulose ether on the variations of various properties of concrete in a wall comparing with two conventional mixes without cellulose ether: one was a mix which has the same mix proportion and about the same strength and slump and the other was a mix which has higher strength and lower slump.

The results are shown in Table 4 ~ 6 and Fig. 3 ~ 5.

Table 4. Variation in compressive strength in a wall.

Mix		H	N	C
n:Nunber		15	15	15
r:Range	(kgf/cm^2)	133	56	34
(Maximum)		476	296	258
$\overline{X}$:Average	(kgf/cm^2)	412	268	238
:Standard deviation	(kgf/cm^2)	38.6	15.0	10.3
V:Coefficient of variation	(%)	9.7	5.8	4.5

Table 5. Variation in tensile strengths in a wall.

	Mix		H	N	C
t_R	n:Number		20	20	20
	r:Range	(kgf/cm^2)	9.1	8.8	8.9
	(Maximum)		41.6	32.4	30.5
	$\overline{X}$:Average	(kgf/cm^2)	37.2	27.8	26.4
	:Standard deviation	(kgf/cm^2)	2.7	2.7	2.3
	V:Coefficient of variation	(%)	7.2	9.9	8.7

	Mix		H	N	C
t_S	n:Nunber		20	20	20
	r:Range	(kgf/cm^2)	10.4	9.7	7.9
	(Maximum)		38.7	30.4	29.2
	$\overline{X}$:Average	(kgf/cm^2)	34.6	26.4	24.2
	:Standard deviation	(kgf/cm^2)	2.7	2.7	2.2
	V:Coefficient of variation	(%)	7.9	10.2	8.8

Table 6. Variation in water absorption in a wall.

Mix		H	N	C
n:Nunber		12	12	12
r:Range	(%vol.)	1.14	1.48	1.15
(Maximum)		8.26	9.92	10.43
$\overline{X}$:Average	(%vol.)	7.86	9.44	9.69
:Standard deviation	(%vol.)	0.30	0.38	0.29
V:Coefficient of variation	(%)	3.9	4.2	3.1

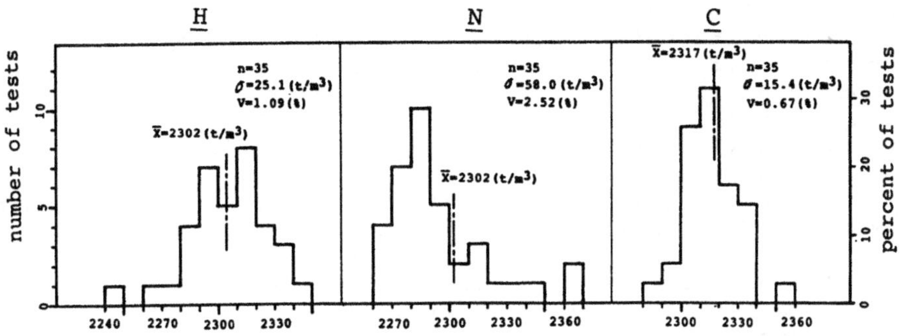

Fig.3. Distribution of density, data from 35 tests.

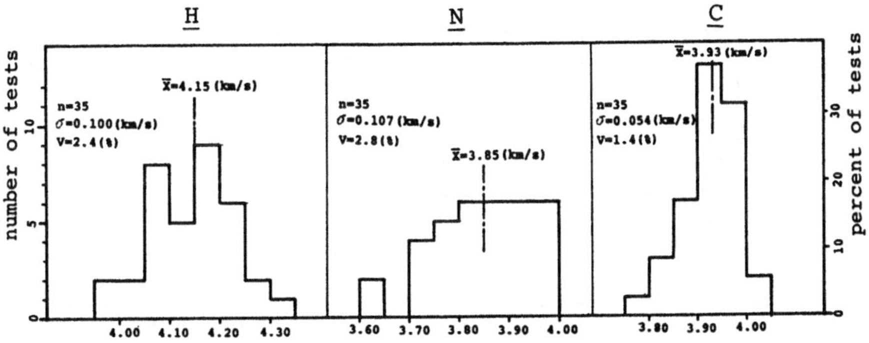

Fig.4. Distribution of pulse velocity, data from 35 tests.

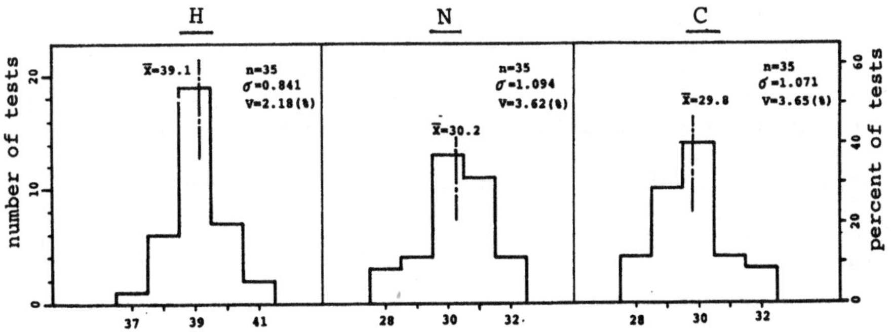

Fig.5. Distribution of rebound number, from 35 tests.

Comparing the three mixes on variations in various properties which indicate concrete qualities in the walls, the variations of the mix C in each properties except rebound number are always smallest and those of the mix H are always smaller than those of the mix N: C < H < N. These are probably due to the differences of stability of concrete during fresh. The stability of fresh concrete influences on mixing, transporting, placing, compacting and movement after casting, consequently on homogeneity of concrete in structures. It may be stated that the addition of cellulose ether was effective signifficantly on the mix C and the less water content with low water cement ratio was effective on the mix H.

Generally speaking the mix H is high quality concrete with high strength and medium consistency. Therefore it can be said that the addition of cellulose ether improves the instability and then heterogeneity in structures of wet consistency concrete which is conventional for building structures in Japan.

Because rebound numbers measured by Schmidt hammer mainly depend on surface hardness of concrete, the variations of them are rather sensitive to existance of coarse aggregate near the surface of concrete. In this experiment the difference in hardness between coarse aggregate and mortar matrix may be the least in the mix H because of its lower water cement ratio. Therefore the variation in rebound numbers of the mix H is smaller than those of the mix N and C which are about the same as shown in Fig. 5.

The difference due to the casting direction in the two tensile strengths in a core was examined by splitting tests on the two ϕ 15 x 15 cm cylinders sawn from a core as shown in Fig. 3. Soshiroda (1972) proposed that the anisotropy of concrete characterized by tensile testing is useful as an indication of change of internal structure brought about from instability of fresh concrete. In this experiment the tensile anisotropy is found in averaged values of all mixes as shown in Table 5. It seems that this was mainly due to defects below the aggregate particles and bleeding channels. Defects below the aggregate particles in concrete were probably caused by bleeding and bridging of aggregates. These might be promoted by the large size of aggregate such as 40 mm maximum and reinforcements.

It is suggested from the result that even a little bleeding possibly brings about bond cracks in concrete in structures.

3. Effect of cellulose ether on strength of construction joint

3.1 Experimental details

Materials and concrete mix data used in the experiment are given in Table 7. The size of specimen, casting direction, the position of joints and splitting tests and testing ages are shown in Fig. 6.

Column Type specimens were 15 x 15 x 45 cm and were cast vertically in steel forms by each 15 cm depth three times at a week interval. Three specimens were made for one condition. Eight kinds of condition for the model of construction joints were prepared combining three kinds of treating method with two kinds of concrete mix as shown in Fig. 7. The treating methods are as follows:

(1) nothing
(2) removing laitance with steel wire brush
(3) making layer of mortar containing cellulose ether just before placing
(thickness of layer ≒ 5 mm, S / C = 1.1, W / C = 0.4)

The strengths of joint were measured by splitting tensile strength.

Table 7. Mix data of concrete.

Mix	W/C (%wt.)	s/a (%vol.)	Water content (l/m^3)	Additive CE (g/cm^3)	Additive DF (g/cm^3)	Slump (cm)
N	60.0	39.0	194	—	—	14 18
C	60.0	39.0	194	642	161	19 21

Materials:	
Cement	Normal Portland cement
Fine aggregate	River sand
Coarse aggregate	Crushed stone maximum size 20mm

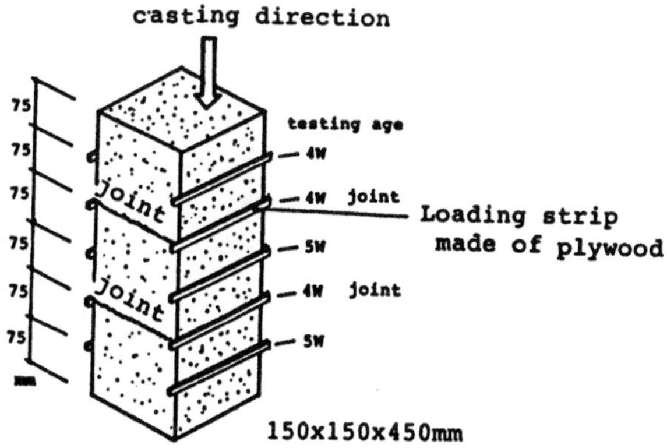

Fig.6. Size of specimen, casting direction, position of joints and loading and testing age.

3.2 Test results

The test results are shown in Fig. 7 and Fig. 8.

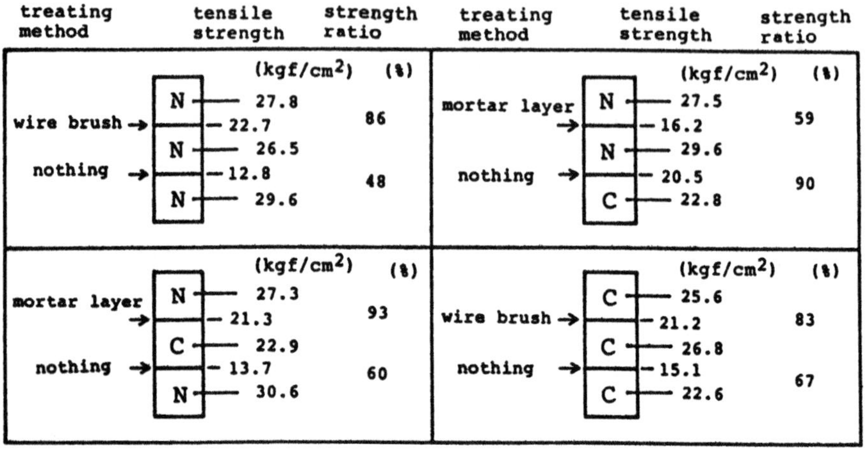

note: Ratio of strength = joint strength / concrete strength.

Fig.7. Splitting tensile strengths of concretes and joints.

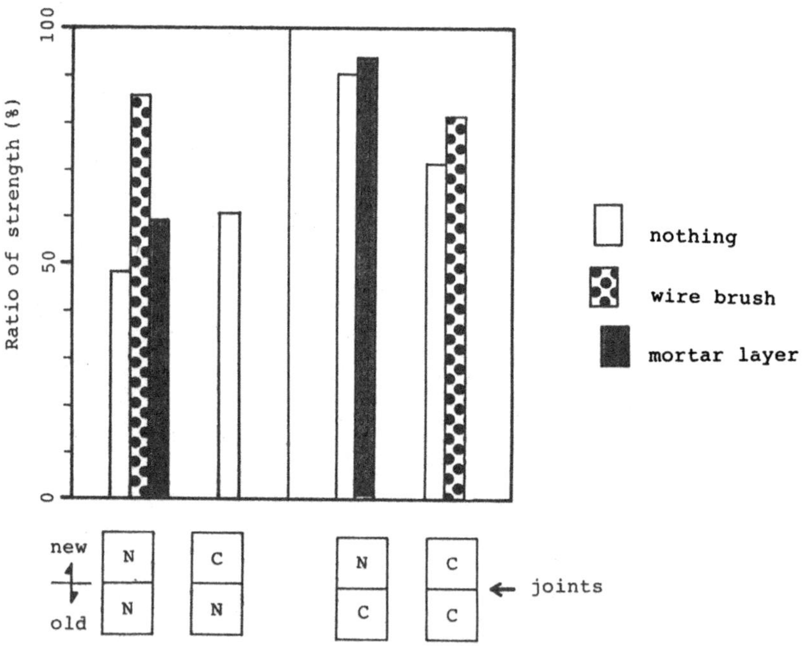

note: Concrete strength is the weaker one compraring new with old.

Fig.8. The ratio of joint strength to concrete strength.

In the case of no treatment: when the old concrete contains cellulose ether, the strengths of joints are considerably higher than those when the old concrete is plain. Comparing the splitted surfaces, in the former case the fracture occurred through the layer near the joint in the old concrete, otherwise in the latter case the fracture occurred through the joint where the laitance layer was observed. This improvement in bond of joint was probably due to decreasing laitance by the effect of the addition of cellulose ether on controlling bleeding.

On the other hand there was little effect when the new concrete contains cellulose ether.

In the case of removal of laitance: when the laitance of surface of joint was removed, the strengths of joints are near the strength of concrete itself and about the same as those when the old concrete contains cellulose ether, regardless of the kind of concrete. The fracture surfaces were through concrete.

In the case of mortar layer: The mortar layer containing cellulose ether improves the strength of joint a little. It seems that the strength of joint mainly depends on the laitance.

4. Conclusions

The main effects of the addition of cellulose ether on heterogeneity characteristics of concrete in structures were as follows:

(1) improving variation in quality of concrete in structures caused by controlling instability of fresh concrete.

(2) improving bond stregth of construction joints without treatment due to reducing laitance by controlling bleeding.

(3) consequently improving heterogeneity characteristics of concrete in structures.

References

Hayakawa, K. and Soshiroda, T. (1985) An experimental study on the concrete containing cellulose ether. Trans. Japan Concrete Institute, 7, 17 ~ 24

Soshiroda, T. (1972) Anisotropy of concrete. Proc. 1st Int. Conf. on Mech. Behav. of Materials, 4, 300 ~ 307

Section 4

Coating and painting

Revêtement et peinture

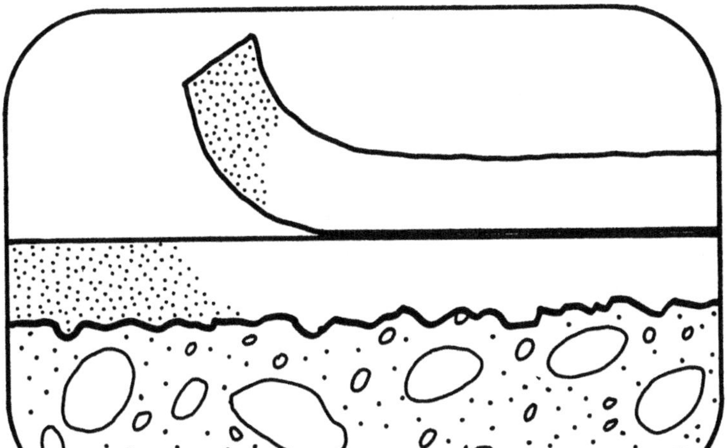

LA MISE EN PEINTURE DES OUVRAGES DE GENIE CIVIL EN BETON - EXEMPLES DE REALISATIONS

D. ANDRE
Section Qualité des Equipements - Laboratoire Central des Ponts et Chaussées

J.P. BENNETON
Section Chimie-Pollution - Centre d'Etudes Techniques de l'Equipement de Lyon

F. MIGLIORI
Section Chimie - Centre d'Etudes Techniques de l'Equipement d'Aix-en-Provence

Résumé
La mise en peinture des ouvrages de génie civil en béton est réalisée soit pour des problèmes d'environnement et d'esthétique, soit pour assurer une protection complémentaire du béton d'un ouvrage ancien venant d'être réparé. Les durées de vie recherchées pour la protection sont de 12 à 15 ans, tant sur les ponts que sur les tunnels. La présente communication traite de l'expérience acquise en France depuis 10 ans pour les systèmes de protection et pour les enduits de ragréages et donne les avantages et les inconvénients des choix réalisés. En particulier les limites de ces revêtements de faible épaisseur sont données avec la protection par peinture de corniches en béton particulièrement gélif et les déboires qui ont suivi.
Mots clés : Peinture, Revêtement, Qualité, Adhérence, Mise en oeuvre, Résine époxyde, Résine polyuréthanne, Béton hydraulique, Réparation, Enduit, Pont.

1. Introduction

La mise en peinture des ouvrages en béton dans le domaine du génie civil s'est développée au cours des dix dernières années en France. Elle concerne principalement trois domaines :
- les ponts neufs ou en entretien ; dans le cas des ouvrages neufs, la protection par peinture permet une meilleure intégration de l'ouvrage dans le site, les peintures étant là à des fins décoratives et esthétiques mais améliorant également la durabilité de l'ouvrage en assurant une protection contre les agents agressifs extérieurs.

Dans le cas des ouvrages en entretien ayant eu par exemple des ragréages et des fissures injectées, la protection par peinture apporte à la fois une amélioration de la protection de surface de l'ouvrage en ralentissant la pénétration de l'humidité et une amélioration esthétique et sécurisante en masquant les zones réparées.

- les tunnels et trémies urbaines ; la protection par peinture apporte un meilleur confort de l'usager, un meilleur rendement des installations d'éclairage et surtout une facilité de nettoyage. On peut citer également les parkings souterrains où la mise en peinture permet en plus des avantages cités ci-dessus une meilleure signalétique et une meilleure sécurité.

- les écrans anti-bruit ; la mise en peinture apporte un meilleur confort par une esthétique améliorée. Les peintures sont en concurrence avec les traitements de surface des bétons, les granulats apparents ou les bétons colorés dans la masse ; certaines peintures peuvent faciliter le nettoyage des graffitis et dégradations diverses.

2. Les durées de vie souhaitées

Lors de la mise en peinture des ouvrages en béton de génie civil, la condition première recherchée est la durée de vie la plus longue possible compte tenu des contraintes d'exploitation liées à la mise en oeuvre et du prix des travaux.

Les durées de vie recherchées sont de 12 à 15 ans, mais bien entendu pour arriver à ces résultats il faut prendre un certain nombre de précautions tant au niveau du choix des produits, que de l'exécution des travaux en particulier pour la préparation de surface du support et le respect des épaisseurs préconisées pour chaque couche.

Pour aider les responsables de travaux, des documents techniques ont été publiés sur le sujet :

- Note d'Information Technique sur la mise en peinture des bétons de génie civil publiée par le LCPC. Cette note donne des informations pour le choix des produits de protection adaptés aux types d'ouvrage et aux buts recherchés, donne des conseils pour la mise en oeuvre et le contrôle des chantiers.

- Annexe technique au fascicule n° 65 ; "Exécution des ouvrages de génie civil en béton armé ou précontraint".

L'observation des préconisations de ces documents permet aux maîtres d'oeuvre d'atteindre l'objectif fixé et d'éviter les déboires trop souvent observés dans le domaine des peintures des bâtiments comme cloquages, décollements, usure prématurée ...

En particulier, une annexe à la note d'information technique comporte des fiches de systèmes de peinture ayant subi avec succès les principaux essais de qualité en laboratoire. Cette liste, remise à jour périodiquement, correspond à une procédure d'évaluation de qualité des produits et non à une procédure de certification de qualité.

A noter que tous ces systèmes de peinture, s'ils assurent une protection effective de la surface du béton, ne peuvent en aucun cas être confondus et assimilés à des produits d'étanchéité. Du fait de leur faible épaisseur et quelquefois de leur rigidité, ils ne peuvent résister en effet à une fissuration même faible des supports.

3. Choix des systèmes de protection

Pour la réalisation correcte des travaux, le responsable doit :

. définir le ou les types de systèmes de peintures adaptés au problème à résoudre,

. utiliser dans la gamme retenue des produits de qualité convenable.

Le choix du système nécessite une analyse fine et une réflexion sur le problème global de la protection et de la durée de vie attendue, liées à l'agressivité de l'ambiance dans laquelle se situe l'ouvrage, à sa structure et à l'état des bétons à protéger.

Le système étant choisi quant à son objectif, il reste à retenir les produits individuels le composant. Pour ce faire, l'utilisation de la liste nominative des produits ayant subi avec succès les essais de laboratoire et ayant le cas échéant des références d'ouvrages reste la meilleure solution.

3.1. Conception générale des systèmes de protection

Comme dans le cas de la protection des aciers, on utilise des systèmes de protection dans lesquels chaque produit individuel a une fonction bien définie.

De façon générale, les systèmes comportent 3 couches :

- la première couche joue un rôle d'accrochage sur le béton en favorisant éventuellement son durcissement. La couche très fluide et peu épaisse doit parfaitement mouiller le support. Les produits utilisés pigmentés ou non, conduisent à une couche d'épaisseur très faible.

- la deuxième couche apporte les fonctions imperméabilité et barrière mécanique. Cette couche doit être suffisamment épaisse et sa fonction sera d'autant efficace qu'elle sera plus riche en résine et que sa pigmentation aura été étudiée pour le critère imperméabilité.

- la troisième couche complète la fonction imperméabilité de la couche précédente et apporte la bonne tenue aux agents agressifs extérieurs, en particulier la tenue à la lumière. Cette couche, très riche en résine, est généralement d'épaisseur plus faible.

3.2. Bilan des systèmes utilisés depuis 10 ans

Les systèmes les plus couramment utilisés sont les suivants :

- systèmes époxydiques ou polyuréthannes ou mixte époxydiques-polyuréthannes,
- systèmes caoutchouc chloré
- systèmes "pliolites".

Les systèmes à base de résines acryliques ou vinyliques de moindre durabilité ne sont pratiquement pas utilisés.

3.2.1. Systèmes mixte époxy-polyuréthanne
Ils concernent plus de 80 % des mises en peinture des ouvrages traités et sont utilisés :

- sur des ponts à des fins esthétiques et protection contre les agents agressifs atmosphériques,

- sur des écrans anti-bruit ou des trémies urbaines de courte longueur pour lesquels le lessivage à l'eau chaude sous pression est réalisé régulièrement.

La constitution du système est la suivante :

. première couche d'impression très fluide souvent non pigmentée à base de résine époxydique à deux composants
. deuxième couche épaisse, pigmentée et chargée, à base de résine époxydique à deux composants
. troisième couche de finition à liant polyuréthanne à deux composants avec un durcisseur non jaunissant de type polyisocyanate aliphatique.

Plusieurs variantes sont utilisées pour la couche de finition en utilisant des liants acryluréthannes à deux composants ou acryliques monocomposants.

Dans presque tous les cas, les résultats obtenus avec ces systèmes sont excellents, la résistance des finitions aux agents atmosphériques, au lessivage à l'eau sous pression et de façon plus générale à toutes les agressions étant bonne.

Les difficultés constatées concernent principalement l'usure des systèmes en raison semble-t-il de couches trop fines, dues d'une part à la conception des systèmes et d'autre part à des quantités de produit réellement déposé trop faibles surtout sur certains bétons poreux.

Aussi, en particulier quand l'entretien par lessivage à l'eau sous pression est prévu, nous avons été amenés à demander un renforcement de l'épaisseur des couches soit au minimum 250 µm sec pour l'ensemble du système. Ceci est également à conseiller pour toutes parties d'ouvrages directement exposées au ruissellement.

3.2.2. Système époxydique phase aqueuse
Ces systèmes sont utilisés principalement dans les tunnels routiers en raison de l'intérêt qu'ils apportent en atmosphère confinée pour l'application du fait de l'absence de solvant.

Le système comporte 3 couches de formulation identique (époxy phase aqueuse), la première étant diluée pour permettre un meilleur accrochage au béton.

Des grandes réalisations, comme le tunnel franco-italien du Fréjus à Modane, ont permis de tester l'applicabilité de ces produits en couches relativement épaisses (350 µm sec).

Actuellement il est prématuré de conclure sur la durée de vie de tels revêtements mais le comportement, en particulier la résistance au lessivage, semblent excellents.

Toutefois, pour l'application de ces émulsions une formation du personnel d'application semble primordiale afin d'éviter les tendances :

- de diluer le produit avec n'importe quoi, ou n'importe comment ce qui induit de forte variation de viscosité donc d'épaisseur des couches

- d'appliquer sur béton humide, voire mouillé, en pensant que l'émulsion "absorbera" l'excédent d'humidité !.

3.2.3. Systèmes "pliolite"
Ils comportent 3 couches (200 µm sec) de formulation identique (copolymère acrylique), la première étant diluée. Ils sont utilisés quand on recherche un aspect mat de la protection, qui ne peut être obtenu avec une finition polyuréthanne. Ces produits conduisent à d'excellents résultats sur le plan de la tenue dans le temps, de l'adhérence. Par contre, la limite des systèmes se trouve dans la résistance au lessivage à l'eau sous pression, ce type de produit étant beaucoup moins performant que les systèmes époxy-polyuréthannes.

3.3. Perspectives
Dans plusieurs cas récents, on a cependant noté les limites d'emploi de ces techniques :

- supports fissurés ou fissurables : si l'on recherche une réelle protection de l'ouvrage, il faut s'orienter vers des enduits plus épais et semi-rigides (épaisseur de l'ordre de 1 mm)

- anciens supports présentant de nombreuses anfractuosités : un boucheporage devient nécessaire

- supports très gélifs ; ces systèmes n'assurent pas une protection suffisante.

4. Enduits de ragréages
Nous ne considérons ici que les enduits de surfaçage destinés à obturer les anfractuosités du béton (quelques mm à 1 cm) et non les matériaux de ragréages en forte épaisseur (mortier).
Un enduit de surfaçage doit avoir :

- une très bonne adhérence au support béton même en faible épaisseur
- une très bonne cohésion interne et ne pas présenter de fissuration au durcissement même en faible épaisseur.

4.1. Enduits hydrauliques
Bien que les enduits hydrauliques soient en principe proscrits, certains ragréages réalisés avec ces mortiers ont conduit à des déboires importants.

En effet le manque d'adhérence entre le béton de l'ouvrage et le mortier hydraulique de ragréage est accentué par les tensions internes de réticulation des peintures bi-composantes (époxydes) qui créent à court terme un craquelage du film de peinture et un décollement ultérieur du ragréage. De plus, l'utilisation des mortiers hydrauliques demande un temps de séchage souvent incompatible avec les problèmes de chantier, ce qui condamne cette technique.

4.2. Enduits aux résines

Dans les enduits aux résines il faut différencier :. les mortiers à haute performance composés d'un mélange de ciment et de charge et d'une résine (en latex) en émulsion . et les mortiers à base de résine époxyde. Ces mortiers sont riches en liant ($>$ 50 %).

Avec ces deux techniques de ragréage lorsque les travaux ont été réalisés dans les règles de l'art - en respectant notamment les conditions de propreté de la surface à traiter et les délais de séchage avant mise en peinture, nous n'avons pas rencontré de problèmes.

Nous avons eu des défauts d'adhérence du film de peinture appliqué trop tôt sur des mortiers haute performance mis en oeuvre en forte épaisseur ($\geq$ 2 cm) dont le temps de séchage n'avait pas été respecté.

Il faut souligner l'importance et la différence du temps de séchage de ces deux techniques de ragréage, en effet :

- pour les mortiers haute performance (mélange ciment, charge et résine en émulsion) il faut que l'enduit soit parfaitement sec

- pour les mortiers époxydes il faut faire l'application de peinture sur un enduit encore poisseux avec un système compatible.

Enfin, il vaut mieux réserver ces enduits à des zones limitées de l'ouvrage comme bouchepore, en particulier après que la préparation de surface de l'ouvrage avant peinture ait fait apparaître des défauts comme bulles, balèvres, défauts de planéité.

5. Limites de la protection par peinture de certains bétons

Les limites d'efficacité des peintures sur béton en faible épaisseur telles qu'elles viennent d'être décrites ont hélas été vérifiées sur les corniches du viaduc des Egratz (Haute-Savoie) situé entre 800 et 1 000 m d'altitude, en partie en vallée encaissée.

Ces corniches préfabriquées en usine sont constituées d'un béton particulièrement gélif ; dès leur pose, après une préparation de surface soignée par décapage à l'abrasif, elles avaient été recouvertes par un système 3 couches mixte époxy/polyuréthanne (150 µm sec). Dès la fin du premier hiver, le décollement du film s'est amorcé entraînant avec lui 1 à plusieurs mm de peau superficielle du béton. Après trois hivers il ne reste plus rien du film sur le haut des corniches et le béton se dégrade très fortement. Il est probable que de l'eau pénètre par quelques discontinuités du film, diffuse dans la partie superficielle du béton et fait éclater celle-ci lors du gel. Ce phénomène est certainement accentué par la rigidité à basse température des films époxy/uréthannes.

Pour tenter de protéger les bétons, une zone d'essais a été effectuée avec une dizaine de produits différents : peintures classiques pliolites, acryliques (perméables à la vapeur d'eau), époxyde/uréthanne, enduit épais polyuréthannes, enduit souple polyuréthannes, enduits épais ciment/polymère.

Actuellement il semble que ce soit les enduits épais (e > 1,5 mm) et très souples (époxy/uréthannes) qui se comportent le mieux mais il convient d'attendre encore 1 ou 2 ans pour se faire une opinion définitive. Bien sûr cette solution est chère (2,5 à 3 fois la peinture) et il convient de protéger l'enduit souple par une protection anti-ultraviolet très compatible chimiquement et mécaniquement avec celui-ci.

On voit donc bien sûr cet exemple particulier les limites de ces systèmes de peinture même de haut de gamme, en raison de la faible épaisseur globale appliquée, limites que l'on retrouve également sur des supports bétons fissurés même réparés en raison de la rigidité des finitions polyuréthannes (cas des extrados d'arcs de ponts).

Références

- Note d'information technique, Laboratoire Central des Ponts et Chaussées, Mise en peintures des bétons de génie civil, avril 1978.
- Fascicule 65, Exécution des ouvrages de génie civil en béton armé ou précontraint - Fascicule spécial n° 85-30 bis.
- SERRES A.M., La mise en peinture des ouvrages d'art en béton, Colloque gestion des ouvrages d'art - Bruxelles Paris - 1981.

ADHESION OF MODERN BARRIER COATS ON CONCRETE MOTORWAY BRIDGES AND TROUGHS UNDER TARMAC

F.J.BUNDIES
Consulting Engineer, Isernhagen/Hannover, Germany

Abstract
Concrete slabs of bridges and motorway troughs under tarmac traffic layers are protected against the effects of water and thawing agents by various sealing systems. The results to date have not been sufficiently satisfactory. Modern systems with crack-bridging properties have been formulated paying particular attention to the adhesion between barrier coat and concrete. Adhesion is greatly influenced by water and moisture in the substrate. Rough estimates give an idea of the quantity of water present at the concrete/barrier coat interface. This has to be taken into account when formulating and applying modern non-bituminous systems.
Key words: Bridges, Troughs, Barrier Coats, Adhesion.

1. Introduction

We all realise the importance of the safe Maintainance of our concrete road bridges as they are the most critical points within the widespread motorway system guaranteeing the uninterrupted flow of people and goods in our countries and across the borders to our neighbours.
E.g., in Germany we have only within the roughly 40.000 km of Autobahn and other motorways under Federal Government administration -leave alone all other roads- 23.000 bridges constituting approx. 1.ooo km by themselves or 40.000.000 m² of concrete surfaces exposed to all kinds of hazards, such as industrial atmosphere, thawing agents, mechanical stresses etc.

According to a systematic survey carried out in Lower Saxony in 1980/81 some

- 18 % of these bridges showed corroded reinforcing steel
- 20 % had cracks in the concrete wider than 0.2 mm
- 25 % were defective in their barrier system between the tarmac and the supporting concrete slab.

In a more recent investigation into the condition of

concrete bridges further evidence has been produced of partly serious damage to the concrete slab due to failures in the various barrier systems used between tarmac and oncrete: in some instances the thawing agents have resulted in completely desintegrated concrete 2 to 5 cm deep, the reinforcing steel in the same areas having diappeared entirely by corrosion.

As a consequence something has to be done -and is being done- to repair the damages and conserve the bridges as long as (and while) this is still cheaper than a complete knock-down and reerection with all its consequences to the traffic.

The facts call for better, more reliable barrier systems to prevent the contact of water and aquous salt solutions with the concrete slab under the tarmac.

At present the existing technical rules are under revision taking past experience into account. The result will be a choice of some 7 different barrier systems, 2 of which as "loose" membranes without adhesion to the concrete slab, while 5 are designed for an intimate adhesive bond to the substrate. Of these one system has crack-bridging properties (up to 2 mm). It is the most modern and promising method, although inevitably requiring somewhat sophisticated application apparatus. This system has further important additional advantages as it alleviates the problems connected with joints between individual concrete sections, between tarmac and concrete and with the adhesion to cast iron or plastic gullies for water drainage.

But whatever is done- one has to realise the presence of water.

2. Water

Water is present everywhere, visible and invisible: as liquid water or in the form of vapour in our surrounding atmosphere with its varying degrees of relative and absolute humidity, as moisture or liquid water on any surface and in the natural soil, and also on and in the concrete we want to -and have to- protect.

The success of barrier systems for concrete structures is greatly dependant on the adhesion of such systems to the substrate. Sealing materials, barrier coats or membranes are basically organic matter, while concrete is inorganic, -and there is little or no physical or chemical affinity between the two.

Therefore, safe adhesion obviously poses some problems. And these problems are complicated additionally by the omnipresent water.

This fact has to be taken into account during the formulation and application of any barrier system, during the application of the hot tarmac (250°C!), and with regard to the service "lifetime" of this heterogenous combination of inorganic concrete, organic barrier coat, and -let say- semiorganic tarmac road covering.

Where is the water? How much water is there? Does it stay in one place or is it migrating?

Water is the smallest molecule of a chemical compound.

- is everywhere,
- permeates practically everything,especially hydrophilic matter like concrete in liquid form or as vapour,
- is "soluble" in air resulting in varying degrees of absolute and relative Humidity, in saturation and condensation processes,
- increases in volume when freezing developing considerable stresses and pressure due to expansion when "trapped", e.g. in concrete,
- is an exellent solvent for many salts resulting under certain circumstances in osmosis and osmotic pressures up to 7 bar,
- is responsible for the corrosive effects of salts on reinforcing steel and other metals used in the building industry,
- interferes with the chemical reactions of organic coating materials.This is of special significance at the concrete/barrier coat interface,
- prevents the adhesion of any coating, unless certain conditions and precautions regarding the substrate and the coating composition are observed.

How much water is in the concrete substrate? Let us look at this question in a very simple practical way (see Table 1). An average example of a concrete composition is

Table 1.

	weight		volume	
Cement	300 kg	= 13,0%	~100 ltr	= 10%
Water	150 kg	= 6,4 %	150 ltr	= 15%
Aggregate	1900 kg	= 80,6 %	~750 ltr	= 75%
	2350 kg		1000 ltr	= 1 m^3

±4/10 of this water is used for the hydration reaction solidifying the cement binder. This leaves
±6/10 of "free" water, i.e. 4% by weight
or 9% by volume
or 90 ltr/m^3 concrete

Under normal circumstances this water is distributed more or less evenly in the bulk of the concrete, occupying voids and "inner surfaces" of pores etc. At the surface of the concrete, however, there is of course contact to the ambient air which may vary in relative humidity. As a consequence various equilibrium degrees of residual "free water"are obtained at the concrete surface. Typical figures for this relation are shown in

Table 2.

Ambient air		Water in concrete	
Temperature	Relative humidity	by weight	by volume
+ 20°C	100 % (immersion in water)	$<$ 9 %	$<$20 %
+ 20°C	85 %	± 4 %	± 9 % = 90 ltr/m^3 =0,09 ltr/m^2/mm
+ 20°C	65 %	2 - 3 %	4,5-7,0 % = 45 - 70 ltr/m^3 $<$0,07 ltr/m^2/mm

Both at 65 and 85 % and at even higher relative humidities the concrete surface would appear to be "dry" although a considerable amount of water is present at its surface. A "wet" appearance (dark or even glossy patches) is only possible on actual immersion into water, or at excessive condensation of moisture when the temperature of the concrete is below the dew point. This can occasionally be observed on bridge soffits in early spring time with quickly rising ambient temperatures (increasing the absolute humidity content in the air) while the temperature adjustment of concrete structures is lagging behind.

The distribution of water and moisture in concrete can vary considerably depending on the respective ambient conditions and the type of structural element concerned. As an example we have a look at a free concete bridge slab and at the retaining wall and bottom slab of a motorway trough (see Figure 1)

Changes in weather conditions cause considerable variations in the moisture/water content of concrete, at least at its surface which to us is of special interest in this context. Coatings of any kind and thickness will interrupt and stop entirely the "natural" equilibrium reactions regarding the "exchange" of water between the concrete surface and the ambient air. The main effect is normally an increase of moisture content

FIGURE 1

DISTRIBUTION OF MOISTURE IN BULK CONCRETE (% BY WEIGHT)

FREE BRIDGE SLAB:

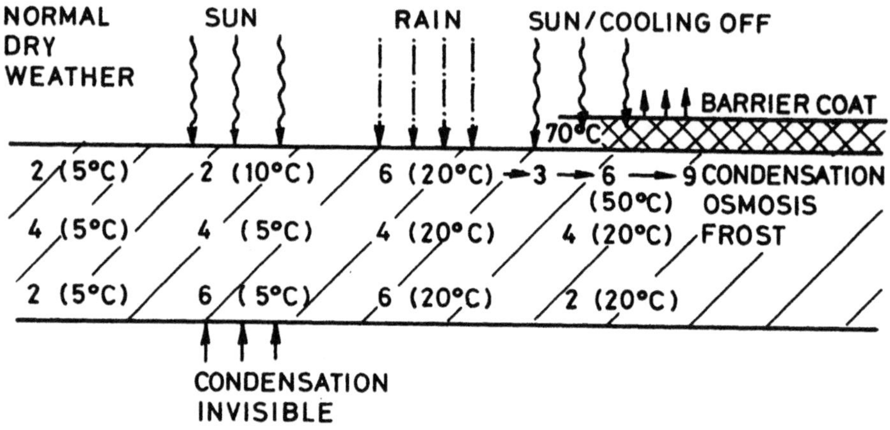

TROUGH

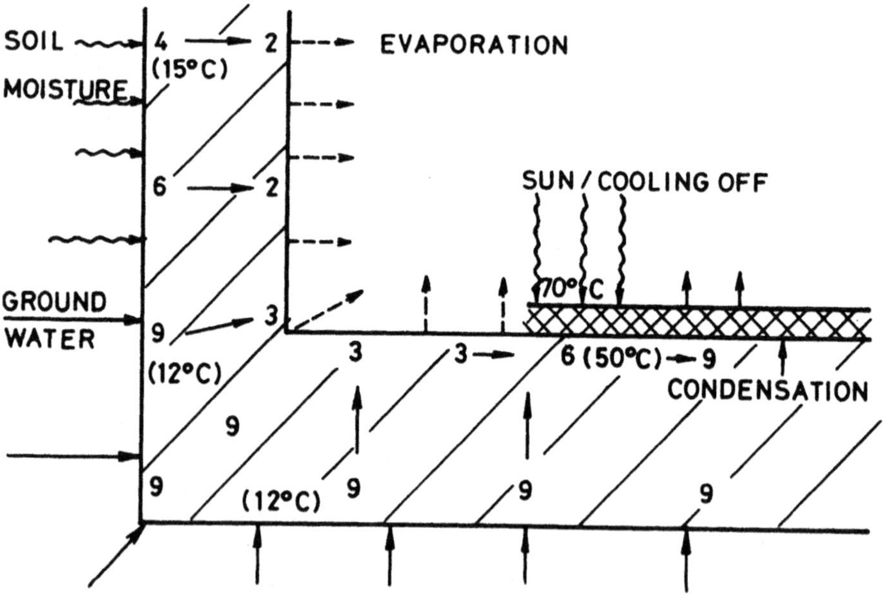

immediately under the coating, i.e. at the contact interface. This can cause inadequate curing of the organic coating at the interface and, as a consequence, loss of adhesion to the concrete substrate.

The significance of water or residual moisture for the success or failure of a barrier coat becomes quite obvious when considering the quantitative relations prevailing at the contact interface, which is decisive for the resultant adhesion(see Table 3).

Table 3.Quantity relations in the contact zone

"Dry" concrete	: 2 - 3 % water	= $<$0,07 ltr/m^2/mm
Wet concrete	: $<$ 9 % water	= $<$0,20 ltr/m^2/mm

<u>2-component epoxy primer,solventfree!</u>

0,3 ltr applied to m^2	= EP in contact	$\sim$ 0,1 ltr/m^2/mm
dry concrete (2 - 3%)	= water	$<$ 0,07ltr/m^2/mm
wet concrete (9%)	= "	$<$0,20ltr/m^2/mm

<u>2-component epoxy primer with solvent</u>, 30-50% solids

0,25ltr applied to m^2	= EP	$\sim$0,075-0,125 ltr/m^2/mm
	solvent	$\sim$0,175-0,125 ltr/m^2/mm
dry concrete (2 - 3%)	=	$<$ 0,07 ltr/m^2/mm

<u>1-component PUR-primer with solvent</u>, 50% solids

0,25ltr applied to m^2	= PUR	$\sim$ 0,125 ltr/m^2/mm (saponifiable!)
dry concrete (2 - 3%)	=	$<$ 0,07 ltr/m^2/mm

Dry concrete with 2 - 3% residual moisture contains up to 0,07 ltr of water per m^2 in its outer 1 mm layer. For wet concrete we can assume some 0,2 ltr of water in this outer layer.

Modern barrier coat systems start with a specially formulated 2-component epoxy primer, <u>solventfree</u>. In spite of the fairly low viscosity of these primers actual penetration into the properly pretreated concrete does not exceed about 0,5 mm. For the sake of simplicity a penetration of even 1 mm is assumed in this context, corresponding to approx. 0,1 ltr epoxy resin being absorbed into the concrete surface. This would mean that approx. 0,10 ltr EP primer solids have to cope with approx. 0,07 ltr of water in the contact zone without seriously disturbing the stoichiometric reaction between the EP main component and hardener. Such interference would prevent a proper chemical curing reaction and the intended safe bond to the concrete surface.

Although these figures are a very rough guess and may not represent the actual ratio at the interface with any degree of accuracy they represent a resin/water ratio of 0 %. From this it is quite clear that only very specially formulated primers can be successful for this purpose bearing in mind that with normal products only some 10 % of water are sufficient to entirely spoil a proper hardening reaction.

One important error which is being encountered every now and again is that some formulators or applicators tend to add diluting thinners to such solventfree primers hoping to improve their penetration into the substrate by reducing the viscosity. This is quite unnecessary and completely wrong, and is the reason for many failures : thinning with organic solvents does not improve the penetration, it only promotes a selective "filtration" of the various primer ingredients upsetting the correct ratios and reactions, at the same time trapping the low molecular solvent in the substrate. This in turn adds to the irregularities in the curing mechanism, let alone other undesirable effects.

Logically the same applies (and is backed by practical experience) to solvent-bourne 2-component and 1-component primers, including those on bituminous basis. With the latter the filtration effect is of lesser importance, but due to their mainly unpolar aliphatic character adhesion to concrete is inferior.

Apart from the above, primers in barrier coat systems have to withstand the heat shock and spontaneous steam pressure developing from the residual water content in the concrete during the application of the hot tarmac (250°C!). Temperatures at the interface concrete/barrier coat rise to up to 130°C within a minute. It is obvious, that this combination of heat, steam and pressure can only be withstood by very special priming and barrier coat systems providing excellent adhesion even under such extreme conditions.

3. Modern non-bituminous barrier coats

The application procedure consists of the following steps:

a) Pretreatment of the concrete to obtain a clean sound substrate; maximum residual moisture content 3 - 4%

b) Repair and levelling-out of damages in the surface using a putty mixed from the primer (see c) and graded dry sand as filler

c) Priming with a specially formulated solventfree 2-component epoxy material. This is a very important step and has to be carried out with great care

d) Application of intercoat adhesive, if required

e) Application of the elastic quick drying crack-bridging barrier coat at a thickness of 2 - 3 mm

f) Application of a bonding agent for spattered aggregate as key for the tarmac road covering.

All these individual layers making up the barrier system have a safe intercoat-adhesion among one another. This is maintained during application of the hot tarmac and during prolonged service.

There are still differing opinions as to the desirability of a firm bond between the elastic but fairly rigid crack-bridging barrier coat and the tarmac layer, or whether a deliberat mechanical separation at this interface, e.g. by the use of a glass fibre mat, is advantageous. Further experience will clear this point.

Due to the short hardening times all the above steps of application from b) to f) can be effected in one or at the most two days. Depending on local circumstances and organisation details a group of 5 - 6 experienced and well equipped men can complete some 500 m² per day.

4. Future development

Although these barrier coats are at present higher in cost than conventional bituminous systems they provide an exellent and reliable barrier for water and aquous salt solutions on concrete bridge and trough structures. Crack-bridging up to 2 mm is effected even for cracks forming <u>after</u> the application of the barrier system, thus giving extra protection to the substrate and its reinforcing iron.

These properties warrant extra costs for new bridges or repair work and will pay in the long run. Naturally, special equipment, skill and sound experience are required for the application of these modern barrier coats.

It is , therefore, to be hoped that in the official tendering and bidding practice for such public engineering work wise decisions are taken on a real long term value-for-money basis.

5. Slides

CRITERION OF CRACKING RESISTANCE OF GLASS FIBER REINFORCED RESINS: A COMPARATIVE STUDY

L. CZARNECKI and J. GRABOWSKI
Department of Civil Engineering, Warsaw University of Technology

Abstract
The paper presents the analysies of the state of knowledge on the problem of cracking resistance of anticorrosion coatings made of glass fiber reinforced synthetic resins (mainly polyester and epoxy). This kind of coatings are most often used to protect the concrete substrate in the chemically active environment. The theoretical bases of relation between the coating and concrete substrate properties have been studied. Different models of the relation have been taken into account. The influence of some coating properties (such as thickness, the tensile strength, the modulus of elasticity) on the value of the limit crack width of the concrete substrate have been analysed. The basic features of the coating which determine its cracking resistance have been found. A review of numerous papers has been presented.
Key words: Bond length, Crack width, Coating, Crack bridging capacity.

I. Introduction

Corrosive resistance and watertightness of reinforced concrete structures are mainly limited by width of cracks which appear in concrete during service time. The suitable standards according to working conditions of structural elements specify the ultimate limit states of cracking. For example Polish standard (PN-76/B-03264) similar to German standard (DIN 1045) requires the maximum width of cracks not to be exceeded 0,1 mm in chemically aggresive enviroment and to be limited 0,2 - 0,3 mm in other conditions. It should be noted, however, that close observation of many existing structures allow to admit that real widths of cracks are very often much higher than permissible values. Also spacings of cracks are different from theoreticaly calculated values, However, that usually has no significant influence on mechanical properties of structures and mainly affects corrosion resistance.

Development of anticorrosion coatings which could resist the formation of cracks in concrete substrate has great practical significance from technical and economical point of view as well.

Although numerous studies have been conducted in the past to develop different types of anticorrosion coatings still there are no theoretical bases of relation between the coating and concrete substrate. Also there is no designing recommendations of crack resistance coatings. Other problems arrive when additionaly the coating should fulfil some physical and mechanical requirments such as impact resistance, temperature and chemical resistance and soon.

The scope of the presented work is the analysis of the state of knowledge on the problem of cracking resistance of anticorrosion coatings made of glass fibre reinforced synthetic resins. This kind of coatings are most often used to protect the concrete substrate in the chemically agressive environment.

2. Types of anticorrosion coatings and basic requirments

As it was mentioned above anticorrosion coatings should fulfil two basic conditions:

- protect the structure against corrosive media,
- quarantee good adherance to concrete, resist the formation of concrete cracks and be capable of taking up movement in the cracks ("cracks bridging capacity").

The hitherto experimental works done by Krak (1967), Mokrzycka (1979), Almahdi (1983), Rieche and Ross (1983), Czarnecki (1985) and others allow to conclude that application of glass fiber reinforced synthetic resins (mainly polyester and epoxy) can satisfy both requirments.

After the extensive study Czarnecki (1985) has described the requirements of protective coatings as follows:

- watertinghtness and chemical resistance,
- adhesion between the coating and the concrete substrate as well as between each two layeres (if any) of a coating,
- heat and thermal shock resistance,
- crack resistance.

The last item is the key-requirement of a good coating-concrete cooperation and the main topic of our consideration in this paper, as well.

There is some contradiction between requirements of high chemical resistance and simultaneously suitable flexibility of coatings. In practice, the balance chemical resistance - elasticity of coatings have been selected in some compr-omise manner. In such a situation authors have recommended multilayer coatings as the resolve. The kind of resin for the top layer is chosen according to the chemical resistance required in a specific environment. According to the type of resin matrix (polyester, epoxide and plasticizers) kind (fabrics and mats) and amount of glass fiber reinforcement the coatings with various tensile characteristics can be obtained.

The ultimate tensile strain of those coatings differ from

$\varepsilon_{max} = 0,3$ % to $\varepsilon_{max} = 2\%$, more frequentyl $\varepsilon_{max} \in (0,5 ; 1,6 \%)$.

There is equivalent to modulus of elasticity, $E_c \in (1;\ 10\ GPa)$ and tensile strength $\sigma_{max} \in (10;\ 150\ MPa)$. Set of tensile charac teristics representative for coatings tested by authors are show in Fig. 1.

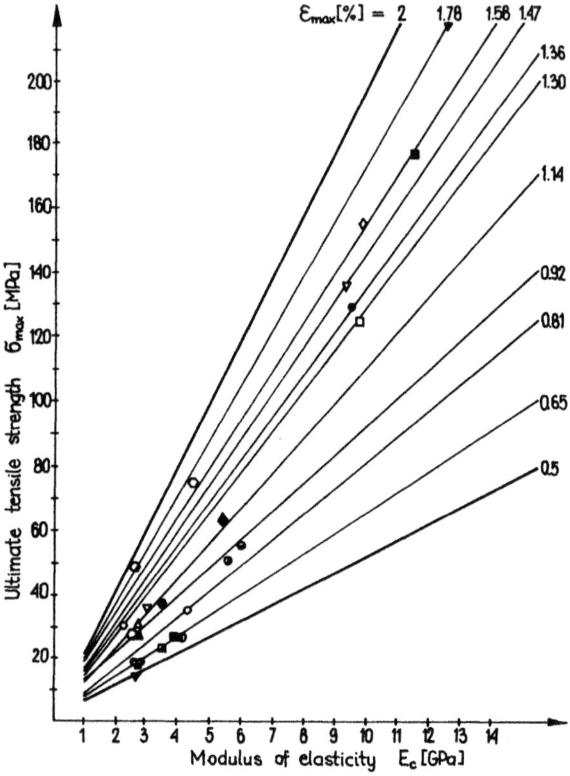

Fig.1. Tensile strength (σ_{max}) vs. modulus of elasticity (E_c) for various ultimate tensile strains (ε_{max}). Set of authors experimental results for various types of coatings.

3. Interface phenomena theoretical bases

A close observation of the failures of coatings applied on concrete has led to several hypothesis of the phenomena taking place at the interface coating-substrate : Teepe (1962), Danilecki et al. (1966), Krak (1967), Wagner and Diecke (1974), Broniewski (1974), Weiss (1974), Menges and Suchenek(1975) Berka (1979), Rieche and Ross (1979, 1983), Gamski (1980), Almahdi (1983), Nemec (1985). For coatings with a relatively high tensile strength as glass fiber reinforced resins, so far the most convinient theoretical approach seems to be presented by Rieche (1979). In the work it was assumed that the thickness of the coating (t) is much bigger than the maximum width of the crack to be covered (w_m) and the narrowing of the coating over the crack is negligibly small. Moreover it was assumed that actual tensile stress in the coating (σ) does not exeed the ultimate tensile strength (σ_{max}) and shear stress (τ) is lower than the ultimate value (τ_{max}). What means no adhesion loss appears.

$$t \gg w_{max} \tag{1}$$

$$\sigma < \sigma_{max} \tag{2}$$

$$\tau < \tau_{max} \tag{3}$$

In such a case nearly infinite strain of the coating in the area over the crack would have to appear. Due to that the last assumption (3) can not be exactly fulfiled. So the local detachment of the coating from the substrate always has to take place. The case when the length of the segment with failed bond was taken into account will be discussed further.

Taking into consideration half of the analysied area(Fig.2) and considering the coating and concrete substrate as a composite material, differential equation of a coating displacement was obtained, as follows :

$$\frac{d^2u}{d\,x^2} = \frac{\tau(u)}{E_c\,t} \tag{4}$$

where: E_c - modulus of elasticity of the coating,
t - thickness of the coating,
u - displacement of the coating.

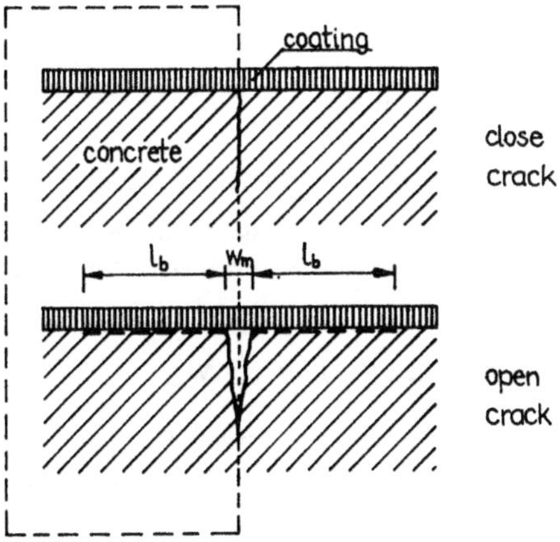

Fig. 2. Crack bridging capacity of coating

To solve this equation it is necessary to find relation between shear stress (τ) and displacement (u) in experimental way or to assume bond stress distribution $\tau(x)$.

The displacements of the coating are the biggest on the edge of the crack and they are reduced to zero at some distance (so called "bond length", l_b) from the crack's edge. The influence of the crack on the coating is limited in the range of the bond length on both sides of the crack (Fig. 2). To simplify the problem it was also assumed, that bond stresses (τ) are constant and equal τ_1. Such assumptions allow to obtain easily the distribution of tensile stresses (σ) and displacements of the coating in the form as it is shown in Fig.3.

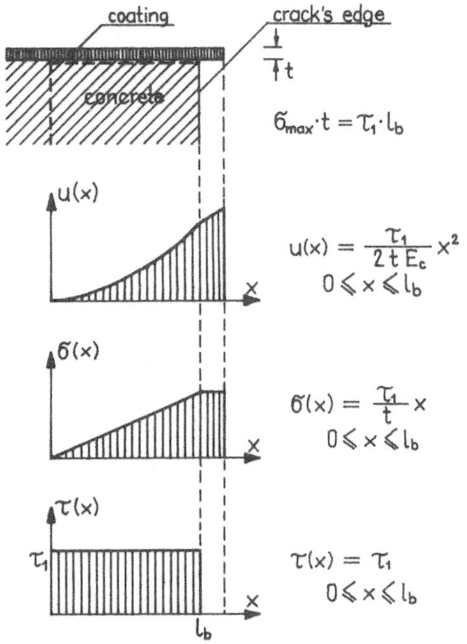

Fig.3. Stresses and displacements distribution.

Because continuity of the coating was assumed, its displacement in the middle of the crack has to be equal one half of crack width, so

$$u(l_b + \frac{1}{2} w_m) = \frac{1}{2} w_m \qquad (5)$$

After some transformations the formula for maximum width of crack which can be covered by the coating without a failure was obtained as follows

$$w_m = l_b \varepsilon_{max}(1 + \varepsilon_{max}) \qquad (6)$$

where: ε_{max} - ultimate strain of the coating.

Presented formula depends on properties of coating - substrate system which under given assumption are defined by the following values: E_c, ε_{max}, t, σ_{max}, l_b. Knowing the range of technical features characterized in practice coatings and concrete substrate the serviceability of the system can be estimate(Fig.4)

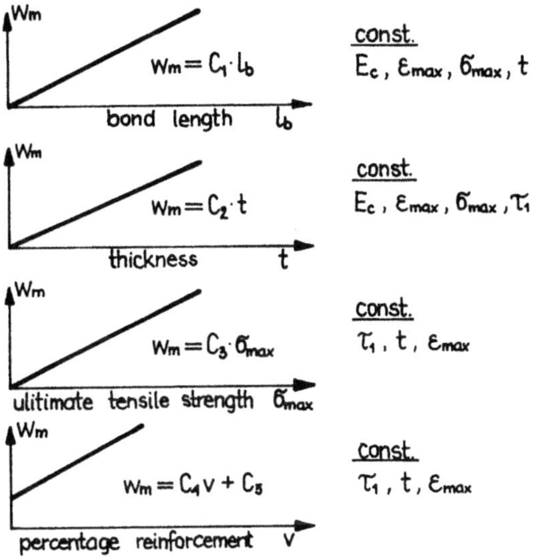

Fig. 4. Maximum crack width (W_m) in function of proper-ties of coating - substrate system (C_1, C_2, C_3, C_4 - constans).

Presented relationships (Fig. 4) have shown that the maximum crack width increases lineary with the increasing value of the bond length, the thickness and the tensile strength of coating and percentage of reinforcement as well. The maximum crack width increases expotentially with increasing ultimate tensile strain (Fig. 5), viz. the max. crack width increases with increasing of coating tensile strength and decreasing of its modulus of elasicity (compare Fig. 1).

It was assumed above that the maximum bond stress was lower than the ultimate value. In fact as it has been already mentioned this assumption is not complitly fulfiled. The case when the length of the segment with failed bond is considerably large should be also analysed. If the segment with failed bond is equal l_o the distribution of shear stresses can be as in Fig. 6.

In such case the formula (5) takes the form

$$u(l_b + l_o + \frac{1}{2} w_m) = \frac{1}{2} w_m \tag{7}$$

Taking into account that the segment $2l_o$ undrgoes additional elastic elongation, formula (6) can be written in following for m:

$$w_m = l_b \cdot \varepsilon_{max}(1 + \varepsilon_{max}) + 2l_o \, \varepsilon_{max} \tag{8}$$

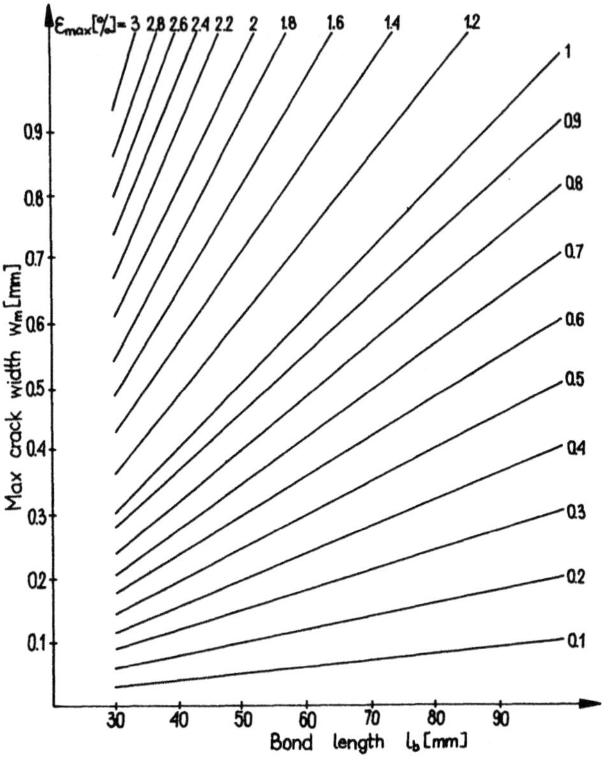

Fig.5. Max. crack width (w_m) in function of properties of coating - substrate system: l_b - bond length, ε_{max} ultimate tensile strain.

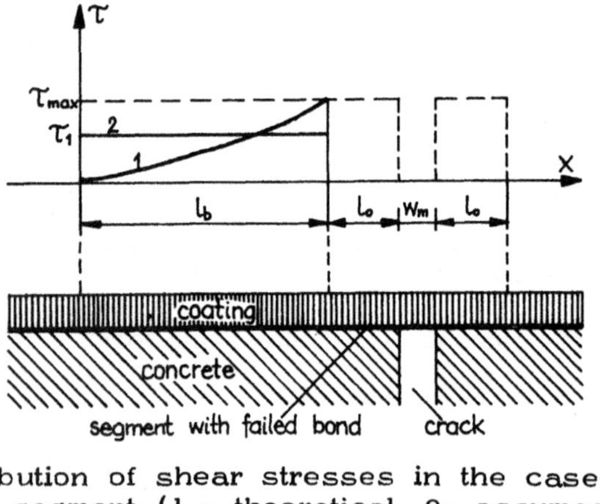

Fig.6. Distribution of shear stresses in the case with failed bond segment (1 - theoretical, 2- assumed).

It is easy to notice, than in case when l_o is equal zero formula (8) takes the form as formula (6)

4. Overview of research studies

Although the problem of cracking resistance of coatings has been the subject of numerous experimental works, the state of the level of information is not yet sufficient. Moreover, when studing this problem, it can be noticed that different investigators working on it introduce various research methods and use different kinds of coatings and substrates (compare e.g. Wagner and Diecke (1974), Krasowskaja (1977), Mokrzycka (1979), Almahdi (1983), Kłosiński and Abramowicz (1986)). This is why the comparison of their results is often very difficult and leads to defferent conclusions. Some basic results are presented in Table 1. Majority of these works were directed to determine the maximum width of the crack which can be covered by the coating. It can be noticed (Table 1) that depending of the thickness of the coating, percentage of reinforcement and kind of resin, the multilayer coating can be designed so, as to be able to cover the crack in the concrete substrate up to 0,9mm according to Wagner and Diecke (1974): 1,30 mm according to Mokrzycka (1979) and on the steel substrate even up to 1,45 mm according to Błaszczyk (1985). The values of the modulus of elasticity of the coatings have been equal: 4,5 : 3,7 : 3,3 GPa respectively.

Krak (1967), Pluta and Ahmed (1982), Almahdi (1983) and others underline the exellent adhesion between epoxy resins based coatings and concrete substrate and also very good intercoats bond property. According to Czarnecki et al.(1984) requirement of good adhesion means: $\tau_{max} > 2,5$ MPa. Usually it is not difficult to obtain this value in the case of epoxy resins (even $\tau_{max} > 6,0$ MPa has been noticed in this case). However, fulfilling that requirement could create troubles in some cases of polyester resins. There has been found satisfactory agreement between the mentioned above experimental values of the maximum crack bridging capacity of coatings with good adhesion to substrate and authors prediction using Rieche formula (Fig. 7a).

If the failed bond segment of the coating appears the max. crack bridging capacity of coatings has been found (Tabl. 1) as follows: 1,9 mm according to Wagner and Diecke (1974) and even above 5 mm according to Mokrzycka (1979): 7 GPa and 2,9 GPa have been the values of suitable modulus of elasticity. The most promising estimation according to the authors using Rieche formula has been presented in Fig. 7b. In this case the result found in Mokrzycka study seems to be too optimistic, It is worth to stress that bond failure has been already observed if adhesion strength level is: $\tau_{max} < 2,0$ MPa (Diecke and Wagner (1974)).

SOME RESULTS OF CRACK RESISTANCE TESTS

TABLE 1

Author(s) (Date)	Method	Coating (Thickness)	Substrate (compresiye) arewnfrh)	Max. crack width W_m[mm]
Danilecki etal. (1966)	Tensile tests	Epoxy resin (0,3 – 1,2 mm) – without rein-forcement – with glass reinforcement	Concrete	 0,01 0,10
Krak (1969)	Tensile tests	Epoxy resin (0,5 – 1,0 mm) with glass reinforcement	Concrete	0,2 – 1,2
Sznejderowa etal. (1969)	Tensile tests	HSPE (0,2 – 0,3 mm)	Concrete	0,1 – 0,3
Broniewski (1974)	Tensile tests	Epoxy resin (0,4 – 0,8 mm) without reinforcement	Mortar	0,4
Wagner, Diecke (1974)	Tensile tests	Epoxy and polyester resins (3 – 12 mm) σ_{max}=2 –70 MPa, E_c = 3 – 17 GPa	Concrete (30 MPa)	0,9 – 1,9*
Krasowskaja (1977)	Bending test (4point Loading)	Epoxy resin with different kind reinforcement σ_{max}= 11–110 MPa, E_c = 3 – 30 GPa	Concrete	–
Mokrzycka (1979)	Tensile tests	Epoxy and polyester resins (1.0 – 3,0 mm) – without reinforcement σ_{max}=9–22 MPa E_c = 2 – 9 GPa – with glass reinforcement σ_{max}= 40 MPa E_c = 3 – 4 GPa	Concrete (~30 MPa)	 0,7 – 4,6* 1,6 – 6,0*
Pluta, Długołęcka (1979)	Tensile tests	Epoxy resin (2,2 – 4,4 mm) with glass reinforcement	Concrete (20 MPa)	0,1
Almahdi (1983)	Bending (3 point loading)	Epoxy and polyester resins (1,2 – 4,2 mm) with glass rein-forcement σ_{max} = 40 – 115 MPa E_c = 6,4–7,5GPa	Concrete (15 – 30 MPa)	0,04 – 0,13
Błaszczyk (1985)	Tensile tests	Epoxy resin with glass reinforcement σ_{max}=33 – 40 MPa E_c = 0,5 – 4, GPa)	Steel	0,03 – 1,45

* The segment with failed bond appears

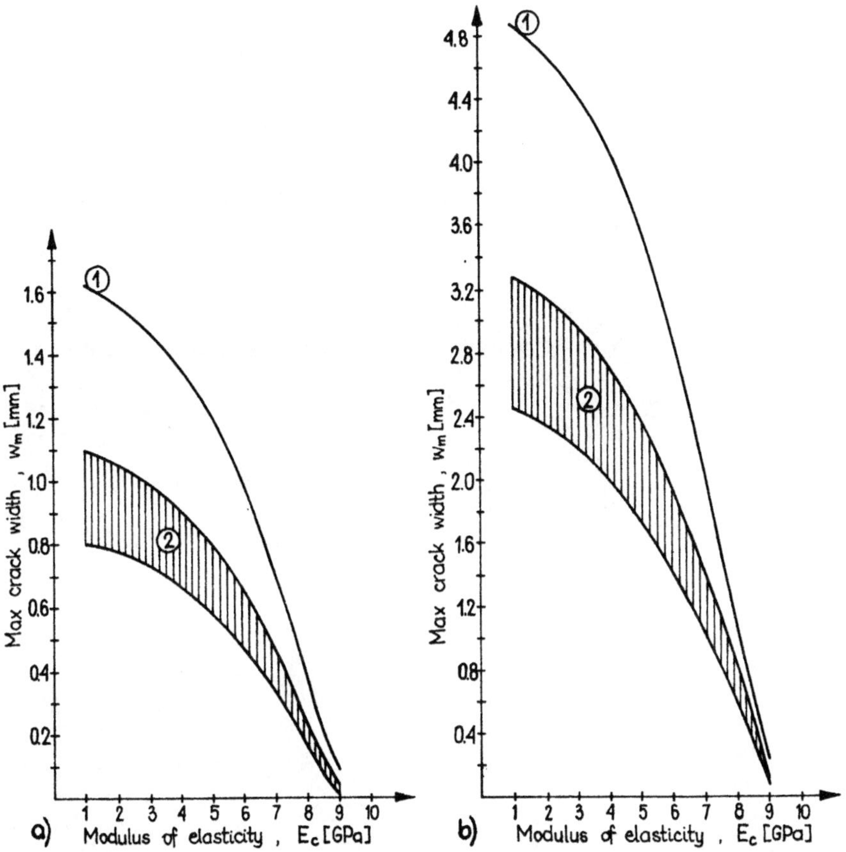

Fig.7. Max. crack width (w_m) of concrete v.s. modulus of elasticity (E_c) of coatings. The functions estimated using Rieche (1979) formula: 1-directly from Rieche, 2-recomended by authors for the concrete substrate (compressive strength 20-40 MPa) and glass reinforced resin coatings with thickness, t > 2 mm; a) no significante loss of adhesion between coating and concrete substrate: τ_{max} > 2,5 MPa, b) the case of failed bond segment of coating: $l_o = l_b$, τ_{max} < 2,5 MPa.

In the case of properly applied coating on concrete, a breakage is usually observed in the top layer of the substrate rather than in-side the coating or at the interface (compare e.g. Krak (1967), Pluta and Długołęcka (1979), Almahdi(1983)) The same phenomena has been observed in properly designed coatings in authors researches (Fig. 8).

The influence of compression strength of concrete on cracking resistance - bridging capacity of the coating have been observed by Pluta and Ahmed (1982). Cracking resistance is increasing with increasing value of concrete strength. The system of coating - concrete substrate can be treated in the category of coupling system with positive feedback.

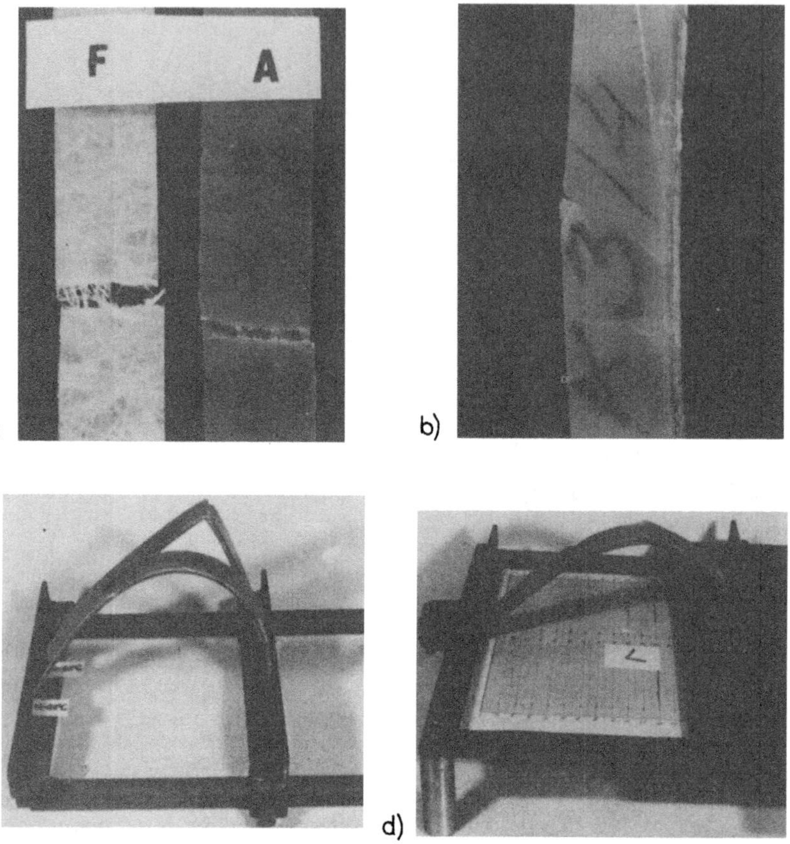

a) b) c) d)

Fig.8. Examples of the top layer breakages of various coatings: a,b) after tensile test, c,d) constant strain test: H,K,L - multilayer coatings, A - two layer coating, F - one layer coating.

The influence of the coating on cracking resistance of concrete substrate - in turn - has been observed by Sołomatow and Szwidko (1969), Krasowskaja (1977), Wei (1985) and others. In some cases increase (even 50 %) of the stress level at the first crack was observed. These results are in good agreement with theoretical ones obtained by Weiss (1974).

Rieche and Ross (1984) have analysed the distribution of strains in the coating in the area where the crack appears in the concrete substrate. These experiments were carried out to verify the theoretical solutions presented in the previous chapter and to measure the bond length (l_b) which is the most characteristic criterion for the crack bridging capacity of the coating applied to a cracked concrete surface. Two different methods were aplied: wire strain gauges and moire technique. Although, the authors notice that they did not obtain useful results appylying the moire method what was probably due to using not proper grating, nevertheless this method seems to be very promising in solving this problem (compare e.g. Wachnicki and Radon (1984), Grabowski et al. (1985). Significant influence of temperature in the range from $-20^{o}C$ to $+20^{o}C$ on the value of bond length was observed. It is in agreement with our previous statement that the relation l_b (T) is similar to that of E(T).

5. Final remarks

The suitable flexible glass fiber reinforced coatings if properly designed as multilayer system can show satisfactory crack bridging capacity in the case of good coating adhesion to concrete as well as if failed bond segment of the coating appears The modulus of the elasticity of the coating has been found the main decesive factor of the crack resistance ability. Using the Rieche formula as the theoretical base some practicaly available estimation of the max. crack width of concrete substrate could be done.

Acknowledgment

The work has been done partially under the Research Program 05.03.C-4 coordinated by Institute of Building Technology and CPBBP 02.L of Institute of Fundamental Technological Research of the Polish Academy of Sciences.

References

Almahdi, A. (1983) Limit state of cracks of reinforced concrete structur-es covered with laminate plastics coatings. D. Sci. Thesis, Warsaw University of Technology (in Polish)

Bares, R.A. (1976) Deformation interaction of soft roofing materials with their base. American Concrete Institute, Fall Convention, Mexico City.

Berka, L. (1979) Eguilibrium states of cracks on a thin plate-rigid basis interface. Staveb Cas., 27, 10, VEDA, Bratislava, 739-751 (in Czech)

Błaszczyk, J. (1985), Study on crack resistance of chemosetting resin coatings on concrete substrate. 2nd Conference on Corrosion of Buildings and Structures, Polish Civil Engineering Society, Warszawa (in Polish)

Broniewski, T. (1974) Crack resistance of anticorrosion coatings on concrete underground. Proceedings on Civil Engineering No 2, Cracow Technical University (in Polish)

Czarnecki, L. (1985) Surface anticorrosion protection of reinforced concrete chimmneys. 2nd Conference on Corrosion of Buildings and Structures, Polish Civil Engineering Society,

Warszawa, (in Polish)
Czarnecki, L (1985), Report on anticorrosion protection of reinforced concrete chimmneys. Instytute of Technology and Organisation of Building Production, Warsaw University of Technology (in Polish)
Danilecki, W., Mączyński, M. and Pluta, J. (1966), The behaviour of plastic coatings on the concrete structures with moving cracks. Proceedings of 12th Scientific Conference KIL PAN and KN PZITB, Krynica , 67-72 (in Polish).
Gamski, A. (1980) Strengthening and repair of concrete structures with epoxy resin. Arch. Inż. Ląd., 26, 1, 169-180 (in French)
Grabowski, J., Jastrzebski, P. Lutomirski, S. and Wichniewicz,S (1985), Application of moire method in the analysis of microcracking of ferrocement. Proceedings of the Second International Symposium on Ferrocement, Banghok, 271-280.
Krak, H.(1967), Cracking of concrete and the influence on epoxy resin based coatings. RILEM Symposium-Synthetic Resins in Building Constructions, Paryż, 401-409.
Krasowskaja, T.A. (1977), Meaning of polymers coatings for reinforced concrete structures, Transportnoje Stroitielstwo, 9, 50-52, (in Russian).
Kłosiński, B. and Abramowicz, M. (1986), Modelling of the work of glass fiber reinforced resin coatings on a cracked surface of concrete. Inetrnational Symposium on Adhesion between Polymers and Concrete.
Menges, G., Suchanek, H.J. (1975) Trial of the explanation of the stress-crack corrosin. 8th Conference of Danaubian Countries on Nature and Artificial Ageing of Plastic Materials (in German), Wurzburg.
Mokrzycka, A. (1979), Analysis of crack resistance of selected polymer coatings on concrete constructions. D.Sci. Thesis, Warsaw University of Technology (in Polish).
Nemec, J. (1985), Some observations about the failure of polymer composites. UTAM-CSAV, Praha (1985).
Pluta, J. and Ahmed, R. (1982), The effect of concrete strength on the unit state of crack's width of concrete structers covered with laminate plastics coatings. Proceedings of 28th Scientific Conference KIL PAN and KN PZITB, Krynica 173-180, (in Polish).
Pluta, J. and Długołecka, E. (1979), Limiting states of laminate plastics coating on the concrete structures with moving cracks. Proceedings of 25th Scientific Conference KIL PAN and KNPZITB, Krynica, 193-198, (in Polish).
Rieche, G. (1979) Crack resistance resin coating on mineral building materials. Farbe+lack 85, 10, 824-831, (in German)
Rieche, G. and Ross, H. (1983), Measurement of bond length of crack bridging coatings of syntetic resins. Proceedings of Collogium of Reinforcing and Repairing of Constructions. Esslingen, 119-121, (in German).
Sołomotow, W.I. and Szwidko, J.I. (1969), Influence of polymer coatings on cracking resistance of reinforced concrete

structures, Bieton i Żieliezobieton, 4, 33-35, (in Russian).
Szniejdierowa, W.W., Miedwiediew, W.M. and Migajewa, G.S. (1969), Cracking resistance of anticorrosion coatings. Bieton i Żeliezobieton 1, 17-18, (in Russian).
Szwidko, J.K. (1972), Influence of polymer coatings on propertis of reinforced concrcte structures. Promyszlennoje Stroitielstwo, 5, 33-34, (in Russian).
Teepe, W. (1962), Mechanical stresser in plastic materials in modythic constructions. Kunstoffe, 52 659-666, (in German).
Wagner, K. and Diecke, W. (1974), Application of synthetic resins for cracking resistance coatings. Bauplanung-Bautechnik, 28,2,74 - 75 (in German).
Wachnicki, C.R. and Radon, J.C. (1984) Fatigue crack growth measurement in a CSM composite using compliance and moire techniques. Butterworth Co (Publishers), Ltd.
Weiss, V. (1974), Failure of brittle materials with tougher surface coating. In Mechanical properties and structure of composite materials (A.M. Brandt - ed), 517-525.
Wei, Z.(1985), A research on glass-fibre reinforced plastic ferrocement (GRPF) and its application in irrigation engineering. Journal of Ferrocement, 15,4, 359-363.

PROTECTION EFFECTS OF POLYMERIC FINISHES ON THE CARBONATION OF CONCRETE AND CORROSION OF REINFORCEMENT

THOSHIO FUKUSHIMA Materials Department, Building Research Institute
FUMINORI TOMOSAWA Materials Department, Building Research Institute
ISAO FUKUSHI Housing and Urban Development Research Institute
Housing and Urban Development Corporation
HITOSHI TANAKA Materials Department, Building Research Institute

Abstract

Protection effects of polymeric finishes on the carbonation of concrete and corrosion of reinforcement were experimentally examined by using accelerated carbonation testing(30°C, 60%R.H., CO_2; 5% for 8 months) and accelerated corrosion testing(50°C, 95%R.H. for 16 months) for test specimens of reinforced concrete with many kinds of polymeric finishes. In order to evaluate the effectiveness of polymeric finishes as repairing materials for partly-deteriorated reinforced concrete, and also to obtain basic data for improvement of the durability of reinforced concrete by polymeric finishes, experimental data were systematically analyzed by the least-square method using parabolic law($\sqrt{t}$ law) involving a constant term, which was theoretically derived for the progress of deterioration of reinforced concrete based upon unsteady state dynamic analysis. It was found that polymeric finishes play the important roles in the retardation of carbonation of concrete and the suppresion of corrosion of reinforcement in neutralized concrete with no chloride ions included under ordinary atmospheric environmental conditions.

Key words: Reinforced concrete, Polymeric finishes, Carbonation, Corrosion, Neutralization, Parabolic law, Repair.

1. Introduction

Recently in Japan, many reports about the damage of reinforced concrete buildings have shadowed the myth of semi-eternal service life of concrete structures. The problem of the deterioration of reinforced concrete is related to the unexpectedly fast progress of carbonation of concrete and corrosion of reinforcement, accompanied with the occurrence of cracking and fall-off of cover concrete even under ordinary atmospheric surroundings. From the viewpoint of the prediction of service life, it is necessary for us to reconsider the durability of reinforced construction, and to find the rational prediction methods for the neutralization of concrete and corrosion of reinforcement(Kishitani et al., 1984). In this context, the progress of neutralization caused by carbonation of concrete is an old, but even now new problem. In order to effectively prevent these deterioration processes of reinforced concrete, we would be able to consider the positive use of polymeric finishes, altogether with the improvement of the quality of concrete itself(Fukushima et al., 1985).

This paper deals with experimental study about the protection effects of polymeric finishes on these deterioration processes for the purpose of evaluation of the effectiveness of ploymeric finishes as repairing materials for partly-deteriorated reinforced concrete, and also to obtain basic data for improvement of the durability of reinforced concrete by use of polymeric finishes.

Table 1. Materials and the designed mix propotion of concrete specimens.

a. Materials.

cement	: Normal portland cement
sand	: River sand (JASS 5 1 grade Fine modulus =2.8)
gravel	: Crushed stone (maximum size : 20mm)
chemical admixture	: Pozolis No.70

b. designed mix proportion.

type		concrete for precasting use	normal concrete for field-mixed use
Slump		8	18
Air cont.(%)		4.0	4.0
Water-cement ratio (%)		55	60
Sand-aggregate ratio (%)		47.9	47.8
Unit water content (kg/m³)		165	185
Absolute vol. (ℓ/m³)	cement	95	97
	Fine agg.	335	324
	Coarse agg.	365	354
Unit Weight (kg/m³)	cement	300	308
	Fine agg.	851	823
	Coarse agg.	967	938
Amount of chemical admixture (cc/m³)		1.050	1.078

2. Experimental methods

2.1 Retardation effects of surface finishes on the neutralization of concrete

Test specimens of concrete(10×10×10cm prism) with various kinds of surface finishes coated in advance were acceleratedly carbonated under the condition of 30°C, 60%R.H., CO_2; 5%. This accelerated condition is expected enough to make fast the progress of the neutraliation of concrete without any change of carbonation mechanisms. In table 1 and Fig. 1 are shown the mix proportions and conditions for preparation and curing of specimen concrete. Types of surface finish materials used in this experiment are summarized in Table 2.

Accelerated carbonation was carried out only from the two side directions with other four surfaces of concrete specimens shielded against carbonation with epoxy resin. The neutralzation depths of acceleratedly carbonated concrete specimens were measured by a 1% phenolphtalein alcohol solution specified by JIS K 8006.

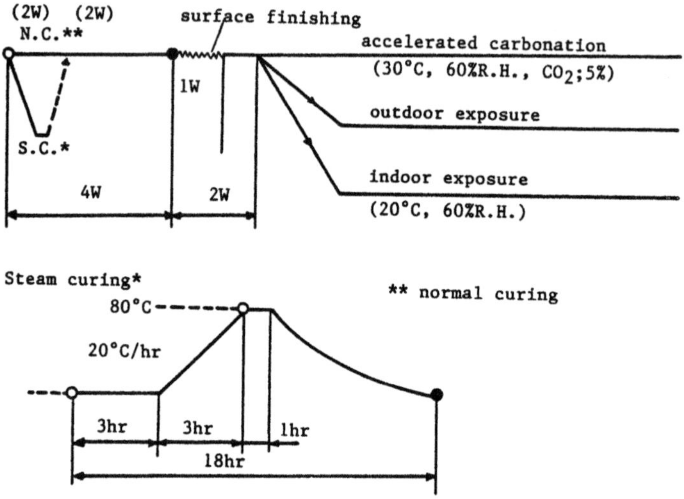

Fig. 1. Conditions for preparation and curing for concrete specimens.

Table 2. Type of surface finish materials.

symbols	
a	no finish
b	mortar
c	vinyl wall paper
d	emulsion paint
e	synthetic resin plaster
f	cement based thin sprayed finish
g	cement based thick sprayed finish
h	acryl based sprayed finish
i	expoxy based sprayed finish
j	surface hardener (lithium silicate)

2.2 Retardation effects of surface finishes on the corrosion of reinforcing bars imbedded in neutralized concrete

Test specimens of reinforced concrete (10×10×40cm prism) with 20 mm concret cover for smooth reinforcing bars (9mmϕ, SGD 3 specified by JIS G 3108) were prepared under the same conditions described Table 1 and Fig. 1, excepting that some concrete specimens including chloride ions added at the concentration of 0.1% NaCl for ovendry sand was also used in this experiment. After confirming by reference samples that cover concrete was completely neutralized, and further that the neutralzation got up to the back side of reinforcing bars, various kinds of finishes shown in Table 2 were coated in the two side surfaces of reinforced concrete specimens with other four surfaces treated with epoxy resin. Accelerated corrosion testing was done under the condition of 50°C, 95%R.H. from two side surfaces with finishes (Fig. 2.).

3. Experimental results

3.1 Results of accelerated carbonation testing

In Table 3 are shown the results of accelerated carbonation testing. The progress of carbonation was effectively delayed by surface finish materials, but the effects depend upon the types of finishes. Fig. 3 shows these situations by plotting the neutralization depths versus real time, and Fig. 4 is the plot of the neutralization depths versus square root of time. It is seen that the progress of carbonation with finishes can be expressed by the parabolic law ($\sqrt{t}$ law) involving a constant term。 Table 4 shows the results of the prediction of the progress of carbonation by least-square methods using this parabolic law.

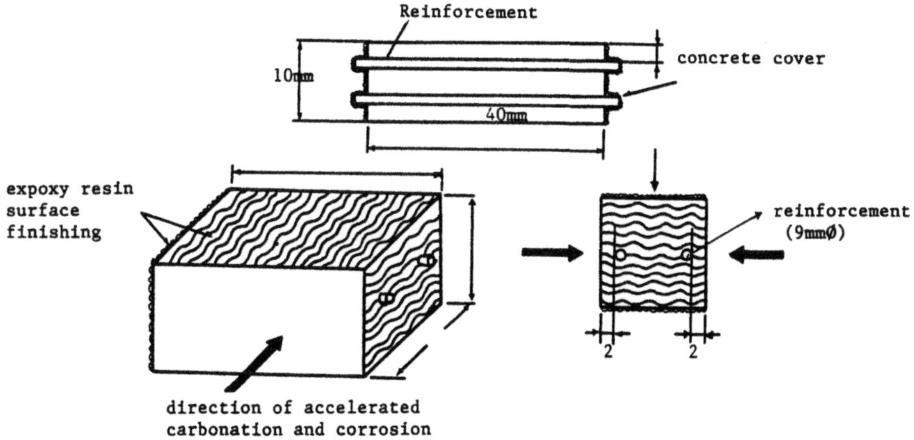

Fig. 2. Specimens of reinforced concrete for testing the retardation effects of finishes against the corrosion of reinforcement imbedded in neutralized concrete.

Table 3. Results of accelerated carbonation testing (30°C, 60%R.H., CO_2;5

type of concrete	type of specimens	mean neutralization depth (mm)			
		1month	2month	4month	8month
normal concrete with river sand for field-mixed use	N - R - a	15.7*	22.3	28.4	34.0
	N - R - b	0.0	1.0	4.5	11.7
	N - R - c	0.0	0.0	0.0	0.3
	N - R - d	8.5	11.2	18.3	25.2
	N - R - e	0.0	2.6	6.4	12.6
	N - R - f	13.1	20.2	25.8	36.0
	N - R - g	13.0	17.6	27.0	37.0
	N - R - h	1.8	3.2	7.7	12.2
	N - R - i	3.8	6.1	8.5	16.9
	N - R - j	13.6	18.7	24.5	31.2
concrete for precasting	P - R - a [1)]	11.7	17.5	22.9	27.3
	P - R - a[*2)]	12.0	17.1	26.1	35.3

1):standard curing 2):steam curing

a~j:surface finishes shown by Table2

able 4. Evaluation of the progress of neutralization by accelerated arbonation testing (30°C, 60%R.H., CO_2;5%).

	type of specimens	equation obtained by least-square method	correlation coefficient
	N−R−a	1) : $X = 0.65 + 14.45\sqrt{t}$	$\gamma = 0.980$
		2) : $X = 7.42 + 9.77\sqrt{t}$	$\gamma = 0.965$
	N−R−b	2) : $X = -7.57 + 6.56\sqrt{t}$	$\gamma = 0.963$
	N−R−c	2) : $X = -0.22 + 0.16\sqrt{t}$	$\gamma = 0.732$
	N−R−d	2) : $X = -1.27 + 9.43\sqrt{t}$	$\gamma = 0.992$
normal concrete for field-mixed use	N−R−e	2) : $X = -7.09 + 6.90\sqrt{t}$	$\gamma = 0.998$
	N−R−f	1)' : $X = 0.60 + 12.72\sqrt{t}$	$\gamma = 0.995$
		2) : $X = 1.79 + 12.14\sqrt{t}$	$\gamma = 0.993$
	N−R−g	1)' : $X = -0.20 + 13.20\sqrt{t}$	$\gamma = 0.998$
		2) : $X = -0.60 + 13.40\sqrt{t}$	$\gamma = 0.996$
	N−R−h	2) : $X = -4.48 + 5.91\sqrt{t}$	$\gamma = 0.989$
	N−R−i	2) : $X = -3.96 + 7.06\sqrt{t}$	$\gamma = 0.959$
		1) : $X = 0.53 + 12.36\sqrt{t}$	$\gamma = 0.994$
	N−R−j	1)' : $X = 1.59 + 11.05\sqrt{t}$	$\gamma = 0.984$
		2) : $X = 4.76 + 9.52\sqrt{t}$	$\gamma = 0.992$
concrete for precasting use	N−R−a	1) : $X = 0.19 + 11.63\sqrt{t}$	$\gamma = 0.996$
		2) : $X = 4.78 + 8.33\sqrt{t}$	$\gamma = 0.956$
	N−R−a	2) : $X = -0.91 + 13.03\sqrt{t}$	$\gamma = 0.996$

X : neutralization depth (mm)
t : time (month)
1) : best fitted for the neutralization by $X = A\sqrt{t} \pm B$ with the initial value setted to be zero using data of 1, 2, 4 months
2) : best fitted using data of 1, 2, 4, 8 months
1)' : best fitted using data of 1, 2, 4, 8 months further by setting the initial neutralization to zero

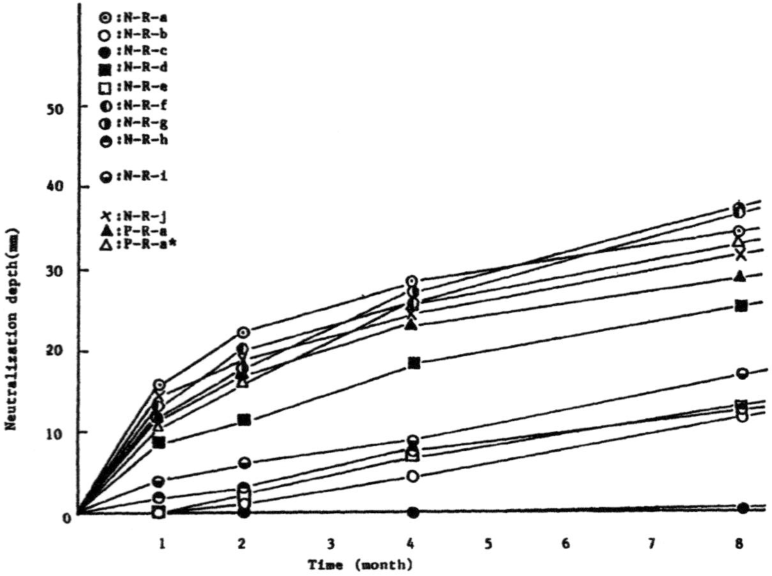

Fig. 3. Results of retardation effects of surface finishes on the progress of neutralization by accelerated carbonation testing(30°C, 60%R.H., CO_2; 5%)-part 1.

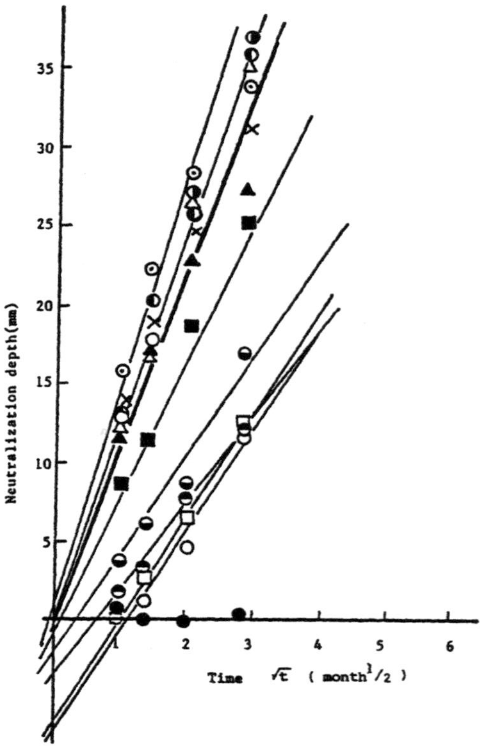

Fig. 4. Results of retardation effects of surface finishes on the progress of neutralization by accelerated carbonation testing (30°C, 60%R.H., CO_2; 5%)-part 2.

3.2 Results of accelerated corrosion testing

In Table 5 are shown the results of measurement of rates of corrosion loss with time of reinforcement imbedded in neutralized concrete. Rates of corrosion loss are calcurated by the following relation:

$$\text{Rate of corrosion loss} = (W_0 - W_1)/ W_0 \times 100(\%) \qquad (1)$$

Here, W_0 is the initial weight before corrosion, and W_1 is the measured weight of reinforcement after the treatment of rust removal by diammonium citrate. Using these data, the progress of corrosion of reinforcement for reinforced concrete specimens with various kinds of surface finishes was evaluated by least-square methods also using parabolic law. The results of evaluation are summarized in Table 6.

4. Discussion

The experimental results obtained by accelerated carbonation testing (Fig. 3 and 4) show the effectiveness of polymeric finishes in the retardation of the progress of the neutralization of concrete, but the types of surface finishes greatly influence the progress of the neutralization. Vinyl wall paper almost completely suppress the progress of the neutralization, and such polymeric finishes as acryl based sprayed finish and epoxy based sprayed finish have the large retardation effects. Emulsion paint and surface hardener have the considerable retardation effects. The progress of the neutralization of concrete with and without finishes can be evaluated by the parabolic law involving a constant term. This law was analytically derived by unsteady state dynamics, assuming the diffusion of CO_2 in wards into concrete from the surface accompanied by the instant conversion reaction of $Ca(OH)_2$ into $CaCO_3$ as follows (Fukushima, 1984):

$$X = A\sqrt{t} - B \quad (t \geqq t_0 = (B/A)^2) \qquad (2)$$

where X means the neutralization depth of concrete,and t indicates age, t_0 indicating the delay factor concerned with the induction period. Making use of this parabbolic law involving a constant term, experimental data were systematically analysed by the least-square method(Table 4). It is found that the retardation effects can be evaluated by the relative comparison of two parameters of A and B. Polymeric finishes are effective in the retardation of the progress of the neutralization of concrete by shutting the diffusion of CO_2 as the direct deterioration agent for the neutralization of concrete.

From the experimental results by accelerated corrosion testing, it is found that polymeric finishes are also effective in the retardation of the progress of corrosion of reinforcement imbedded in neutralized concrete, as long as chlorides ions are not included in concrete. Even if the neutralization has already reached to reinforcement, the shielding by surface finishes against the simultaneous diffusion of water and oxygen from the surface enables us to expect the retardation of the progress of corrosion of reinforcement for a temporary time. In that sense it may be said that the use of polymeric finishes is the most convenient repairing method for the deterioration of reinforced concrete. If the deterioration of surface finishes or the bond failure between finishes and concrete occur after the longer time, however, the retardation effects can be hardly expected because of the reduction of the shielding fuction against external deterioration agents. On the other hand, the progress of the corrosion

Table 5. Results of measurement of rates of corrosion loss with time of reinforcement imbedded in neutralized concrete by accelerated corrosion testing (50°C, 95%R.H.).

type of concrete	curing condition of concrete	type of specimens	calculated rates of corrosion loss (%X10^{-2})				
			1 month	2 month	4 month	8 month	16 month
concrete with river sand	normal concrete for field-mixed use	N - R - a	72	79.5	98	54.5	66.5
		N - R - b	51	85	113	77	88
		N - R - c	50	43	51.5	60.5	69
		N - R - d	13	36.5	72	73	48.5
		N - R - e	20	100	86	66	78.5
		N - R - f	15	43.5	78	67.5	61.5
		N - R - h	14.5	36	77.5	73	82.5
		N - R - i	10	36.5	99	73.5	69
		N - R - j	58.5	80.5	108.5	61.5	74
	concrete for precasting	P - R - a	49.5	86	87.5	76.5	55.5
		P - R - c	56	51	65	75	89
		P - R - e	14	103.5	81.5	72	71
concrete containing chloride ions (0.1%NaCl)	normal concrete for field-mixed use	N - S - a	56.5	109.5	120	127	178
		N - S - b	79	123.5	134.5	106	159
		N - S - c	70	74	83.5	114	145
		N - S - e	30	106.5	125	116	160
	concrete for precasting	P - S - a	49.5	121	128	121	176
		P - S - b	71.5	122	126.5	152	233
		P - S - c	73	67	91	129.5	185
		P - S - e	37	103	125.5	136.5	218.5

Table 6. Evaluation of the progress of corrosion of reinforcement imbedded in neutralized concrete by accelerated corrosion testing (50°C, 95%R.H.).

	type of specimens *	equation obtained by least-square method	correlation efficient	
concrete with river sand	N－R－a1 *	$Y= 0.44+0.26\sqrt{t}$	$\gamma=0.980$	2)
	N－R－b1	$Y=-0.07+0.61\sqrt{t}$	$\gamma=0.976$	3)
	N－R－c1	$Y= 0.45+0.02\sqrt{t}$	——	2)
	N－R－d1	$Y=-0.47+0.59\sqrt{t}$	$\gamma=1.000$	3)
	N－R－e1	$Y=-0.20+0.60\sqrt{t}$	$\gamma=0.498$	3)
	N－R－f1	$Y=-0.47+0.63\sqrt{t}$	$\gamma=0.998$	3)
	N－R－h1	$Y=-0.51+0.64\sqrt{t}$	$\gamma=0.993$	3)
	N－R－i1	$Y=-0.73+0.83\sqrt{t}$	$\gamma=0.938$	3)
	N－R－j1	$Y= 0.09+0.50\sqrt{t}$	$\gamma=0.933$	2)
	P－R－a1	$Y= 0.09+0.36\sqrt{t}$	$\gamma=0.999$	2)
	P－R－b1	$Y= 0.02+0.10\sqrt{t}$	$\gamma=0.693$	2)
	P－R－c1	$Y= 0.43+0.10\sqrt{t}$	$\gamma=0.501$	2)
	P－R－e1	$Y=-0.23+0.61\sqrt{t}$	$\gamma=0.425$	3)
concrete containing chloride ions (0.1 % Nacl)	N－S－a1	$Y= 0.20+0.41\sqrt{t}$	$\gamma=0.905$	2)
	N－S－b1	$Y= 0.38+0.33\sqrt{t}$	$\gamma=0.700$	2)
	N－S－c1	$Y= 0.18+0.33\sqrt{t}$	$\gamma=0.933$	2)
	N－S－e1	$Y= 0.16+0.40\sqrt{t}$	$\gamma=0.822$	2)
	P－S－a1	$Y= 0.22+0.41\sqrt{t}$	$\gamma=0.834$	2)
	P－S－b1	$Y= 0.16+0.40\sqrt{t}$	$\gamma=0.953$	2)
	P－S－c1	$Y= 0.09+0.44\sqrt{t}$	$\gamma=0.972$	2)
	P－S－e1	$Y= 0.32+0.54\sqrt{t}$	$\gamma=0.949$	2)

* 1) : accelerated corrosion testing (50℃, 95%R.H.) of neutralized concrete
2) : indicate that corrosion has already occurred before accelerated corrosion testing
3) : indicate that corrosion has temporarily delayed by finishes
Y : corrosion loss ratio t : time (month)

of reinforcement in neutralized concrete with chlorides ions in advance added is not suppressed or retarded by surface finishes. The retardation effects of surface finishes on the progress of corrosion were evaluated also by the least-square method(Table 6).

5. Conclusion

In order to examine the protection effects of polymeric finishes on the carbonation of concrete and corrosion of reinforcement, accelerated carbonation testing and accelerated corrosion testing were carried out for test specimens of reinforced concrete with various kinds of surface finishes. Experimental data were systematically analysed by the least-square method by using the parabolic law involving a constant term. Although we should appreciate the limitations of the effectveness of polymeric finishes as repairing materials due to their deterioration and bond failure to reinforced concrete, polymeric finishes were considered to be effective enough in the retardation of carbonation of concrete and the suppresion of corrosion of reinforcement in neutralized concrete with no chloride ions included under under ordinary atmospheric environmental conditions.

References

Kishitani, K., Shiire, T., Tomosawa, F., Fukushi, I., Kasami, H., Izumi, I. (1984) Carbonation of concrete in existing structures up to 55 years old. Proceedings of 3rd International Conference on the Durability of Building Materials and Components, Espoo, Finland vol. 3 pp. 250-266.

Fukushima, T. (1984) Chemical behaviour of calcium hydroxide in cement paste and concrete. Proceedings of 27th Japan Congress on Materials Research, Kyoto, Japan pp. 225-232.

Fukushima, T. (1985) Deterioration processes of polymeric materials and their influence on the durability of reinforced concrete. American Chemical Society Symposium Series No. 287: Polymer Wear and Its Control (ed. Lieng-Huang Lee) pp. 347-361

REVETEMENTS ANTI-COROSION IKAS POUR LES STRUCTURES EN BETON ARME

Z.KAMAITIS, A.SNIUKSTA, A.KUDZYS
Institut du Génie Civil de Vilnius

Résumé
La présente communication rend compte des résultats des études des revêtements IKAS pour les structures en béton armé. On décrit les copositions et leur propriétés principales, on propose un modèle de calcul des compositions et des paramètres des revêtements anti-corrosion. Des applications pratiques les plus intéressantes sont présentées.
Mots clés: Revêtement, Résine indène-coumarone, Béton armé, Durabilité, Application.

1. Introduction

La dégradation du béton et des armatures est le phénomène bien connu dans les bâtiments et les ouvrages d'art en béton armé ou précontraint. La cause la plus fréquente des désordres constatés est la corrosion chimique et physique des matériaux. L'expérience montre que la vitesse de corrosion du béton armé est étroitement liée à la qualité des structures d'une part et aux conditions de service d'autre part. Ajoutons que les millieux agressives présentent toujours une très grande variétée.

Le moyen efficace pour assurer la protection et la conservation des structures consiste à revêtir ces structures d'une couche de revêtement è base des mortiers, mastics et bétons ou d'un film de peinture principalement à base de polymères. Pour la protection des bâtiments industriels et des ouvrages en béton armé on fait principalement appel aux résines époxydes, polyesters ou polyuréthanes. Nous avons élaborées des compositions de la résine indène-coumarone, de l'acrylate polyesterique et du styrène (IKAS) qui présentent un ensemble de bonnes propriétés.

L'efficacité d'un revêtement ou d'une peinture à base des polymères est caracterisée par les qualités suivantes:

stabilité chimique
perméabilité
adhérence au support (béton ou acier)

résistance à l'usure
tenue à la fissuration
prix de revient.

Donc, pour choisir le système efficace de protection d'une structure, il faut disposer des données quantitatives de ces qualités ansi que le mécanisme de détérioration du béton et du revêtement en fonction du temps et des actions de service.

La présente communication rend compte de resultats des travaux et recherches systématiques effectuées à l'Institut du Génie Civil de Vilnius depuis 20 ans. Dans celle-ci nous décrivons les compositions IKAS et leur propriétés principales, nous analysons le modèle de calcul des systèmes de protection et nous présentons les certaines exemples de l'application pratique des revêtements IKAS.

2. Compositions IKAS

Les compositions proposées sont des produits poli-composants se polymérisant à température ordinaire (15-20° C) sous l'action d'un catalyseur (peroxydes) et d'un accélérateur (sels de cobalt). Du fait du durcissement les copolymères de la résine indène-coumarone se forment.

La viscosité des résines pures est de l'ordre 2-5 P. De plus, les résines possèdent un pouvoir mouillant considérable permettant d'obtenir les mélanges extrêmement chargés, mais en même temps fluides et autolissants. Une fois le durcisseur ajouté aux résines époxydes ou polyesters la viscosité augmente progressivement jusqu'à la prise. Par contre, les résines IKAS gardent à peu près une viscosité constante jusqu'à le temps de prise. La DPU vers +20° C est de 4 à 5 h pour les résines pures et de 2 à 3 h pour les résines chargées.

Le retrait en cours de polymérisation des résines pures est toujours plus important que celui des résines chargées. Les tensions internes dues au retrait ne sont pas grandes à cause de répartition du phénomene de retrait sur une durée de la réaction exothermique prolongée. Ce qui permet de revêtir de grandes surfaces sans risque de fissuration du revêtement.

Les résines IKAS possèdent des résistances mécaniques plus faibles que celles des époxydes ou polyesters. Mais c'est la résistance chimique qui est la qualitée de préférence pour un revêtement anti-corrosion. De nombreux essais nous ont permis d'établir les coefficients de tenue chimique des compositions aux agents chimiques. Notons, que les revêtements IKAS présentent une bonne résistance chimique aux solutions acides et basiques des concentrations habituelles, aux solutions salines, aux hydrocarbures, aux atmosphères industrielles, etc ...

La méthode de calcul de compositions IKAS pour les revêtements anti-corrosion est basée sur le principe à obtenir des compositions de résistance mécanique ou chimique maximale avec le minimum du liant en gardant une maniabilité désirable.

Pour les mastics de protection la quantité de filler

$$G_1 = \frac{\gamma_m V_m}{1+\beta A_{rel}^{0,2}}$$

et celle du liant

$$G_p = \beta A_{rel}^{0,2} G_1 .$$

Pour les mortiers de résine de protection le poids du sable

$$G_2 = \frac{\gamma_{0,2}}{1+(\alpha-1) V_{0,2}}$$

et celui du mastic

$$G_m = \frac{\alpha V_{0,2} \gamma_m}{1+(\alpha-1) V_{0,2}}$$

Dans ces formules: γ_m et V_m la dancité et le volume du mastic; $\gamma_{0,2}$ et $V_{0,2}$ la masse volumique et le volume des vides du sable; A_{rel} la surface spécifique du filler; β le module de richesse en liant a une valeur de $11{,}5 \cdot 10^{-2}$ pour les compositions optimales en résistance mécanique et de $11{,}5 \cdot 10^{-2}$ à $14 \cdot 10^{-2}$ pour les compositions optimales en résistance chimique; α le coefficient d'enrobage des grains du sable (pour les melanges autolissants $\sim 1{,}5$).

3. Durabilité des revêtements

La durabilité d'une structure en béton armé est caractérisée par la fonction probabiliste

$$Z(t) = R(t) - S(t),$$

où $R(t)$ est la résistance et $S(t)$ est la charge (fig. 1).

La fiabilité d'une structure s'exprime par l'expression:

$$P\{T \geqslant t_{ser}\} = P\{Z(t) > 0\}.$$

Il est évident que dans le cas des millieux agressives les revêtements anti-corrosion assurent la protection, augmentent la durée de vie et la fiabilité d'une structure. Le problème est donc de déterminer le temps de vie et ses caractéristiques stochastiques. Dans le cas général, pour une structure en béton armé (fig. 2)

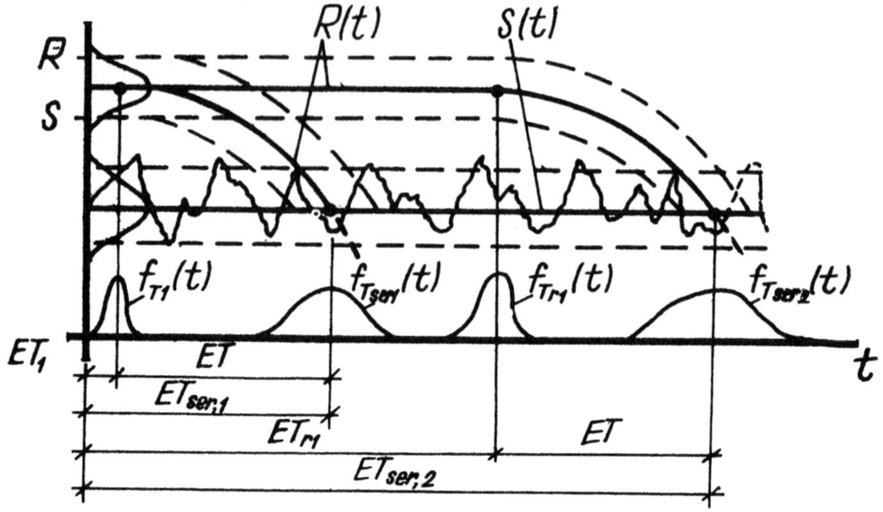

Fig. 1. Durabilité d'une structure sans (1) et avec (2) un revêtement anti-corrosion.

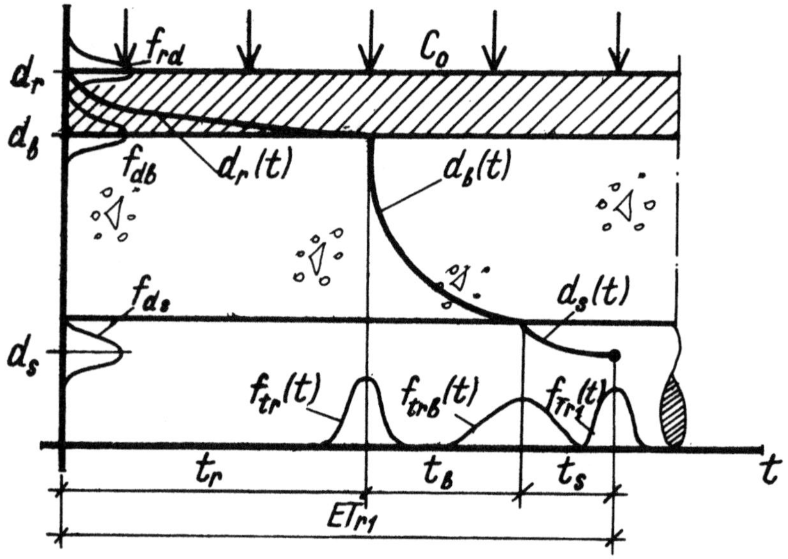

Fig. 2. La durée de vie du système de protection (t_r+t_b).

$$T_{r1} = t_r + t_b + t_s \ ,$$

où: t_r et t_b la durée de vie du système de protection (revêtement + recouvrement du béton); t_s le laps de temps écoulé entre le commencement de la corrosion des armatures et la valeur admissible de celle-ci. Donc la durée de vie du revêtement $T_{r1} - (t_b + t_s)$. Dans présente étude nous n'avons pas de possibilité d'arreter sur les modèles de calcul de t_b et t_s . La durée de vie t_r est la fonction des conditions de l'environnement et des paramètres du revêtement lui-meme (type du liant, composition, épaisseur). Un revêtement peut perdre ces propriétés protectrices soit à cause de dégradation chimique ou physique, soit à cause de pénétration des agents à travers du revêtement, soit enfin à cause de décollement du revêtement.

L'épaisseur du revêtement:
dans le cas de dégradation

$$d_r = d_0 e^{-\lambda t_r}$$

dans le cas de pénétration

$$d_r = \sqrt{\pi \beta t_r} \ (1 - \frac{C_x}{C_0})$$

où: d_0 l'épaisseur admissible de la couche du revêtement; λ la constante de corrosion; β le coefficient de diffusion; C_0 et C_x la concentration du millieu agressif sur la surface et dans une profondeur X .

La durée de vie d'un revêtement est étroitement liée également à l'adhérence de celui-ci au support

$$\sigma_r + \sigma_{tem} + \sigma_{ch} \leq R_{bt} \ ,$$

où: σ_r , σ_{tem} et σ_{ch} les contraintes dues au retrait de polymer, la variation de température et des charges; R_{bt} la résistance du béton à la traction.

4. Applications

Les applications sont les suivantes:

terrasses et sols industriels

reservoirs et cuves en béton armé ou em acier pour les eaux acides

étanchéité des superstructures des bâtiments (facades, murs, poteaux, dalles, etc...) en présence des atmos-

phères agressives
revêtements des installations agricoles (citernes, fosses de neutralisation, silos)
chapes d'étanchéité des surfaces en béton ou pierre (sous-sols, fondations, ponts, parkings, etc.).

Les revêtements multicouches IKAS doit être appliqués sur un support sain, propre, sec et le plus régulier possible (pour les peintures). Il est nécessaire d'appliquer sur le support une ou deux couches d'accrochage avec le mélange sans charge. Grâce à la viscosité basse et le temps de copolymérisation prolongé la résine s'impregne fortement dans le béton, renforce la surface en profondeur et l'adhérence du revêtement. L'application de revêtements se fait à la brosse, au rouleau, au pistolet.

Pour choisir un système de protection le plus économique, il faut tenir compte des coûts de construction (C) et des frais de l'entretien et de réparation pendant le temps de service t_{ser}. A condition que la durée de vie du revêtement remonte à t_r années, le coût total E est égal approximativement à:

$$E = C\left(1 - \frac{t_{ser}}{t_r}\right).$$

La comparaison de E pour différents types de revêtements montre que les revêtements IKAS dans la majorité de cas présentent une économie considérable. Par exemple, les coûts des peintures à base de résines perchlorvinyliques, époxydes, polyuréthanes, de caoutchouc chloré et IKAS interviennent dans un rapport suivant: 2:4:9:3:1.

Bibliographie

Kamaitis, Z., Sniuksta, A. (1986) Revêtements anti-corrosion IKAS pour les structures en béton armé, L'institut de l'information technique et des recherches techno-économiques de Lithuanie, Vilnius, 40.

Kudzys, A. (1985) Reliability estimation of reinforced concrete structures, Vilnius Mokslas publichers, 156.

MODELLING THE WORK OF GLASS FIBER REINFORCED RESIN COATINGS ON A CRACKED SURFACE OF CONCRETE

W.M. KŁOSINSKI, M.W. ABRAMOWICZ, W. GLINKA
Institute of Building Constructions, Warsaw University of Technology

Abstract
The use of glass fiber reinforced resin coatings for protection of concrete tanks for chemical aggressive industrial liquids depends on whether they are crack-resistant enough. During the work of the tank the width of the cracks can change. The appropriate equipment enabling the research into resin coatings crack-resistance has been built. It allows to test different coatings over changeable widths of cracks in the modelled surface with fixed frequency, number of changes and dimentions of the created chinks.
Key words: Glass fiber reinforced resin coating, crack-resistance, surface, width of cracks

1. Introduction

In concrete building constructions the possibility of cracking on the surface of concrete and reinforced concrete has to be taken into consideration. In practice in a majority of cases the forming of cracks is beyond our control but we can control their width e.g. in terms of the percentage amount and the way of steel reinforcement. The problem of the work of laminates on concrete occurs in those situations where a crack appears under the resinous glass layer.

A great number of researches /refered to separetaly in the paper by Czarnecki, Grabowski/ have been carried out on the work of resinous fiber reinforced and fiberless coatings on the concrete surface which can be subjected to cracking.

2. General description

The research in question has been concerned with observations of the behaviour of glass fiber reinforced resin coatings used for anticorrosion protection of reinforced concrete tanks /for liquids aggressive toward concrete/ in which the volume of liquid may be changed manifoldly. As a result of that, changing extending forces in the tank construction must take place; in consequence, the width of cracks in concrete will be changing. Therefore, fundamental research has been concerned with observations of coatings on the so modelled base, in which the width of crack can change repeatedly. That crack may be opened and closed and its width may increase from the initially fixed width to practically any width which might be observed in concrete constructions. Initial observations consisted in recognising crack-resistance of different kinds of plast-materials by a single dilatation of the crack. This crack-resistance made it possible to eliminate those materials and laminates which turned out to be not crack-resistant or were damaged over the crack the width of which was less than 0.1 mm. Those materials have been recognized to be useless in covering concrete surfaces on which cracks can be developed.

Up to now plasticised epoxy and polyester resin laminates reinforced with glass fiber have been tested.

3. Testing equipment

The equipment has been designed and made in the Institute of Building Constructions, Warsaw University of Technology.

It consists of special swing dimerous forms and a device which stretches and tightens them /Fig. 1,2/ with a controlled force and over a fixed distance.

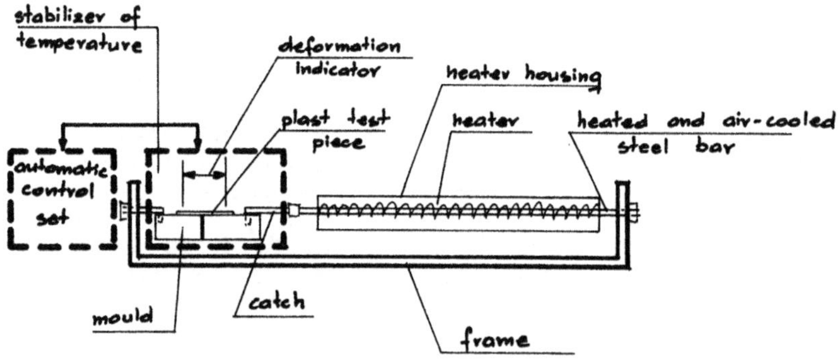

Fig. 1. Scheme of research equipment.

Test pieces are placed in these forms. Each form consists of a steel frame /Fig. 3./ with horizontal slides in which two steel plates used as a ground for the coating piece are placed.

Fig. 2. Test piece during research cycle.

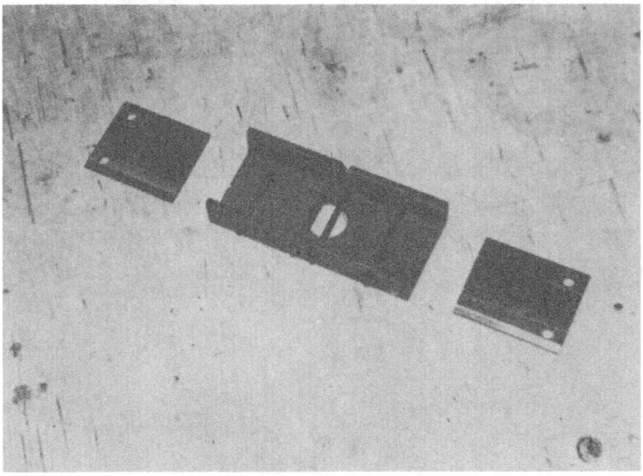

Fig. 3. Decomposed steel frame-form.

The sliding of both plates along their rails models the rising of crack under the coating piece. Test pieces were 120 mm long and 40 mm wide.

Mutual displacement of both plates of dimerous ground has been obtained through a thermal expansion of a steel bar and through the medium of catches fixed to the plates. The plates may be pulled and pushed together at a fixed speed and frequence. The test piece was observed by means of a microscope. The equipment recorded automatically:

- the width of crack /between plates/ in time
- the force needed to move both plates
- the temperature of the test piece

4. Course of tests

Research has been carried out with an assumption that every test piece should be tested at least 200 times over a crack of a width of between 0 and 0.1 mm, at least 150 times over a crack of a width between 0.1 and 0.2 mm, at least 100 times over a crack of a width between 0.2 and 0.3 mm, and finally at least 50 times over a crack of a width between 0.3 and 0.4 mm. Altogether this gives at least 500 cycles. The number may be related to the real work of tanks. Test pieces had a temperature of +30 °C. The examination of each piece was completed by making a 0.5 mm wide crack and this chink was left for one week. Finally, the width of the crack was increased until the test piece had been damaged. Fig. 4. shows a test piece of glass fiber reinforced resin coating after damage. The destruction of a piece usually was a consequence of its break, sometimes it happened as a result of the coming off of the coating.

Fig. 4. Test piece after damage.

5. Results of tests

Laminates with plasticised epoxy resin:
- 50 pieces were tested

- 15% of them were damaged over the crack of a width up to 0.2 mm

- 7% of them were damaged over the crack of a width up to 0.4 mm

- in all the other pieces the crack of a width up to 0.5 mm was brought about by the static load

Laminates with plasticised polyester resin:

This type of resin turned out to be far less crack-resistant. Only 10% of pieces withstood the crack of a width up to 0.4 mm. All the others were crack-resistant over the crack of a width up to 0.2 mm, but almost instantly were damaged over the crack of a width up to 0.3 mm.

6. Conclusions

The research into glass fiber reinforced resin coatings on a modelled cracked surface of concrete has made it possible to describe their relative resistance and durability. Therefore an adequate material can be chosen depending on the anticipated width of cracks in concrete, or the concrete construction must be designed so that the crack is appropriate for the used resin, generally by means of steel reinforcing.

STUDY ON THE USE OF CRACK RESISTANT POLYESTER MORTAR AS ANTICORROSION COATING FOR OUTDOOR CONCRETE

LOU XIAN-NENG
Central Research Institute of Building and Construction Ministry of Metallurgical Industry, People's Republic of China

Abstract
The paper presents preparation of a low shrinkage polyester mortar coating, its physico-mechanical properties and corrosion resistance in various acid media. The coating is made of polyester resin as binder and hollow micropellets in fly ash as powder material. It features extraordinary low shrinkage approaching zero, and low tensile modulus of elasticity. Applied on the surface of portland cement concrete, it forms a resin mortar coating that is highly adhesive, impermeable, waterproof as well as abrasion-, impact-, weathering- and chemical resistant. It has been successfully used in southen China under intense heat of summer when temperature on the ground is as high as 70-80 c as well as temperature of several degrees below zero. It has been used with success as a isolating for large-sized waste acid reservoirs.
Key Words: Polyester mortar coating, Corrosion resistance, Fly ash, Contraction.

1. Introduction

Polyester mortar and its use as a protective coating for concrete have been studied for over 20 years. Plenty of studies have been conducted with breakthroughs by a number of countries on issues such as cracking and delamination when using polyester mortar as a protective coating for large-sealed concrete floors as well as the properties of polyester mortar and construction of coating. To meet the needs of polyester mortar as an anticorrosion coating for outdoor concrete, small amounts of rubber type elastic additive are incorporated into a mixture of fly ash and barium sulfate powder to reduce contraction and elastic modulus of polyester mortar and improve its crack resistance. Crack and corrosion resistant polyester mortar with bulk density 1.4-1.5 has been prepared with success as anticorrosion coating for outdoor concrete at a smelting works in the south China's hot and damp region.

2. Fundamental

When polyester mortar is used as an anticorrosion coating for outdoor concrete, the coating has to resist at least two kinds of failure due to curing contraction stress and thermal stress. The internal stress of a coating is equal to the sum of contraction stress and thermal stress. $\sigma_G = \sigma_s + \sigma_t = E(\varepsilon_s + \alpha \Delta t)$ where σ_G - internal stress; E-elastic modulus; α - coefficient of expansion; ε_s - contraction percentage; Δt - temperature difference. Besides, an autdoor coating is exposed to strong sunshine and may be damaged by ultraviolet light, a destructive factor which should not be underestimated. Ultraviolet light causes breakdown of the resin's molecular chain and macroscopic structure, discoloration of coating surface as well as cracking and pulveritation also occur.

In order that a polyester mortar coating may be used in the open, an elastic additive is incorporated and a suitable powdered mixture selected to simultaneously reduce hardening contraction and elastic modulus with coefficient of thermal expansion constant or reduce so that internal stress does not exceed tensile strength of coating. On the other hand, FRP barrier is provided between coating and concrete substrate. During construction thickness of coating should be controlled and expansion joints provided at intervals. To prevent ultraviolet damage fly ash with low bulk density is used, which floats on coating surface and prevent polyester resin from oxidation; fine sand is spread on coating surface for protection. With these measures taken, polyester mortar coating can be prevented from cracking, delamination, weathering and other damage during its construction, curing and service.

3. Materials and preparation

199A polyester resin is used in crack resistant polyester mortar; 50% dibutyl cyclohexanone peroxide paste, 8% cobalt octoate styrene solution and N,N-dimethyl aniline are used as initiator, accelerator and co-accelerator respectively, mixture of fly ash and barium sulfate is used as powdered material, quartz sand as sand, and rubber powder as elastic additive. Mix proportion of mortar is shown in table 1.

Table 1. Referense mix ratio

Polyester resin(g)	Quartz sand(g)	Powdered mixture(g)	Rubber powder(g)
100	400	90	7

4. Selection of powder and elastic additive

Powder considerably affects contracting and elastic modulus of polyester mortar. Comparison tests were conducted on quartz powder, precipitated barium sulfate power, fly ash etc. It is found that fly ash-based mixture with small amount of precipitated barium sulfate powder causes polyester mortar to have minimum hardening contraction, which polyester mortar prepared with quartz powder has maximum contraction percentages are in the following order: Quartz powder > cast stone powder > lightweight calcium carbonate powder > barium sulfate powder mixture of fly ash and barium sulfate powder. Contraction percentages of polyester mortars are compared prepared with barium sulfate powder and mixture of fly ash and barium sulfate (without elastic additive incorporated). Data are shown in Table 2.

Table 2.

Types of powdered material	Contraction (%)					Elastic modulus in tension
	1 day	3 day	7 day	14 day	28 day	($N/mm^2 \cdot 10^5$)
Barium sulfate	.206	.246	.257	.279	.303	.118
Fly ash + barium sulfate	.106	.127	.136	.159	.168	.122
contraction reduction factor %	48.5	48.3	47	43	44	

As seen from Table 2, mixture of fly ash and barium sulfate causes polyester mortar to have contraction reduction factor 43-48.5%, compared with barium sulfate powder, it reveals advantage in using fly ash as powdered material.

Elastic additive has a significant effect on contraction and elastic modulus in tension. Comparision tests were carried out on the effect of elastic additive on contraction and elastic modulus in tension by employing PVC powder, polyethylene powder, PVC fiber and rubber powder. It is found that for the first three additives despite satisfactory effect of reduction in contraction and elastic modulus the quantity required is too large that the coefficient of thermal expansion of mortar increases, especially when temperature reaches 80 c, thermal deformation increases significantly. The Rubber-type additive has no such drawbacks, dosage under 10% has a significant effect on contraction reduction with elastic modulus in tension decreased accordingly. Results are shown in Table 3.

Table 3. Effect of rubber-type additive on contraction and elastic modulus

Dosage of rubber powder	Contraction %					Elastic modulus in tension
%	1day	3day	7day	14day	28day	N/mm² · 10^5
0	0.106	0.127	0.136	0.159	0.168	0.122
5	.0469	.0625	0.563	.0583	.064	0.103
Contraction reduction factor %	55.75	50.78	58.6	63.3	61	
7.5	0	.0031	.0025	.0029	.003	0.111
Contraction reduction % factor	100	97.6	98	98	99.8	

As seen from Table 3, incorporation of rubber-type additive causes significant decrease in contraction of mortar. At dosage of 7.5% maximum contraction reduction factor reaches 100% and minimum factor 97.6%, reduction factor of elastic modulus in tension reaches 15.6% and 9% respectively. It reveals that rubber-type additive has significant effect on contraction reduction and a certain effect on reduction in elastic modulus. Mixed with fly ash, it results in satisfactory crack resistance of polyester mortar.

5. Properties

5.1 Physico-mechanical properties

Test procedures follow "Test and inspection instruction in metallurgical capital construction" Ye-Ji-Gui 103-77. Data are shown in Table 4.

Table 4. Physico-mechanical properties

Coefficient of expansion 1/°C	Compressive strength N/mm²	Tensile strength N/mm²	Flexaral strength N/mm²	Impact strength N.mm/mm²	Abrasive resistance
$1.67 \cdot 10^{-5}$	42.8	9.9	21	13	6.1

Bond strength with cement mortar N/mm²	Freezing and thawing resistance (150 cycle) %
1.6	-14

5.2 Corrosion resistance

Results are shown in Table 5.

Table 5. Corrosion resistance	
Media	Assessment
H_2SO_4 1 - 30 %	satisfactory
HNO_3 15 %	satisfactory
HCl 15 %	satisfactory

6. Application

6.1 Service portion and conditions

Crack resistant polyester mortar is used as anticorrosion coating for outdoor concrete in the following portions:

(a) Floor for waste acid treatment	756 m^2 (36 x 21 m)
(b) Floor for waste water treatment	200 m^2 (10.3 x 19.5 m)
(c) 2 Wastewater basins	each 198 (9 x 22 m) total 396 m2
(d) Water purifying tank	70 m^2 (7 x 10 m)
(e) Small wastewater basin	16 m^2 (4 x 4 m)

Service condition: Corrosion medium: 1 - 15% H_2SO_4. Temperature: winter -10 °C, summer 70 - 80 °C, outdoor use subjected to sunshine and rain.

6.2 Construction produre

Preparation ob substrate → Painting with epoxy varnish → Laying Polyester FRP → Coating with 5 m/m thick polyester mortar → Painting with polyester varnish and spreading fine sand → Curing and service.

6.3 Cases Histories

During construction and curing of polyester mortar coating cracking and delamination have not been observed, the coating is of even thickness and smooth. Since over one year service coating has withstood sunshine and rainwater, year round temperature change -10-70 °C, ultraviolet radiation, and corrosion of corrosive media including waste acid and waste water discharged from production processes; coating, delamination, weathering and corrosion have not been observed, service performance is satisfactory.

7. Conclusions

(a) Practical use of polyester mortar coating in large-scale construction and as anticorrosion coating for outdoor concrete shows that crack resistant polyester mortar coating has satisfactory crack resistance, it can resist hardened contraction stress and thermal stress without cracking and has shown satisfactory service performance.

(b) In polyester mortar incorporation of rubber type elastic additive and use of fly ash-based powdered mixture result in simultaneous significant decrease in contraction and elastic modulus in

tension as well as certain decrease in coefficient of thermal expansion, significant crack resisting effect is achieved. Also, fly ash mixture improves considerably construction properties. Fly ash with low bulk density, floating on the surface, can isolate air and prevent polyester mortar surface from oxidation and sticking.
The use of fly ash mixture can accelerate hardening of coating and shorten curing period and construction cycle. Meanwhile, the use of fly ash reduces bulk density of polyester mortar coating to 1.4-1.5, ie by 25%-30%. The use of polyester mortar for making construction elements can reduce its weight by 25-30% and transportation cost as well. When it is used as floor coating, the quantity per sq meter and consequently the cost is decreased. Fly ash is a waste material of soud fuel power plants, its successful use in polyester mortar opens up a new way to comprehensive use of fly ash.

(c) Construction of crack resistant polyester mortar coating is appropriate. Polyester FRP barrier improves crack resistance and impermeability of coating.

(d) Crack resistant polyester mortar coating has achieved significant technical econommical results. Its cost is three times less compared with granitic floor, construction rate 6 times faster, lower labour intensity, safety in construction, ease of ensuring construction quality. Its cost is half of that of epoxy mortar floor. It can be seen that crack resistant polyester mortar is a monolithic coating material for indoor and outdoor concrete easily to be popularized.

References

H.T. Peschke, Stress and strain analysis between cementitious concrete and polymer concrete, proceedings voli pp 477-489

T.Okada and T.Konaka, Low shrinkage mechanisms in single-phase and two-phase unsaturated polyester resin mortar system, Third international congress on polymers in concrete proceedings, Vol 1 pp 447-461.

D.W.Fowler, A.H. Meyer and D.R.Paul, Low temperature curing of polymer concrete, proceedings Vol 1 pp 421-434.

R.S.Ayyar and S.N.Deshpand, Creep studies on polymer mortars, Third international congress on polymers in concrete, proceeding Vol pp 504-523.

UTILISATION DES RESINES SYNTHETIQUES DANS L'ENTRETIEN DES BARRAGES D'ELECTRICITE DE FRANCE

Daniel RENIER
Adjoint au Chef de la Division Génie Civil,
Service de la Production Hydraulique.

RESUME

Des informations sont données sur les matériaux et les procédés utilisés actuellement sur les ouvrages EDF pour l'étanchéité des parements amont de barrage ainsi que pour le traitement des fondations et les protections contre l'abrasion des eaux. L'expérience acquise sur quelques revêtements superficiels d'étanchéité permet d'estimer approximativement la pérennité de ces interventions.

Introduction

Les matériaux de synthèse ont fait leur apparition sur les chantiers de génie civil, il y a une vingtaine d'années environ. Depuis cette époque, leur usage n'a cessé de se développer, notamment en ce qui concerne la protection ou la réparation des bétons dégradés. Pour l'entretien de ses barrages, EDF ne pouvait manquer d'expérimenter ces nouveaux matériaux et ces nouvelles techniques.

Leur emploi le plus ancien et le plus fréquent concerne l'étanchéité des parements amont de barrage par revêtement superficiel. Il arrive aussi que, pour faire écran aux percolations, des résines soient mises en oeuvre par injection, soit en fondation, soit dans le corps même de l'ouvrage. Quelquefois enfin, des chapes synthétiques sont appliquées sur les radiers pour les protéger contre l'abrasivité de l'écoulement. Il s'agit toujours de chercher à utiliser le type de produit répondant le mieux au problème posé.

Ce document présente des exemples relatifs à ces emplois caractéristiques.

1. Etanchéité superficielle des parements amont de barrage

Certains parements amont de barrages en béton vieillissent mal. Après quelques temps de service, des fuites croissantes peuvent apparaître à travers le corps de l'ouvrage. Depuis plus de 20 ans maintenant, EDF est intervenu sur un certain nombre d'entre eux pour enrayer les risques de dégradation qui s'ensuivent, en réalisant notamment des revêtements étanches en résine synthétique.

Le tableau ci-dessous donne les caractéristiques principales des ouvrages cités dans ce qui suit.

	LA GIROTTE	MIGOELOU	LANAU	PORTILLON
Situation	Savoie	Hautes Pyrénées	Truyère	Haute Garonne
Année de mise en eau	1950	1958	1962	1951
Altitude	1 750 m	2 300 m	660 m	2 566 m
Type	18 voûtes + contreforts et 1 barrage poids	9 voûtes + contreforts et 1 voûte annexe	1 voûte	Enrochements masque maçonnerie + gunite
Hauteur sur fondations	48 m	29 m	30 m	23 m
Longueur en crête	510 m	225 m	178 m	205 m
Epaisseur (voûte ou masque)	1,70 m	1 m	1,3 m à 3 m	1,2 m à 0,6 m + 0,08 env.
Surface de parement	400 à 1 400 m2 par voûte	5 200 m2	3 500 m2	3 100 m2
Ciment	CPA du Teil	CPB 250/315	CPAL et CLK 210/325	CPA 250/315
Dosage	325 et 380 kg/m3	350 kg/m3	300 kg/m3	600 kg/m3
Année d'intervention sur les parements	à partir de 1963...	1970/1980	1972	1985

La médiocre qualité de certains constituants du béton, une exécution peu soignée de celui-ci, peuvent être à l'origine d'une dégradation rapidement aggravée par les conditions de vie très difficiles de l'ouvrage, le gel particulièrement, ou l'agressivité des eaux.
A la Girotte, le barrage a été construit dans l'immédiat après guerre et n'a certes pas bénéficié des meilleurs matériaux ; des fissures sont apparues dès la fin de la réalisation et le phénomène s'est accru après la mise en eau ; l'effet d'encastrement lié à la forme complexe de la géométrie des voûtes -qui comprennent de haut en bas une portion de tore, une partie cylindrique inclinée vers l'aval, une portion de sphère et pour certaines encore une portion de cylindre vertical-, est sans doute aussi à l'origine des désordres : les fissures affectent surtout la jonction des différentes portions et la clé de voûte.

A Migoëlou, la construction des bétons de ce barrage de haute altitude a nécessité 2 campagnes de travaux d'été. Tout le soin a été apporté pour avoir un matériau de bonne tenue au gel ; les zones de faiblesse sont les reprises de bétonnage entre levées successives, qui ont constitué dès la mise en eau des plans de cheminements de fuite importants. De plus, les voûtes de Migoëlou sont relativement minces, et le PH des eaux pyrénéennes est très acide...

Au barrage de Lanau aussi, les nombreuses fissures horizontales correspondent pour la plupart à des reprises de bétonnage. Au Portillon, le liant de la maçonnerie et la gunite du masque, détériorés par le gel, se dégradent chaque année davantage.

L'augmentation des fuites à l'aval des barrages ou l'évolution des suintements marquant le parement aval des voûtes, sont quelquefois telles qu'il apparaît nécessaire d'étancher les parements amont.

Les premières tentatives faites dans ce sens avec des revêtements superficiels en brai ou autres produits bitumineux ont en général échoué. C'est l'emploi de résine de synthèse à partir de 1960/65 environ qui permît pour la première fois d'espérer des résultats positifs et durables, spectaculaires en tous cas : la preuve en était la disparition quasi totale des tâches humides sur les parements aval des ouvrages.

Le choix du produit a évolué et la technique mise en oeuvre s'est perfectionnée au fil des années. Les paragraphes qui suivent donnent une brève information sur les réalisations actuelles.

1.1. Réparation des zones de béton dégradé

Suivant l'importance du volume concerné, la reprise est effectuée soit avec un mortier de résine époxy, soit avec un mortier dosé à 500 kg de ciment CLK additionné d'acétate de polyvinyle ou de styrène-butadiène.

Elle a lieu après démolition des zones altérées, jusqu'au béton sain, soufflage et lavage à haute pression. Le meilleur collage est assuré au moyen d'une couche d'imprégnation en résine époxy appliquée après séchage du support (lorsque la surface est importante et pour un mortier de ciment, l'époxy peut être remplacé par une barbotine à l'acétate ou au styrène-butadiène). Quel qu'il soit, le mortier est mis en oeuvre sur une couche d'imprégnation encore poisseuse (soit 10 à 20 minutes après celle-ci).

Ce type de réparation est utilisé sur nos ouvrages depuis de nombreuses années et donne entière satisfaction. L'un des premiers exemples concerne le traitement d'une écaille de béton à La Girotte en 1965. En 1985, les zones dégradées du masque amont du Portillon ont aussi été repris de la même manière.

1.2. Traitement des joints ou des fissures

Dans les années 1960/70, plusieurs types de produits ont été mis en oeuvre, notamment :

- des silicones
- des polysulfures d'éthylène (type Thiokol)
- des brais polyuréthannes
- des polyuréthannes.

Des difficultés ont été rencontrées en ce qui concerne l'adhérence des silicones. Les premiers mastics polyuréthannes n'ont pas toujours répondu à notre attente (à Migoëlou, en 1970, par exemple). Finalement, les élastomères les plus satisfaisants, et qui continuent d'être utilisés actuellement sont les Thiokol (à la Girotte depuis 1962 et à Lanau en 1972) ainsi que les brais polyuréthanne (au Portillon en 1985 comme à la Girotte en 1965). Il faut aussi signaler l'utilisation à Migoêlou, d'un polyuréthanne plus rigide, appliqué, en 1972 à titre d'essai, et en 1980, dans tous les joints pile-voûte et de clé : après cinq années une fine fissure apparaît à la surface des joints, en raison de la trop grande raideur du matériau.

En ce qui concerne le Thiokol, la qualité des produits proposée sur le marché peut être très variable : il convient pour notre usage de choisir parmi les bicomposants ceux dont les performances sont les meilleures. Et ils doivent être appliqués sur un primaire "spécial" : les décollements constatés sur certains joints de Lanau résultent probablement d'une application effectuée directement sur le polyuréthanne utilisé en revêtement de parement.

La préparation propre au traitement de joints et de fissures comporte si nécessaire, la réalisation d'une mortaise de 2 à 4 cm de largeur et de 1 à 2 cm de profondeur par sciage, ou bien l'enlèvement du vieux produit de joint par sablage et la reconstitution éventuelle des lèvres du joint avec un mortier d'époxy, la pose d'une mousse ou d'un papier adhésif en fond de mortaise. L'expérience a montré qu'il était utile de coller aussi du papier de part et d'autre des lèvres du joint de façon que les débordements du mastic, inévitables lors du bourrage, ne constituent pas l'amorce ultérieure d'un décollement.

Cette préparation est le plus souvent réalisée en même temps que les travaux préliminaires au revêtement général du parement.

S'il n'est pas prévu de revêtement général, le traitement décrit ci-dessus est assorti d'un revêtement partiel concernant une bande de parement de 1 m environ de largeur de part et d'autre du joint ou de la fissure.

1.3. Préparation du support

Cette phase de travaux est très importante, et le soin apporté à son exécution conditionne l'efficacité du traitement.

Les surfaces à revêtir doivent être saines et propres. Elles peuvent être bouchardées, mais le risque existe alors de s'appuyer sur un béton microfissuré et désorganisé superficiellement. La méthode la plus sûre, la moins "traumatisante", habituellement employée dans nos ouvrages, est le sablage suivi d'un dépoussiérage par brossage énergique ou lavage à haute pression · un chantier récent (Escoubous en 1983) a expérimenté, en remplacement de la sableuse traditionnelle, une pompe à haute pression munie d'un dispositif permettant la projection simultanée du sable (Pression 200 bars) ; il en résulte un travail moins "poussiéreux" mais le rendement est inférieur.

La technique du sablage est également utilisée pour enlever complètement l'ancien revêtement - usé ou cloqué - (La Girotte).
Dans certaines zones de forte adhérence, l'opération est plus difficile et délicate.

Depuis 1972 (Lanau), la préparation du support est souvent complétée par un masticage ou "bouche-porage" de toutes les cavités (reprises de bétonnage, nids de cailloux, bulles de coffrage) avec un mortier de même nature que la résine appliquée en revêtement. Cette opération est effectuée à la spatule, en général après mise en oeuvre de la couche d'imprégnation. Elle diminue le nombre des bulles d'air qui peuvent se trouver enfermées derrière le revêtement et constituent les points sensibles de celui-ci.

1.4. Le revêtement

Deux voûtes du barrage de La Girotte possèdent encore le revêtement d'hypalon en 4 couches appliqué en 1969 sur une imprégnation en 2 couches d'époxy-buton. Mais il est très usé : l'époxy apparaît par endroit. Ces 2 voûtes sont un peu l'exception. La tenue de l'hypalon appliqué sur d'autres voûtes du même ouvrage n'a guère excédé 10/12 ans.

Depuis 1970, la résine mise en oeuvre est le plus souvent un polyuréthanne (à la Girotte comme à Lanau, à Migoëlou comme à Escoubous et au Portillon). Appliquée tantôt à la brosse ou au rouleau, tantôt au pistolet bi-composant, la résine comporte en général 5 à 10 % de solvant pour la première couche, ce qui facilite l'imprégnation du béton, elle est sans solvant pour les deux ou trois couches appliquées ensuite. Rappelons que le solvant non évaporé, enfermé entre couches, a longtemps été à l'origine des déboires (formation de cloques) constatés avec les premiers produits de synthèse utilisés. La résine employée actuellement en revêtement est donc sans solvant et a deux composants : la résine de base et le durcisseur ; le polyuréthanne monocomposant, souvent rencontré dans le bâtiment, a, jusqu'à présent, échoué dans les conditions de travail plus difficiles de nos ouvrages.

La quantité de résine mise en oeuvre varie de 1,2 à 2 kg par m2 de parement (boucheporage non compris). Des contrôles effectués après travaux sur le barrage de Lanau (en 1972) ont vérifié la bonne adhérence du revêtement, et montré que, pour une consommation moyenne de 2 kg/m2, l'épaisseur totale était partout au moins égale à 1 mm pour les 3 couches, l'épaisseur de chaque film étant elle-même très variable.

Depuis 1978, à la Girotte, le revêtement est complété par l'application au pistolet Airless d'une couche de résine spéciale destinée à empêcher le farinage superficiel que provoque l'action des rayons UV. Il semble que cette protection se détériore assez rapidement, l'écaillage de la couche étant général au bout de 5 années : on peut se demander si l'application d'un film supplémentaire du revêtement lui-même ne serait pas préférable à "l'anti UV".

De toute manière, le revêtement polyuréthanne n'est pas "éternel". Ou bien des cloques apparaissent peu à peu, au fil des années s'amplifient et risquent de se déchirer. Ou bien il s'agit d'un phénomène d'usure ; c'est le cas le plus fréquent à la Girotte où les glaces du plan d'eau raclent la surface inclinée des voûtes pendant les marnages de printemps. Sur ce dernier ouvrage, comme à Migoëlou et à Lanau, la durée du revêtement polyuréthanne peut être estimée à 15/20 ans.

En 1983, le cylindre inférieur d'une voûte de la Girotte a été revêtu, à titre d'essai, par un copolymère d'époxy-uréthanne (2 couches sur 1 primaire époxy + 1 couche anti UV). Lors de la visite de 1985, le comportement de cet essai s'avérait satisfaisant.

2. Traitement des fondations de barrages

La fondation des barrages constitue une zone sensible des ouvrages. Des dispositifs d'auscultation permettent d'en suivre les évolutions en contrôlant notamment le débit des fuites ou (et) la piézométrie existant sous le barrage. Si des variations trop importantes apparaissent, il peut s'avérer nécessaire d'intervenir, par exemple en renforçant le voile étanche et en augmentant les drainages. Une intervention de ce type effectuée récemment sur le barrage de Saint-Etienne Cantalès a fait appel à des résines de synthèse.

Saint-Etienne Cantalès est un barrage poids-voûte de 70 m de hauteur, 270 m de longueur en crête, 35 m de largeur maximale au niveau des fondations, et 5,5 m de largeur au couronnement. Il a été construit entre 1939 et 1945 sur la Cère, affluent rive gauche de la Dordogne. Il retient 133 hm3 d'eau. La vue d'aval et le profil du plot central HI sont schématisés sur la figure n° 1.

A partir des années 1970-1975, les débits de fuites et les sous-pressions mesurés sous les plots centraux du barrage ont augmenté façon sensible. En 1981, le total des fuites dans cette zone atteignait 300 l/min., et certain piézomètre (PZ6), installé en fondation (à la cote 442) dans le 1/3 amont du barrage mais à l'aval du réseau drainant, indiquait la cote 500 (pour un plan d'eau amont à 516).

Une campagne d'injection fut entreprise au début de l'année 1982, pendant la période la plus froide de l'année et alors que le plan d'eau amont était à la cote haute, de façon à bénéficier de "l'ouverture" maximale de la fondation.

Des forages ont été réalisés à partir de la galerie basse et du fond de la galerie rive droite, à 4 m environ du plan du parement amont, jusque 5 mètres sous le niveau du contact béton-rocher. D'abord implantés à la maille de 3 m, ils ont été doublés, les premiers essais d'eau ayant montré peu de communication entre eux. Après exécution des trous supplémentaires par contre on obtenait d'un forage à l'autre une percolation plus systématique de l'eau, injectée sous une pression à peine supérieure à la charge d'eau amont (6 bars).

Les essais d'eau préalables montraient que le volume des vides étaient rélativement peu importants. Le choix du produit à injecter s'est donc porté sur un coulis à base de résine acrylique additionné de phosphates disodiques et trisodiques et de copolymères vinyliques.

Près de 9 000 litres de coulis ont été injectés. Les pressions d'injection ont varié de 6 à 14 bars ; dans certains cas, la communication s'établissait avec les forages voisins, avec formation de gel.

Le traitement avait été précédé d'une obturation des drains de la zone concernée. Il fut complété par la réalisation sous les deux plots centraux d'un nouveau réseau de drainage, constitué de sept forages espacés de 5 m environ et de 30 mètres de profondeur.

L'ensemble des travaux a durée deux mois et demi.

L'intervention s'avère efficace : en effet, la piézométrie sous le barrage a été divisée dans certains profils par 2 ou par 3 (voir la figure n° 1) et le débit des fuites mesurées dans la galerie basse est maintenant inférieur à 3 l/mn.

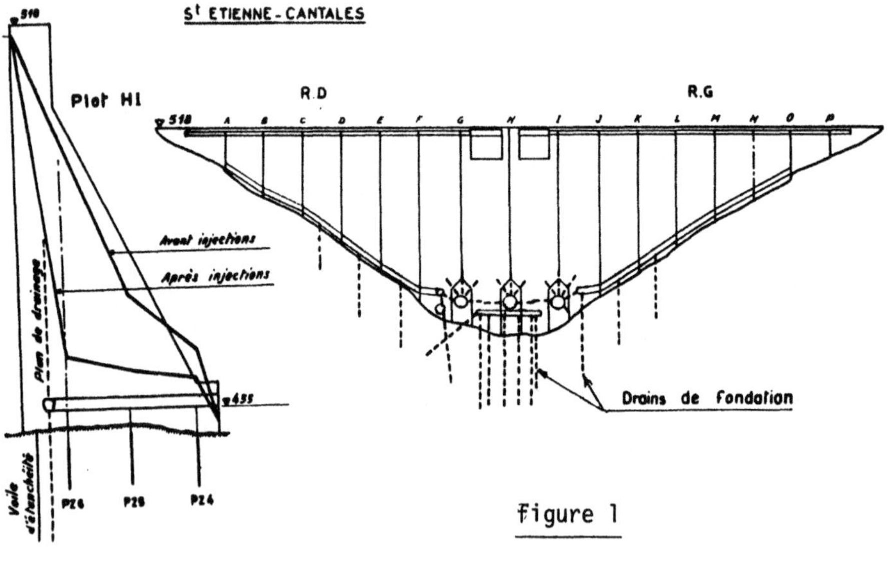

Figure 1

3. Traitement d'un parement amont de barrage par injection du corps de l'ouvrage

L'année qui a suivi les travaux effectués à St Etienne Cantalès, la même résine acrylique a été injectée dans le corps en maçonnerie du barrage de Combes (Usine de Confolent sur la Creuse).
Cet ouvrage est assez ancien. Mis en service en 1919, il a été surélevé en 1928 avec adjonction de 4 contreforts ; sa hauteur est de 31 m pour une longueur en crête de 140 mètres. L'amont est protégé par un enduit ciment de 2 500 m2 de surface.

Les travaux entrepris en 1983 concernent à la fois le parement amont et le contact de l'ouvrage avec la fondation. Il y avait, en effet, des fuites importantes, des suintements sur le parement aval et, sur un ou deux profils, une piézométrie relativement élevée à l'intérieur de l'ouvrage.

Les travaux ont concerné 1700 m2 de parement, soit environ les deux tiers de la surface totale. Ils ont comporté une injection de la partie amont de la maçonnerie sur 50 cm à 1 m environ de profondeur. Des forages ont été réalisés (25 mm de diamètre, 50 cm de profondeur) à raison de trois trous par m2. Un calfeutrement a été effectué sur les zones où l'enduit était fissuré. De l'eau colorée a d'abord été injectée dans les trous pour piloter les injecticns de remplissage elles-mêmes : pour les faibles vides, le coulis injecté était une résine acrylique (pression maximale 2 bars) ; les vides plus importants étaient traités avec un mélange acrylique-ciment-CPMF, ou bien avec une résine pôlyuréthanne gonflante et durcissant en présence d'humidité en 15 minutes à température normale - l'addition de durcisseurs permettait d'accélérer éventuellement la polymérisation -.

Cette méthode avait pour avantage de ne polluer ni les piézomètres, ni les drains qui se trouvaient à proximité. Le temps de polymérisation était réglé par les absorptions d'eau, mesurées au préalable, lors des essais d'eau. Le cheminement de la résine était limité aux premiers mètres d'épaisseur de la maçonnerie. L'absorption moyenne a été de 26 l/m2, avec un maximum de 46 l/m2.

La fondation a été traitée un peu de la même façon après exécution de forages obliques, pénétrant de 2 à 3 mètres dans le rocher, tous les 2 m 50, quelquefois tous les 1 m 25 ; les produits injectés ont été dans l'ordre, par absorption décroissante :

- d'abord un coulis de ciment CLK à la pression de 1 bar,
- puis un coulis acrylique-ciment (40 % de résine acrylique, 60 % de ciment) à la pression de 2 bars,
- enfin une résine acrylique pure.

La figure 2 donne une élévation aval, la coupe de l'ouvrage et le schéma des travaux. Sur le graphique sont représentés les relevés piézométriques avant et après traitement de la maçonnerie. Il montre que, d'une façon générale, une baisse très sensible de la piézométrie a été obtenue. Le traitement a en outre diminué considérablement les fuites : elles étaient de 18 l/minute avant travaux, elles sont maintenant de l'ordre de 5 l/minute et le parement aval est complètement asséché.

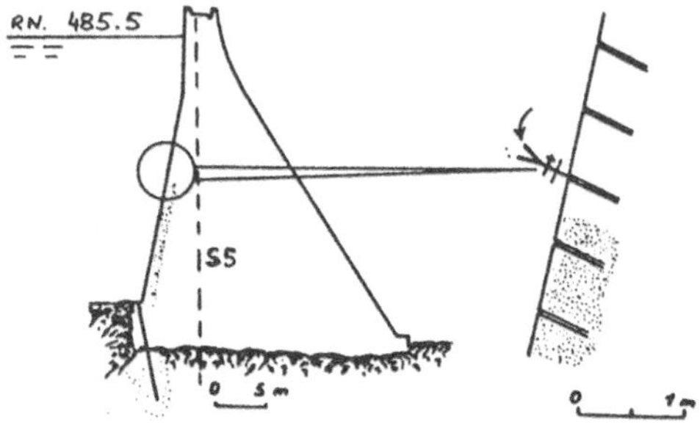

figure 2

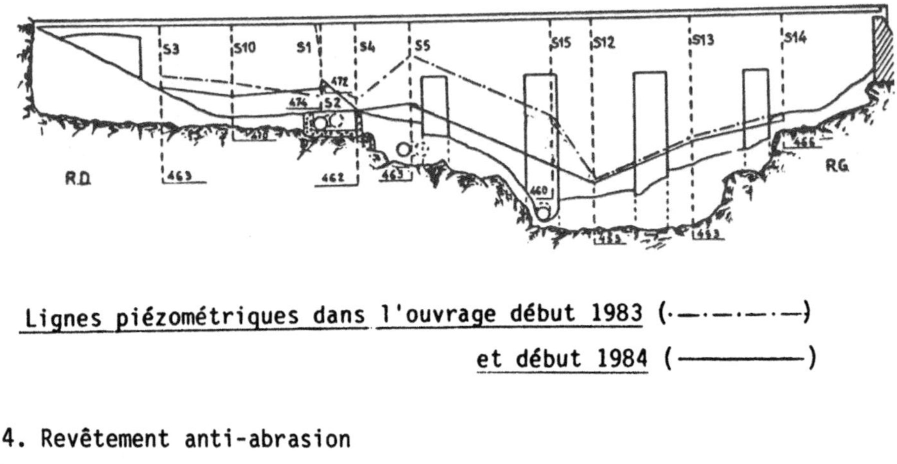

Lignes piézométriques dans l'ouvrage début 1983 (·—·—·—·—)

et début 1984 (————)

4. Revêtement anti-abrasion

Dans des régions montagneuses comme les Alpes, les radiers des ouvrages hydrauliques sont souvent soumis à de fortes sollicitations résultant des grandes vitesses de l'eau, des transports solides ou de l'impact des lames déversantes ; la tenue des bétons peut en être rapidement affectée, d'autant plus que les conditions climatiques y sont difficiles. Différentes protections contre l'abrasion ont été imaginées et utilisées depuis longtemps sur nos ouvrages, avec plus ou moins de réussite : revêtements en bois, chapes renforcées au corindon, blindages métalliques, revêtement de granit,

Fig. 3

BASSIN DU CLAPIER

Revêtement anti-usure

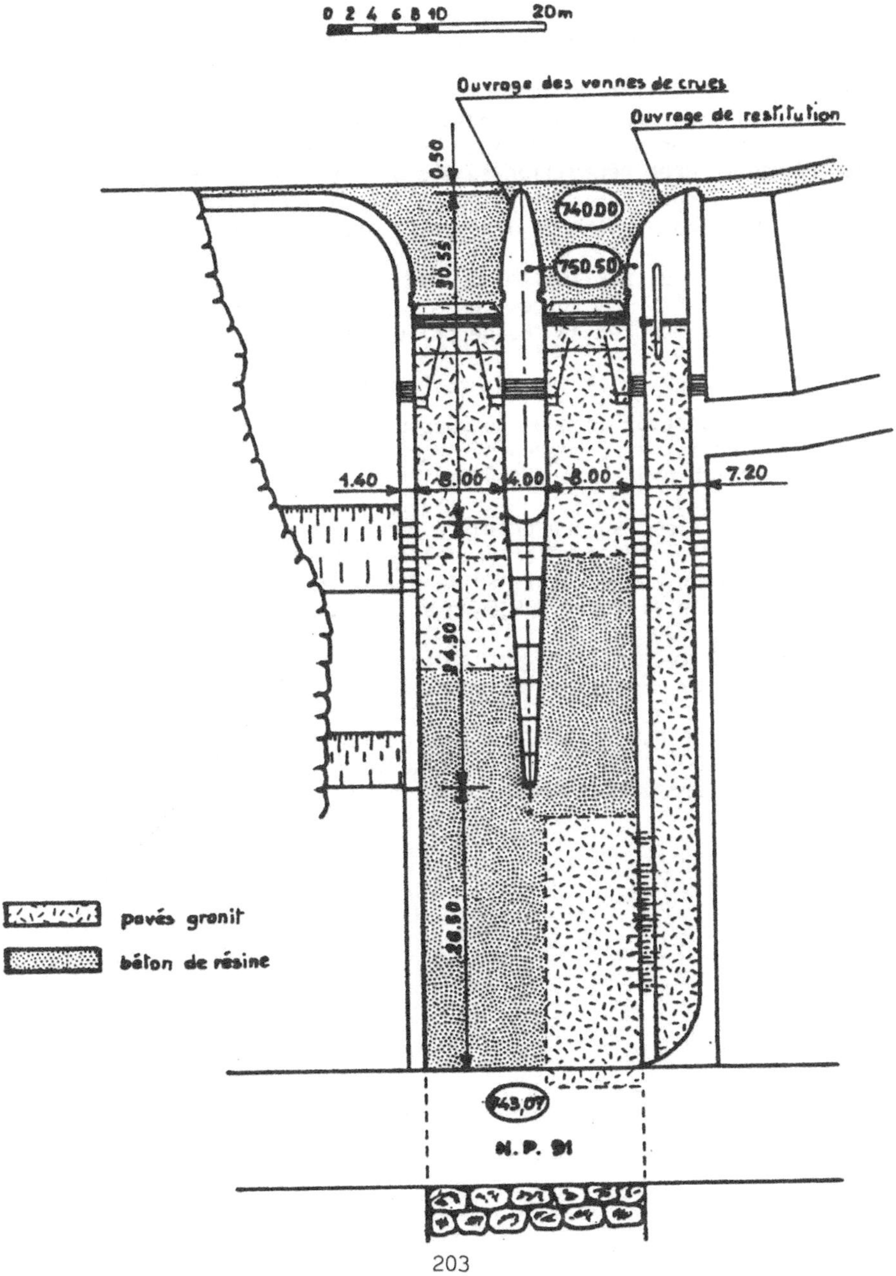

pavés en fonte, etc... Les deux dernières solutions, qui sont souvent les plus satisfaisantes, sont aussi parmi les plus onéreuses.

Dans la dernière décennie, quelques essais de protection ont également été effectués avec des revêtements utilisant des matériaux synthétiques : soit époxy, soit polyuréthannes.

L'une de ces expériences a été tentée en 1982 pour protéger le radier de l'évacuateur de crue du bassin du Clapier sur la Romanche. Cet ouvrage fait partie de la chute de Saint Guillerme, réaménagée cette année-là. Il comporte une passe de restitution, et deux passes d'évacuation des crues capables de laisser passer le débit de pointe de la crue décamillénale évalué à 676 m3/s. Chaque pertuis d'une largeur de 8 m est équipé d'une vanne segment de 7 m de hauteur. La vitesse de l'eau peut atteindre 10 à 11m/s.

Les parties les plus sollicitées du radier ont été protégées avec des pavés de granit, les autres zones (soit 900 m2) ont été revêtues avec un mortier de résine polyuréthanne (voir figure n°3).

Les travaux concernant ce type de revêtement ont comporté :

. une préparation soignée du support par sablage et lavage à haute pression et un reprofilage des bétons dégradés avec un mortier hydraulique prédosé

. l'application, sur un primaire expoxydique frais - bicomposant- (250 g/m2), d'un mortier de résine polyuréthanne de 20 mm d'épaisseur, comportant par m2 9,7 kg de résine et 19,5 kg de sable de silice 0,2/0,8 mm et 3/6 mm.

Sur les bajoyers, une plinthe de 0,20 m de hauteur a été exécutée avec un mortier de même type, thixotropé.

Ce revêtement a de bonnes performances en laboratoire : (dureté Shore A de 60 à 80 - résistance aux tractions de 35 bars). Il lui reste à subir l'épreuve du temps et des crues sur le site de Saint Guillerme.

Section 5

Repair of concrete surfaces

Réparation des surfaces en béton

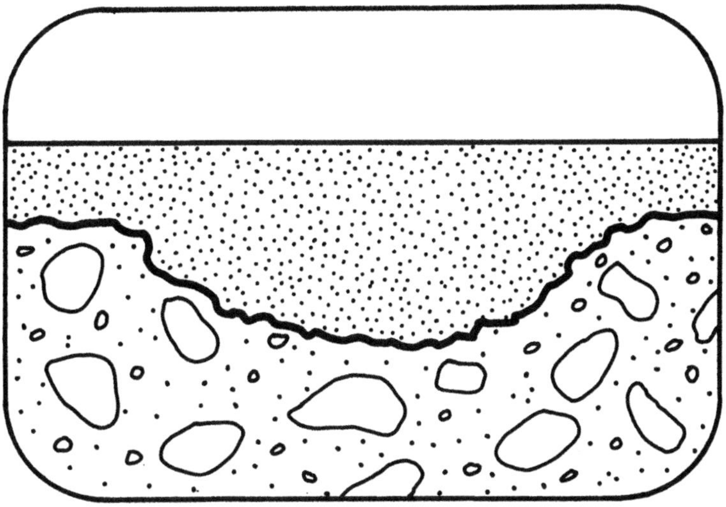

EXPERIMENTS ON THE USE OF GUNITE WITH RESINS IN THE MAINTENANCE OF DAM FACINGS

M. BERRA and S. VENESIA
ENEL, Research Center on Hydraulics and Structures(CRIS), Milano

Abstract
The replacement and repair of deteriorated surfaces with gunite is one of the most common operations of facing maintenance for hydraulic works (dams, canals, etc.). However, a certain number of cracking and deteriorated phenomena occurred during these last years, mainly due to temperature and shrinkage problems, to a poor adhesion to the base and, generally, to an objective difficulty of quality control, has lead the Research Center on Hydraulics and Structures (CRIS) of the Italian Electricity Board (ENEL) to study any possible improvement for gunite. For this purpose, the possibility was considered of adding acrylic resins in water dispersion to the gunite, during spraying, and its effectiveness was tested through a series of experimental investigations. Besides a small experimental application on a dam facing, a set of specimens of gunite, with and without resins, was prepared on site, under normal operating conditions. Laboratory tests concerned the following properties: compression and bending strength, adhesion, water absorption, porosity and water permeability. The results obtained do not seem to indicate, however, in the specimens with resins, improvements in the physical-mechanical characteristics such as expected.
Key words: Gunite, Acrylic Resin, Adhesion Test, Dam Facing.

1. Introduction

The problem of early deterioration of concrete structures, even those built recently, has become so important, in terms of the costs incurred in repairing and maintenance of the works, that significant progress in cement and concrete technology is required.
Particularly regarding hydraulic works, one of the most frequently used techniques for repairing the concrete facings of dams and canals, deteriorated over the years, is the sprayed concrete or

gunite, for both economic and technological considerations.
Despite its high cost per cubic metre, in fact, gunite is less expensive than normal concrete for thin layer applications. It is also an extremely versatile material, requires rather simple plant and equipment and can be used effectively wherever normally prepared concrete would present problems of transportation, application, compaction and finishing.
If correctly applied, gunite is the best cement-based material for thin layers and repairs on old and new concrete. In addition it provides good physical and mechanical characteristics, being theoretically able to ensure, thanks to the high impact velocities and consequent compaction of the mixture, good adhesion between the underlying base and the new concrete.
The greatest difficulties, however, are obtaining a uniform, high quality material in the field, especially where the sections are considerably reinforced, the thicknesses exceed certain limits and are applied in more than one layer and, perhaps the most important condition, when the gunning crew is not particularly experienced.
These difficulties, and recent cases of gunite cracked and deteriorated after only a few years, often with adhesion problems, convinced the Research Center on Hydraulic and Structures (CRIS) of ENEL, the Italian Electricity Board, to study the possibility of improving gunite performance through the use of resins in water dispersion.

2. The problems with gunite

The most common causes of failure may be summarised as follows:

- the presence of cavities due to gunite rebound; this defect occurs particularly when the gunite thickness is placed in more than one layer.
  Observable damage: poor adhesion between the various layers with lack of uniformity, presence of laminations and detachment of the gunite layers.
- insufficiently wet curing conditions, due to operational difficulties and to frequent adverse environmental conditions (the plants are available mainly during the summer months).
  Observable damage: crumbling of the gunite due to the impossibility of normal setting and rapid shrinkage crackings.

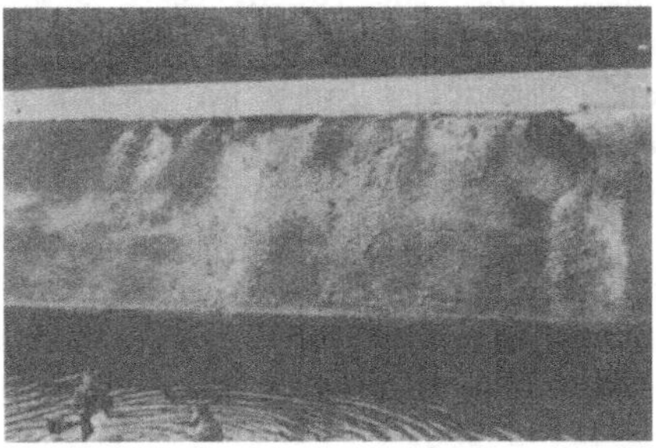

Fig. 1 - Bank of a canal only 8 months of normal service after gunite application.

- inadequate cleaning of surfaces: the remaining particles make adhesion difficult.
  Observable damage: detachment of gunite sections.
- presence of congested or too large reinforcement bars; incorrect placing chairs, gunite nozzle too far away from the surface to be coated, which causes the formation of non-compacted zones under the reinforcing and weak sections for subsequent shrinkage.
  Observable damage: gunite detachment and formation of cracks which follow the pattern of the reinforcing network.
- ununiform nozzle feeding: this makes it difficult to hydrate the mixture correctly and produces gunite that is too rich or too lean, or in any case not uniform.

The difficulties associated with the first three causes of failure could be overcome by improving the characteristics of the material.
For this reason it was thought to modify the gunite by adding resins in water dispersion which had proven, both in laboratory experiments and in practical applications, to be suitable for improving the characteristics of traditional concrete mixes.
In addition, since with the use of resins dry curing is required, the problems associated with wet curing, even with high ambient temperatures, could be avoided.

3. Experimental programme

The purpose of the experimental programme was to verify, through an evaluation of the physical-mechanical characteristics of gunite with and without resins, whether that addition would provide an overall improvement in the gunite, especially regarding the principal problems related to its placing listed above.
The experimentation consisted of a small application of about 15 m^2

of gunite with resins onto the facing of a dam and, utilizing the same materials and equipment of the job site, preparing a series of gunite panels for laboratory testing, with and without resins, curing them in both dry and wet environments.
Since the gunite was sprayed with the method of dry-mixing sand and cement, with the addition of hydration water in the nozzle, the resin was added by dispersion directly into the hydration water.
This, in effect, is the only difference in the use of the resin compared to a normal concrete in which the addition and mixing is performed before application.

3.1 Materials

The materials used and the general weight compositions of the mixtures examined are shown in Table I.

Table I - Weight composition of gunites examined (kg/m^3).

Components	Gunite	
	Normal	With resin
Pozzolanic cement R 325	500	500
Sand max.s. = 1.2 mm	1350	1350
Powered water repelling agent	10	10
Water	150	130
Acrylic lattice dispersion (% of dry = 60)	--	50

The applied gunites are, in effect, richer in cement due to the inevitable presence of rebound.

3.2 Experimental application

The application of modified gunite on the upstream facing of a dam involved about 15 m^2 of surface on the left abutment; it was performed in a single layer of about 8 cm thick on a reinforcing network of 10x20 cm of ∅ 4 mm iron rod (Fig. 2). In that test area the gunite cured under dry conditions, despite the high ambient temperature due to the summer season.

Fig. 2 - Experimental application of gunite with resins on a small part of the dam facing.

3.3 Preparation of the test panels

At the same as the experimental application, a series of gunite panels was prepared alongside the dam. For support we utilized reinforced concrete elements 6 cm thick, with dimensions of (50x60) cm^2 for the special panels for adhesion testing and (40x50) cm^2 for the normal panels from which to take core samples for the other tests.

The support elements were positioned almost vertically on scaffolding planks and separated by wooden spacers.

Fig. 3 illustrates the gunite placing, up to a thickness of 7-8 cm in a single layer.

Fig. 3 - Preparation of the test panels.

The special panels, shown in the lower portion of the Figure 3, were prepared to evaluate, after 28 days of curing, the adhesion of the gunite layer to the concrete support element.

The special panels were prepared in the following steps:

- Cutting of a concrete slab to obtain (50x60) cm^2 elements;
- Anti-adhesion coating of the surface with bituminous emulsion in paste form. The coating was left off a circular area 22.5 cm in diameter located in the center of the element.
- Chiseling of the circular area to roughen the surface and uncover

the aggregates.

- Application of a reinforcing network with square mesh 5 cm on the side of Ø 3 mm iron rod. Four threaded rods were secured to the network, arranged in a 30-mm square, to permit the attachment of a plate for the later adhesion test.

The normal panels were prepared by simply spraying gunite onto the support elements (40x50) cm^2.
Table II summarises the meaning of the symbols.

Table II - Meaning of prepared panel symbols.

GUNITE	PANELS NORMAL	PANELS SPECIAL
Normal	G1-G2	G3-G4
With resin	R1-R2-R3	R4-R5

3.4 Curing

The gunite panels were left to cure in loco for the first two weeks and then transferred to the laboratory for testing at the age of 28 days.
Panels G1,G2,G3,G4 and R1 were kept continually wet for the first 14 days with sprayed water. Panels R2,R3,R4 and R5, on the other hand, were cured in the air at ambient conditions.

4. Test procedures and evaluation of results

Core samples and prisms of the prescribed dimensions for the testing programme were cut from the panels prepared and cured as described above, both those of simple gunite and those of gunite with resin.
The tests may be classified into three groups:

- Tests to determine the mechanical characteristics of the gunite material, such as compression, bending and tensile strength and compression elasticity moduli.
- Tests to verify the adhesion characteristics of the gunite to the concrete reference support.
- Tests to define some physical properties of the gunite such as water absorption, water permeability and porosity.
  These, together with the adhesion tests, permit us to evaluate indirectly the durability of the gunite, as its capacity to resist external physical and chemical attacks.

Table III - Compression, bending strengths and elastic moduli of the gunites examined after 28 days of curing (average of 3 or more specimens).

GUNITE	CYLINDERS σc(MPa)	CYLINDERS E(MPa)	PRISMS σc(MPa)	PRISMS σb(MPa)
G(1-2)	36.3 +/- 3.1	22,820 +/- 2040	36.4 +/- 3.7	6.4 +/- 1.5
R(2-3)	33.2 +/- 1.6	20,840 +/- 2780	41.6 +/- 3.0	7.4 +/- 1.9
R(1)	29.2 +/- 3.5	22,830 +/- 1020	41.4 +/- 1.7	7.8 +/- 1.4

Table IV - Direct tensile strength of gunites examined.

GUNITE	DIRECT TENSILE STRENGTH (MPa)
G(1-2)	0.59 1.30
R(2-3)	1.47 1.20
G(1)	0.84 1.34

4.1 Mechanical strength

The mechanical strength and compression elastic modulus were measured after 28 days of curing on cylinders approximately 5 cm in diameter and 7 to 8 cm in height, according to the thickness of the panel from which they were bored. Since the length-to-diameter ratio of these cylinders was less than 1.94, the values were properly adjusted to obtain the cylindrical resistances, according to ASTM C42 standard. The elastic modulus was referred to stress values up to 30% of maximum load.

Other values of compression resistance were obtained by (4x4x16) cm^3 prism sections, broken beforehand by bending.

The results, reported in Table III, show that the gunites examined

have a fairly similar mechanical behaviour. The dispersion, and sometimes apparent contradiction, of the data obtained is surely due to the lack of homogeneity of material, even macroscopically visible on the test specimens.
The direct tensile tests, finally, conducted after 28 days on two cylinders for each gunite (dimensions Ø 4 cm x h 8 cm), provided the dispersed results shown in Table IV.

4.2 Adhesion

Particular care was given to preparing a test which would provide indications on the adhesion of the gunite (panel adhesion test).
We worked on the specimens described above, suitably arranged so that a tensile action could be exerted on the gunite and cause its detachment from the support.
The tests were conducted after 28 days of curing by securing the reference support of the gunite with epoxy adhesive to a fixed surface and exerting the tensile action on the gunite-support interface by means of a 40-ton hydraulic jack and a counterframe (Fig. 4).
The load application velocity was 0,05 MPa.s.

Fig. 4 - Adhesion test on panels.

The tests conducted on the specimens of normal gunite (G3-G4) and with resin (R4-R5) provided low values of adhesion resistance similar for both gunites (Table V), even though the detachment in some cases took place completely at the gunite-support interface and in others also involved the gunite (Fig. 5).
Direct tensile tests on cylinders obtained by boring the entire thickness of the panels composed of support and gunite (dimensions Ø 4 cm, h 15 cm), provided further information on the adhesion properties. The load velocity was the same 0.05 MPa.s. The failure did not in all cases involve the interface, located at about the centre of the cylinder height, but in some cases the gunite, the tensile strength of which thus proved to be less than the adhesion

strength to the support.
The values obtained, also shown in Table V, cover a rather wide range which also covers the values of direct tensile strength shown in Table IV.

Table V - Values of resistance to adhesion obtained with the adhesion test on panels and cylinders (*failure that involved the gunite and not the interface).

GUNITE	ADHESION RESISTANCE (MPa) PANELS	CYLINDERS
G(3-4)	0.30-0.33	-
G(1-2)	-	0.52-1.00-0.70
R(4-5)	0.34-0.31	-
R(2-3)	-	1.10-1.33*
R(1)	-	0.97-1.08*

Fig. 5 - Gunite panel after adhesion test.

The adhesion values of the cylinders are higher than those provided by the adhesion test on the panels: this may be justified in light of the lack of homogeneity of the gunites and the different sections involved in the tests 13 cm^2 in the case of the cylinders and 400 cm^2 in the panels).
In addition, the probable stress concentration all around the edge of the resistant circular section of the adhesion test on the panels may have contributed to reducing the detachment values. In general, however, the values obtained from this last test may be considered the most reliable, since they are representative of a broader surface which takes the lack of homogeneity of material more into account.

4.3 Physical characteristics

The characteristics of water absorption, porosity and water permeability were determined, operating on cylindrical specimens ∅ 10 cm and h 7–8 cm. For water absorption, we referred to Italian UNI standard 7699, which prescribes reaching in sequence a phase of saturation and then complete drying of the specimen until constant mass is obtained.

Porosity was determined in accordance with U.S. standard ASTM D 3203. The water permeability test, finally, was performed on specimens suitably covered laterally by a rubber membrane and subjected to a water load of 0.4 MPa, with a lateral containment of 0.5 MPa. The permeability values were calculated according to Darcy. Table VI shows the results obtained.

Table VI - Values of gunite absorption, porosity and permeability.

GUNITE	WATER ABSORPTION % of mass	WATER ABSORPTION % of volume	POROSITY % of volume	PERMEABILITY K m/s
G(1-2)	9.8	19.7	19.3	4.61×10^{-11}
R(2-3)	10.0	19.5	21.8	5.33×10^{-11}
R(1)	10.6	20.6	22.3	5.05×10^{-11}

4.4 Observations

The overall analysis of the results of the characteristics of the gunites examined lead us to affirm that the addition of resins to gunite prepared by the dry mix process provides no appreciable improvements.

The normal reference gunite, examined through the specimens prepared and cured under real jobsite conditions, was not of excellent quality: the compression strength remained around 35 MPa; the elasticity modulus around 20,000 MPa; the tensile strength and adhesion strength to a reference concrete proved lower than 1 MPa; porosity, absorption and permeability proved generally high.

Despite this, the quality of this gunite was not in any way improved in the properties examined through the addition of resins.

One possible interpretation for explaining this unexpected behaviour is the difficulty of obtaining, at least for gunite sprayed with the dry mix process, a through mixture of the resin into the cement mortar, perhaps because of the low mobility of the polymer macromolecule with respect to that of water.

5. Conclusions

Reports of problems observed on rather recent gunites and the lack of engineeringly-reliable data on its properties and performance, prompted a series of tests aimed at verifying the effectiveness of using resins to improve the characteristics of the gunite material.
In addition to the mechanical characteristics and the physical properties, particular attention was given to a test which could evaluate the adhesion properties of the gunite to a concrete reference support, directly comparing the performance of normal gunite and gunite with resins. Despite the optimistic expectations, however, the results proved disappointing, indicating that the addition of resins to gunite prepared by the dry mix process does not substantially improve the quality, even in the case of the particular type of gunite examined, which did not present excellent characteristics.
In effect, even with the presence of resin, the resistance to adhesion was below 1 MPa, and the values of mechanical resistance, absorption, porosity and permeability did not change significantly, as though the resin played no role in the gunite sprayed by the dry mix process, perhaps because of the application procedure, which makes it difficult to thoroughly mix the water dispersion of the resin with the cement and sand.

References

Internal Report. (1983) Indagine su malta cementizia additivata con una emulsione a base di acetato di polivinile. ENEL-CRIS Report n° 3134.

The Concrete Society.(1975) Polymers in Concrete. Proceedings of the First International Congress on Polymer Concretes.

Ryan T.F. (1973) Gunite, a handbook for Engineers. Cement and Concrete Association.

Special Pubblication, SP-14. (1966) Shotcreting. ACI Pubblication.

L'EMPLOI DES MATERIAUX POLYMERES EN TECHNIQUES DE REPARATION ET DE CONSOLIDATION DES CHAUSSEES AERONAUTIQUES EN BETON HYDRAULIQUE DES AEROPORTS D'ORLY ET DE ROISSY-CHARLES DE GAULLE

R. CHANUT et R. LORIN
AEROPORTS DE PARIS, Laboratoire

Résumé
Les chaussées en béton hydraulique qui constituent les pistes, voies de circulation pour avions des aéroports d'Orly et de Roissy-Charles de Gaulle finissent, sous le trafic sévère qu'elles subissent, par présenter localement certains désordres. Dans le cadre des travaux d'entretien, les matériaux polymères de type époxydique sont couramment employés depuis une quinzaine d'années en technique de réparation ou, plus récemment, en technique de consolidation de ces chaussées.

Ces techniques sont des injections, des réparations de surface, des collages de béton frais sur béton ancien et des scellements de dispositifs de transfert de charge. Leur comportement est pleinement satisfaisant. Le résultat, qui peut être en partie attribué à la sélection attentive des produits employés, reste toujours très dépendant de la qualification de la main d'oeuvre affectée à ces travaux et du soin apporté à leur exécution.
Mots clés: Chaussées aéronautiques en béton, Techniques de réparation, Résines époxydes, Evaluation des produits de réparation.

1. Introduction

AEROPORTS DE PARIS, Etablissement public autonome, a pour mission la conception, la construction et l'aménagement d'aéroports en France et à l'étranger. Les aéroports d'Orly et de Roissy-Charles de Gaulle, classés au 7e rang mondial par leur trafic passager, ainsi que 12 autres aéroport et aérodromes de la région parisienne, constituent son domaine propre qu'il gère, exploite, développe et entretient.

Les chaussées aéronautiques (pistes, voies de circulation, aires de trafic) des aéroports d'Orly et Charles de Gaulle sont essentiellement réalisées en béton hydraulique. Elles couvrent sur chaque plateforme une superficie supérieure à 2 millions de mètres carrés. Elles ont été conçues et construites en grande partie avant que les effets de l'exploitation des avions gros porteurs - dont la charge totale peut atteindre 380 tonnes - n'aient encore été pleinement ressentis.

2. Principaux désordres survenant sur les chaussées aéronautiques

Les principaux désordres subis sont ceux affectant habituellement tout revêtement en béton. Ce sont:

- . des épaufrures se présentant sous l'aspect de ruptures, le long de joints ou fissures, dont la profondeur ne dépasse pas la mi-épaisseur des dalles.
- . des fissures longitudinales, transversales ou d'angles de dalles.

L'évolution de l'assise des dalles, sous l'effet du trafic, est à l'origine de la plupart des fissures. La présence d'eau dans le corps de chaussée, l'érosion provoquée par la mise en pression instantanée de cette dernière, à l'interface des couches, lors du passage des avions, produisent cette évolution. Un phénomène dit de "pompage" apparaît qui se signale par des rejets d'eaux chargées de produits d'érosion par les joints et fissures. (fig. 1).

Fig. 1

Un entretien précoce doit être effectué pour préserver la bonne qualité des chaussées et différer la réfection de certaines parties d'ouvrages. Il consiste principalement à rétablir:

- . l'intégrité des dalles par des réparations partielles
- . des transferts de charge de part et d'autre des fissures et des joints.

3. Principales techniques de réparation et de consolidation des revêtements

Depuis une vingtaine d'années, AEROPORTS DE PARIS procède à la réparation des revêtements en faisant usage, presque exclusivement, de matériaux polymères de type époxydique qui entrent dans la composition de produits d'injection, de réparation de surface, de collage de béton frais sur béton ancien et de scellement.

3.1 Application des produits d'injection
Deux genres d'injection se pratiquent sur revêtements en béton:

- les injections de fissures pour tenter de réparer certaines dalles
- des injections sous dalles pour combattre le phénomène de pompage et en limiter les effets.

Des essais d'injection de fissures ont été effectués, à maintes reprises, à l'aide de résines époxydes spéciales à très faible viscosité. La réparation a échoué, dans la grande majorité des cas, soit par mauvaise adhérence de la résine provoquée par l'état d'encrassement de la fissure, soit par non-rétention de la résine qui tend à s'écouler par gravité, en cheminant sous les dalles.

Les injections sous dalles ont fait, ces dernières années, l'objet de campagnes d'intervention, en particulier, à l'aide de brai-époxy spécialement formulé à cet effet. Cette technique (fig. 2) a donné des résultats encourageants mais encore insuffisants en raison de la réapparition du phénomène de pompage au bout de 2 à 3 ans.

Fig. 2

3.2 Application des produits de réparation de surface
La réparation de surface la plus fréquemment pratiquée consiste à substituer, sur une épaisseur le plus souvent inférieure à 10cm, un matériau à base de résine époxyde à une couche de béton en cours de désorganisation.

Les figures 3 à 7 illustrent un exemple de ce type de réparation effectuée de part et d'autre d'une fissure.

Fig. 3

Fig. 4

Fig. 5

Fig. 6

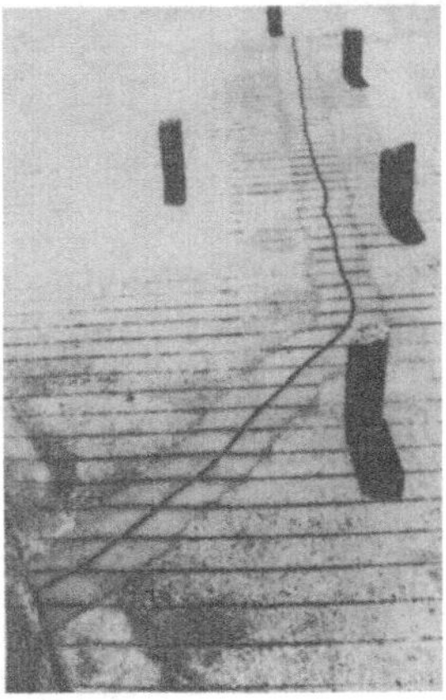

Fig. 7

Après délimitation de la zone à réparer, marquage (facultatif) des fissures à l'aide d'un colorant (fig. 3), la démolition du béton se pratique à l'aide de pics et burins pneumatiques puis s'achève par un bouchardage fin (fig. 4). Un joint est constitué par une bande de caoutchouc mousse (fig. 5) adhérant au béton, au droit de la fissure, par l'intermédiaire d'une colle au néoprène. Une couche d'accrochage de résine époxyde pure (≃ 200 grammes par m2) est ensuite appliquée, suivie de la mise en oeuvre de mortier de réparation (Fig. 6 et 7).

La composition habituelle du mortier est la suivante: 4 volumes de sable siliceux sec de granularité 0/2mm pour 1 volume de résine époxyde.

Le comportement de telles réparations est pratiquement sans défaillance, dès lors que les travaux sont pratiqués avec des produits sélectionnés, par des techniciens compétents, dans des conditions climatiques convenables, et qu'un transfert de charge peut être assuré désormais par un goujonnage oblique (couture de la fissure) (cf 3.4.1.1).

3.3 Application de produits de collage de béton frais sur béton ancien

Le collage de béton frais sur ancien béton est une pratique désormais courante. Elle est retenue comme mode de réparation dans le cas de fissures transversales ou d'angle, tendant à provoquer la fragmentation partielle de dalles.

Les figures 8 à 11 illustrent un exemple d'application

Fig. 8

Fig. 9

Fig. 10

Les différentes phases de cette réparation sont les suivantes:

Fig. 11

- délimitation et démolition (fig. 8) de la partie de dalle en voie de dégradation
- taille puis préparation par bouchardage fin d'un plan de collage formant avec la surface un angle maximal de 60° (fig. 9)
- réfection avec, si possible, renforcement de la couche de fondation
- application d'un produit bitumineux sur la tranche des dalles avoisinantes et placement d'un feuil isolant sur la couche de fondation
- application d'un produit de collage (fig. 10) selon les prescriptions de la notice technique du fabricant (moyens, quantité, précautions) en veillant particulièrement à respecter avec suffisamment de marge de sécurité la durée pratique d'utilisation et le temps d'assemblage
- mise en oeuvre de manière traditionnelle du béton frais nécessaire pour reconstituer complètement la dalle (fig. 11)

Les produits de collage employés sont des résines époxydes spécialement formulées à cet effet. Leur thixotropie doit les rendre aptes à se maintenir sur une surface non horizontale. Leur durée pratique d'utilisation après mélange des constituants, leur temps d'assemblage après application sur le plan de collage, doivent être suffisamment longs, dans toutes les conditions prévisibles sur chantier, pour ne pas risquer d'être dépassés.

La tenue des collages ainsi effectués ne pose généralement pas de problème dès lors que le chantier est bien organisé, l'équipe chargée des travaux hautement qualifiée, et le produit employé soigneusement sélectionné.

3.4 Application de produits de scellements
Les produits de scellements sont employés principalement pour:

- consolider les revêtements en plaçant des dispositifs de transfert de charge entre dalles ou parties de dalle
- encastrer des équipements spéciaux.

3.4.1 Application aux techniques de consolidation
Trois types de technique de consolidation sont employées ou en cours d'expérimentation par AEROPORTS DE PARIS.

3.4.1.1 Le goujonnage oblique
Il permet de solidariser les 2 parties d'une dalle fissurée par des goujons en acier de 32mm de diamètre et de 40cm de longueur pour une épaisseur de dalle de 40cm.
Les figures 12 à 16 illustrent cette technique.

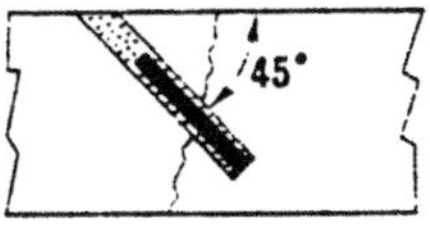

Fig. 12

Fig. 13

Fig. 14

Fig. 15

Fig. 16

Les différentes phases de cette opération sont:

. perforation de la dalle avec un angle d'attaque de l'ordre de 45° (fig. 12 et 13), en dépassant suffisamment la fissure existante sans traverser la dalle. Diamètre de la perforation: 50mm
. soufflage de la perforation à l'air comprimé et enduction de sa paroi à l'aide de résine époxyde pure (fig. 13)
. versement dans la perforation d'un volume déterminé de produit de scellement
. introduction du goujon (fig. 15)
. cachetage du trou

Les goujons sont répartis le long de la fissure tous les 0,50m. Le produit de scellement employé est un mortier de consistance fluide dans la formulation duquel entrent respectivement 1 et 1,5 volume de résine époxyde pure et de sable siliceux 0/2mm. La résine doit pouvoir s'accommoder de la présence d'eau. Le produit de cachetage est également un mortier de résine époxyde tel celui employé pour les réparations de surface (cf 3.2).

Le succès de cette technique (960 goujons placés en 1981, 1936 en 1982, 1644 en 1983, 2192 en 1984 et 1972 en 1985) témoigne de son excellent comportement et de son efficacité.

3.4.1.2 Le post-goujonnage horizontal

Le post-goujonnage horizontal placé à cheval sur un joint, introduit à mi-épaisseur de dalles contigües existantes, est une technique employée à titre expérimental depuis octobre 1984 par AEROPORTS DE PARIS. Cette technique rétablit un transfert de charge entre deux dalles (joint de retrait flexion non initialement goujonné) lorsque l'état du revêtement, par l'apparition de décalage entre dalles, en signale la nécessité. Elle est illustrée par les figures 17 à 20.

Fig. 17

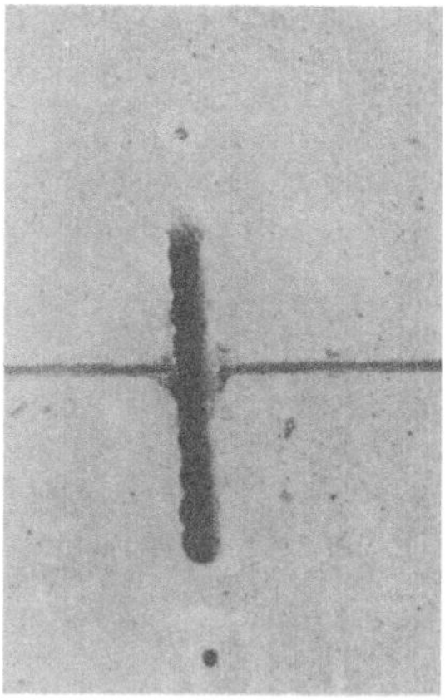

Fig. 18

Fig. 19

Fig. 20

Exception faite de la découpe du logement du goujon par juxtaposition de trous sécants (figures 17 et 18), la technique de scellement du goujon horizontal à mi-épaisseur de dalle ne diffère pratiquement pas de celle employée pour le goujonnage oblique (cf 3.4.1.1) et fait appel aux mêmes mortiers de scellement et de cachetage.

Le goujon à paroi lisse, revêtu de produit anti-adhérent sur la moitié de sa longueur, permet les mouvements de dilatation des dalles.

Une excellente tenue des scellements, après 18 mois de service sur voie de circulation pour avions, est constatée. (3 joints)

3.4.1.3 Le clavetage

Le clavetage est également une technique expérimentale dont l'adaptation et la première application aux chaussées aéronautiques datent d'octobre 1984. Sa fonction, comme le goujonnage horizontal, est de rétablir, si nécessaire, un transfert de charge entre deux dalles, de part et d'autre d'un joint de retrait flexion non goujonné initialement.

Le dispositif introduit dans le revêtement (fig. 21) est un produit manufacturé, protégé par des brevets déposés par le Laboratoire Central des Ponts et Chaussées et la Société FREYSSINET International. Ce dispositif est placé à cheval sur le joint entre deux dalles. Il comporte deux demi-coquilles scellées aux dalles. Une clavette prend appui sur celles-ci et assure le transfert de charge. Un joint en caoutchouc sépare les coquilles (non présenté sur la figure).

Fig. 21

La technique de mise en place de ces clavettes est illustrée par les figures 22 (percement et nettoyage du logement de clavette), 23 (déversement du produit de scellement), 24 (introduction de la clavette) et 25 (joint équipé).

Fig. 22

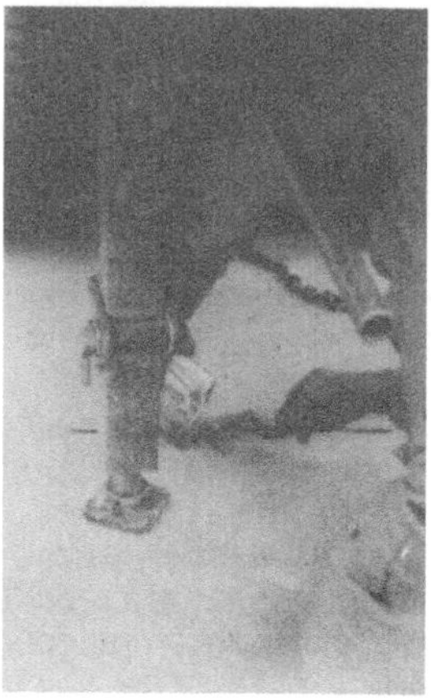

Fig. 23

Fig. 24

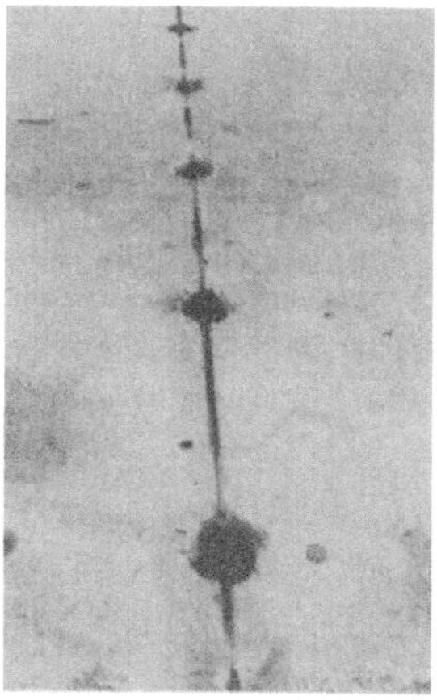

Fig. 25

Le produit employé pour le scellement des clavettes est une colle époxydique spécialement étudiée par la société développant cette technique. Sa rapidité de polymérisation est adaptée aux conditions climatiques.

Trois joints expérimentaux équipés de clavettes sur une voie de circulation de l'aéroport de Roissy-Charles de Gaulle, il y a 18 mois, ont également un excellent comportement.

3.4.2 Application à l'encastrement d'équipements spéciaux

Les équipements encastrés dans les pistes et voies de circulation pour avions sont des feux de balisage principalement implantés dans l'axe ou au voisinage de l'axe de ces ouvrages.

Fig. 26

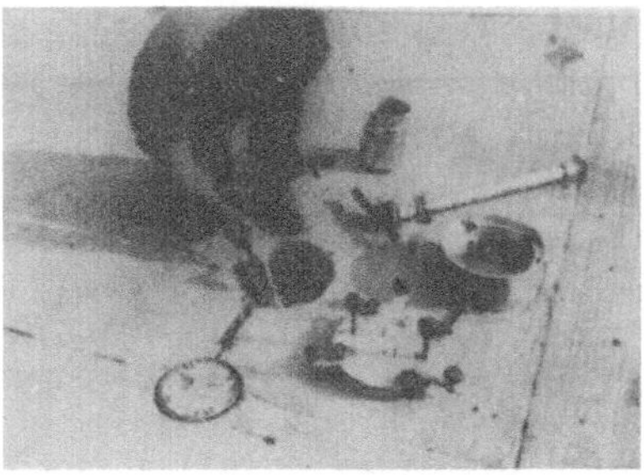

Fig. 27

La technique de pose de ces équipements est illustrée par les figures 26, 27 et 28

Les produits initialement employés, voici une vingtaine d'années, étaient des mastics polyesters. Des produits à base de résine époxyde leur ont été rapidement substitués et donnent entièrement satisfaction.

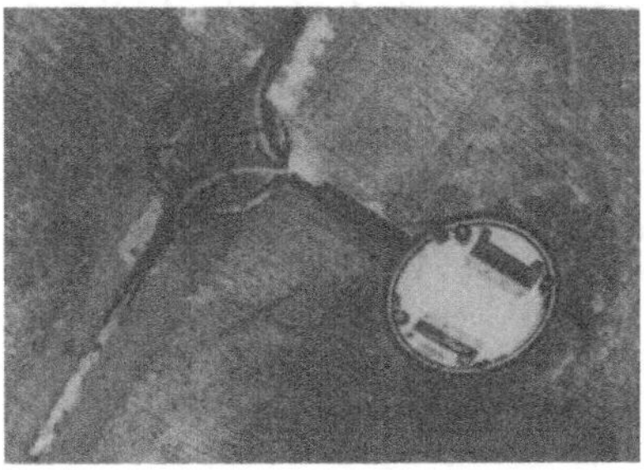

Fig. 28

4. Sélection des produits

Jusqu'en 1984, la sélection des produits dans le cadre des marchés passés par AEROPORTS DE PARIS a été faite par des procédures particulières s'appliquant aux produits de réparation de surface, de collage de béton frais sur béton ancien, et de scellement. Ces procédures prévoyaient le contrôle de la durée pratique d'utilisation, de la perméabilité et du retrait (lorsque cela se justifiait), et s'attachaient plus particulièrement aux essais d'efficacité.

Le principal essai d'efficacité concernait les produits de collage et de réparation. Il consistait à vérifier l'adhésion des produits polymères au béton par un essai de résistance à la flexion sous moment constant qui sollicitait le plan de collage d'une éprouvette reconstituée (Fig. 29).

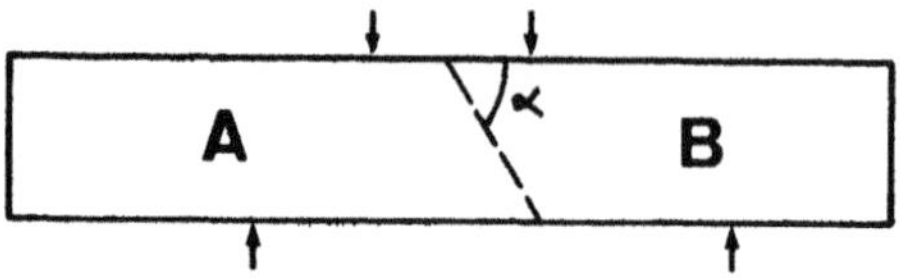

α = 60° angle formé par le plan de collage avec la surface

A = béton ancien

B = produit de réparation

Fig. 29

L'essai se pratiquait avec et sans vieillissement conventionnel.

Cette situation a évolué lorsque référence a pu être faite dès 1984 à la PROCEDURE D'EVALUATION DES PRODUITS DE REPARATION, élaborée sous l'égide du LABORATOIRE CENTRAL DES PONTS ET CHAUSSEES avec la participation de nombreux organismes officiels ou professionnels et laboratoires, parmi lesquels celui d'AEROPORTS DE PARIS.

Cette procédure s'attache à définir par nature de produits (de scellement, de réparation de surface, de collage structural, d'injection) des essais sélectionnés dont le mode opératoire a été minutieusement discuté avant d'être officiellement approuvé.

Ces essais permettent l'identification du produit, l'appréciation de son aptitude à l'emploi et l'évaluation de ses performances dans des conditions standardisées.

AEROPORTS DE PARIS a approuvé cette procédure et s'est engagé à y faire référence, dans ses marchés, en vue de l'agrément de produits.

5. Conclusions

5.1 Tenue des réparations

L'expérience acquise par AEROPORTS DE PARIS dans le domaine des réparations des chaussées aéroportuaires en béton à l'aide de produits époxydiques permet de dire, exception faite des injections de fissures de dalles, qu'une très bonne tenue dans le temps des réparations sous trafic est possible. Mais il a été nécessaire pour atteindre un tel objectif que les produits soient sélectionnés avec soin, ce qui était le cas, déjà, avec la procédure d'agrément particulière mise en place depuis une quinzaine d'années.

Il faut aussi que les entreprises chargées de réaliser de tels travaux disposent de personnels suffisamment compétents et très sensibilisés aux particularités de ces travaux.

En ce qui concerne la consolidation des chaussées existantes par introduction et scellement de dispositifs de transfert de charge entre dalles, l'expérimentation entreprise est prometteuse mais un peu trop récente (18 mois) pour que soient tirées des conclusions définitives. De nouvelles expérimentations, plus étendues, sont prévues.

5.2 Coût des réparations

Réparer une dalle permet de la conserver plus longtemps, voire très longtemps, si techniques de réparation et de consolidation peuvent être associées.

Par rapport au coût de la réfection complète d'une dalle représentant 22,5m3 de béton et de la fondation qui lui est associée, celui de réparations d'importance moyenne peut être estimé comme il suit :

. Réparation d'épaufrures : 9%
. Réparation d'une cassure d'angle par collage de béton frais: 8%
. Goujonnage oblique d'une fissure :10%
. Goujonnage horizontal d'un joint complet :13%

Si nécessaire, plusieurs de ces opérations peuvent être pratiquées sur une même dalle.

Le pourcentage du nombre de dalles concernées par ces réparations, sur un aéroport comme celui d'Orly dont les chaussées aéronautiques ont souvent entre 15 et 30 ans d'âge, est voisin de 1%.

REPAIR OF CONCRETE FLOORS WITH POLYMER MODIFIED CEMENT MORTARS

DR. ERHARD GF. CHORINSKY
Chemotechnik Abstatt, Germany

Abstract
The long-term experience with the formulation of repair mortars for concrete, which are exposed to weather and trafic loads, has shown, that the adhesion of modified cement mortars can be explained as follows:

Adhesion of cement mortars, which contain oligomeric additives only, is based on cristalline bonding of the cement paste to the old concrete. This bonding cannot be broken by the exposure to water.

Adhesion of polymer cement mortars, formulated with thermoplastic polymer dispersions is based on physical interaction, the chemical adhesion is prohibited by the polymer cover on the cement particles.

Adhesion of epoxy-modified cement mortars on old concrete is based on physical interaction too. But in contrary to the thermoplastic polymer dispersions the resistance of the epoxy-resin to saponification can be so high, that the bonding to the old concrete can be water resistant over a long period of time.
Key words: Concrete floor, Repair mortars, Oligomer modified cement mortars, Polymer modified cement mortars, Epoxy modified cement mortars.

Requirements

Since 1950 several systems of polymer modified cement mortars for the repair of demaged concrete were developped by the chemical industry. Although a lot of informations are published about the advantages and disadvantages of the different systems, there is a significant lack of knowledge about the influence and functioning of polymeres in cement mortars on the mode of bonding to old concretes.

To understand the bonding mechanisms we first of all have to define the concrete surface as clean and not contaminated by any dirt like oil, grease and other chemicals. We also have to consider whether the concrete surface is carbonated or has reacted with other acid agents. Therefore the surface of a concrete floor, which has to be repaired with cement mortars should meet the following requirements:

1. The old concrete should have a sufficient compression strength and the weak surface has to be removed either by steel blasting, milling or other mechanical means.

2. The concrete surface to be coated should be alcaline with a ph-value above 10, to secure, that no carbonated products weaken the bond ing of the repair mortars.

3. The concrete should not contain any water repelling agents, which could diminish the absorbtion properties on the subsurface and the adherance of the repair materials.

Unmodified cement mortars

The workability and the compactability of cement mortars increase with the cement content. Because the agregates, which are used on the jobsite for repair purposes have a maximum grain diameter of 8mm, which means that the optimum of workability is given at a cement content of 400 kg/m^3. If the granulometric composition is unfavourable for example because of a high content of fine particles, the repair mortars are even made with up to 500 kg of cement per m^3.

Mortars with that high cement contents naturally show high shrinkage and cracking.

If an unmodified cement mortar is applied on a dry concrete surface a part of the mixing water will be sucked off into the subsurface before any solvable and reactive components in the cement paste are formed. In this case a chemical reaction of calcium hydroxyde and soluble silicates is restricted on the outer areas of the agregates. Penetration of reactive components into the capillary fringe is prohibited and the repair mortars will not adhere firmly to the old concrete.

If the surface of the old concrete is saturated with water before the repair mortar is applied, we can assume, that the capillary fringe is closed for penetrating of hydration products out of the cement paste. The access water from the capillars will raise the water to the cement ratio in the boundary of the fresh mortar and lower its mechanical properties.

The result of too dry or too wet surfaces of the old concrete on the adhesion of an unmodified cement mortar is always the same: the adhesion strength is weakened in the interface and we can assume only scattered cristalline adhesion of the hardened cement paste. In other words the repair mortar is not rooted tightly in the capillars of the old concrete and will break away if it will be exhibited to dynamic loading.

In the last years another problem appeared with the usage of cement mortars formulated on the jobsite and will become even more severe in the next years: to reduce costs in the production of cement, in Germany about 5 % of inert raw material is grinded to the burned clinker. This material cannot contribute to the adhesion of the cement paste. If part of the cement clinker is replaced by fly-ash, the pH-value of the cement paste is lowered by the reaction of the contained silicium dioxide with formed calcium hydroxide. This could have a negative effect on the bond ing if the silicia particles are too big to fit in the capillary spaces of the old concrete.

Oligomer modified cement mortars

At the end of the sixties a new group of synthetic products for the improvement of concrete were offered to the construction industries on the basis of sulfonated naphtaline and melamine formaldehyde resins. Because of their oligomeric structure these products are water soluble and able to reduce the water to cement ratio quite significantly. Unfortunately they tend to increase bleeding and are therefore not suitable in unformulated form as mortar additives. In combination with other polymers however the oligomeric salts of several sulfonated resins can be used as auxiliary systems in cement mortars, which can be used as repair materials on concrete floors.

The advantage of these oligomer products in the cement paste is that they don t influence the crystalline growth as much as polymer dispersions. Incorporated in a high alkaline system, which must be stabilised against bleeding and separation, they increase the adhesion strength on old concrete quite impressively.

The functioning of oligomer modified cement mortars is better on concrete surfaces, which are not carbonated, the adhesion is reduced on calcitic surfaces. We therefore can assume that their mode of operation is based mainly on a crystalline growth on the silicatic phases of the old concrete. As the result of this bonding the interface cannot be disturbed by exposure to alkaline water.

The practical experience with this kind of bonding principle over several years shows that the oligomer modified cement paste grows into the capillary fringe of the old concrete and form a connection with an adhesion strength of more than 2,5 N/mm^2. This is usually more than a good concrete and more than an unmodified cement mortar can reach.

The oligomeric additives and the other components of the additive system remain as calcium salts in the hardened cement paste, but their contribution to the properties of the repair mortar is negligible, probably because of their low concentration of less than 1 % of the cement content.

Polymer modified cement mortars

Polymer dispersions as bonding additives for cement mortars have a great market in Germany. The begin of their technology was in the early fifties, when the chemical industry offered thermoplastic polymers dispersed in water to formulate paints and adhesives. Combined with cement the polymers improve the adhesion of mortars on old concrete as long as the interface is not exhibited to water. For cementicious flooring materials as levelling compounds the polymer dispersions can be applied as primers without the addition of cement, because these materials are used to reach surface smoothness under floor covers only.

Although the development of polymer dispersions, which cannot be saponified in an alkaline milieu, has made great progress in the last years, the sensitivity of polymer modified cement mortars is still a problem in some cases. This has the following reason: the polymer particles of a dispersion are coated with hydrophil monomers and stabilizing agents. As ingredients of a cement mortar many formulations entrain air to the cement paste and this improve its workability. The entrainment of air however weakens the adhesion-strength of the repair mortars. At the same time the tensile strength can be increased , which is of great importance for the durability of a repair material.

The polymer particles form a cover around the cement grains if the dispersion is added in sufficient concentration to the mortar to improve its flexibility. By this way the hydration products of the cement cannot contribute in an appreciable manner to adhesion on the old concrete and the bonding is based mainly on the properties of the used polymer system. The polymer particles on the other hand are not able to penetrate into the capillary fringe of a dense concrete, because they are too large. They cannot act as impregnating agents and stabilize the outer surface of the old concrete only. The binding of dust and loose particles to the subsurface is of course a desirable effect but not sufficient for the repair of industrial floors.

The problem of the remaining auxiliary system, which is incorporated in a polymer and a cement matrix can be neglected if the repair mortar is not attacked by water in the interface. But if a concrete floor is built directly on soil without a moisture barrier, all repair systems basing on polymer modified cement mortars are in danger because the waterborne auxiliary materials in the dispersions can deteriorate the adhesion.

Fortunately the bonding dispersions are used less concentrated for preparing the subsurface of old concrete than the formulators of these materials recommend. Therefore the crystalline bonding of the cement in the repair mortars can contribute to the adhesion. The applicators usually esteem highly the good workability of polymer modified cement mortars and do not add more than 5 % polymer to cement for repair purposes.

Epoxy modified cement mortars

This group of materials still has a very restricted but growing position under the repair materials for industrial concrete floors. Inspite of the outstanding properties of some products, which are since the early seventies on the German market, a considerable reluctance against the epoxy polymer emulsions can be observed. The reason for this might be the high price and the application problems with the two component resins. The advantages of these polymers for the improvement of cement mortars are not generally known.

Epoxy modified cement mortars can be formulated with a higher compression strength than normal polymer modified cement mortars. They adhere on carbonated concrete as well as on dense and alkaline concrete surfaces, even if the cement paste contains water repelling agents. The experience with these systems shows that the main difference between a non saponificable thermoplastic polymer dispersion and the duroplastic epoxy emulsion in the cement paste is remarkable: the possibility to apply repair mortars in less than 5 mm thickness even for highest loading can only performed by ECC-systems.

It can be expected that epoxy modified cement mortars will substitute even pure epoxy coatings because of their outstanding characteristics as repair systems. The mode of bonding to the old concrete, which is obviously very simular to the bonding of polymer dispersions but less sensitive to water attack has a definite advantage against the pure resins: ECC-systems adhere excellent to wet concrete surfaces, that means that repair work can be done immediately after high pressure jet water cleaning of the old concrete, which is without doubt the most effective preparation method after mechanical milling of the old concrete.

BOND STRENGTH OF PATCH REPAIR MORTARS FOR CONCRETE

D.J. CLELAND Department of Civil Engineering,
M. NADERI The Queen's University of Belfast.
A.E. LONG

Abstract

When diagnosing a faulty concrete structure, the cause of the fault or deterioration should be established so that the most appropriate solution to the problem can be selected. For example if corrosion of the reinforcement is the main cause of deterioration, then one might call for an impermeable or less permeable repair system which may be found among the resinous system. However one should always bear in mind the compatibility of the system with the old or deteriorated concrete structure.

No matter what repair system is chosen one of the requirements of a repair system is the ability to provide an adequate bond between the repair and the old concrete. Ideally the bond strength should be such that the composite structure behaves as monolithial.

The factors which in the authors' view plays an influential role in creation of bond between the repair and old concrete, has been examined in this paper and some comparison is made between the different repair systems.

Keywords: Bond Strength, Patch Repair, Durability, Surface Treatment, Repair Materials, Bonding Agent, Spalled Concrete, Shear and Tensile Bond.

1. Introduction

Assessment and the repair of deteriorated concrete structures has been examined by many researchers (1,2,3,4). Practising engineers are frequently faced with a wide range of causes of deterioration. Some of these are as follows:

1. Poor concrete mixes or placing techniques.
2. Design errors.
3. Construction errors.
4. High external and accidental forces.
5. Fire damage.

6. Weathering

7. Chemical attack.

8. Corrosion of reinforcement.

It is important to establish the cause of the problem before attempting a solution. Indeed it is also necessary to establish the objective of the repair because both of these will influence the decisions, relating to the repair. If it is decided to opt for a "patch type" repair, then a range of systems are available both in terms of material and supplier.

As well as the cause of failure and the objective of the repair, other factors such as location, area, volume, colour of the main structure, accessability, etc., will influence the engineer's choice of a repair system. For example if carbonation is known to be the dominating factor causing deterioration, then one might select an impermeable or less permeable system such as a resineous or a polymer modified cementitious system (7,8). Different strengths and properties may be obtained by choosing different repair systems, but one of the most important factors which must be inherent in every repair system is its adhesion to the original concrete. This must have adequate strength so that the composite structure can function in the desired manner. Some of the factors that play an important role in forming a strong bond between the repair system and the original concrete are as follows:

(a) Characteristics of the original concrete and its surface conditions and preparation.

(b) Characteristics of the repair system.

(c) Workmanship.

(d) Environmental condition.

Work is on-going at The Queen's University of Belfast to evaluate different repair systems in relation to their bond strength between the repair and original concrete. Some of the findings of the first series of tests are discussed in this paper.

2. Test Procedures

To produce a surface that resembles the actual spalled concrete surfaces which occur in practice, concrete specimens of 600 x 300 x 200mm were cast with three holes along the mid-plane (parallel to 300mm side). At seven days, three wedges were driven into the three holes, and the block was split into two specimens of 600 x 300 x 100mm dimensions with a fractured/rough surface of 600mm x 300mm. To produce saw cut surfaces, some of the split specimens were sawn to produce two specimens of 600 x 300 x 50mm size. The concrete strength at 28 days was about 65N/mm^2.

These slabs were repaired using nine different repair materials. The slabs were cored down into the old concrete at the required age, using diamond tip coring drill producing 50mm diameter cores. These cores were tested applying tensile or shear force (9).

To assess the tensile bond strength, steel or aluminium probes of 50mm diameter were glued on top of the cores. Then using "limpet" apparatus (6), the core was fractured. Then having the failure taken place along the bonded plane, the average tensile strength at failure was calculated simply by dividing the failure tensile force by the cross sectional area of the probe. From inspection of the failure, the type i.e., concrete, repair or bond, was also noted.

To measure the shear strength of the bond between the repair and the original concrete, "Friction Transfer" method (5,9) was used which consist of fitting a specially devised apparatus on top of the core, then applying a torque and recording the maximum shear bond strength at failure.

All the above tests were carried out under dry laboratory conditions and the detail of the original concrete mix proportions is shown in Table 1.

To see the mechanical keying effect and increased contact area between the repair and the concrete, one of the repair material (fibre re-inforced acrylic modified cementitious mortar) was applied on to a brushed and washed (exposed aggregate) surface of newly cast concrete, i.e., to produce a highly roughened and relatively strong surface. Reference 9 gives the details of the application of the repair material.

3. Results and Discussion

The average tensile strength and the maximum shear strength of the bond between the repair system and original concrete are shown in Figs.1 and 2, belonging to saw cut and rough (chisel hammered) surfaces. It is worth mentioning that no attempt has been made to improve the bond strength of the 'CONTROL' system which is orinary sand/cement mortar. It may be possible to achieve better bond strength using ordinary S/C mortar with different W/C, S/C ratio, but at the time of writing, the results of this optimization are not complete.

From the results of Figs. 1 and 2 it is apparent that a good correl-ation exists between the tensile and shear bond strength, and Table 3 indicates that the age of 14 days chosen here is appropriate.
It can be seen from the results of 'split rough' and 'saw cut smooth' surfaces that:

TABLE 1 : MIX PROPORTION OF THE ORIGINAL CONCRETE

CEMENT (kg/m^3)	WATER (kg/m^3)	FINE AGG. (kg/m^3)	COARSE AGG. (kg/m^3)	28 DAY CO[illegible] STRENGTH (N[illegible]
560	190	578	1278	68

TABLE 2 : COMPARISON OF BOND STRENGTH ON ROUGH AND SLIGHTLY ROUGH SURFACES

REPAIR SYSTEM	SURFACE TEXTURE	TENSILE BOND STRENGTH (N/mm^2)
FIBRE REINFORCED ACRYLIC MODIFIED CEMENTITIOUS MORTAR	SAWN	1.63
	SLIGHTLY ROUGHENED	1.81
	ROUGH	FAILURE OF ORIGINAL CONCRETE (2.96)

TABLE 3 : THE "AGE-BOND STRENGTH" RELATIONSHIP OF 3 REPAIR MATERIALS APPLIED ONTO SAWN SURFACE

REPAIR MATERIAL	AGE (DAY)	SHEAR BOND STRENGTH (N/mm^2)	TENSILE BOND STRENGTH (N/mm^2)
SAND/CEMENT MORTAR + ACRYLIC BONDING AGENT	14	4.00 (9)	2.4 (2)
	30	5.66 (7)	2.44 (2)
SAND/CEMENT MORTAR	14	2.52 (5)	-
	17	-	1.24 (6)
	21	2.37 (5)	1.26 (6)
	28	2.78 (6)	1.11 (6)
	50	2.54 (5)	1.23 (5)
SAND/CEMENT MORTAR + EPOXY BONDING AGENT	14	4.22 (13)	3.06 (1)
	31	6.27 (2)	2.54 (1)
	81	5.31 (6)	2.54 (3)

( ) SHOWS THE NUMBER OF BOND FAILURE

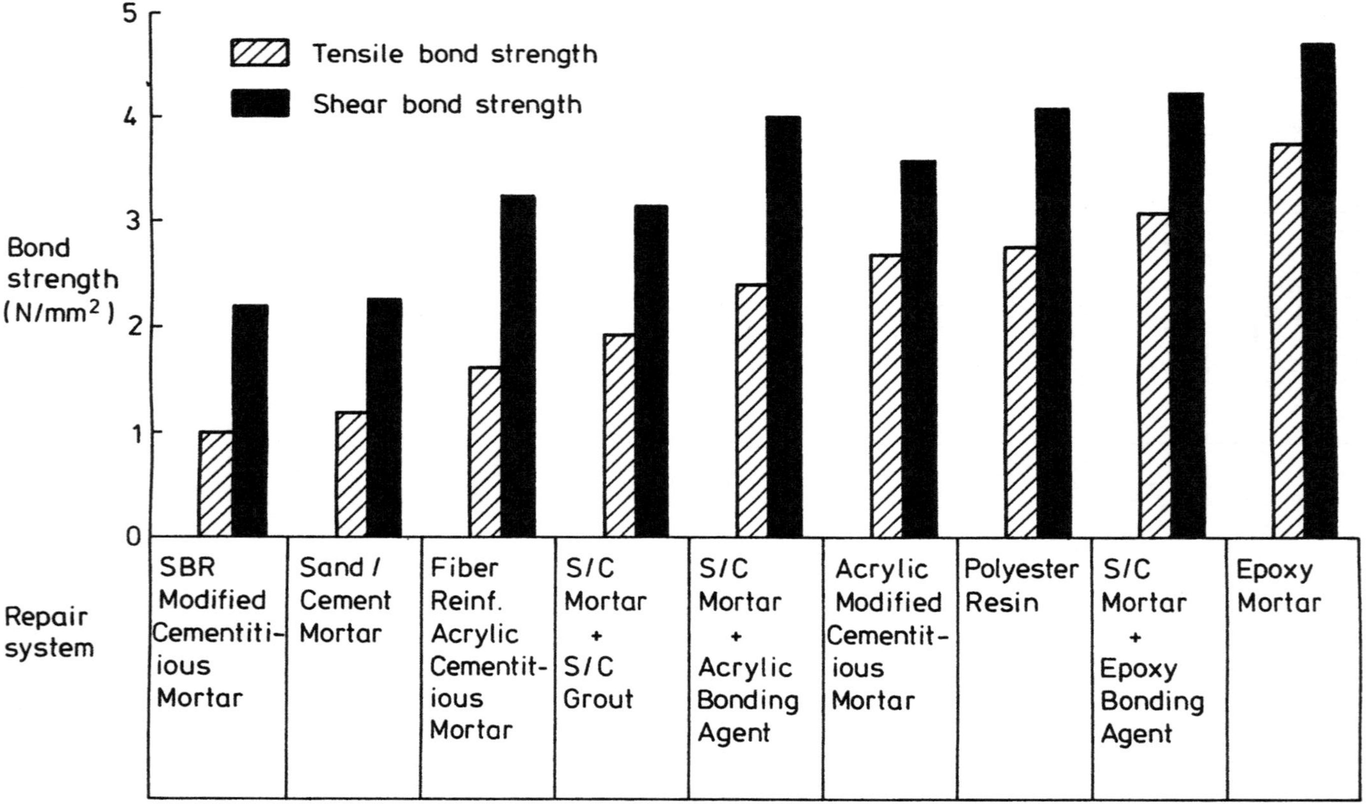

Fig. 1. SHEAR AND TENSILE BOND STRENGTHS OF DIFFERENT REPAIR SYSTEMS AT 14 DAYS, APPLIED TO SAW CUT SURFACES

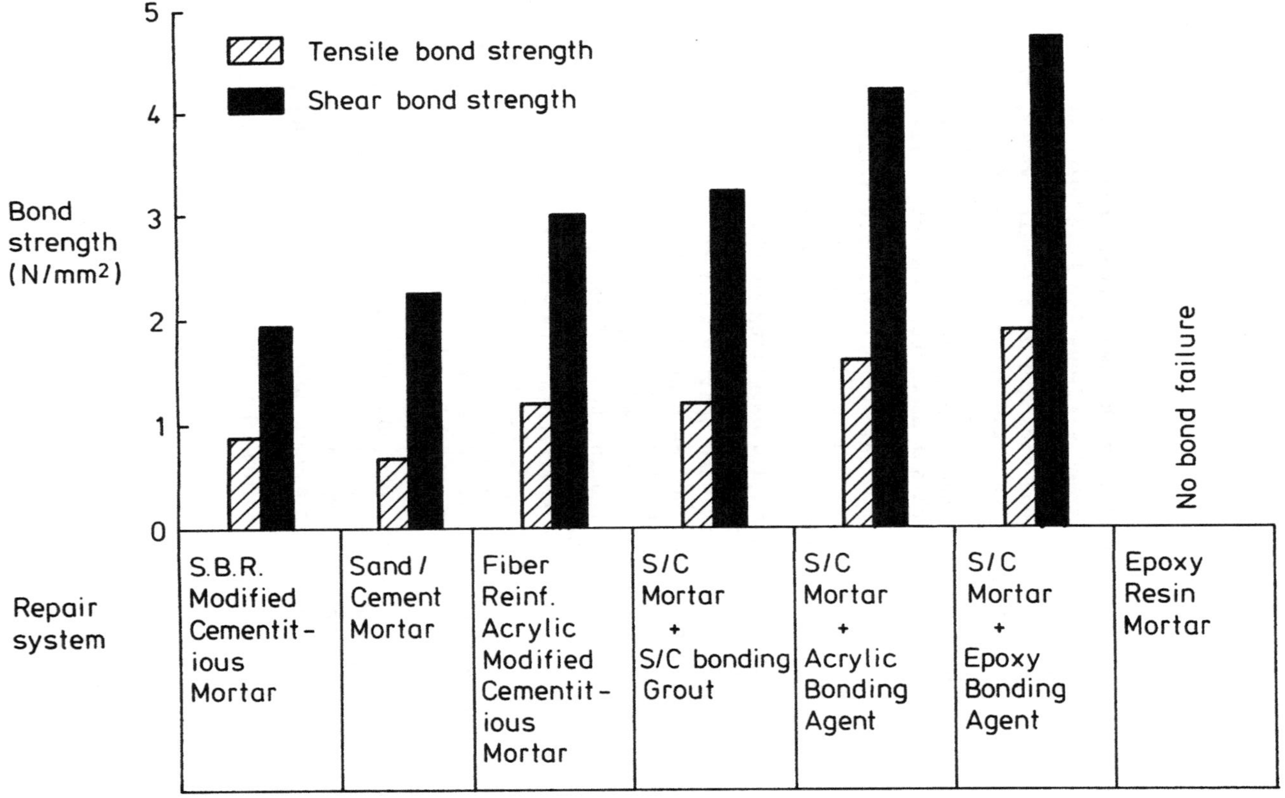

Fig. 2. SHEAR AND TENSILE BOND STRENGTH OF DIFFERENT REPAIR SYSTEM AT 14 DAYS APPLIED TO SPLIT/ROUGH (CHIZEL HAMMERED) SURFACES

(a) The saw cut values are generally higher.

(b) The difference is greatest for the epoxy resins.

This may be explained as follows:

When dealing with a spalled or split concrete surface, spalling or splitting stresses leaves most of the aggregate and some mortar particles loose and therefore even after surface treatment (such as chiseling, bush hammering, sand blasting, water jetting, etc.). These weak surface layers form the exposed part of the deteriorated or damaged concrete. Thus no matter what system of repair is employed if the bond strength between the repair and original concrete is higher than the strength of this weak layer, the fracture will take place along the weak layer of the original concrete which can produce misleading results, both at research and quality control level. Therefore when testing the bond strength of a repair system, the failure stress values should be noted as well as the fracture planes. In practice this layer will exist and can only be minimised by using an appropriate method of surface preparation. When a repair material is applied to rough (split, chisel hammered) concrete surfaces the roughness provides more contact area and mechanical key which results in an increased bond strength. This effect was proved when one of the repair materials (fibre reinforced acrylic modified cementitious mortar) was applied to a strong rough surface (aggregates were exposed during setting and hardening of the concrete). However with rough surfaces, the compaction and expelling of the trapped air from the depressions and protrusions is more difficult. Another point worth mentioning here is that when dealing with rough surfaces pure quantitative measurement of bond strengths in tension and shear will not be as simple as it is when dealing with a saw cut smooth surface. Needless to say that the viscosity and fluidity of the repair system can play a major role in overcoming the foregoing mentioned difficulties. By considering the performance of epoxy resin that gave no bond failure when applied to split/chisel hammered surface, it might be noted that above a certain repair and bond strength the failure will take place in the weak surface of the damaged concrete. Therefore depending on the type of surface preparation as indicated in earlier paragraphs, this point should be borne in mind as far as the bond strength is concerned. To avoid the complication of the bond strength results with the factors inherent in the weak surface layer of damaged and exposed concrete surface as explained in the foregoing paragraphs, the authors believe that comparison of repair systems is best achieved by using strong sawcut concrete surfaces. (However the interpretation of results and the actual performance of the repairs should be examined with the points about the weak surface layer in mind).

From the graphs of Figs. 1 and 2 it can be seen that S.B.R. modified cementitious mortar produced the weakest bond strength both in shear and in tension. As it was expected epoxy resin gave the highest bond strength. It is interesting to see that a prepacked fibre reinforced acrylic modified cementitious mortar gave bond strength values similar to that of sand/cement mortar used with S/C bonding grout.

When comparing sand/cement mortar used with different bonding agents such as acrylic emulsion and epoxy resin, there seems to be an increase in bond strength up to about 25% and 50% respectively.

One of the difficulties encountered in using some of the prepacked systems was proper proportioning and mixing of the component of the systems, even with some of the systems such as polyester resin we found that it did not set after about seven days, so the manufacturer's instructions has to be followed with great care. The other point was that, when coring or testing at the age of 14 days most of the modified and prepaked mortars gave off the smell of their chemical constituants which maybe is due to not having complete polymerization taking place. Compaction problems existed when dealing with some of the polymer modified cementitious materials. Further tests to examine the effect of different curing conditions and service condition on the bond strength are being carried out at The Queen's University of Belfast. However from these initial results it would appear that the bond strength of epoxy systems and modified cementitious system is only marginally higher than that for sand/cement mortar. Further in real situations where the quality of the surface preparation is necessarily suspect it may not be possible to mobilise this increased strength.

4. Conclusion

Two methods of testing the bond strength between a repair and the original concrete have been used to evaluate various repair systems and three different surface preparations.

The test methods measure the tensile and shear bond strengths of the repair system.

Some difficulties were encountered with interpretations of the manufacturer's instructions.

Epoxy resin system gave the highest (about twice the bond strength of the sand/cement mortar + S/C grout) and the S.B.R. modified cementitious mortar gave the lowest (about half of the bond strength of sand/cement mortar + S/C grout) bond strengths.

Bonding agents used with sand/cement mortar improved the bond strength by 20% - 50% over that obtained using a grout alone.

If repair is applied to a sound rough surface, e.g., brushed or lightly sand blasted, it will give a better bond strength mainly due to increase in bond area and mechanical keying effects, providing the air pockets are not formed around the protrusions.

However, when a repair applied to a more natural rough surface obtained by fracturing the concrete and subsequent chiselling to remove loose sections most of the tested repair materials produced bond strengths lower than that for smooth sawn surfaces. This was particularly true for the tensile bond strength and appears to be due to entrapped voids and air pockets around the protrusions.

References

1. Rasheedozzafar, Fahd H. Dakhil, Ahmad S. Gathani, "Corrosion of reinforcement in concrete structures in the Middle East", Concrete International, September 1985, P.48.
2. Tabor, J.L. "The evaluation of resin systems for concrete repair", Magazine of Concrete Research, Vol. 30, No. 105, December 1978.
3. Mangat, P.S., Gurusamy, R., "Steel fibre reinforced concrete for structural repair", The 2nd International Conference on Structural faults and Repairs, 30th April - 2nd May 1985, organised by the Institution of Civil Engineers, London.
4. Lloyd, J.P., Heidersback, R.H., "Use of the Scanning Electron Microscope to study cracking and corrosion in concrete", Concrete International, May 1985, P.45.
5. Naderi, M., Internal research report, Civil Engineering Department, The Queen's University of Belfast, Oct. 1985.
6. Long, A.E., "Early Detection of Structural Faults in Concrete by using partially Destructive Tests", The 2nd International Conference on Structural Faults and Repair, 30th April - 2nd May, 1985, organised by The Institution of Civil Engineers, London.
7. Robery, P.C., "Structural Repairs", Concrete, February 1983, P.23
8. Hewlett, P.C., Hurley, S.A., "Repair Materials - Selecting and Testinf", The 2nd International Conference on Structural Faults and Repair, 30th April - 2nd May, 1985, organised by The Institution of Civil Engineers, London.
9. Naderi, M., Cleland, D., Long, A.E., "Insitu test methods for repaired concrete structures", International Symposium on Adhesion between Polymers and Concrete, September 16th - 19th, 1986, France.

STUDY OF 3200 VINYL ESTER RESIN MORTAR AND ITS APPLICATIONS

LIN BAOYU, LU ANQI AND CENG RUIYU
Department of Material and Structural Engineering
Nanjing Hydraulic Research Institute

Abstract
This paper describes briefly a new type of high molecular material—vinylester 3200 resin mortar, including the characteristics of its formulation, the varieties of performance and its applications in the field. Experiments show that the vinylester 3200 resin mortar has excellent resistance to cavitation-erosion, abrasion-erosion, chemical-corrosion and freeze-thaw cycling, with long-term durability to weathering and soaking in water, good deformability and strong adhesion to the surface of concrete and steel as well as low permeability. It is proved 3200 resin mortar is a new type of abrasion-erosion-resistant material for hydraulic structures and chemical-corrosion resistant material for construction after practical applications.
Key words: Resin mortar, Adhesive strength, Abrasion-erosion resistance, Cavitation resistance, Corrosion resistance, Sluiceway.

1. Introduction

Vinyl ester resin, also known as epoxy acrylic resin, is a new type of polymer developed in the sixties, which is formed by a ring opening addition reaction between a low-molecular weight epoxy resin and an unsaturated carboxylic acid. This type of resin not only retains the basic properties of the skeletal polymer epoxy resin but also possesses the room-temperature setting behavior of the unsaturated polyester. In fact, it embodies the properties of both epoxy resin and unsaturated resin, hence yielding a more perfect product which exhibits better performance in field applications.

3200 vinyl ester resin is the vinyl ester resin modified with a small amount of fumarate, which, containing a lot of active double bonds, allows free polymerization in the presence of a peroxide initiator and can cure at room temperature, enabling its application to be free from seasonal effects. This type of resin is characterized by high mechanical strength, relatively great elongation, and good adhesiveness. Using the resin, a resin mortar has been formulated and investigated in respect of various properties. This type of resin mortar has come into use as a new type of covering material resisting scouring abrasion and cavitation erosion in hydraulic structures or as a new type of coating material resisting corrosion in buildings. A brief description of its properties and applications is given as follows:

2. Features of the formulation of 3200 vinyl ester resin mortar and its properties

Recently resin mortars have been widely used as anti-corrosion, protective-covering and bonding materials. The behavior of resin mortars varies with resin type, addition agent type, filler type, aggregate type, and application method in use. In order to prepare a good type of vinyl ester resin 3200 mortar (called hereafter 3200 mortar for short), mixtures were made using various types of aggregates of different gradings according to the densest resin mortar mix design and porosities measured so as to find out their composition in a state of the greatest density. And then the aggregate and cementing material were mixed in a state of the greatest filling density. From the characteristics, i.e., density, segregation characteristic and strength, measured prior to and following hardening, dosages of cementing material were found, thereby enabling determination of the optimum mix proportions. As indicated by systematic comparison tests, the features of the developed formulation are as follows:

(a) Addition of a shrinkage-reducing admixture can cause a reduction in shrinkage during the process of curing improving the bond stength to the concrete substrate. The resin in 3200 mortar will disply extremely high shrinkage in the presence of an initiator and a promoter while being transformed from the viscous-state linear structure into the solid-state network structure. This sort of shrinkage involves thermal shrinkage and chemical shrinkage, the former being due to a differential between the room temperature and polymerization exothermic peak, and the latter due to the migration, i.e., the transformation of the van der waals distance into the chemical bond distance, which occurs when some of double bonds in the resin macrochains are copolymerized with micromolecular monomers to form three-dimensional construction macromolecules. A suitable type of shrinkage-reducing admixture and its proper dosages have been found out through repeated experimental studies, thus making it possible to reduce shrinkage shear stresses in the mortar and concrete substrate, to improve bond strength, and to reduce the possibility of cracking.

(b) The properties of the resin mortar can be improved by adding a silane coupler to enhance the resin-aggregate bond strength. The silane coupler is a low-molecular compound in which two different active groups are bonded to the same silicon atom, R - Si - X, where X denotes the reactive group that tends to react with an inorganic filler whose surface has an affinity for water to form chemical bonds, Si - o - Si, and R represents the reactive carbon functional group that tends to react with an organic polymer to turn into an effective part of it. The silane coupler when added to resin mortar can serve as a bridge at the interface between the resin and aggregate and couple up them by forming chemical bonds so as to enhance the resin-aggregate bonding strength and thereby to improve the properties of the resin mortar, particularly its water resistance. An appropriate coupler and its optimum dosages have been found out by investigations.

The microstructure of 3200 mortar specimens with and without a coupling agent was observed by means of a scanning electron microscope, as shown in photos 1 and 2. From the photos it is clear that the bond of the resin to the aggregate is improved by adding the coupling agent, as indicated by comparison of the sand shown at the upper left in photo 1 with that at the lower left in photo 2.

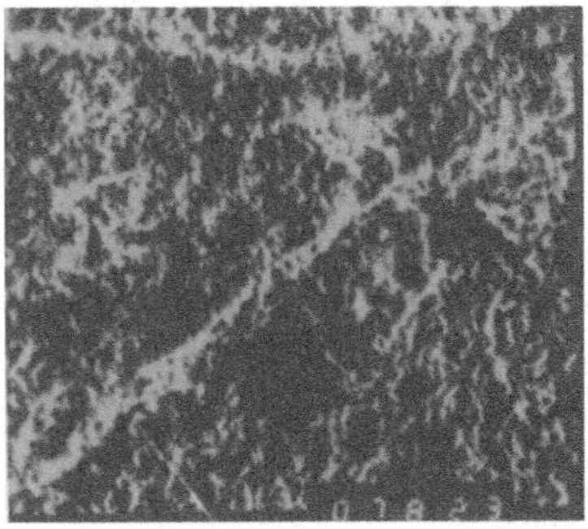

Photo 1 -- Microstructure of 3200 mortar containing no coupling agent, multiplied 1500 times.

The principal properties of 3200 mortar are listed in Tables 1 and 2. As can be seen in these tables, the 3200 mortar has excellent cavitation resistance, scouring-abrasion resistance and frost resistance, good resistance to corrosion by chemical fertilizer, and a certain level of ageing resistance, long-term water resistance and deformation resistance. In the case of a cavitating flow with a velocity of 48m/sec and cavitation numbers in the range $0.1 \leq \sigma \leq 0.136$, it shows a 60-fold increase in cavitation erosion resistance over the high-strength concrete, and in case of a sediment-laden flow with a velocity of 14m/sec and a sediment concentration of 2% its resistance to scouring abrasion is increased two or three fold. Results of quick soaking and baking and long-term soaking tests of specimens in urea, ammonium nitrate and ammonium sulfate

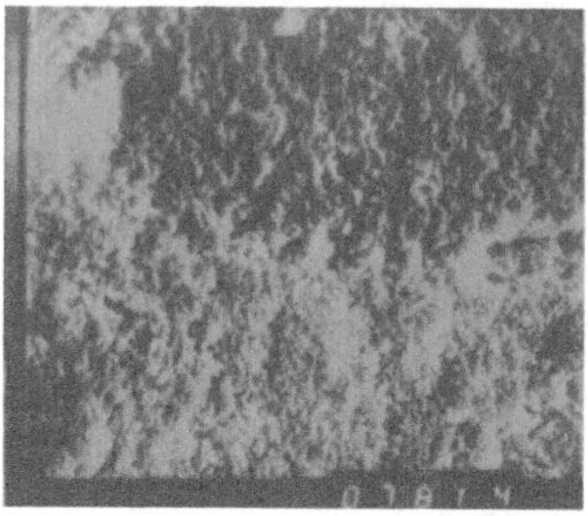

Photo 2 -- Microstructure of 3200 mortar containing a coupling agent, multiplied 1500 times.

saturated solutions indicated that there were no changes in appearance and that fluctuations in strength and weight fell in the range of allowed values for corrosion resistant materials. Furthermore, the 3200 mortar had a higher adhesive bonding strength to the substrate no matter whether it was a dry cement substrate or a steel plate, while the comparison ordinary cement mortar showed separation the instant the forms were removed. This is due to the fact that 3200 vinyl ester resin contains a relatively large number of polar groups, -O- -$\underset{OH}{CH}$-, which enable formation of molecular bond forces between the resin and the adjacent boundary surface, producing a denser bond.

Table 1. Principal Physicomechanical Properties of 3200 Vinyl Ester Resin Mortar

Properties	3200 mortar	Comparison set	Remarks
Resistance to cavitation erosion (hr/kg/m^2)	56.16/6380	0.88/100	R-500 high-strength concrete was used for the comparison set.
Field abrasion resistance at Danjiangkou (hr/cm)	727/313	232/100	do.
Lab resistance to scouring abrasion (hr/kg/m^2)	3.64/433	0.84/100	R-600 high-strength mortar was used for the comparison set.
Ultimate elongation (x 10^{-6})	1053/462	228/100	do.

Tensile elastic modulus (x 10^5kg/cm^2)	2.25/87	2.60/100	do.
Linear expansion coefficient (X 10^{-6}/^oC)	36/225	16/100	do.
Compressive strength (kg/cm^2)	923/153	605/100	do.
Tensile strength (kg/cm^2)	180/333	54/100	do.
Bending strength (kg/cm^2)	316/224	141/100	do.
Impact ductility (kg-cm/cm^2)	46.4/1079	4.3/100	do.
Variation in bending strength after 2000 hrs of ageing (%)	+8	+10	do.
Bonding strength to dry cement mortar substrate (kg/cm^2)	62	Separation took place at the time of form removal.	do.
Tensile bond strength to dry steel plate (kg/cm^2)	225	do.	do.
Variation in strength after 300 cycles of freezing and thawing(%)	+9.0	Collapse, no strength	R-600 high-strength air-entrained mortar was used for the comparison set.
Shrinkage (x 10^{-6})	414/31 (4 days)	1271/100 (28 days)	do.
Water resistance (retained percentage of strength after 2.5 years of immersion in water)	95%	71%	The comparison set was made with 3200 mortar containing no coupling agent.

Note: In the table the numerators are absolute values, while the denominators are percentages relative to the comparison sets.

Table 2. Corrosion-Resistance of 3200 Vinyl Ester Resin Mortar in Saturated Fertilizer Solutions

Fertilizer type	Testing method	Strength variation (%)		Weight variation (%)		Appearance change	
		Comparison mortar	3200 mortar	Comparison mortar	3200 mortar	Comparison mortar	3200 mortar
Urea	36 cycles of quick soaking and baking	+20	+3	-6.89	-0.15	Surface peeling, sand exposed throughout	no change
	Long-term soaking for 5.5 months	-21	-10	+0.69	-0.21	no change	no change
Ammonium nitrate	36 cycles of quick soaking and baking	-98	-3	-27.00	+0.57	Severe peeling	no change
	Long-term soaking for 5.5 months	-71	-7	-5.44	+0.19	no change	no change
Ammonium sulfate	36 cycles of quick soaking and baking	-48	-3	-26.00	+1.72	Severe peeling	no change
	Long-term soaking for 5.5 months	-25	+2	-14.00	+0.31	Surface peeling, sand exposed throughout	no change

3. Application of 3200 vinyl ester resin mortar to the sluiceway of the Liujiaxia Hydropower Station

The Liujiaxia sluiceway located on the left river bank has a total length of 241 m and consists of sluices through the dam and a ski-jump chute downstream of the dam with a width of 8 m and a plane turning radius of 296 m. There is a continuous deflector bucket at the tail end of chute. The maximum discharge velocity is 38m/sec. The concrete mixes had a design strength grade 350, but in reality a grade 400 concrete was ultimately produced. This structure, however, has sustained over a number of years severe damage resulting from the abrasive effects of water borne debris and from pitting due to cavitation at high water velocities. According to the overall inspection carried out in the first ten-day period of April, 1983, the area of severe damage amounted to 770 m^2, including 235 m^2 for the floor, 230m^2 for the right wall and 305m^2 for the left wall. There was an area of general stone exposure of 1300 m^2 and an area of casual stone exposure of 1800 m^2 .

Nine repairs have been successively made to the sluiceway since 1975. Epoxy coal tar mortar and dry high-strength cement mortar were used as repair materials, but they failed to afford a satisfactory solution. Therefore, during the period from April 1983 to April 1984, 3200 mortar was used in connection with the overhaul of the sluiceway to repair the high-velocity section. At the same time, low-toxicity epoxy mortar and 6196 unsaturated polyester resin mortar were applied, serving as comparison mortats. In 1983, the total amount of mortar applied was 18.8 m^2 for 3200 mortar, 30.8 m^2 for low-toxicity epoxy mortar and 21.4 m^2 for unsaturated polyester mortar; in 1984, the total amount of mortar application was 172.1 m^2 for 3200 mortar, and 38.2 m^2 for low-toxicity epoxy mortar. Besides, six abrasion-resisting materials were applied to the left steel-lined wall at the outlet of Sluice No.2 on a trial basis. They went through the trials of practical operations during the flood seasons of 1983, 1984 and 1985, which involved approximately 2600 hrs of operation, discharging about $91 x 10^8$ m^3 of water, and passing more than 14 x 10^6 t of sediment. They showed essentially good performance in high-velocity sediment-laden flows. The 3200 mortar bonded soundly to the concrete substrate and exhibited only signs of very slight abrasion erosion. It was found that the 3200 mortar applied to the portions of steel lining which had originally substained relatively severe cavitation damage remained unaltered after passing water and sediment over one flood season while the others exhibited signs of distress.

Test results from samples taken in the field, as shown in tables 3 and indicated that the properties of the resin mortars were for the main similar to those originally obtained in the laboratory.

Table 3. Tests Results on the Properties of Samples Taken in the Field in 1983

Material	Resistance to cavitation erosion ($hr/kg/cm^2$)	Resistance to scouring abrasion ($hr/kg/cm^2$)	Bonding strength (kg/cm^2)	Tensile strength (kg/cm^2)	Bending strength (kg/cm^2)	Compressive strength (kg/cm^2)
3200 mortar	43.76	1.85	61.3	174.3	333.0	1047
6196 mortar	22.46	1.81	82.4	168.1	347.1	1116
6101 mortar	58.18	1.62	98.0	170.9	357.2	1014
6101 coal tar mortar *1	31.41	1.64	73.6	139.5	312.8	942
3200 concrete	41.30	1.67				1078
6101 concrete	47.06	2.01				1034
*2 Dry high strength mortar	1.22	0.67	Fracture took place the instant the forms were removed.	41.0	107.9	526 *3

Notes:
1. 6101 epoxy coal tar mortar is the repair formulation originally used for the Liujiaxia Hydropower Station.
2. The dry high-strength mortar had a cement-sand ratio of 1:2.5 and a water-cement ratio of 0.36 and contained 0.75% water-reducing admixture UNF.
3. The compression specimens of mortar had the dimensions of 4 cm by 4cm by 6.25cm.

4. Use of 3200 Vinyl Ester Resin Mortar as Corrosion Resistant Material for Buildings

The problem of corrosion in buildings is encountered in all the trades and industries where there is the possibility of making contact with corrosive media. According to the statistics from some western countries, the economical loss due to building corrosion accounts for 6% of the gross national product. Not only does damage due to building corrosion cause economical losses but it brings about frequent subsequent maintenance and repairs, which have influence on production continuity and threaten safety, constituting a far-reaching concern.

According to the technical specifications for Anti-corrosion Design of Industrial Buildings GBJ46-82 in China, anti-corrosion measures corresponding to corrosion grades 11 and 111 should be undertaken to protect fertilizer packaging areas, fertilizer storages and bag fertilizer storehouses. Field investigations also show that serious building corrosion damage exists in fertilizer factories.

Seven materials including 3200 mortar were investigated with a view to developing corrosion preventive materials for the concrete of fertilizer wharves and storages. With the aid of laboratory long-term soaking tests and quick soaking and baking tests in combination with field comparison tests conducted in the Qixiashan Fertilizer Factory over half a year, it was found that 3200 mortar was capable of resisting the chemical and physical corrosion effects of ammonium nitrate, urea and ammonium sulfate. And then trial large-area applications were made, with basically satisfactory results.

The reasons why 3200 mortar performs fairly well in corrosion prevention can be summarized as follows:
This kind of vinyl resin tends to bring in a diphenolic type a epoxy resin structure of larger dimensions whose molecular weight is much higher than that of unsaturated polyester. Little possibility exists for hydrolyzing in the presence of corrosive media because of no repeated occurrence of ester bonds in the main chain. In addition, modifying the resin with fumarate enhances steric hindrance among molecules, thereby reducing the hydrolytic reaction of ester bonds. On the other hand, controlling reaction temperature makes it possible to avoid the syn- and anti-isomerism of divinyl ester and to keep the anti-type double bond content at a relatively high level. Furthermore, not only does 3200 mortar contain intramolecular double bonds but its end groups contain double bonds, which facilitates free radical polymerization, leads to a relatively high degree of solidification and enables the resin to gain a dense network structure after solification, resulting in higher corrosion resistance.

5. Concluding Remarks

(a) It is shown that the bonding strength of 3200 vinyl ester resin mortar to concrete substrates is greatly enhanced by adding a shrinkage-reducing admixture together with a coupling agent. In view of its excellent properties of cavitation resistance, scouring abrasion resistance and corrosion resistance and its fairly good properties of long-time water resistance, freezing resistance and ageing resistance, it may be advisable to popularize this type of mortar as a new type of coating material.

(b) Experience with the application of 3200 vinyl ester resin mortar to the Liujiaxia Sluiceway, which has successfully withstood three years of operation for passing water, silt and debris in high-velocity sediment-laden flows, shows that this type of mortar can be popularized as a protective covering material to resist scouring abrasion and cavitation erosion or as a repair material for hydraulic structutes.

(c) When using 3200 vinyl ester resin mortar as a coating material for large-area corrosion prevention strict control of construction conditions and operations is necessary, and special attention should be paid to the treatment of concrete substrates so as to yield satisfactory workmanship.

References

Lawrence E. Kukacka. (1981) Applications of Polymer Concrete. ACI Publication SP-69.

People's Republic of China National Standards. (1983) Technical Specifications for Anti-corrosion Design of Industrial Buildings GBJ 46-80, Beijing.

Lin Baoyu, Ceng Ruiyu and Lu Anqi. (1983) Preliminary Investigation of Three Types of Resin Mortars for Use in Hydraulic Structures. J of Nanjing Hydraulic Research Institute, No.4, 35-46

Lin Baoyu. (1985) Summary of Field Tests on Cavitation-Erosion and Scouring-Abrasion Resistant Materials Conducted in Liujiaxia Sluiceway, Water Conservancy and Hydropower Technology, No. 5, 20-22

ADHERENCE DES MORTIERS HYDRAULIQUES MODIFIES PAR DES RESINES MISCIBLES : APPLICATION A LA REPARATION DES BETONS ARMES

Martine MASSON
Service Technique, Société SIKA

Abstract
Le problème des réparations est abordé sous l'angle des critères de qualité auxquels doivent répondre les produits de réparation.
Ces critères sont déterminants dans le choix d'une dispersion de polymères utilisés en addition aux mortiers de liants hydrauliques.
Les dispersions de copolymères acryliques dont les particules sont de très petite taille, apportent aux mortiers des qualités d'adhérence élevée et une excellente protection contre la carbonatation et la corrosion des armatures en milieu agressif.
Keys words : Réparation, Béton armé, Résine acrylique, Mortier, Adhérence, Carbonatation, Corrosion.

1. Introduction

La dégradation des bétons armés est souvent liée à une augmentation de la porosité du béton par suite d'erreurs de conception, fabrication ou mise en oeuvre.

Le passage d'agents agressifs (acides, gaz carbonique...) en milieu humide, favorise la diminution du pH du béton, entraînant la dépassivation puis la corrosion des armatures. La formation de rouille a pour conséquence le décollement puis l'éclatement des bétons.

La réparation doit donc reconstituer le monolithisme du matériau et les armatures doivent être protégées d'une nouvelle progression de la corrosion.

Depuis plusieurs années, les liants hydrauliques ont été utilisées pour des réparations avec plus ou moins de succès.

Fabricants et chercheurs ont tenté d'améliorer les propriétés des mortiers de ciment par un apport de résine miscible pour en adapter les qualités à la réparation des bétons.

C'est dans cette voie que s'est engagée notre recherche avec la démarche suivante :
. Quelles qualités sont recherchées pour un mortier de réparation ?
. Quels polymères peuvent apporter ces qualités ?
. Quelles sont les caractéristiques du mortier modifié par le polymère?

2. Qualités des produits de réparation

Le principe même de la réparation est la reconstitution du monolithisme de l'ouvrage, tant sur le plan mécanique que sur le plan de son comportement dans son milieu, c'est-à-dire la corrosion. Il s'agit dont de redonner au matériau béton ses propriétés initiales mécaniques et physico-chimiques.

Un mortier de ciment traditionnel ne peut répondre à l'ensemble des critères de qualité d'un bon produit de réparation et c'est dans ce sens qu'il va falloir l'améliorer. La résine miscible utilisée doit donc apporter les qualités suivantes :

2.1. Adhérence élevée au support

C'est certainement le critère déterminant puisque la réparation du béton suppose une parfaite cohésion du matériau reconstitué. D'une manière générale, on prévoit que l'adhérence du mortier doit être supérieure à la cohésion du béton.

Il ne faut cependant pas ignorer que c'est le critère sur lequel la mise en oeuvre va introduire un facteur d'incertitude élevé. L'adhérence du mortier sera liée à la méthode de mise en oeuvre. La conception du produit doit donc tenir compte de ce critère déterminant.

2.2. Imperméabilité à l'eau

La plupart des dégradations du béton sont liées à une pénétration d'agents agressifs véhiculés par l'eau dans les pores et capillaires (eau de pluie, condensation...). La réparation doit présenter une durabilité élevée et le produit utilisé doit donc être insensible à ces agressions extérieures. Au même titre que l'adhérence, l'imperméabilité est un facteur de durabilité de la réparation et est étroitement liée aux techniques de mise en oeuvre (serrage, talochage...).

2.3. Coefficient de dilatation thermique équivalent à celui du béton support

En effet, pour tous les ouvrages soumis à des chocs thermiques ou de larges amplitudes thermiques, l'utilisation d'un produit à coefficient de dilatation trop élevé ou trop bas pourrait entraîner des fissurations ou même des décollements.

2.4. Protection des aciers

Si le béton, grâce à son alcalinité naturelle protège par passivation les armatures, il est indispensable que le produit de réparation ait le même comportement. C'est pour cette raison qu'il est préférable d'utiliser un produit à base de ciment et c'est dans cette voie que les essais du produit fini ont été réalisés.

2.5. Résistances mécaniques supérieures ou égales à celles du support
Module d'élasticité inférieur ou égal à celui du support

Le mortier de réparation, comme il a été souligné plus haut, doit permettre à l'ouvrage de retrouver ses caractéristiques mécaniques. L'excès de résistances mécaniques ou de flexibilité du produit n'est donc pas indispensable en terme de réparation des bétons.

3. Choix du polymère d'addition

Depuis plusieurs décennies, l'amélioration des mortiers a pu être obtenue par addition de résines thermoplastiques en poudre ou en dispersion. Ces dispersions sont des suspensions dans l'eau de polymères auxquels s'ajoutent des émulsifiants et/ou des colloïdes, le tout étant dispersé de façon régulière et stable. Ces particules en suspension ont des diamètres variant entre 0,1 et 5 microns, leur taille étant liée à la taille du polymère d'une part et à la taille des additifs (colloïde ou émulsifiant) d'autre part.

Les principales résines utilisées dans les dernières décennies peuvent se diviser en trois groupes comme indiqué dans le tableau 1.

Tableau 1

Les différentes familles de polymères utilisés dans les mortiers de ciment

	1re génération	2e génération		3e génération	
Type de polymère	PVA Acétate de polyvinyle	PVP Propionate de polyvinyle	PVCO Copolymères vinyliques	SBR Styrène butadiène	AC Copolymères acryliques
Additif à la dispersion aqueuse	Colloïde			Colloïde + émulsifiant	Emulsifiant

Notre choix s'est donc orienté vers un copolymère acrylique de la 3e génération. En effet, la nature même des molécules et l'utilisation d'un additif émulsifiant entraînent des tailles de particules très faibles, de l'ordre de 0,1 micron. Ces particules de polymère dont la taille est très inférieure à celle du ciment (voir schéma 1) s'intègrent dans le réseau sable/ciment et donnent au mortier de réparation une forte cohésion interne.

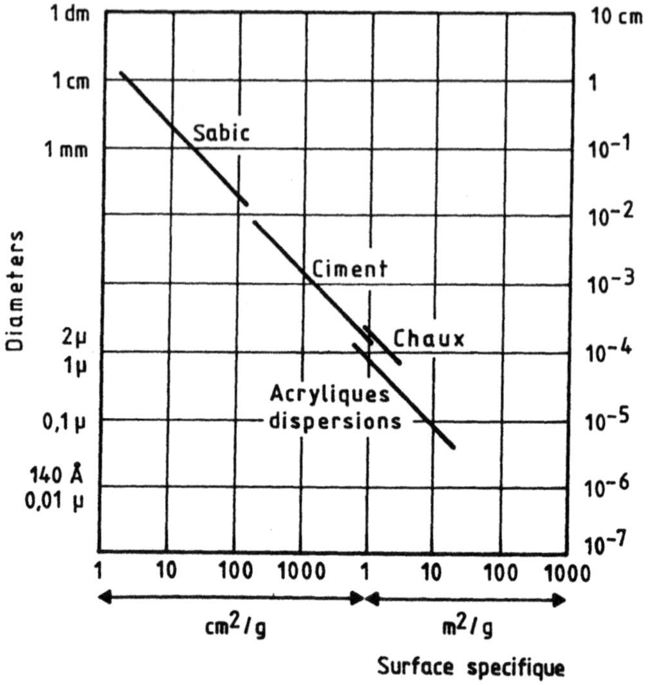

Tailles des particules de polymère comparées à celles du ciment

Schéma 1.

Cette faible dimension va également intervenir dans le mécanisme d'adhérence, les particules pénétrant dans les pores et capillaires du support béton. L'humidification préalable du support va donc jouer un rôle essentiel, l'eau véhiculant ces particules du mortier vers le support.

Ce mécanisme d'intégration du polymère dans le mortier va donc apporter :

. Une très forte adhérence (voir schéma 2),
. La suppression des incertitudes liées à la mise en oeuvre puisqu'il suffit d'humidifier le support après nettoyage. On utilise ni primaire ni barbotine d'accrochage, éliminant ainsi tous les problèmes de succession de plans de collage que pose ce type de mise en oeuvre.

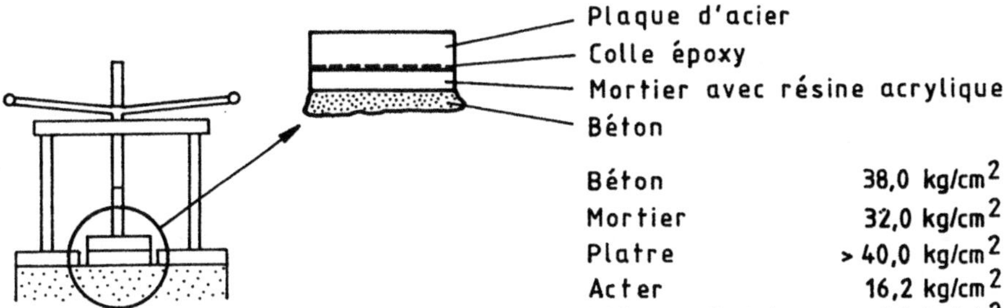

Béton	38,0 kg/cm^2
Mortier	32,0 kg/cm^2
Platre	> 40,0 kg/cm^2
Acter	16,2 kg/cm^2
Béton cellulaire	6,9 kg/cm^2

Schéma 2 - Essais d'adhérence

Ces performances élevées sont obtenues avec un ciment parfaitement dosé et une granulométrie précise et régulière du sable. C'est pourquoi ces produits sont formulés sous forme de mortiers prédosés offrant ainsi le maximum de sécurité sur chantier.

4. Les autres propriétés apportées par la résine

4.1. Module d'élasticité

Avec la résine acrylique choisie, le module d'élasticité est inférieur à celui des mortiers traditionnels et se situe à 25 000-27 000 MPa. Il faut cependant noter qu'en utilisant le large éventail de copolymères acryliques, il est possible de formuler des mortiers plus flexibles avec des modules d'élasticité de l'ordre de 19 à 22 000 MPa. Ces mortiers ont d'autres utilisations dans l'imperméabilisation ou la protection des bétons.

4.2. Imperméabilité

Elle est directement liée à la taille des molécules de polymères et à leur mode d'action décrit plus haut. L'absorption capillaire est de trois à quatre fois inférieure à celle d'un mortier témoin 1/3. Il faut également souligner que ces mortiers ont également une excellente imperméabilité en sous pression jusqu'à 2 MPa (essais LCPC).

4.3. Résistance mécanique

Les dispersions synthétiques des générations précédentes présentaient l'inconvénient d'augmenter la résistance à la traction par flexion au détriment de la résistance à la compression.
Au contraire, les copolymères acryliques, apportent une modification sensible, tant de la traction par flexion que de la compression.
Par ailleurs, la montée en résistance est également rapide (Schéma 3).

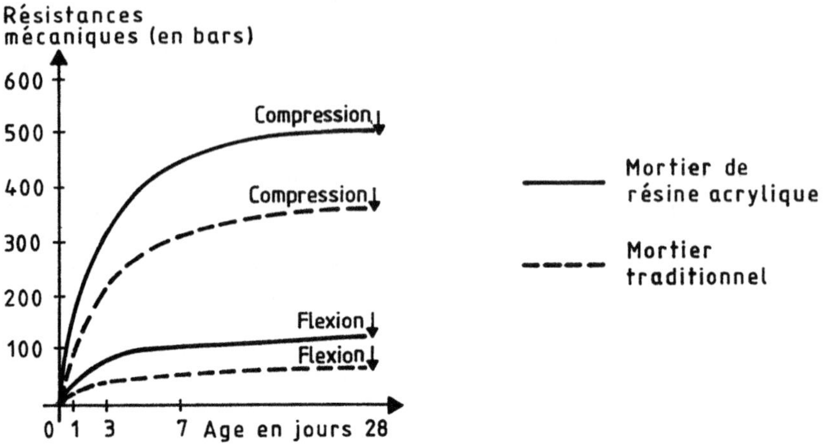

Schéma 3. : Résistances mécaniques des mortiers de résine acrylique

5. Protection des aciers

Avec des mortiers présentant une telle cohésion, il était important de vérifier, dans l'optique de la réparation, le comportement de ces produits aux agressions chimiques atmosphériques.

Les essais de carbonatation et de corrosion accélérée ont été réalisés en collaboration avec le CERILH (Centre d'Etudes et de Recherches de l'Industrie des Liants Hydrauliques).

5.1. Carbonatation accélérée

Le principe de ces essais consiste à comparer, dans le temps, les profondeurs de carbonatation d'un mortier témoin (400 kg de CPJ 45 par mètre cube de sable) et de mortiers additionnés de résine acrylique conservés dans une ambiance contenant 70 % de gaz carbonique CO2. On mesure régulièrement la profondeur de carbonatation.

La carbonatation est évaluée par pulvérisation d'une solution de phénolphtaléine sur la section rompue (coloration rose de la zone non carbonatée). (Schéma 4.).

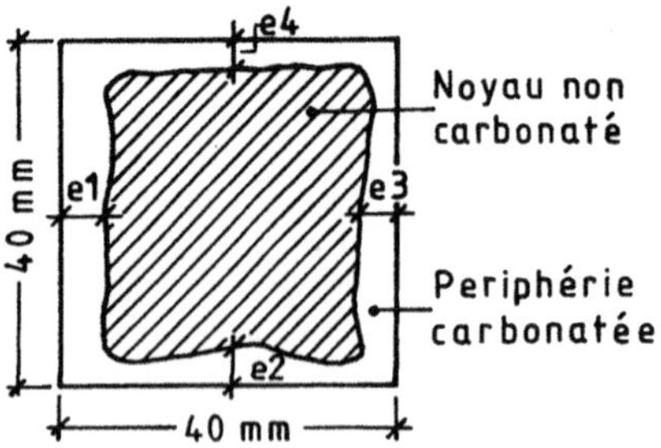

Schéma 4 - Evaluation de la profondeur de carbonatation sur une section

On constate, après 28 jours d'essais dans ces conditions que, du point de vue de la carbonatation, 1,5 mm à 3,5 mm d'épaisseur (selon la granulométrie) de mortier additionné de résine synthétique sont équivalents à 20 mm de mortier traditionnel (voir tableau 2.).

Tableau 2

Carbonatation accélérée

	1 jour	3 jours	7 jours	14jours	28jours
Témoin CPJ 45	11	20 Carbonatation à coeur	20	20	20
Mortier additionné de résine acrylique Sable : 0-0,5 mm	0	1	2	3	3,5
Mortier additionné de résine acrylique Sable 0-3 mm	1	2	3	4,5	5,5

5.2. Essais de corrosion des armatures

La mesure de corrosion accélérée des aciers est effectuée sur éprouvettes prismatiques de 2 x 2 x 16 cm confectionnées avec une armature centrale en acier doux de 5 mm de diamètre (Schéma 5.).

Les éprouvettes sont conservées 28 jours dans l'air puis 28 jours dans l'armoire à carbonatation accélérée. Elles sont ensuite entreposées verticalement dans un bac contenant une solution très agressive (25 g/l de sulfate, 25 g/l de chlorure) jusqu'au milieu de l'éprouvette (test semi-immersion).

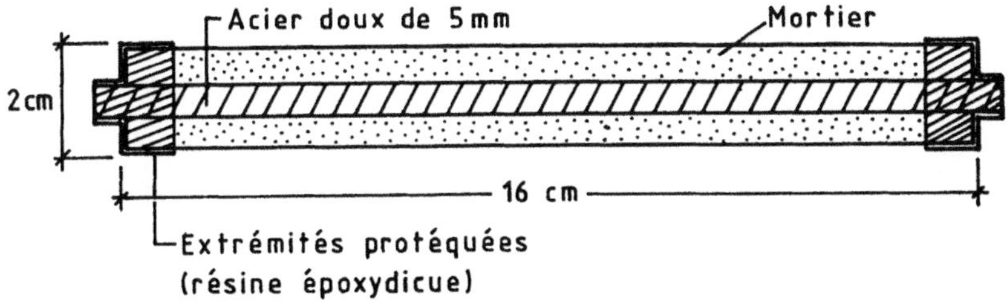

Schéma 5. - Essai de corrosion des armatures

En matière de protection des armatures, les produits testés retardent de façon significative le début de corrosion des armatures.

Alors que l'armature du témoin est corrodée à partir de 22 jours d'immersion, celles protégées par des mortiers additionnés de résine acrylique sont intactes après 4 mois d'immersion.
Ces deux essais mettent donc en évidence une des propriétés essentielles apportées par le copolymère acrylique dans les mortiers de réparation.

6. Conclusion

Si l'addition de résines acryliques permet de formuler des mortiers de réparation possédant des performances élevées, il est essentiel que les formulations soient adaptées à la spécificité des applications.

Sur ces principes et en utilisant différents copolymères acryliques et différentes granulométries de sables, a été élaborée une technique de réparation. Cette technique est basée sur une gamme de produits prédosés adaptés à chaque cas de réparation, en fonction des épaisseurs, des contraintes de fonctionnement ou d'application et allant du traitement de l'armature jusqu'à la protection finale du béton.

CARACTERISTIQUES D'ADHESION DU MORTIER EN MONOCOUCHE EPAISSE POUR LA REPARATION DES SURFACES EN BETON

R. NANIWA
Département d'Architecture, Faculté des Sciences Appliquées,
Université Kogakuin

Résumé
La tendance actuelle au sujet de la finition extérieure dans la construction est l'emploi de revêtements monocouches minces en vue d'éviter le décollement dû aux tensions internes. Cette réduction d'épaisseur simplifie les travaux et économise les matériaux. Toutefois, le revêtement épais améliorait la durabilité de la construction, apportait une excellente isolation thermique et protégeait la structure contre les facteurs agressifs. D'autre part, le mortier en monocouche épaisse est très efficace lors de la réparation de décollements, éclatements, écornures et d'autres creux apparaissent sur des surfaces extérieures particulièrement exposées aux intempéries. Pour cette raison, nous avons élaboré une nouvelle technique de cimentage en monocouche de 10 à 50 mm d'épaisseur grâce à un nouveau mortier de ciment-polymère dont les agrégats principaux sont le sable siliceux, le sable de laitier granulé de haut-fourneau et le sable de microsphères creuses en verre. Nous avons constaté de plus, que pour la monocouche épaisse, la résistance d'adhésion de ce mortier est aussi nettement supérieure par rapport aux mortiers classiques.
Mots-clés : Mortier en monocouche épaisse, Mouvement différentiel, Sable de laitier granulé de haut-fourneau, Ouvrabilité de truellage, Rhéomètre de charge oblique, Force de la pression de la truelle, Facteur de résistance au glissement de la truelle, Moment produit à la poignée.

1. Introduction

En ce qui concerne la finition extérieure en gunitage ou en carrelage, la tendance actuelle est de rendre le support en mortier de plus en plus mince afin d'éviter les fissurations et le décollement partiel ou total, dus à la haute tension interne dans le cimentage; tension interne provoquée par les mouvements différentiels (thermique et hygrométrique) entre la structure en béton armé et le revêtement. Les tensions internes s'accroissent exponentiellement avec l'augmentation de l'épaisseur totale du cimentage. De plus, une épaisseur mince réduit la durée des travaux par la suppression des périodes de repos nécessaires au séchage et à l'obtention d'une stabilité dimentionelle suffisante, et d'autre part, permet une économie. Toutefois, n'oublions pas qu'un revêtement épais améliore fortement la durabilité de la construction ainsi que son isolation thermique, tout en la protégeant des facteurs agressifs.

Au sujet de l'épaisseur des revêtements, il faut satisfaire à deux impositions contradictoires :

- Lorsque le cimentage est mince, bien que les tensions internes soient faibles, sa durabilité est insuffisante suite à la carbonatation du béton.
- Lorsque le cimentage est épais, la durabilité est insuffisante suite aux phénomènes de fissuration et de décollement.

De sorte que le mortier idéal devrait avoir :

peu de tension interne (couche mince), pas de carbonatation (couche épaisse), et pas de fissuration ni de décollement (couche mince).

Cependant, bien plus, le mortier épais doit se maintenir à l'état durci avec ses excellentes performances, pour la finition extérieure de nouveau, après la réparation de défauts comme écornures et d'autres creux (figure 1).

En conséquence, nous avons développé un mortier de ciment-polymère pour la monocouche épaisse qui répond mieux aux exigences énoncées ci-avant. Ce mortier épais a été mis au point par l'auteur et ses collaborateurs au Japon et en Belgique, et adopté aux chantiers dans la réparation de défauts de la construction tels que fissuration, décollement, ainsi que pour des travaux de finition de nouveaux bâtiments etc..., en utilisant sa capacité d'être appliqué par truellage en monocouche épaisse. Pour ces travaux, nous avons élaboré une nouvelle technique de cimentage en monocouche de 10 à 30 mm d'épaisseur pour les travaux de finition et pouvant aller jusqu'à 50 mm ou à 70 mm pour les travaux de réparation partielle.

Les principaux agrégats utilisés dans ce mortier sont : le sable siliceux, le sable de laitier granulé de haut-fourneau ainsi que des microsphères creuses en verre.

2. Composition du mortier en monocouche épaisse

Les fonctions de chaque agrégat constitutif du mélange sont les suivantes :

(a) Sable siliceux de carrière - La présence d'un grand angle de frottement interne dû à sa forme angulaire aiguë engendre l'effet d'arche avec comme conséquence une légère résistance à l'affaissement. Il ne provoque pas de retrait au séchage, lors de présence du sable de laitier. Dans le cas où cet agrégat est incorporé dans le mortier en grande quantité, le truellage en monocouche épaisse est rendu impossible.

(b) Sable de laitier - En plus de sa basse densité, son absorption d'eau idéale, le grand angle de frottement interne dû à la grande surface individuelle est inégale, il contribue à la cohésion interne du mortier et de ce fait diminue le risque d'affaissement. Par sa réaction chimique avec les alkalis, il durcit lentement. En réduisant l'eau absorbée, il devient compact et augmente ses caractéristiques mécaniques. Toutefois, une trop grande quantité de sable de laitier a pour conséquence la formation de microfissures dans le ciment durci au cours des six premier mois d'age.

(c) Microsphères creuses en verre - L'extrême légèreté, leur forme parfaitement sphérique permettent d'obtenir une haute ouvrabilité par l'effet "ball bearing". En outre, elles augmentent la thixotropie du mortier et permettent de former une monocouche très épaisse conservant l'eau incluse. Comme pour le sable de laitier, une trop grande proportion de microsphères peut provoquer des microfissures.

Nous considérons que le mortier épais doit être obtenu en sélectionnant les combinaisons optimales des fonctions des agrégats de la façon dite ci-avant.

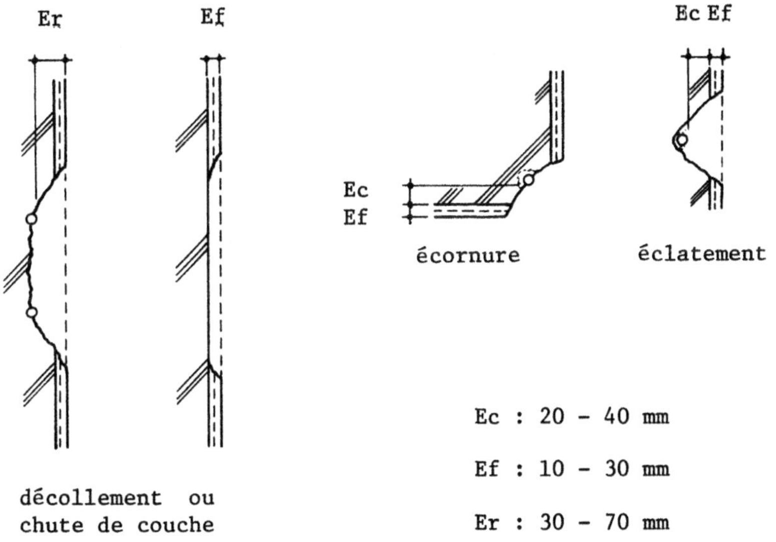

Figure 1 : Classification de sortes de défauts pour la réparation

3. Caractéristiques d'adhésion du mortier en monocouche épaisse

Outre que le mortier en monocouche épaisse répond à des besoins actuels aux chantiers, il est essentiel que ce mortier possède de meilleures caractéristiques d'adhésion au support en béton comparées avec celles de mortier classique, ce qui évite des effets excessifs de tensions internes dus aux mouvements différentiels entre le mortier et le béton. De sorte que ce but soit atteint les caractéristiques suivantes devraient être exigées du mortier durci :

(a) Une bonne déformabilité
(b) Un bas module d'élasticité
(c) Un bas retrait dimensionnel au séchage
(d) Une bonne imperméabilité
(e) Une excellente résistance au gel-dégel
(f) Une bonne isolation thermique

Notre mortier peut répond aux performances mentionnées ci-dessus. D'autre part, une excellente ouvrabilité serait requise du mortier en monocouche épaisse à l'état frais. On peut supposer qu'il y ait du principe de causalité entre l'adhésion et l'ouvrabilité sur le revêtement en mortier épais.

3.1 Ouvrabilité du mortier à l'état frais
Un rhéomètre de charge oblique, conçu par l'auteur met d'aspects d'ouvrabilité en évidence qui distinguent le mortier épais de mortier ordinaire.

La conception du rhéomètre est illustrée à la figure 2 ainsi que l'image de l'appareil est présentée à la photo 1.

La valeur de F = -Fr (Fr : force de réaction, résultant de Fvr et de Fhr) représente la force de la pression de la truelle.

La valeur de φ = Fhr/Fvr, représente le facteur de résistance au glissement de la truelle.

La valeur de M = -Fr·e, représente le moment produit à la poignée. La figure 3 montre les résultats d'essais d'ouvrabilité.

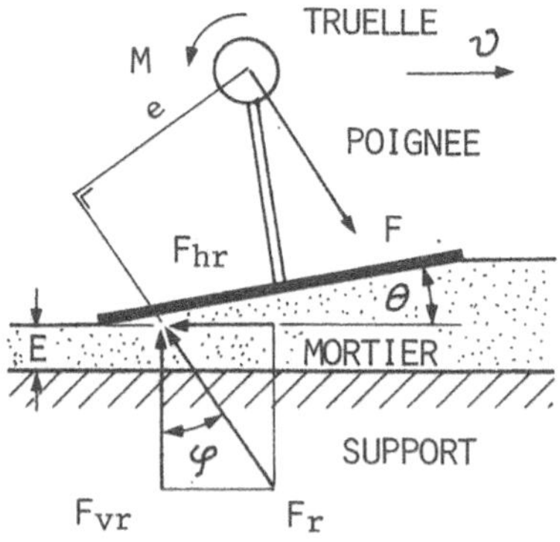

Figure 2 : Schéma de travail lors d'un truellage

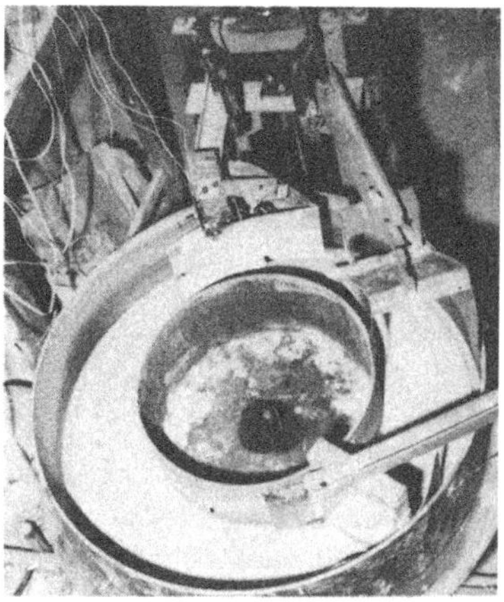

Photo 1 : Aspect du rhéomètre

Les conditions d'essais étaient : θ = 10(degré), E = 5 et 20 (mm), et υ = 55 (cm/sec), et 20°C, 65 % H.R. de l'environnement.

La quantité d'eau ajoutée de chaque type des mortiers employés dans ces essais est déterminé de telle façon que les plâtriers - collaborateurs considérant comme optimal au point de vue de l'ouvrabilité. Les deux cadres enclorés par la ligne pointillée à la figure 3 indiquent les champs optimals ainsi déterminés dont l'un pour le truellage mince (5 mm d'épaisseur, ceci en haut à gauche de cadre) tandis que l'autre pour le truellage épais (20 mm d'épaisseur, en bas à droite). Pour une ouvrabilité optimale du mortier, le point sur les φ - F coordonnées doit se localiser à l'intérieur de chacun des cadres.

Toutes les deux valeurs de φ et de F provenant de deux types de mortier épais, soit de SBR 2-E1(Styrène-Butadiène Rubber) et de MC(Methyl Cellulose) sont situées parfaitement à l'intérieur de ces deux cadres, ce qui signifie que ces deux mortiers sont convenables comme enduit en monocouche épaisse grâce à leurs caractéristiques optimales pour le glissement avec un bon étalement et pour la force de pression de la truelle.

Malgré le fait que les valeurs de φ sont optimale (en cas de 5 mm) et aussi très petite (en cas de 20 mm),les valeurs de F de mortier de ciment ordinaire sans polymère se trouvent bien à l'extérieur des cadres, soit, plus hautes parmi des mortiers employés. Cela signifie, donc, que le truellage en monocouche épaisse par ce mortier ne peut se faire qu'à l'occasion d'application partielle au chantier mais que dû à l'étalement extrêmement pauvre, il en résulte d'une chute de couche à l'état frais ou d'une très faible résistance d'adhésion éventuellement après durcissement. La raison pour laquelle ces résultats se produisent est que la charge due à la force de pression de truelle se transmet jusqu'au niveau de la surface de contact avec le support tout en y provoquant un mouvement déformable à l'interface.

Si la quantité d'eau est élevée de telle sorte que ce problème soit réglé, ce mortier se comporte comme de la boue ("slurry-like"). Par conséquent, il n'est jamais approprié à l'enduit épais.

Dans le cas du mortier de ciment-polymère ordinaire avec du sable siliceux, la valeur de F diminue par rapport au mortier ordinaire sans polymère ci-avant mentionné et ceci malgré une petite quantité d'eau ajoutée. Cet avantage provient du fait que ce type de mortier

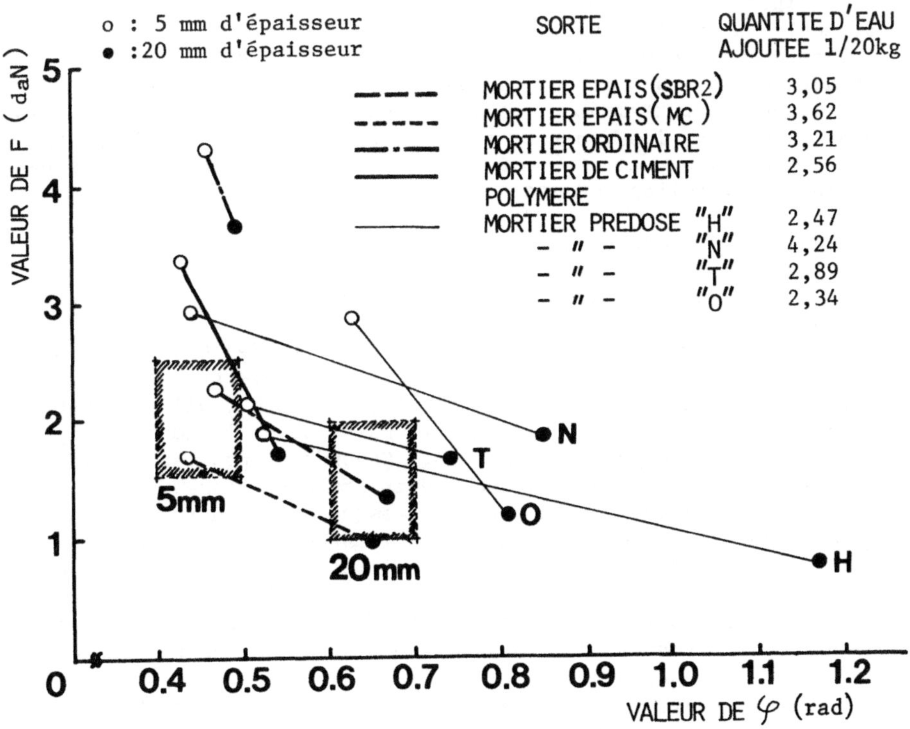

Figure 3 : Ouvrabilité de mortiers variés

profite des polymères en dispersion additionnels. L'ouvrabilité est ainsi bien améliorée par rapport aux mortiers ordinaires. Cependant, même ce mortier n'est pas approprié comme mortier épais, car la valeur de F est légèrement trop haute et celle de φ est trop basse au dehors du cadre en cas de 20 mm d'épaisseur.

D'ailleurs, nous avons constaté que les quatre mortiers prédosés en vente au Japon se diffèrent d'un à l'autre tel qu'indiqué à la figure 3.

Le mortier H montre sa valeur de φ la plus élevée, ce qui veut dire qu'il est très visqueux sans bon étalement lors d'application. Nous pouvons, donc, en conclure qu'il est impossible d'utiliser ce mortier pour l'enduit épais.

Le mortier N s'est situé également en dehors des limites appropriées en tant que mortier épais. Il ne convient qu'au truellage mince.

Le mortier T ressemble le plus à notre mortier épais parmi les mortiers éprouvés dans ces essais. Notons, cependant, que la force de pression de truelle se transmet jusqu'à la surface de contact avec le support lors de truellage épais, car la valeur de F est légèrement plus haute que le seuil indiqué par les cadres.

Autrement dit, la partie de la couche de mortier touchant à la surface de support en béton est sujet au déplacement répétitif provoqué par le mouvement de truelle et par la suite des forces de cisaillement et de pression s'y produisent, ce qui en résulte l'apparition d'eau immigrée et de bulle d'air. A cause de ce mécanisme extrêmement complex, nous voyons, en fin, le phénomène de chute de couche partielle ou totale appelé "peeling down".

Finalement, le mortier O est surtout convenable en truellage à l'épaisseur de 1 à 3 mm qui est une sorte de manière de truellage très mince en frottement. Ce mortier caractérise comme le plus visqueux parmi tous les types de mortier autant pour le truellage épais que pour celui mince. Son ouvrabilité est, en général, la plus pauvre lors de truellage en monocouche épaisse.

3.2 Adhésion du mortier à l'état durci

La figure 4 montre les résultats d'essais d'adhérence-traction au support en béton utilisant des éprouvettes de 50 x 50 mm de diverses sortes de mortier.

Les essais ont été effectués à la température de 20°C et l'humidité relative a été fixée à 60 %. L'âge des éprouvettes a été de 28 jours. La vélocité de tension était 19,6 daN/sec (pull-off test).

En ce qui a trait du truellage, l'épaisseur de 10 mm a été appliquée par une seule passe à une fois. Cependant, pour atteindre l'épaisseur de 30 mm, nous avons enduit par deux passes, 15 mm d'épaisseur chaque fois mais sans repos presque aucun entre les deux passes. En cas des deux mortiers de ciment-polymère ordinaires (SBR 1-ORD et SBR 2-ORD) contenant seulement du sable siliceux comme agrégat, une très faible pression de truelle a été observée pour l'épaisseur de 30 mm tout en produisant la résistance d'adhésion presque nulle. Par contre, en cas des trois sortes de mortier épais de ciment polymère, elles ont présenté, en général, de meilleures performances.

Parmi les trois types, nous avons constaté que la SBR 2 est supérieure à la SBR 1 mais l'EVA (Ethylène - Acétate de Vinyle) possède la plus haute résistance.

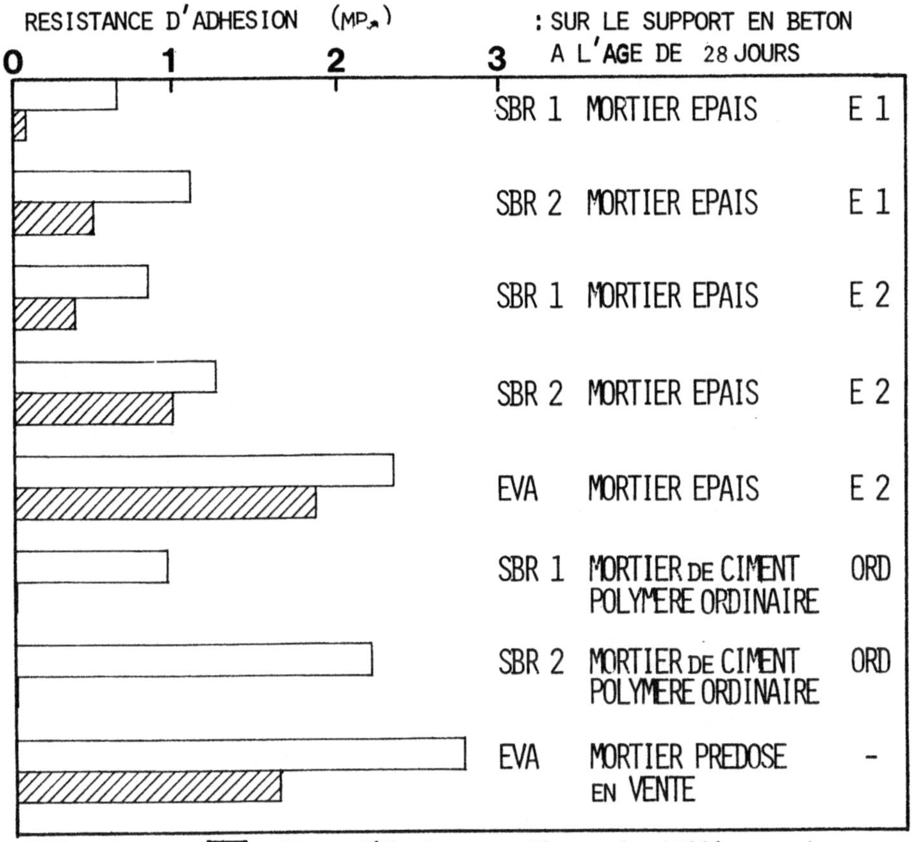

(blanc) : 10 mm, (hachuré) : 30 mm d'épaisseur, il y a la différence de dosage entre E 1 et E 2 (sable mélangé) et ORD est sable siliceux.

Figure 4 : Efficacité de mortier épais au point de vue d'adhésion

Bien que la résistance d'adhésion se rapetisse, en général, avec l'augmentation d'épaisseur d'éprouvettes, les mortiers épais développés par l'auteur (SBR 2-E1, SBR 2-E2 et EVA -E2) gardent un niveau suffisamment satisfaisant par rapport au mortier de ciment ordinaire.

4. Mise en oeuvre

La technique du truellage en monocouche épaisse est la suivante :

(a) Gratter suffisamment la surface du support (béton armé ou couche de mortier), nettoyer et dégraisser la surface.

(b) Préparer le mortier riche de ciment-polymère pour le primer d'accrochage. Régler la quantité d'eau à ajouter suivant la différence et l'état du support ainsi que de la température et de l'humidité relative de l'environnement. Mettre en oeuvre le mortier en une couche très mince de 1 à 3 mm en se servant d'une truelle et en appuyant bien la truelle sur le mortier en fort frottement. Peigner ou brosser la surface de couche mince.

(c) Composer le mortier de ciment-polymère pour la monocouche principale et déterminer la quantité d'eau à mettre en considérant la température et l'humidité relative de l'environnement.

(d) Réaliser le truellage en monocouche épaisse, soit immédiatement après la mise en place du primer d'accrochage, soit après avoir laissé le temps de repos (de quelques dizaines de minutes à 3 jours).

(e) Effectuer d'abord un cimentage à la truelle à l'épaisseur mince (5 à 10 mm) avec la pression de truelle très forte. Réaliser la répétition immédiate de la mise en place du mortier à la truelle jusqu'à l'obtention d'une monocouche d'épaisseur prévue.

Au cours de ces travaux, appuyer fortement sur la truelle pour les premières passes (jusqu'à ce que l'épaisseur du cimentage atteigne 15 mm environ d'épaisseur), puis à partir de cette épaisseur voulue, réduire graduellement la pression exercée sur la truelle. N'appliquer jamais de forte pression répétive par la truelle afin d'éviter l'apparition de tension interne à l'interface du support.

5. Cas pratiques

5.1 Réparation des défauts

Les défauts suivants peuvent par exemple être observés :

(a) éclatements dus à la rouille des armatures métalliques;
(b) décollement partiel ou total avec chute de matière;
(c) relief et creux accidentés de surface irrégulière observés lors de l'enlèvement des éléments du coffrage;
(d) éclatement dus au "gel-dégel".

Dans ces cas, la technique de réparation est la suivante :

- application d'une couche mince de mortier d'accrochage;
- remplissage en une fois par plusieurs passes successives de mortier monocouche;
- couche de finition.

5.2 Application d'un revêtement pour une surface de 1 à 3 m^2 environ

La technique d'application est identique à celle décrite ci-avant, en ce qui concerne la couche mince d'accrochage. Toutefois, le mortier est appliqué par couches successives jusqu'à l'obtention de l'épaisseur voulue.

5.3 Application d'un revêtement pour une plus grande surface

Après la mise en place d'une mince couche de mortier d'accrochage, l'application est réalisée par une première couche sur toute la surface des murs comme dans le cas de la technique traditionnelle mais, grâce aux caractéristiques du mortier, on peut immédiatement continuer le travail et appliquer une deuxième couche et ainsi de suite jusqu'à l'obtention de l'épaisseur voulue. Il faut, cependant, peigner

Photo 2 : Protection d'armature par résine époxyde

Photo 4 : Epaisseur de 30 mm

Photo 3 : Epaisseur de 50 mm

Photo 5 : Ecornure d'un pilier,épaisseur maximum, 50 mm

préalablement la surface de chaques couches.

Ces trois applications mettent en évidence la diversité d'emploi du mortier et surtout le gain de temps appréciable qu'il permet de réaliser, et, par suite, de l'économie qui en résulte. Les photos de 2 à 6 montrent des exemples de réparation aux chantiers et d'application à un modelage de masque en mortier épais sur un mur de béton.

6. Conclusion

Il est indispensable que l'on considère l'ouvrabilité du mortier de ciment-polymère à l'état frais qui excercera une grande influence sur les performances d'adhésion entre le béton et le mortier à l'état durci. En ce qui concerne ce mortier, nous avons constaté que l'addition de quelques matériaux (matière fibreuse, billes et morceaux de mousse de polystyrène, sable naturel et artificiel de faible densité etc.., ajoutés dans des proportions judicieuses, en améliore les ca-

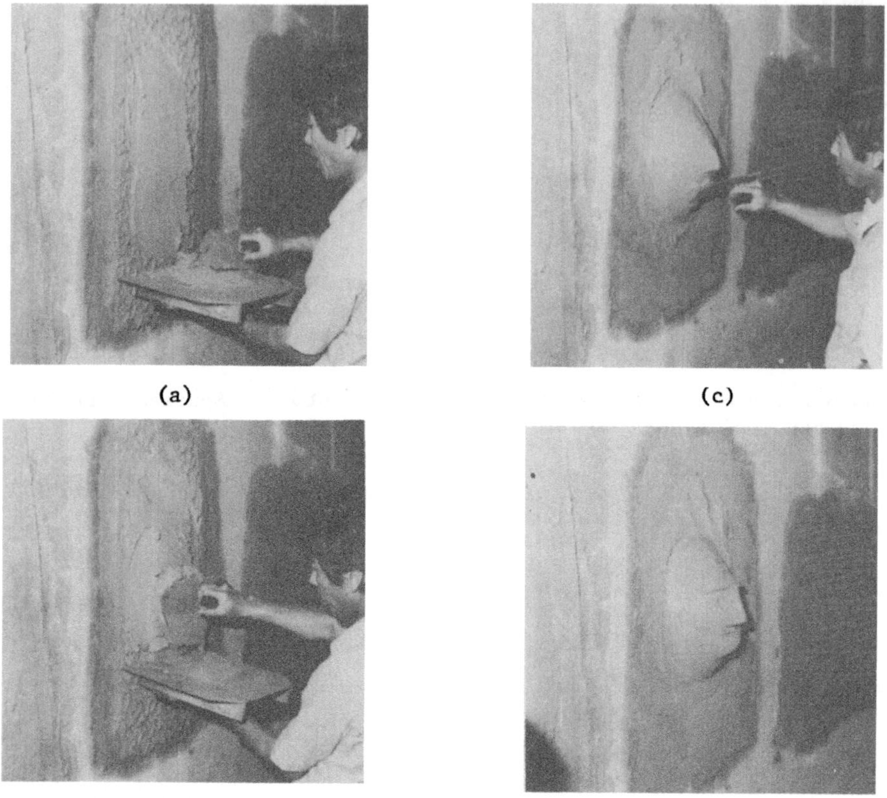

(a) (c)

(b) (d)

Photo 6 : Durée de 5 minutes pour faire un dome avec plusieurs passes, épaisseur de 110 mm au sommet de dome et hauteur jusqu'au nez de masque, 135 mm

ractéristiques. Ces améliorations ont été vérifiées par des essais.

Remerciements

Cette étude a été réalisée grâce à une collaboration belgo-japonaise. L'Auteur exprime ses remerciements à Monsieur le Professeur K. GAMSKI et à Monsieur L. HENNO, Administrateur à l'Institut du Génie Civil de l'Université de Liège - Belgique.

Bibliographie

Naniwa, R. (1984)Mortiers en monocouche épaisse. Journal du CIB "Building Research & Practice", Mai/Juin, pp.164-169.

Maruyama, E. et al. (1984)Workability and application of thick setting polymer-modified cement mortar in finishing and repairing works. Proc. Int. Sym. on Future for Plastics in Bldg. & Engineering.

PROTECTION CONTRE LA CORROSION DES ARMATURES APPORTEE PAR LES MORTIERS DE REPARATION DE SURFACE A BASE DE POLYMERES

A.M. PAILLERE LABORATOIRE CENTRAL DES PONTS ET CHAUSSEES PARIS
D. COCHET LABORATOIRE REGIONAL DES PONTS ET CHAUSSEES DE STRASBOURG
J.J. SERRANO LABORATOIRE CENTRAL DES PONTS ET CHAUSSEES PARIS

Résumé

Lorsqu'une réparation de surface est effectuée, le béton dégradé est éliminé et remplacé par un mortier de réparation qui se trouve en contact du béton et de l'acier mis à nu.
L'étude a consisté à simuler des réparations comportant des aciers en contact en partie avec du ciment (béton durci non dégradé) et en partie avec du polymère ou du liant hydraulique mixte ciment-polymère.
Après réparation et durcissement du mortier appliqué, les éprouvettes ont été placées en ambiance agressive, eau potable, eau pure, eau sulfatée, eau de mer, eau magnésienne, avec immersion totale ou semi-immersion alternée, et en atmosphère naturelle.
La progression de la corrosion a été contrôlée par :

- mesures électrochimiques
- examens visuels
- détection des ions agressifs.

Les résultats sont les suivants :

- immersions totales : aucune corrosion des aciers
- semi-immersion alternées :
  - . très forte corrosion des aciers en contact avec le mortier de polymère sans primaire,
  - . forte corrosion des aciers en contact avec le mortier de polymère avec primaire,
  - . corrosion nulle pour les aciers en contact avec les mortiers à liant hydraulique mixte quel que soit le milieu agressif, exception faite de l'eau de mer où l'on a noté une oxydation faible des aciers.

Mots clés : Corrosion, Armatures, Mortiers, Réparations, Liants hydrauliques, Polymères.

1 - INTRODUCTION

Lorsque le matériau béton se dégrade soit à cause de facteurs internes (gel-dégel, alcali réaction, etc...) ou externes (milieux agressifs par exemple), soit à cause de facteurs accidentels (incendie, chocs, etc...) on est amené à procéder, lors de la réparation, à l'élimination du béton dégradé jusqu'à la partie de l'ouvrage considérée comme saine, et à la reconstitution, du matériau éliminé, avec des mortiers ou bétons à base de polymère, ou de liant hydraulique modifié avec des polymères, ou de liant hydraulique traditionnel (ciment avec ou sans adjuvants).

Si la stucture est en béton armé ou en béton précontraint on se trouve en présence, dans la majeure partie des cas, d'armatures apparentes, l'acier ayant été mis à nu par la dégradation ou par l'élimination du béton dégradé.

Dans ces conditions, cet acier est enrobé d'une part, par le béton de la structure d'autre part, par le mortier ou le béton de réparation. Dès lors, la question se pose, de la protection contre la corrosion de ces armatures, dont une partie est en contact avec le milieu passivant connu : le béton hydraulique, et l'autre partie avec un milieu dont on ignore exactement son pouvoir de protection.

En effet, les mortiers polymères sont réputés étanches et adhérents mais non passivants, et en cas d'infiltrations d'eaux agressives la protection des aciers est remise en question.

C'est pour apporter une réponse à ce problème, que nous avons entrepris la présente recherche, qui a consisté à simuler des réparations comportant des aciers en contact, à la fois, avec du béton hydraulique et du mortier de polymère ou à base de liant hydraulique modifié avec un polymère.

2 - CONDUITE DE LA RECHERCHE

L'expérimentation a porté sur des éprouvettes de béton et de mortier armées avec des aciers Ø 7 mm et soumises à des milieux environnants agressifs.

2.1) Béton armé

On a confectionné des demi-éprouvettes prismatiques 10x10x40cm avec un béton de qualité (résistance à 28 j de l'ordre de 50 MPa) dosé à 400 Kg de ciment par mètre cube, et avec un rapport $\frac{E}{C}$ (eau/ciment) de 0,47.

Dans chaque prisme on a disposé 4 armatures au niveau des angles, à des épaisseurs d'enrobage de 1 ou 2 cm, comme indiqué dans la fig. 1

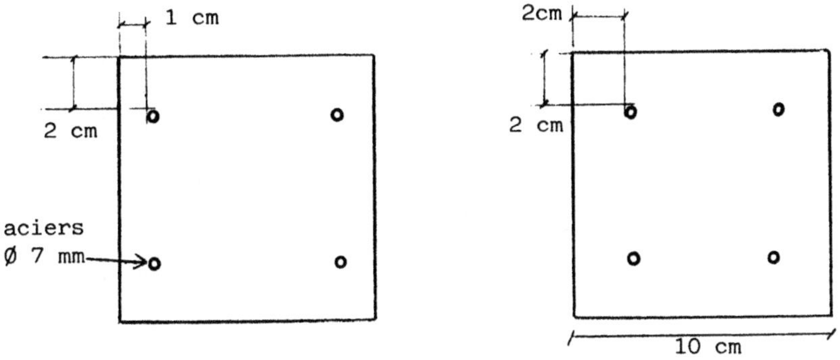

Fig. 1 : Positionnement des aciers Ø 7 mm dans les prismes 10 x 10 x 40 cm de béton.

A 20 jours d'âge du béton hydraulique on a complété ces demi-prismes en coulant le mortier de réparation comme indiqué dans la fig. 2.

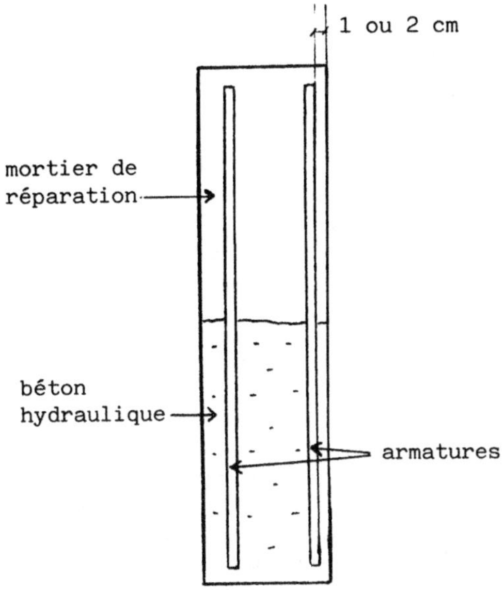

Fig. 2 : Schéma d'une éprouvette prismatique 10 x 10 x 40 cm armée avec 4 aciers Ø 7 mm et reconstituée avec un mortier de réparation.

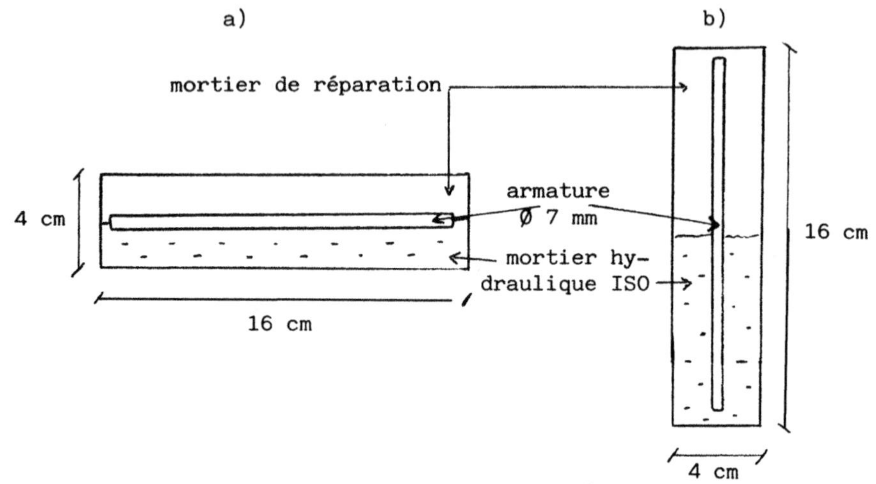

Fig. 3 : Eprouvettes de mortier ISO reconstituées avec le mortier de réparation.
a) Le mortier de réparation a été appliqué tout le long de l'acier sur la face 4x16 cm de l'éprouvette prismatique.
b) Le mortier de réparation a été appliqué sur la face 4x4cm de l'éprouvette prismatique.

A 28 jours d'âge du mortier de réparation on a soumis les éprouvettes ainsi reconstituées, pendant 1 an, aux milieux agressifs suivants :

. eau de mer artificielle dosée à 30 g/l de NaCl, 5 g/l SO_4 Mg, 6 g/l $MgCl_2$, 1,5 g/l SO_4 Ca,
. eau magnésienne dosée à 30 g/l de SO_4 Mg,
. eau pure déminéralisée,
. atmosphère de SO_2,
. intempéries (exposition en atmosphère naturelle),
. eau potable.

A l'exception des éprouvettes exposées à une atmosphère riche en SO_2 et aux intempéries, les autres éprouvettes ont été posées verticalement sur une face 4x4 cm et à demi immergées dans l'eau agressive pendant une semaine, puis retournées et placées sur la face opposée pendant la semaine suivante. Cette immersion alternée a duré 1 an.

Cette méthode a été utilisée dans le but de permettre une absorption capillaire et une évaporation alternatives, par le mortier hydraulique et le mortier de réparation.

Les mortiers de réparation étudiés étaient les suivants :

. Mortier avec résine époxydique A mélangée à un sable silico-calcaire dans le rapport 1/4.

. Mortier avec résine époxydique B mélangée à un sable silico-calcaire dans le rapport 1/4.

. Mortier avec résine époxydique C mélangée avec une charge dans le rapport 1/3.

. Mortier avec résine époxydique D mélangée à un sable siliceux dans le rapport 1/4.

. Mortier avec un polyméthacrylate de méthyle avec un fin granulat siliceux à raison de 15 Kg pour 2 litres de résine.

. Mortier hydraulique modifié avec une résine styrène butadiène ayant la composition suivante :

1 partie de ciment
½ partie de résine en émulsion
3 parties de sable normal ISO.

A 28 jours d'âge de la réparation les éprouvettes ont été immergées totalement dans l'eau de mer artificielle (à 30 g/l de NaCl) pendant 4 mois.

22) Mortier armé

La composition du mortier hydraulique était celle du mortier AFNOR ou ISO. Le ciment était, pour les mortiers soumis aux eaux agressives, un ciment résistant aux eaux à forte teneur en sulfates (type V ASTM), pour les autres mortiers un CPA 55 à 11 % de C_3A.

On a confectionné des ½ prismes 4x4x16cm en mortier hydraulique avec une armature Ø 7 mm placée au centre du prisme.

Le mortier de réparation reconstituant le prisme a été coulé comme indiqué dans les figures 3a et 3b sur les demi prismes de mortier hydraulique âgé de 28 jours.

Les mortiers de réparation expérimentés ont été :

- le mortier avec résine époxydique A,
- le mortier hydraulique modifié avec styrène butadiène (latex).

3 - PRINCIPAUX RESULTATS

Le contrôle de la corrosion des armatures a été effectué soit par :

- La mesure de pénétration des chlorures, par la méthode de dosage potentiométrique (argentimétrie).

- L'examen visuel

  A la fin de la période d'immersion totale, ou de semi-immersion alternée, les éprouvettes ont été rompues et on a procédé à l'extraction et examen des armatures.

31) Bétons armés

311) Pénétration des chlorures :

Le tableau récapitulatif n° 1, fait apparaître une pénétration et dosage des ions en Cl^- pratiquement nuls dans le cas des mortiers de résine époxydique, une pénétration légèrement supérieure dans le cas de résine polyméthacrylate de méthyle. Enfin une plus grande pénétration dans le cas des mortiers à base de liant hydraulique modifiés ou traditionnels, les valeurs les plus élevées correspondant à ces derniers.

TABLEAU RECAPITULATIF N° 1 PENETRATION ET DOSAGE DES CHLORURES

MORTIER DE REPARATION A BASE DE	PENETRATION DES CHLORURES EN MM			PROFONDEUR DOSAGE DES CL EN %	
	A 6 MOIS	A 8 MOIS	A 12 MOIS	A 12 MOIS *	
RESINE EPOXYDIQUE	PAS DE PENETRATION	IDEM	IDEM	DE 0 A 10 MM DE PROFONDEUR 0,011 %	
	PAS DE PENETRATION	IDEM	IDEM	/	
	PAS DE PENETRATION	IDEM	IDEM	/	
	PAS DE PENETRATION	IDEM	IDEM	/	
METHACRYL	PAS DE PENETRATION	IDEM	IDEM	DE 0 A 15 MM DE PROFONDEUR 0,0064 %	
LIANTS HYDRAULIQUES POLYMERES	1 A 2MM	3 A 4 MM	3 A 4 MM	PROF 0 A 10 MM 0,108 %	PROF 10 A 15 0,0060 %
	0 A 15 MM	22 A 25 MM	22 A 25 MM	PROF 0 A 15 MM 0,346 %	PROF 15 A 25 0,111 %
LIANTS HYDRAULIQUES	20 MM	25 MM	25 MM	PROF 0 A 15 MM 0,394 %	PROF 15 A 25 0,143 %
	20 A 23 MM	25 A 30 MM	25 A 30 MM	PROF 0 A 15 MM 0,343 %	PROF 15 à 25 0,136 %

* PAR LA METHODE DU DOSAGE POTENTIOMETRIQUE (ARGENTIMETRIE)

312) Examen visuel :

Aucune corrosion d'armature n'a été décelée. Ceci montre le pouvoir passivant des liants hydrauliques qui protègent bien les aciers de l'attaque par les ions Cl^-.

32) Mortiers armés

A la fin des cycles de semi-immersion alternée, compte tenu de l'aspect extérieur des éprouvettes (traces de rouille ou coloration brunâtre), il nous a paru inutile de procéder à la détection de la profondeur des ions agressifs. Les résultats présentés concernent donc uniquement ceux issus d'une observation visuelle des éprouvettes rompues.

Le tableau synoptique 2 présente une synthèse de l'ensemble des résultats.

Les éprouvettes se classent en deux familles distinctes : celles confectionnées avec des liants hydrauliques (support et produit de réparation) et celles comprenant un mortier de résine (support avec liant hydraulique et produit de réparation à base de résine époxydique).

Cas du mortier de réparation avec liant hydraulique modifié

On constate, quelle que soit la nature du milieu agressif environnant, qu'il n'y a pratiquement pas de corrosion des aciers. On remarque tout au plus une légère amorce de formation de rouille sur l'acier, soit à son extrémité, soit au niveau de la surface de contact mortier support - mortier de réparation (fig. 4, 5 et 6).

Cas du mortier de réparation à base de liant époxydique

Ce mortier a été appliqué systématiquement sur l'acier de deux façons :

. avec un primaire, composé de la même résine que celle du mortier sans charge, badigeonné sur la surface apparente de l'acier et sur le mortier support.

. sans primaire, le mortier époxydique a été coulé directement sur l'acier et le mortier support.

Seule les éprouvettes exposées pendant un an aux intempéries (atmosphère de Paris) n'ont présenté qu'une légère corrosion dans la partie de l'acier située au centre du mortier de réparation (fig. 7).

Toutes les armatures placées, avec ou sans primaire, dans le mortier époxydique présentent une corrosion importante et pour certaines, très avancée. C'est le cas particulier des conservations en eau magnésienne et eau de mer (fig. 8, 9, 10 et 11).

Il est à noter, que même le milieu le moins agressif qu'est l'eau potable, conduit à la corrosion de l'armature protégée par de la résine époxydique (fig. 12).

Par contre, quel que soit le milieu environnant, la partie de l'acier noyée dans la demi éprouvette en mortier de ciment est restée parfaitement protégée et ne présente, dans les cas de l'eau magnésienne et de l'eau de mer, qu'une très légère corrosion. On observe ainsi des armatures corrodées par moitié (contact avec le mortier époxydique) et intactes sur l'autre moitié (contact avec le mortier hydraulique).

TABLEAU 2

mortier de réparation
armature métallique
mortier support

pas d'oxydation
légère oxydation
forte oxydation

MORTIER DE REPARATION	TYPE DE REPARATION	MODE DE CONSERVATION					
		EAU POTABLE	EAU PURE	ATMOSPHERE DE SO_2	ATMOSPHERE NATURELLE (intempéries)	EAU DE MER	EAU MAGNESIENNE
LIANT HYDRAULIQUE MODIFIE AVEC STYRENE BUTADIENE		pas d'oxydation	légère oxydation	pas d'oxydation	pas d'oxydation	pas d'oxydation	légère oxydation
		pas d'oxydation	pas d'oxydation	légère oxydation	pas d'oxydation	légère oxydation	légère oxydation
MORTIER EPOXYDIQUE AVEC RESINE A SANS PRIMAIRE		forte oxydation	forte oxydation	forte oxydation	légère oxydation	forte oxydation	forte oxydation
		forte oxydation	forte oxydation	forte oxydation	légère oxydation	forte oxydation	forte oxydation
MORIER EPOXYDIQUE AVEC RESINE A AVEC PRIMAIRE		forte oxydation	forte oxydation	forte oxydation	légère oxydation	forte oxydation	forte oxydation
		forte oxydation	forte oxydation	légère oxydation	légère oxydation	forte oxydation	forte oxydation

4 - CONCLUSIONS

Cette recherche a confirmé les règles de l'art de la construction à savoir : lorsque les armatures en acier sont suffisamment recouvertes par un béton ou mortier hydraulique de qualité, il n'y a pas de risque de corrosion par les milieux agressifs environnants.

L'immersion totale d'un élément de béton armé de qualité, même dans une eau agressive, n'entraîne pas de corrosion des armatures.

Par contre l'immersion alternée qui fait appel aux phénomènes de succion capillaire, d'évaporation et absorption répétées, est la plus dangereuse pour la longévité des aciers, lorsqu'ils ne sont pas placés dans leur totalité dans le milieu passivant qu'est le ciment Portland. Malgré sa bonne adhérence à l'acier et faible perméabilité, le mortier époxydique ne parvient pas à empêcher les infiltrations d'eau et ions agressifs. L'acier se trouve ainsi sans protection, exposé à leur attaque.

Il importe donc, dans tous les travaux de réparation où il y a des armatures mises à nu, de protéger systématiquement les aciers avec des produits passivants, avant l'application de mortier de résine.

MORTIER DE REPARATION AVEC LIANT HYDRAULIQUE MODIFIE
(ciment + polymère)

liant hydraulique modifié(P.C.C.) | liant hydraulique pur

FIG 4:éprouvettes conservées enimmersion alternée dans l'eau pure.Absence de corrosion de l'acier dans les deux éprouvettes ;quelques traces de rouille à l'extrêmité de l'acier.

FIG 5:éprouvettes conservées en atmosphère SO_2 absence de corrosion.

MORTIER DE REPARATION AVEC LIANT HYDRAULIQUE MODIFIE
(ciment + polymère)

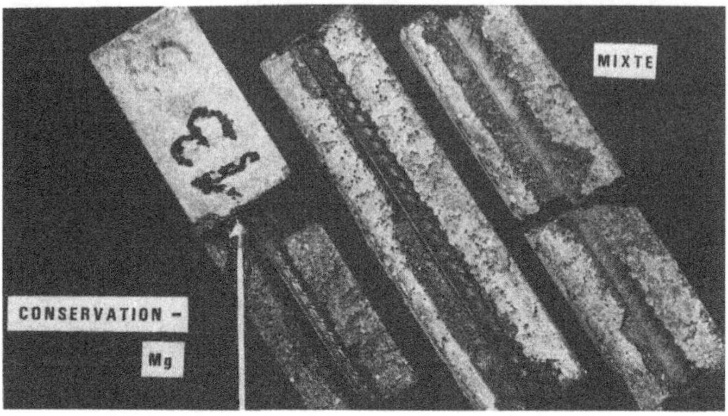

surface de contact
mortier hydraulique,
mortier P.C.C.

FIG 6: éprouvettes conservées en semi-immersion alternée dans l'eau magnésienne.Absence de corrosion de l'acier au coeur de l'éprouvette .Légère corrosion au niveau de la surface de contact du mortier hydraulique et du mortier avec ciment-polymère

MORTIER DE REPARATION AVEC LIANT POLYMERE

(résine époxydique)

(sans primaire)

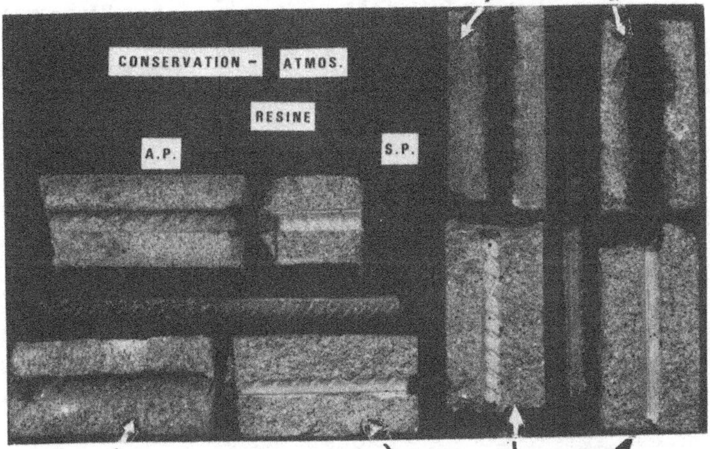

MORTIER POLYMERE
(avec primaire)

MORTIER HYDRAULIQUE

FIG 7 :éprouvettes exposées aux intempéries .Légère corrosion de l'acier enrobé de mortier polymère,absence de corrosion du côté du mortier hydraulique.

MORTIER DE REPARATION AVEC LIANT POLYMERE

(résine époxydique)

FIG 8
éprouvettes conservées en semi immersion alternée:
EAU PURE

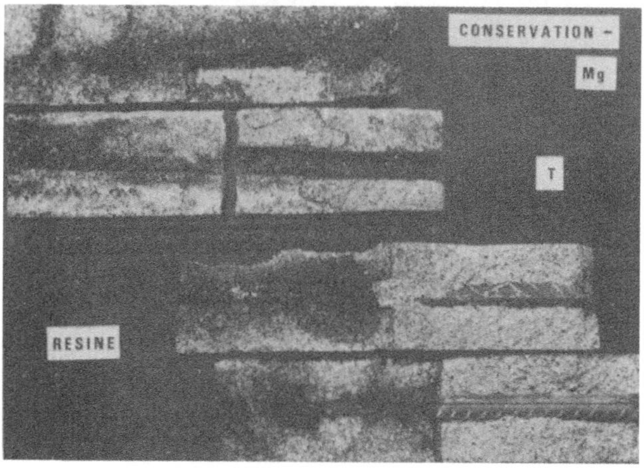

FIG 9
éprouvettes conservées en semi immersion alternée:
EAU MAGNESIENNE

On constate une notable corrosion dans la partie de l'acier en contact direct avec le mortier polymère et une absence totale de corrosion dans la partie de l'acier en contact avec le mortier hydraulique.

MORTIER DE REPARATION AVEC LIANT POLYMERE
(résine époxydique)

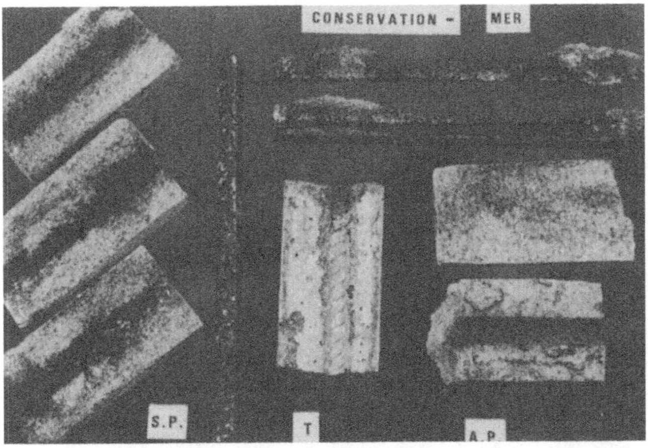

FIG 10

éprouvettes conservées en semi immersion alternée:
EAU DE MER

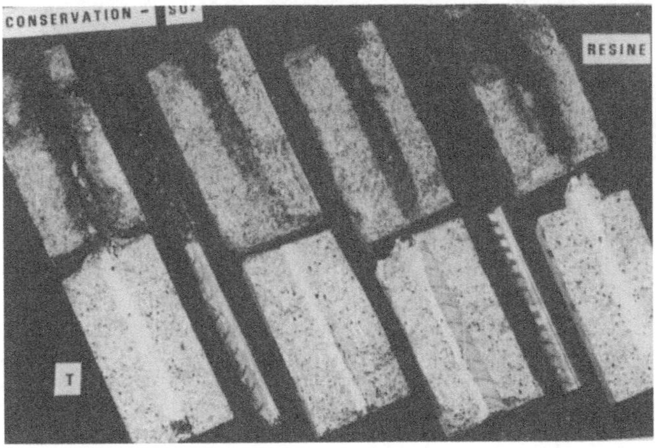

FIG 11

éprouvettes conservées en semi-immersion alternée :
atmosphère de SO_2

On constate une notable corrosion dans lapartie de l'acier encontact direct avec le mortier polymère et une absence totale de corrosion dans la partie de l'acier en contact direct avec le mortier hydrauli que

MORTIER DE REPARATION AVEC LIANT POLYMERE
(résine époxydique)

FIG 12 :éprouvettes conservées en semi immersion en EAU POTABLE

(même processus de corrosion que dans les autres milieux agressifs)

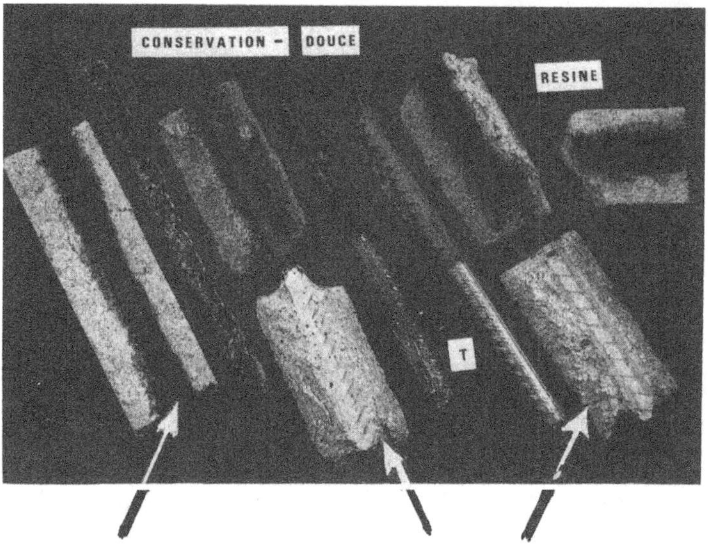

MORTIER POLYMERE   MORTIER HYDRAULIQUE

BEHAVIOUR OF TOUGHER COATING WITH CONCRETE UNDER DIFFERENT TYPES OF LOADING

V. WEISS
Department of Concrete Structures and Bridges, Faculty of Civil Engineering, Czech Technical University

Abstract
The beneficial effect of relatively thin tougher coating on the tensile strength and extensibility of concrete and other quasi-brittle materials was discovered by our ageing . short-term static tests and explained by the energy absorption due to the deformation of the coating spanned over a critical crack. Afterwards, this approach was used in U.S.S.R. but another energy condition was considered, namely the improvement of the energy balance due to the debonding. Under this assumption, no improvement of mechanical properties of the brittle core can occur for the cases of the long-term static loading and of the repeated loading because the energy is consumed during the whole loading process and is then not available at the moment of the sudden final failure. Nevertheless, our recent tests have confirmed our first statement, i.e. that the increase in tensile strength and in extensibility is similar like under short-term static loads. Therefore, the energy absorption due to the deformation of the coating prevails that-one due to the debonding. This conclusion is also in accordance with our theoretical results and will be used for the formulation of the new unified Czechoslovak-Soviet recommendation for so-called layered structures.
Key words: Concrete, Coating, Brittle, Tough, Strength, Energy, Critical crack, Bond.

The mechanical behaviour of quasi-brittle coarse heterogenous materials like concrete is rather influenced by co-action with tougher materials. If strains of the different co-acting materials are fully or greatly equalized at least at their contact planes then the quasi-brittle material becomes more "plastic" (better: pseudo-plastic), more strong and more tough, related to its original properties. Two examples of such a cooperation are here presented:

(a) Substantial increase in deformability of concrete specimens tested in a parallel arrangement with stiff elements (made e.g. of steel). In this case, complete stress-strain diagrams with descending branches up to great strains can be gained. (The same effect is obtained at testing of concrete specimens in a very stiff testing machine.)

(b) Significant increase in the crack resistance of concrete cast in situ in composite structures with prestressed concrete elements.

These phenomena, even if less pronounced, occur also in concrete elements combined with a very weak reinforcement like a protective surface coating. (A similar situation is in the concrete reinforced by closely-spaced wires or thin fibres.) Of course, such reinforcing elements cannot carry substantial forces due to their low stiffness and low bond strength to concrete (this bond strength being limited by the shear strength of concrete). Nevertheless, such weak reinforcement can substantially improve the stability of microcracks (or, more exactly, of critical structure cracks) in concrete. Therefore, the crack resistance, tensile strength and extensibility of concrete are significantly increased and the first information about it was given at the RILEM Symposium Paris 1967 by Weiss (1971).

The mentioned improvement of mechanical properties of the more brittle concrete core in cooperation with a tougher but weak coating is caused by the increased total energy absorption due to the action of this coating, particularly due to its local straining, in the region of a critical structure crack. This principle is clearly demonstrated in Fig. 1 where absorbed energies A and released strain energies U are plotted against the depth c of a critical surface crack. It is obvious that a more steep released strain energy curve U_2 yielding from a higher nominal stress σ_2 in the core must correspond to a higher fictive critical energy release rate $G_{c2} = (dA_1 + dA_r):dc$ of the system core + coating (where A_1 denotes the fracture energy absorbed in the core by development of the critical crack and A_r is the energy absorbed by the deformation of the coating over the crack) than the curve U_1 for the basic value $G_{c1} = dA_1:dc$ of the core only. Thus, the change of a critical structure crack in the core into a running crack is delayed and can occur at the higher nominal stress σ_2 only.

These facts together with adopted assumptions were proved just previously by extensive tests of beams and prisms made of various heterogenous materials (cement concrete and mortar, resin concrete, polymer-cement concrete, coarse ceramics, fused basalt) provided with thin coatings of glass-fibre-reinforced plastics (GRP), resin

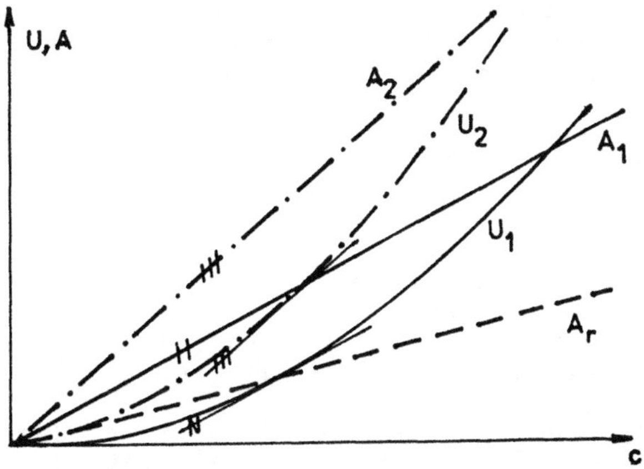

Fig. 1. Improvement of the energy balance of a crack due to the deformation of the coating

concrete or epoxy resin only, respectively, and subjected to bending or to simple tension under static short-term loading. Some pilot tests in bending on concrete beams and hourdis with GRP coating were also performed under sustained loading and the processes connected with the development of surface structure cracks were more elucidated with the aid of reflexive photoelasticity as reported by Weiss (1971) and Kocian and Weiss (1982).

The described energy action of the strained tougher coating on the quasi-brittle core has been expressed by explicite formulae; more detailed results were presented by Weiss (1973). For the simplest case of a planar crack under pure tension (plane stress state), the energy dA_1 absorbed by the core due to the growth of a critical crack by dc at the instant of its change into a running crack is to be superimposed with the energy dA_r absorbed in this region by the deformation of the adjacent front coating due to the corresponding increase in the opening of this crack. This improvement of the energy balance is expressed by the formula

$$\frac{dA + dA_r}{dA} = 1 + \frac{2R_b \sqrt{c}}{\sqrt{\pi E G_c}} \tag{1}$$

and the corresponding improved tensile strength of the core by the expression

$$R_2 = R_1 \sqrt{\frac{dA + dA_r}{dA}} \tag{2}$$

where E denotes the Young's modulus of the core and R_b is the shear bond strength between the coating and the core, the other symbols being defined yet above. Another important case for the evaluation of performed tests is represented by a planar crack under pure bending while other conditions from the previous case remain unchanged. A similar but more complicated formula

$$\frac{dA + dA_r}{dA} = \frac{2R_b \sqrt{e}}{\sqrt{\pi E G_c}} \frac{\sqrt{\frac{c}{e}} \left(1 - \frac{c}{e}\right)}{\sqrt{1 - \frac{3c}{2e} + \frac{2c^2}{3e^2}}} \tag{3}$$

was here established for the improvement of the energy balance with e = distance of the neutral axis from the tensioned edge of the cross-section, whereas the same formula 2 is valid for the calculation of the improved bending strength.

Our energy approach described above was later generally used in U.S.S.R. by Davydov et al. (1977) for design of so-called layered structures but a different governing energy condition was adopted, namely the improvement of the energy balance of a critical surface crack due to the debonding between the tougher coating and the more brittle core in the vicinity of this crack according to Potapov et al. (1973). Further, with regard to our work, the Soviet solution was in several other assumptions different:

Development of a critical crack of a parabolic shape at the tensioned face of the beam (provided with a coating) towards its sides.
Consideration of the plane-strain state.
Assumption of the plastic strain energy (corresponding to the rectangular stress distribution in the tensioned zone of a concrete beam according to the Soviet standard instead of the elastic-one.
A numerical error in the integration of the energy absorbed by the failure of the bond between the coating and the core was made (as a matter of fact, this energy is only half of the calculated-one).
Assumption for the strain in the coating over the debonded region to be equal to $R:E_r$ where R is the tensile strength of the core and E_r is the Young's modulus of the coating. (The introduction of the tensile strength for bending instead of a greater value of the bending strength corresponds to the adopted rectangular tensile stress distribution for concrete.)

Thus, the recommended Soviet formula for the increase in the tensile strength (see above) of the core by

$$R_2 - R_1 = \frac{2{,}55 G_c E_r}{R_1^2} \tag{4}$$

is quite different from our formulae 3,2 and gives greater values of the increased strength.

Let us solve the problem of the improvement of the energy balance of a critical surface crack by debonding more exactly.

Beginning this work, it is necessary to subject the mentioned further assumptions in the Soviet solution to a critics. The first of them, the growth of the critical crack towards the sides of the beam instead to the neutral axis, is realistic only for such structural elements where the quasi-brittle part made e.g. of concrete under the tougher coating has only a limited thickness and is followed again by a tougher material reaching further to the neutral axis. This case is not too usual in the practice where mostly compact concrete elements coated by a thin protective surface layer occur. The second assumption concerning the plain-strain state causes only a reduction in the formulae for the energy balance by the factor 1: $\sqrt{1-\nu^2}$ where ν denotes the Poisson's ratio. The third of these assumptions is more formal, but, as a matter of fact, it means that the task of bending is changed into the task of tension. The error in the calculation of the debonding energy does not need to be commented. The last Soviet assumption about the strain in the coating over the debonded area can be fulfilled only in very special cases. Usually, the coating should be sufficiently strong to assure a safe spanning of individual cracks, and, therefore, its stresses and strains are then far below the values needed for its fracture. Thus, the assumption used in our previous solution and considering the equilibrium of the tensile force in the coating with the shear force determined by the bondstrength and by the size of the unloaded region at the critical crack is here much more pertinent.

Now, it is possible to perform the derivation of the formula for the improved energy balance of a critical crack due to the debonding in its vicinity. For comparison with our old results and with the Soviet formula 4, the case of simple tension will be investigated.

As known, the critical planar crack of semi-elliptical cross-section is bound by the condition

$$\pi \varepsilon_1^2 E c_1 = G_{c1}$$

wherefrom the nominal strain in the core is

$$\varepsilon_1 = \sqrt{\frac{G_{c1}}{\pi E c_1}} .$$

The opening of this crack at the surface of the core is determined by

$$s_1 = 4 c_1 \varepsilon_1 = \sqrt{\frac{G_{c1} c_1}{\pi E}} .$$

The strain of the coating over this crack according to the equilibrium condition mentioned above is given by the expression

$$\varepsilon_{r1} = \frac{R_b c_1}{t_r E_r}$$

where t_r is the thickness of the coating. Hence, the length L_1 of the debonded area between the core and the coating is

$$L_1 = \frac{s_1}{\varepsilon_{r1}} = 4\sqrt{\frac{G_{c1}}{\pi E}}\,\frac{t_r E_r}{R_b\ \sqrt{c}} .$$

Now, if the critical crack c_1 undergoes to the unstable growth and increases to $c_2 = c_1 + dc$, then the debonded length will be

$$L_2 = \frac{s_2}{\varepsilon_{r1}} = 4\sqrt{\frac{G_c}{\pi E}}\ \frac{t_r E_r}{R_b\sqrt{c_1}}\ \frac{c_2}{c_1} .$$

The absorbed energy corresponding to this change in the debonded length is

$$dA_b = G_{cb}(L_2 - L_1) = G_{cb}4\sqrt{\frac{G_{c1}}{\pi E}}\ \frac{t_r E_r}{R_b c_1\sqrt{c_1}}\ dc$$

(where G_{cb} = critical energy release rate for debonding) and the energy

$$dA_1 = G_{c1}dc$$

absorbed in the core by the increase dc in the length of the critical crack is in this way changed according to the formula

$$\frac{dA_1 + dA_r}{dA_1} = 1 + \frac{G_{cb}}{g_{c1}}\,\frac{4t_r E_r}{R_b c_1}\sqrt{\frac{G_{c1}}{\pi E c_1}} . \qquad (5)$$

The increase in strength in this case can be computed again using the formula 2.

The calculated improvements in the energy balance according to the formula 5 for the debonding mechanism are substantially smaller than those according to the formula 1 for the straining mechanism. Consequently, the same statement is qualitatively valid for corresponding strengths according to the formula 2 as so as for some further mechanical properties. Further, there is still another important factor to be respected: if the energy consumption by debonding was decisive for the improvement of the tensile (or bending) strength of the core, then this mechanism would be inefficient for the cases of long-term static loading or of cyclic loading because energy would be consumed in a great extent continously or step-wise during the whole loading process and would not be available at the moment of the sudden final failure of the loaded structural member.

To more elucidate these questions, we have undertaken new extensive tests of concrete beams with a resin concrete coating under long-term static loading and with a GRP coating under cyclic loading (i.e. in fatigue). Obtained results in the cases of a good bond between the co-acting parts were very close to those of our previous tests under short-term static loading and in accordance with the formulae 3,2 (see Fig. 2), and, in this way, they have unambigously confirmed our first hypothesis that the energy consumption due to deformation of the tougher coating spanned over a critical crack in a more brittle core is the most important process governing the improvement of the tensile (or bending) strength of this core and that the effect of energy absorbed by destruction of the bond between the coating and the core is only of second order and can be neglected.

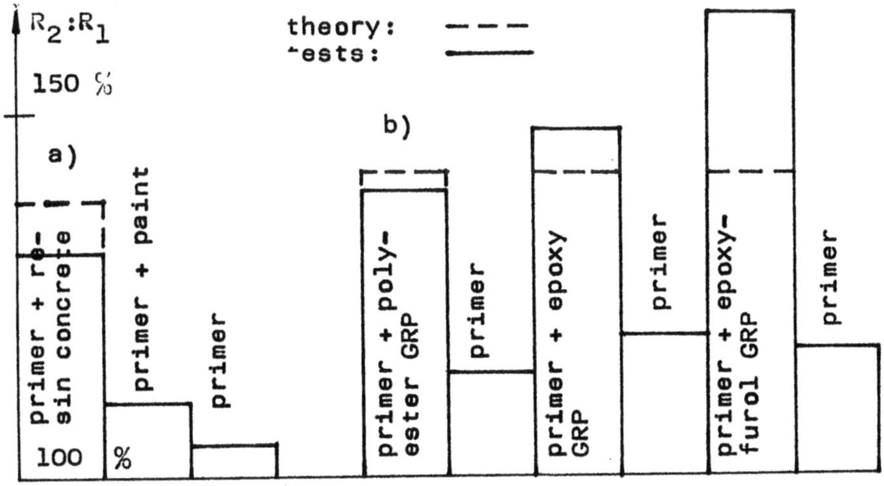

Fig. 2. Increase in the bending strength due to the action of a tougher coating (average values):
a) under long-term static loading
b) under cyclic loading

These conclusions are rather important for all cases in the building practice where tough coatings with a sufficient bond to concrete or similar heterogenous materials, respectively, are used as a protection of such structures. Therefore, the obtained theoretical and experimental results will be also exploited in the future for the work on new unified Czechoslovak-Soviet recommendation for design of layered structures.

References

Davydov, S.S. and Solomatov, V.I. and Shvidko, Ya.I. and Krasnovskaya, T.A. (1977) *Armoplastbetonnye konstruktsyi. Zhelezobetonnye konstruktsyi s polymernymi pokrytiyami*. MIIT, Moscow

Kocian, V. and Weiss, V. (1982) The use of photoelastic foils for stress and strain investigation of concrete beams. *Cement and Concrete Research*, 12, 497-510.

Potapov, Yu.B. and Solomatov, V.I. and Seliayev, V.P. (1973) *Polymernye pokrytiya dlya zhelezobetonnych konstruktsyi*. Stroyizdat, Moscow.

Weiss, V. (1971) The behaviour of concrete with reinforced plastics coating, in *Les résines de synthèse dans la construction. Synthetic resins in structures*. Eyrolles, Paris, tome 2, pp. 569-580.

Weiss, V. (1973) Failure of brittle materials with tougher surface coating, in *Dritte Internationale Tagung über den Bruch*, VDE, Düsseldorf, tome VIII, pp. 335 I-6

Section 6

Bonding of fresh to old concrete

Adhérence du béton frais au béton ancien

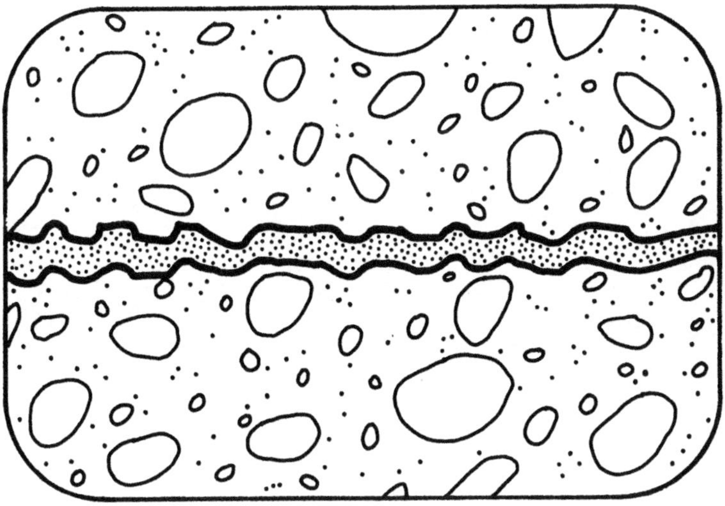

ASPECTS DES RECHERCHES ET REALISATIONS PRATIQUES DANS LE DOMAINE DE LA REPARATION DES ELEMENTS PREFABRIQUES ET MONOLITHES EN BETON

ION IONESCU et TRAIAN ISPAS
Institut des Recherches et Projection pour Matériaux de Construction, Bucarest

Résumé
L'ouvrage présente condensé les suivants aspects: quelques résultats des recherches de laboratoire concernant l'adhérence béton de diverses classes-polymères, l'adhérence béton-béton aux polymères, résultats obtenus aux travaux effectués pour la réparation de divers éléments préfabriqués et monolithes en béton. Les observations sur le comportement en exploitation des éléments préfabriqués et monolithes en béton, réparés, pendant 10-25 années; quelques conclusions pour la prochaine activité.
Mots clés: Béton, Dégradations, Polymères, Réparations, Ségrégations, Fissures, Pellicules.

1. Introduction

Les causes qui peuvent produire les dégradations pendant l'exécution des éléments préfabriqués et monolithes en béton, au transport et au montage des éléments préfabriqués sur les chantiers et aussi dans l'exploitation des travaux de constructions sont nombreuses. Parce que une analyse détaillée n'est pas possible dans l'ouvrage présent on mencionnera par la suite quelques causes, plus fréquentes, des dégradations, comme: erreurs de conception (inclusivement erreurs concernant la choix des matériaux adéquants, projection technologique, projection des compositions de béton etc.); erreurs d'exécution, erreurs au transport, magasinage sur les chantiers et montage des éléments préfabriqués, entretien défectueux; erreurs concernant le changement de la destination ou des programmes initials d'exploitation de certaines constructions et d'adoption des mesures techniques qui s'imposent; sollicitations accidentales au caractère particulier. L'analyse des dégradations des éléments préfabriqués et monolithes doit être effectuée toujours avec une attention particulère parce que les différentes causes peuvent produire des effets similaires ou les mêmes catégories de causes peuvent avoir de nombreux effets. En général, les dégradations des éléments préfabriqués et monolithes en béton peuvent être groupées aproximativement en dégradations de surface (grâce aux certaines erreurs d'exécution, choc au décoffrage, transport et montage, actions de l'eau et du gel sur le béton, corrosion de l'armature etc.); dégradations dans la masse du béton (grâce aux erreurs d'exécution, inclusivement l'utilisation des agregats de faible qualité grâce aux certaines réactions chimiques entre liants et agregats ou liants et eaux agressives etc.) et dégradations d'origine mécanique (à cause de la non-réalisation des performances du matériel ou de la sursollicitation du matériau permanent ou accidental etc.). L'analyse technico-économique concrete dans chaque situation dans laquelle sont produits les dégradations indique dans la plupart de cas la nécessité et la possibilité du rétablissement de la capacité d'utilisation (réparation) des éléments préfabriqués ou monolithes en béton simple, béton armé ou béton précontrait. Dans l'ouvrage présent on analyse seulement les dégradations qui se produisent pendant l'exécution des éléments préfabriqués et monolithes, aussi comme au transport et montage des éléments préfabriqués en béton, on mencionne:

(a) Ségrégation grâce aux certaines erreurs de projection soit de dosage des composants du béton soit à cause de la vibration nonadéquante du béton, nonétanchement des moules ou décoffrages etc.;
(b) Vides en section, ruptures aux bords et aux coins au décoffrage, ruptures au transport, magasinage défectueux des éléments préfabriqués;
(c) Fissures grâce au traitement thermique nonadéquant, sollicitations défectueuses au transport, magasinage et montage des éléments préfabriqués.

2. Quelques résultats des recherches effectuées au laboratoire

Les recherches de laboratoire pour la détermination de l'adhérence ont contenu, en général, essais au compression sur eprouvettes prismatiques 10 x 10 x 55 cm fig. 1.a-b, essais au flexion fig. 2.a-c, essais après la méthode brasilienne sur cylindres 16 x 32 cm fig. 3.a-b, essais au cisaillement fig. 4.a-b, essais aux cycles de gel-dégel répétés (un cycle égal 4 heures dans l'eau à 20 $\pm$ 2°C et 4 heures à -20 $\pm$ 3°C) aussi comme quelques sollicitations répétées 100-300 fois à la compression et flexion pour des charges de 0,30 - 0,36 de l'effort de rupture des eprouvettes témoin. L'essai de certains planchers et poutres en béton armé et précontrait qui

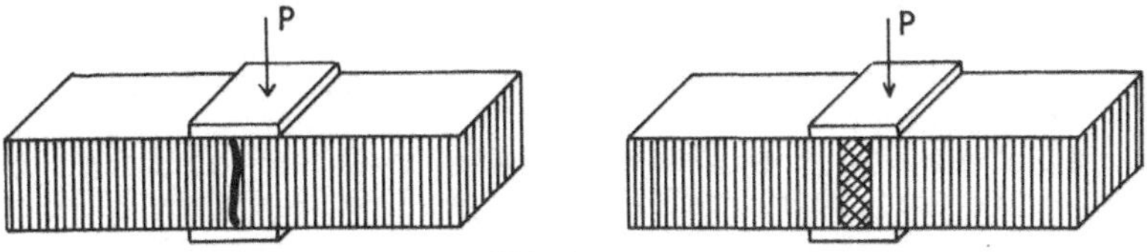

FIG. 1

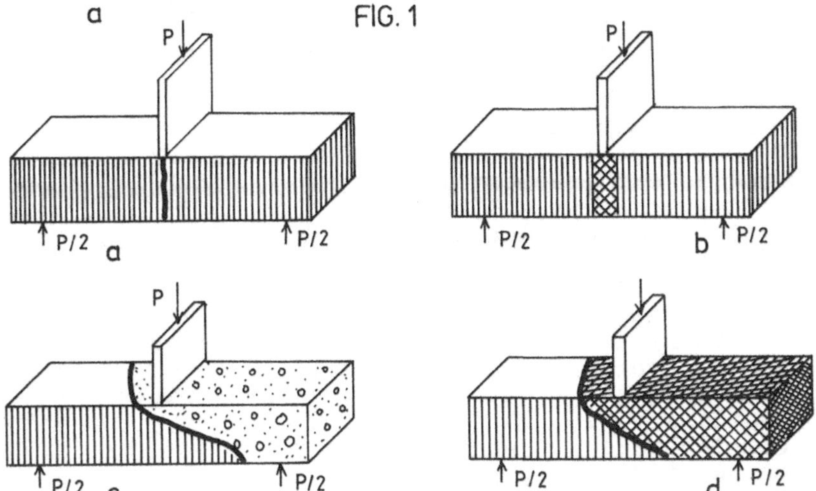

FIG. 2

FIG. 3

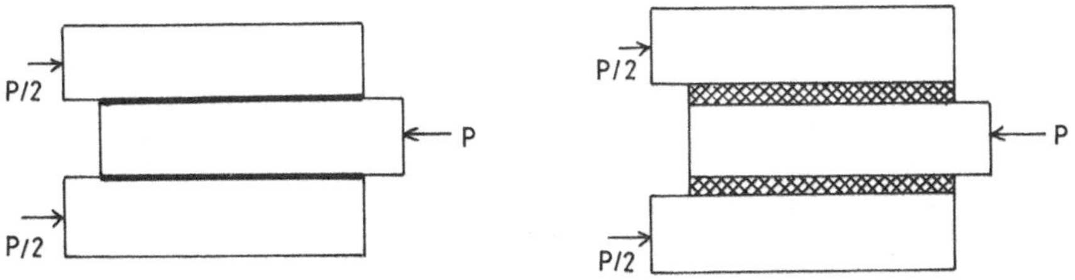

FIG. 4

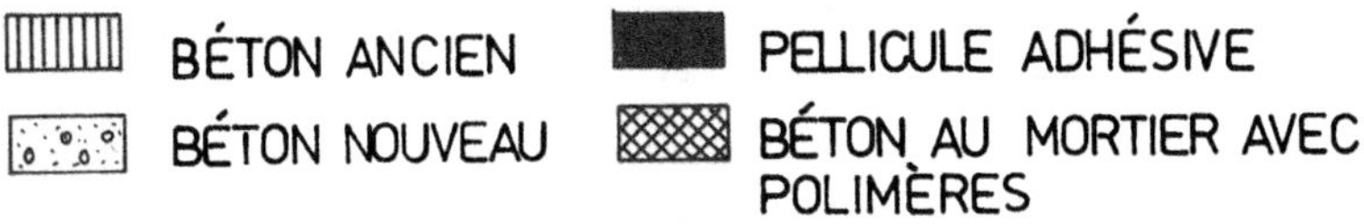

ont souffert dégradations en diverses étapes du processus de fabrication après 7-28 jours de la réparation. A la réalisation des recherches ont été utilisé les suivants matériaux:

- mélanges d'agregats naturels o-1, o-5, o-1o, o-16, o-31 mm qui accomplissaient les exigences des normes roumaines pour le béton et des récommandations pour la réparation des bétons aux polymères;
- matériau de remplissage broyé à une surface spécifique (Blaine) de 25oo - 28oo cm²/g (filler obtenu par la mouture des calcaires, quartz, andésites basaltiques et coke de haut fourneau);
- ciments portland normales P4o - P5o (PN), ciments portland au durcissement rapide P4o - P5o (PR), ciments portland avec maximum 15-2o% additions Pa35 (PA), ciments portland résistants aux sulfates SR 4o (SR);
- divers types de produits sur base des polymères (résines sinthétiques) des groupes acetate de polyvinyle (P.V.A.), émulsion E5o mélamino-formaldéhydes, polyesters et époxides. De tous les produits mencionnés, après les essais préliminaires ont été retenu pour amples recherches et utilisation en production: E5o, superplastifiant VIMC 11 (de la groupe des résines mélamino-formaldéhydes) et aussi quelques produits de la groupe des résines époxides Dinox o4oA, Dinox olo5, Dinox oloC;
- cendre des électrofiltres de centrale thermique type A ou B.

Toutes les recherches ont commencé de l'idée que le rétablissement de la capacité d'utilisation (réparation) des éléments préfabriqués et monolithes en béton doit être réalisé aux matériaux qui assurent une classe de résistance supérieure au béton utilisé initialement à la réalisation des éléments préfabriqués et monolithes en béton. Par exemple, pour les bétons des marques B2oo - B25o utilisées à la réalisation des éléments préfabriqués pour constructions de logements le béton (au ou sauf addition des polymères) utilisé pour les réparations doit assurer une résistance minimum conforme à la marque B3oo, à la réparation de certains éléments en béton précontrait de marque B6oo, le béton fin aux polymères utilisés pour les réparations doit assurer une résistance, au moins, à la marque B8oo etc.
Aussi on a eu en vue les conditions réelles d'exploitation des éléments préfabriqués et monolithes en béton auxquels étaient nécessaires les réparations. Quelques résultats des recherches sont présentés dans les tableaux 1-3. Chaque serie d'eprouvettes pour des essais (excepté les essais au cissaillement) a contenu trois éléments témoin (sauf réparation, pour comparaison) et trois éléments réparés. Les essais se sont effectués au 3, 7, 28 et 365 jours après la réalisation des

réparations. Les eprouvettes en béton utilisées à l'exécution des réparations avaient à la date de l'opération l'âge de 14-28 jours. L'essai des eprouvettes témoin a été effectué de la même manière que l'essai des eprouvettes réparées.

Tableau 1. Quelques résultats des recherches concernant l'adhérence des pellicules et bétons au ciment et polymères

Serie	Eprouvette	Essai	Ciment	Polymère	Résistance daN/cm² par jour			Mode de rupture
					7	28	360	
1.	Témoin	Compress.	P.N.	-	178	261	304	Normal
2.	Réparée	1.a	P.N.	E50	196	272	316	Normal
3.	Réparée	1.b	P.N.	E50	209	280	329	Normal
4.	Réparée	1.b	P.N.	VIMC 11	181	258	309	Normal
5.	Témoin	Compress.	P.R.	-	227	262	304	Normal
6.	Réparée	1.a	P.R.	E50	216	270	291	Normal
7.	Réparée	1.b	P.R.	E50	246	285	327	Normal
8.	Témoin	Flexion	P.N.	-	23,8	27,2	28,6	Normal
9.	Réparée	2.a	P.N.	E50	19,5	26,1	27,9	Dans la zone de joint
10.	Réparée	2.b	P.N.	E50	24,0	28,1	29,4	"
11.	Réparée	2.c	P.N.	E50	23,5	28,1	29,7	"
12.	Réparée	2.d	P.N.	E50	25,0	30,4	35,6	"
13.	Témoin	Brasil.	P45	-	17,5	18,2	18,8	Normal
14.	Réparée	3.a	P45	E50	17,3	18,6	19,5	En dehors de joint
15.	Réparée	3.b	P45	E50	18,2	18,9	20,7	"
16.	Réparée	4.a	SR	E50	26,8	29,7	30,5	A côté de joint
17.	Réparée	4.b	SR	E50	28,4	30,9	33,6	En dehors de joint

Les dates du tableau 1 demontre que, dans les conditions de la préparation correcte des surfaces et protection adéquate des travaux réparés pendant 3-7 jours, après l'exécution de la réparation, les éléments préfabriqués et monolithes en béton simple et béton armé de marque maximum B300 (inclusivement B300-CT) ont été soumis en exploitation à l'humidité permanente de l'ambiant de maximum 7o% et aux sollicitations de compression ou compression et flexion comme les grands panaux préfabriqués pour les murs de constructions de logements, industrielles, poteaux des bâtiments industriels etc., peuvent être réparés en conditions techniques avantageuses utilisant les solutions indiquées dans la fig. 1-4 et tableau 1.

Tableau 2. Les résultats des recherches concernant l'adhérence des pellicules et bétons aux résines epoxides

Serie	Eprouvette	Essai	Polymère	Materiau	Résistance daN/cm² par jour 3	28	365	Lieu de rupture
1o1	Témoin	Compress.	-	-	464	682	746	Normal
1o2	Réparée	1.a	E4o	Pellicule	486	7o2	759	Normal
1o3	Réparée	1.b	E4o	Béton	5o2	734	786	Normal
1o4	Témoin	Flexion	-	-	32,7	36,6	38,1	Normal
1o5	Réparée	2.a	E4o	Pellicule	32,1	37,5	38,4	En dehors de joint
1o6	Réparée	2.b	E4o	Béton	34,7	38,2	39,6	"
1o7	Réparée	2.c	E4o	Pellicule	32,3	35,8	38,4	"
1o8	Réparée	2.d	E4o	Béton	36,2	39,7	4o,5	"
1o9	Témoin	Brasil.	-	-	28,5	34,6	35,8	Normal
11o	Réparée	3.a	E4o	Pellicule	28,8	35,2	36,o	En dehors de joint
111	Réparée	3.b	E4o	Béton	32,8	36,6	38,1	"
112	Réparée	4.a	E4o	Pellicule	38,9	4o,2	4o,8	"
113	Réparée	4.b	E4o	Béton	42,6	43,5	44,1	"
114	Réparée	4.b	oloS	Béton	44,5	45,8	43,3	"
115	Réparée	4.b	oloC	Béton	41,9	42,7	43,2	"

Les éléments préfabriqués en béton simple et armé de marque maximum B3oo (inclusivement B3oo-CT) aussi comme les poutres et planchers sollicités en exploitation au flexion et exploité à l'humidité permanente de l'ambiant sous 7o% peuvent être réparés en conditions technico-économiques avantageuses utilisant les schémas indiquées dans la fig. 2.b, c, d, à l'obligation de la préparation correcte des surfaces et l'assurance de la protection des réparations effectuées pendant minimum 3-7 jours après l'exécution. Les compositions des matériaux indiqués pour la réparation des éléments préfabriqués et monolithes en béton simple et béton armé de marque maximum B3oo (inclusivement B3oo-CT) en conditions mencionnées sont indiquées dans le tableau 5.

Tableau 3. Les résultats obtenus après l'exécution des cycles gel-dégel répétés

Serie	Eprouvette	Essai	Polymère	Matériau	No cycles	daN/cm² après cycles de gel	Mode de rupture
1o1	Témoin	Compress.	-	-	1oo	627	Normal
1o3	Réparée	1.b	E4o	Béton fin	1oo	685	Normal
1o4	Témoin	Flexion	-	-	1oo	32,8	Normal
1o5	Réparée	2.b	E4o	Béton fin	1oo	34,5	En dehors de joint
1o7	Réparée	2.c	E4o	Pellicule	1oo	33,6	"
1o9	Témoin	Brasil.	-	-	1oo	32,7	Normal
111	Réparée	3.b	E4o	Béton fin	1oo	34,5	En dehors de joint

Les dates des tableaux 2-3 demontrent que par l'utilisation des résines époxides on peut effectuer les réparations des éléments préfabriqués et monolithes en béton armé et précontrait, dans quelque situation dans laquelle l'analyse technico-économique préalable justifie l'exécution de tels travaux parce que les résines époxides et les travaux qui s'exécutent avec tels résines ont un coût plus élevé et demande une equipe specialisée en exécution et une surveillance technique pendant toute la durée de l'exécution. Aussi dans le cas de l'utilisation des résines époxides on demande: assurer une préparation conforme aux surfaces et protéction des travaux de réparation effectués pendant 6-24 heures en fonction de la temperature de l'ambiant.
Entre les années 1961-1984 on a effectué de nombreux essais sur des éléments préfabriqués et monolithes, en béton armé et précontrait, qui ont présenté diverses dégradations d'exécution, transport et montage etc., en résultant les suivants:
- tous les planchers en béton armé avec ségrégations ou ruptures qui ont été réparé aux pellicules de E5o

Tableau 4. Le comportement en exploitation de certains éléments préfabriqués et monolithes en béton et béton armé et précontrait réparés.

Eléments réparés	Défaut réparé	Matériau	Conditions exploitation	Année réparation	Comportement en exploitation
Plancher bâtiment	Ségrégation Rupture aux bords et coins Fissure	Pellicule E5o et ciment Béton Béton avec E5o	Edifice de logement	1961-1984	Très bon
Poutres en béton armé	Vides en section Ségrégation Ruptures aux bords et coins	Pellicule E5o et ciment Béton Béton avec E5o	Divers Types de bâtiments	1963-1968	Très bon
Poutres en béton armé	Vides en section Ségrégation Ruptures aux bords et coins	Pellicule en ciment et VIMC 11 Béton avec VIMC 11	Divers Bâtiments de divers types	1979-1982	Très bon
Poutres pour ponts-groue en béton précontrait de 12 m ouverture	Fissures	Injection avec E4o dilué	En dépôt ext.de produits finis Ponts-groue 12,5t au programme travail 16h/jour	Avril 1972	Très bon
Divers types poutres en béton précontrait	Fissures	Injection avec E4o	Divers edifices industriels	1968-1982	Très bon

et ciment et béton de ciment ou réparés au béton fin avec addition de E5o, soumis au moins de 7 jours après la réalisation de la réparation aux sollicitations égales au 1,25 charge d'exploitation, se sont comportés très bien;

- tous les poutres en béton armé réparés avec béton fin avec addition de E5o soumis aux sollicitations égales de 1,25 - 1,3o charge d'exploitation se sont comportés aussi très bien;
- tous les poutres en béton précontrait réparés avec béton fin aux résines époxides ou injectés aux résines époxides se sont comportés très bien aux sollicitations de 1,3o - 1,4o charge d'exploitation;
- à deux poutres en béton armé qui, soumis initialement aux sollicitations de rupture, concernant les essais d'agréer, sont rompus aux charges concentrées situées au milieu d'ouverture de 1o,6 t (poutre no 1) et 11,8 t (poutre no 2) après trois jours de l'injection des fissures, surgies au premier essai, au résines époxides, on a obtenu aux essais charges de rupture de 12,3 t et réspectivement 14,2 t. On demontre que par injection on a obtenu une bonne consolidation de la structure des poutres fait qui a conduit à la croissance de la capacité portante.

Les observations effectuées jusqu'à 25 années concernant le comportement en exploitation de certains éléments préfabriqués et monolithes en béton armé et précontrait qui ont été réparé, ont conduit à l'obtention des résultats du tableau 4, en résultant un très bon comportement en temps des éléments réparés, en diverses conditions d'exploitation, c'est-à-dire un élément préfabriqué ou monolithe qui a été réparé n'a pas souffert dans l'exploitation des dégradations dans les zones lesquelles on a exécuté les réparations. Certaines compositions de pellicules, bétons au ciment et addition de E5o ou VIMC 11 et pellicules en base des époxides ou bétons aux époxides utilisées aux travaux des réparations effectués sont indiquées dans le tableau 5.

3. Quelques conclusions.

Pendant 25 années des recherches et exécution des travaux pratiques et par observations effectuées dans le domaine de la réparation des éléments préfabriqués et monolithes en béton on a accumulé une importante experience qui permet, au présent, le rétablissement de la capacité d'utilisation pratique de divers types d'éléments préfabriqués ou monolithes, en base d'une analyse technico-économique préalable.

Les dates que nous disposons, au présent, concernant le comportement et la durabilité des éléments préfabriqués

Tableau 5. Compositions de pellicules et de bétons utilisées aux travaux de réparations éléments préfabriqués et monolithes en béton

Pellicules		Bétons fins	
	sur base de E5o		
E5o	= 25-3o %	E5o	= 7-9 %
Ciment	= 55 %	Ciment	= 22-28 %
Eau	= 15-2o %	Sable o-3mm	= 54-66 %
		Eau	= 5-9 %
	sur base de VIMC 11		
VIMC 11	= 8-1o %	VIMC 11	= 3-5 %
Ciment	= 75-8o %	Ciment	= 18-22 %
Eau	= 5-1o %	Agregat o-16	= 64-7o %
		Eau	= 5-8 %
	sur base de E4o		
E4o	= 7o-8o %	E4o	= 16-25 %
Diluant	= 11-18 %	Diluant	= 3-5 %
Durcisseur	= 7-12 %	Agregat o-1 ou o-3 mm	= 7o-8o %

et monolithes auxquels on a effectué les réparations aux résines synthétiques produites couramment en différentes conditions d'exploitation demontrent que une réparation correcte exécutée peut durer plusieurs décennies et assurer une exploitation en bonnes conditions des constructions.

Bibliographie

Ionescu, I. et collaborateurs, (1966) Les composants macromoléculaires dans les constructions hydrotechniques et travaux spécials. Ed. Technique Bucarest, chap. IV.

Ionescu, I., (1973) Quelques remarques sur les propriétés des bétons aux résines synthétiques et liants minéraux. Revue Roumaine de Chimie no 6, pag. 1o41-1o65.

Ionescu, I., (1974) Quelques remarques sur les propriétés des bétons aux résines synthétiques sans liants minéraux. Revue Roumaine de Chimie no 1 pag. 97-1o7 et no 2 pag. 292-32o.

Ionescu, I. et Ispas T., (1986) La Pratique actuelle des Bétons. Ed. Technique Bucarest, chap. 3, 9.

BONDING BETWEEN OLD AND NEW CONCRETE IN STRUCTURES LOADED BY STATIC AND TIME-DEPENDENT LOAD

J. SILFWERBRAND
Department of Structural Mechanics and Engineering,
Royal Institute of Technology in Stockholm

Abstract

Concrete bridge decks often show extensive damages caused by wear, de-icing salts and freeze-thaw cycles. The damaged concrete has to be removed and replaced by a new-cast concrete topping.

The bonding between old and new concrete has been studied by tests with composite concrete beams. The beams were composed of old concrete with a stripped, rough interface against a new concrete topping.

In one investigation composite concrete beams were loaded by a single load at the center of the beam. Both static and fatigue tests were carried out. The results show good bonding in both cases.

In a second investigation the effect of differential shrinkage was studied. The composite concrete beams in the tests could resist the residual stresses caused by differential shrinkage without cracks along the rough interfaces or in any of the parts of the beams. The stress relief caused by the creep, which is simultaneous to the shrinkage, is the main cause of the resistance of the concrete beams to residual stresses caused by the differential shrinkage.

Key words: Concrete bridge deck, New-cast topping, Composite structures, Bonding, Differential shrinkage, Fatigue.

1. Introduction

Concrete bridge decks often show extensive damages caused i.a. by de-icing salts and freeze-thaw cycles in combination with heavy traffic loadings. The damage starts at the surface and progresses down in the structure with carbonation and increase in cloride content. The damaged concrete has to be removed and replaced by a new-cast concrete topping. The structural interaction between the old concrete and the new-cast topping is dependent on the bonding between them and consequently studies of the bond strength are of great interest.

In one investigation composite concrete beams were loaded by a single load at the center of the beam. Both static and fatigue tests were carried out.

In another investigation differential shrinkage was studied, i.e. the shrinkage difference between the new-cast concrete topping and the old concrete. Investigations referred in /1/ have shown that the bonding between pre-cast and in situ concrete is disrupted more often at the boundary of the structure than in the rest of it. The main cause is considered to be the differential shrinkage which can cause high stresses at the end of the beam or slab. The aim of this investigation was to study this case of loading closer.

Detailed reports dealing with the two investigations are published in /5/ and /6/.

2. Beams loaded by a single load

2.1 Tests specimens

In the first investigation two homogeneous and 16 composite concrete beams were simply supported and loaded by a single load at the center. Twelve composite beams consisted of old concrete cut out of a concrete bridge deck and a new-cast concrete topping. The age of the topping was at time of loading about four months. The interfaces were rough, stripped by a pneumatic hammer. Four composite beams consisted of old and new concrete cast with an interval of two months with smooth, steel-grinded interfaces. The interface was in all cases at mid height of the composite beam. The cross section of the specimens is shown in Fig. 1. The beam span was 1020 mm. Data of the specimens are shown in Table 1. As can be seen in the table some composite beams were strengthened with stud bolts crossing the interface.

The specimens were divided in two equal series, one for static tests and one for fatigue tests.

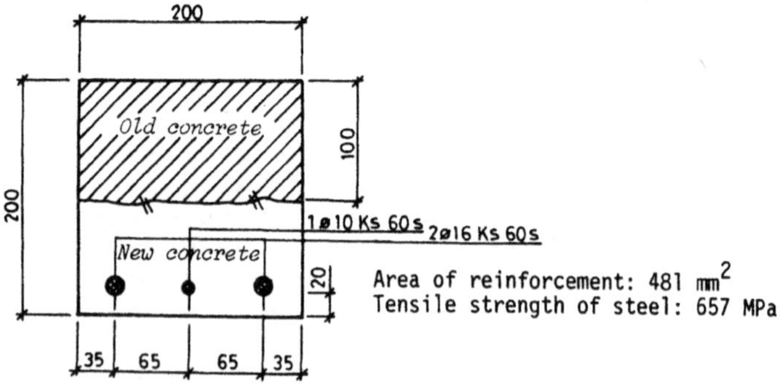

Fig. 1. Cross section of the specimens.

Table 1. Data of the specimens.

Specimen No.	Old concrete f_{ccu} (MPa)	New concrete f_{ccu} (MPa)	Interface	Stud bolts Number x diam. (mm)	f_{st} (MPa)
1	85	57	Rough	-	-
2	85	63	Rough[1])	-	-
3	85	62	Rough	8.7	466
4	85	63	Rough	8.10	436
5	85	63	Rough	8.12	362
6	85	64	Rough[2])	-	-
7	63	63	Homogeneous	-	-
8	85	61	Rough	-	-
9	85	61	Rough[1])	-	-
10	85	61	Rough	8.7	466
11	85	62	Rough	8.10	436
12	85	60	Rough	8.12	362
13	85	61	Rough[2])	-	-
14	62	62	Homogeneous	-	-
15	68	64	Smooth	-	-
16	71	62	Smooth	-	-
17	68	64	Smooth[3])	-	-
18	71	62	Smooth[3])	-	-

[1]) The interface was epoxy-treated before casting.
[2]) The fresh concrete was vacuum-treated after casting.
[3]) A special mortar (Compacta) for better adhesion was used.

Note: f_{ccu} is the ultimate compressive strength of concrete and f_{st} is the tensile strength of steel

2.2 Static tests

Nine specimens were statically loaded to failure. The ultimate loads are shown in Table 2. Failure at the interface occured only for the two beams with smooth interfaces. The specimens with rough interfaces reached at least as high ultimate loads as the homogeneous specimen. The utlimate loads of the specimens with stud bolts were increased above the ultimate load of the homogeneous specimen because of the effect of shear reinforcement.

In Table 2 the measured ultimate loads are compared with calculated ultimate loads. These are calculated according to CEB /2/.

The measured ultimate loads were in all cases greater than the calculated loads at failure at the interface.

Table 2. Measured and calculated utlimate loads.

Specimen No.	Load at failure at the interface (kN)	Ultimate load (kN)	Mode of failure	Calculated load at flexural failure (kN)	Calculated load at shear failure (kN)	Calculated load at failure at the interface (kN)
1	-	130	Shear	195	140	65
2	-	150	Shear	195	155	70
3	-	200	Shear	195	185	65
4	-	235	Flexural	195	225	80
5	-	235	Flexural	195	235	125
6	-	130	Shear	195	140	65
7	-	130	Shear	190	150	-
15	40	70	Shear	65[1]	85[1]	0
17	17	69	Shear	65[1]	85[1]	0

[1]) The loads are calculated for the lower (reinforced) half of the cross section of the beam.

The deflections at the centers of the beams were measured. The relationships between deflection and load for some of the specimens are shown in Fig. 2. The solid lines are culculated for a cracked section with a linear stress-strain distribution and a neglected tensile concrete zone. There is a good correspondence between calculated and measured values.

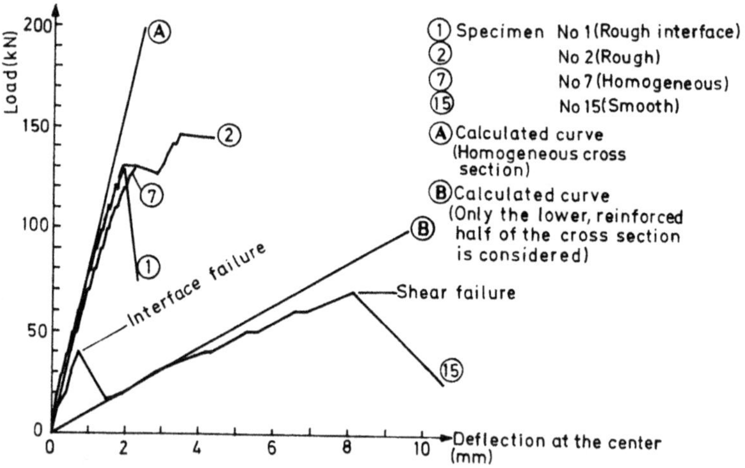

Fig. 2. Calculated and measured relationships between deflection and load.

2.3 Fatigue tests

The nine specimens in the second series were loaded to failure or to one million cycles of repeated loading. The relationship between the maximum deflection at the center of the beam and the number of cycles of repeated loading for the homogeneous specimen and for the composite specimens with rough interfaces, but without stud bolts is shown in Fig. 3. The load was sinusoidal with a maximum value of 90 kN and a mean value of 50 kN.

As shown in Fig. 3 the ratio between the increase of the deflection and the logarithm of the number of cycles was of the same magnitude for the composite specimens and the homogeneous specimen.

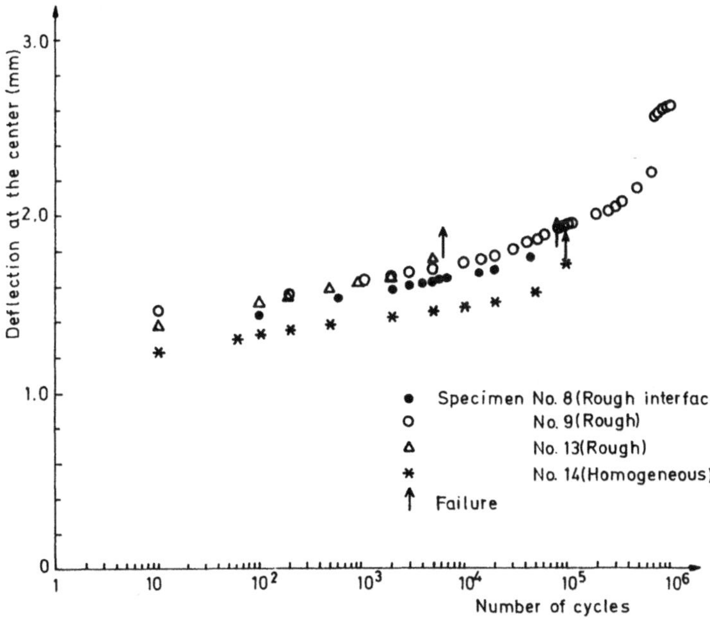

Fig. 3. Relationship between number of cycles of repeated loading and deflection.

The failures were in all occuring cases shear failures. Failure occured at 85000 cycles for specimen No. 8, at 6700 for specimen No. 13 and at 110000 cycles for the homogeneous specimen No. 14. Failure did not occur for specimen No. 9.

The interfaces were strengthened by studs in specimens Nos. 10, 11 and 12. Failure did not occur for specimens Nos. 10 and 12, but failure occurred at 813000 cycles for specimen No. 11.

According to the fatigue tests there is no reason to suppose that a composite concrete beam with a rough, stripped interface has a less resistance to fatigue than a homogeneous beam.

3. Differential shrinkage in composite concrete beams

The investigation treated tests with four composite concrete beams. The cross section was in all tests composed of old concrete with the height 250 mm and a new concrete topping with the height 100 mm. The old concrete was at the casting of the new topping more than two years old. Before casting of the new topping the interfaces were stripped by a pneumatic hammer and carefully cleaned. Two beams were supported on a soft elastic foundation in form of an air-bag and two beams were simply supported on three rigid supports. Data of the specimens are shown in Table 3.

Table 3. Data of the specimens.

Specimen No.	Beam length (m)	Beam width (m)	Beam depth (m)	Depth of the topping (m)
1 [1])	1.1	0.27	0.35	0.10
2 [1])	2.1	0.27	0.35	0.10
3 [2])	6	0.27	0.35	0.10
4 [2])	8	0.27	0.35	0.10

[1]) Simply supported on three rigid supports
[2]) Elastic foundation in form of an air-bag

The tests were carried out outdoors in a tent between september 1983 and november 1984.

It was natural to treat the effects of differential shrinkage in un-cracked beams, since no cracks were observed in the tests. Assume a constant shrinkage ε_{cs} in the concrete topping and a constant modulus of elasticity E in the whole beam. The strain ε in the beam can be calculated according to the simplest linear-elastic theory for full structural interaction between old and new concrete. The calculated strain is based on the free shrinkage measured on three reference prisms. Fig. 4 shows examples of measured and calculated strain as functions of time for some measuring points. (All 52 measuring points are shown in /6/).

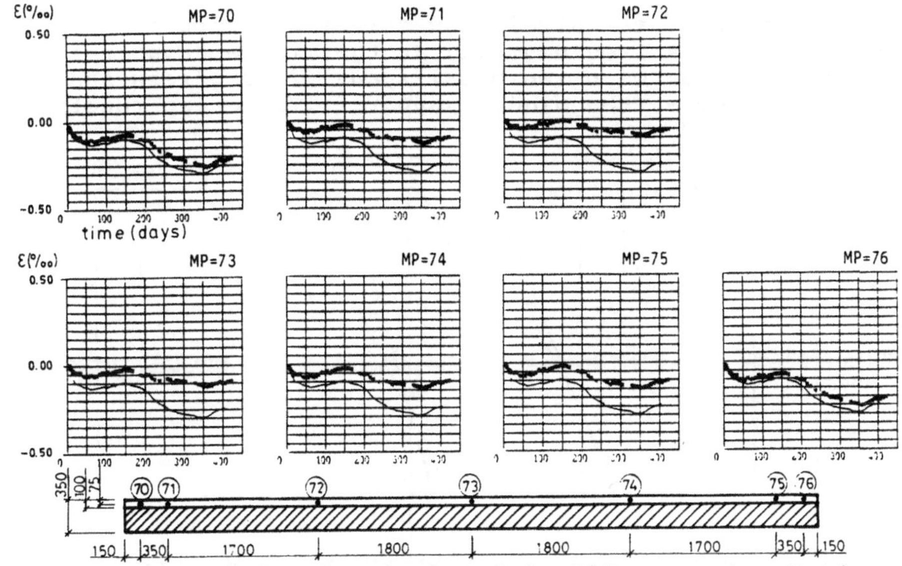

Fig. 4. Measured and calculated (solid curves) relationships between time and concrete strain for specimen No. 4.

As shown in Fig. 4 the measured and calculated functions are affine, but the magnitude of the calculated strain is lower than the magnitude of the measured strain. There are two main causes. Standard prisms from whitch data are taken shrink more and faster than the new-cast topping because of their smaller dimensions. The second explanation is that in the simple theory described above the creep is neglected.

The concrete creep ε_{ccr} can be written as (See /3/):

$$\varepsilon_{ccr}(t) = \int_0^t \frac{\varphi(t,\tau)}{E_c(\tau)} \frac{d\sigma(\tau)}{d\tau} d\tau \tag{1}$$

or if the integral is substituted by a sum and the modulus of elasticity E_c is assumed to be independent on time t:

$$\varepsilon_{ccr}(t) = \frac{1}{E_c} \sum_k \varphi(t,[k-\frac{1}{2}]\Delta t)\{\sigma(k\Delta t) - \sigma([k-1]\Delta t)\} \tag{2}$$

were φ is the creep coefficient and σ the stress.

The free beem and the continuous beam on three supports have been divided into elements. The time is divided into time steps Δt. At every time step the effect of creep in a particular element upon the strain and stress distributions is studied. The creep is calculated according to Eq. (2). The effects of creep in all elements are summed and added to the effect of differential shrinkage. With a known stess distribution in the beam, new values of the creep can be calculated according to Eq. (2). The problem is solved numerically by direct iteration.

Fig. 5 shows the strain and stress distributions in the inner part of a free beam and at the middle support in a continuous beam with two equal spans. (The stress has its maximum value at the middle support). The creep function in /3/ is used.

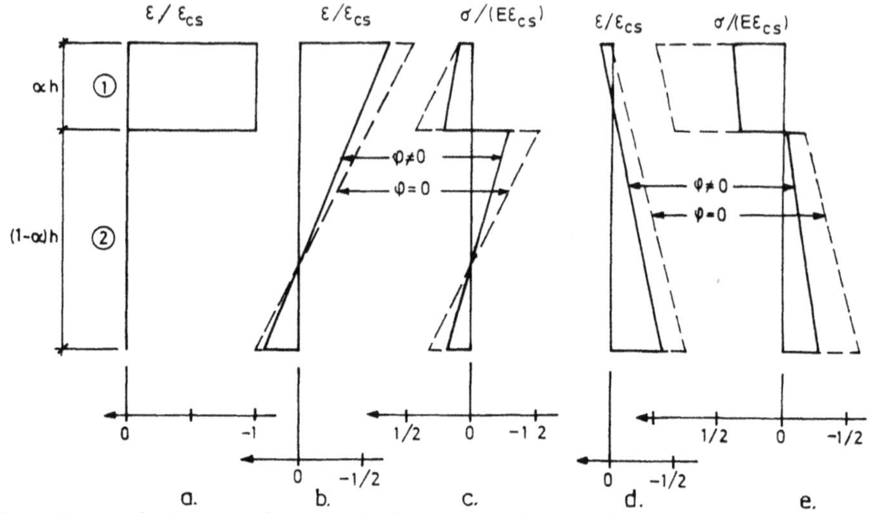

Fig. 5. Distributions of free differential shrinkage (a), resulting strain and stress in a free beam (b and c) and resulting strain and stress at the middle support in a simply supported continuous beam with two equal spans (d and e).

As can be seen from Fig. 5 the calculated strain is reduced with a factor about 0.8 if the creep is considered and the calculated stress is reduced to less than the half. This relief caused by creep is the main cause of the resistance of the concrete beams to residual stresses caused by the differential shrinkage. Calculations predict maximum tensile stresses lower than the tensile strength of concrete.

Calculations according to FIP /4/ predict maximum shear stresses lower than the tensile strength of concrete.

At the tests the deflections of the beams supported on air-bags were measured in five equidistant points for each specimen. The measured curvature κ was estimated as:

$$\kappa(x) \approx \frac{d^2w(x)}{dx^2} \approx \frac{w(x+\Delta x) - 2w(x) + w(x-\Delta x)}{(\Delta x)^2} \tag{3}$$

where w(x) is the deflection at the distance x from the beam end.

The distance between consecutive points is ΔL. By using Eq. (3) with Δx=ΔL and Δx=2ΔL and Richardson extrapolation the calculated value of the curvature at the center of the beam can be improved. This value is below called 'measured curvature'.

The measured curvature is compared with the calculated according to the simplest linear-elastic theory. The curvature is also obtained as an intermediate result at the calculation of the effect of creep. In Fig. 6 the measured curvature is compared with the calculated curvature for the cases of neglected and considered creep. If the creep is considered the measured curvature corresponds fairly well to the calculated.

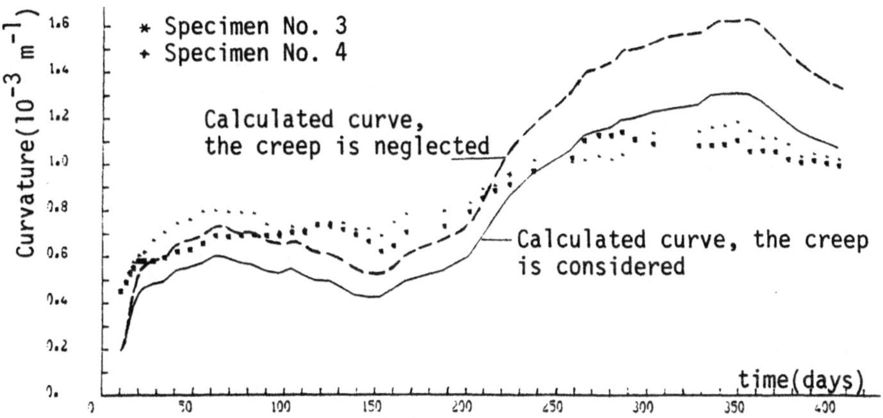

Fig. 6. Measured and calculated relationships between time and curvature at the center of the beam.

After terminated tests the tensile strength between old and new concrete was determined. Cylinders were drilled through the topping into the concrete under the interface. The force at tensile failure was registered. The mean value of the tensile strength was 1.9 MPa.

4. Further tests

Very recently hydrodemolition, i.e. concrete removal by use of a high pressure water jet, has become a technique to be used besides the conventional pneumatic hammer. Studies at the Department show that the roughness of a surface stripped by a water jet has a greater magnitude than the roughness resulting from stripping by a pneumatic hammer. The aim of a third investigation is to compare interfaces stripped by hydrodemolition with interfaces stripped by a pneumatic hammer by tests with composite concrete slabs.

Since November 1985 tests with square concrete slabs are carried out at the Department. Five composite concrete slabs and two homogeneous concrete slabs are supported at the corners. The load cases will be differential shrinkage and loading by a central single load.

5. Concluding remarks

Composite concrete beams of old and new concrete with a stripped, rough interface have been tested. Simply supported composite beams have shown as high ultimate loads as homogeneous concrete beams.

There is no evidence to assume that a composite concrete beam with a stripped, rough interface has less resistance to fatigue than a homogeneous concrete beam.

Furthermore, a composite concrete beam of old concrete with a stripped, rough interface against a new concrete topping can, with certain precautions, resist the residual stresses caused by differential shrinkage without cracks along the interface or in any of the parts of the beam.

The stress relief caused by the creep, which is simultaneous to the shrinkage, is the main cause of the resistance of concrete beams to residual stresses caused by the differential shrinkage.

Tests with beams indicate that a smooth interface is detrimental to the loading capacity of composite concrete beams.

To obtain as good results in situ as in the laboratory tests, it is necessary to do all works with removal and casting of concrete carefully and to demand that dust and all loose particles are removed from the surface before casting.

6. Acknowledgements

The tests described in this paper were carried out at the Department of Structural Mechanics and Engineering at the Royal Institute of Technology in Stockholm in co-operation with the National Swedish Road Administration. The research programme was initiated by professor H. Ingvarsson at the National Swedish Road Administration. The author is working at the Department since 1982 under the guidance of professor S. Sahlin. The author wishes to express his thanks to the laboratory staff under the supervision of K. Lindgren.

References

Bellander, U. and Thorsén, A. (1972) Samverkan mellan TT-element och pågjuten överbetong - en fältundersökning. (Composite action between TT-units and topping - a field study). Cement and Concrete Research Institute, CBI Report 7242, Stockholm.

Comité Euro-International du Béton (CEB). (1978) Model Code for Concrete Structures.

Comité Euro-International du Béton (CEB). (1984) Structural Effects of Time-Dependent Behaviour of Concrete.
FIP Comission on Prefabrication, Guide to Good Practice. (1982) Shear at the Interface of Precast and In Situ Concrete. FIP/9/6.
Silfwerbrand, J. (1984) Samverkan mellan delvis nedbilad betongplatta och pågjutning. Balkförsök. (Bond between a partly stripped concrete bridge deck and a new-cast concrete topping. Tests with beams.) Bulletin No.142, Department of Structural Mechanics and Engineering, Royal Institute of Technology, Stockholm.
Silfwerbrand, J. (1986) Differenskrympning i samverkanbalkar av gammal och pågjuten betong. (Differential shrinkage in composite concrete beams of old concrete and a new-cast concrete topping). Bulletin No.144, Department of Structural Mechanics and Engineering, Royal Institute of Technology, Stockholm.

EFFETS DE L´APPLICATION DE L´ENDUIT EPOXYDE SUR LA RESISTANCE L´ADHERENCE LORS DE LA CONTINUATION DU BETONNAGE

B. STANIVUKOVIC et N.KNEZEVIC-VUKSANOVIC
Institut de Máteriaux et Constructions de la Faculté de Génie Civil, Sarajevo

Résumé
La réalisation des constructions en béton présentant une complète continuité est rare dans le bâtiment. Le plus souvent, le bétonnage est interrompu aux endroits déterminés de la construction, et la continuation du bétonnage représente un point potentiellement faible. Un tel point peut être amélioré si le béton déjà durci avant de la continuation du bétonnage est enduit d´une sorte de résine époxyde adéquate. Dans le travail on a présenté les données concernant les recherches expérimentales effectuées au moyen de modèles sous forme de cylindres et de prismes. Les recherches ont porté sur les cas des modèles bétonnés en continuité, ensuite des modèles où la continuation du bétonnage a eu lieu 24 heures ainsi que des modèles, comme dans le cas précédent, où l´on a appliqué l´enduit époxyde à l´endroit de la continuation. Les modèles on été exposés à la traction, au cisaillement et à la flexion jusqu´à la rupture. On a analysé les caractéristiques de la rupture pour les cas particuliers et on a présenté les valeurs et rapports des résistances obtenues.
Mots-clés: Epoxyde, Continuation du bétonnage, Traction, Cisaillement, Flexion.

Introduction

Il est rare d´avoir dans la pratique de la construction des cas où l´édification des constructions en béton ou en béton armé se réalise sans interruption, c´est-à-dire de façon continue. Il est possible d´éviter des interruptions du bétonnage quand il s´agit des formes de constructions ou des éléments de constructions de dimensions moins importantes ou présentant une forme simple, par exemple lors de la production des éléments de montage préfabriqués. Cela peut se réaliser dans certains cas même quand il s´agit d´importants ou de très importants ensembles de construction, ce qui exige de considérables efforts de caractère tehcnologique et d´exécution, par exemple en cas de construction des tables turbo-génératrice des centrales thermo-électriques.

Dans la majeure partie des cas, et particulièrement lors du bétonnage sur place, il est pratiquement impossible d´éviter des interruptions et continuations du bétonnage, ce qui résulte de certaines raisons de caractère technologique et constructif. De tels points représentent les points les plus délicats de la construction et, potentiellement, ses points les plus faibles en ce qui concerne la résistance, l'étanchéité et autres caractéristiques. Ces points peuvent être particulièrement dangereux quand il est question des constructions dans les zones exposées aux contraintes dues à la traction, au cisaillement ou à la flexion, auxquelles le béton de par sa nature n'est pas suffisamment résistant.

Les essais expérimentaux effectués sur les modèles d'essai ont permis d'obtenir certaines données informatives concernant l'influence de l'interruption et de la continuation du bétonnage sur l'affaiblissement des points de coupe par raspport aux points de coupe correspondants lors du bétonnage continu. Les essais ont été effectués à l'Institut des matériaux et constructions de la Faculté du Génie civil à Sarajevo.

Données sur les modèles et les procédés appliqués

Les eesais ont été effectués sur trois sortes de modèles qui ont été exposés à la traction directe, au cisaillement et à la traction par flexion. Pour chacune des trois sortes de contraintes on a fait des modèles correspondants présentant les caractéristiques suivantes:

(a) modèles bétonnés de façon continue, c'est-à-dire sans interruption du bétonnage (étalons),

(b) modèles bétonnés jusqu'à la mi-hauteur, c'est-à-dire jusqu'à la mi-portée, avec l'interruption et la continuation du bétonnage jusqu'à la pleine hauteur, c'est-à-dire de la portée, effectué 24 heures après,

(c) modèles identiques à ceux sous (b) mais où la surface de contact du béton ancien a été recouverte, avant la continuation du bétonnage, par un enduit de résine synthétique.

La surface de contact du béton ancien a été nettoyée au moyen d'une brosse en acier, en vue d'obtenir un meilleur lien avec le béton nouveau.

Dans le cas des essais des modèles exposés à la traction directe on a utilisé des modèles en béton cylindriques, présentant le diamètre de 15 cm et hauts de 30 cm. L'interruption du bétonnage a éte faite à la mi-hauteur du modèle, perpendiculairement à la direction de l'action de la force de traction, appliquée jusqu'à la rupture du modèle.

En vue d'etablir la résistance au cisaillement on a utilisé des prismes de 12 x 12 x 36 cm. L'interruption du bétonnage a été effectuée dans ce cas au milieu de la longueur du prisme, dans le sens vertical, c'est-à-dire parallèlement avec la direction de l'action de la force de cisaillement, appliquée jusqu'à la rupture du modèle.

Pour obtenir la résistance à la flexion on a utilisé des prisme de 10 x 15 x 70 cm, présentant la continuation du bétonnage au milieu de la portée, qui était de 60 cm.

Pour la fabrication des modèles d'essais on a utilisé le béton présentant la composition suivante:

- agrégat en calcaire broyé et lavé, avec les granulations suivantes:

0/4 mm ... 20 %
4/8 mm ... 20 %
8/16 mm ... 20 %
16/31,5 mm .. 40 %

- ciment Pc 35 de production yougoslave, en quantité de 300 kg/m^3 de béton utilisé,
- facteur eau-ciment de 0,55.

La résistance à la compression moyenne, constatée sur 10 cubes de béton, présentant un coté de 20 cm et vieux de 28 jours, était de l'ordre de f_{bm}= 44,3 MPa.

La fabrication des modèles d'essais a été faite au moyen de machines, et la mise en place du béton au moyen de la pervibration.

En vue d'améliorer le lien entre le béton récent, au point de la continuation, on a utilisé une résine d'époxyde bicomposite, de production yougoslave.

La forme des modèles et les procédés de la réalisation des essais sont représentés sur le schéma suivant:

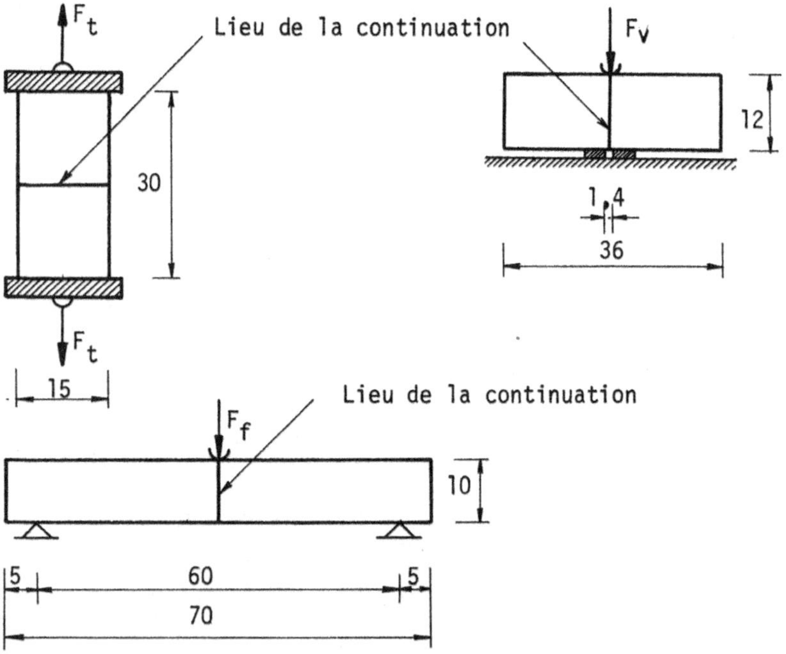

Les essais concernant les trois sortes de contraintes, c'est-à-dire pour chacune des formes des modèles, ont été effectués en trois séries, chacune comprenant 10 modèles (étalons, continuation du bétonnage sans résine et continuation avec enduit résineux). Lors des essais on a utilisé le béton vieux de 28 jours (béton récent) et de 29 jours (béton ancien).

Résultats des essais et commentaires des résultats obtenus

Les résultats des essais effectués sont représentés sur la table n° 1.

Table n° 1.

Résistance (MPa)	sorte d'essai		
	étalon	continuation béton-béton	continuation béton-résine-béton
traction directe f_t			
minimale	2,50	1,43	1,52
maximum	2,79	1,60	1,68
moyenne	2,65	1,52	1,60
cisaillement f_v			
minimale	4,83	1,15	4,50
maximum	5,96	2,19	4,98
moyenne	5,41	1,57	4,86
traction par flexion f_f			
minimale	5,14	0,17	2,70
maximum	5,68	0,70	3,44
moyenne	5,30	0,26	2,98

Lex caractéristiques de la rupture pour les cas examinés particuliers sont représentés sur la table n° 2.

Lors des essais effectués sur des étalons on a constaté les relations suivantes des valeurs moyennes pour les particulières formes des résistance:

$$f_{bm} : f_{tm} : f_{vm} : f_{fm} = 1 : 0{,}060 : 0{,}122 : 0{,}120.$$

Il faudrait accepter ces valeurs comme valeurs d'orientation avec la possibilité de certaines oscillations, en fonction du choix de la forme et des dimensions des modèles examinés. Des écarts plus considérables peuvent apparaître en fonction de la composition et de la sorte du béton utilisé.

Table n° 2

Résistance	Caractéristique de la rupture		
	étalon	continuation béton-béton	continuation béton-résine-béton
traction directe f_t	à travers le béton au milieu de la mi-hauteur	au lieu de la continuation	au lieu de contact résine-béton récent
cisaillement f_v	à travers le béton au-dessous de la force	au lieu de la continuation	au lieu de contact résine-béton récent
traction par flexion f_f	à travers le béton au-dessous de la force	au lieu de la continuation	au lieu de contact résine-béton récent

Dans le cas de la traction directe les relations des résistances moyennes des étalons (f_{tm1}) par rapport aux résistances moyennes de la continuation du bétonnage béton-beton (f_{tm2}) et béton-résine-béton (f_{tm3}) ont été comme il suit:

$$f_{tm1} : f_{tm2} : f_{tm3} = 1 : 0{,}574 : 0{,}604.$$

Partant de ces relations on peut constater que l'influence de l'enduit résineux au lieu de la continuation est pratiquement insignifiante et que la résistance à la traction directe représente au lieu de la continuation environ 60 % de la résistance correspondante au lieu où le bétonnage a été réalisé de manière continue (étalon).

Les valeurs moyennes de la résistance au cisaillement, constatées sur les étalons (f_{vm1}) au cas de la continuation béton-béton (f_{vm2}) et béton-résine-béton (f_{vm3}), ont témoigné des interdépendances suivantes:

$$f_{vm1} : f_{vm2} : f_{vm3} = 1 : 0{,}290 : 0{,}897.$$

Dans ce cas-là l'application de l'enduit de résine synthétique au lieu de la continuation produit un effet très favorable, car la résistance d'un tel lien se rapproche de la résistance propre à l'étalon, c'est-à-dire elle lui est inférieure de seulement environ 10 %. Lors de la continuation du bétonnage habituelle, sans application d'enduit d'époxyde, la résistance au cisaillement, au lieu de la continuation, est inférieure de plus de trois fois à celle du même béton sans interruption du bétonnage.

Les relations respectives des résistances à la flexion moyennes sont les suivantes:

$f_{fm1} : f_{fm2} : f_{fm3} = 1 : 0{,}049 : 0{,}562.$

Dans ce cas la résistance à la flexion au lieu de la continuation, sans enduit d'époxyde, est pratiquement insignifiante et représente environ 5 % de la résistance correspondante du béton en dehors de la continuation. L'enduit d'époxyde, au lieu de la continuation, s'est montré très efficace et a considérablement amélioré la qualité du lieu de la continuation, exposée à la contrainte de flexion.

Conclusions

Les essais ont été effectués dans un volume réduit, vu les éléments suivants:

- le nombre d'essais du même caractère est relativement petit,
- lors des essais on a utilisé une seule sorte de béton, de composition bien précisée,
- aux lieus de la continuation on n'a utilisé qu'une seule sorte de résine d'époxyde.

C'est pour ces raisons-là que l'on peut considérer les résultats obtenus uniquement à titre informatif, sans la possibilité objective de leure application générale. Il serait utile d'effectuer ces essais à une échelle plus élevée et en variant les paramètres cités, c'est-à-dire en combinant différentes sortes et qualités de béton, différentes sortes d'enduits synthétiques, en étudiat également d'autres sortes de continuations et non seulement celle perpendiculaire à la coupe etc.

Les essais effectués permettent quand même de formuler certaines conclusions de caractère général:

- La qualité du lien au lieu de la continuation du bétonnage depend de maints facteurs, tels que les qualités du béton, la sorte de l'état de contrainte au lieu de la continuation, la position de la continuation (horizontale, verticale, oblique),la direction du bétonnage, la façon dont est traitée la surface de contact etc.
- Dans certains cas, surtout celui des continuations exposées au cisaillement et aux contraintes de flexion, la qualité de la continuation peut considérablement être améliorée si l'on recouvre la surface du béton ancien par un enduit d'époxyde, immédiatement avant la continuation du bétonnage.

Nous estimons qu'un programme de recherches plus vaste permettrait d'obtenir d'utiles données qui devraient trouver leur place dans les correspondants reglements concernant le domaine du béton et du béton armé.

BOND BEHAVIOR OF POLYMER-IMPREGNATED CONCRETE PERMANENT FORMS FOR PRESTRESSED CONCRETE BEAMS

K. TSURUTA and M. OZAWA Technical Research Laboratory, Ozawa Concrete Industries, Ltd.
T. NAITO Technical Research Institute, Taisei Corporation

Abstract
The permanent forms of polymer impregnated concrete (PIC) panels have been developed to protect prestressed concrete beams from chloride ions.

This paper discusses bond property between this form and placed concrete. First, PIC panels with highest bond property were selected with various surface configurations. Second, post-tensioned prestressed concrete beams using these panels were manufactured. The behavior of these beams was studied by an experimental analysis; prestressing, static flexure and dynamic flexure. PIC panels were found to include effective sectional area.
Key words: Bond Behaviour, Polymer-impregnated Concrete, Permanent Forms.

1. Introduction

Recently the rapid deterioration of various reinforced or prestressed concrete structures by the chloride-induced corrosion of reinforcing steel has been a widely recognized problem in the construction industry. For the purpose of preventing the chloride-induced corrosion, polymer-impregnated concrete permanent forms (PIC permanent forms) for prestressed concrete beams (PC beams) were developed by the authors. This corresponds to the "Method of Using Panel Excelling in Property to Shut Off Corrosion Factors as Permanent Form" (Ref. 1) in the Type 2 Corrosion Protection Method of the JCI Recommended Practice for Corrosion Protection of Offshore Concrete Structures. In such case, 1 capacity to shut off chloride ions, etc., 2 durability as concrete structure, 3 integrity with placed concrete proper, 4 good working properties, etc. are demanded. Leaving 1 and 2 to separate investigation, this paper deals with the bond behavior of PIC permanent forms to the concrete placed in them to mold PC beams.

Firstly, the results of physical property tests of PIC are summarized in Table 1.

With regard to integrity, PIC permanent form with six types of surface configuration were fabricated to study the properties of the bond with placed concrete, and flexural and shear tests were performed.

Next, based on the results of the fundamental experiments, forms were fabricated with PIC panels of good bond properties and post-tensioned prestressed concrete beams were made. Static and dynamic loads were applied to the prestressed concrete beams and comparison studies of their behaviors were carried out with ordinary prestressed concrete beams.

Table 1 Summary of Properties of PIC

Description of Properties	Specification of Test	Test Results
1. Mechanical Properties	Compressive strength, ϕ10x20cm	1500kg/cm^2
	Modulus of rupture, 10x10x40cm	240kg/cm^2
	Tensile strength, ϕ15x15cm	120kg/cm^2
	Modulus of elasticity, ϕ10x20cm	4.5x10kg/cm^2
	Poison's ratio, ϕ10x20cm	0.18
	Density	2.4
	Coefficient of thermal expansion	1.1x10/°C
	Creep coefficient	0.08
	Fatigue limit (10^{7}cycle)	63%
		Compressive Strength
2. Resistance to freezing and thowing	(1) in Air -40°C +60°C 30 cyclel	No change
	(2) in Water -30°C +60°C 230 cyclel	4% decrease
3. Resistance to chemical attach	(1) 2% H_2SO_4 1000 hours	40% decrease
	(2) 2% NaOH 1000 hours	No change
4. Resistance to Ultra-Violet radiation	Weather meter 3000 hours	No change

2. Materials and preparation of PIC permanent form

Concrete materials of PIC permanent form for flexural and direct shear tests and permanent forms used of PC beams were prepared with the water-cement ratio of 0.37, the cement content per unit volume of 450 kg/m^3, the slump of 8±1 cm and the air content of 2 to 3 percent. Normal portland cement, river sand, a crushed stone and a superplasticizer were used.

The impregnated materials were methyl methacrylate (monomer) and Azobisisobutyronitrile (catalyst) and its formulation was 100:1 by weight.

The manufacturing process of PIC panels is illustrated in Fig. 1.

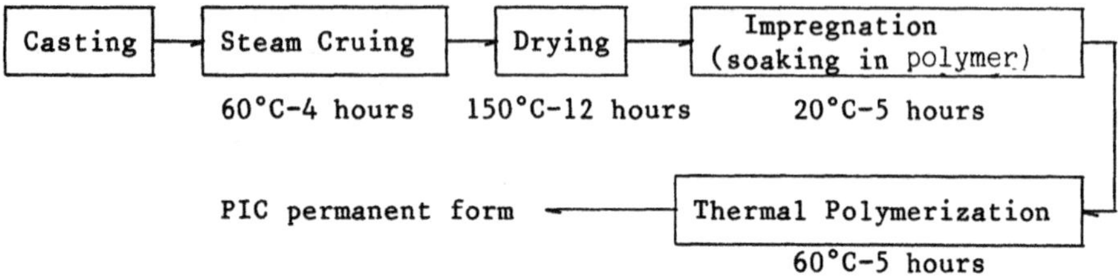

Fig. 1 Fabrication process of PIC permanent form

The polymer-loading is from 4.5 to 5.5 percent by weight. The mechanical properties of PIC panels such as the compressive strength, the modulus of rupture and the elastic modulus were 1350 kg/cm^2, 240 kg/cm^2, and 4.5×10^5 kg/cm^2 respectively.

3. The bond properties between PIC permanent form and placed concrete

The PIC permanent form of bond surfaces such as shown in Fig. 2 were fabricated to study the bond properties between PIC permanent form and placed concrete. The flexural tests and direct shear test were carried out.

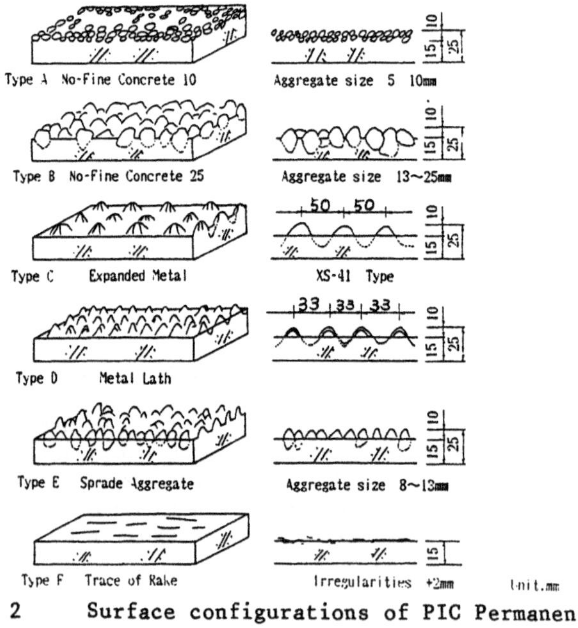

Fig. 2 Surface configurations of PIC Permanent Forms

3.1 Flexural tests

The details of test specimen are shown in Fig. 3 and a bond failure was expected to occur in the notch. The specimens were tested under three-point loading. Electrical wire strain gauges were fixed in the 5-cm distance from the notch. Fig. 4 shows the typical relationship between load and strain. By increasing the load, the strain increases; and then decrease when the bond failure between PIC permanent form and placed concrete starts. The strain change point was judged to be the startof the bond failure. The bond strength τ_i is calculated by the following equation: (Ref. 2)

$$\tau_i = \frac{Ec \cdot \Delta\varepsilon \cdot a \cdot b}{b \cdot c}$$

Where

- Ec: modulus of elasticity of PIC permanent form (kg/cm^2)
- $\Delta\varepsilon$: strain at start of bond failure
- a: depth of notch (cm)
- b: width of specimen (cm)
- c: distance from notch to measureing gauge (cm)

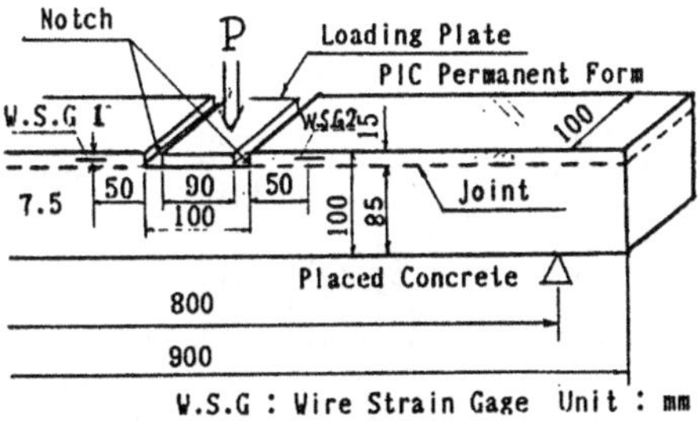

Fig. 3 Setup for Flexural Test

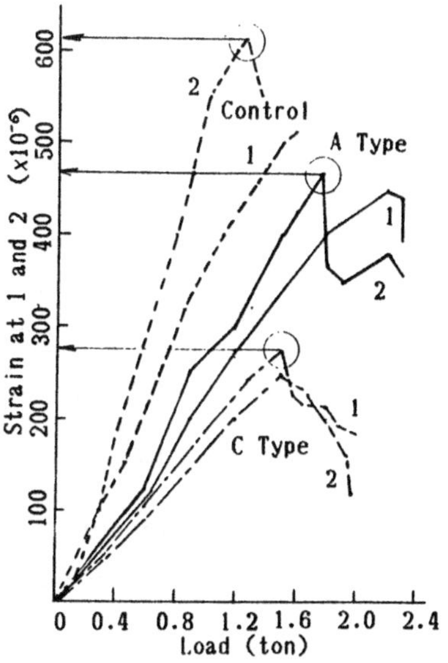

Fig. 4 Typical relationship between load and strain for flexural test

The calculated results are shown in Table 2. The A type excells in bond strength.

Table 2 Bond Strength τ_i between PIC panels and concrete

Specimen Type	Depth of Notch (mm)	τ_i (kg/cm^2)
A	25	34
	15	51
A'	15	38
B	25	24
C	15	27
D	15	11
E	15	31
F	15	23
G	15	37

Where

A': Total thickness 15mm,

Compressive strength of placed concrete, (ϕ10 x 20 cm) 400 kg/cm^2

Modulus of rupture of placed concrete, (10 x 10 x 40 cm) 50 kg/cm^2

3.2 Direct shear test

The thickness of PIC permanent form were 25 mm and the loaded areas were $2x180cm^2 = 360cm^2$. Fig. 5 shows the direct shear test setup.

The test results are shown in Fig. 6. The cracking and ultimate shear strength was obtained by dividing cracking and ultimate shear load by loaded areas. The A and B types have superior bond strength.

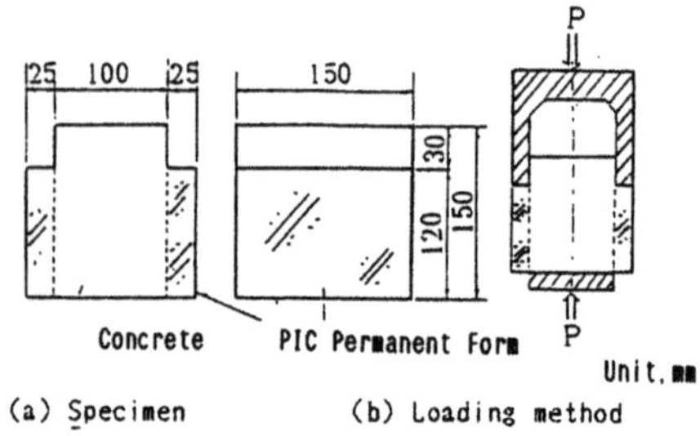

Fig. 5 Setup for Shearing Test

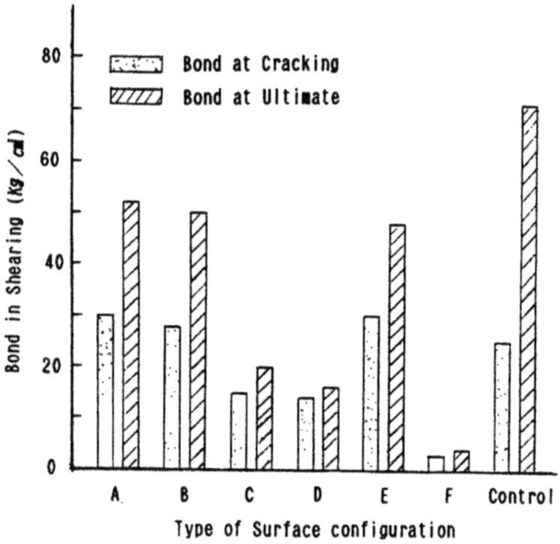

Fig. 6 Bond in Shearing between PIC Permanent Forms and Placed Concrete

4. Behavior of post-tensioned beams with PIC permanent forms

From flexural and shear test results, the A type PIC permanent form was adopted in permanent forms of PC beams.

4.1 FABRICATION OF POST-TENSIONED BEAMS

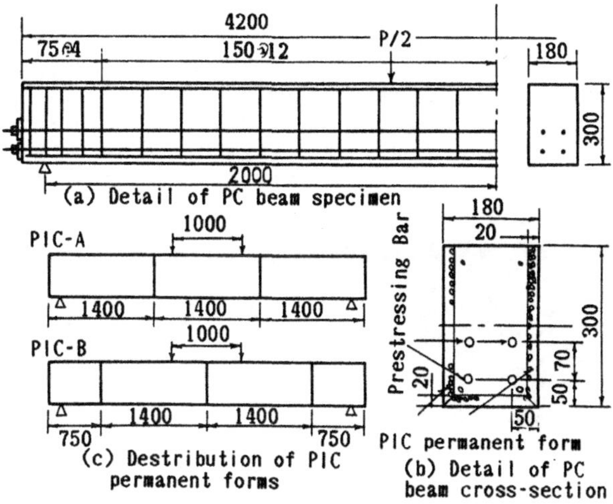

Fig. 7 The post-tensioned PC beam and loading method

The details and loading method of posttensioned beams are shown in Fig. 7. The location of PIC panels were jointed in two ways for loading point as shown in Fig. 7 (c). An ordinary posttensioned beam was made for comparison purpose. The post-tensioned beams were fabricated as follows: 1 fabrication of PIC panel 2 formwork of PIC panel jointing with epoxy resin (Photo 1) 3 tendon and reinforcing bar placement 4 placing of concrete 5 steam curing 6 prestressing 7 grouting 8 curing in air. The specified concrete strength was 500 kg/cm^2 and the mix design was for maximum size aggregate of 15mm, the water cement ratio of 0.40 and a unit cement content of 420 kg/m^3, with slumps to be 8±1cm and air contents 4±0.5 percent. High-early-strength portland cement, SBPD 130/145 prestressing steel, D-10 reinforcing bar and ∅6 stirrups were used. The prestressing with 41.76 ton was carried out when compressive strength get to more than 450 kg/cm^2, and the following grout was injected. The mix design of the grout was the water-cement ratio of 0.40 and addition of aluminum powder at the rate of 0.0075 percent by weight of cement, while compressive strength was 300 to 350 kg/cm^2 at the time of loading. The beams were cured in air until loading.

The jointing of PIC permanent form was simple butt joining. The epoxy resin was a bisphenol A type epoxy resin and a modified aliphatic amine, and its formulation was 1:1 by weight.

4.2 Behavior at prestressing

The strain distribution immediately after prestressing is shown in Fig. 7. Initial tensioning force was 41.76 ton with compressive stress intensities at bottom fibers 178 kg/cm^2 for the standard type and 157 kg/cm^2 for the PIC type as a composite material. The tensile stress intensities at top fibers were 24 and 20 kg/cm^2, respectively.

Calculation of stress for the PIC permanent form is made using the transformed sectional area and transformed section modulus obtained from the ratio of modulus of elasticity of the PIC permanent form and the placed concrete.

These modulus of elasticity, at time of prestressing, were assumed to be 3.0×10^5 and 4.5×10^5 kg/cm^2 for the placed concrete and the PIC permanent form, respectively. The results of calculation are also given in the Fig. 8. According to the strain distribution and the calculated values in the Fig. 8, it appears that the PIC panels and concrete were acting integrally against tensioning force.

Photo 1 PIC panels

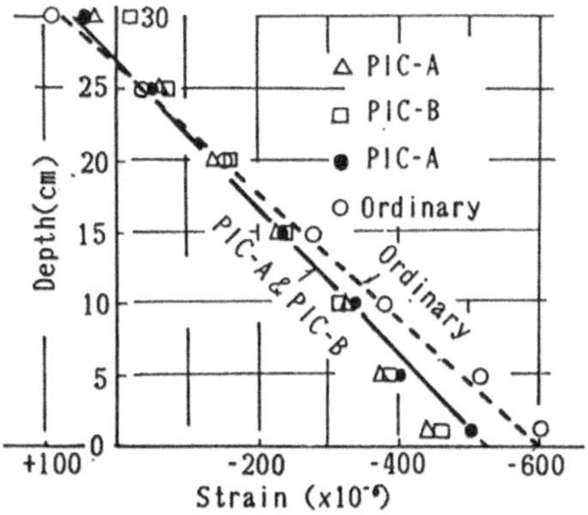

Fig. 8 Strain distribution immediately after prestressing

4.3 Loading tests

Loading Method: Each of ordinary, PIC A, and PIC B types were subjected to static flexural loading tests (Nos. 1, 2, 3), while similarly, specimens were subjected to 2 million cycles of repetitive loading at service loads, and followed by static flexural loading tests (Nos, 4, 5, 6). The load conditions in repetitive loading were selected (considering the amount of reduction in prestress) so that a tensile stress of 30 kg/cm^2 would act on the bottom fiber of the ordinary type prestressed concrete beam. The upper limit load determined from this condition was 6.5 ton and the lower limit load 0.9 ton.

The testing machine was a pulsator type fatigue testing machine. The repeated speed was 300 r.p.m. Deflections and strains were measured while receiving the load of between 0 and 6.5 ton. The measurement was taken after the testing machine was stopped at every 10^n cycles.

Test Results: The measured values of initial cracking loads and failure loads are shown in Table 3.

Table 3 Summary of Flexural Test of PC Beams

No.	Specimen Type	Concrete Strength (kg/cm^2) Compressive	Flexural	Initial Clacking Flexural Load (ton)	Ultimate Flexural Load (ton)
1	Ordinary	557	62	8.0	16.0
2	PIC-A	567	63	10.0	15.1
3	PIC-B	596	63	9.8	15.4
4	Ordinary	591	63	8.4	17.2
5	PIC-A	616	62	10.5	16.8
6	PIC-B	546	51	10.8	16.6

The cracking loads by calculation are 7.7 ton for the ordinary type and 10.4 ton for the PIC type. In case of PIC, the flexural strength of the PIC permanent form cannot be ignored. Thus, it was assumed that the plane of the composite cross section would be kept, the apparent modulus of rupture of composite cross section were calculated from the balance between the tensile stress of the PIC permanent form and the compressive stress of the placed concrete. The initial cracking load was then determined. The apparent flexural strength with this cross section was 120 kg/cm^2. This calculated initial cracking load value agreed roughly with measured values and it appears that initial cracking load can be evaluated by this method. From the test results the initial cracking load with this cross section is 20 to 30 percent higher for the PIC type than for the ordinary type.

The relations between load and deflection are shown Fig. 9. PIC permanent form types had greater rigidities than the ordinary types in the elastic range before the crack. During the process from crack initiation to failure, the trend was the same for the PIC as for the ordinary type. The residual deflections at midspan when 6.5 ton were loaded every 10^n cycles are shown in Fig. 10, and PIC permanent form types were less affected by repetitive loading. Regarding cracking and failure properties, the number of cracks was half that of the ordinary type, while crack spacing was approximately double and crack width was the same. All failures were compressive failures of concrete after steel had yielded, while the PIC permanent form was not spalled off up to failure.

Further, with respect to jointing locations for PIC permanent form, since the modulus of elasticity of epoxy resin was 1/10 smaller than that of concrete, the PIC B type had deflection and strain about 10 percent larger than A type, but as shown in Fig. 9, these did not seem particularly to constitute weak points.

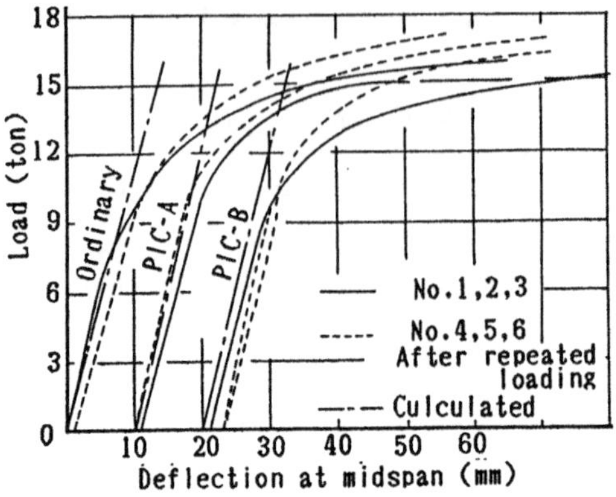

Fig. 9 Load-Deflection of PC beams

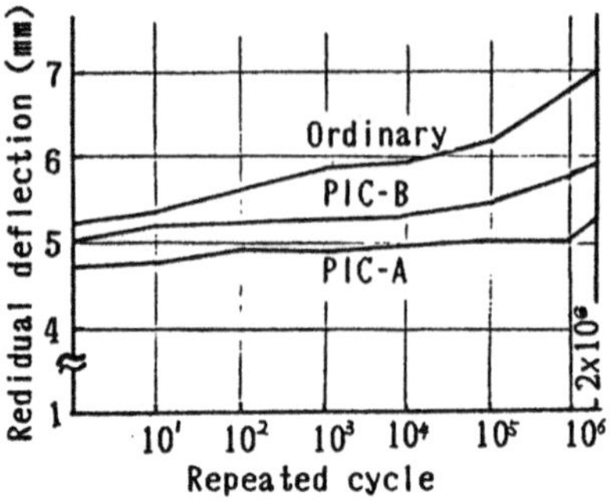

Fig. 10 Redidual deflection at midspan when 6.5 ton were loaded

5. Conclusion

As a result of studies concerning integrity in case of using PIC panels as permanent forms for prestressed concrete beams, it was learned that integrity with placed concrete could be obtained by innovations with regard to surface configurations, with no problems concerning prestressing and actions of static and dynamic loads. It was further learned that application could be done in the same way as with ordinary prestressed concrete beams. Studies are now being made on the sealing-off properties and durabilities of PIC permanet form, and good results are obtained.

References

1. Kobayashi, K.: Lecture Series, "Deterioration of concrete Structures from Corrosion of Steel and Countermeasures," Public Works, vol. 25, No. 7- Vol. 26 (in Japanese).

2. Akai, K., Okawa, S., and Fujita, S.: Effect of Construction Joint Surface Treatment on Bond Strength Between New and Old Concretes. CAJ Review of the 33rd General Meeting, Technical Session, 1979 (in Japanese).

TESTING OF BOND BETWEEN FRESH AND HARDENED CONCRETE

J.S. WALL, N.G. SHRIVE and B.R. GAMBLE
Department of Civil Engineering, The University of Calgary

Abstract
The bond between fresh and hardened concrete can be subjected to various stress states depending on the application and service conditions. Four test methods, reflecting different stress states at the bond between the fresh and hardened concrete, are evaluated. The four methods are a slant shear test, an indirect tension test and two different flexure tests. The slant shear test is shown to be consistent and the most sensitive test in showing the strength of the bond, and was used to evaluate the effect of the following parameters on that strength:

(1) the water cement ratio of a portland cement mortar;
(2) the thickness of the bond layer;
(3) the effect of various curing conditions;
(4) the effect of wetting the surface of the hardened concrete before application of the portland cement mortar bonding agent; and
(5) the effect of delay between mixing a copolymer Polyvinyl acetate (PVA) bonding agent, and its application to hardened concrete.

We conclude that copolymer PVA is a poor bonding agent over a wide range of curing conditions and that the thickness of a portland cement mortar layer has a strong influence on bond strength. The other factors did not affect strength significantly.
Key words: Bond tests, Cold joints, Concrete bond, Flexure test, Indirect tension test, Polyvinyl acetate, Portland cement mortar, Slant shear test.

1. Introduction

Fresh concrete is laid on top of hardened concrete in many instances. Often, specific attempts are made to bond the layers together. The service stress state at the subsequent bond can vary considerably depending on the type and use of the structure. For example, a bridge deck overlay may have shear with tensile or compressive stresses induced by shrinkage or thermal effects in addition to compression and shear from service loads. On the other hand, the bond between precast concrete segments can be exposed to very high compressive stresses due to axial loads (i.e. arch components) or high tensile stresses caused by thermal contraction. The stresses in bonds used to secure repair concrete to a wide range of structures will vary with each repair situation.

A single method of test should involve a stress state fairly typical of service, while also being sensitive to the strength of the bond. Four test methods were evaluated, each reflecting a different stress state at the bond. The tests were a slant shear test, an indirect tension test and two flexure tests. The slant shear test was consistent and the most sensitive test with respect to strength of the bond. Thus, this test was used to evaluate certain different aspects of chosen bonding agents.

The bonding agent should provide a bond which is as strong as the materials being joined. Failure will then initiate in one of the latter, rather than the bond itself. A wide variety of bonding agents exist, including: epoxy resin, acrylic latex, SBR latex, Polyvinyl acetate (PVA) and portland cement. The latter two are widely used in North America as relatively inexpensive bonding aids and were thus of primary interest. A copolymer PVA bonding agent subjected to adverse curing conditions was also used as a source of a relatively weak bond.

2. Evaluation of Bond Test Methods

2.1 The tests

The four different tests selected for evaluation are depicted in figure 1. The slant shear test was chosen because various versions have been used previously, for example by Kreigh (1976). The stress state at the bond face involves compression and shear macroscopically. Specimens with a height:width ratio of 3:1 were used to reduce the effects of platen restraint in the bond region of the prism.

The flexure tests, with bond planes at 45° and 60° relative to the horizontal (Figure 1), were chosen to determine if exposing bonds to (flexural) tension and shear stress states is more rigorous, providing greater sensitivity to bond strength. Third point loading was used on specimens 12" (305 mm) long with 3" (76.2 mm) square cross-section. Thus, the bond was contained in the zone of constant moment between the two centre loading points. A similar test, with the bond plane at 30° to the horizontal was used by Moss and Batchelar (1975) to determine the effectiveness of epoxy resins in bonding concrete.

The indirect tensile prism test was used to subject the plane of the bond to an essentially biaxial principal stress state with tension directly across the bond. A prism was used (rather than a cylinder as in the standard 'Brazilian test') to facilitate casting. In order to maximize the tension stresses across the bond line, and to minimize tearing stresses adjacent to the loading points, finite element analysis was used to assess various arrangements of plywood packing and steel loading strips.

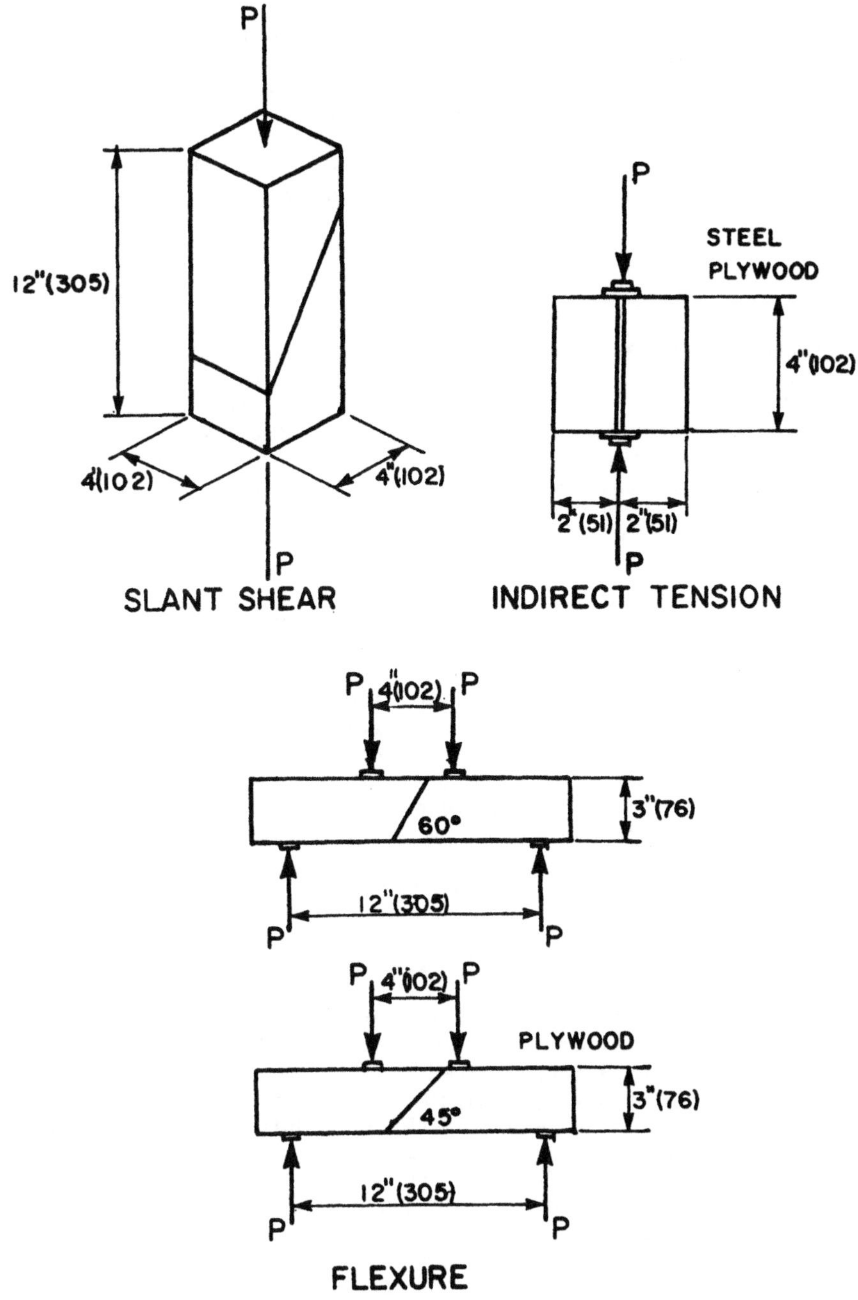

Fig. 1. Configuration of bond tests
Lengths in inches (mm)

2.2 The program for assessing test methods

The test methods were evaluated by examining their sensitivity to the bond strength of a typical portland cement mortar and a PVA agent. The materials were used to bond a concrete mix designed with a low water/cement ratio to have a strength around 45 MPa. The intention was to ensure that the bond strength, as assessed by the test, would not be obscured by failure in the concrete. Also, the fresh concrete used in typical applications is often high strength - for example in high density overlays on bridge decks.

The bond mortar consisted of equal parts by weight of dry sand and portland cement with a water cement of ratio of 0.35. A PVA based bonding agent was used in the 'paint' form (without the addition of aggregate or cement) and was subjected to moist curing to ensure a relatively weak bond.

Six composite prisms were made for each combination of test type and bonding agent. The base components were cast in steel moulds with oiled wooden inserts. The open ends of the moulds were covered with plastic immediately after the concrete was cast. After 24 hours, the prisms were demoulded and placed in a curing room with 100% relative humidity. At an age of seven days, the prisms were sandblasted on the bond plane surface, replaced in the moulds, and the bond coat and overlay concrete applied.

The bond surface was kept dry when the portland cement bonding agent was used, but was wetted immediately prior to the application of the PVA bond agent.

Specimens were again cured for 24 hours in the moulds, under plastic, and a further 27 days at 100% relative humidity before testing. Six monolithic prisms of each test type were used as "controls". Between batch variation of the concrete was monitored through casting six 3 x 6" (76.2 x 152.4 mm) cylinders from each batch. It was found that the strengths obtained from these cylinders matched the strengths obtained from the monolithic control prisms in the slant shear test.

The slant shear tests were performed in a 2000 kN testing machine subject to stroke control. The loading rate was kept in the range 0.14 - 0.34 MPa per minute as specified by ASTM C39. Load/stroke plots were obtained for half the specimens and stress/strain plots for the other half. Strains were measured with Linear Variable Differential Transducers (LVDT's) mounted on a frame attached to the specimen via measuring points glued on the specimen sides. The gauge length was 5 1/2" (140 mm), centred ± 1/8" (3 mm) on each specimen.

The flexure tests were performed in accordance with ASTM C78 at a loading rate between 1.25 and 1.75 kN per minute. Central deflection versus load was recorded for each specimen. Casting procedures left a 1-3 mm offset between the surfaces of the base and overlay components, and half of each of the top and bottom surfaces of the specimens as cast faces. To compensate for the differences, strips of plywood were placed between the loading points and the specimen to level the specimen and to provide for a relatively uniform distribution of the load across the specimen width.

The indirect tension tests were carried out at a constant loading rate of 1 kN per second. Load was transferred to the specimen through 1/2" (12.7 mm) wide, 1/4" (6.4 mm) deep steel strips contacting plywood packing 30 mm wide by 3 mm deep.

2.3 Results and comparison of the four test methods

The average strength for each bonding agent and each test is given in Table 1. One obvious conclusion to be drawn is that the PVA bond was much weaker than both the portland cement and the control specimens with no bond - as desired. The reduced strength was detected with varying degrees of success by the tests.

The slant shear test was the most sensitive to the weakness of the PVA bond. In this test the PVA bonded specimens had an average strength of only 22% of the average control prism strength. The prisms with the portland cement mortar bond had an average strength 72% of the control strength. Strains measured with the LVDT's in this test (Table 2) indicate the PVA bond was the more flexible of the two bonds, and that both bonds were more flexible than the control prisms with no bond. However, strains measured in later tests for most portland cement mortar bonded specimens were closer to strains in the control prisms than in Table 2.

The flexure tests also indicate a significant difference between the strengths of the two types of bond. However, neither test is as sensitive as the slant shear test and there is an additional disadvantage in that many of the results are more variable than with the slant shear test. The central deflections of the beams varied considerably, but nevertheless indicated the bonded beams were more flexible than the control beams. The variability in deflection can not be attributed to varying compaction of the plywood shims at the

Table 1. Strengths from the bond tests.
Each value is the average of six specimens.

	Control (MPa)	PC Mortar (Strength as a % of Controls, MPa)	PVA (Strength as a % of Controls, MPa)
Slant Shear (Prism Compressive Strength)	41.3	29.8 (72%)	9.03 (22%)
Flexure-45° Bond (Modulus of Rupture)	5.76	5.17 (90%)	2.66 (46%)
Flexure-60° Bond (Modulus of Rupture)	5.76	4.87 (85%)	2.14 (37%)
Indirect Tension (Tensile Strength)	3.13	2.99 (96%)	2.44 (78%)

Table 2. Strains at specific stresses (slant shear test).

	Control			PC Mortar			PVA		
Stress (MPa)	5	8	20	5	8	20	5	8	20
Strain (μ)	140	237	672	175	312	925	217	-	-

load points; the shims were precompressed to minimize their effect on measured deflection. The offset between the two materials at the bond interface must clearly affect the strength since it will act as a stress raiser. This latter effect could well be part cause of the variability observed with the flexural tests.

The results of the indirect tension test showed the least variation in strength between the bond types. Nevertheless, the specimens failed at the bond line in almost every case, indicating that the test does provide some indication of relative bond strength. Of course, none of the tests provides a value for the direct tensile bond strength; all the tests provide relative data in this context.

The slant shear test used here is the most sensitive to the varying strengths of the bonds, and also provided results with a low coefficient of variation. Thus, this test was selected for examining various parameters in relation to portland cement mortar and PVA bonds.

3. Investigation of Various Bond Parameters

The slant shear test described above was used to investigate the effects of the following parameters on bond strength:

(1) the water cement ratio of the portland cement mortar;
(2) the thickness of the bond layer;
(3) the effect of various curing conditions;
(4) the effect of wetting the surface of the hardened concrete before application of the portland cement mortar bonding agent; and
(5) the effect of delay between mixing a copolymer PVA bonding agent, and its application to hardened concrete.

Minor changes to the procedure included replacement of the wooden mould inserts with steel inserts, and use of steel templates with a plastering trowel to control the thickness of the bond layer.

The details of the bonds, curing conditions and strengths are provided in Table 3. The data have been compared statistically using a modified version of Student's t-test from Walpole and Myers (1978). The results are shown in Table 4. A significant difference occurs at the 95% confidence level whereas a highly significant difference is deemed to occur at the 99% confidence level.

As in the first series of tests, the portland cement mortar bonds were stronger than the PVA bonds - for all but two PVA bonds, where the results are statistically similar to two portland cement bonds.

In the portland cement mortar bonds, thickness can be seen to be important by comparing the strengths of bonds B11, B12 and B13. All these bonds contained the mortar described earlier. The 1/4" (6.4 mm) thick joint is highly significantly weaker than either the 3/16" (4.8 mm) or 1/8" (3.2 mm) thick joints. The 1/4" specimens all failed at the bond line whereas for other thicknesses the bond did not dominate the failure, indicating bond strength as good as or better than the concrete (Table 5). This result is similar to masonry where the thickness of mortar joints is highly significant with respect to both the compressive strength (eg: Francis et al

Table 3. Details of bonds, curing conditions and bond strength using slant shear tests (average of eight tests).

Bond (Designation)	Active Ingredient	Thickness (ins (mm))	Water Cement Ratio of Bond Mortar	Curing Period after 24 hrs (Days) 100%RH/ 50%RH	Avg. Compr. Strength (MPa)	Stand. Dev. (MPa)
B10	No Bonding Agent			27/0	43.5	2.90
B11	PC	3/16 (4.8)	0.35	27/0	45.1	0.75
B12	PC	1/4 (6.4)	0.35	27/0	34.9	5.57
B13	PC	1/8 (3)	0.35	27/0	44.4	1.79
B14	PC	1/8 (3)	0.40	27/0	41.7	1.23
B15	PC	1/8 (3)	0.32	27/0	31.5	1.05
B16+	PVA	Paint*	-	13/14	20.2	1.87
B17+	PVA	Paint	-	13/14	17.4	1.31
B18+	PVA	1/8 (3)	0.20	13/14	31.9	4.77
B19+@	PVA	1/8 (3)	0.20	13/14	30.2	4.34
B20+	PVA	1/8 (3)	0.22	13/14	26.1	2.03
B21+@	PVA	1/8 (3)	0.22	13/14	26.4	2.36
B22+	PVA	1/8 (3)	0.22	27/0	20.9	1.84
B23+	PC	1/8 (3)	0.40	27/0	46.0	3.36

PC portland cement
+ substrate was wetted
* 2 coats
@ PVA pot life expired

Table 4. Test for statistical significance.

Bonds Compared	Parameter Varied	Mean Strength of First Bond (MPa)	Mean Strength of Second Bond (MPa)	Significant Difference (Y = Yes, N = No)	Highly Significant
B11:B12	3/16":1/4"*	45.1	34.9	Y	Y
B13:B11	1/8":3/16"*	44.4	45.1	N	N
B15:B13	0.32:0.35+	31.5	44.4	Y	Y
B13:B14	0.35:0.40+	44.4	41.7	Y	Y
B14:B23	Dry:Wet†	41.7	46.0	Y	N
B16:B17	Coats of PVA	20.2	17.4	Y	Y
B18:B19	No Delay:Delay	31.9	30.2	N	N
B20:B21	No Delay:Delay	26.1	26.4	N	N
B18:B20	0.20:0.22+	31.9	26.1	Y	Y
B20:B22	Curing	26.1	20.9	Y	Y

* thickness
+ water-cement ratio
† surface moisture condition

Table 5. Mode of failure and batch control strengths.

Bond	At Bond (# of Prisms)	Not at Bond (# of Prisms)	Ductility of Failure* Brittle (# of Prisms)	Ductile (# of Prisms)	Strength of Control Cylinders (MPa) Base	Overlay
B10	7	1	6	2	44.4	41.5
B11	1	7	1	7	45.0	41.0
B12	8	0	8	0	43.5	44.7
B13	4	4	4	4	42.5	44.2
B14	0	8	0	8	41.5	36.6
B15	4	4	0	8	41.0	27.9
B16	8	0	8	0	53.3	49.6
B17	8	0	8	0	56.0	52.2
B18	8	0	8	0	55.3, 53.3	51.2, 48.6
B19	8	0	8	0	55.3, 53.3	51.2, 48.6
B20	8	0	8	0	51.3, 51.1	43.6, 52.4
B21	8	0	8	0	51.3, 51.1	43.6, 52.4
B22	8	0	8	0	48.4	45.0
B23	8	0	8	0	48.6	48.6

* Any specimen which produced a sudden loud noise and experienced a rapid reduction in load within approximately one second of the ultimate load was considered brittle. Any other specimen was considered ductile.

(1971)) and the bond strength (Sise (1984)). As observed here for bonded concrete, thicker mortar joints are weaker in masonry also. This is true for both clay brickwork and concrete blockwork. There has been no explanation of this phenomenon to date.

The effect of water cement ratio is not as clear. Failure at the bond was not the majority occurrence for this series (Table 5). Bond B15, with a mortar water-cement ratio of 0.32 appears much weaker than bonds B13 and B14 with mortar water-cement ratios of 0.35 and 0.40, respectively. However, the strength of the batch control cylinders (Table 5) indicates that the overlay concrete for the B15 specimens had unusually low strength. Nevertheless, failure occurred at the bond four times. This was different to set B14 where prisms with a slightly higher overlay strength always failed in the concrete. It is difficult to draw conclusions with respect to the effect of the water-cement ratio of the bond mortar when so few specimens failed at the bond. However, it appears that a very low water cement ratio can cause a reduction in bond strength.

Wetting the concrete surface prior to applying a portland cement mortar appears to be slightly beneficial, as may be seen by comparing series B23 with series B14. Using no bonding agent at all (series B10) did not appear to be particularly detrimental under the controlled laboratory conditions. The strength results were similar to the better portland cement mortar results, although the brittle failure mode was not desirable. High strengths have been noted before (eg: Tabor (1979)) for composite specimens prepared under laboratory controlled conditions.

Examination of the PVA bond results reveals a number of interesting points. The PVA used without mortar, and cured at 100% relative humidity for 13 days then at 50% relative humidity for the next 14 days before testing, produced weak bonds (series B16 and B17). The strength however, was much higher than with the previous test series (9.0 MPa, Table 1) where the prisms were cured in 100% relative humidity for the complete 27 days before testing. When two coats of PVA were applied in a span of about 20 minutes (bond B17) the strength was reduced from 20.2 to 17.3 MPa, a highly significant drop statistically. A more drastic drop in strength was recorded by Dixon and Sunley (1977) for an SBR bonding grout applied in a similar manner.

PVA modified mortars yielded bonded prism strengths of 31.9 and 26.1 MPa for water-cement ratios of 0.2 (B18) and 0.22 (B20), respectively. The effect of delaying the application of the overlay until the pot-life of the PVA (as stated by the manufacturer) had expired, was negligible (B19 and B21). This suggests the manufacturer's pot-life is conservative or the PVA may have contributed little to the bond strength.

The effect of curing the 0.22 water-cement ratio mortar, in 100% relative humidity from age 1 day until testing 27 days later was to reduce the bond strength significantly to 20.9 MPa (B22). Some of the strength reduction can be attributed to the difference in specimen moisture content during testing. Neville (1981) shows data which indicate that moisture content caused by long exposure to relative humidities between 50% and 100% does not affect recorded mortar strengths significantly. For confirmation, three 3 x 6" concrete cylinders were cured and tested in each condition. The difference in strength was less than 10%. The results therefore suggest that copolymer PVA bonds are sensitive to moisture over time periods as short as 27 days. The fact that the PVA bonds exposed only 13 days to 100% relative humidity are weaker than most of the portland cement bonds, suggests moisture may damage copolymer PVA bonds over an even shorter period of time. Shaw (1983) has noted previously that homopolymer PVA is known to be vulnerable to alkaline moisture but that copolymer PVA was developed to perform better in this regard.

4. Conclusions

(1) Of the four tests evaluated, the most sensitive and least variable method of testing for bond strength and flexibility, is the slant shear test.

(2) For portland cement mortar bonds, the thickness of the mortar layer has a highly significant effect on bond strength. An exces-

sively thick bond will cause a considerable reduction in bond strength.

(3) The treatment of the bond surface of the substrate concrete by pre-wetting appears to have a small beneficial effect on bond strength.

(4) Copolymer PVA is a poor bonding agent over a wide range of curing conditions and mortar mix designs. Under the laboratory conditions employed, the use of PVA produced consistently weaker bonds than using no bonding agent at all.

Acknowledgements

The authors gratefully acknowledge the financial support of the Natural Sciences and Engineering Research Council of Canada and the University of Calgary. The technical assistance provided by Mssrs. Harry Pollard, Don McCullough, Don Anson, Paul Houle, Erich Damson and Ron Begert, under the guidance of Mr. Howard Johnson, is sincerely appreciated.

References

Dixon, J.F. and Sunley, V.K. (1977) Use of bond coats in concrete repair. Concrete (London), 17, 8, 34-35.

Francis, A.J., Horman, C.B. and Jerrems, L.E. (1971) The effect of joint thickness and other factors on the compressive strength of brickwork, Proc. 2nd Int. Conf. on Brick Masonry, B. Cer. Res. Ass., Stoke-on-Trent, England.

Kreigh, J.M (1976) Arizona slant shear test: a method to determine epoxy bond strength. J. ACI, 73, 7, 372, 373.

Moss, P.J. and Batchelar, M.L. (1975) Structural bonding using epoxy resin adhesives. New Zealand Engineering, 30, 2, 45-52.

Neville, A.M. (1981) Properties of Concrete 3rd. Ed. Pitman Publishing, Inc., Marshfield, MA.

Shaw, J.D.N. (1983) The use of polymers in concrete repair, Civil Engineering (London), June, July 1983.

Sise, A. (1984) Flexural bond strength of masonry, M.Sc. Thesis, University of Calgary.

Tabor, L.D. (1979) Concrete repair. Precast Concrete, 10, 2, 65-68.

Walpole, R.E. and Myers, R.J. (1978) Probability and Statistics for Engineers and Scientists 2nd Ed. Macmillan Publishing Co., New York, N.Y.

Section 7

Crack injection

Injection de fissures

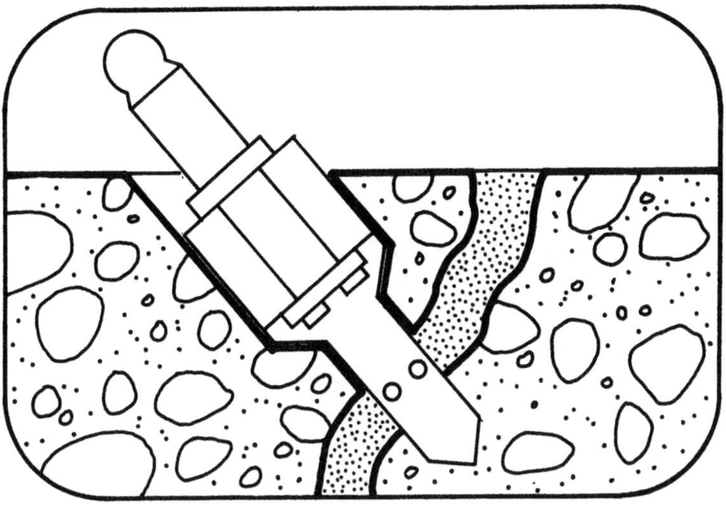

INJECTIONS POUR COMPLEMENT D'ETANCHEITE A L'AIR DU BETON DES ENCEINTES DE REACTEURS NUCLEAIRES PW 1300

B. BAUDOUIN   E.D.F - C.E.M.E.T.E SECTION PRODUITS SPECIAUX
QUARTIER VALCROS CD 65 13090 AIX EN PROVENCE

Résumé
L'étanchéité de l'enceinte de confinement des centrales nucléaires PWR du Palier 1300 MW est assurée par une double paroi : interne en béton précontraint et externe en béton armé.

Pour obtenir un débit de fuite de l'enceinte interne inférieur au debit spécifié en cas d'accident, les dispositions suivantes sont réalisées :

- confection et mise en oeuvre d'un béton étanche à l'air,
- traitement des discontinuités telles que reprises, fissures, traversées des fourreaux, par l'injection de produits appropriés.

Ce sont, en effet, ces discontinuités qui sont généralement le siège des fuites.

La sélection des produits d'injection doit tenir compte des particularités de ces fuites :

- les études ont porté entre autres sur :
  . l'évolution de la viscosité, en fonction du temps,
  . le pouvoir mouillant des produits,
  . les caractéristiques mécaniques (résistance à la traction, dureté Shore A) des résines.

C'est ainsi qu'une gamme importante de produits a été examinée, surtout en fonction de leurs possibilités d'emploi sur chantier.

En parallèle, ont été menés des essais d'application in situ, afin de vérifier le comportement des produits sélectionnés.

Nous disposons après cette étude, de deux types de solutions pour le traitement des singularités de bétonnage :

- des résines époxydiques super-fluidifiées,
- des silicates de soude de diverses provenances.

Mots-clés : Etanchéité, Béton, Centrales nucléaires, Résines, Silicates, Injection.

1. Introduction

Le bâtiment réacteur des centrales nucléaires PWR doit résister aux conditions accidentelles de référence dont l'hypothèse consiste en la rupture d'une des tuyauteries d'évacuation de l'énergie du réacteur. Dans cette hypothèse, le bâtiment réacteur doit résister mécaniquement à la pression engendrée par l'accident et doit confiner par une parfaite étanchéité les matières radio-actives qui pourraient être libérées.

Pour le palier 1300 MW, le bâtiment réacteur est entièrement réalisé en béton et l'étanchéité est assurée par une double paroi :

interne en béton précontraint et externe en béton armé. L'espace inter-paroi est maintenu en dépression et l'air extrait est traité par filtration.

Pour obtenir un débit de fuite de l'enceinte interne inférieur au débit spécifié en cas d'accident, les dispositions suivantes sont réalisées :

- confection et mise en oeuvre d'un béton étanche à l'air,
- traitement des discontinuités telles que reprises, fissures, traversées des fourreaux par l'injection des produits appropriés.

Ce sont, en effet, ces discontinuités qui sont généralement le siège des fuites.

Dans ce cadre, il a été prévu, dès la construction, d'incorporer au niveau des discontinuités de la structure béton, un dispositif de contrôle et d'injection constitué d'un tube spiralé métallique (type protection de câbles électriques), d'un diamètre de 25 mm environ, entouré d'un feutre non tissé (BIDIM) destiné à empêcher la laitance de colmater les espaces entre les spires du tube métallique.

Ce dispositif permet par injection d'air de contrôler l'étanchéité de l'ouvrage et, éventuellement, le traitement du défaut ainsi équipé.

2. Méthodes de sélection des résines

A la lumière d'une part d'essais antérieurement effectués sur maquettes, d'autre part des caractéristiques définies pour l'étanchéité, les produits devant être utilisés pour améliorer de façon sensible l'étanchéité des singularités de bétonnage doivent avoir les propriétés suivantes :

- une faible viscosité permettant au produit de cheminer dans des fissures très fines (certaines fissures des maquettes, quasi invisibles à l'oeil nu, présentaient des débits de fuite prohibitifs),
- un bon pouvoir mouillant, pour les mêmes raisons, avec en outre pour les meilleurs produits la possibilité de remplir les fissures par capillarité,
- une relative insensibilité à l'eau, en raison de l'humidité subsistant en permanence dans une paroi de béton épaisse,
- pas trop de retrait de polymérisation, ce qui risquerait d'aboutir à une fuite résiduelle,
- un temps de polymérisation suffisamment long pour permettre l'injection de sections assez étendues, et la pénétration du produit jusqu'à l'extremité fine des fissures avant que celui-ci ne polymérise dans le canal de répartition,
- un coût final raisonnable relativement à la satisfaction obtenue.

2.1 Moyens mis en oeuvre pour une 1ère sélection

2.1.1 Mesure de la viscosité

La viscosité des produits est appréciée au moyen d'un viscosimètre (type DRAGE ou BROOKFIELD).

2.1.2 Mesure de la consistance
La mesure de la consistance à la coupe AFNOR (norme T 30 014) est une méthode permettant d'apprécier également la viscosité du produit. C'est un moyen d'identification sommaire mais rapide du produit sur chantier.

2.1.3 Montée capillaire dans une colonne de sable
Nous avons noté la hauteur de la colonne de sable que les produits envoyés avaient atteint par montée capillaire avant de polymériser. Cet essai a été effectué sur sable sec et humide.

Nous avons utilisé du sable siliceux obtenu par tamisage du sable SP de "Silice et Kaolin", la granulométrie retenue pour sélectionner les produits correspond à la fraction comprise entre les tamis 0,315 mm et 0,160 mm. Ce sable est enfermé dans un tube plexiglass dont la partie inférieure est obturée par un tamis métallique de 80 μm ; il constitue une colonne de 10 cm.

Les essais sur sable humide se font sur colonne de sable préalablement humidifiée par capillarité.

2.1.4 Pénétration dans la farine de silice
Une goutte d'eau, déposée sur la farine de silice (référence C 400 de la Société SIFRACO) mouille celle-ci immédiatement.

Une goutte de produit a été déposée sur cette silice, pour apprécier la vitesse de pénétration de celui-ci.

La même goutte déposée sur la silice humidifiée a en général un comportement différent.

2.1.5 Durée pratique d'utilisation
Cette notion est liée à la vitesse de polymérisation du produit, qui elle même est influencée par de nombreux facteurs, dont le volume de l'échantillon de mesure.

La détermination du pic d'exothermie comme moyen d'appréciation de la durée pratique d'utilisation n'est réalisable que sur des produits à réaction relativement rapide, aussi nous nous sommes limités à l'observation visuelle de la gélification, sur un échantillon de 100 grammes.

2.1.6 Injection d'une fissure type
Pour modéliser le comportement de chacun des produits à l'intérieur des singularités de bétonnage il a été décidé d'injecter avec chacun d'eux une fissure type créée à partir d'une éprouvette de béton sciée puis reconstituée, chaque partie ayant été rectifiée et séparée de l'autre par deux lamelles de clinquant de 0,02 mm formant génératrice.

Ce béton ayant une perméabilité connue il est possible de déterminer dans des conditions définies le débit de fuite d'une éprouvette de béton avant sciage, et dans les mêmes conditions le débit de fuite de l'éprouvette avec la fissure type, vide et injectée de produit.

Ces mesures sont effectuées sous une pression de 4 bars d'azote, sur éprouvettes séchées à poids constant.

L'amélioration de l'étanchéité apportée par chaque produit essayé a été appréciée sur béton injecté à sec et humide.

2.2 Produits soumis aux essais

Nous avons sélectionné parmi quelques formulateurs de résine ceux qui disposaient de résines d'injection dont les caractéristiques répondaient selon eux à nos objectifs.

Par ailleurs, les essais réalisés sur maquettes avaient utilisés des résines très fluides qui devaient être testées en laboratoire pour comparaison.

Nous avons donc soumis aux essais mentionnés plus haut les produits suivants :

Société A :
. Produit 1 = Résine époxydique sans solvant

Société B :
. Produit 2 = Résine polyester très fluide

Société C :
. Produit 3 = Résine styrénique. Cette résine a été mise en oeuvre sur les maquettes de TRICASTIN et de PALUEL

Société D :
. Produit 4 = Résine époxydique sans solvant

Société E :
. Produit 5 = Résine époxydique sans solvant

Société F :
. Produit 6 = Résine époxydique sans solvant

Société G :
. Produit 7 = Résine époxydique sans solvant

Société H :
. Produit 8 = Résine époxydique sans solvant

Société I :
. Produit 9 = Résine époxydique sans solvant

2.3 Résultats

Les résultats concernant les produits soumis aux essais figurent dans les tableaux A et B ci-après.

Nous avons eu quelques problèmes lors de l'essai d'injection de fissures avec les résines Produit 2 et Produit 3 qui se sont injectées très rapidement, mais dont on n'a pas pu arrêter la progression. Un essai de polymérisation rapide à la flamme molle a enflammé l'appareillage.

Nous avons rempli la fente de façon à affleurer le haut de l'éprouvette mais la résine a disparu avant prise (évaporation ou capillarité dans le béton ?) et la fente s'est vidée en partie. Les éprouvettes séparées dans le plan de la fente après l'essai ont d'ailleurs montré le mauvais remplissage et la mauvaise adhérence de ces résines.

On constate en outre qu'aucune résine ne chasse l'eau par simple capillarité. Par contre toutes pénètrent dans la silice mouillée. Cette facilité de pénétration sur support humide est confirmée par tous les autre essais : la pellicule d'eau enrobant les grains semble servir de lubrifiant.

Si on prend pour principaux critères la DPU suffisamment longue jointe à une viscosité basse et un comportement satisfaisant à l'étanchéité de fissure, hormis les deux résines citées plus haut, on remarque les trois produits n° 1, n° 5, n° 7.

Cependant il faut noter que le dernier produit (n° 7) ne mouille pas les surfaces sèches (test à la farine de silice et temps d'injection de fissures très long).

Tableau A : Résultats des premiers essais de sélection

			Montée capillaire		Pénétration farine de silice 15 min.			
	Viscos. Drage à 22°C (cP)	Consist. coupe Afnor n°6 (s)	Haut. sable sec (mm)	Haut. sable humide (mm)	Sèche	Humide	DPU annoncée	DPU mesurée sur 100g
A Produit 1	300	21	45	0	OUI	OUI	2h10	1h45
B Produit 2	1	5	100	0	OUI	OUI	10h	8h
C Produit 3	2,2	5	100	0	OUI	OUI	2h30	7h
D Produit 4	270	20	32	0	OUI	OUI	0h50	0h30
E Produit 5	110	11	62	0	OUI	OUI	3h	1h30
F Produit 6	440	26,6	25	0	OUI	OUI	0h20	0h30
G Produit 7	470	26,2	30	0	NON	OUI	1h30	1h30
H Produit 8	202	14,6	32	0	MOYEN	OUI	4h	1h05
I Produit 9	950	72	25	0	OUI	OUI	4h	2h25

Tableau B : Résultats des premiers essais de sélection.

Essais d'injection de "fissures types"

	Débit éprouvette béton 1/h	Débit éprouvette fissure 1/h	Débit éprouvette injectée 1/h	Temps injection mn.s	Pression injection bar
A Produit 1	S 1,8	S 12	S 2,3	S 13.35	4'30":0,6 9'05":1
	H 3,3	H 15	H 3	H 2.55	0,6
B Produit 2	S 1,8	S 47	S 45	S 2s	0,6
	H 1,8	H 47	H 40	H 1.00	Capillarité
C Produit 3	S 1,5	S 15	S 13	S 6s	0,6
	H 1,5	H 30	H > 12	H 35s	0,6
D Produit 4	S 1,8	S 5	S 2	S 20.00	1'30":0,6 puis :1,5
	S 1,8	H 25	H 12,5	H 7.35	0,6
E Produit 5	S 3,3	S 20	S 0,94	S 6.30	1'30":0,6 3'45":1,2
	H 3,3	H 7,5	H 6	H 31s	0,6
F Produit 6	S	S	S	S	
	----- D.P.U. TROP COURTE -----				
	H	H	H	H	
G Produit 7	S 2,5	S 8	S 1,8	S 22.00	3' :0,6 3'45":1,2
	H 2,9	H 18	H 13	H 7.06	0,6
H Produit 8	S 1,6	S 10	S 8,2	S 34.00	0,6
	H 1,6	H 12	H 12	H 33.00	0,6
I Produit 9	S	S	S	S	
	----- RESINE TROP VISQUEUSE -----				
	H	H	H	H	

2.4 Essais sur chantier

Les produits sélectionnés, n° 1 (300 centipoises) et n° 5 (100 centipoises) ont été injectés sur une maquette grandeur nature, réalisée sur l'aménagement de PALUEL.

Ces essais ont montré la nécessité de s'orienter, compte tenu de la configuration des réseaux d'injection, vers la recherche de produits plus fluides et ayant une durée pratique d'utilisation plus longue.

Tenant compte des informations ainsi obtenues, il a été élaboré un mode opératoire d'injection à partir de cette expérience : description du matériel, pression d'injection, etc...

Cette phase d'essais a permis l'injection du radier du bâtiment réacteur de la tranche 1 de Paluel, avec des résines époxydes fluidifiées et de durée pratique d'utilisation allongée conçues par les sociétés A et E.

2.5 Seconde série d'essais

L'expérience de l'injection du radier de Paluel nous a orientés vers des définitions plus précises des qualités à attendre des produits d'injection :

- valeur d'extrait sec élevée,
- viscosité durablement basse,
- mouillabilité bonne vis-à-vis du béton,
- résistance mécanique suffisante (mais pas forcément importante).

Des essais plus complets ont donc été entrepris en liaison avec les formulateurs, sur différentes familles de produits :

a/ Résines polyesters

Les résines polyesters peuvent être diluées par un solvant réactif, et donc avoir une très grande fluidité ; de plus le temps de prise est réglable par adjonction de plus ou moins d'accélérateur de prise. Des résines formulées dans cet esprit nous ont été proposées, mais de nombreux inconvénients se sont manifestés :

- un échantillon n'a jamais durci,
- un autre a pris dans le récipient d'emballage, avec ébullition du produit et débordement...,
- l'inflammabilité est importante,
- l'odeur est très désagréable (ce qui peut poser un problème dans les endroits mal ventilés),
- le produit subit un léger retrait à la prise,
- il est incompatible avec l'eau qui subsiste toujours un peu dans les fissures.

Les résines polyester sont des produits envisageables mais délicats et dangereux d'emploi dans le cadre de nos installations.

b/ Résines acryliques

Une résine d'injection à base de résine acrylique nous a été proposée par la Société J : Produit 10.

La mesure de ses caractéristiques (figure 1) montre une durée pratique d'utilisation beaucoup trop courte. De plus, cette résine est acide (pH : 3) et donc corrosive vis-à-vis du béton et des fers.

c/ Résines époxydes

Différentes résines, répondant à nos souhaits, nous ont été proposées par les formulateurs. Il faut noter que ces résines sont expérimentales, et que nous avons exprimé des souhaits légèrement variables selon les formulateurs, ce qui nous permet de disposer d'un éventail des possibilités pour sélectionner le meilleur profil répondant à nos besoins.

C'est ainsi que nous avons mesuré les caractéristiques des résines suivantes :

Société E : E00 SF 2 (produit 11)
E00 SF (produit 12)
E00 SF/R (produit 13)
Société B : IN 51 F (produit 14)
IN 52 F (produit 15)
IN 2 F (produit 16)
In 4 F (produit 17)
Société A : 03C 14 (1) correspondant à 2 (produit 18)
03C 14 (2) livraisons différentes (produit 19)
03C 19 (produit 20)
Société I : J x 1435 (produit 21)
Société K : RE 06 (produit 22)

2.5.1 Mesure de la viscosité en fonction du temps

Les mesures des viscosités, effectuées au moyen d'un viscosimètre DRAGE sur 150 g de produit, sont effectuées à la température ambiante de 20°C, en fonction du temps écoulé depuis le mélange des composants.

Les résultats sont portés sur les figures 2, 3, 4, 5 et 6.

2.5.2 Mesure de la montée capillaire en fonction du temps

La montée du produit par capillarité est mesurée en fonction du temps à la température ambiante de 20°C sur 100 grammes de produit.

Le support de montée est constitué par un tube de plexiglass rempli de sable. Celui-ci est obtenu par tamisage du sable SP de "Silice et Kaolin", la granulométrie étant celle comprise entre le tamis 0,315 mm et 0,160 mm. La partie inférieure du tube de plexiglass est obturée par un tamis de 80 µm.

Les résultats sont portés sur les figures n° 7, 8, 9 et 10.

2.5.3 Mesure de la résistance mécanique

La résistance mécanique des produits est évaluée sur éprouvettes haltères moulées, avec un durcissement réalisé à l'air libre, ou en milieu confiné par l'application d'une lame de verre sur les moules. En effet les caractéristiques obtenues dans l'un et l'autre cas sont parfois notablement différentes.

Les résultats des mesures effectuées sur les produits 11 (Société E) et 16 (Société B) sont portés sur les figures n° 11, 12, 13 et 14. Les résistances mécaniques des autres produits fluides n'ont pas pu être appréciées (faute de temps pour le produit 21).

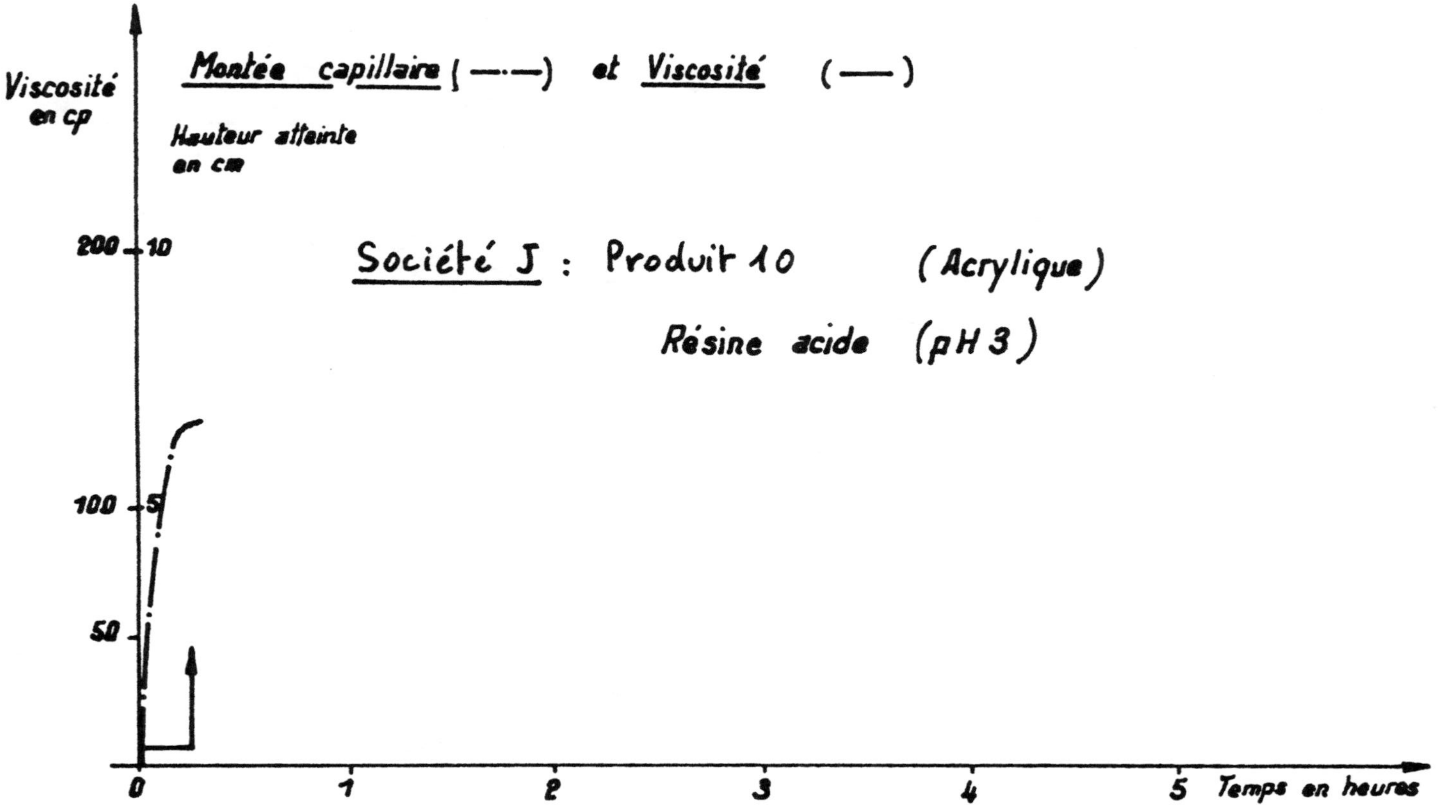
Montée capillaire (—·—) et Viscosité (—)
Viscosité en cp
Hauteur atteinte en cm
Société J : Produit 10 (Acrylique)
Résine acide (pH 3)
200
10
100
5
50
0
1
2
3
4
5
Temps en heures

Figure N° 1

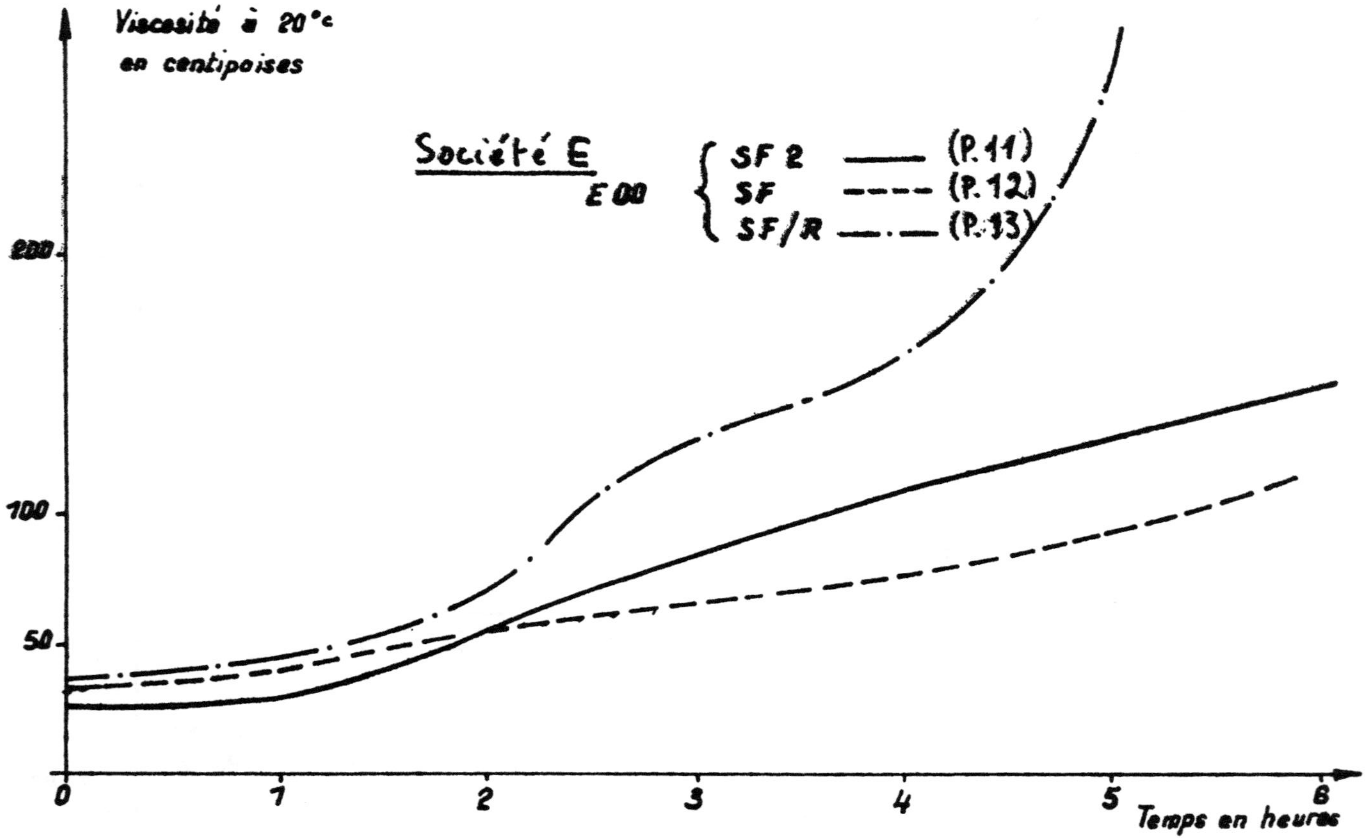

Viscosité à 20°c
en centipoises
Société E
E 00
SF 2 (P.11)
SF (P.12)
SF/R (P.13)
200
100
50
0
1
2
3
4
5
6
Temps en heures

Figure N° 2

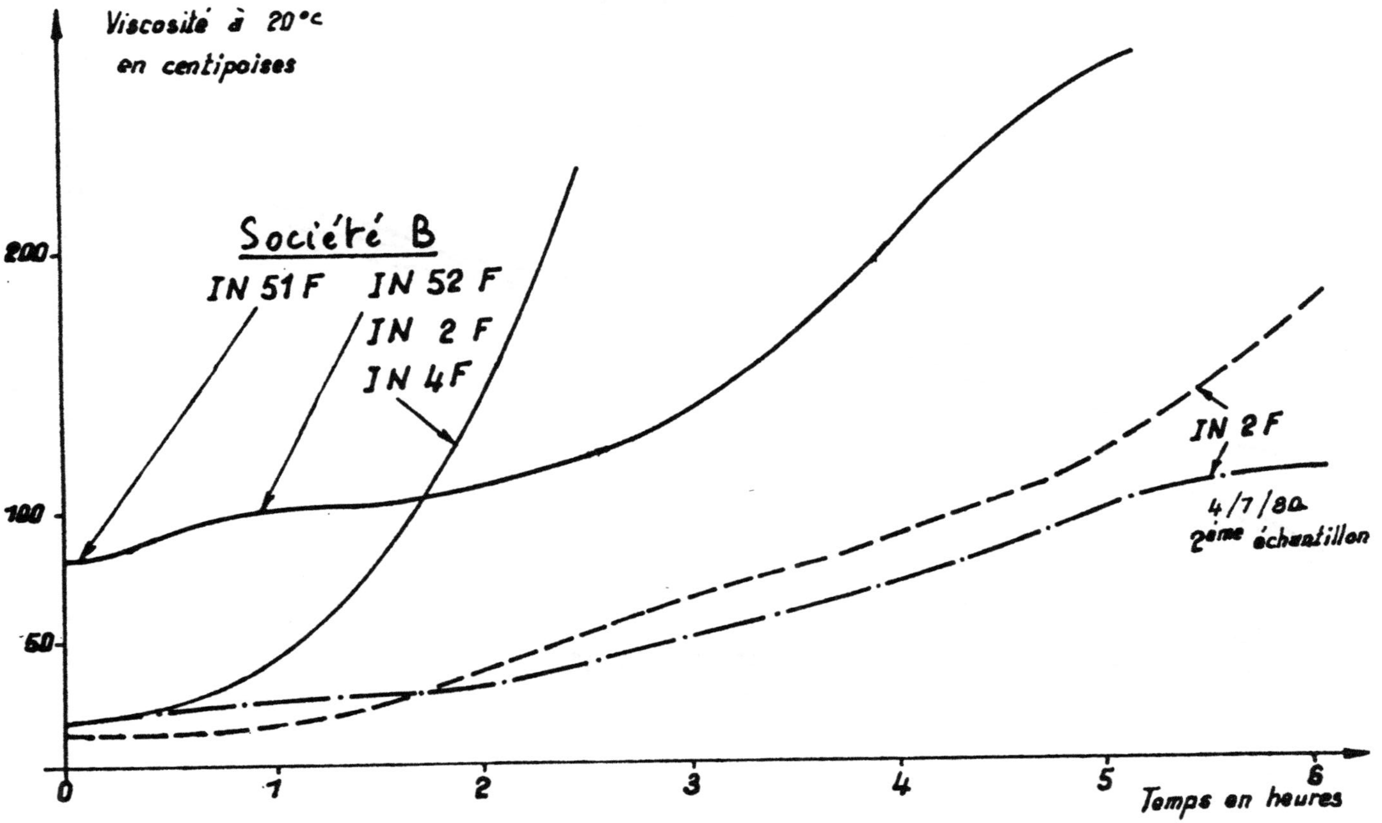
Viscosité à 20°c
en centipoises
Société B
IN 51 F
IN 52 F
IN 2 F
IN 4F
IN 2 F
4/7/80
2ème échantillon
200
100
50
0
1
2
3
4
5
6
Temps en heures

Figure N° 3

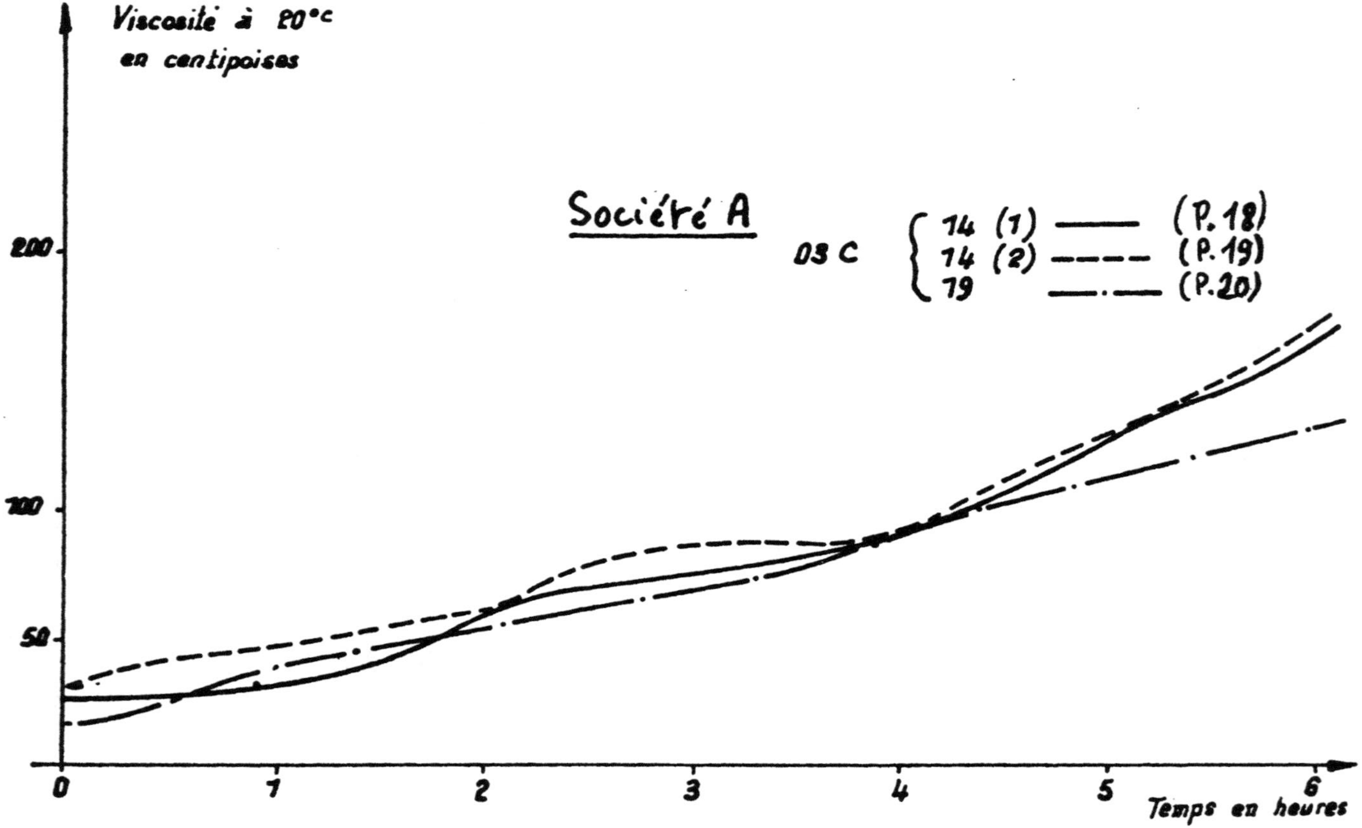

Viscosité à 20°c
en centipoises
Société A
03 C
14 (1) (P.18)
14 (2) (P.19)
79 (P.20)
200
100
50
0
1
2
3
4
5
6
Temps en heures

Figure N° 4

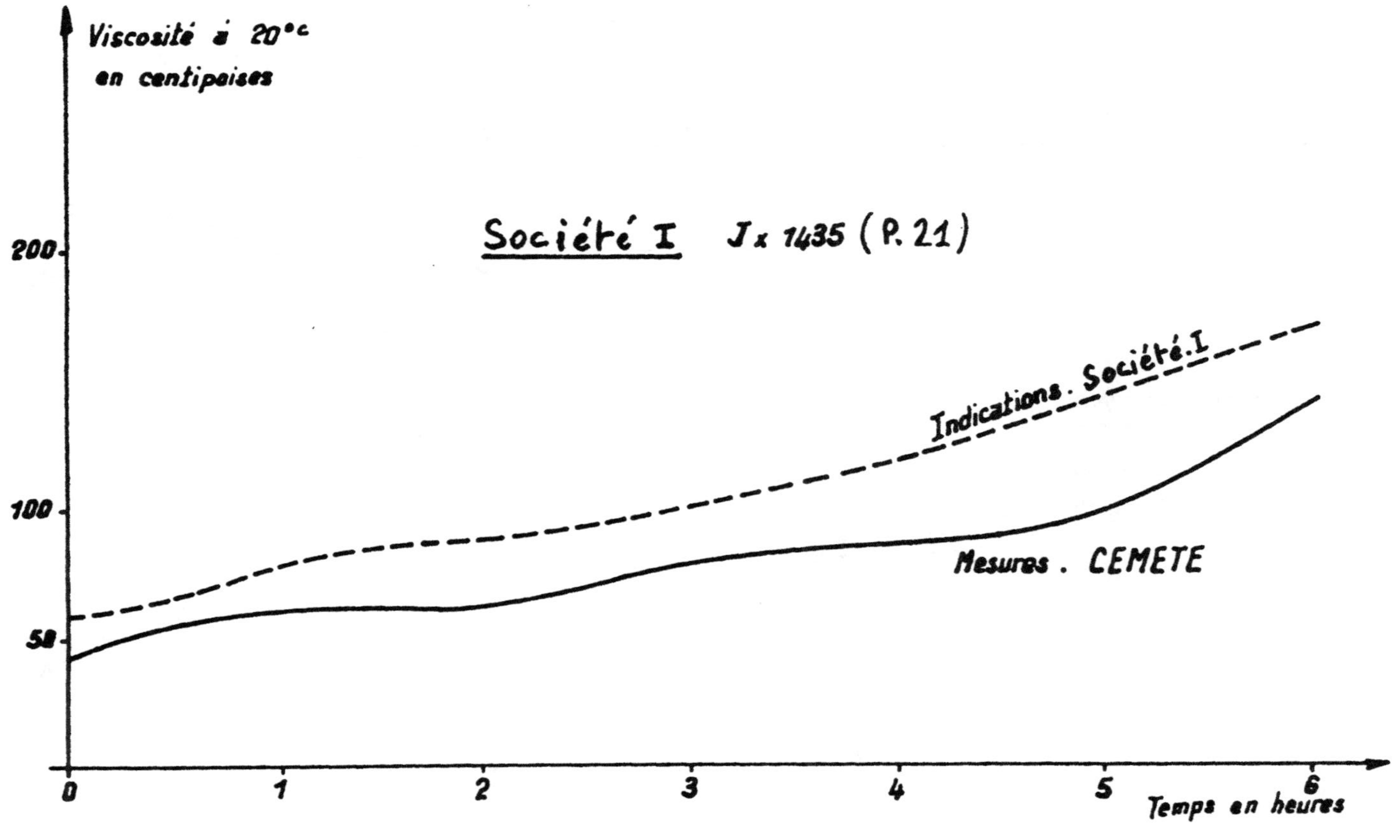
Viscosité à 20°c
en centipoises
Société I Jx 1435 (P. 21)
200
100
50
Indications. Société.I
Mesures. CEMETE
0
1
2
3
4
5
6
Temps en heures

Figure N° 5

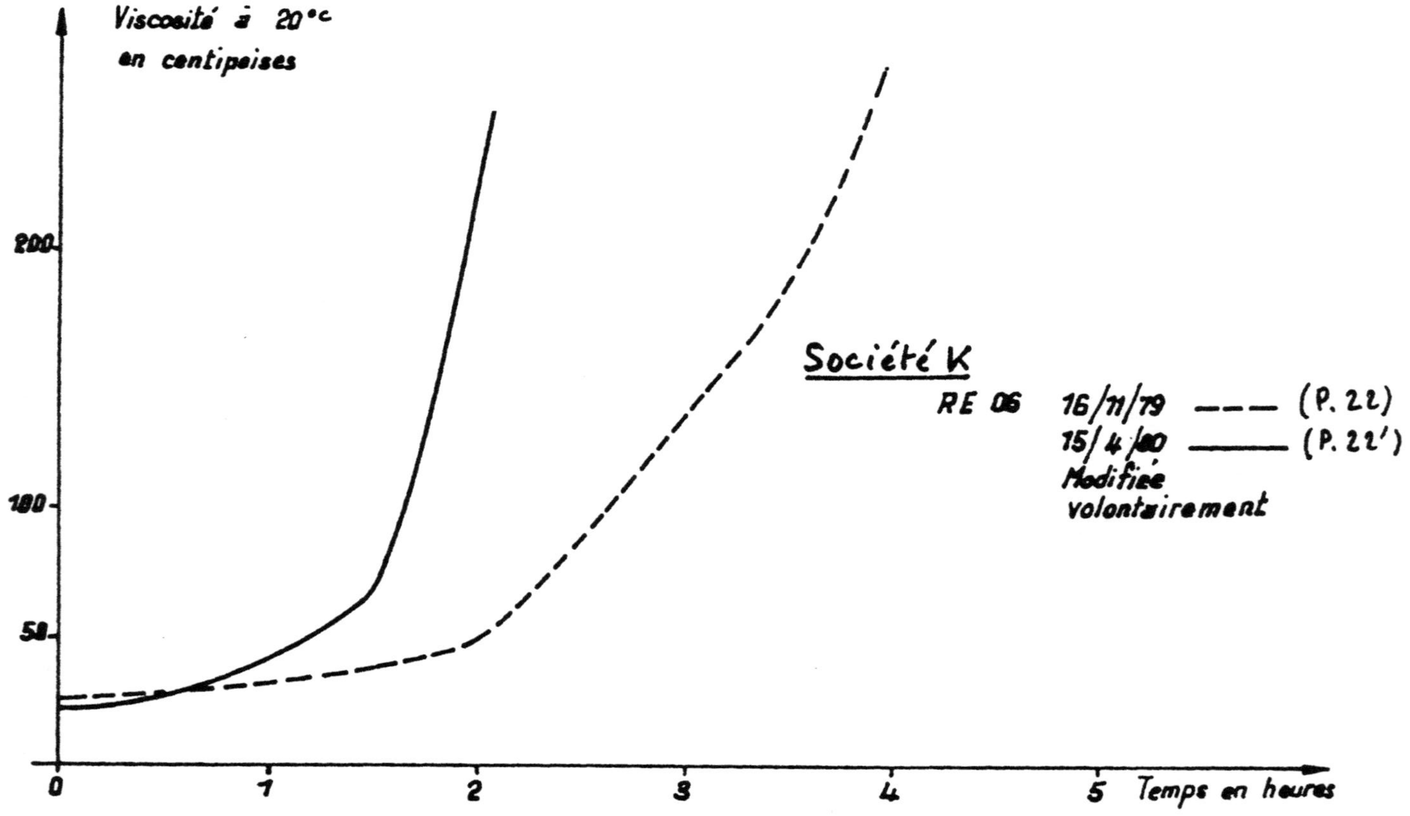

Viscosité à 20°c
en centipoises
200
100
50
0
1
2
3
4
5
Temps en heures
Société K
RE 06
16/11/79 --- (P. 22)
15/4/80 ——— (P. 22')
Modifiée
volontairement

Figure N° 6

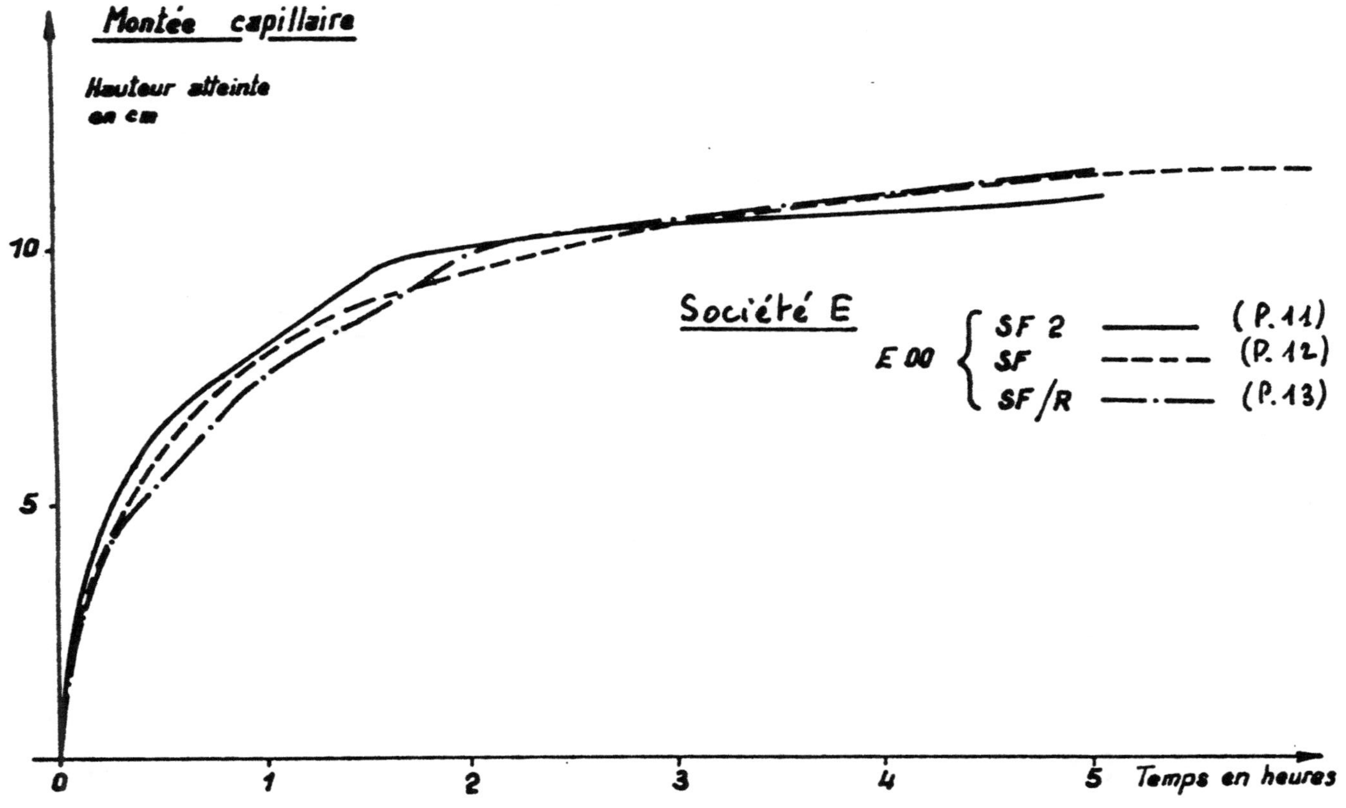
Montée capillaire
Hauteur atteinte en cm
10
5
0
1
2
3
4
5
Temps en heures
Société E
E 00
SF 2 (P. 11)
SF (P. 12)
SF/R (P. 13)

Figure N°7

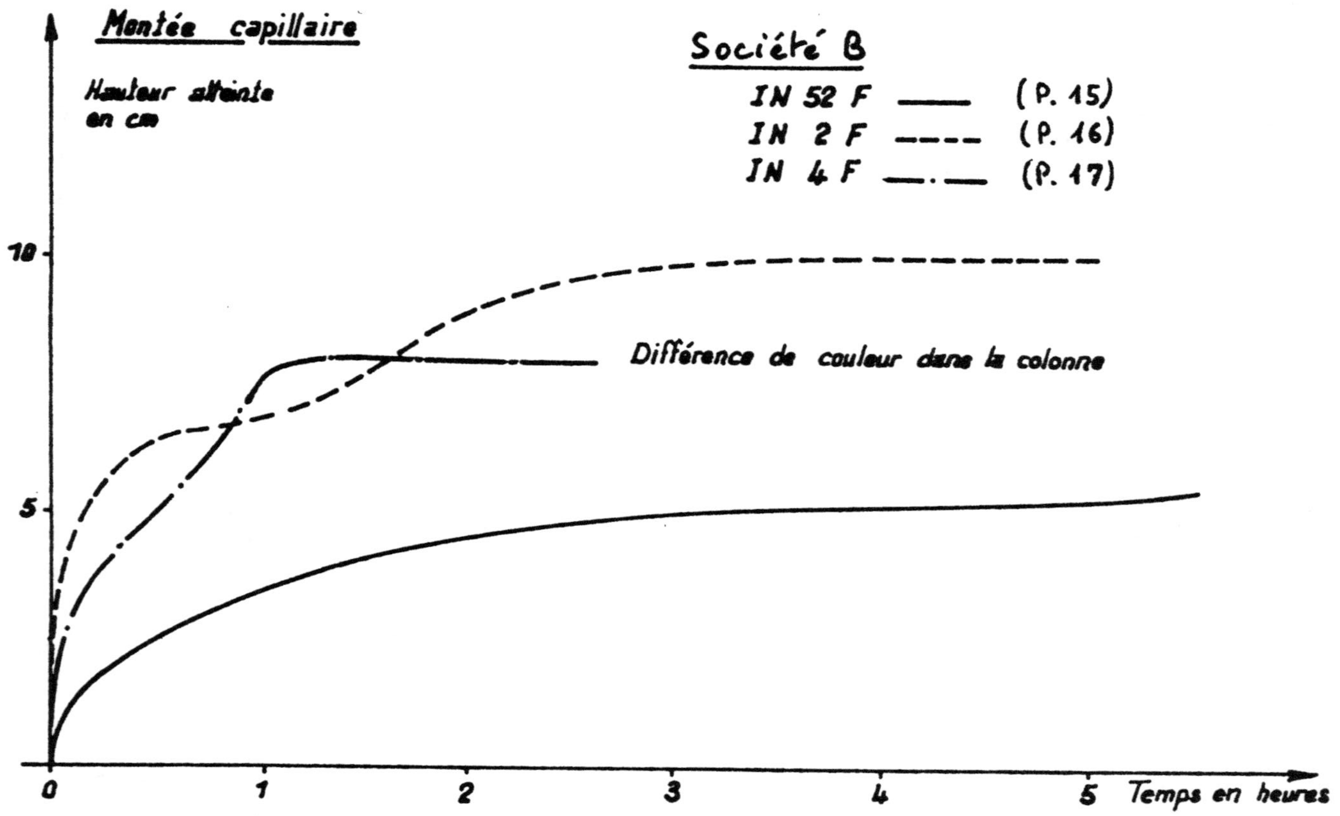
Montée capillaire
Hauteur atteinte
en cm
Société B
IN 52 F (P. 15)
IN 2 F (P. 16)
IN 4 F (P. 17)
10
5
Différence de couleur dans la colonne
0
1
2
3
4
5
Temps en heures

Figure N° 8

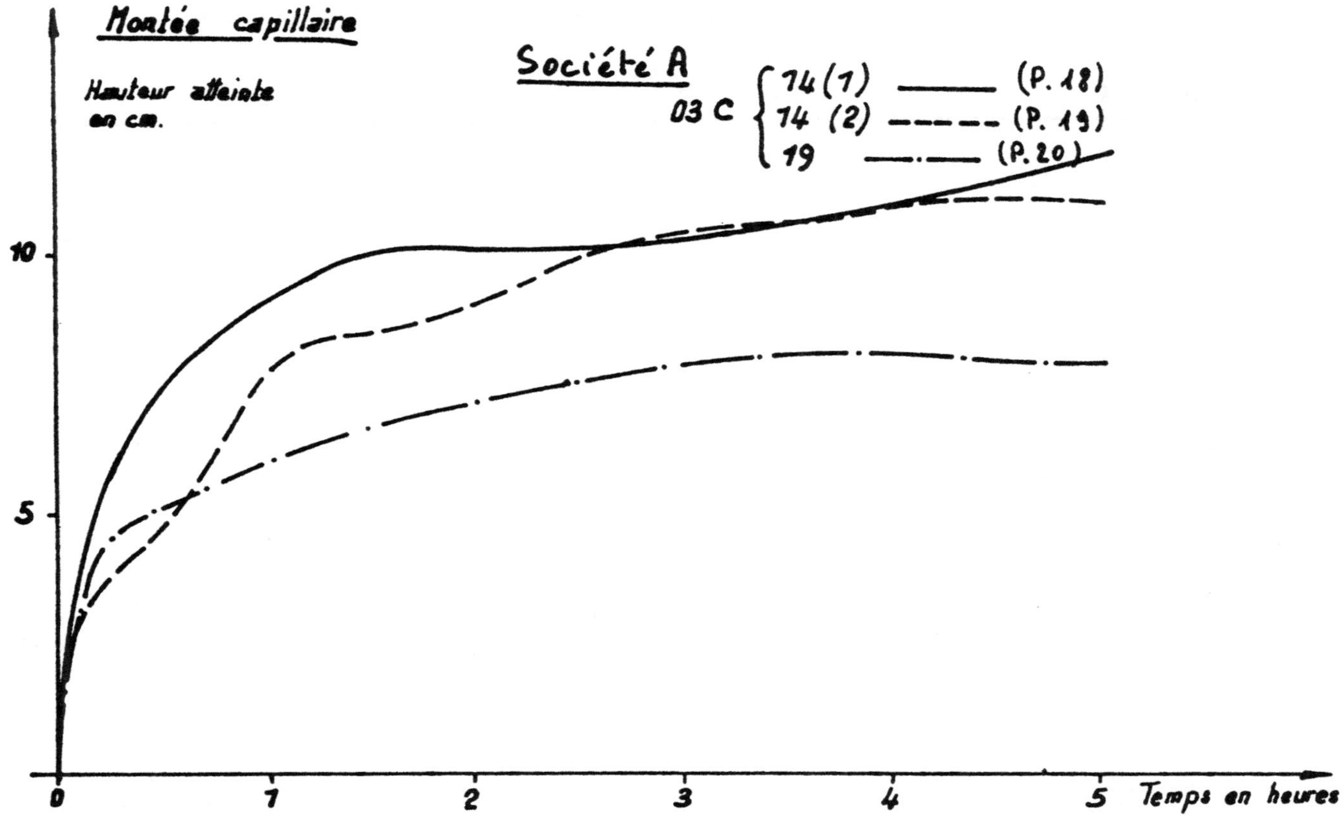
Montée capillaire
Hauteur atteinte en cm.
Société A
03 C
14 (1) (P. 18)
14 (2) (P. 19)
19 (P. 20)
10
5
0
1
2
3
4
5
Temps en heures

Figure N° 9

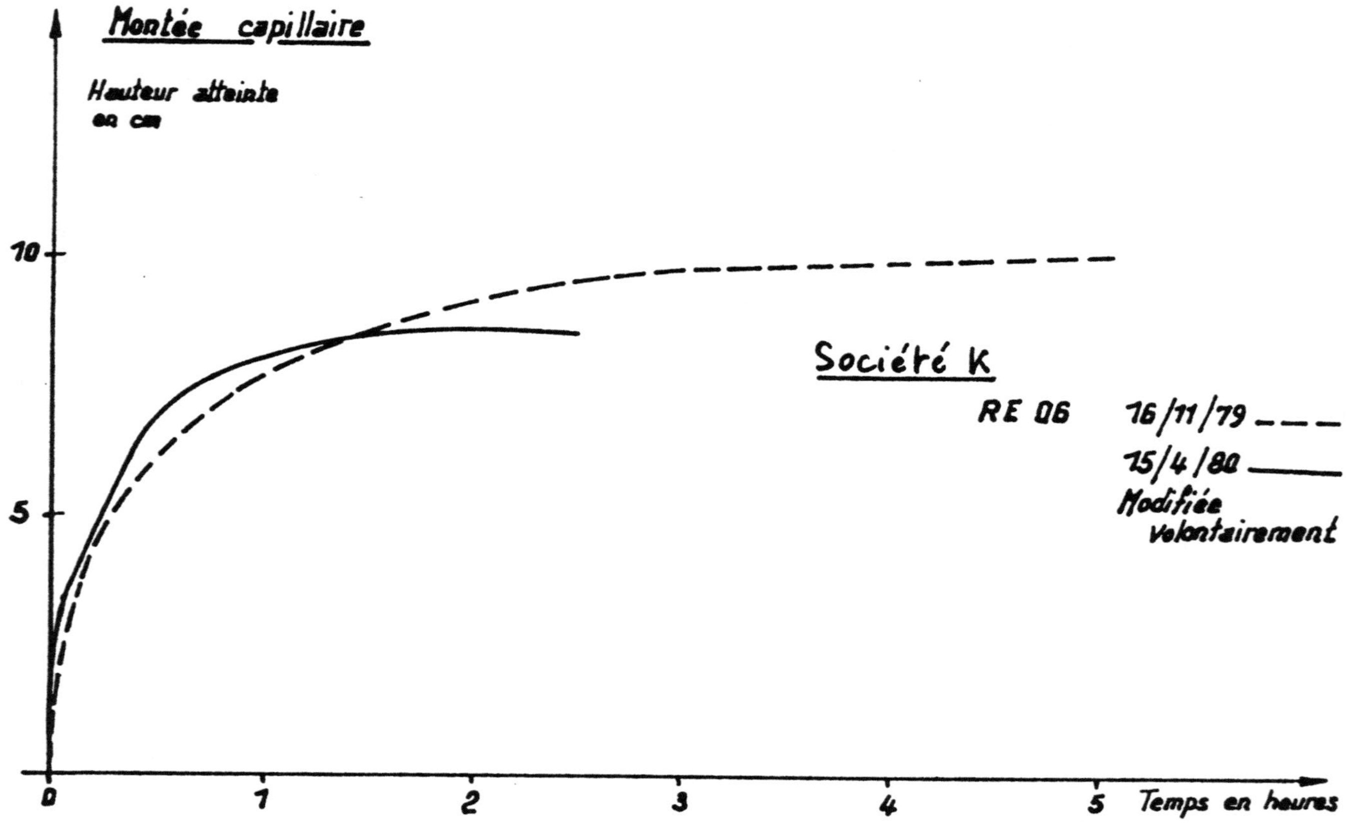
Montée capillaire
Hauteur atteinte en cm
10
5
0
1
2
3
4
5
Temps en heures
Société K
RE 06
16/11/79 ---- (P.22)
15/4/80 —— (P.22')
Modifiée volontairement

Figure N°10

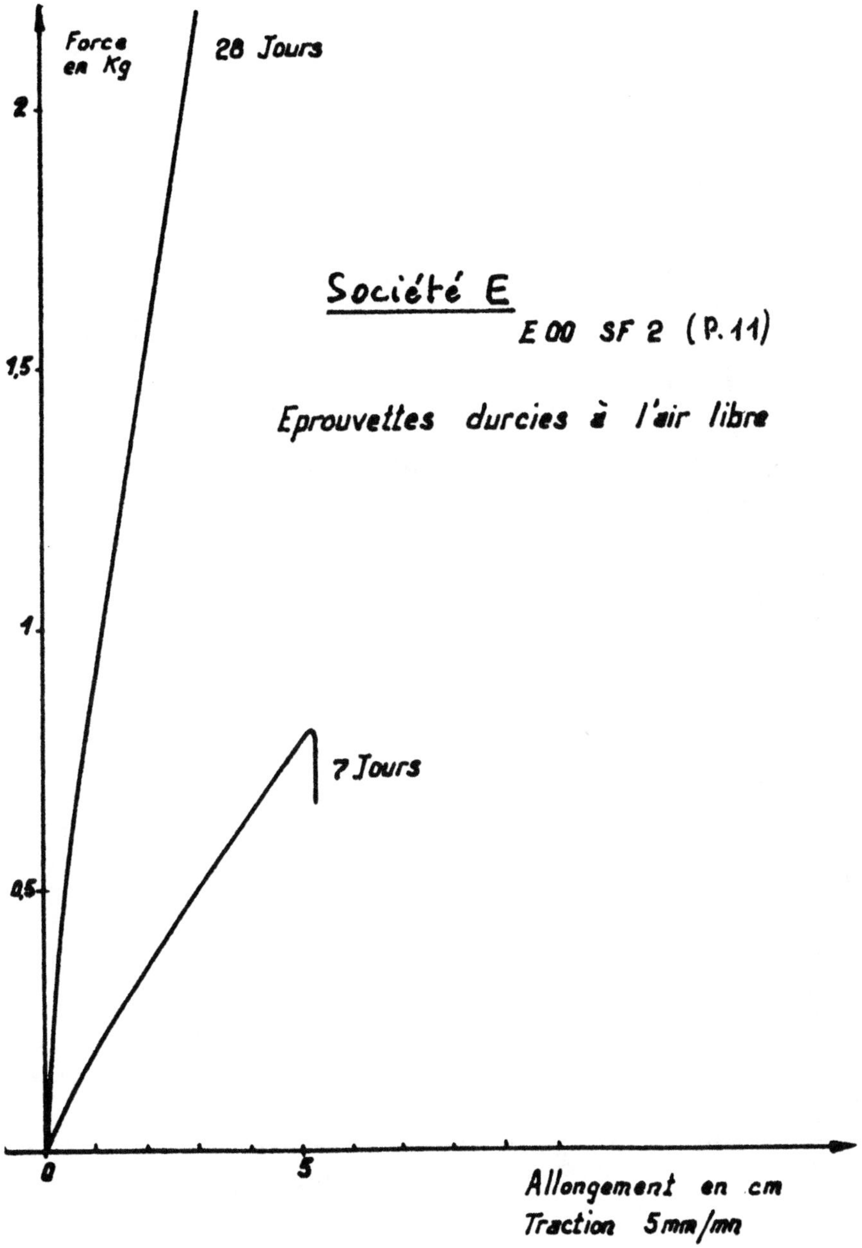
Force en Kg
28 Jours
2
1,5
1
0,5
0
5
Société E
E 00 SF 2 (P. 11)
Eprouvettes durcies à l'air libre
7 Jours
Allongement en cm
Traction 5 mm/mn

Figure N° 11

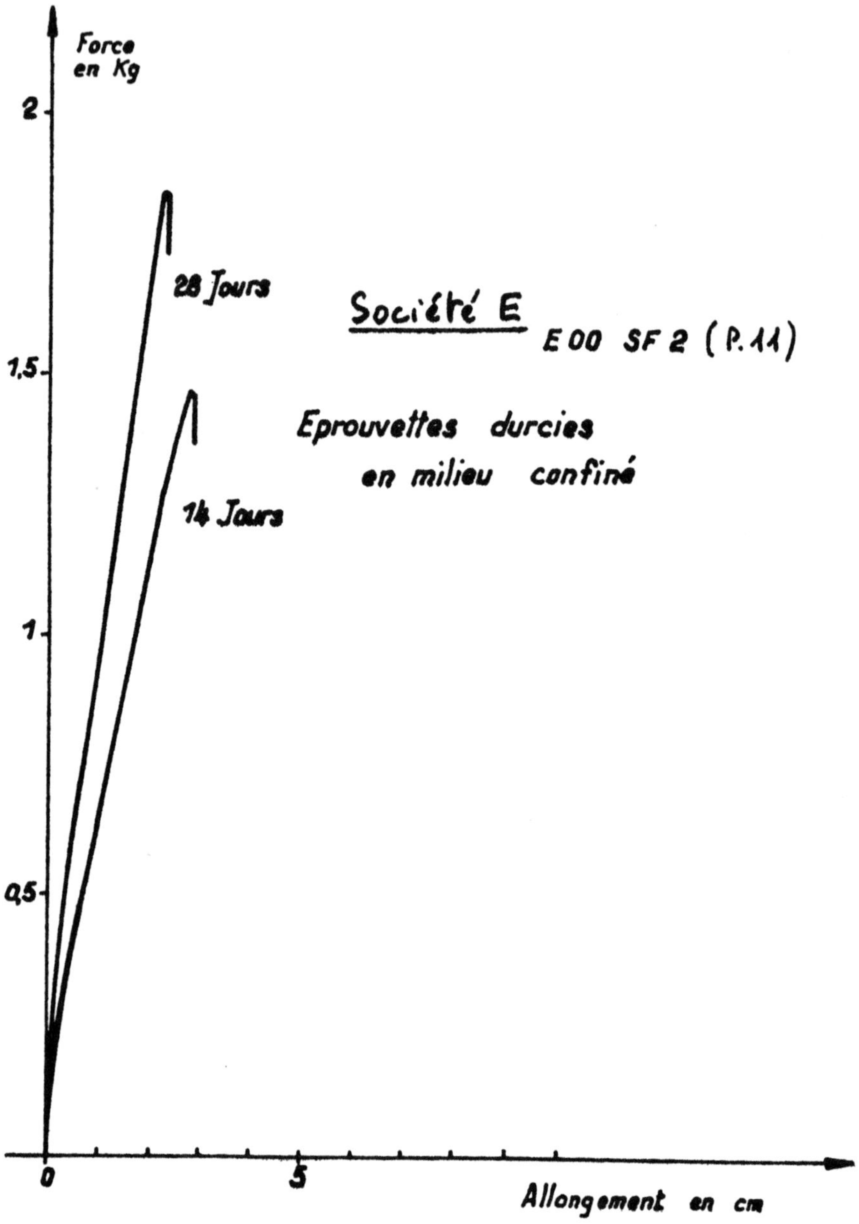
Force en Kg
2
1,5
1
0,5
0
5
Allongement en cm
28 Jours
14 Jours
Société E
E 00 SF 2 (P.11)
Eprouvettes durcies en milieu confiné

Figure N° 12

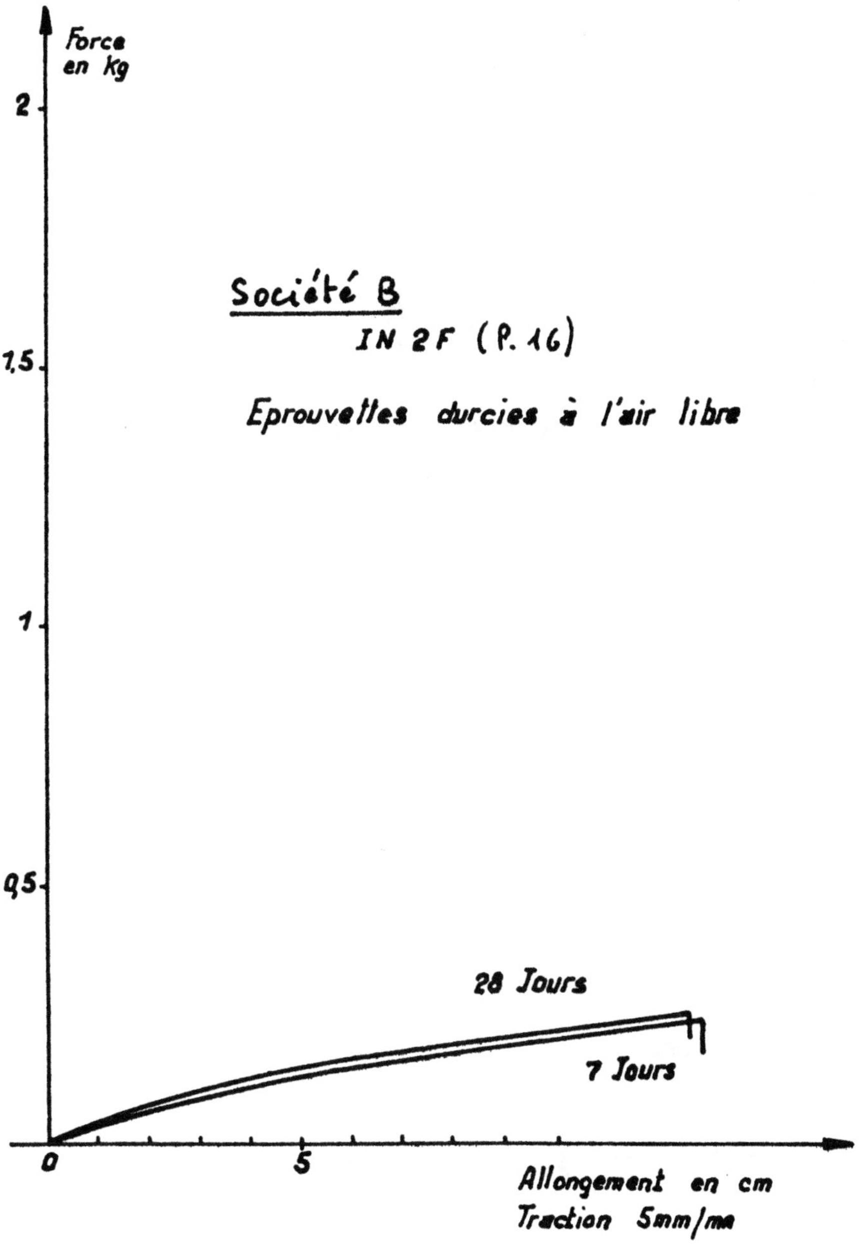

Force en kg
2
1,5
1
0,5
Société B
IN 2F (P. 16)
Eprouvettes durcies à l'air libre
28 Jours
7 Jours
0
5
Allongement en cm
Traction 5mm/mn

Figure N° 13

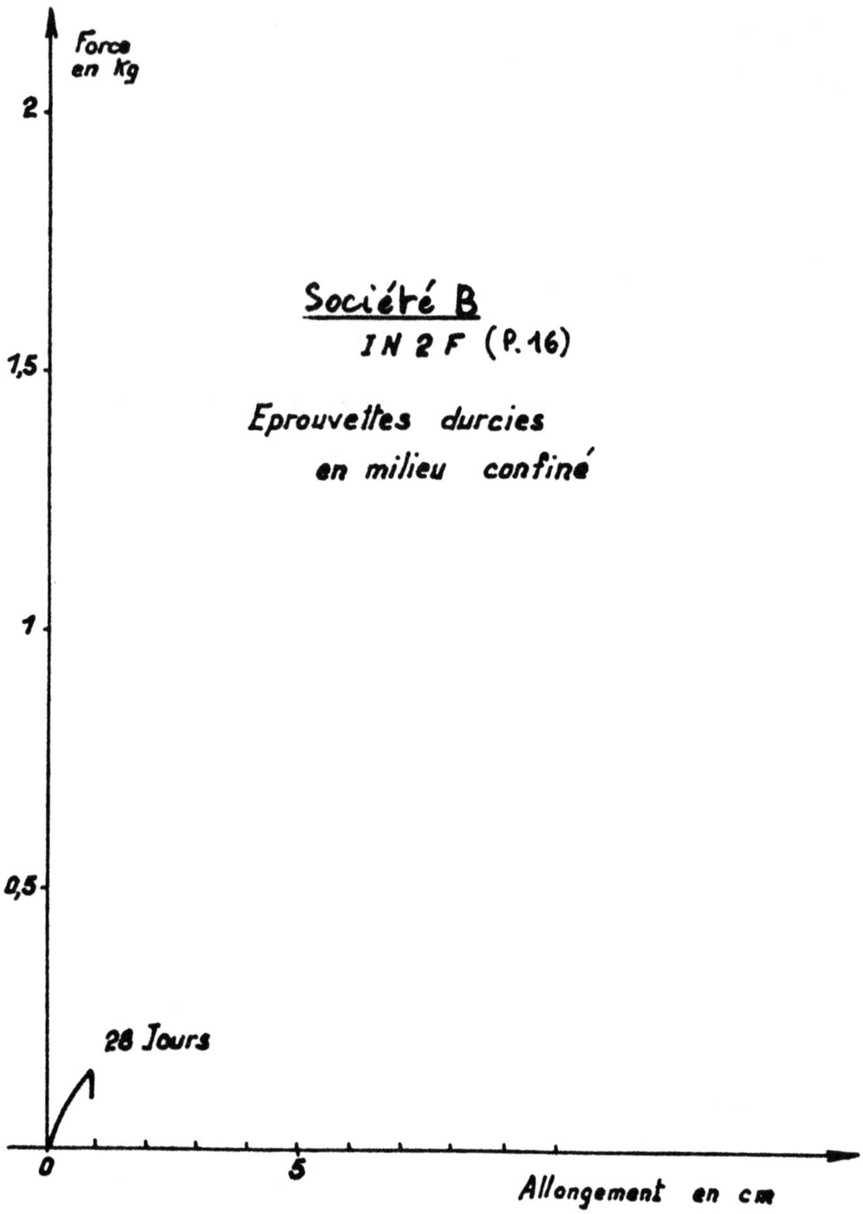
Force en Kg
2
1,5
1
0,5
0
5
Allongement en cm
Société B
IN 2 F (P.16)
Eprouvettes durcies en milieu confiné
28 Jours

Figure N° 14

DURETE SHORE A - EXTRAIT SEC

Polymérisation en milieu confiné

Eprouvettes haltères

Produit	Echéances				Extrait Sec %	Viscosité initiale cp
	7 Jours	14 Jours	21 Jours	28 Jours		
Société I Produit 21	Non polymérisé	Poisseux Shore A:28	Shore A:30	Shore A:27	96,6	40 à 50
Société K Produit 22	Non polymérisé	Poisseux Shore A:0	Poisseux Shore A:0	Poisseux Shore A:0	84,8	20 à 30
Société E Produit 12	Poisseux	Shore A:40	Shore A:46	Shore A:48	90,2	20 à 30
Société B Produit 16	Friable	Friable Shore A:0	Friable Shore A:0	Friable Shore A:0	52,0	10 à 20

Figure N° 15

2.5.4 Dureté Shore A
La dureté shore A a été mesurée sur les talons des éprouvettes fabriquées ci-dessus en milieu confiné.

Les résultats sont notés dans le tableau de la figure n° 15.

2.5.5 Extrait sec
L'extrait sec conventionnel des produits a été mesuré sur des échantillons polymérisés en milieu confiné, puis séchés à l'étuve à 105°C pendant 3 heures.

Les résultats sont notés dans le tableau de la figure n° 15.

2.6 Conclusions
Nous pouvons constater, au vu des figures 2 à 6, que la viscosité initiale n'est pas significative de la viscosité à 1, 2 ou 3 heures.

L'expérience acquise lors des essais sur maquette ou pendant l'injection du radier de Paluel nous a montré l'importance de cette notion.

Nous pourrons donc définir un fuseau réaliste dans lequel doit s'inscrire la courbe viscosité/temps.

Il est également nécessaire de connaître l'influence de la température sur les variations de viscosité. Il est souhaitable qu'elles soient aussi faibles que possible.

Cependant le désir de gagner dans ce domaine ne doit pas conduire à faire baisser l'extrait sec des produits, par adjonction de solvants. Certaines formules sont tombées dans ce travers, comme le montrent les extraits secs du tableau de la figure n° 15.

Nous avons également observé sur une formule, une instabilité du plastifiant utilisé causant une hétérogénéité dans la colonne de sable (par un phénomène de chromatographie). La présence exagérée de plastifiants ou de solvants dans les formules influe également sur les résistances mécaniques : en empêchant ceux-ci de s'évaporer nous avons mis en évidence la perte de résistance mécanique, voire la non-polymérisation lorsque les produits durcissent en milieu confiné (cf. figure n° 15).

La résine à haut extrait sec conventionnel EOO SF 2 (produit 11) a des courbes de résistances mécaniques très concordantes à 28 jours (figures 11 et 12) après durcissement à l'air libre et en milieu confiné.

La résine IN 2 F (produit 16) dont l'extrait sec conventionnel est relativement faible, a par contre des courbes de résistance très dissemblables dans l'une ou l'autre condition (figures 13 et 14).

2.7 Caractéristiques finales
Suite à ces divers essais, tant en laboratoire que sur chantier, nous pouvons définir les caractéristiques auxquelles doivent satisfaire les résines pour répondre à nos conditions d'emploi.

2.7.1 Viscosité fonction du temps
La viscosité en fonction du temps doit être à l'intérieur du fuseau ci-après :

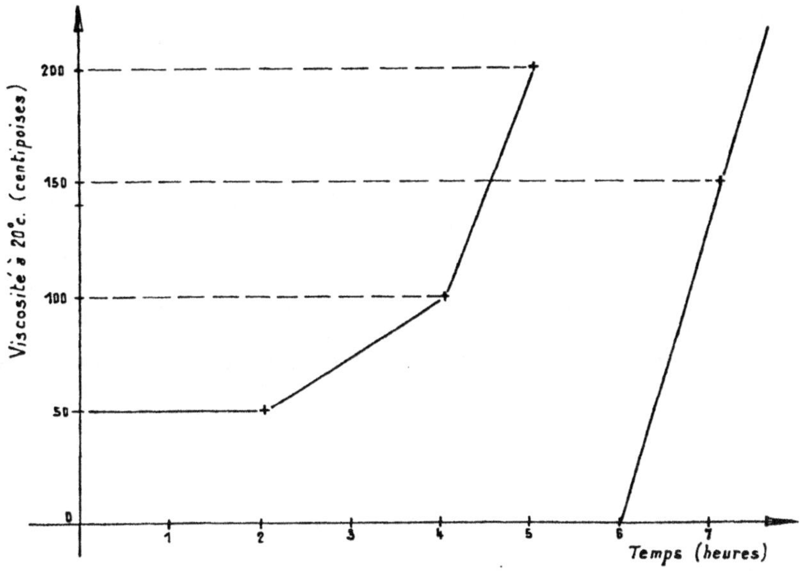

La mesure de cette caractéristique peut se faire à l'aide d'un viscosimètre BROOKFIELD, sur 500 g de produit, à la température ambiante de 20°C. Les temps sont mesurés à partir de la fin d'introduction du dernier composant.

2.7.2 Montée capillaire
La montée capillaire doit être supérieure à 10 cm.

Le support de montée est constitué par un tube de plexiglass rempli de sable. Celui-ci est obtenu par tamisage du sable SP de "Silice et Kaolin", la granulométrie étant celle comprise entre les tamis 0,315 mm et 0,160 mm. La partie inférieure du tube de plexiglass est obturée par un tamis de 80 µm.

2.7.3 Extrait sec
L'extrait sec de la résine doit être supérieur à 90 %.

L'extrait sec conventionnel est mesuré sur un échantillon de résine que l'on a fait polymériser en milieu confié à 20°C (coupelle de mesure couverte). Les caractéristiques de séchage sont 105°C pendant 3 heures.

2.7.4 Dureté shore A
La dureté shore A, doit être, après polymérisation de 28 jours en milieu confiné, supérieure à 25.

2.7.5 Produits existants
Il existe à ce jour, trois produits répondant à ces caractéristiques en provenance des Sociétés I, K, E.

Il s'agit, malgré tout, de produits quasi expérimentaux, dont il est primordial de s'assurer de la conformité avec l'échantillon ayant servi à déterminer les caractéristiques.

Il est donc indispensable, avant toute utilisation, d'effectuer des contrôles de temps de prise, d'extrait sec et de densité au moins sur chaque lot de fabrication.

3. Essais sur les silicates

Les silicates, produits minéraux en solution aqueuse, sont des composés qui sont susceptibles de réagir avec les ions calcium du béton, en formant des cristaux insolubles. L'avantage de cette formule est que ces cristaux ne se forment qu'au sein du béton et que le canal d'injection noyé dans la paroi reste disponible pour des contrôles d'efficacité du traitement.

Par contre, les essais d'efficacité du produit doivent se faire sur des fissures réelles du béton. Les seuls essais de laboratoire possibles sont des mesures de viscosité des produits.

3.1 Produits essayés

Les produits essayés ont été les suivants :

- silicate de soude 35/37°B de la Société J (produit 23)
- silicate de soude 40/42°B de la Société J (produit 24)
- silicate de soude 38/40°B de la Société J (produit 25)
- silicate de potassium de la Société J (produit 26)
- produit 27 de la Société L
- produit 28 de la Société M
- produit 29 de la Société N

Ces deux derniers produits se sont avérés par la suite provenir de la même source.

- produit 30 de la Société M.

3.2 Essais en laboratoire

3.2.1 Viscosité, fonction de la dilution

Les viscosités des produits, dilués dans différentes proportions d'eau, sont portées sur les figures n° 16 et 17.

3.2.2 Viscosité, fonction de la température

Les viscosités des produits, diversement dilués, en fonction de la température sont portées sur les figures n° 18, 19, 20, 21 et 22.

3.2.3 Montée capillaire

a/ Le support de montée est un tube en verre de diamètre intérieur 10 mm, rempli de sable. Celui-ci est obtenu par tamisage du sable SP de "Silice et Kaolin", la granulométrie étant celle comprise entre le tamis 0,315 mm et 0,160 mm. La partie inférieure du tube est obturée par un tamis de 80 µm.
Le volume des vides peut être estimé à 40 % par pesée de ce sable rempli d'eau.

b/ Le support de montée est un feutre de polyester (BIDIM). La notice du fabricant donne les caractéristiques suivantes :

- diamètre des filaments 27,3 µm.
- porométrie : 94 à 59 µm.

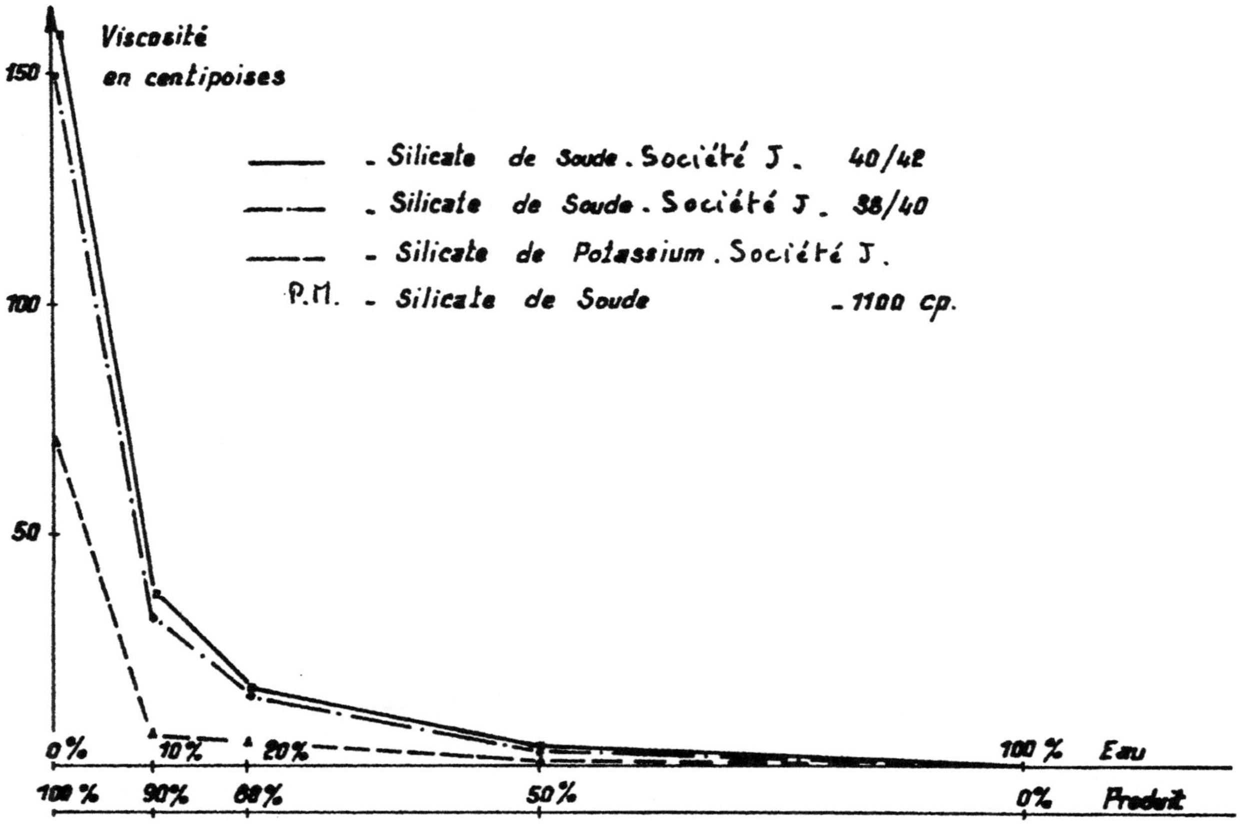
Viscosité
en centipoises
150
100
50
- Silicate de Soude . Société J. 40/42
- Silicate de Soude . Société J. 38/40
- Silicate de Potassium . Société J.
P.M. - Silicate de Soude - 1100 cp.
0%
10%
20%
100 %
Eau
100 %
90%
80%
50%
0%
Produit

Figure N° 16

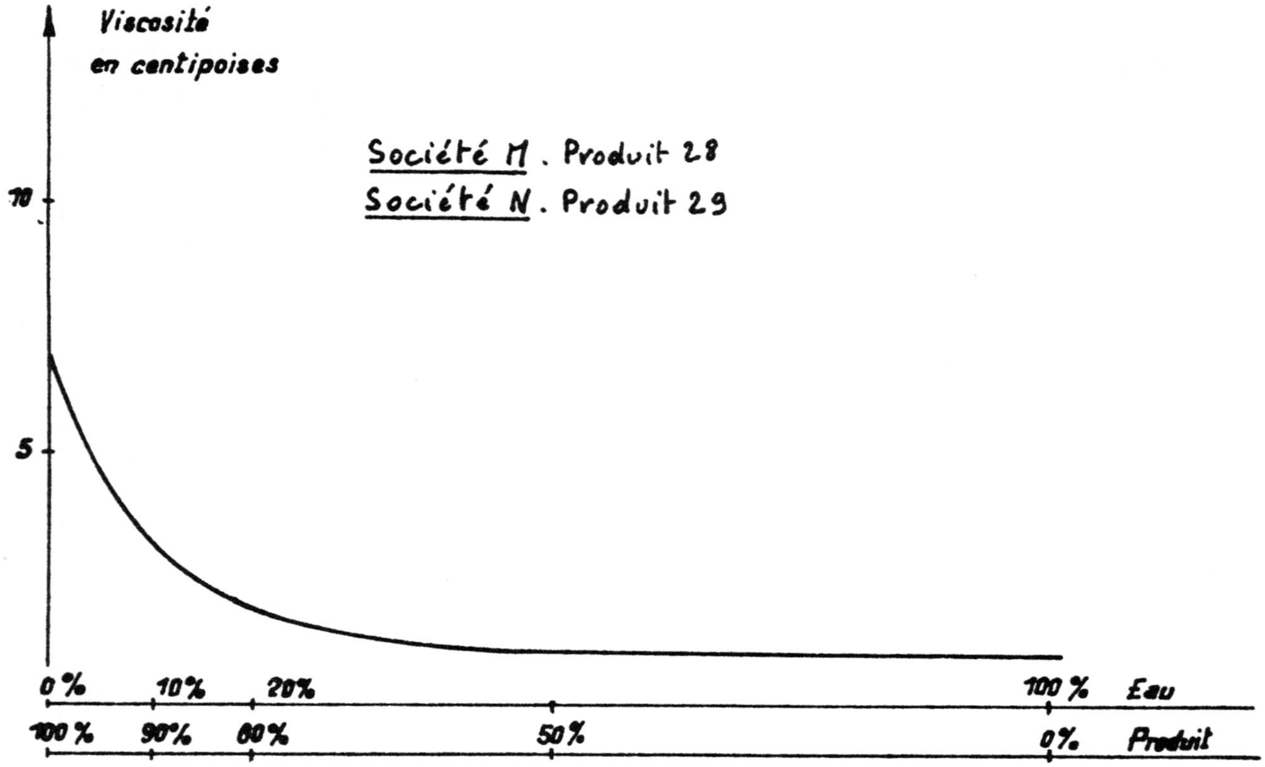
Viscosité
en centipoises
Société M. Produit 28
Société N. Produit 29
10
5
0 %
10%
20%
100 %
Eau
100 %
90%
80%
50%
0%
Produit

Figure N° 17

SILICATE DE SOUDE 40/42 (P.24)

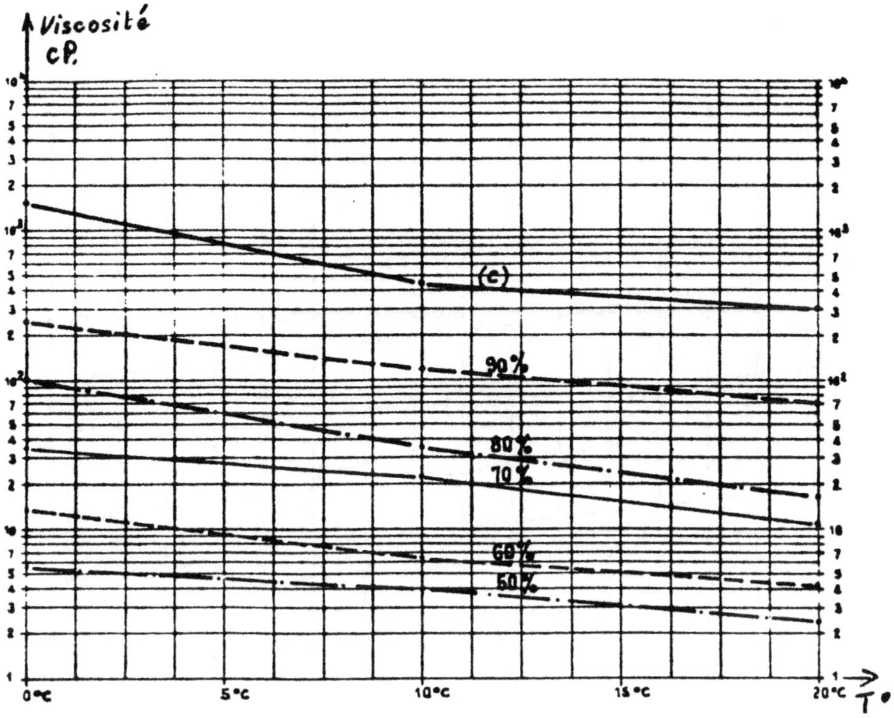

Figure N° 18

SILICATE DE SOUDE 38/40 (P.25)

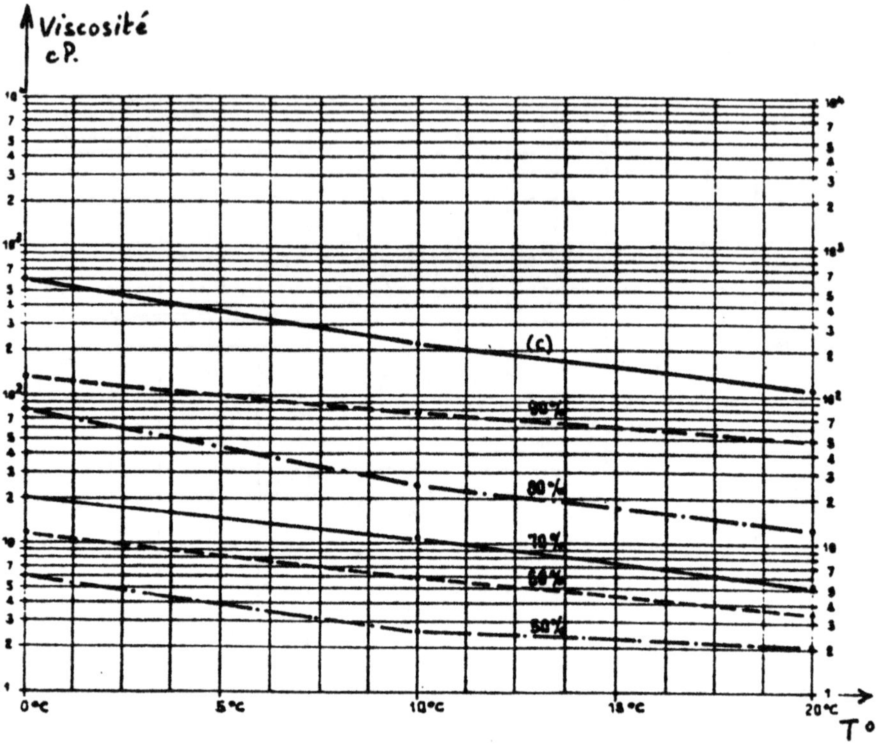

Figure N° 19

SILICATE DE POTASSIUM GK 23 (P. 26)

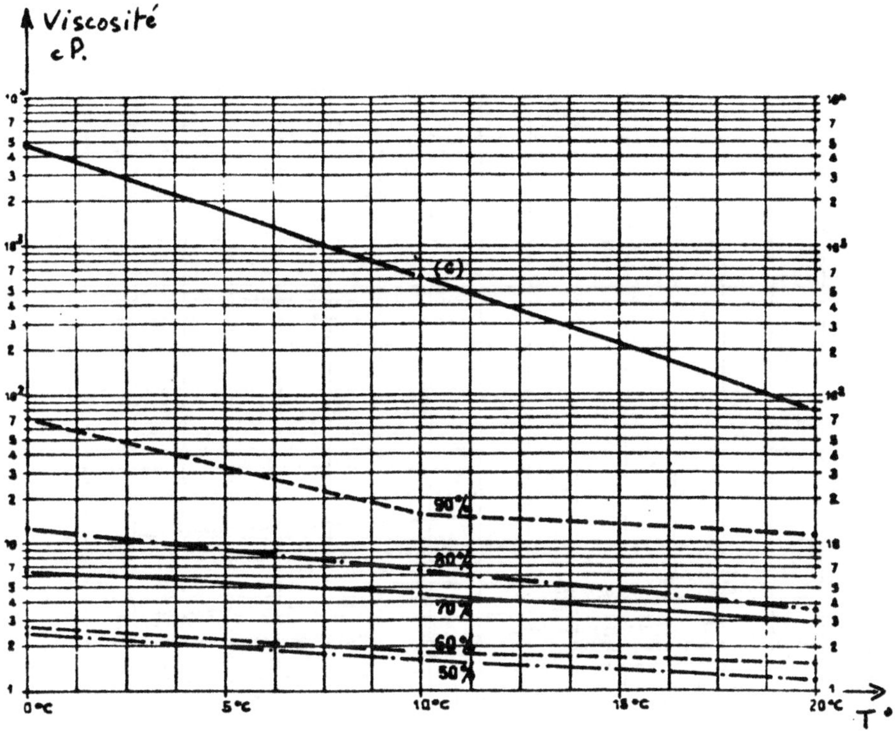

Figure N° 20

SOCIETE M. Produit 30

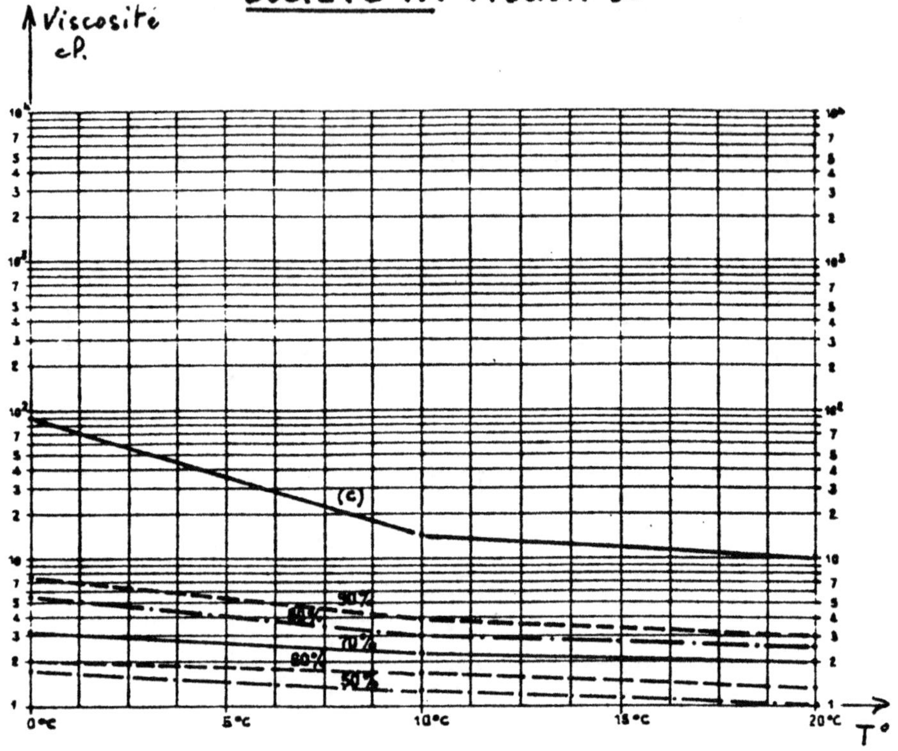

Figure N° 27

SOCIETE N. Produit 29

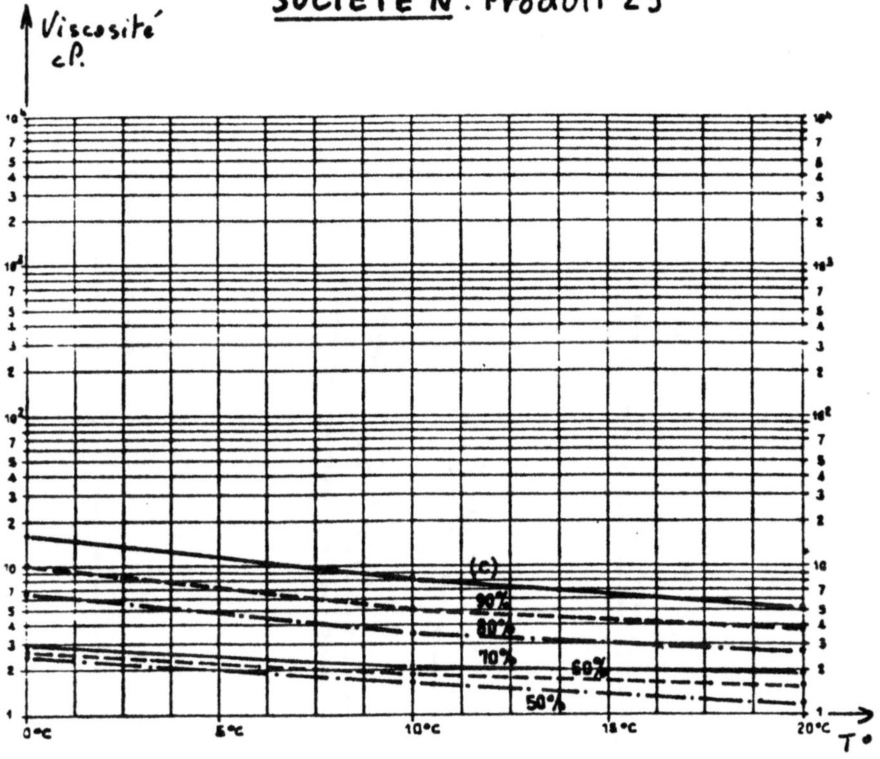

Figure N° 22

MONTÉE CAPILLAIRE . PRODUITS PURS

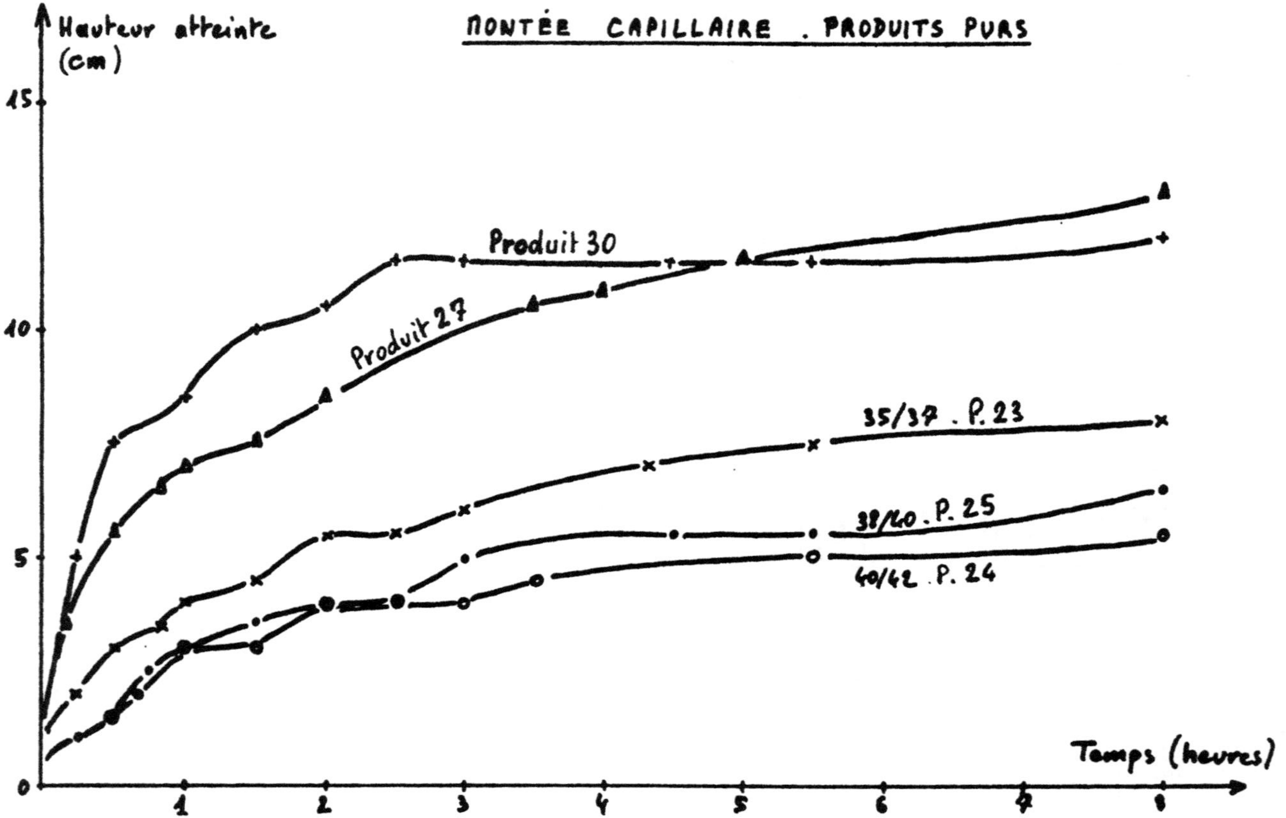

Figure N°23

MONTEE CAPILLAIRE . PRODUITS DILUES à 12 cP.

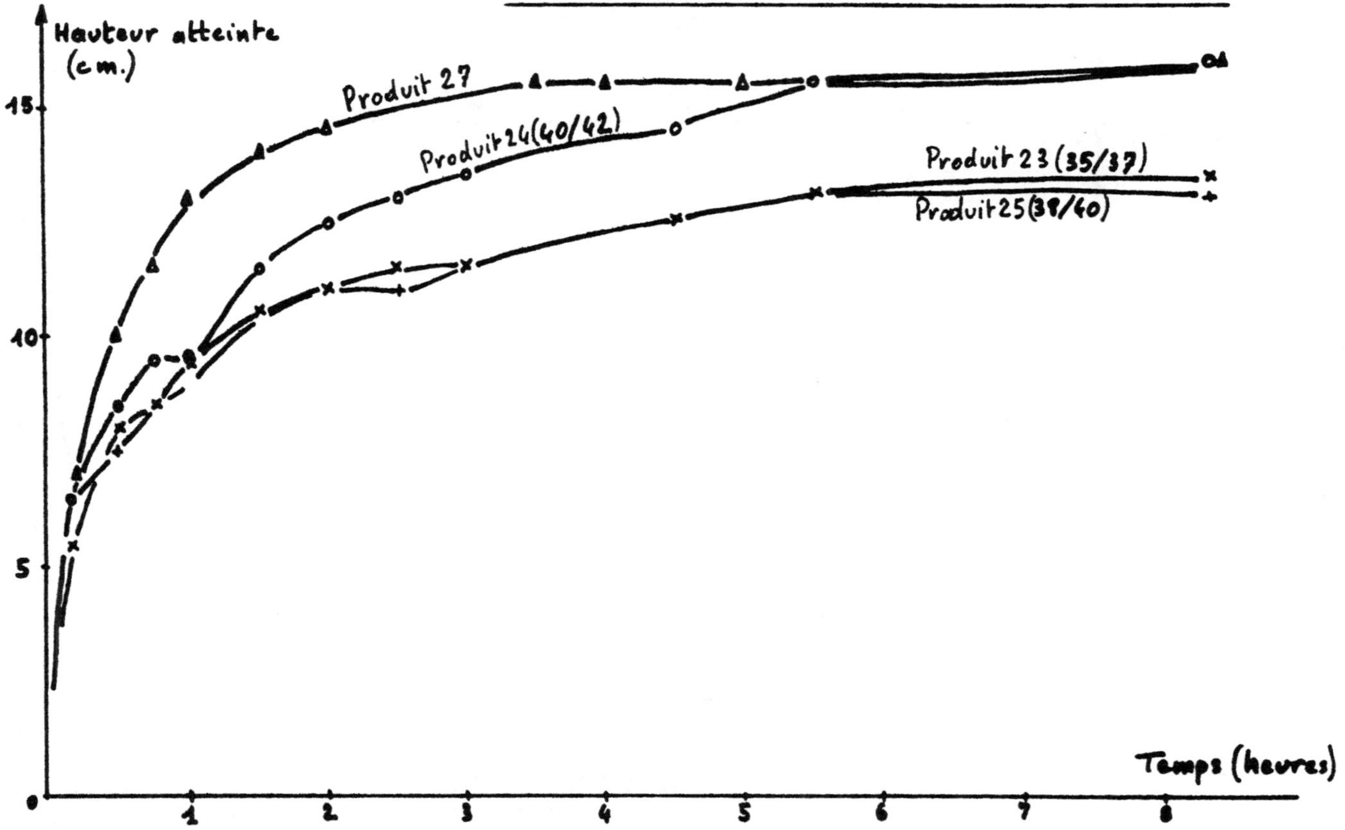

Figure N° 24

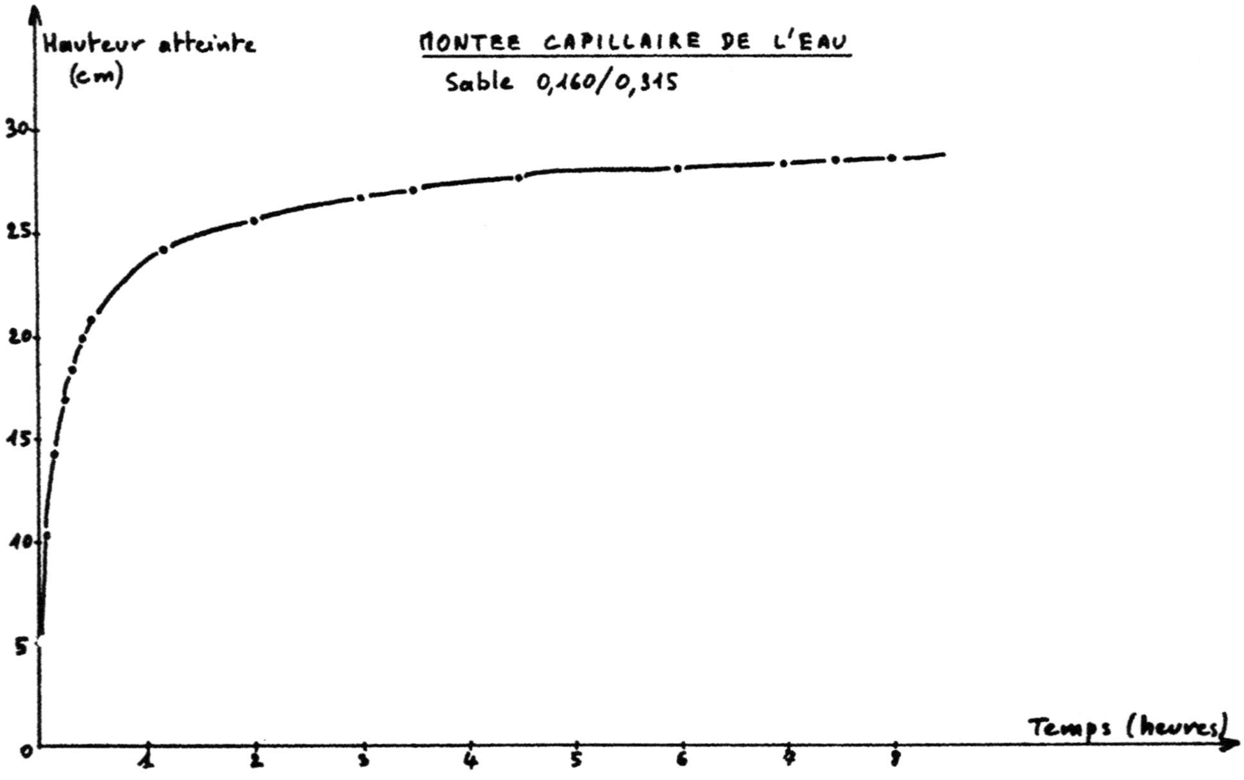
MONTEE CAPILLAIRE DE L'EAU
Sable 0,160/0,315
Hauteur atteinte (cm)
30
25
20
15
10
5
0
1
2
3
4
5
6
Temps (heures)

Figure N°25

Les résultats de la vitesse de montée capillaire sont portés sur les figures n° 23 et 24 respectivement pour les produits utilisés purs et pour les produits dilués jusqu'à une viscosité de 12 centipoises favorable à une injection correcte des fissures fines du béton.

Pour comparaison, la vitesse de montée capillaire de l'eau dans le même support est donnée figure n° 25.

Les produits ne progressent pas du tout sur le support de BIDIM.

3.3 Essais sur maquette d'épaisseur identique à celle des enceintes

Une maquette d'épaisseur identique à celle des enceintes (90 cm) équipée de systèmes de contrôle de fuite et d'injection a été construite sur le site du C.E.M.E.T.E, en comptant sur la masse du béton pour provoquer des fissures de retrait thermique (voir figure 26).

En fait, il a été nécessaire de chauffer le béton, à l'aide de résistances autour des armatures, pour obtenir ce résultat.

Les résultats des débits de fuite à l'air mesurés sont rassemblés sur la figure n° 27. Sauf indication contraire, ces débits sont obtenus sous 0,5 bars de pression.

Ce massif de béton a été injecté de différents produits, dont les silicates de soude suivants :

- Section 1 : Silicate de soude 38/40°B (produit 25) de la Société J.
- Section 2 : Produit 28 de la Société M, dilué de 25 % d'eau.
- Section 3 : Silicate de soude 40/42°B (produit 24) de la Société J, non dilué.
- Section 7 : Silicate de soude 38/40°B (produit 25) de la Société J, dilué de 50 % d'eau.
- Section 8 : Produit 29 de la Société N, dilué de 20 % d'eau.
- Section 9 : Produit 29 de la Société N, pur injecté sous pression.

Les débits de fuite, mesurés à diverses échéances après ces injections, sont notés dans le tableau de la figure n° 28.

3.4 Essais sur le chantier de Belleville

Afin de tester sur chantier la validité de la méthode, l'enceinte externe du réacteur de Belleville sur Loire a été équipée de dispositifs d'injection.

Ces dispositifs, après mesure du débit de fuite, ont été injectés avec des silicates.

Les résultats de ces opérations sont récapitulés dans les tableaux des figures n° 29 et 30.

3.5 Conclusions

La mesure des débits de fuite après traitement, à l'aide du système mis en place pour l'injection au centre de la paroi, est possible avec les silicates, qui ne réagissent qu'au sein des fissures.

Ces mesures montrent toute l'efficacité de ce type de produit.

L'exception présentée par la section 9 provient de la trop forte pression d'injection du produit (> 10 bars) qui a vraisemblablement ouvert des passages. D'autres avantages apparaissent à l'usage :

- possibilité d'ajuster la viscosité par dilution en fonction des dimensions présumées des fissures ou de la température.

Schéma du mur maquette

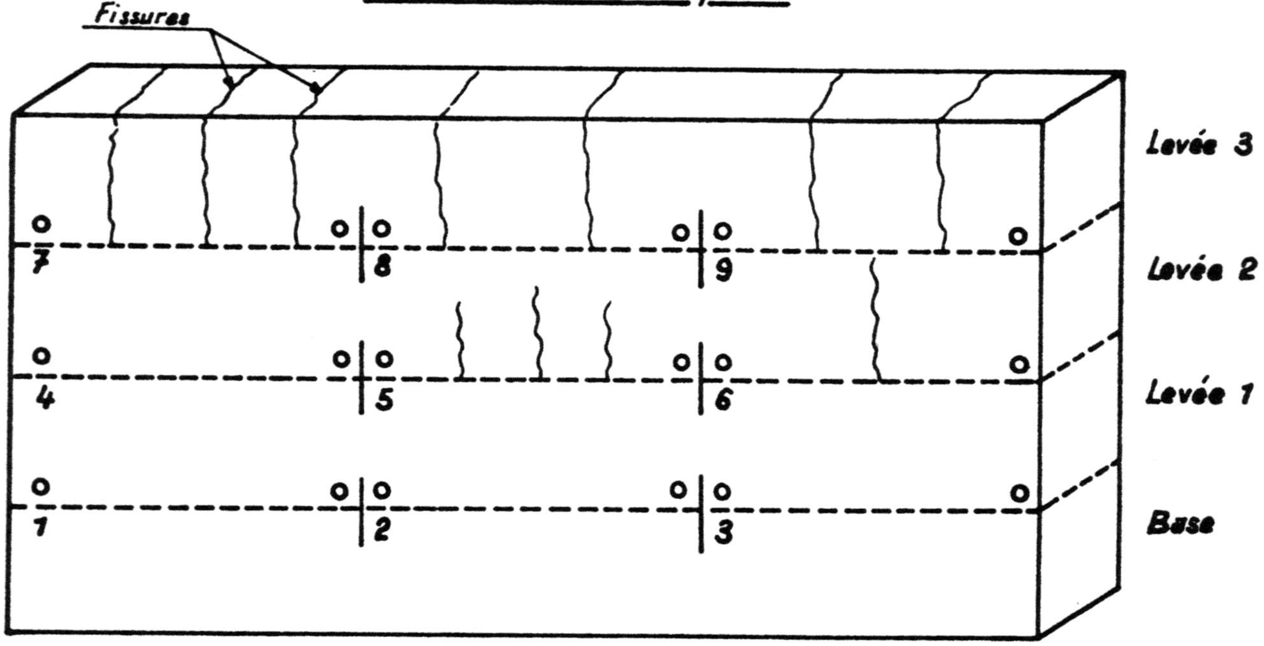

Figure N° 26

DEBITS DE FUITE AVANT INJECTION

Mesures sous 0,5 bar - Exprimés en nl/h/m

	N° de la section								
	1	2	3	4	5	6	7	8	9
INITIAL	1 117 sous 0,1b	10 sous 0,1 b	> 3 000 sous 0,1 b	10	4,33	0,5	11	25	22
15 jours			> 3 000 sous 0,1 b				7	22	15
120 jours			> 3 000 sous 0,1 b				7	16	20

Figure 27

DEBITS DE FUITE APRES INJECTION

Mesurés sous 0,5 bar - Exprimés en nl/h/m

	N° DE LA SECTION								
	1	2	3	4	5	6	7	8	9
1 jour	5	1	-				-	-	9
8 jours	3,6	1	> 3 200				0,5	0,25	0,56
15 jours	-	-	> 3 200				0	0	-
30 jours	1	1	> 3 200				-	-	25
38 jours	0,3	0,4	> 3 200				-	-	-
60 jours	-	-	-				-	-	30
70 jours	0,20	0,4	> 3 200				0	0	-

Figure 28

BELLEVILLE S/ LOIRE

DEBITS DE FUITE AVANT INJECTION

Pression d'essai 0,5 bar, valeur de débit en Nl/h/m

/	N° DE LA SECTION			
LEVEE	1	2	3	4
3 - 4	4,5	555	147,72	colmatée
6 - 7	24,83	73,7	200	168,7

Figure 29

BELLEVILLE S/ LOIRE

DEBITS DE FUITE APRES INJECTION

Pression d'essai 0,5 bar , valeur des débits en Nl/h/m

/	LEVEE 3 - 4 N° DE LA SECTION			
Délai après injection	1	2	3	4
30 jours	0	92 (sous 1 bar)	0	
60 jours	0	39,8 (après injection)	0	

	LEVEE 6 - 7 N° DE LA SECTION			
Délai après injection	1	2	3	4
45 jours	0	0	Gaine colmatée	1,19

Figure 30

- durée d'intervention non limités donc possibilité d'injecter à basse pression pendant un temps long.
- les produits, bien que légèrement caustiques, ne présentent pas de dangers importants.

Par contre, un point a nécessité des études complémentaires : le gel formé peut se désssécher et se rétracter, permettant ainsi une réapparition de fuite.

En fait, deux raisons nous montrent que cette occurence est peu probable.

a/ S'agissant de parois de béton épaisses (0,90 m à 1,20 m) l'expérience montre que l'humidité interne ne diminue pas, sauf sur les quelques centimètres proches des parements.
De plus, les conditions de fonctionnement de l'enceinte ont toutes les chances de provoquer des condensations de la vapeur d'eau interne dans les parois de celle-ci.

b/ Des études complémentaires menées avec l'aide du CERILH (Centre d'Etudes et de Recherches de l'Industrie des Liants Hydrauliques), ont montré que le gel formé évoluait avec le temps vers un composé cristallin insoluble, analogue aux composés formé par le ciment (silicate hydraté C-S-H). On peut donc parler de "fermeture" de la fissure.

4. CONCLUSION GENERALE

Nous disposons, après cette étude, de deux types de solutions envisageables pour le traitement des singularités de bétonnage :
- des résines époxydiques super fluidifiées, pour lesquelles des caractéristiques générales sont définies,
- des silicates de soude de diverses provenances.

INFLUENCE DE L'EAU SUR LES LIANTS EPOXYDES UTILISES EN REPARATION PAR INJECTION

A. BONONI, J. BORREILL, Y. MOUTON
Laboratoire Central des Ponts et Chaussées, Paris, France

Résumé

L'étude de l'influence de l'eau sur la réticulation isotherme de quelques liants époxydiques utilisés en réparation par injection a permis de mettre en évidence les points suivants :

- la miscibilité initiale des systèmes d'époxydes avec l'eau dépend de la nature chimique des durcisseurs et de la teneur en diluant réactif du mélange ;
- l'étude des cinétiques de réticulation isotherme montre que la température agit non seulement sur la vitesse de réticulation, mais également sur le taux de réticulation. La cinétique de réticulation isotherme en masse suit une loi de type Arrhénius ;
- l'étude des cinétiques de retrait volumique isotherme a permis d'évaluer leur miscibilité finale avec l'eau et la modification de leur vitesse de réaction ;
- la variation du module élastique en compression est abaissée de 16 % pour le système le moins miscible à l'eau à 70 % pour le plus miscible pour une teneur en eau de 6 %.

Mots clefs : Miscibilité, Eau, Liants époxydes, Réticulation, Cinétique, Injection, Réparation, Béton.

Introduction

Pour rétablir le monolithisme des structures fissurées en béton et protéger leurs armatures de la corrosion, il est couramment procédé à l'injection de liants d'époxydes de faible viscosité dans des fissures de l'ordre du millimètre ou en dessous. On obtient une "bonne réparation" lorsque le liquide réactif injecté crée, après sa réticulation, une liaison mécanique entre les parties disjointes de la fissure et restaure l'étanchéité des éléments endommagés.

Dans une précédente étude, nous nous sommes préoccupés des propriétés de l'assemblage béton-liant d'époxyde à partir d'un modèle de poutre soumis à fissuration. Cette expérimentation a soulevé un certain nombre de questions, notamment en ce qui concerne l'influence de l'eau sur la réticulation. Il s'agissait d'évaluer l'évolution des propriétés intrinsèques du liant lorsqu'il est injecté dans une fissure contenant de l'eau. Cette question est essentielle car, entre la surface en béton saturée d'eau et le système d'époxyde non perturbé, il se forme une couche liant + eau dont les propriétés conditionnent la qualité de la réparation.

Résultats
Cette étude a porté sur 4 liants (tableau I). Le liant de référence est numéroté 0. Les liants numérotés 1, 2 et 3 correspondent à des produits commerciaux.

Tableau I : Composition du liant de référence (0) et des trois liants commerciaux formulés pour l'injection (1, 2, 3)

Référence du système	Partie Base B = R + d		Partie Durcisseur D
	Résine R	Diluant réactif d	Polyamines
0	DGEBA	sans	DETA pure
1	DGEBA	PGE	à base de TMHMDA*
2	DGEBA	DGEBD	à base de TMHMDA*
3	DGEBA	BGE	cycloaliphatiques

* TMHMDA = triméthylhexaméthylène diamine

Nous nous sommes proposés de préciser la capacité de ces systèmes en réaction à absorber l'eau par diffusion. La figure 1 représente les quantités d'eau absorbées en fonction du temps.

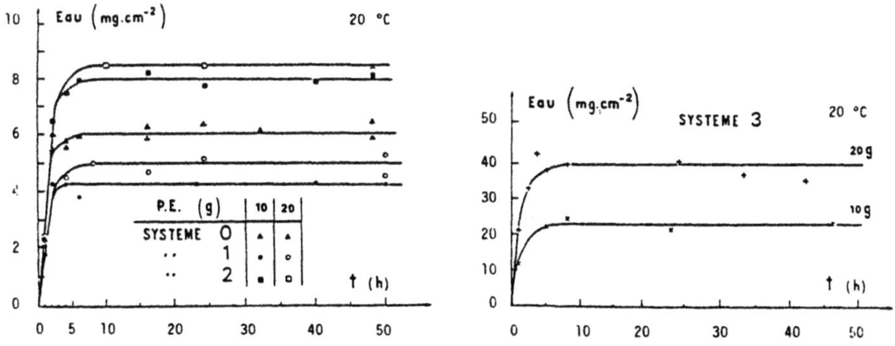

Fig. 1 : Eau absorbée par les systèmes 0, 1, 2 et 3.

On constate que l'eau est surtout absorbée dans les deux premières heures de la réaction quel que soit le système. Pour les systèmes 0, 1 et 2 ayant une DPU inférieure à 45 min, il se forme rapidement une couche qui protège le reste du liant de la pénétration de l'eau, alors que le liant 3 (DPU = 110 min) est perturbé par l'eau dans toute sa masse.

La miscibilité initiale avec l'eau des systèmes commerciaux a été mesurée : 1,8 % pour le système 1, 12,7 % pour le 2 et 15,4 % pour le 3.

Pour tenter de comprendre le mécanisme d'action de l'eau, nous avons alors construit les diagrammes ternaires base + durcisseur + eau pour le système 0 dans plusieurs cas de figure :

- Lorsque la base est diluée par un diluant réactif, soit de type aliphatique, le butylglycédyléther (BGE), soit de type aromatique, le phénylglycédyléther (PGE), les courbes de solubilité (fig. 2 et 3) montrent que la miscibilité de l'eau croît avec la teneur en diluant réactif.

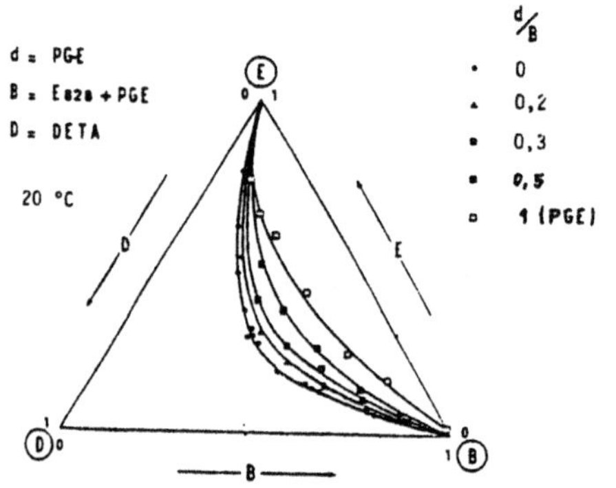

Fig. 2 : Diagrammes ternaires des mélanges (E 828/BGE + DETA + eau).

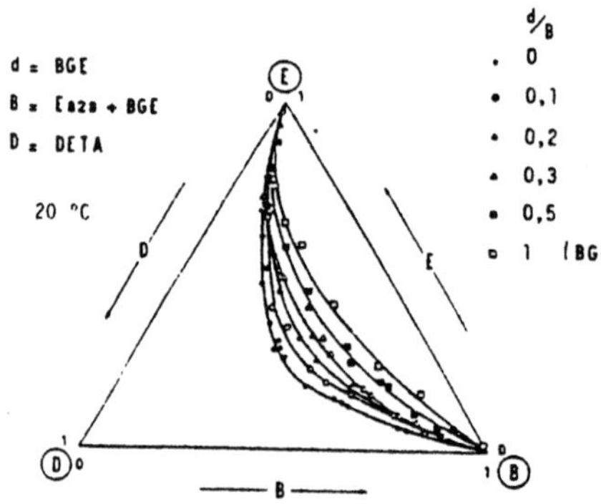

Fig. 3 : Diagrammes ternaires des mélanges (E 828/PGE + DETA + eau)

Un diluant glycédyléther aromatique rend le système moins sensible à l'eau qu'un diluant glycédyléther aliphatique (fig. 4).

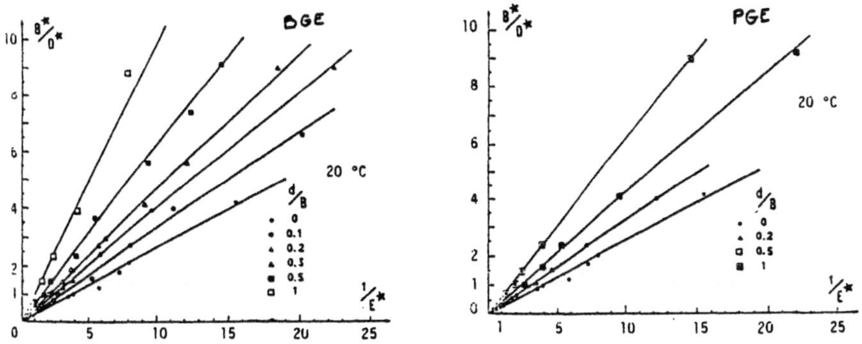

Fig. 4 : Droite de miscibilité de l'eau dans les systèmes (E 828 + BGE ou PGE) + DETA

- Lorsqu'on utilise différents durcisseurs. Les courbes de solubilité pour deux séries de durcisseurs, l'une de type polyamines aliphatiques (fig. 5) et l'autre de type monoamines primaires et secondaires (fig. 6) montrent que la miscibilité de l'eau est d'autant plus faible que le nombre de motifs $-NH-(CH_2)_2-$ est plus élevé.

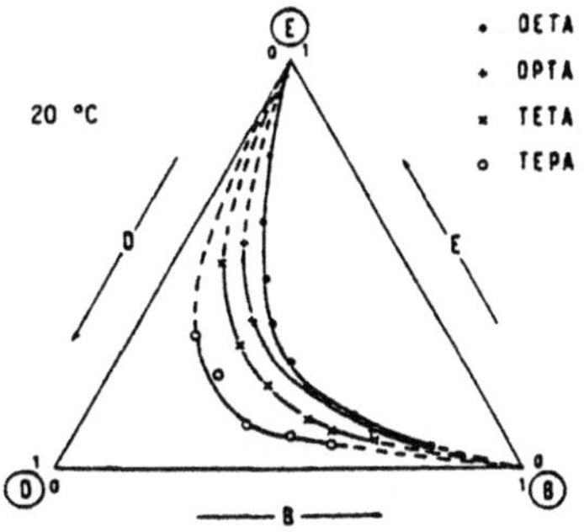

Fig. 5 : Diagrammes ternaires E 828 + eau + série des polyamines aliphatiques

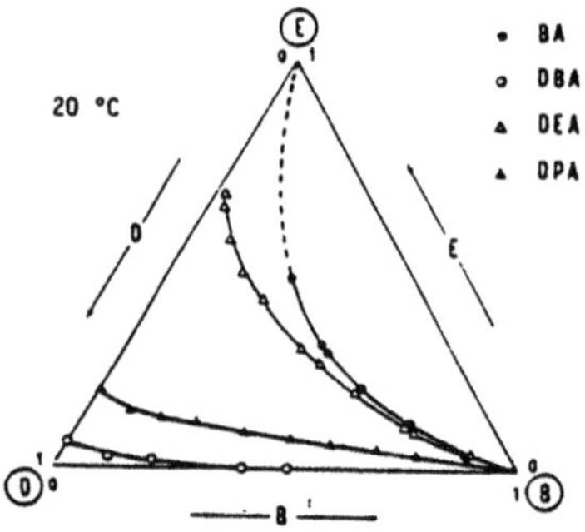

Fig. 6 : Diagrammes ternaires E 828 + eau + série de monoamines primaires et secondaires.

L'étude des cinétiques de réticulation isotherme (fig. 7) montre que la température agit non seulement sur la vitesse de réticulation, mais également sur le taux de réticulation (85-88 % à 20 °C pour 68-78 % à 5 °C). La cinétique de réticulation isotherme en masse suit une loi de type Arrhénius (fig. 8).

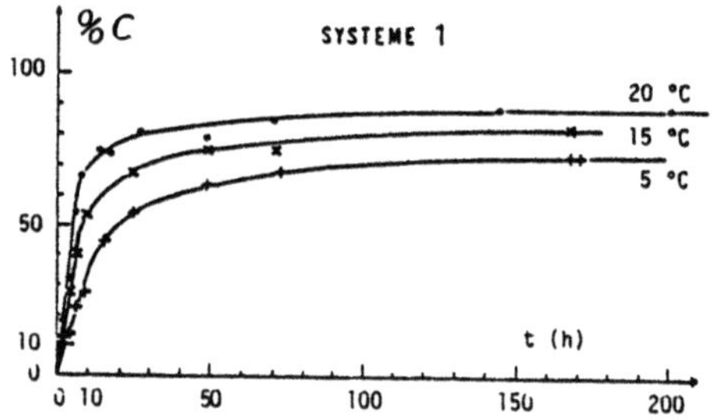

Fig. 7 : Cinétique de conversion des cycles oxirannes au cours de la réticulation isotherme du système 1 à 5 °C, 15 °C et 20 °C.

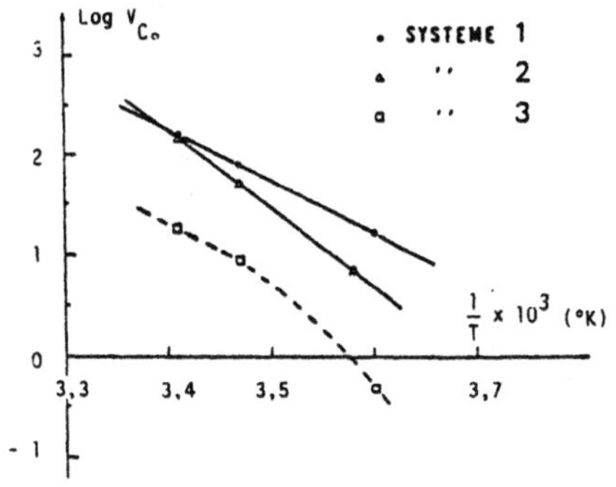

Fig. 8 : Diagramme d'Arrhénius des systèmes 1, 2 et 3 obtenus à partir de la vitesse initiale V_{Co} de conversion des cyles oxirannes.

Les cinétiques de retrait volumique isotherme des liants d'époxydes contenant de 0 à 10 % d'eau (fig. 9) font apparaître d'importants changements du comportement de chaque système en fonction de l'inclusion d'eau.

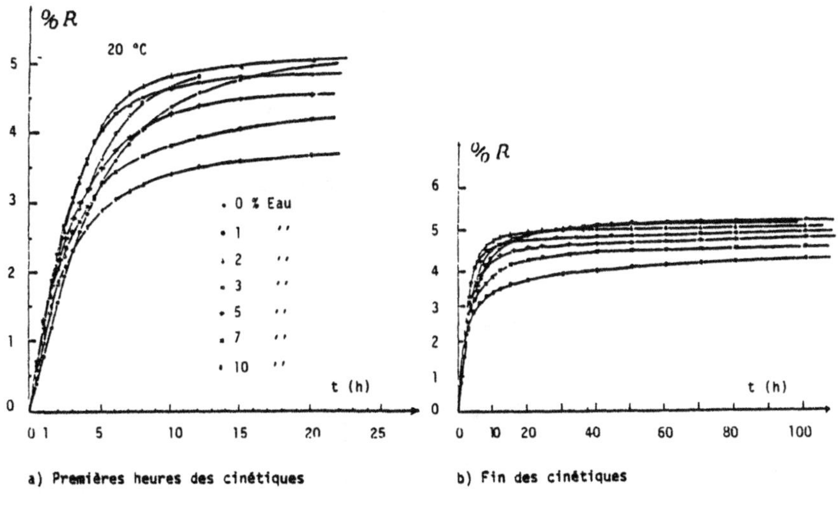

Fig. 9 : Cinétique de retrait volumique du système 1 en fonction de sa teneur en eau.

La vitesse initiale de retrait volumique est fonction croissante de la teneur en eau tant que le mélange liant + eau reste homogène (fig. 10).

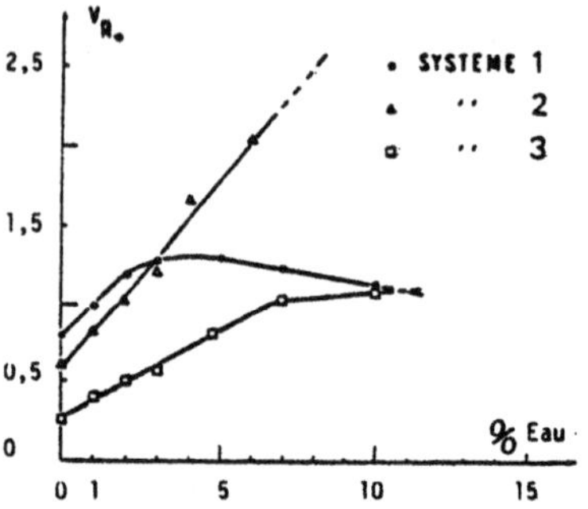

Fig. 10 : Vitesse initiale de retrait volumique, à 20 °C, des systèmes 1, 2 et 3 contenant de 0 à 10 % d'eau.

Pour vérifier si l'incorporation d'eau dans les liants modifie les propriétés mécaniques nous avons mesuré le module élastique en compression des systèmes d'époxydes réticulés à 20 °C et contenant de 0 à 10 % d'eau (fig. 11).

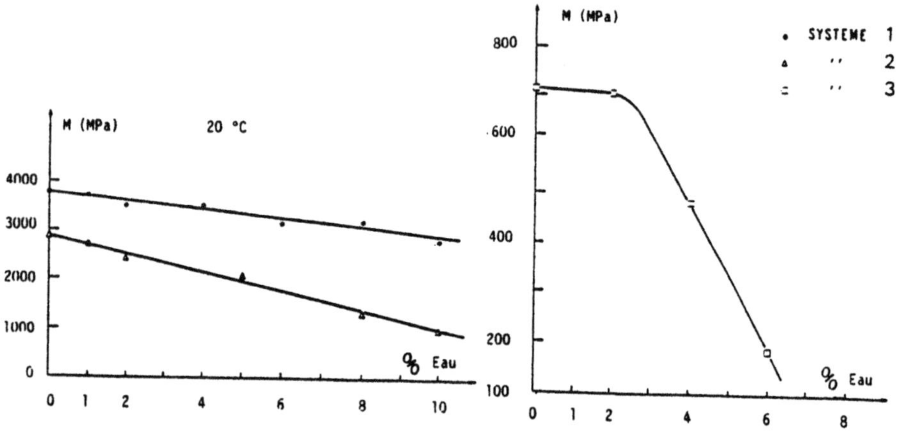

Fig. 11 : Variation du module élastique en compression (M) des systèmes 1, 2 et 3 en fonction de leur teneur en eau.

Ces mesures mettent en évidence l'incidence de la miscibilité des systèmes d'époxydes sur leur cohésion lorsqu'ils réticulent en présence d'eau : l'addition de 6 % d'eau aux liants 1, 2 et 3 (classés par ordre de miscibilité croissante) provoque une chute du module élastique en compression respectivement de 16 %, de 38 % et de 70 %.

Conclusion

Au cours de cette étude nous avons mis en évidence les points suivants :
- la miscibilité initiale des systèmes d'époxydes avec l'eau dépend étroitement de la nature chimique des durcisseurs et est une fonction linéaire de la teneur en diluant réactif, de pente d'autant plus faible que ce dernier est plus aromatique ;
- l'étude des cinétiques de réticulation isotherme montre que la température agit non seulement sur la vitesse de réticulation, mais également sur le taux de réticulation (85-88 % à 20 °C pour 68-75 % à 5 °C). La cinétique de réticulation isotherme suit une loi de type Arrhénius ;
- l'étude des cinétiques de retrait volumique isotherme des liants d'époxydes contenant de 0 à 10 % d'eau a permis d'évaluer leur miscibilité finale avec l'eau et surtout la modification de leur vitesse de réaction ;
- la variation du module élastique en compression montre l'effet catastrophique que peut avoir l'addition de l'eau sur la cohésion de certains liants après réticulation (6 % d'eau provoque une chute du module de 16 à 70 % pour les liants étudiés).

Par ailleurs, lorsqu'un système d'époxyde réticulé est placé au contact de l'eau, il réagit en fonction de sa structure macromoléculaire et de sa nature physico-chimique.

Il est très important, lorsque l'on envisage qu'une structure réparée avec des liants d'époxydes soit mise en contact avec de l'eau, de prendre en compte ces phénomènres dans le choix du produit à injecter.

Bibliographie

Mouton Y. Contribution à l'étude de la réparation des structures en béton fissurées, par injection de liants époxydiques. Rapport de Recherche LPC n° 86, juin 1979.

Mouton Y. et Prin D., L'injection de résines : étanchéité ou restitution du monolithisme, Colloque Internat. Gestion des Ouvrages d'Art, Bruxelles-Paris, avril 1981, 473-478.

Mouton Y., Durrieu F. et Prin D., Efficiency of cracked concrete repair by époxy binder injection, Symposium Internat. ICP/RILEM/IBK, Prague, juin 1981, thème 3/B-91, 527-532.

Borreill J., Etude de la réticulation isotherme, en présence et en absence d'eau, de quelques liants époxydiques utilisés pour la réparation des bétons hydrauliques, Rapport de Recherche LPC n° 124, sept. 83.

PRODUITS SPECIAUX POUR L'INJECTION DES BETONS FISSURES OU MICROPOREUX

C. CARON
Laboratoire Géatec - Venelles - France

Résumé
Lorsque se présente un cas d'injection de béton fissuré ou microporeux, on pense instantanément "époxyde". Or un époxyde, du fait de sa viscosité élevée, n'est pas forcément le produit idéal pour injecter un massif peu perméable. A titre d'exemple nous verrons dans ce texte comment on peut étendre le domaine d'utilisation des époxydes (par fluidification et par allongement du temps de polymérisation) et quels autres types de résines, plus fluides, peuvent être envisagés.
Mots clés : Injection, Viscosité, Fissure, Béton poreux, Epoxyde, Furanique, Phénolique, Résine aquaréactive.

1. Introduction

Il y a forcément des grandes analogies entre la micro-injection des bétons fissurés ou poreux et l'injection classique des sols :

analogie dans la diversité des milieux à injecter qui peut aller de la fissure franche et propre à la microporosité.
analogie dans le but recherché qui peut aller de la simple étanchéité au renforcement structural.

Alors que l'injecteur ne connaissait au XIXe siècle que le ciment, ce qui limitait grandement son champ d'application, il dispose actuellement d'une palette de quelques centaines de coulis nettement différenciés sur le double plan de l'injectabilité et de la résistance. Au XXe siècle le micro-injecteur est nettement moins riche. Il fut un temps où il ne connaissait que l'époxyde, puis sont venus les acrylamides. Mais cela ne représente que les deux extrêmités des travaux de protection du béton : haute restructuration dans un cas et simple étanchéité dans l'autre.

2. Critère de choix pour une résine d'injection

Pour qu'une injection remplisse son rôle il faut que :

(a) Le coulis chemine parfaitement dans tous les vides et interstices qu'il a à bloquer, ce qui implique un choix de la viscosité optimale du coulis.

(b) Le coulis ait, après durcissement, des caractéristiques mécaniques suffisantes, lesquelles dépendent de sa composition chimique.

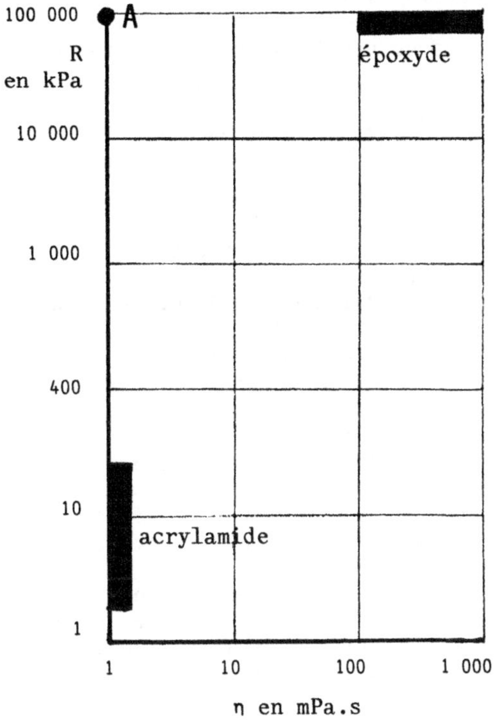

On peut représenter ces deux critères par le diagramme ci-contre. Le coulis idéal serait défini par le point A caractérisé par une viscosité aussi faible que celle de l'eau et une résistance supérieure à celle du béton.

Les deux classes normales de résine : les époxydes et les acrylamides aqueux sont évidemment bien loin de ce point idéal A. Sans espérer arriver jusqu'à ce point utopique en l'état actuel des connaissances, on peut tout au moins essayer de s'en approcher soit en minorant la viscosité du prépolymère époxyde (§ 4), soit en augmentant la résistance des résines aqueuses (§ 5).

3. Rappel des formules d'injection. Conséquences

En injection radiale le débit Q est :

pour une fissure d'épaisseur e : $Q = \frac{\pi e^3 P \alpha}{6\,\eta \ln R/r}$

pour un massif poreux de coefficient de perméabilité K :

$$Q = \frac{2\pi K l \eta_0 P}{\varpi \eta \ln R/r}$$

avec α = coefficient de rugosité (0,3 à 0,7 en moyenne)
P = pression d'injection
l = longueur de la passe injectée
R = rayon d'action de l'injection
r = rayon du forage
η_0 = viscosité de l'eau
η = viscosité du coulis
ϖ = poids volumique du coulis

Dans les conditions pratiques : r = 1 à 2 cm, R = 1 à 2 m, on a

R/r = 50 à 200 et $l_n R/r$ = 4 à 5. Les formules précédentes deviennent approximativement :

pour une fissure d'épaisseur e : $Q = \frac{\pi e^3 P}{54\eta}$

pour un massif poreux de coefficient de perméabilité K :

$Q = \frac{\pi K l \eta_o P}{2,25\bar{\omega}\eta}$. Comme $\eta_o = 10^{-3}$Pa.s et $\bar{\omega} \simeq 10^4$ N/m3, on a en U. I. :

$Q = \frac{\pi K l P}{2,25 \times 10^7 \eta}$.

Comme en injection des sols P, de même que e et K, sont fixés par le terrain on choisira un coulis dont la viscosité η soit telle que la cadence d'injection Q puisse être maintenue dans la fourchette réaliste de 0,5 à 2 m3/h (pour des raisons économiques, il est extrêmement rare qu'une cadence inférieure à 0,5 m3/h soit utilisée).

On opérera de même en micro-injection avec cependant une différence fondamentale : comme les quantités à injecter sont très faibles, la notion économique de cadence d'injection disparaît et on a donc, dans le choix des paramètres, une variable complémentaire qui est Q. Encore faut-il qu'en diminuant cette cadence d'injection la résine ne soit pas prise avant d'avoir traité la zone prévue.

Pour avoir un rayon d'action de 1 m :

il y aura à injecter dans une fissure d'épaisseur e : πe. Comme le béton de part et d'autre de la fissure est quelque peu poreux et microfissuré, il y a lieu de doubler, voire tripler, cette valeur. Il en résulte un temps d'injection de $t = \frac{135\eta}{e^2 P}$.

il y aura à injecter dans un massif perméable de porosité 0,15 : 0,15 πl, ce qui conduit à un temps d'injection de $t = \frac{35 \times 10^4 \eta}{KP}$.

La limite d'injectabilité est alors fixée par les conditions suivantes

épaisseur minimale de la fissure : $e = \sqrt{\frac{135\eta}{Pt}}$,

exemple avec P = 10^5Pa, η = 0,5 Pa.s et t = 3 600 s,

$e = \sqrt{\frac{135 \times 0,5}{3,6 \times 10^8}} = 4 \times 10^{-4}$m = 0,4 mm.

perméabilité minimale du massif poreux : $K = \frac{35 \times 10^4 \eta}{Pt}$,

exemple avec les mêmes données,

$K = \frac{35 \times 10^4 \times 0,5}{10^5 \times 3\ 600} = 5 \times 10^{-4}$m/s.

4. Perfectionnement dans le domaine des époxydes

L'époxyde a l'avantage d'une résistance fort élevée et d'un retrait nul. Inversement il a l'inconvénient, en injection fine, d'une viscosité assez élevée et d'un temps de polymérisation relativement court. En conséquence son domaine d'application est limité à l'injection de fissures franches, propres et sans ramifications secondaires.

Pour accroître le champ d'application des époxydes et aborder avec succès celui de la régénération des bétons poreux, une première voie consiste à abaisser fortement la viscosité de l'époxyde. Ainsi que nous l'avons exposé à un précédent Congrès Rilem (Liège 1984), cette fluidification est possible jusqu'à une vingtaine de mPa.s, ce qui permet de traiter à basse pression (100 kPa) des bétons poreux dont la perméabilité est de l'ordre de 10^{-5}m/s.

Une autre façon d'augmenter la profondeur de pénétration dans les bétons poreux ou microfissurés consiste à accroître le temps de polymérisation de l'époxyde. Jusqu'à présent on pensait que seul les systèmes base polyesters permettaient de repousser au lendemain le début de la polymérisation ; nous savons maintenant le faire avec des systèmes base époxyde sans que cela nuise aux qualités du polymère. Par cette voie on peut aussi traiter à basse pression des bétons poreux de perméabilité 10^{-5}m/s.

Enfin le couplage de deux méthodes (fluidification + allongement du temps de polymérisation) permet de repousser à $K = 10^{-6}$m/s la limite des époxydes.

Il est rare que l'on ait à régénérer des bétons ségrégés dont la perméabilité soit inférieure à 10^{-6}m/s. Si cela était, il faudrait soit travailler à haute pression (si l'état du béton le permet) soit rapprocher les forages d'injection.

5. Résines de résistance intermédiaire

L'époxyde d'une part, l'acrylamide d'autre part représentent les deux extrêmités de l'échelle des résistances actuellement connues. Entre une époxyde de résistance 100 000 kPa et un gel d'acrylamide de résistance 10 kPa il y a un hiatus à combler. Il est de nombreux cas où une résistance intermédiaire est souhaitable, surtout si cette modification s'accompagne d'une meilleure injectabilité. Par exemple il est parfaitement légitime de bloquer les fissures d'un massif travaillant en compression avec une résine dont la résistance est inférieure à celle du béton (cf. formule de Prandtl).

Ces résines aux résistances intermédiaires peuvent être obtenues par les classiques associations époxyde-brai mais celles-ci ne permettent pas de gagner en injectabilité puisque le brai et l'époxyde ont sensiblement la même viscosité. Il est alors bien plus performant de passer à des résines aqueuses formulées à partir de monomères. De tels systèmes ont en effet la fluidité de l'eau quelle que soit leur concentration en résine. La résistance est réglable par le dosage en en résine.

Concentration du phénolique (%)	Résistance en kPa (normalisation AFTES)
25	2 100
40	4 000
50	5 800
60	7 800
70	10 000

Comme ces monomères aqueux base phénoplaste ou furanique ont une viscosité égale à celle de l'eau et que leur temps de polymérisation peut être supérieur à 24 heures, leur limite d'injectabilité peut être repoussée à $K = 10^{-7}$m/s.

6. Injection sous courant d'eau

L'injection d'une fissure parcourue par un courant d'eau est une opération fort délicate. Certes on y parvient en accélérant le temps de prise de la résine mais c'est au prix d'un grand gâchis. Aussi la découverte par les Japonais, experts en la matière, des systèmes aquaréactifs (résines qui polymérisent instantanément en présence d'eau en décuplant de volume) a permis un progèrs décisif dans ce type de traitement. Devant l'intérêt de ce coulis nous avons mis au point une formule purement française.

Ce bref aperçu, limité à trois classes d'application, montre que le micro-injecteur, comme son grand frère l'injecteur classique, doit et peut disposer d'une gamme de coulis fort étendue, lui permettant de résoudre au mieux chaque cas spécifique d'étanchéité ou de consolidation dans chaque type de perméabilité. Selon l'écrivain japonais Yoshida Kenndoo le vrai maître samouraï doit et peut se contenter d'une seule flèche ; n'ayant pas atteint ce stade de maîtrise, nous pensons que le micro-injecteur se doit prudemment, en plus de sa flèche classique, d'en avoir de nombreuses autres en réserve.

IMPROVEMENT IN THE DURABILITY OF CRACKED CONCRETE ELEMENTS INJECTED WITH SYNTHETIC RESIN BY OPTIMIZING THE BOND BEHAVIOUR

K.P. GROSSKURTH and W. PERBIX
Institut für Baustoffe, Massivbau und Brandschutz der Technischen Universität Braunschweig, Federal Republic of Germany

Abstract
Cracked building elements are increasingly being repaired by the injection of reactive resins. The success of a repair not only depends on the choice of the right injection material but rather it is also influenced in a complex way by many practical construction conditions. This includes the non-positive injection of wet crack surfaces. At present very little information is available about the influence of processing and material-related parameters on the bond behaviour. The cause of premature bond failure must be sought in the microregions of the adhesive areas, wherefore the application of scanning electron microscopy is indispensable. The influence of different boundary conditions on the bond behaviour of the injected resin systems was investigated in several tests. In the first test unit with building elements, it could be shown that dry crack injection at a temperature generelly presents little difficulty. The investigation of 25 °C carbonated injected crack surfaces also presented positive results. However, the presence of water during the injection and the curing phase resulted in a reduction in the bond tensile strength. In cooperation with the chemical industry, the adhesive properties of some wet injected resins have been improved considerably.
Key words: Crack injection, Reaction resins, Non-positive injection, Positive injection, Adhesive behaviour, Epoxy resins, Polyurethanes, Acrylates.

1. Introduction

Reinforced concrete constructions require uncracked concrete building elements. Structurally this requirement can only be fulfilled at high costs by either choosing prestressed building elements or using a high degree of reinforcement. It is more economical to accept the risk of cracks and seal any occuring cracks positively or in some cases non-positively by taking suitable action. Using a positive injection, only the protection against corrosion of the reinforcement is guaranteed, whereas a non-positive connection of the crack surfaces is advantageous in so far as the behaviour of the building element is additionally restored to its former condition.

Appropriate crack injections with reactive resins are increasingly carried out by numerous special firms with various resin formulations. The success of a repair using resin crack injection depends, in a complex way, on the resin composition and on the processing conditions. Usually it can be determined by means of mechanical testing methods. However, the cause of premature bond failure must be sought in the microregions of the adhesive areas and is scarcely found with the aid of macroscopic tests. In these cases, a clear correlation between the variety of influencing parameters and the cause of the bond failure can be expected from the application of scanning electron microscopy. Moisture and temperature conditions, the influence of carbonated crack surfaces and the loading of the building elements during the injection and the curing phase of the reaction resin were taken into special consideration.

2. Execution of the research task

2.1 Tests with building elements and results

In the first test section three reinforced concrete beams (concrete grade B 35) with the dimension b/d = 30/60 cm and a span of 4.00 m were chosen as specimens. In each of the test beams an air-entraining agent was added to the concrete of one half of the beams.

The cracks, which were later to be injected, were induced by a static loading which caused at least 12 cracks with a crack-width of 0.2 to 0.5 mm. For the simulation of further loading conditions one side of the bending beams was heated with infrared radiators up to approximately 35 ^{o}C. The temperature on the opposite side was 25 ^{o}C. After the reappearance of normal moisture conditions cracks with alkaline surfaces (subsequently referred to as "dry") as well as those with carbonated crack surfaces were injected. In order to include the influence of moisture conditions, cracks filled with water and emptied shortly before injection as well as cracks filled with water, were injected under static and pulsating loading conditions.

In conjunction with the chemical industry three compositions of epoxy resin were chosen. One resin (subsequently referred to as resin system 1) was noticed to be especially balanced and to have a positive spectrum of characteristics. Regarding the two other compositions (resin system 2 and 3) the manufacturers claimed improved moisture compatibility. The resin composition 3 showed a higher reactivity than the other two.

The reaction resin, mixed within the package, was injected by low pressure injection into the previously sealed cracks under static as well as pulsating loading. For the pulsating loading the crack motion was set to 0.05 mm at a frequency of 0.6 per second. This loading was sustained even during the curing phase. Subsequently the test specimens were statically overloaded.

In investigating the quality of the bonding area between injection resin and crack surface, on which the main interest is focused, scanning electron microscopy proved to be particularly suitable. Figure 1 shows on the basis of resin system 1 an example of a complete bond of both crack surfaces. The dry crack was injected under static loading at a temperature of 25 ^{o}C. Comparable positive results have been obtained for the injection of similar cracks under pulsating loading. Even the injection of cracks with carbonated surfaces gives no cause for complaint concerning the bonding behaviour.

For wet crack surfaces however, regions with predominately disturbed bonds occur (see Fig. 2); moreover, the resin shows a distinct porosity, which is caused by chemical reaction of the still uncored polymerized resin with the adhering water of the crack surfaces. The entirely unsatisfactory injection of a wet crack under pulsating loading at an elevated temperature can be seen in Fig. 3.

The most important aim of further investigations is to improve the adhesive bond of wet crack surfaces.

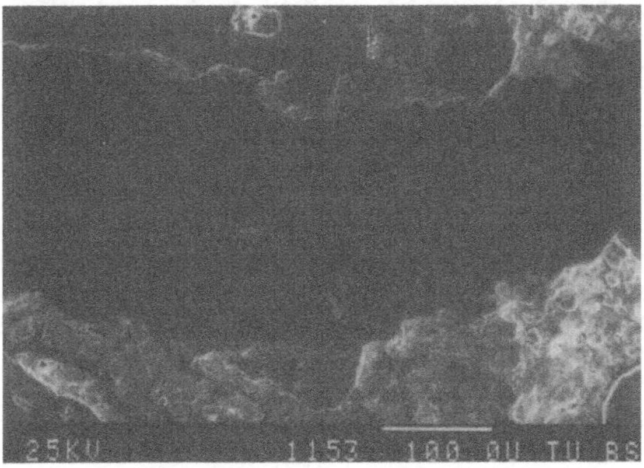

Fig. 1. Static loading, resin system 1
Concrete: dry, 25 ^{o}C

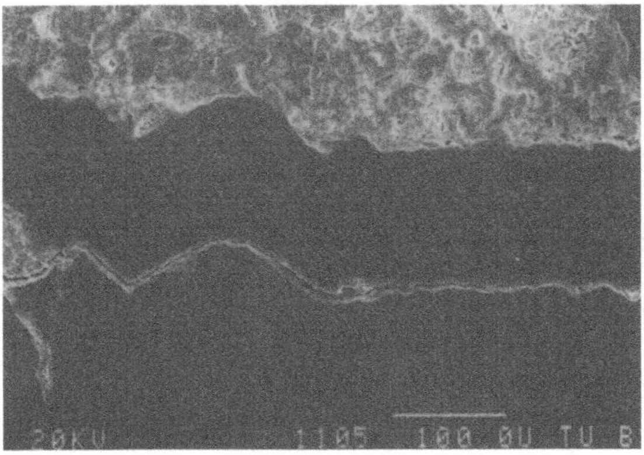

Fig. 2. Static loading, resin system 1
Concrete: wet, 25 ^{o}C

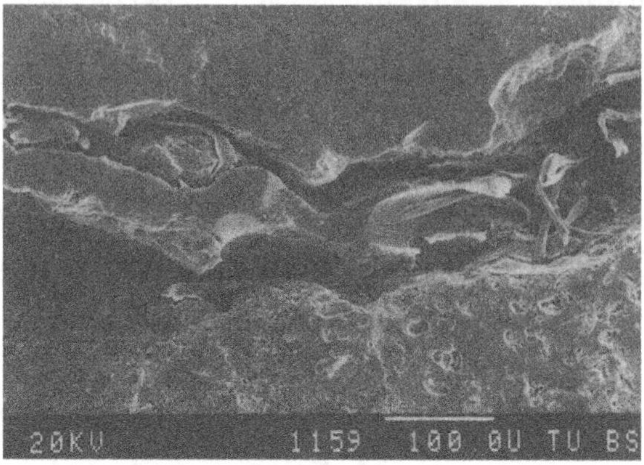

Fig. 3. Pulsating loading, resin system 1
Concrete: wet, 35 °C

In order to give a clear summary of the test results of the scanning electron microscopy structure examination in a quantified form, a numerical assessment of the contacting behaviour between the crack surface and the reactive resin as well as the observed tendency to form pores is given in table 1.

crack surface	load	resin system 1 bond	resin system 1 pores	resin system 2 bond	resin system 2 pores	resin system 3 bond	resin system 3 pores
dry	stat. puls.	2.0 2.5	2.0 4.0	1.0 2.0	1.0 3.0	3.0 3.0	3.0 4.0
carbon-ated	stat. puls.	2.0 2.5	3.0 3.0	1.0 1.0	1.0 4.0	2.0 2.5	2.0 3.0
wet	stat. puls.	3.0 4.0	5.0 4.0	3.0 3.0	3.0 4.0	3.0 3.0	5.0 5.0
mean	stat. puls.	2.3 3.0	3.3 3.7	1.7 2.0	1.7 3.7	2.7 2.8	3.3 4.0

Table 1. Assessment of contacting behaviour and forming of pores. Numerical rating between 1 (very good) and 5 (unsatisfactory)
stat. = static loading
puls. = pulsating loading

It is evident that dry as well as carbonated cracks fundamentally enable far better contacting than wet cracks. Altogether, resin system 2 is superior to the other two resin systems. Nevertheless, the knowledge obtained so far does not justify characterizing the resin system 2 as a reactive resin, which is insensitive to moisture. For resin system 3, the high reactivity, which makes normal use virtually impossible, and a corresponding short pot-life at higher temperatures proved to be so detrimental, that practical use does not seem worth considering.

For beams under pulsating loading the crack-width measurements during the injection and the subsequent curing phase showed, at the end of the curing phase, a high reduction in the crack movements, which means that the behaviour of a nearly uncracked cross-section was achieved. Fig. 4 shows the change in crack-width of a dry crack under pulsating loading, in the case of the highly reactive resin system 3.

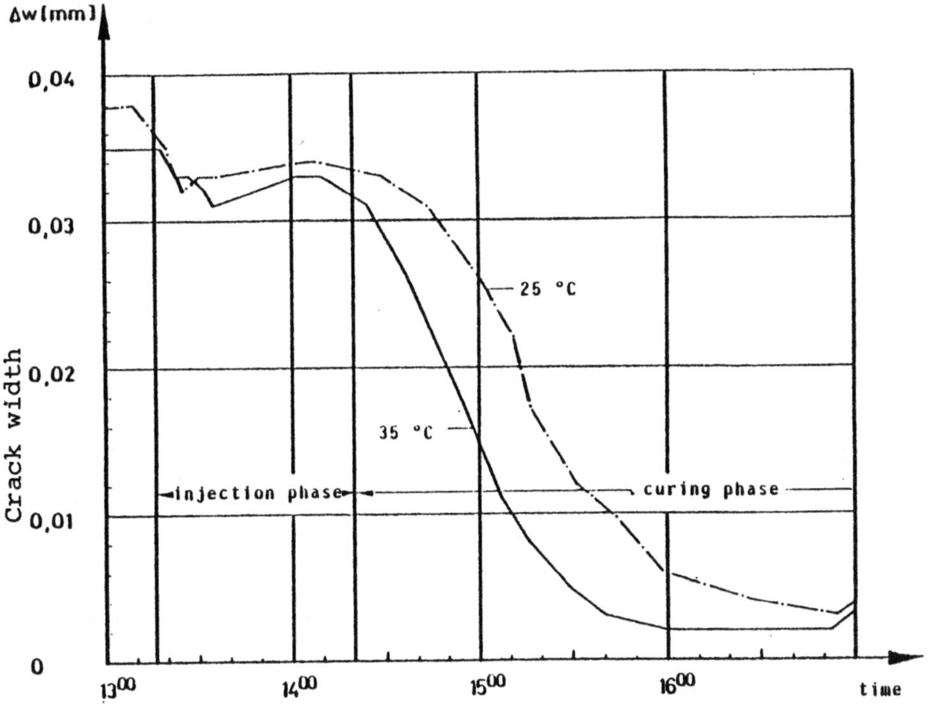

Fig. 4. Development of the change in crack-width dry crack-injected under pulsating loading

With respect to the influence of the different crack surface states, the results of the electron scanning investigations correspond closely with the results of tension tests on the specimens taken from cores. Whereas for injected dry and carbonated cracks the fractured surface was exclusively situated within the concrete, injected wet cracks were able only in rare cases to resist the recurring adhesive stresses.

2.2 Model experiments and results

Considering these results, the dependence of the adhesive bond between the concrete and various synthetic resin compositions on the moisture state of the jointing edge surfaces should be tested in a second test phase by further mechanical-technological investigations and scanning electron microscopy structure examinations. The test would seem to be necessary in order to obtain information on the suitability of the injected resin at climatic boundary conditions and in order to optimize the adhesive behaviour of the resin formulations in collaboration with the chemical industry.

The results of the central tension test with injected specimens should serve as criteria for the estimation of the influences on the bond between resin and concrete. Toward this end the bending beams used in the test were substituted by concrete prisms with the dimension 4x4x16 cm^3 (B 35, W/Z = 0.5), which were halved normal to the longitudinal faces by means of fine sawing.

After fixing the crack-width the packers were glued diagonally staggered on two opposite faces and the "crack" was sealed. Regarding the test specimen conditioning, it was differentiated between dry and wet crack surfaces. For the conditioning of wet crack surfaces the sealed cracks were filled with distilled water, afterwards emptied and injected.

In addition to the resin systems EP 1 and EP 2 of the tests with the building elements, four epoxy resins, specially developed for these investigations were used (EP 3...6). Furthermore, two compositions on the basis of polyurethane with and without admixtures were considered as well as acrylic resins. The scope of these systems comprises so far exclusively non-drip crack injection. The application of the resins was realized by using contact pressure moulding with a pressure of 2 bar.

The results of the tension tests with dry and wet injected specimens are shown in figure 5, where the bond tensile strength is normalized with respect to the concrete tensile strength. A proportionality factor of 100 % means that the specimen broke in the concrete because the bond tensile strength was higher than the tensile strength of concrete.

The epoxy resins, which were specially composed for this application, show a uniform behaviour. Analogous to the test results with the building elements, the adhesion strength of all tested epoxy resins exceeds the tensile strength of the concrete in the case of dry crack surfaces. Only the presence of water during solidification results in negative effects with regard to the bond behaviour in some cases. Nevertheless, three epoxy resins show no measurable reduction in bond strength.

The attendant scanning electron microscopy structure investigations show that the epoxy resins in case of dry injection completely covered the crack surfaces and cured to a homogeneous resin matrix. The reaction process of wet injected cracks is disturbed by cavitation. However, the strength tests exhibit only an insignificant reduction in the adhesive behaviour.

The polyurethanes possess an unbalanced relation of the bond tensile strength (Fig. 5). Acceptable bond strength with the composition PUR 1 requires the addition of hydrated lime for linkage carbon dioxide gas, which results from the reaction of the resin and the moisture of the crack surfaces. PUR compositions with water-absorbing zeolite additives are not suitable for non-positive crack injection. Injected gel systems on the base of acrylic resin also have no effect on wet crack surfaces.

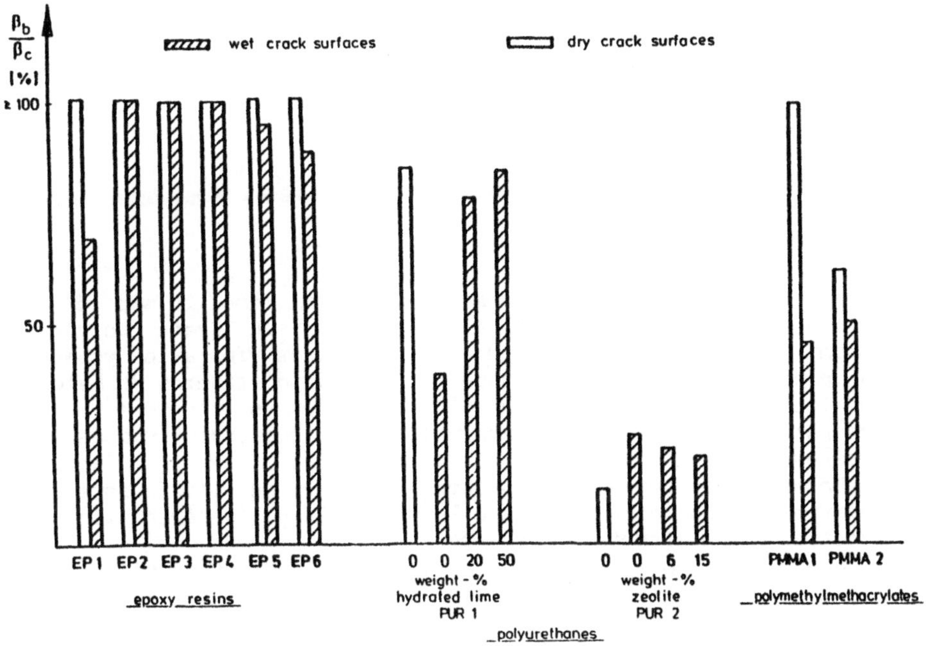

Fig. 5. Bond strength (β_b) related to the concrete tensile strength (β_c); β_c = 4.5 N/mm^2 in average

In conclusion it can be summarized that at present only modified epoxy resins are suitable for non-positive crack injection of dry as well as wet crack surfaces. Up to now resins on the basis of polyurethanes and acrylates have been used only for positive crack injection. However, these systems may possibly be suitable for non-positive crack injections after adequate modifications. In collaboration with the chemical industry improved products are in the process of development.

MODIFICATIONS APPORTEE PAR LES VARIATIONS DE TEMPERATURES SUR L'ADHERENCE DES RESINES EPOXYDES INJECTEES DANS LES FISSURES DE STRUCTURES EN BETON

A. KALLEL Ecole Nationale d'Ingénieurs de Tunis TUNISIE
A.M. PAILLERE Laboratoire Central des Ponts et Chaussées PARIS FRANCE
J.J. SERRANO Laboratoire Central des Ponts et Chaussées PARIS FRANCE

Résumé
On peut restituer une grande partie du monolithisme d'une structure en injectant dans les fissures un liant polymère adéquat. Seulement le matériau utilisé pour la réparation possède des propriétés physico-mécaniques tout à fait distinctes du béton. Sous l'effet des sollicitations extérieures couplées à la température on assiste à l'apparition d'autocontraintes dans les fissures.

La présente recherche concerne l'étude de l'influence des variations de la température sur la stabilité de la réparation par injection d'une structure fissurée.

Dans ce but on a étudié la rupture d'éprouvettes de béton 10 x 10 x 40 cm, fissurées artificiellement et injectées, après qu'elles aient subi des cycles à basse température (+ 20° C, -15° C) et des cycles à température élevée proche de celle rencontrée dans les ouvrages tunisiens (+ 20° C, + 60° C).

Les premiers résultats montrent qu'il n'y a pas d'influence notable des cycles froids, sur l'adhérence du polymère injecté. Par contre le passage de 20° à 60° C montre qu'au bout de deux cycles, l'adhérence de la résine epoxyde au béton est modifié de telle sorte que la rupture devient du type adhésif au lieu d'avoir lieu dans le béton, comme dans le cas des éprouvettes conservées à 20°C.

Des essais de traction directe, effectués à différentes températures, comprises dans la plage 20° C à 60° C, sur des éprouvettes de résines époxydes durcies, ont montré un comportement de ces matériaux variable avec la température.
Mots Clés : Polymère, Béton, Injection, Fissure, Température, Résistance, Flexion.

1. Introduction

Lorsque les polymères sont choisis judicieusement il est possible d'obtenir, par injection dans les fissures d'une structure en béton, une adhérence excellente de ce matériau aux lèvres desdites fissures. On peut donc ainsi assurer l'étanchéité et un certain monolithisme dans la structure. Toutefois les polymères, ou résines, utilisés ont des propriétés physico-chimiques très distinctes du béton, en particulier, du point de vue du coefficient de dilatation linéique (6 à 13 10^{-6} dans le cas du béton contre 50 à 85 10^{-6} dans le cas des polymères) et du module d'élasticité (30 000 à 40 000 MPa pour les bétons et 1 000 à 4 000 MPa pour les polymères).

Ainsi donc, on peut en déduire qu'à température égale, le polymère aura une déformation nettement supérieure à celle du béton, et que du fait de son module plus faible, il pourra "encaisser" une certaine déformation. Cependant on peut penser que, comme une autre partie des déformations de la résine est empêchée par le béton, il va y avoir naissance à des auto-contraintes qui risquent de se manifester lors d'une élévation de température. De même, compte tenu de la nature chimique du polymère, il n'est pas évident qu'au cours de cycles thermiques, celui-ci ne subisse des transformations pouvant affecter son adhérence au béton.

C'est pour approfondir ces divers points que nous avons étudié l'influence des variations de la température sur la stabilité de la réparation par injection.

2. Choix et principe de l'essais

Les fissures des éprouvettes, prismatiques 10 x 10 x 40 cm, de béton ont été créées artificiellement avec une entaille, effectuée par sciage jusqu'à une hauteur bien déterminée (fig. 1), puis on a injecté cette fissure avec le polymère époxydique à étudier.

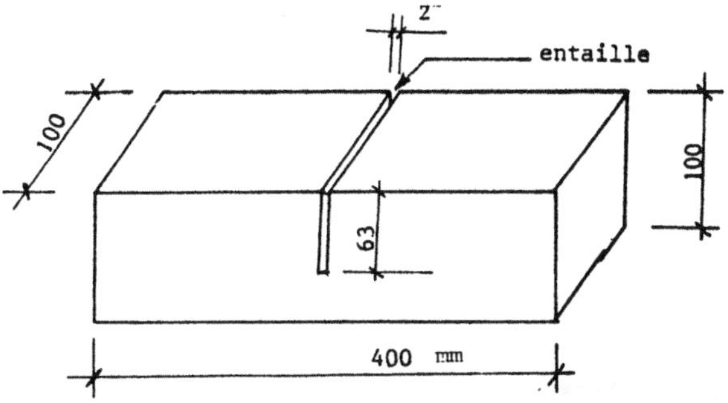

Fig. 1. Dimension de la fissure des éprouvettes

Ces polymères étaient les suivants :

S 52 Base époxydique et durcisseur polyamine
F 105 Base époxy-polyuréthanne et durcisseur mélange de polyamine
F 108 Base époxydique et durcisseur mélange de polyamines
RM 27 G4 polyméthacrylate de methyle

S 276 polyméthacrylate de methyle

Les éprouvettes ainsi injectées et les témoins correspondants ont été conservées à 20 ± 2° C et 50 ± 5 % HR pendant 15 jours.

Après ces échéances une partie des éprouvettes ont été soumises à des cycles froids (+ 20° C, - 15 ° C) et une autre partie à des cycles chauds (+ 20° C, + 60° C) la figure 2 donne le schéma des montées et descentes en température

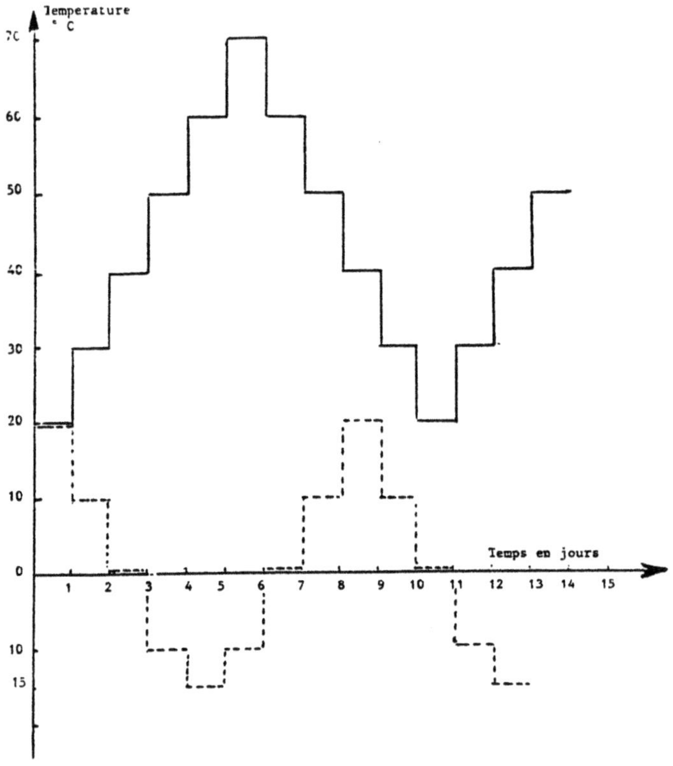

Fig. 2. Cycles thermiques

Des sondes thermométriques indiquaient les températures atteintes dans les éprouvettes. De même les mouvements des fissures (ouverture et fermeture) ont été contrôlés à l'aide d'extensomètres mécaniques (figure 3).

Après les cycles thermiques les éprouvettes réparées et les témoins ont été maintenus 24 h à 20° ± 2° C et soumises ensuite à l'essai de flexion centrée.

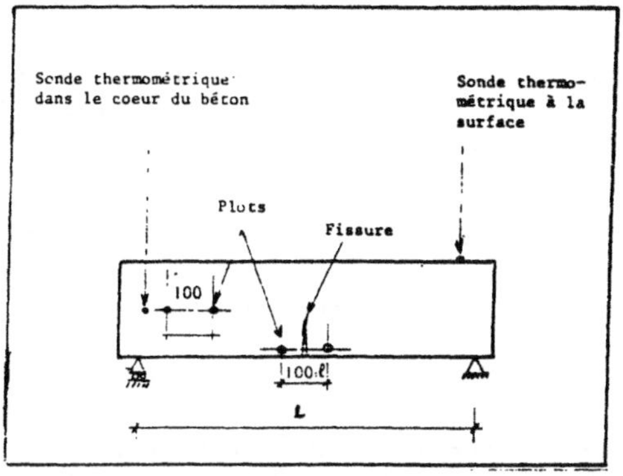

Fig. 3. Instrumentation des éprouvettes soumises aux cycles thermiques lors du chauffage de l'éprouvettes

3. Principaux résultats

3.1 Dilatation de la fissure injectée

La figure 4 présente les principales valeurs de dilatation mesurées pendant les cycles thermiques. On constate ainsi que lors de la montée en température,

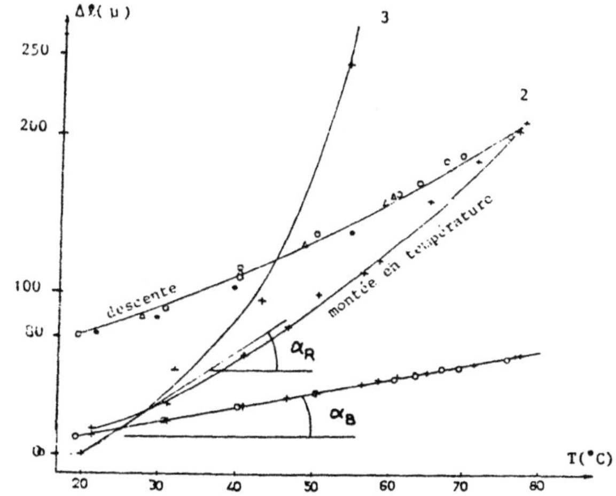

Fig. 4. Dilatation du béton de la fissure sur une éprouvette en béton. Dilatation du polymère en fonction de la température

- le béton se dilate linéairement avec la température $\alpha_B = 11.10^{-6}$.

- La résine S 52 se dilate très rapidement avec l'augmentation de la température ($\alpha_R = 62\ 10^{-6}$) pour atteindre une direction asymptotique aux environs de 60 degrés.

- la dilatation de la fissure injectée, est plus accentuée que celle du béton tout seul, tout en restant inférieure à la dilatation de la résine pure S 52. Ainsi la fissure s'ouvre lorsque la température augmente.

Lors du cycle de refroidissement des éprouvettes (passage de 60 à 20° C) on remarque que :

- La courbe de dilatation et de contraction du béton témoin sont confondues.

- La courbe de contraction ou fermeture de la fissure injectée suit, lors de son refroidissement, un trajet différent de celui du chauffage : On relève l'apparition d'une déformation permanente $\Delta l = 80\ \mu m$. L'ouverture de la fissure est donc plus accentuée qu'au début de l'essai.

3.2 Essais de flexion

Les essais de flexion réalisés sur des éprouvettes fissurées et réparées par injections (tableau 1 et figure 5) montrent que les éprouvettes réparées par les polymères epoxydiques (S. 52, F 108 et S 276), et <u>conservées à température ambiante</u>, présentent les meilleures performances, la rupture est obtenue systématiquement dans le béton. Par contre, pour les deux autres produits à base de méthacrylate de methyle, F 105 et RM 27G4 la rupture est obtenue dans le joint de collage.

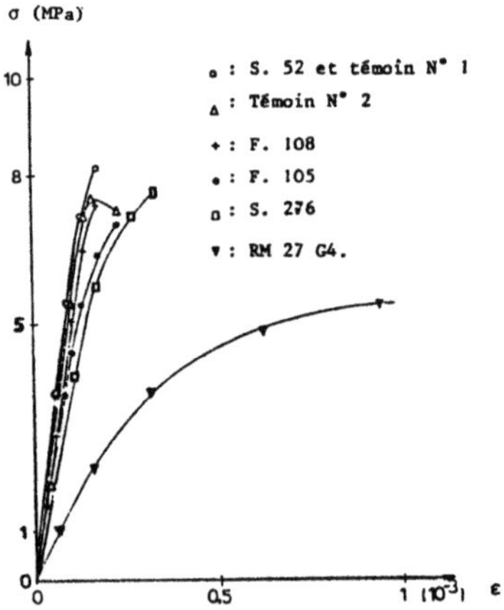

Fig. 5. Résultat des essais de flexion effectués sur des éprouvettes témoins et réparées conservées à température ambiante (0 = 20° C)

Tableau 1. Résultats des essais de flexion effectués sur les éprouvettes réparées et traitées à différents cycles thermiques.

Nature du traitement	Nombre de cycle	Type de liant utilisé	Charge à la rupture (daN)	Résistance à la flexion (bars)	Type de rupture
Conservation à température ambiante à 20° C	Béton	Eprouvettes témoin	1425 1373 1301	75,4 72,6 72,3 68,8	
	Témoins	S. 52	1484	78,5	Rupture dans le béton
		F.105	1275	67,4	Rupture dans le joint
		F.108	1354	71,6	Rupture dans le béton
		RM27G4	968	51,2	Rupture dans le joint
		S.276	1400	74,0	Rupture dans le béton
à Chaud	2 cycles (2 C.C.)	S. 52	981	51,9	Rupture dans le joint
		F.105	615	32,5	Rupture dans le joint
		F.108	876	46,3	Rupture dans le joint
		RM27G4	811	42,9	Rupture dans le joint
		S.276	1236	65,4	Rupture dans le joint
	6 cycles (6 C.C.)	S. 52	451	23,9	Rupture dans le joint
		F.105	216	11,4	Rupture dans le joint
		F.108	157	8,3	Rupture dans le joint
		RM27G4	255	13,5	Rupture dans le joint
		S.276	491	25,9	Rupture dans le joint
à Froid	2 cycles (2 C.F.)	S. 52	1361	71,9	Rupture dans le béton
		F.105	1302	68,8	Rupture dans le joint
		F.108	1550	82,0	Rupture dans le béton
		RM27G4	955	50,5	Rupture dans le joint
		S.276	1589	84,0	Rupture dans le béton
	8 cycles (8 C.F.)	S. 52	765	40,5	Rupture dans le joint
		F.105	1164	61,6	Rupture dans le joint
		F.108	1518	80,2	Rupture dans le béton
		RM27G4	780	41,1	Rupture dans le joint
		S.276	1256	66,4	Rupture dans le joint

En ce qui concerne les variations de température, on constate que si la structure réparée est soumise à des cycles froids (cas des climats tempérés), le comportement de la fissure injectée avec polymère est le même qu'à la température voisine de 20° C.

L'adhérence ne se trouve pas affectée (figures 6 et 7 tableau 1) lors des cycles + 20 C, - 15° C.

Il en est tout autrement si la réparation se trouve soumise à des variations thermiques atteignant 60° C. Déjà après 2 cycles + 20° C, + 60° C les résistances diminuent et les ruptures ont lieu, même dans le cas des polymères époxydiques, dans la résine injectée (tableau n° 1 figures 6 et 7). Ce phénomène s'accentue au fur et à mesure que le nombre de cycles augmente.

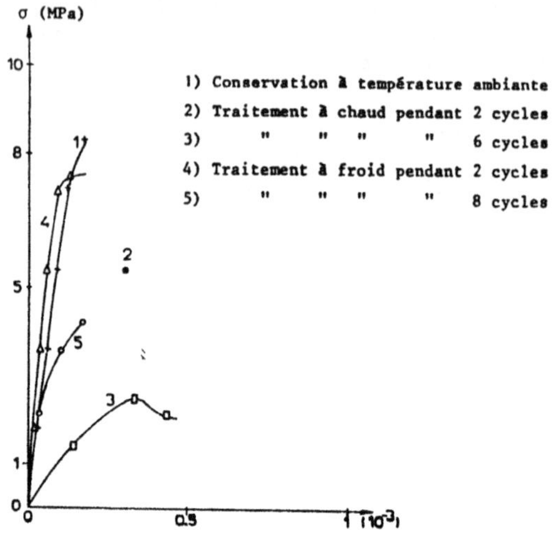

Produit de réparation S.52.

Fig. 6. Résultats des essais de flexion effectués sur des éprouvettes réparées et soumises à différents types de cycles

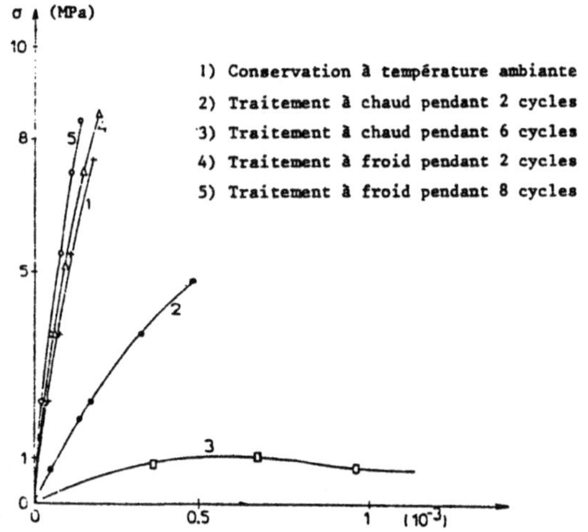

Produit de réparation F.108

Fig. 7. Résultats des essais de flexion effectués sur des éprouvettes réparées et soumises à différents types de cycles

Ainsi donc une élevation de température dégrade l'adhérence des polymères au béton.

On peut relier cette dégradation aux variations constatées dans la résistance à la traction directe des polymères lorsque les températures s'élèvent (fig. 8). Des études en cours sur la température de transition vitreuse des polymères permettront certainement une meilleure compréhension des phénomènes constatés

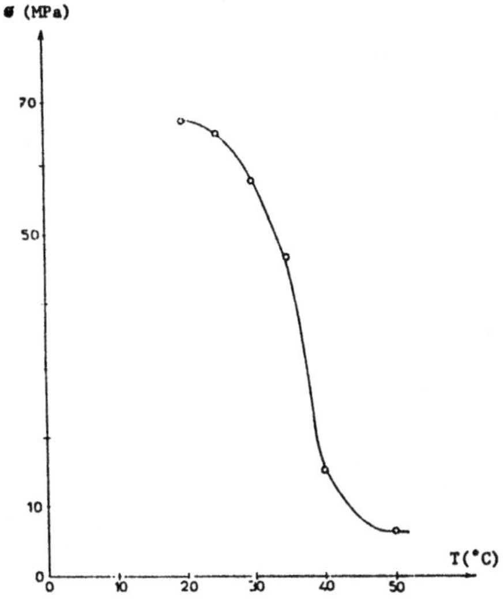

Fig. 8. Variation de la résistance à la traction du polymère S. 52 en fonction de la température (Eprouvettes I.S.O.)

4. Conclusions

Les résultats de la présente recherche font apparaître qu'il importe de bien connaître les variations thermiques auxquelles seront soumis les ouvrages dégradés lors des réparations, par injection de polymères.

En effet, si le milieu climatique environnant est tempéré ou froid, il semble que l'adhérence du polymère injecté ne soit pas mise en cause. Par contre dans le cas des pays chauds où les structures peuvent atteindre périodiquement 50 ou 60° C, l'adhérence se dégradera au fur et à mesure et la réouverture de la fissure aura lieu à plus ou moins brève échéance. Cette dégradation de la réparation peut être reliée, soit à des transformations dans le polymère lui-même soit à la formation d'auto-contraintes dans la fissure, le polymère injecté se dilate lorsque la température s'élève et exerce une pression sur les lèvres de la fissure, puis se contracte quand la température baisse, soit encore à la combinaison des deux phénomènes précédents. Des études approfondies actuellement en cours devraient permettre de mieux cerner le problème du comportement des strucutres reparées avec des polymères et soumises à des cycles thermiques.

Reférences

Mouton, Y. Réparation des structures en béton fissurées par Injection de liants époxydiques, Rapp. Rech. L.C.P.C. n° 86, 1979

Lattouf, M. Collage des structures en béton armé, utilisation de résines synthétiques, thèse Université Claude Bernard de Lyon 1974.

Kallel A. Réparation des structures en béton fissurées par des liants polymères, mémoire de D.E.A. en mécanique apliquée au Génie Civil. Ecole National d'Ingénieurs de Tunis (ENIT), Mars 1984.

Guide pour le choix et l'application des produits de réparation des ouvrages en béton -LCPC - SETRA PARIS.

Paillère A.M., Rizoulières Y., Lazzri Y., possibilités d'application des résines époxydiques - Réparation des Bétons Hydrauliques -Bull. Liaison Laboratoire des Ponts et Chaussées n° 74 nov. déc. 1974.

Paillère A.M., Rizoulières Y., Réparation des structures en béton par injection de polymères Bull. Liaison Labo Ponts et Chaussées n° 96 Juillet août 1978.

A THEORETICAL AND EXPERIMENTAL STUDY OF CONCRETE MEMBERS REPAIRED BY RESIN INJECTION

J.G. KEER and N.K. EMBERSON

Department of Civil Engineering, University of Surrey

Abstract
Simple analytical models are presented which are used to examine the theoretical behaviour of an axially reinforced concrete prism under tension pre and post-repair by resin injection. The effect of repair on the steel strain at the crack and the crack movement has been derived in terms of the resin modulus and the original crack width. The load sustained post-repair before recracking has been shown to be dependent upon the load at which repair is carried out and on the previous load history of the member. The experimental work was conducted using axially reinforced concrete prisms loaded in tension and monitored for strain. Plain and ribbed reinforcements were used. A novel part of the test program was that the repair and resin curing period of some members was undertaken while they were held under load, rather than at zero load. Most members tested were able to support, following repair, substantial proportions of the original cracking load before recracking occurred, although for ribbed bars, the post-repair load increment is sensitive to the success in repairing secondary cracking.

Key words: Reinforced concrete, Resin injection, Repair, Cracks, Analysis.

1. Introduction

The injection of a resin adhesive into cracks in a concrete structure is a potentially valuable technique for restoring the strength and durability of the uncracked concrete, as ACI Committee 244 (1984) and The Concrete Society Report 26 (1984) record. A number of laboratory investigations e.g. Chung (1975), Mouton (1979) and Hewlett (1982) have confirmed the effectiveness of the technique in restoring a large proportion of the uncracked stiffness of the repaired member. This paper reports results from an analytical study of a resin-injected crack zone in a concrete member and from an experimental programme of tensile tests on axially reinforced concrete prisms which included specimens repaired whilst held under load - a simulation of the practical situation.

2. Analysis of the behaviour of resin-injected cracks in concrete members.

The objective is to assess theoretically the behaviour of a tensile crack zone in a reinforced concrete member which has been repaired by resin injection. In particular, the stress and strain generated in resin and steel and the associated crack movement under load are to be investigated. Assuming bond is maintained between resin and concrete, the dominant influence on the behaviour is the direct stiffness of the resin layer, which is controlled principally by the modulus of elasticity of the resin and the crack width.

As a preamble, it is useful to record two points. Firstly, concrete is a brittle material with an ultimate tensile strain of about $50\text{-}100 \times 10^{-6}$ corresponding to a tensile strength in the region of 2-3 MN/m^2. Secondly, a feel for the magnitude of crack movement due to factors other than imposed load is important. The Concrete Society Technical Report No. 22 (1982) contains useful data on this. For example the total strain due to seasonal temperature and drying shrinkage (6 months to 30 years) for an external member assuming 50% restraint is estimated as 239×10^{-6}. If it is assumed that a 300mm length of member contributes to the crack movement, then the value of that movement is $300 \times 239 \times 10^{-6}$ or about 0.07mm.

2.1 Simple analysis of a resin layer in a plain concrete member.

Injection resins are comparatively low modulus, flexible materials in the hardened state. Their modulus of elasticity, E_r, may be 1-3 GN/m^2, one-tenth that of concrete and one-hundredth that of steel. They are present, however, as very thin layers (of thickness t say) in cracks and, because of this, the axial stiffness of the resin member, $E_r A/t$ is very high. (A is the area of resin normal to the direction of stress). A simple one-dimensional analysis of two sections of a concrete prism joined by a thin layer of resin (Fig. 1) is valuable in demonstrating this, with interesting results.

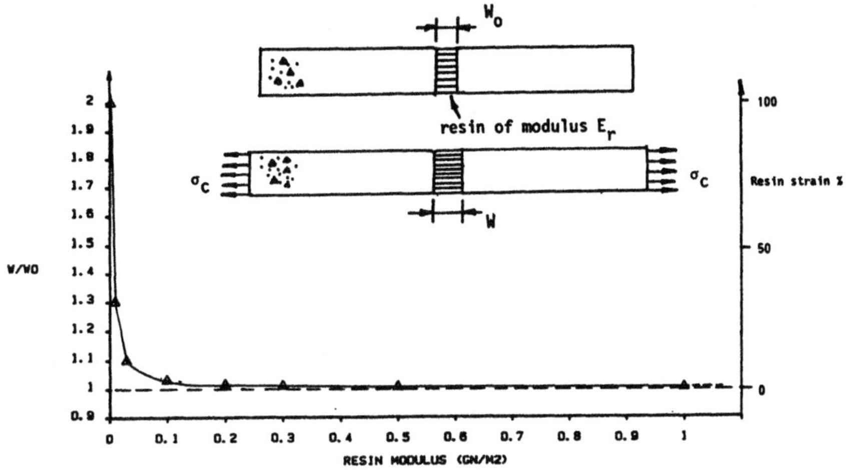

Fig. 1 W/Wo, resin strain v resin modulus

In Fig. 1, the width of the resin layer is Wo at zero load. A uniform stress of 3 MN/m^2 (σ_c) is applied to the concrete section, a reasonable value for the maximum tensile stress that could be sustained without cracking. The resin is assumed to be stressed uniformly and the crack width increases to W. Fig. 1 illustrates the relationship between W/Wo and E_r, the resin modulus. Significant opening of the crack only occurs for resins of very low modulus (<0.1 GN/m^2). This simple analysis illustrates the effectiveness of an injection resin in restricting crack opening and hence restoring the stiffness of concrete members. The corollary of this argument, however, explains why recommendations are made that the techniques should only be used to repair dormant cracks (ACI Committee 244 (1984), Concrete Society (1984)). A 0.5mm wide crack filled with a resin of modulus 1 GN/m^2 would only accommodate a movement of 0.0015mm before the stress level generated in the resin would, when transferred to the bonded concrete, cause the concrete to re-crack.

2.2 Analysis of a resin-filled, preformed crack in a reinforced concrete member

The model comprises a concrete prism axially reinforced by a single plain reinforcing bar and containing a pre-formed crack of uniform width Wo (Figure 2(a)). At zero load the stresses throughout the steel and concrete are assumed to be zero. Further assumptions are:

(a) Plane sections remain plane, thus on loading the stress in the concrete is uniform over the whole concrete section. Similarly the stress in a resin layer in a crack is uniform;

(b) It follows from (a) that there is slip between the concrete and the steel (and between resin and steel) and force will be transferred between the steel and concrete by bond stresses acting at the interface. The transfer of stress into the thin resin layer from the reinforcing bar by bond stresses is considered to be negligible;

(c) The bond stress has a constant value along the interfacial zone. This is a simplification of the bond stress distribution assumed in other one-dimensional analyses of cracking in reinforced concrete (Thomas (1936), Bennett (1974)) but its use should not invalidate the general conclusions drawn from the analysis;

(d) The stress transfer mechanism is unaffected by the injection process, which results simply in the opposite faces of a crack being bonded by the resin layer. Stress transfer across the crack takes place via the steel and resin.

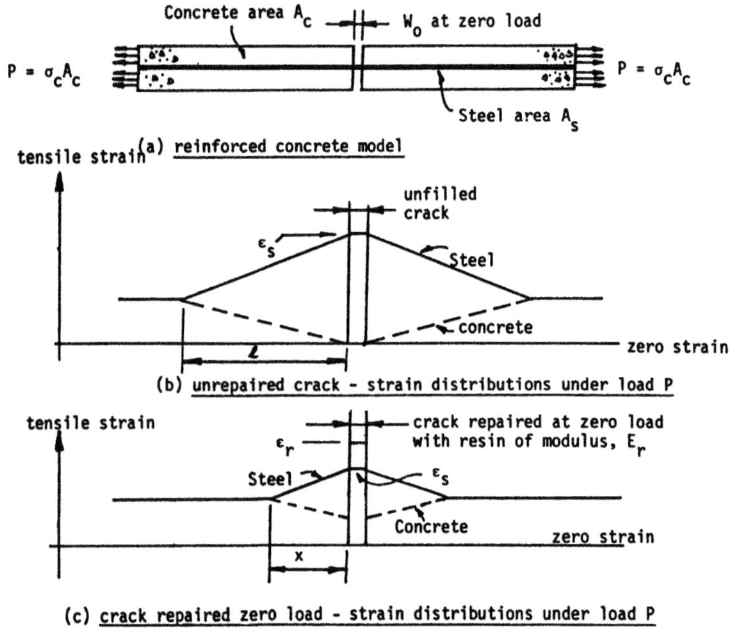

Fig. 2 Reinforced Concrete Member with Preformed Crack

A tensile load P is applied to the ends of the member remote from the pre-formed crack (Figure 2(a)). Figures 2(b) and (c) illustrate the assumed strain distributions in concrete and steel close to the crack under load P for an unrepaired crack and for a crack repaired at zero load. Particular attention was paid to the stress and strain in the steel and resin at the crack and the crack opening caused by loading. These are dependent upon:

(i) the load applied, P. In Figure 2(a) the load P is applied at the ends of the member remote from the crack as a stress σ_c on the concrete area A_c such that $P=\sigma_c A_c$;

(ii) the moduli of elasticity of the steel (E_s), concrete (E_c) and resin (E_r). In the discussion which follows constant values of E_s (= 200 GN/m^2) and E_c (= 30 GN/m^2) have been assumed. The effect of varying E_r was investigated;

(iii) the reinforcement ratio $p = A_s/A_c$ where A_s and A_c are the cross-sectional areas of the steel and concrete respectively (Figure 2(a));

(iv) the initial crack width at zero load, W_0. The effect of varying W_0 was investigated;

(v) the stress transfer capability between the concrete and the steel bar. In the discussion which follows, a notional value of the stress transfer capability is given by the length parameter ℓ (Figure 2(b)), which is the transfer length under load in the unrepaired specimen. The effect of improved stress transfer (ℓ reducing) or poorer stress transfer (ℓ increasing) was investigated.

Examples of some of the relationships which were derived (Keer, 1986) are now given.

Figure 3 shows the relationship between the steel strain at the repaired crack and the resin modulus for different assumed transfer lengths, ℓ, and for a stress σ_c (=P/Ac) of 3MN/m^2, a reinforcement ratio ρ of 0.01 and an original crack width at zero load of 1mm. For very flexible resins, the steel strain approaches the value for an unrepaired crack under the same load - 1500×10^{-6} (stress 300 MN/m^2). The effect of a 'stiff' resin is to reduce markedly the steel strain and stress at the crack. For a typical injection resin with E_r =3 GN/m^2, steel strain and stress at the crack are about 250×10^{-6} and 50 MN/m^2 respectively. The stress and strain induced in the resin are about 2.7 MN/m^2 and 900×10^{-6} respectively, well below the tensile capacity of a typical injection resin which may be 60 MN/m^2 and $20{,}000\times10^{-6}$. Figure 4 illustrates the effect of the original unrepaired crack width, W_0, on the steel strain at the crack, assuming ℓ = 100mm. In fact Figures 3 and 4 can be combined into Figure 5 since the steel strain at the crack is dependent on the term $E_r\ell/W_0$ for a given stress σ_c. Figure 6 illustrates the relationship between increase in repaired crack width, i.e. crack movement due to load and the resin modulus, for different original crack widths, W_0 (and assuming ℓ = 100mm). It shows how effective a typical injection resin can be in restricting movement at a crack and restoring stiffness to the member, movement being reduced by two orders of magnitude for an original crack width of 1mm.

Some preliminary work has been carried out using a finite element analysis of a 'no-slip' model of an axially reinforced concrete prism in tension, in which the concrete and steel are assumed to remain bonded together. For resin properties typical of an injection resin the 'slip and 'no-slip' models give similar indications of the magnitude, under load, of the crack movement and steel stress following repair of the crack.

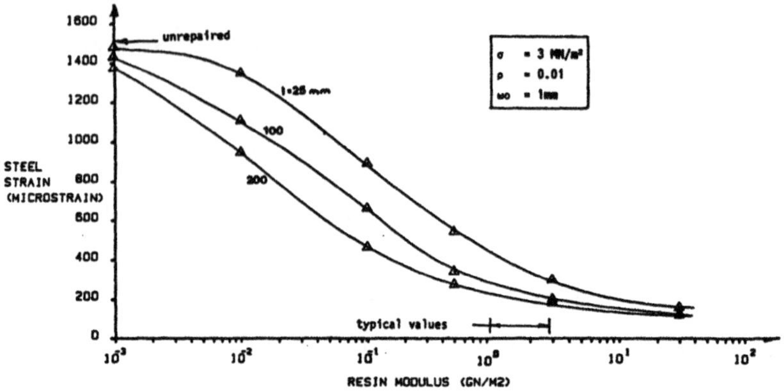

Fig.3 Theoretical relationship between steel strain at a repaired crack and resin modulus

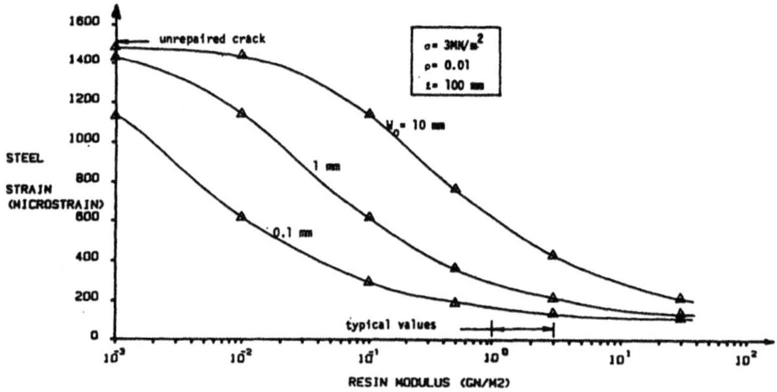

Fig.4 Theoretical relationship between steel strain at a repaired crack and resin modulus for different original crack widths

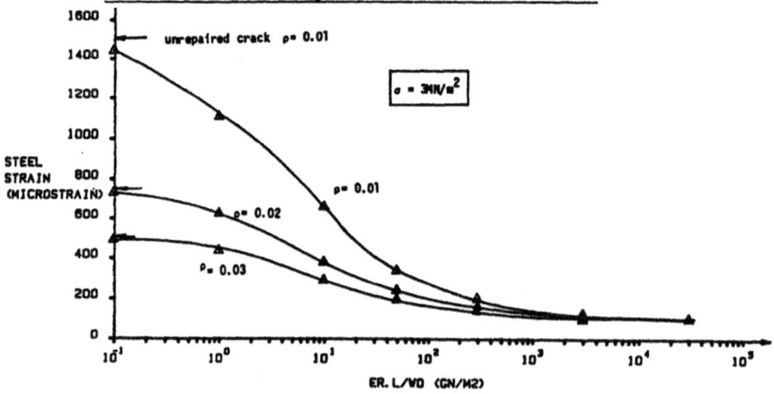

Fig.5 Theoretical relationship between steel strain at a repaired crack and the factor $E_r \ell / W_o$ for different reinforcement ratios, ρ

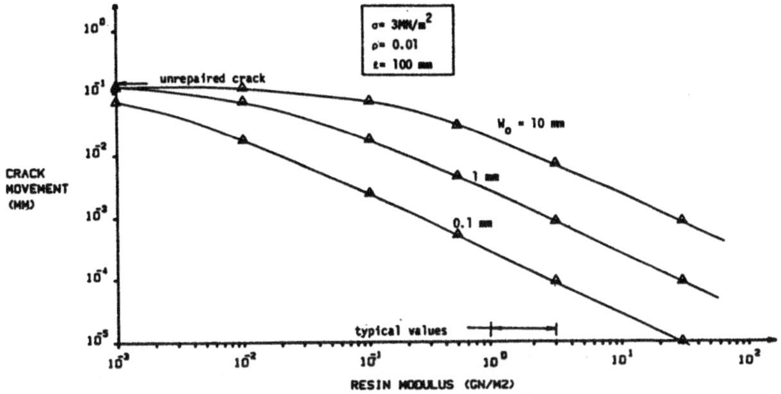

Fig. 6 Theoretical relationship between crack movement and resin modulus at a repaired crack for different original crack widths

2.3 Analysis of members cracked by loading and subsequently repaired by resin injection.

The model consists of a concrete prism axially reinforced by a single, plain reinforcing bar and loaded in tension. Cracks are induced in the member by an initial loading. The behaviour of the member is examined following repair after a number of different load histories. The same assumptions as in Section 2.2 are made, but in this case stresses/strains are present in steel and concrete at the repair load, even if this is at zero load.

The theoretical behaviour under tensile load of a 100mm square concrete prism reinforced centrally with a 12mm diameter steel bar was examined. The reinforcement ratio was 0.011. The following parameters were assumed in the analysis:

Ultimate concrete tensile strain ε_{ct} = 100×10^{-6}
Modulus of elasticity of concrete E_c = 30 GN/m²
Modulus of elasticity of steel E_s = 200 GN/m²
transfer length ℓ, before repair = 100mm (see fig. 2b)

Theoretically, the crack spacing in the member will lie between ℓ and 2ℓ. In the analysis it was assumed that cracks form at the upper bound of the spacing i.e. 2ℓ. In Figure 7 the path OABD represents the initial loading path of the specimen. The average strain is the steel strain averaged over the length 2ℓ, which constitutes the transfer length each side of the crack. The concrete cracks at A, whereupon the strain increases at constant load as the member becomes fully cracked with cracks at spacing 2ℓ. Unloading (and reloading) paths from B and D are shown. The reloading behaviour after repairs at points C,E,F,G and H is shown by dashed lines. The resin modulus was taken as 3 GN/m². The member was assumed to recrack on loading after repair where the concrete tensile strain again reached its ultimate value, ε_{ct}, of 100×10^{-6}. These positions are labelled J to N on Figure 7.

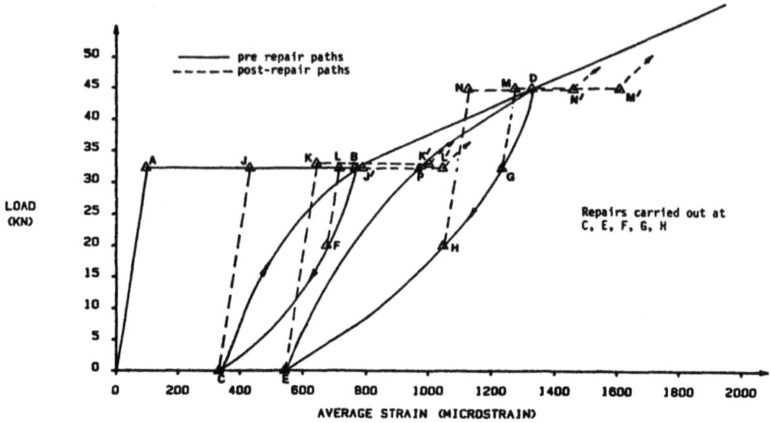

Fig.7 Load-average strain relationship of a reinforced concrete member in tension

The following points emerge from the analysis. The reloading paths after repair are virtually as stiff as the original uncracked path OA. The increase in load sustained by the member after repair is dependent upon the load at which the repair is carried out and the previous loading history. For example, increasing the load after repair at B results in immediate re-cracking since the concrete midway between the cracks is already at its ultimate failure strain. Reloading from the same load at G after repair, however, results in load increment GM before re-cracking since the residual tensile strain in the concrete at G is less than the ultimate failure strain. Similarly paths FL and HN result from repair at the same load, but with different load history. The position of new cracks is dependent upon the load at which the repair is carried out and the previous loading history. Following repair at points C,F,G and H, theoretically new cracks would form midway between the repaired cracks, whilst following repair at E, new cracks at K would form close to the original cracks, due to the shape of the residual strain distribution in the concrete at E.

Figure 7 illustrates that theoretically the load that can be sustained before re-cracking following repair under load depends upon the residual stress state in the concrete, which itself depends upon the load at repair and the load history of the member. The tensile stress sustained by concrete after cracking of a reinforced member is acknowledged in design procedures by allowing for 'tension stiffening' effects. In Figure 7 the maximum concrete stress between cracks at points F and G is about 1.8 MN/m^2. At point H the value is about 0.7 MN/m^2, which accounts for the higher load sustained post-repair in this case. (The cracking stress of concrete was assumed to be 3MN/m^2).

In reality, residual concrete stresses in members at particular loads and after often complex loading histories cannot be easily defined. Peak values of about half the tensile strength have been suggested (Beeby, 1972). Furthermore, tensile stresses between cracks will reduce with time under permanent loading and under cyclic loading (Stevens, 1973) and may be very small at the time of repair. The post-repair load increment before re-cracking may therefore be close to the original cracking load. If this is the case, whether or not a member cracks post-repair will be governed by the relative values of the load causing cracking and the superimposed load which the structure must sustain post-repair.

3. Experimental Work

The specimens used were 100mm square concrete prisms one metre in length, reinforced axially by a single reinforcing bar. Plain round bars and deformed bars with helical ribs were used. After manufacture, the specimens were kept in a fog room at 20 °C for 28 days, after which time they were stored in laboratory air until test. Cube strengths at 28 days were about 65 MN/m^2.

Each prism was tested in an Avery Universal testing machine. Protruding lengths of steel bar at each end were attached to the test machine via a universal joint coupling to reduce bending effects. The first loading induced cracks in the prisms suitable for subsequent repair by resin injection. Strain readings were taken at regular load intervals over 100mm gauge lengths between demec studs. In order to achieve crack widths which were injectable and which covered a range between narrow (~0.2mm) and wide (~4mm), the load was sometimes increased to a level sufficient to yield the steel locally at some of the crack locations. Prisms were either repaired after unloading to zero load, or repaired in the test machine after some unloading to an intermediate load, which was held while the prism was repaired and the resin cured. Following repair and curing, the prisms were reloaded and strain readings retaken.

The injection resin generally used was a proprietary epoxy injection resin (of modulus 2.8 GN/m^2). Three specimens were repaired with a very low modulus resin ($E \sim 0.1$ GN/m^2) which was more viscous than a typical injection resin. A manually operated cartridge gun was used for injection. With some narrow cracks, resin in an inlet tube was injected with the aid of a pneumatic gun.

4. Results

Table 1 summarises the results of tests on 15 prisms.

Series A prisms were reinforced with plain round bars and repaired at zero load with the typical injection resin. Generally crack repairs were successful in that the prisms were able to sustain between 63% and 107% of the original first crack load before further cracks formed. In a typical test, an increase in load of 30 KN prior to cracking resulted in average strain increases of 60-70 $\mu\varepsilon$ ($\mu\varepsilon$ = strain x 10^{-6}) over the 100mm gauge length. After cracking and on reloading from zero load, an average strain increase measured across a crack was about 3000$\mu\varepsilon$. The corresponding crack width increase was, therefore, about 0.3mm. Following repair of a crack, the average strain increase was about 90με for a 30 KN load increment. Allowing for the extension of the concrete either side of the crack within the gauge length, this suggests that the crack movement was reduced to about 0.003mm by the resin repair.

Series B prisms were reinforced with plain bars and repaired at zero load with the specially formulated, very low modulus resin. Repaired prisms were able to sustain between 80 and 128% of the original first crack load before further crack formation. New cracks formed close to the original repairs in this case. Possibly the higher viscosity of this resin formulation resulted in fine fissures associated with the original cracks not being penetrated successfully. The crack movement post-repair was considerably reduced by the very low modulus resin and a stiff repaired zone resulted.

Series C prisms were reinforced by ribbed bars and repaired at zero load by the typical injection resin. Figure 8 illustrates the crack pattern pre and post-repair for specimen 9 and the load-average strain relationship measured across the first crack to form. A new crack (crack 11) formed within the same gauge length post-repair. The pattern of cracking for the ribbed bar specimens was different from the plain bar specimens and similar to that observed in other tests on axially reinforced prisms (Beeby, 1972), being characterised by the presence of longitudinal cracks and secondary tranverse cracks. This type of crack network is more difficult to inject successfully and this probably accounts for the lower loads at which new cracks formed post-repair compared to plain bar specimens. The locations of new cracks were frequently at positions where secondary tranverse cracks had occurred or existed internally and which had not been fully penetrated by resin.

Series D and E specimens were reinforced with plain and ribbed bars respectively and were repaired under load by the typical injection resin. Figure 9 shows the load-average strain relationship recorded over the gauge length in which crack 3 of prism 15 formed on the first loading. The prism was repaired whilst held under a load of 30 kN and subsequently reloaded, when crack 12 formed within the same gauge length at a load of 55 kN. It is interesting to note that on the crack pattern of Figure 9 the resin was considered by observation to have successfully penetrated secondary cracks associated with crack 3. The high load increment of 25 kN before formation of crack 12 reflects this success.

5. Discussion

Most members tested were able to support, following repair, substantial proportions of the original cracking load before recracking occurred (Table 1). Theoretical considerations suggested that the load sustained, post-repair before cracking, depended on the load history of the member and that in certain circumstances, this load could be negligible when there were substantial residual tensile stresses in the concrete at repair. Experimental evidence suggests, however, that residual tensile stresses in the concrete at repair, even when carried out under load, are small. There is some

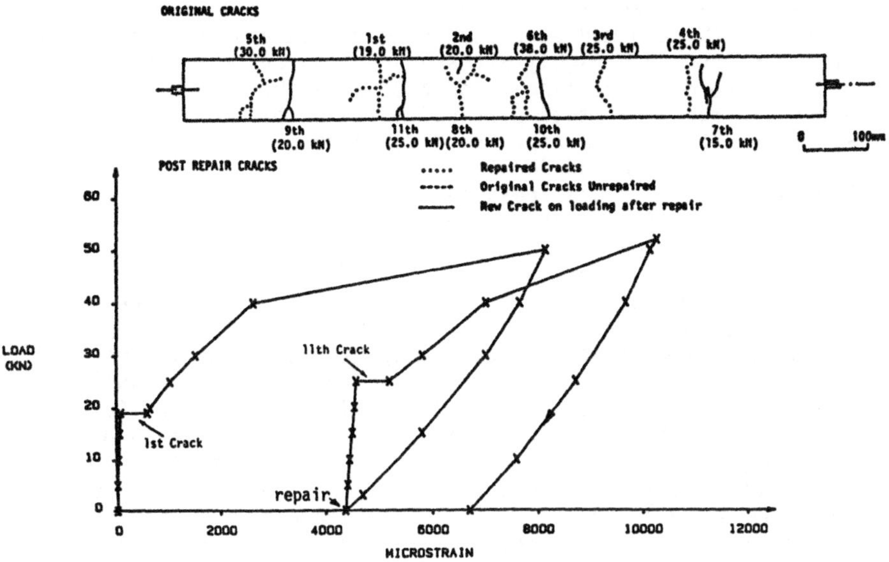

Fig. 8 Crack pattern and load-average strain relationship for prism 9

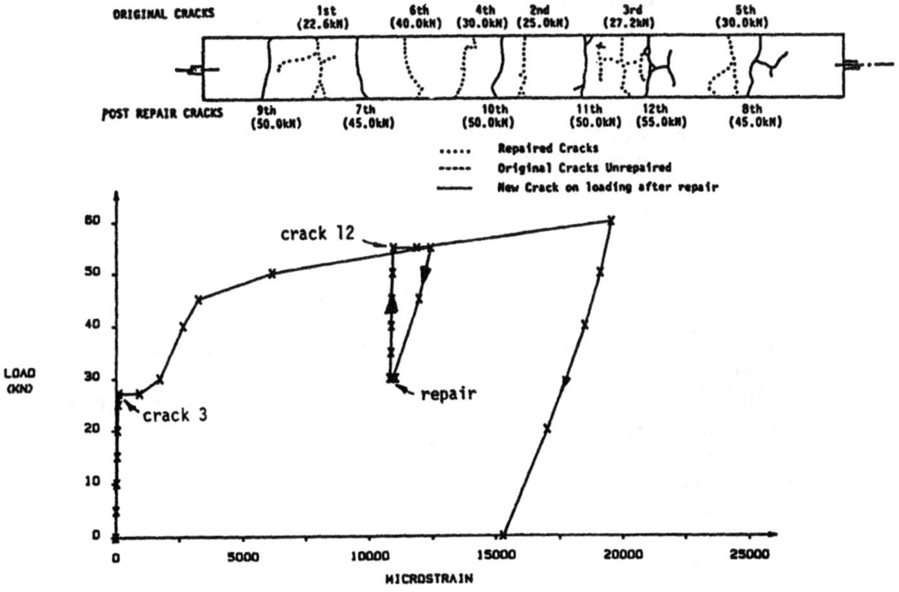

Fig. 9 Crack pattern and load-average strain relationship for prism 15

evidence that substantial post-repair capacity of specimens reinforced with deformed bars cannot be relied upon (particularly those which have been highly stressed) because of the difficulty of repairing secondary cracking associated with the mechanism of stress transfer via ribs. If this difficulty can be overcome, it would appear that even members repaired under load may not re-crack under a subsequent superimposed load.

The restoration of stiffness in a cracked member following repair was evident, even for a resin with a very low modulus of elasticity of 0.1 GN/m^2. On the other hand, theoretical and experimental results emphasise that active cracks, even if movement is as small as 0.01mm, should not be repaired by a typical injection resin of modulus 2-3 GN/m^2 (or indeed by one with a modulus an order of magnitude lower) without accepting that recracking of the member is likely if the processes causing movement recur. A crack which may have been acting usefully as an undesigned movement joint (perhaps compensating for the lack of provision or poor performance of pre-formed joints) by accommodating movements of perhaps 0.1mm will, if sealed by an injection resin, be able to accommodate a movement of the order of 0.001mm only. The situation prior to original crack formation is therefore restored and further cracks are likely to occur.

6. Acknowledgements

The work was supported by the Transport and Road Research Laboratory, Crowthorne, Berks and the authors would like to acknowledge the help of Dr. D.M. Thompson from that establishment.

References

ACI Committee 224 (1984). Causes, evaluation and repair of cracks in concrete structures. J.Am. Conc. Inst. 211-230.

Beeby A.W. (1972) A study of cracking reinforced concrete members subjected to tension. Cem. and Conc. Assoc. Technical Report. Cem. and Conc. Assoc.

Bennett E.W. (1974). A theory for the movement of cracks in concrete members with plain bar reinforcement or plain wire tendons. Mag. Conc. Res. 26(87), 83-92.

Chung H.W. (1975) Epoxy-repaired reinforced concrete beams. J.Am. Conc. Inst. 72(5) 233-234.

Concrete Society (1982) Technical Report No.22. Non-structural cracks in concrete. The Conc. Society.

Concrete Society (1984) Technical Report No. 26. Repair of Concrete damaged by reinforcement corrosion. The Conc. Society.

Hewlett P.C. and Morgan J.G.D. (1982). Static and cyclic response of reinforced concrete beams repaired by resin injection. Mag. Conc. Res. 34 (118) 5-17.

Keer J.G. (1986) Analytical study of concrete members repaired by resin injection. Univ. of Surrey, Dept. of Civil Engineering Report.

Mouton Y. (1979). Reparation des structures en beton fissurees par injection de liants epoxydiques. Rapport de recherche LPC No.86 Laboratoire central des ponts et chaussees.
Stevens R.F. (1973) Deflection of reinforced concrete beams. Proc. Inst. of Civ. Engrs. 53, 207-224.
Thomas F.G. (1936) Cracking in reinforced concrete. The Struct. Engineer. 298-320.

SERIES	A				B			C		D			E		
Prism No.	1	2	3	4	5	6	7	8	9	10	11	12	13	14	15
Bar dia / type *	12P	12P	12P	16P	12P	12P	12P	12R	12R	12P	12P	12P	12R	12R	12R
Resin **	x	x	x	x	Y	Y	Y	x	x	x	x	x	x	x	x
Loads at which cracks form 1st loading (kN)	30 34 39.5	30 30 37.5	30 30 30 37	35 35 35 35	24.5 24.5 25 27.5 27.5	24.5 25	25 25 30	20 20 23 30 34	19 20 25 25 30 38	25 25 30 33 33.5	25.2 28 36.5	20 20 28.5	18 18 22 24 24 30	20 23 26 29 27	22.6 25 27.2 30 30 40
Load at repair (kN)	0	0	0	0	0	0	0	0	0	22	22	15	30	30	30
No. of cracks repaired	2	2	3	3	2	2	3	5	6	3	3	3	6	5	6
Loads at which cracks form post-repair (kN)	24 29 30	24	32	22 28 35	32	25 25	20 25	10 10 10 15 25 25 30	15 20 20 25 25		48	35	35 40 40 40 40 45 50	45 50 50 55	45 45 50 50 50 55
cracking load increment post-repair ÷ 1st crack load (%)	80	80	107	63	131	98	80	50	79		103	100	28	75	66

* p = Plain bar, R = ribbed bar
** x = typical injection resin $E \sim 2.8$ GN/m²
y = very low modulus resin $E \sim 0.1$ GN/m²

Table 1 Cracking loads of prisms pre-and post-repair

RESIN INJECTION OF CRACKS WITH CHANGING WIDTH

R.H. LETSCH
Baustoffinstitut, Technical University Munich

Abstract
Cracks in concrete usually have to be closed either to make the structure water tight again or to protect the reinforcement against corrosion. As many cracks are not stable in their width, a repair is only successful, when the injected material can change its volume according to the volume-change of the crack. The volume of gases can be changed to a great amount by applying small external forces. To achieve volume compressibility of an injecting resin combined with watertightness, plastic micro-balloons or micro-spheres (average diameter 0,04 mm) are inserted into the resin. Using a rubberlike epoxy resin, volume deformations of more than 50 % are possible. If the epoxy resin is stiff (modulus of elasticity about 2000 N/mm^2) forces can be transmitted still allowing a volume change of 10 to 20 %. The applied deformations are fully reversible.
Key words: crack injection, width change, volume compressibility, plastic micro-balloons.

1. Introduction

Cracks in concrete structures occur, when the tensile strain of the concrete is overcome. In many cases the cracks have to be closed e.g. to restore the load-carrying capacity of the structure, to protect the reinforcing steel against corrosion or to make the structure watertight again /1, 2, 3/. In general the width of a crack does not remain constant, but is changing due to varying loading conditions or temperatures. To allow a change in the width of a filled crack the injecting material must be able to change its volume, when external forces are applied.

2. Crack injection

Up till now crack-injection is done in general with epoxy-, polyurethane- or acrylic resins /4, 5/. A crack injection of this type is only successful, if the crack stays at constant width. Any movement of the sides of the crack will cause a change in the volume of the crack and therefore in the injected resin under the condition that the adhesion to the concrete is not overcome /6/. As the injected materials are nearly volume-incompressible a crack either opens again or - as in many cases - a new crack forms in the neighbourhood of the original crack.

For the sealing of cracks, which are not stable, resins are used, which are foaming in the crack. The foam is thereby in most cases produced by the chemical reaction of a polyurethane with water. Due to this wet cracks can also be filled. This method of crack-injection can be problematic if the foam developed is open-pored - this means that it would not be watertight - and of uncontrollable pore-volume. The resin used has to show rubberlike elasticity, as brittle foams will disintegrate at the deformations possible, which can reach more than 50 %. These foams cannot transfer stresses, therefore their application is limited to a purely sealing function.

3. Volume compressible injecting and sealing material

To overcome the before-mentioned problems a material for crack-injection has to be used, which is volume-compressible, can be injected into narrow cracks (width smaller than 0,1 mm), has good bonding properties to concrete, is watertight and can - if it is required - transfer stresses.

The only materials whose volume can be changed within the engineering means are the gases. Gases can be compressed or expanded to a large amount without using great forces. This compressibility is used in soft foams or in any air-cushioned device.

An injecting material, which fulfills the above mentioned requirements, has to contain a gas in the form of very small bubbles, which are distributed equally and separated from each other. During the injecting-process the bubbles must neither be damaged nor their distribution significantly changed.

One way to produce a material with the above named qualities is to insert plastic micro-balloons into a resin either showing rubber elasticity, when the deformations are big (up to 50 %), or with relatively high modulus of elasticity, when stresses have to be transferred and the deformations are smaller (appr. 20 %).

Any modification in the properties of the resin should be possible to meet the requirements optimally.

The diameters of the plastic micro-balloons must be significantly smaller than the width of the crack to be injected. The micro-balloons should show only a little stiffness.

4. Tests

4.1 Materials

Tests were conducted using three different types of epoxy resin - the modulus of elasticity at 23 °C corresponding to 2000 N/mm^2, 500 N/mm^2 and 0,2 N/mm^2 - and plastic micro-balloons of an average diameter of 0,04 mm.

The volume-ratio of resin to micro-balloons varied between 1 : 1,5 and 1 : 2. Figure 1 shows a cut through a rubber elastic epoxy resin (E = 0,2 N/mm^2) with plastic micro-balloons after repeated compression to 50 %.

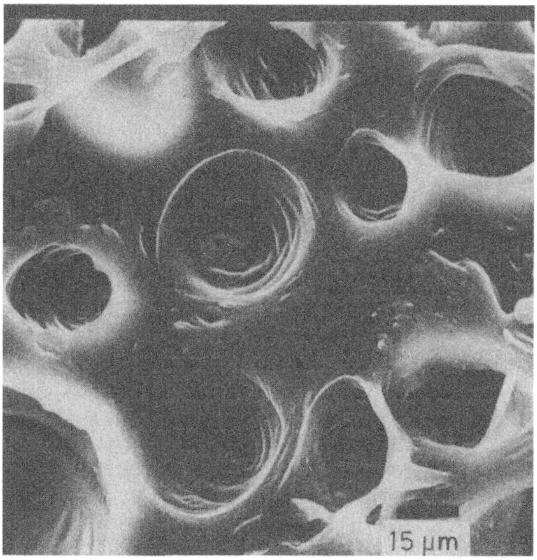

Fig. 1. SEM-picture of micro-balloons embedded in a rubber elastic epoxy resin after repeated deformation up to 50 %.

4.2 Test procedure
Specimen for the compression tests were cast in a stiff ring of 100 mm diameter with a thickness of 2 and 50 mm. The tests were conducted at a deformation speed of 50 %/min.

Compression was obtained by pressing the compression platen (100 mm diameter) into the ring with the specimen.

Due to the lateral restraint a volume compression was obtained.

The sides of the ring were treated with releasing agents to allow slipping.

Tensile tests were conducted at joints with a cross section of 20 mm x 20 mm cast between prism of cement mortar.

4.3 Test results
Depending on the stiffness of the resin volume compression between 12,5 and 50 % was possible with full recovery of the material within a short period of time (0,5 to 2 hours). The results are summarized in table 1. Figures 2 to 4 show the mechanical behaviour of the materials tested.

Table 1. Material properties of epoxy resins filled with plastic micro-balloons.

Modulus of elasticity of the resin (N/mm^2)	Volume ratio of resin to micro-balloons	Stress (N/mm^2)	Strain (%)	Lateral restraint
2000	1 : 2	-8,72	12,5	yes
500	1 : 2	-2,47	35	yes
0,2	1 : 2	-1,65	50	yes
0,2	1 : 1,5	-1,64	45	yes
0,2	1 : 1,5	-0,53	45	no 1)
0,2	1 : 2	-0,22	50	no 2)
0,2	1 : 2	+0,16	50	no 2)
0,2	1 : 1,5	-0,43	45	no 2)
0,2	1 : 1,5	+0,39	45	no 2)

1) plate with 100 mm diameter, 50 mm thick
2) joint with cross section, 20 mm x 20 mm and 50 mm long

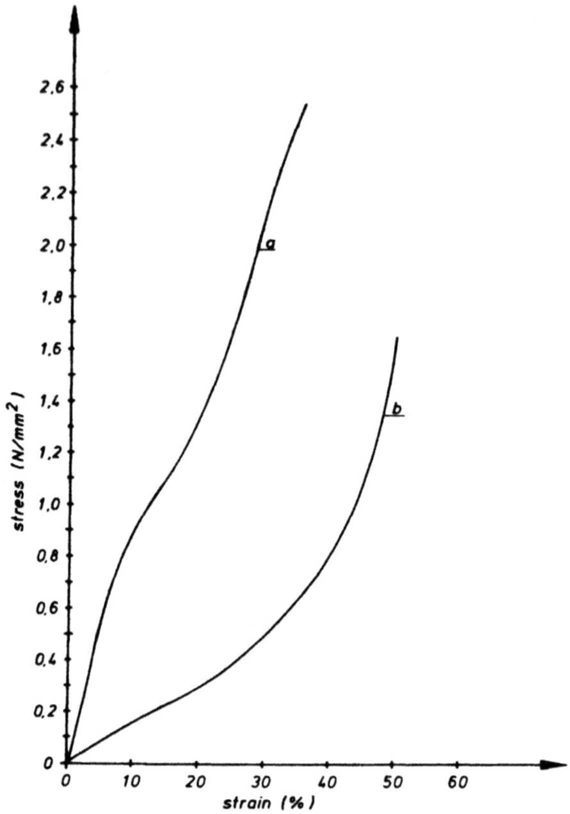

Fig. 2. Stress-strain curves (compression with lateral restraint) of epoxy resins of different moduli of elasticity filled with micro-balloons (volume ratio 1 : 2)
a) E = 500 N/mm^2 b) E = 0,2 N/mm^2.

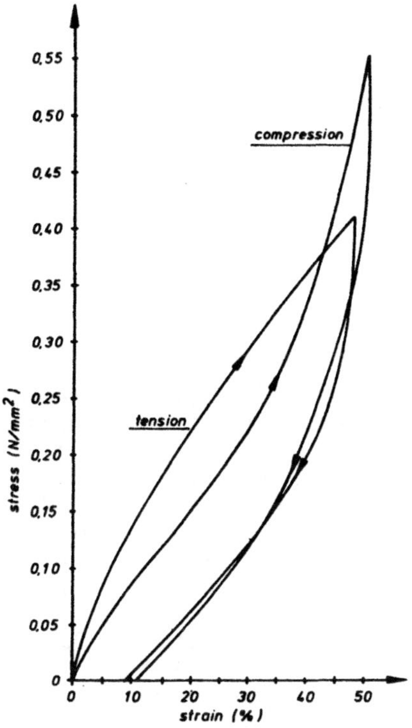

Fig. 3. Stress-strain curves of a rubber-elastic epoxy resin filled with micro-balloons (volume ratio 1 : 1,5) without lateral restraint. Joint with cross section 20 mm x 20 mm and 50 mm long.

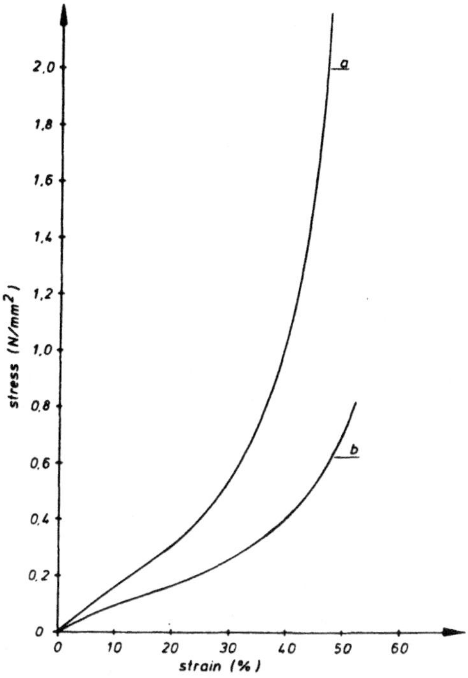

Fig. 4. Stress-strain curves (compression) of a rubber elastic epoxy resin filled with micro-balloons (volume ratio 1 : 1,5)
a) with lateral restraint
b) without lateral restraint

References

/1/ Iványi, G.: Sanierung von Rissen durch Epoxidharzinjektionen. Haus der Technik, Essen, 15.11.1985.

/2/ Kordina, K.; Blume, F.; Günther, J.: Schadensbehebung durch Schließen von Rissen. Institut für Baustoffe, Massivbau und Brandschutz der Technischen Universität Braunschweig, Heft 65, 1984.

/3/ Engelke, P.; Iványi, G.: Kraftschlüssiges Verpressen von Rissen in Überbauten von Massivbrücken mit Epoxidharzen. Beton- und Stahlbetonbau 80 (1985) 2 + 3, S. 29-35, S. 79-82.

/4/ Kern, E.: Dichten von Rissen und Fehlstellen im Beton durch Injektion von Kunststoffen. VDI-Berichte Nr.384, 1980.

/5/ ACI Comittee 224: Causes, Evaluation, and Repair of Cracks in Concrete. ACI Journal 81 (1984) 5-6, pp. 211-230.

/6/ Asendorf, K.: Prüfverfahren zur Beurteilung von Injektionsharzen. Tunnel (1985) 2, S. 107-110.

USE OF HIGH MOLECULAR WEIGHT METHACRYLATE FOR REPAIRING CRACKS IN CONCRETE

DAVID W. FOWLER
Department of Civil Engineering, University of Texas at Austin

Abstract
High molecular weight methacrylate (HMWM) is a relatively new monomer that has many of the advantages of methyl methacrylate (MMA), namely low viscosity, potential for curing over a wide range of temperatures and good wetting characteristics. However, the HMWM has low odor and a high flash point which are signigicant improvements over MMA.

HMWM has been used to produce polymer concrete for overlays and other specialty applications. It has very good mechanical and durability properties, and bonds very well to portland cement concrete.

One of its most important uses, however, is to seal cracks and restore structural integrity of portland cement concrete. Due to its excellent wetting characteristics, it penetrates even very fine cracks. It has been used for a wide range of crack repair applications in the United States, particularly for bridges.
Key words: Concrete, Cracks, Repair, Monomer, High Molecular Weight Methacrylate.

1. Introduction

The use of high molecular weight methacrylate monomer has been successfully used to seal cracks in concrete bridges and floors. The monomer has the ability to penetrate most of the crack depth even for narrow cracks. The monomer is easy to apply and provides a significant restoration of strength.

2. Monomer system

There are several monomers which are categorized as high molecular weight methacrylate. The primary difference in the resulting polymer is the modulus of elasticity.

Some of the initial monomers produced positive results when subjected to laboratory toxicity tests. The use of these monomers should be in accordance with strict safety procedures, particularly the use of protective clothing, by personnel trained in the use of the monomers. Variations of these monomers are being developed which do not exhibit the toxicity problems and will simplify the use of the materials in the field.

Three basic monomers were used in the study, indicated by numbers 200, 1100, and 1300. Monomer 200 produces the polymer with the highest modulus and monomer 1300, the lowest. The monomer system consisted of:

100 weight parts monomer
4 weight parts benzoyl peroxide (40% solution)
1 weight part cobalt napthenate (12% solution)

Monomer 400 was the same as monomer 200 except that cumene hydroperoxide was substituted for benzoyl peroxide.

One of the most difficult tasks in the repair of portland cement concrete is the repair of cracks. Epoxy injection has proven effective for many applications, but it is time consuming and expensive and requires specialized personnel and equipment. There have been many materials and sealers sold for the purpose of repairing cracks with only varying success. Fine cracks are especially difficult to seal. High molecular weight methacrylate (HMWM) has been found to be an effective sealer and to restore at least some of the structural integrity.

3. Laboratory testing

A laboratory test program was conducted to determine the effectiveness of the materials for sealing cracks (1, 2). Portland cement concrete blocks (30 cm x 14 cm x 36 cm), were cast with four strands of 10 ga wire. The wire was used to prevent complete separation of the concrete when cracked. The slabs were cured 24 hours at room temperature and then cured in a moist room for seven days for Type III cement and 30 days for Type I cement.

The concrete blocks were sawed down the length parallel to the wire to create two blocks 15 cm x 14 cm x 36 cm. The specimens were cracked using centerpoint flexural loading with a 25 cm span (Fig. 1). Crack widths were varied from 0.2 mm to 2 mm. Polyester putty was used to prevent the monomer from escaping during the treatment.

The monomer system was poured onto the slabs and brushed over the surface until the cracks were full. Monomer was ponded on the surface. Most of the slabs were treated in laboratory conditions (24°C) although four slabs were treated outside at room temperatures of about 32°C.

3.1 Effect of crack width

After curing for a minimum of 24 hours the specimens were sawed perpendicular to the cracks to expose the crack for determining the percentage of the crack length filled. It was found the polymer generally filled the cracks full depth with some discontinuity present along the crack length, apparently due to air pockets. The percentage of the crack length filled was made for each specimen using sawed sections exposing the crack in each specimen. Figure 2 summarizes the findings for the four monomer systems. There was little difference in

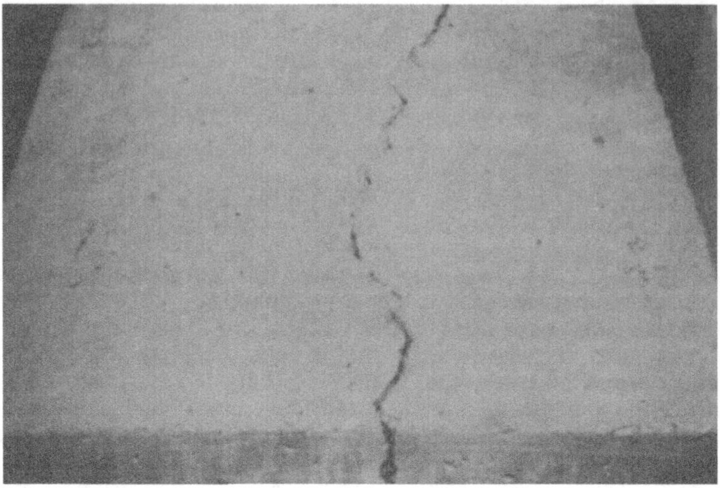

Fig. 1 Typical Laboratory Cracked Slab for Sealing Study

Fig. 4 Strength of the Repair is Shown in Cross Section of Cracked Slab Which Was Repaired and Recracked

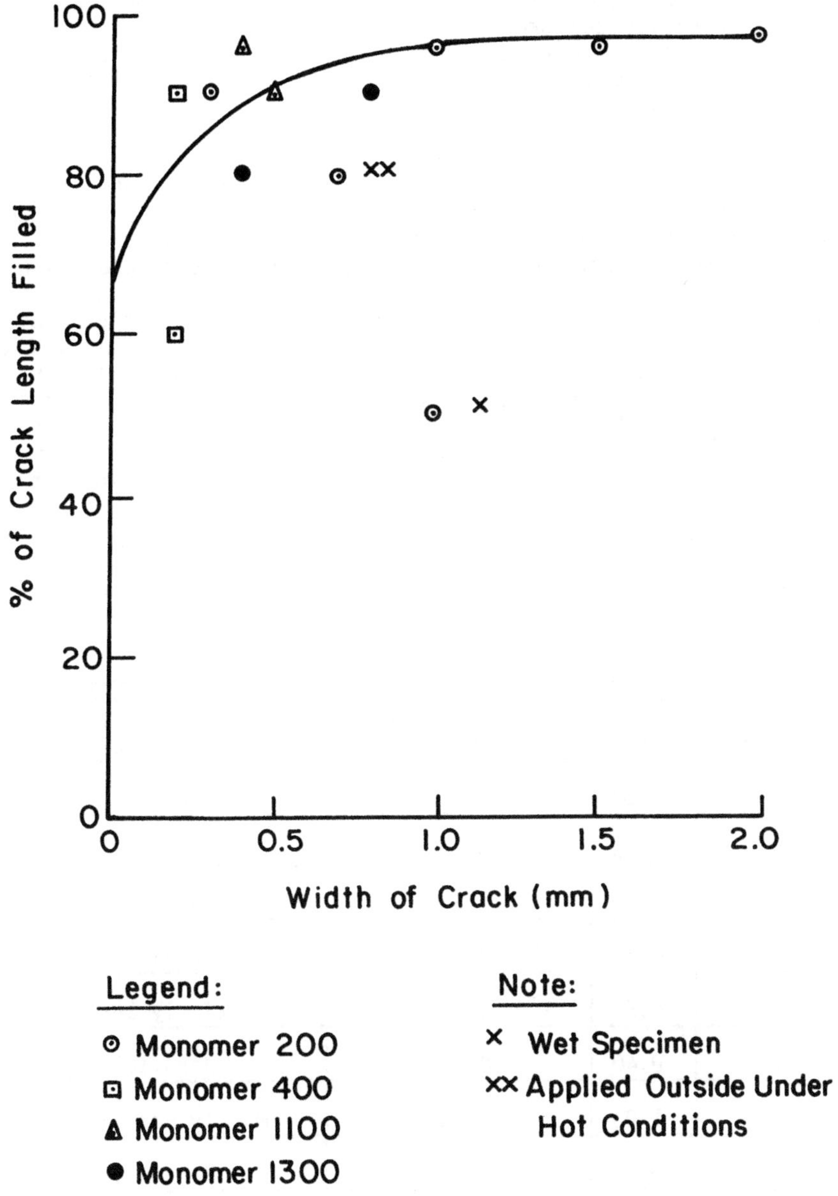

Fig. 2 Width of Crack vs. Percentage of Crack Length Filled

the resins based on monomer type. Later, monomer 1100 was chosen because of its slightly lower modulus. Figure 3 shows the results for this material. Generally, the percentage of crack length filled with increasing crack width for widths up to 1 mm. The specimens treated at 32°C generally performed about as well as those treated at 24°C. In all but one specimen, at least 80 percent of the crack length was filled which would usually be more than sufficient filling to prevent water intrusion.

3.2 Recracking strength

Although the primary objective is usually to prevent water from penetrating the concrete, a secondary objective is to restore strength to the cracked slab. The treated specimens, before sawing to expose the cracks, were reloaded in flexure following the same procedures used to develop the initial cracks to determine (1) the location of the crack and (2) the flexural stress for the second loading. Table 1 indicates that only two of the ten specimens failed partially inside the repaired crack; the other specimens failed parallel or nearly parallel to the original crack. Figure 4 shows a section through a recracked slab. Figure 5 shows the recracking stress as a function of the initial stress required to crack the concrete. The lowest ratio of recracking stress to initial stress was 76 percent. Even when the failure was outside the repaired crack, the ratio was less than 1.0 in all but one case, most likely due to the formation of microcracks during the initial loading. Figure 6 indicates that the cracking stress ratio did not exhibit a strong relationship to the crack width.

Table 1. Results of recracking tests

SPECIMEN	CRACK WIDTH (mm)	RECRACKING STRESS / INITIAL CRACKING STRESS	POSITION OF RECRACKING CRACK IN RELATION TO ORIGINAL CRACK
1	0.60	0.86	Parallel
2	0.35	0.79	Parallel
3	1.40	1.28	Mostly Parallel
4	0.25	0.98	Parallel
5	0.50	0.76	Partially Inside
6	0.40	0.98	Parallel
7	1.00	1.00	Mostly Parallel
8	0.65	0.80	Parallel
9	1.00	0.89	Partially Inside
10	2.00	0.73	Parallel

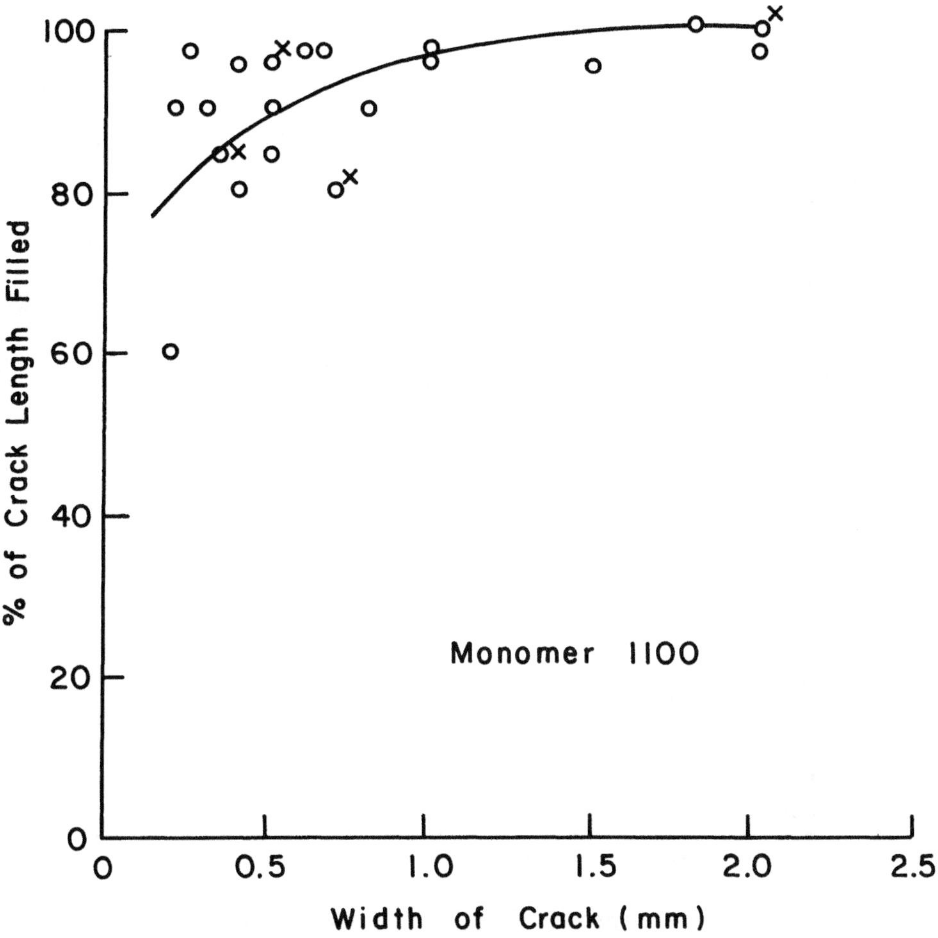

NOTE: x Applied Outside Under Hot Conditions

Fig. 3 Width of Crack vs. Percentage of Crack Length Filled

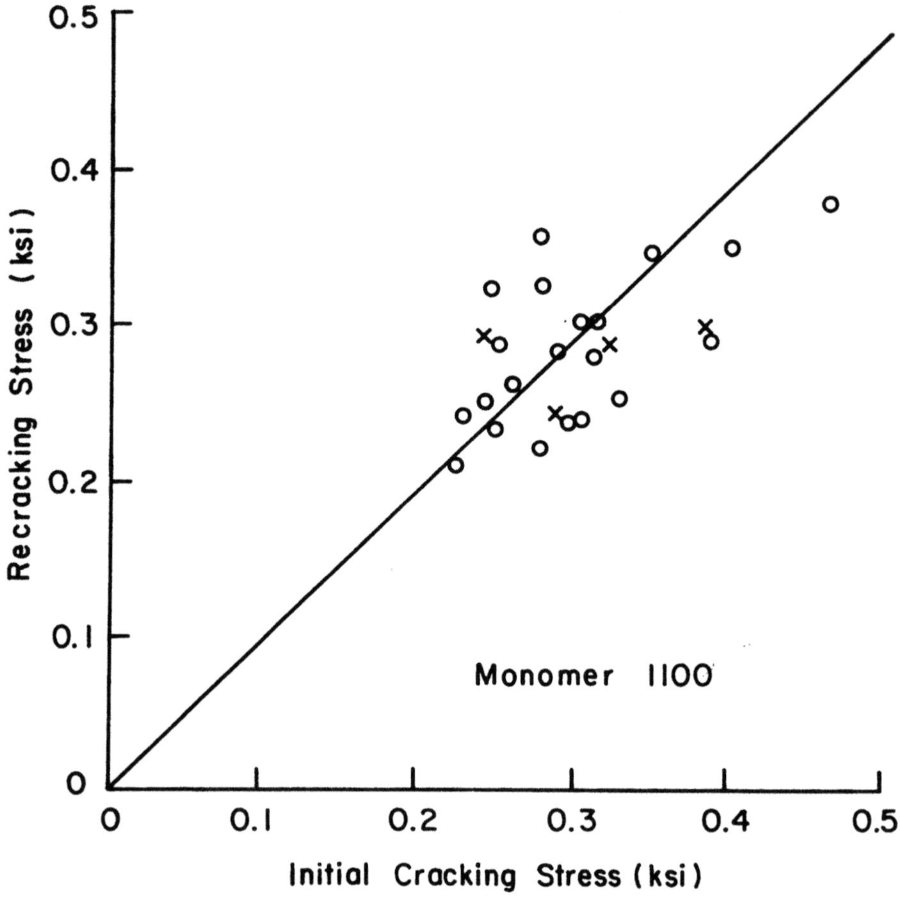

NOTE: x Applied Outside Under Hot Conditions

Fig. 5 Recracking Stress vs. Initial Cracking Stress

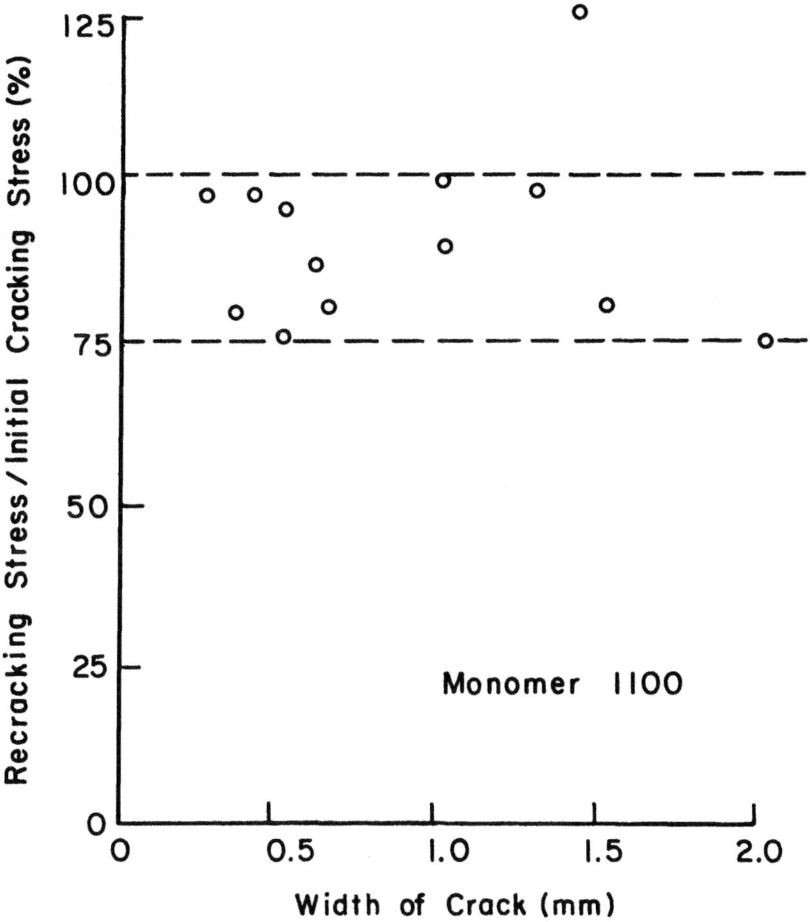

Fig. 6 Ratio of Recracking Stress to Initial Cracking Stress vs. Crack Width

3.3 Effect of moisture

The presence of moisture has the potential for interfering with the penetration of the monomer. Two sets of five specimens each were prepared, cracked and placed into a moist room for two weeks until they were thoroughly saturated and the cracks were filled with water. Five of the specimens were removed from the moist room (100 percent relative humidity) and placed outside under direct sunlight at a temperature range of 24°C to 35°C and a relative humidity of about 50 percent (referred to as "dry conditions"). Four specimens had medium crack widths (0.7 to 1.1 mm) and one had a small crack (0.2 to 0.6 mm). The specimens were allowed to dry for 8, 16, 24, 48, and 72 hours. The other five specimens were placed in an environmental chamber at 18°C and 90 percent relative humidity (referred to as "wet conditions"). These specimens contained small cracks (0.2 to 0.6 mm) and were also dried for 8, 16, 24, 48, and 72 hours. One other specimen was treated immediately after removal from the moist room.

After drying the specimens were treated with monomer. After curing, the specimens were recracked. The cracks were examined for the percentage of the crack length filled with polymer. The results are plotted in a ratio of the percentage of the crack length filled for those partially dried specimens to the percentage of crack length filled for fully dried specimens (Fig. 7). The two curves represent two extremes: moderate widths cracks under good drying conditions and small cracks under very damp conditions. Over 80 percent of the crack lengths were filled for the fully dried slabs in this part of the study and the ratio of cracks filled or partially dried to fully dried specimens was over 60 percent for 24 hours or more of drying. This indicates that at least 50 percent of the cracks would be filled even for small cracks under damp drying conditions.

4. Field studies

Several field studies have been conducted to determine the effectiveness of the HMWM in sealing cracks in actual structures. The procedure used is very simple. Monomer is mixed on site and then poured onto the concrete bridge deck or floor slab. Brooms are used to work the monomer into the cracks. It is important as the monomer seeps into the crack and the monomer level in the crack is lowered that additional monomer be worked into the crack to insure maximum filling. Excess monomer should be swept onto untreated areas or off the deck to prevent a slick polymer film. For bridges it has been found that a very light application of sand broadcast over the surface after monomer has been applied results in a good skid-resistant surface. Generally, even without sand, the surface has about the same skid resistance as the untreated concrete although the surface has the appearance of being slick.

4.1 Fort Worth bridge

A concrete bridge, on Interstate 20 near Ft. Worth, Texas, 130m in length, developed significant random cracking due to plastic shrink-

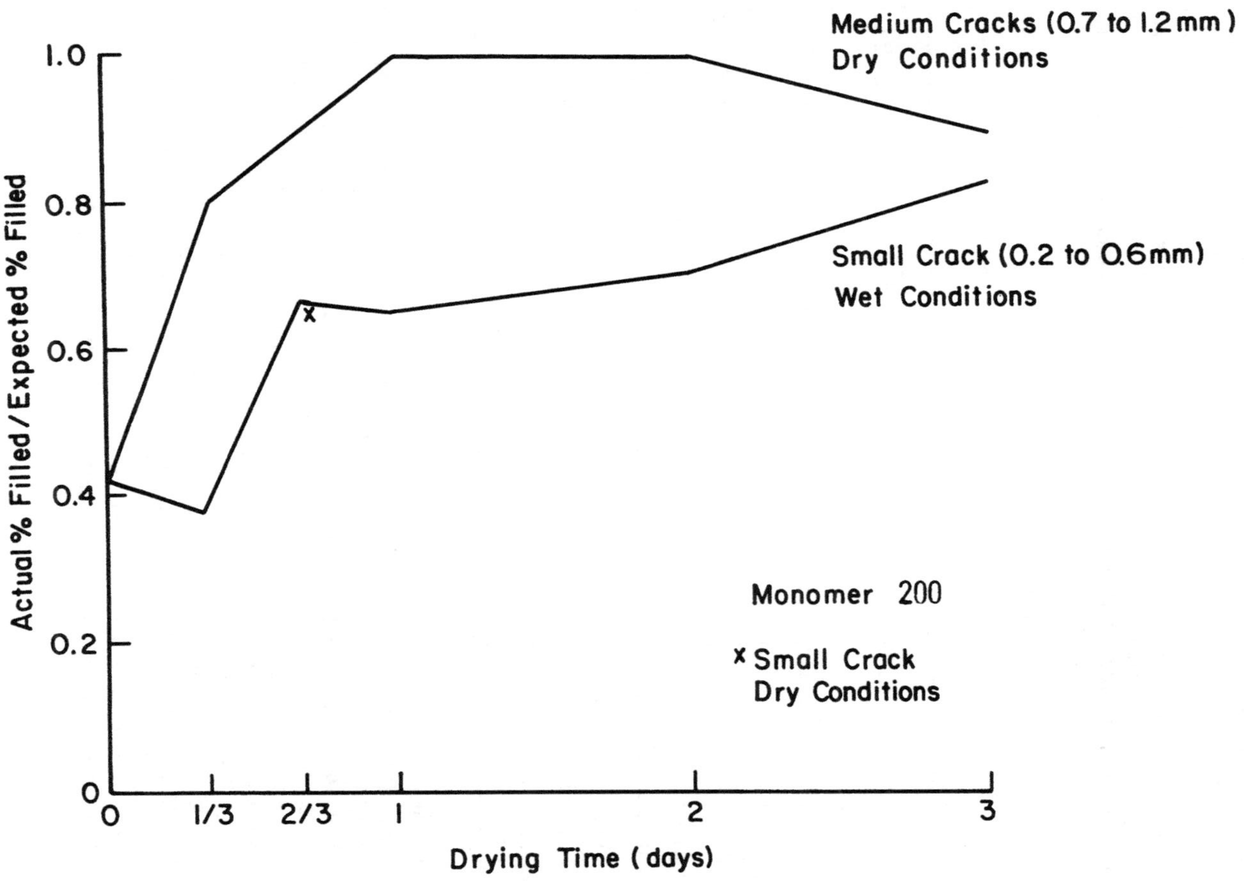

Fig. 7 Ratio of Actual Percent Filled to Expected Percent Filled vs. Drying Time

age. An initial field test was performed to investigate methods for repairing the cracks. Three repair methods were considered: (1) brooming the monomer into the cracks (Fig. 8); (2) injecting the monomer into the cracks via injection ports by means of an airless electric paint sprayer; and (3) epoxy injection. Monomer 200 was used for the first two methods. A commercially-available epoxy was used for the epoxy injection. The monomer injection was accomplished by using the same type of ports and surface sealing of the cracks as in epoxy injection. Cores taken from the bridge had crack widths at the surface ranging from 0.9 to 1.3mm. The results of the field study are shown in Table 2. All three methods performed equally well. Since the application of monomer with brooms was much easier, it was selected for the treatment on the full bridge in June 1984. Only the cracks were treated; the entire bridge surface was not sealed with monomer. The treatment was successful.

Table 2. Fort Worth bridge results

Core No.	1	2	3
Sealant	Epoxy	Monomer 200	Monomer 200
Type of Application	Injection	Pour	Injection
Crack Width at Top (mm)	1.30	0.95	1.10
Depth of Crack Crack (in.)	6-1/2	8-1/4	5-1/8
Percent of Crack Filled	95	95	95
Percent of Depth Filled	90	92	97

4.2 Monahans bridge

A bridge deck on Interstate 20 near Monahans in west Texas developed significant cracking. The bridge, 13m wide x 49m long, was constructed in 1981 using the wax bead process. In this process very small wax beads were mixed into the portland cement concrete. After the concrete was cured, the surface of the concrete was heated to melt the wax which blocked the pores, preventing intrusion of water and salt into the concrete. However, the high temperature (85°C) required to melt the wax apparently caused the cracking. (A companion bridge constructed at about the same time without wax beads has very little cracking.) The cracks were in evidence on the bottom of the bridge, and water penetrated through the cracks.

The deck was treated with monomer 1100 in July 1985. The deck

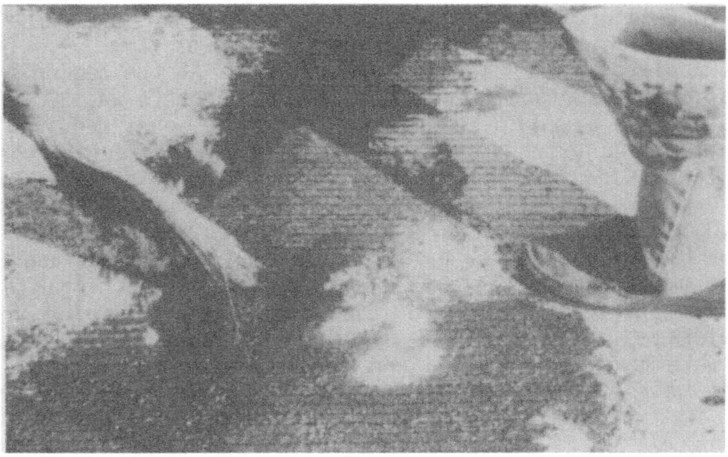

Fig. 8 Brooming Monomer into Cracks on I-20 Bridge Deck in Fort Worth, Texas

Fig. 9 Monomer Being Applied With Push-brooms and Squeegees on Bridge Deck in Monahans, Texas

was first swept with brooms. The monomer was then mixed in approximately four-liter batches and poured onto the surface. The monomer was applied with push-brooms and rubber squeegees (Fig. 9). The monomer was applied at the rate of 2 sq.m/liter. At the beginning of the application the surface temperature of the concrete was 38°C, and the temperature increased to 50°C during the three hours required for the application. An extender was added to the monomer during the application to provide more time before polymerization began. Two cores taken several days after the treatment showed that the cracks ranged from 1.0 to 5.0mm in width at the surface, but that the width reduced to about 0.1 mm or less at a depth of 2 cm from the surface. In one core the monomer penetration stopped at the reinforcing steel at a depth of about 7.5 cm and, in the other, debris in the very fine crack width at a depth of 7.5 cm stopped the monomer. However, the cracks were completely sealed and, after 9 months, there is no evidence of water leaking through the cracks on the bottom of the deck.

5. Safety

Some of the monomers used in this study have been shown to be toxic. Care should be exercised in the mixing, handling, and application of the materials. Protective shoes, clothing, and eye goggles should be worn. Disposable chemically-resistant suits have been found to be useful.

New monomer analogs have been developed which are less toxic. They are now being tested to determine if they seal cracks as effectively.

6. Conclusion

High molecular weight monomer has been shown by laboratory tests to be an effective sealer of cracks from the standpoint of length of crack filled and from the restored strength of the concrete. Monomer can be applied within one or two days after exposure of the concrete to rain or other sources of moisture. Field tests confirm the simplicity of the applications method and the effectiveness of the treatment.

References

Bermudez-Goldman, Alejandro J., "Mechanical and Durability Investigation of a Modified High Molecular Weight Methacrylate Polymer Concrete," Masters' Thesis, The University of Texas at Austin, May, 1986.

Mangum, Wayne D., "Improved Methods for Sealing Joints and Cracks in Portland Cement Pavement," Masters' Thesis, The University of Texas at Austin, December, 1984.

Section 8

Glueing of precast elements
Collage d'éléments préfabriqués

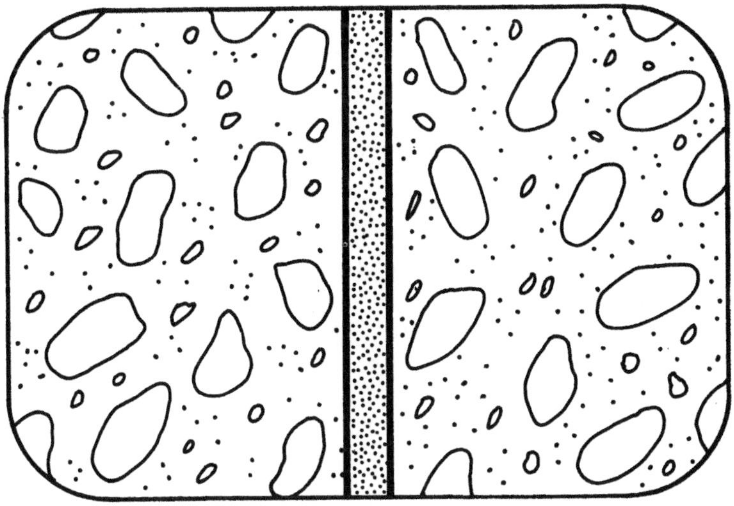

GLUEING OF REINFORCED AUTOCLAVED AERATED CONCRETE SLABS TO FORM ROOM SIZE WALL UNITS

D. BRIESEMANN and E. FREY
Hebel GmbH Holding, D 8089 Emmering-Fürstenfeldbruck

Abstract
By means of suitable synthetic resin adhesives, storey-high slabs of reinforced autoclaved aerated concrete (AAC) can be joined to form transportable room size wall units within very short time. By means of the adhesive joints, the slabs are combined to form monolithic wall panels, which can be transported to the building site as soon as about 30 minutes after the glueing. Tests have been made with adhesives based on epoxide resins, polyurethane and polyester resins. All three kinds can be used in principle and the polyester adhesive has proved to cure faster than the others. The adhesives may also be used to glue AAC to reinforced normal-weight concrete and to steel. This permits lintels made of reinforced normal-weight concrete to be glued into the units over window and door openings. The safety in handling is improved in that steel rods extending horizontally at right angles through the surfaces defining the joints are fixed by nuts to the units. Said steel rods are intended only to prevent a fracture of the units in case of unforeseen shocks and of a possible failure of the adhesive joints as the units are handled and installed in the site.
Key words: Reinforced autoclaved aerated concrete, Synthetic resin adhesive, Wall units.

1. Synthetic resin adhesive

1.1 Requirements

To be suitable for glueing components of autoclaved aerated concrete (AAC) in a factory, an adhesive must meet the following requirements:

- It must cure within short time to such a degree that it can carry load; the units must be capable of being handled not more than 1 hour after they have been glued;
- the curing must be sufficiently fast even at relatively low temperatures of 10 to 15°C;
- the adhesive must be suitable for making joints to form monolithic wall units. For this reason the compressive strength and the flexural tensile strength of the adhesive must be higher than the corresponding strengths of AAC 1 hour after the glueing;

- the adhesive must be capable of joining AAC also to reinforced normal-weight concrete and to steel;
- the adhesive joints must resist moisture and alkalies and must be durable;
- the joints must be thin and the consumption of material must be low.

1.2 Properties of polyester resin adhesive

In principle, adhesives made of polyurethane (Lechler), epoxide resin (Ciba) and unsaturated polyester resins (BASF, Bayer) have proved to be suitable for glueing AAC. It has finally be found that the polyester resins available under the names Palatal from BASF and Leguval from Bayer have the best properties as regards short curing time and very high initial strength. Most glueing tests were carried out with Palatal having the following composition:

Palatal P 4	80 parts by weight
Benzoyl peroxide powder	2% of resin
Palatal P 65	20 parts by weight
Quartz powder W4 (Quarzwerke Frechen)	225 parts by weight

The most important parameters consisting of the compressive strength, flexural tensile strength and bulk density of Palatal and, for comparison, of reinforced AAC of the strength class GB 3,3 have been compiled in Table 1.

Table 1. Properties of Palatal polyester resin and of GB 3,3 Autoclaved aerated concrete

	Compressive strength (MPa)	Flexural tensile strength (MPa)	Bulk density (kg/dm^3)
Palatal (after 28 days)	120 - 140	30 - 40	1.7 - 1.9
GB 3.3	4.5 - 5.5^{+}	1.2 - 1.5^{+}	0.53 - 0.57^{++}

\+ at a moisture content of 3-10 vol%

++ when oven-dried at 105°C

To determine the long-term strength of the polyester resin, standard prisms (4 cm x 4 cm x 16 cm) of Palatal were stored in the air and others were stored in saturated water that contained particulate AAC in suspension. The compressive strength and the flexural tensile strength were measured after 10, 100 and 1000 hours. Some test specimens were stored in water that contained particulare AAC because that storage simulated conditions (high moisture content, alkaline environment) which were more severe than must be expected in practice. The results of the long-term tests have been compiled in Table 2.

Table 2. Compressive strength and Flexural tensile strength of Palatal under continuous load

Time	Storage in air		Storage in water	
	Compressive strength	Flexural tensile strength	Compressive strength	Flexural tensile strength
(h)	(MPa)	(MPa)	(MPa)	(MPa)
10	103	28.2	143	46.0
100	106	29.8	113	39.7
1000	131	33.7	115	34.5

The adhesive had almost attained its final strength after only 10 hours. During air storage, the strength increased with time whereas it decreased with time during storage in water which contained particulate AAC (pH value 9 to 10). In case of a linear extrapolation, the elements stored in water containing AAC would have a compressive strength of 90.5 MPa and a flexural tensile strength of 24 MPa after 10^5 hours (about 11-1/2 years). These values are higher by a factor of about 25 than those for AAC.

2. Composites of polyester resin and AAC

2.1 Durability of adhesive joints

In addition to the tests with the prisms, tests were conducted with test specimens which had been glued with Palatal. They were stored in the open air on the weathering stand and tested at different times under loads applied as shown in Figure 1. A control specimen consisted of an AAC block having the same size but no adhesive joint. The adhesive joint was perfectly intact after one year. The fracture under load occurred always in the AAC close to the adhesive joint. The adhesive joints had not been weakened after three years. It was interesting to note that the flexural tensile strength of the blocks having adhesive joints was only one-half of the flexural tensile strength of the control specimen having no adhesive joints although the composite fractured only in the AAC. Similar results have been obtained in compressive strength tests of test specimens consisting of a plurality of adhesively bonded pieces and of monolithic control specimens.

2.2 The influence of the moisture in the AAC

Because owing to its production process AAC may still have a relatively high moisture content when it is placed, the influence of the surface moisture on the bond to the adhesive was investigated. By means of the test set-up shown in Figure 2, the bond strength of the adhesive was measured 3-1/2 hours after AAC specimens having different moisture contents had been glued. In case of a moisture content in excess of 15 vol%, the fracture occurred in most cases in the adhesive joint whereas it always occurred 1 to 2 cm from the adhesive joint in which AAC contained less than 15% moisture. The variation of the measured values of the ultimate stripping stress was distinctly larger than with the specimens containing less than 15% moisture.

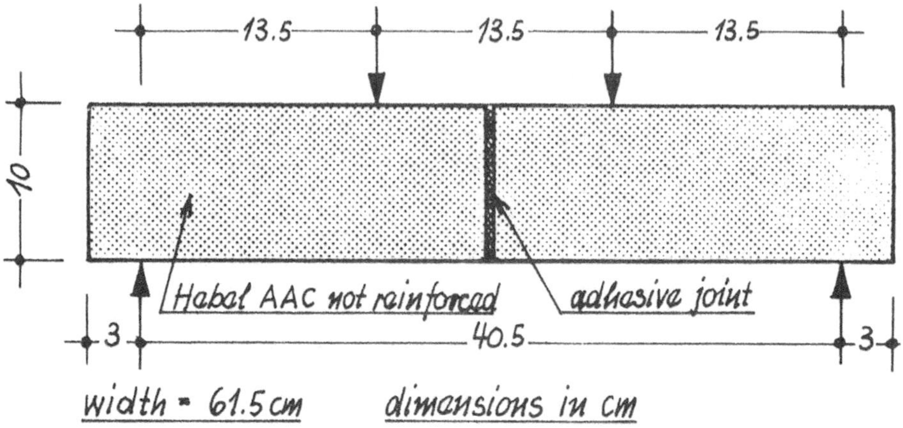

Fig. 1. Application of load for testing the flexural tensile strength

2.3 The influence of a silicone-treated AAC surface
The test set-up shown in Figure 2 was used also to determine the influence of a potential treatment of the AAC with silicone on the bond strength of the adhesive. It has been found that the bond strength of the polyester resin is reduced by a treatment of the AAC surface with silicone. The test specimens treated with silicone (BS 15 of Wacker, diluted 1:20 with water) were glued with Palatal after one day and tested after additional 3-1/2 hours. One-half of the silicone-treated specimens fractured at the adhesive joints; the control specimens fractured in the AAC close to the adhesive. The silicone-treated specimens had an ultimate stripping stress of 0.27 MPa, which was only one half of the ultimate stripping stress of 0.52 MPa of the control specimens.

3. Manufacture of units

3.1 Preparatory work
To ensure a satisfactory bonding throughout the surface of the AAC, that surface must be rubbed off with the grinding plate so that small surface irregularities will be removed, and must then be cleaned with a broom or a vacuum cleaner to remove loose material. Faultless joints were always obtained when the adhesive was applied to both surfaces to be joined. In case of adhesive joints having a thickness of 1 to 1.5 mm, the consumption of adhesive amounted to 1.1 to 1.3 kg pure polyester resin or about 4 kg filled resin per m².

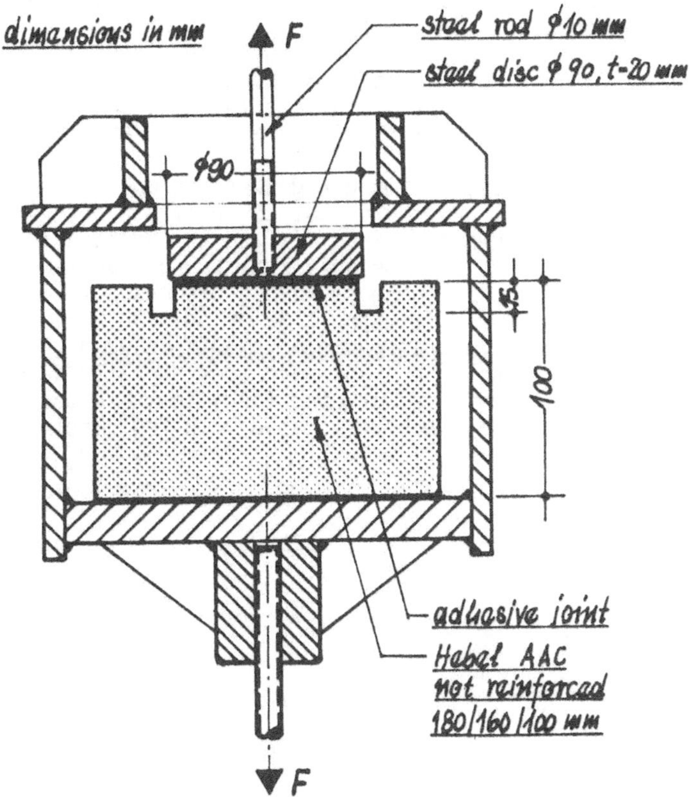

Fig. 2. Test equipment for measuring the bond strength of adhesives glued to AAC

Joints in which the adhesive had been applied only to one surface sometimes failed because the adhesive had not bonded to part of the initially uncoated surface. But an application of adhesive to only one surface will be sufficient if the other surface defining the joint is primed with a dilute polyester resin solution (without a filler). That practice will require a shorter working time than the application of adhesive to both surfaces and will result in somewhat thinner joints.

3.2 Assembling the AAC slabs

Simple wall units are made from 2.5 m long AAC wall slabs. The horizontally disposed slabs are coated with the adhesive at their edge faces, e.g., with the aid of a serrated spatula or trowel. The slabs are then properly positioned and forced against each other on a frame, in which the justaposed slabs are pushed against each other and held under compressive stress be means of a pressure-applying screw until the adhesive has cured. The slabs are also connected by two steel rods, which are 10 mm in diameter and extend through the joints at right angles to the longitudinal axis of the AAC slabs, and at right angles to the surfaces defining the joint and are fixed by nuts at their ends. Figure 3 shows the arrangement of the rods, which are fixed to the nuts only to increase the safety in case of unforeseen shocks as the units are handled and are placed on the site.

When the slabs have been clamped together for about 1 hour the adhesive has sufficiently cured to that the unit can be released from the clamping pressure and can be lifted at the handling loops and be erected. At that time the unit can also be handled while suspended from a crane. In the most favourable cases it was possible to lift units after a clamping time of only 30 to 40 minutes. Whereas this may result in measurable movements between the slabs of a unit at the joints, when the slabs have been properly adhesively joined throughout the surfaces defining the joint such movements will be so slight that no cracks will appear in the coating to be applied to the units.

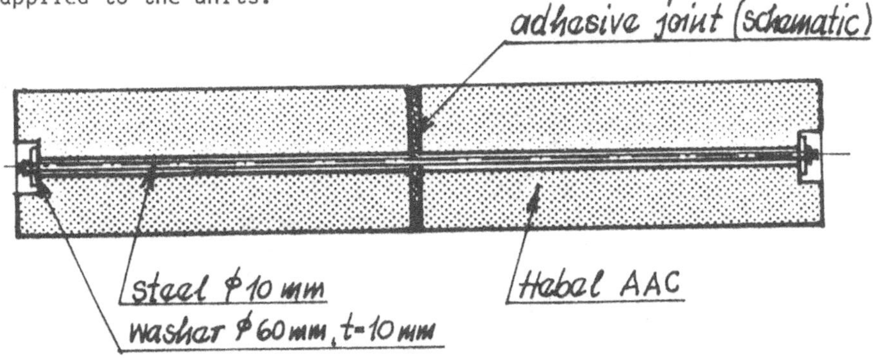

Fig. 3. Arrangement of the steel rods within the glued AAC panels (schematic); horizontal section through a standing wall

3.3 Handling loops

Experiments have been conducted with different handling loops and have resulted in the selection of the loops shown in Figure 4, which have been successfully used. The loops are made of I-section reinforcing steel having a thickness of 8 or 10 mm and owing to the small radii of bend shall be bent at elevated temperature. The vertical legs of the loops have a length of 60 cm and the horizontal arms have a length of 18 cm. Before the AAC slabs are joined, the short arms are struck into the AAC slabs and will provide a mechanical interlock in case of a failure of the adhesive joint. Recesses for receiving the long legs must be milled into the surface of the AAC slab to be joined. The adhesive bonding of said vertical legs of the loops will improve the ultimate stress of the handling loops.

3.4 Wall units

In addition to simple wall units made of identical AAC slabs, it is possible to make units provided with window or door openings in

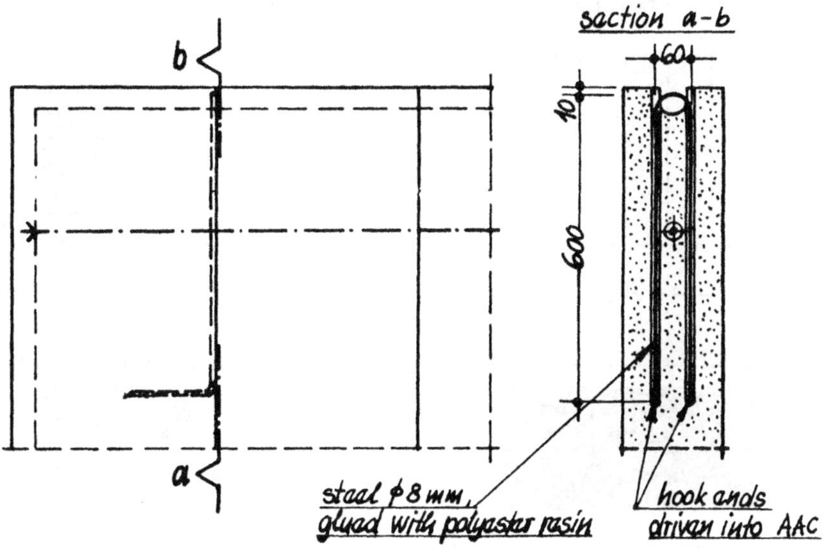

Fig. 4. Details of the handling loops

numerous variations. Figure 5 shows examples of each of said elements. The lintels over the window and door openings are also adhesively joined and may be made of AAC or of reinforced normal-weight concrete, as may be required.

For a fixation of the slabs which adjoint a door opening and for taking up stresses arising in transport, said slabs are interconnected at their lower edges by means of flat bars or T sections of steel, which are adhesively joined and nailed to the AAC. The largest wall units made thus far had a length of 5.38 m and the largest inside width of a window opening amounted to 3.6 m.

By means of the glueing process which has been described, wall units can be made which are in accordance with the Approval Z 17.1-43 "Storey-high, Load-carrying Hebel Wall Units Made of Unreinforced Autoclaved Aerated Concrete" and in accordance with the Approval Z. 17.1 - 43.1 "Reinforced, Storey-high, Load-carrying Hebel Wall Units Made of Autoclaved Aerated Concrete of Strength Classes GB 3,3 and GB 4,4".

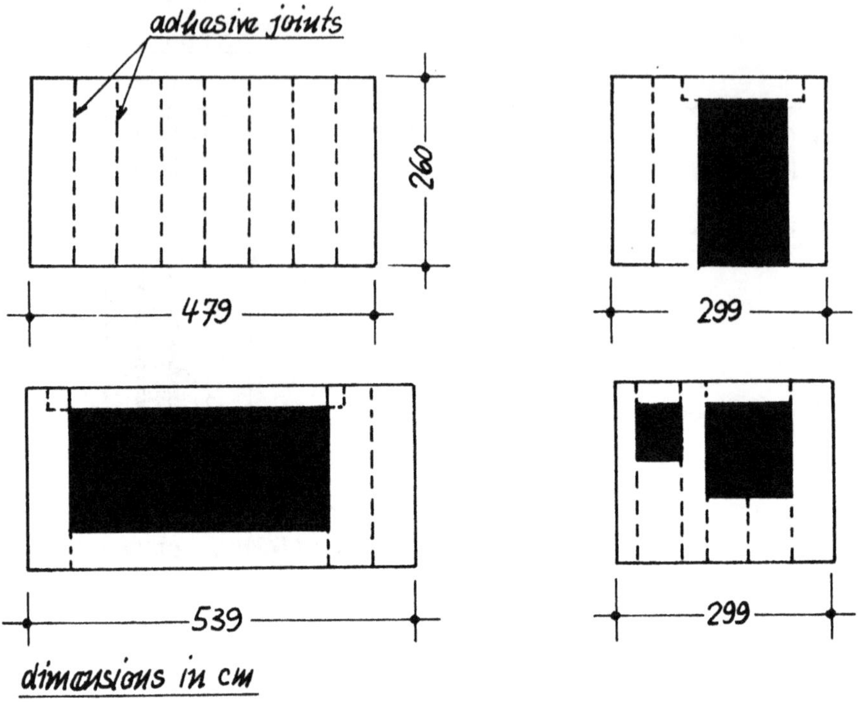

Fig. 5. Design examples of wall units made out of glued AAC panels

THE DIMENSIONING OF ADHESIVE-BONDED JOINTS IN CONCRETE BUILDING COMPONENTS

L. Franke
Institute for Building Physics and Bulding Material Science
Technical University Hamburg-Harburg, Germany

Abstract

In order to be able to predict the durability of adhesive-bonded (glued) joints made in concrete construction using epoxy resin adhesives, it is necessary to know the change undergone in the adhesion term under a variety of conditions. The first part of the present work thus deals with investigations of the performance and sustained load strength of adhesive-bonded joints under damp conditions and of the sustained load strength at increased temperatures, which were carried out by the author over a period of several years. On the basis of these results and of the results of investigations carried out by other authors a series of simple dimensioning equations are provided, taking the strength of the concrete into consideration. In addition to this, simple tests are suggested with which the suitability of epoxy resin mortars for adhesive-bonded joints in concrete construction can be checked.

1. Introduction

To date the technique of joining building components with the help of epoxy resin adhesives has only been used to a relatively small extent for construction work using prestressed concrete segments and for building component reinforcement. The type of joints which are, among others, meant by this, are outlined in Fig. 1. One of the main reasons for the reticence with which this building technique has been treated up to the present is the fact that there is still uncertainty with regard to the permissible long-term loading to which such joints may be subjected. In what follows indications for the safe dimensioning of adhesive-bonded joints in concrete construction shall be provided, developed on the basis of the results of investigations carried out both by the author of this work and by other authors.

2. Investigation of the strength of adhesive-bonded joints

Adhesive-bonded joints may be provided as predetermined design features in precast concrete structures or be used for the strengthening or the renovation of existing structures. In the former case the joints will, as a rule, be at right angles to the main loadbearing or spanning direction of the members concerned. In the latter, the joints will generally extend in the direction of spanning, more particularly when existing structural sections are subsequently enlarged or strengthened by means of bonded-on plates, slabs or strip-type units made of steel, reinforced concrete or prestressed concrete (Fig. 1).

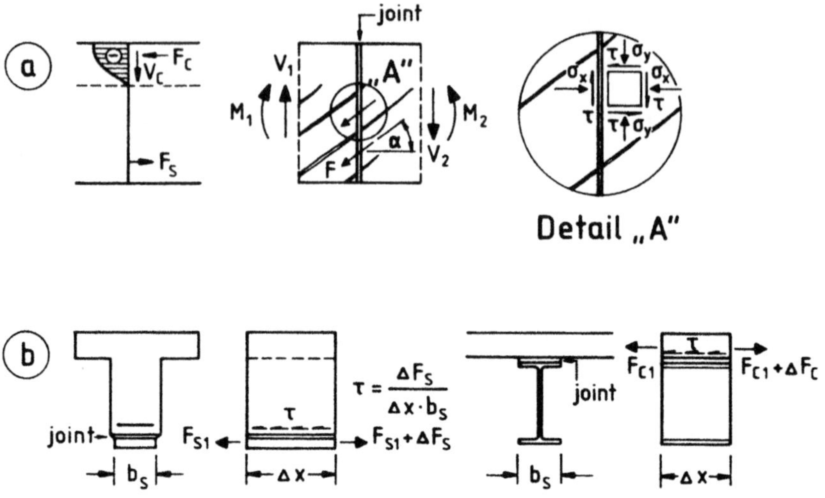

Fig. 1. Examples of the stress conditions in adhesive joints of composed concrete members

In the last-mentioned cases, however, the boundary face of the joint is subjected mainly to a shear-and-compressive stress combination or to compressive stresses alone or predominantly to shear stresses. The investigations that were carried out accordingly concentrated more particularly on the question how an adhesive-bonded joint, or a concrete member comprising such a joint, behaves under such loading situations when exposed to a number of different environmental conditions.

It is now necessary to verify, wether even a joint extending at right angles to the axis of the specimen causes some weakening. Now the question as to the strength of a prismatic concrete specimen comprising a joint extending perpendicularly to the direction of the force is in fact relevant, because the modulus of elasticity of reaction resin mortars for structural adhesive jointing is only about one-third of that of concrete, so that a certain amount of transverse tensile stressing of the concrete occurs in the vicinity of the joint. This has, for example, been confirmed by means of a finite-element analysis carried out in the context of the present research. The transverse tensile effect is less, however, the thinner the joint is formed.

For verification, the present authors carried out compressive tests on prismatic concrete specimens ($fc = 47$ N/mm^2) with 1.5-2 mm thick epoxy resin-bonded joints at temperatures ranging from 20° to 80° C. It was found that, under the given conditions, these concrete specimens always had a higher compressive strength than the monolithic reference prisms made of concrete with the same cube strength (Fig. 2). The reason for this behaviour must be sougth more particularly in superior strength of the reaction resin bonding mortar which exerts a crack-restraining effect in the state of failure of

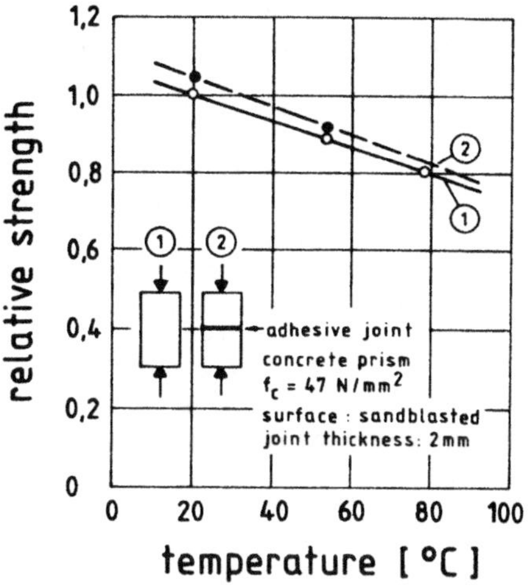

Fig. 2. Compressive strength of monolithic and jointed concrete prisms at increasing temperature (mean values of five measurements)

the jointed prism specimen and thus increases the strength thereof.

The shear stress that a joint can transmit is expressed with sufficient accuracy by the Coulomb failure criterion:

$$\tau_u = c - \mu \cdot \sigma_n \qquad (1)$$

Here c denotes the portion assignable to adhesion, while the portion associated with friction and dependent on the normal stress is $\mu \cdot \sigma_n$.

Tests carried out to date and theoretical considerations lead to the conclusion that the relative danger of failure in the joint is the greatest at an angle of approximately 30° between the loading direction and the direction in which the joint runs.

In order to make a reliable assessment of the behaviour of adhesive-bonded joints in concrete under an wide range of different conditions such as long time exposure to wet environment, long term or fatigue loading, it is necessary continuously to observe the behaviour of the strength of the joint (or the bond strength developed at the interface) over a specific period of time so as to obtain well-founded data on which, if necessary, extrapolitions can be based with some confidence. From this point of view the results of the behaviour of adhesive-bonded joints hitherto reported were unsatisfactory because in most cases it was possible only to arrive at a "yes/no" decision as to whether the concrete or the interface (boundary face of the joint with the adhesive) was responsible for failure of the joint during the test period. No function of time was available, and it was not possible to predict any changes in properties with sufficient certainty. Hence a major objective of the investigations was to study the expected time-dependent change in the adhesion strength of joints over as a long a period of time as practicable and under various conditions, as already mentioned.

Verification tests showed quite definitely that reliable results for the actual magnitude of the adhesion strength can be obtained only if very high-strength concrete or mortar is used in the tests and if the specimens are so designed that the joint is subjected to conditions which approximate as closely as possible to those in actual practice and the stress distribution in the vicinity of the joint is as streightforward and well-difined as possible.

Thus the specimens chosen for the tests were high-strength cement mort r prisms of 4 x 4 x 16 cm^3 (fc = 86 N/mm^2) and high strength concrete prisms of 10 x 15 x 70 cm^3 (fc = 80 N/mm^2). The adhesive-bonded joints in these prisms were approximately 2 mm thick and ran at an angle of 30° to the direction of force.

In the cement mortar prisms the inclined surfaces were formed by sawing the prisms with special diamond-studded cutting discs. The concrete specimens were each produced from two separately cast halves which were sandblasted after removal from the moulds. The joints were formed with adhesive applied to the surface-dry concrete faces (moisture content approx. 4 per cent by weight). After jointing, the specimens were subjected to various environmental conditions, as already stated. By means of roughness measurements, microscopic examinations and comparative shear/compression tests it was first ascertained that the results obtained with sawn mortar prisms were valid also for sandblasted concrete joint faces, and vice versa.
For comparison, adhesive mortars based on epoxy resins and polyester resins were included in the investigations.

The mix proportions of resin to aggregate were about 1: 2 (by weigth). The epoxy resin hardeners used were Triethylenetetramine (adhesive EP I) and the polyamine HY 2992 from Ciba (adhesive EP II).

The results of the investigations that are important to assessing the behaviour of epoxy adhesive-bonded joints under various environmental and loading conditions, especially the long-term behaviour, will be briefly outlined here. Shear/compression tests at elevated temperatures confirmed that with all the adhesive systems under investigation the temperature at testing has a distinct effect on the adhesion strength. Thus, the shear/compressive strength of the jointed prism specimens is found to decrease at 50° C to 70-90 per cent of the value at at 20° C. Though these values are still to be regarded as relatively high, it is not possible to utilize them in a sustained-load situation. In sustained-load shear/compression tests extending over an period of more than a year, with loads of varying magnitude applied to respective EPI-bonded specimens at a constant temperature of 50° C, the sustained-load adhesive strength was found to be about 60 per cent of the short-term shear/compressive strength at that temperature (Fig. 3).

The desired information on the quantitative change in adhesion strength after prolonged exposure to the action of moisture was obtained from shear/compression tests of four years' duration with the various adhesive systems. Fig. 4 shows the behaviour of the epoxy resin bondings.

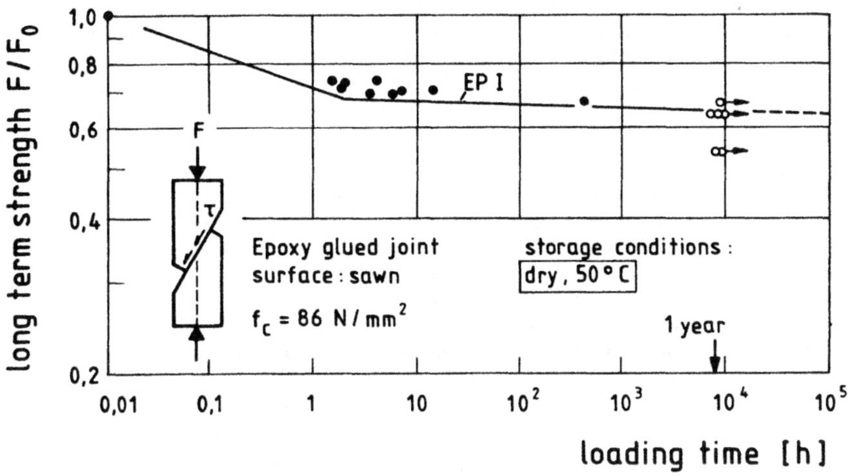

Fig. 3. Shear-compressive strength under sustained load for storage at 50° C

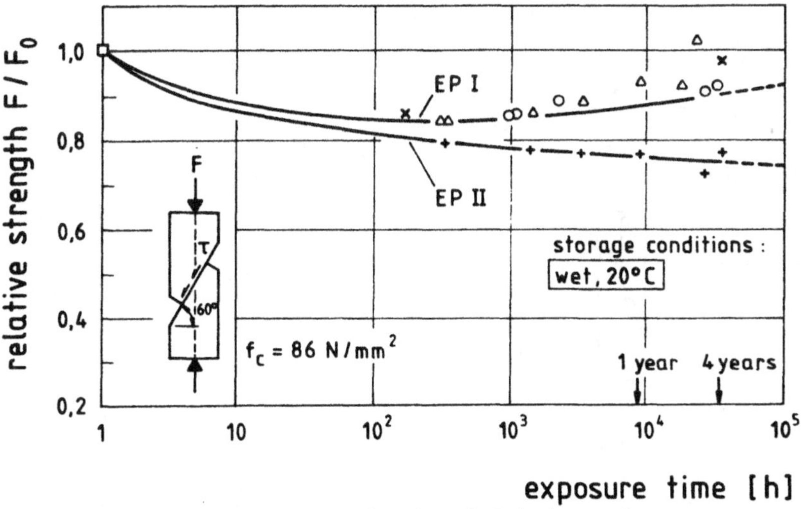

Fig. 4. Strength of epoxy adhesive jointed specimens after long time wet storage

It emerges that all the adhesive bondings undergo something very like a spontaneous decline in adhesion strength in the first 14 days of moist storage. After passing through a minimum, this strength undergoes an increase in the case of the adhesives EP I, whereas with EP II there is a further, though slight, decrease in adhesion strength during the period of observation.

In the course of the entire test period, which lasted for four years, the strength tests carried out on the prisms bonded with EP I demonstrated a reduction of strength of a maximum of 15 % caused, according to plan, by failure in the boundary face between the adhesive mortar and the concrete; in the case of EP II, however, there was an increasing rate of cohesion failure in the adhesive mortar itself.

Results of storage and sustained load tests which Jagfeld carried out over a period of eight years on adhesive-bonded joints using EP I and low-strength cement mortar prisms (fc = 47 N/mm) are now available. Over the entire test period there was an increase in the joint shearing strength; in each individual case the cement mortar was responsible for the failure.

Of major importance with regard to the assessment of the load-transmitting capacity of adhesive-bonded joints in concrete under practical conditions is their behaviour under sustained load in conjunction with continuous exposure to moisture (Fig. 5).
In all the sustained load tests the occurrence of failure was, as planned, due to adhesion strength being exceeded at the interface between the adhesive and the concrete. The initial values indicated in Fig. 5 are the shear/compressive strengths obtained after 3 days' storage in standard climate followed by 7 days in water. It emerges that the long-term strength in wet environment is about 50 per cent of the initial value.

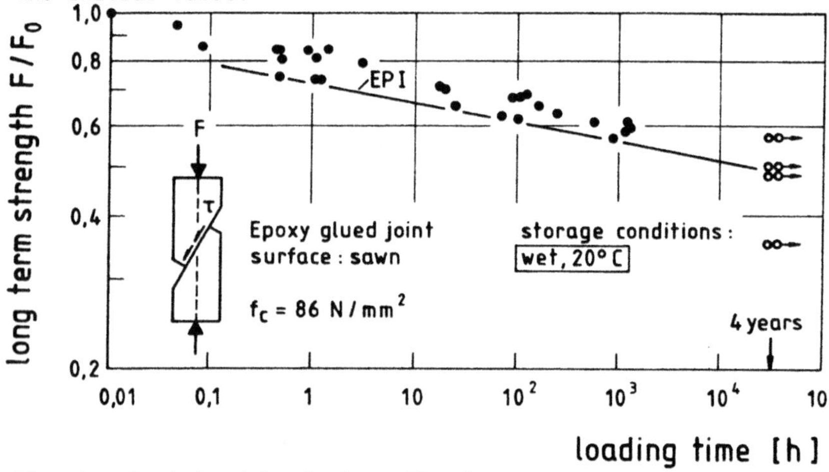

Fig. 5. Sustained load strength of epoxy-adhesive jointed specimens in wet environment

In cases where adhesive-bonded joints are subjected to non-static loading, the change in adhesion strength and in the instrinsic strength of the contact materials themselves due to fatigue loading cycles will have to be ascertained. Comprehensive investigations relating to this set of problems were therefore also carried out. To characterize the fundamental behaviour, Fig. 6 shows a stress-number (S-N) diagram which was obtained with EP I-bonded specimens which were continuously stored in water (also during the test itself). The ordinates show the upper stress values of the cycles; the lower stress was zero. It is notable that the diagram, plotted in a single-logarithmic co-ordinate system, becomes horizontal already from about $N = 4 \cdot 10^4$ cycles onwards. The results obtained with specimens which had been previously stored in water for periods of 6 weeks and 2 years, respectively, were found to belong to the same collective of results. The fatigue strength ist about 55 per cent of the short-term shear/compressive strength and thus approximately corresponds to the fatigue strength of ordinary concrete. However, it was ascertained - inter alia, by means of scanning electron microscopy - that failure always occurred at the interface of concrete and adhesive, as had indeed been presumed.

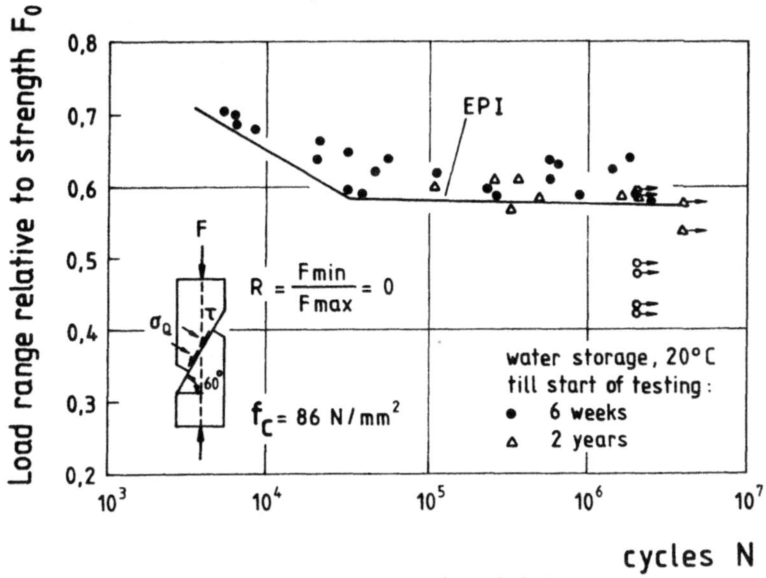

Fig. 6. Behaviour of epoxy-adhesive jointed specimens under fatigue loading in wet environment

3. Dimensioning of adhesive-bonded joints in practice

For describing the efficiency of the joints completely it is necessary to know c and μ as functions of the surface condition of the joint faces and the loading and/or storage conditions. For this purpose, besides the measurements already mentioned, a fairly large number of tests with various angles of inclination of the joint, various τ/σ_n stress ratios and various types of adhesive mortar were carried out.

The effect of various conditions of storage, e.g. continuous moist and of loading, such as sustained load or fatigue load, therefore manifests itself solely in a parallel shift of the shear/compression relationship "downwards" to lower values as compared with the initial behaviour. The above-mentioned influences thus bring about a reduction of the adhesion term c, namely to 50 per cent of the initial values. The roughness coefficient μ ist evedently governd only by the surface texture or roughness of the concrete and not by the storage conditions, at any rate in the case of adhesive mortars whose composition corresponds approximately to that of the adhesive mortars used in these tests.

The efficiency of adhesive-bonded joints in concrete construction is, on the other hand, restricted by the efficiency of the concrete itself or by the strength of the surface of the concrete. In the case of multi-axial loading the strength of the concrete is limited by the well-known failure criteria. The shear stresses to which one may, theoretically, subject the concrete, and in particular the surface of the concrete, can be obtained from Mohr's envelope (Fig. 7). According to this, the shearing strength τ_0 corresponds to the tensile strength f_{ct}, even though one often obtains higher shearing strengths in the course of usual shearing tests carried out on adhesive-bonded joints. The results (both those of the author and all other available results) of shear/compression tests and shearing tests carried out on adhesive-bonded joints

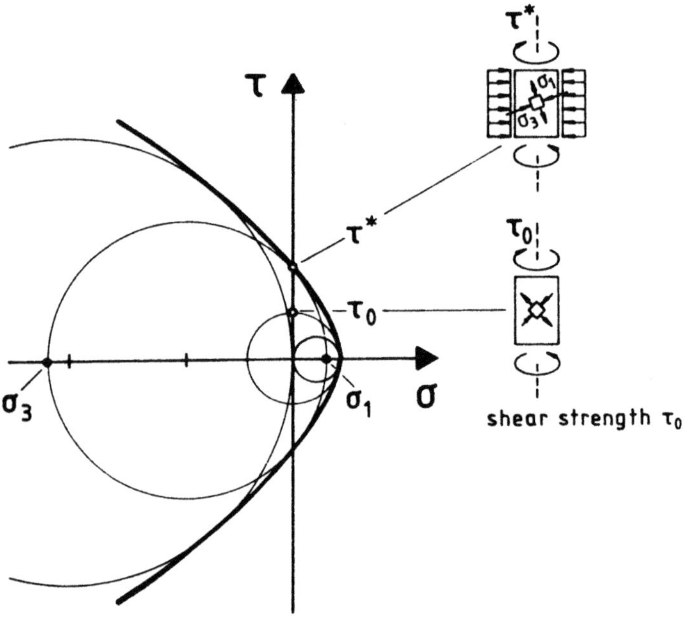

Fig. 7. Mohr envelope for concrete and concrete shear strength τ_0

in concrete were evaluated and combined in a τ-σ_n-diagram, related to the tensile strength of the concrete f_{ct} (Fig. 8). The associated data can be found in Table 1. The evaluation demonstrates that the strength of adhesive-bonded joints in concrete made using EP adhesives, when they are subjected to long-term loading, can be expressed as a function of the concrete strength (σ_n negative as ceompressive stress):

$$\tau_u = f_{ct} - 0.7 \cdot \sigma_n \tag{2}$$

The permissible stresses τ_s and σ_{ns} in service follow from this to

$$\tau_s \cdot \gamma \leq f_{ct\ 5\%} - 0.7 \cdot \sigma_{ns} \cdot \gamma \tag{3}$$

where γ = safety coefficient, and
$f_{ct\ 5\ \%}$ = 5%-fractile of the centric tensile strength

In accordance with Eurocode No. 2 the centric tensile strength here is

$$f_{ct\ 5\ \%} = 0.71\ f_{ctm} \text{ and } f_{ctm} = 0.27\ f_c^{2/3} \tag{4}$$

The influence of adhesive-bonded joints with any joint inclination upon the magnitude of the tolerable main stresses is shown in Fig. 9.

In the case of a uniaxially loaded prismatic concrete specimen with an inclined adhesive bonded joint running at the most disadvantageous angle ($\alpha \approx 60°$) the strength of the joint as set by the adhesion coefficient c and the friction coefficient μ gives us an minimum compressive strength of

$$f_{c,min} = c \cdot \frac{1+(\mu + \sqrt{\mu^2 + 1})^2}{\sqrt{\mu^2 + 1}} \tag{5}$$

Where $c = f_{ct\ 5\%}$ and $\mu = 0.7$ (from equation 3), it follows that (cf. Fig. 9)

$$f_{c,min} = f_{ct\ 5\%} \cdot 3.84 \tag{6}$$

Using equation (4) this then results, with regard to a concrete prism made of, for instance, concrete with a cylinder strength of $f_c = 35\ N/mm^2$ and with an inclined joint, in a minimum compressive strength of

$$f_{c,min} = 7.7\ N/mm^2 = 0.22 \cdot f_c \tag{7}$$

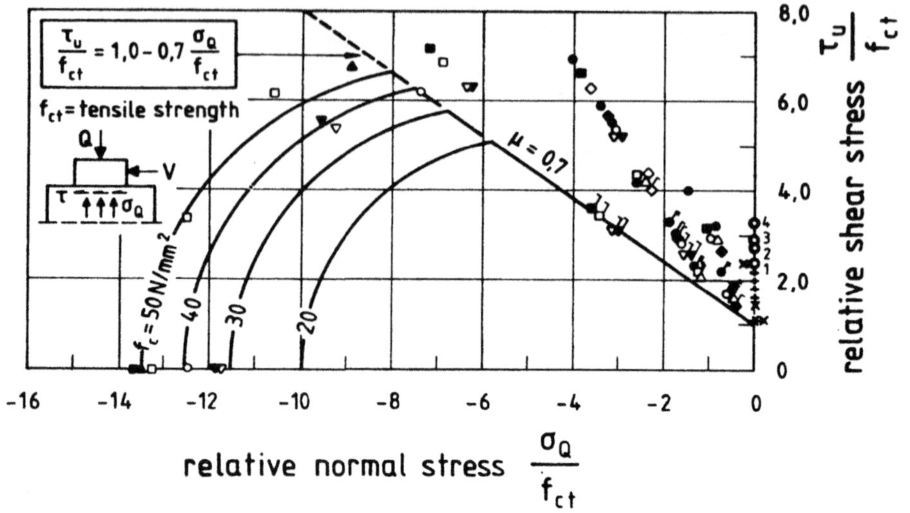

Fig. 8. Available results of shear-compression tests and shearing tests on epoxy adhesive jointed concrete specimens related to the tensile strength of the concrete f_{ct} (symbols are mean values, cf Tab. 1)

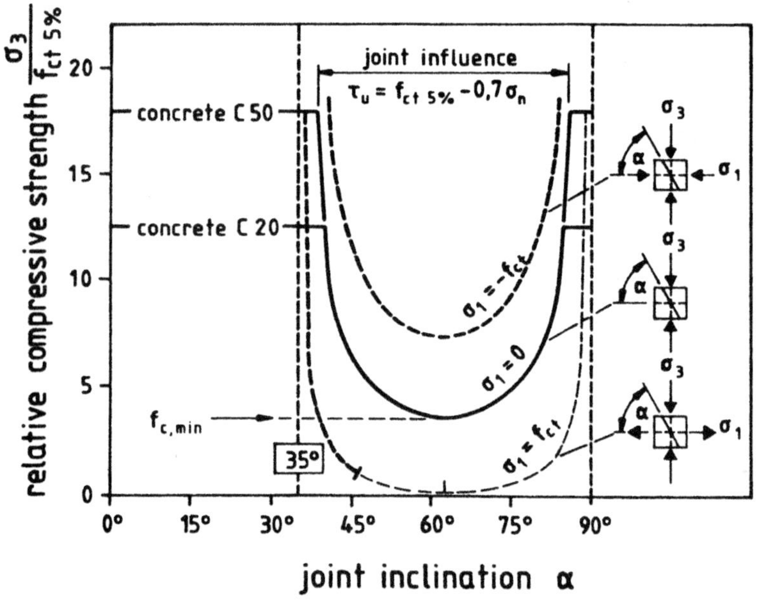

Fig. 9. Influence of the joint bearing behaviour on the bearing capacity of uni- or two-axially stressed specimens comprising an inclined adhesive-bonded joint

4. Requirements upon the concrete surface and selection of the adhesive mortar

Before an adhesive-bonded joint is made the surface of the concrete must be properly treated. Sand-blasting, water sand-blasting, treatment with needle equipment and flame cleaning are suitable methods. However, after flame cleaning another procedure must also be carried out in order to remove loose particles, e.g. by means of rotating brushes, and the surface must then be primed with EP resin.

At the time when the adhesive mortar is applied, the concrete should be at least surface-dry. Reliable limiting values for the moisture content are to be tested using the adhesive concerned. At present it is considered that the moisture content at the surface of the concrete to be bonded must not exceed 6 per cent. In addition, pull-off tests should be done. The measured pull-off strength of the concrete surface should exceed the value of $f_{ct\ 5\%}$.

In order to test the suitability of an epoxy resin motar for use in adhesive-bonded joints in concrete construction one must, basically, carry out two series of tests: one series to test the temperature behaviour under load and another series to test whether the long-term strength of the joint under damp conditions corresponds, as a minimum requirement, to equation (2).

The disadvantageous assumption should be made that in conditions of continuous damp storage the sustained load strength is 40 % of the short-term strength of the adhesive-bonded joint when it has been stored wet. On the basis of this assumption the adhesive-bonded joint is sufficiently strong when the compressive strength $f_{c,min\ (0)}$ of a concrete prism with a bonded joint running at $\alpha = 60°$ has at least the following value after being stored in water for 28 days:

$$f_{c,min\ (0)} \geq 2.5 \cdot f_{c,min} \qquad (8)$$

If one takes $c = f_{ctm}$ and $\mu = 1.0$ as the upper value of the dispersion range, then one obtains the following as the minimum value to be demonstrated

$$f_{c,min\ (0)} \geq 2.5\ f_{ctm} \cdot 4.8 = 12 \cdot f_{ctm} \qquad (9)$$

This corresponds approximately to the compressive strength of the concrete which is to be adhesive-bonded.

In order to characterize the temperature stability of the adhesive-bonded joint one should use the limiting temperature at which a failure occurs in a concrete prism with a 60° bonded joint and which is subjected to a load of $f_{c,min}$ and a continuously increasing temperature.

References

Büchting, F., Moosbrugger, P. (1976) Brücke über den Europakanal aus Fertigteilen mit Verbindungsfugen aus Epoxidharz (Bridge over the European Canal made of Prefabricated Sections Bonded with Epoxy Resin), Bauingenieur, p. 137

Burkhardt, P., Hertig, P.; Aeschlimann, H.V. (1975) Expériences sur les poutres mixtes en acier-béton liées à l'aide d'adhesifs epoxydes (Investigations on Composed Reinforced Concrete Beams Jointed by Epoxy Adhesives), Matériaux et Constructions, p. 261

Diaz, B.E., (1975) The Technique of Glueing Precast Elements of the Rio-Niterol Bridge, Matériaux et Constructions, p. 43

Eibl, J., Franke, L., Hjoith, O., (1972) Versuche mit Kunstharzmörteln (Experiments with Synthetic Resin Mortars), Bautechnik, p. 348

Eurocode No. 2, (1984) Common General Rules for Concrete, Reinforced Concrete and Prestressed Concrete Structures. Draft.

Hallquist, H., (Dec. 1967) An Investigation of Epoxy and Polyester Resin Mortars as a Jointing Material, Bulletin Rilem, p. 233

Jagfeld, P., (1984) Langzeitverhalten von epoxidharzverklebten Betonproben (Long-term Behaviour of Concrete Samples Bonded with Epoxy Resin), Werkstoff und Konstruktion, Bericht des Otto-Graf-Inst., Stuttgart

Ladner, M., Flüeler, P., (1974) Versuche an Stahlbetonteilen mit geklebter Bewehrung (Experiments on Reinforced Concrete Building Components with Adhesive-Bonded Reinforcement), Schweizerische Bauzeitung, Issue 19.

Rehm, G., Franke, L., (1982) Kleben im konstruktiven Betonbau/ Adhesive-Bonded Construction in Concrete (German/English), Betonwerk + Fertigteiltechnik, Issue 10, p. 632, Issue 11, p. 670 and Issu 1 (1983), p. 31

Roubin, E., (1976) Die Segmentbauweise mit geklebten und vorgespannten Fugen im österreichischen Brückenbau (The Segmental Construction Technique using Adhesive-Bonded and Prestressed Joints in Bridge-Building in Austria), Zement und Beton, Issue 5, p. 218.

Schorn, H. (1978) Langzeitverhalten von Epoxidharzen im Betonbau (Long-Term Behaviour of Epoxy Resins in Concrete Construction), Bericht F 1466 des Instituts für Bauforschung, RWTH Aachen

Symbol	Reference	fc N/mm²	Surface treat-ment	Storage/Test Conditions	Testing
■ ■′	Schorn	70	sawn	room climate	■ s-c[1] strength ■′ s-c long term strength,[2]
□ □′	Diaz	57	sand-blasted	room climate	□ □′ "
▲	Buchting/ Moosbrugger	46	"	"	s-c strength
▼ ▼′	Roubin	33	sawn	"	▼ s-c strength ▼′ s-c long term strength,[2]
▽ ▽′	Roubin	33	"	"	▽ ▽′ "
△	Eibl/Fran-ke/Hjorth	32,54	wire brushed	dry and wet	s-c strength
○ ○′	Rehm/ Franke	39	sand-blasted	4 years wet	○ s-c strength ○′ s-c long term strength,[2]
◆	"	80	"	28 days wet	s-c strength
◆′	"	80	"	wet + {long term load	s-c long term strength
◇	"	86	sawn	room climate	s-c strength at 50 °C
◇′	"	86	"	50 °C + long term load	s-c long term strength at 50 °C
●	"	86	"	dry (2 series) 4 years wet (2 series)	s-c strength
●′	"	86	"	4 years wet + long term load	s-c long term strength
●″	"	86	"	wet + dyn. loading	s-c fatigue strength ($N \geqq 2 \cdot 10^6$)
+	Ladner/ Flueler	18-26	sand-blasted	dry	shear strength
1 **O**	Jagfeld	47	sawn	7 days room climate	"
2 **O**	"	47	"	8 years room climate	"
3 **O**	"	47	"	8 y.wet with/ without load	"
4 **O**	"	47	"	8 y. weather-ing with/ without load	"
*	Burkhardt/ Hertig/ Aeschlim.	30	sand-blasted	room climate	torsion test
×	Hallquist	29	wire brushed	"	shear strength

Table 1 1) shear compressiv 2) calculated

ETUDE EXPERIMENTALE DE L'ADHESION ENTRE UN BETON ET UNE RESINE EPOXYDIQUE LORS DE L'ASSEMBLAGE PAR COLLAGE DE VOUSSOIRS PREFABRIQUES DANS LES PONTS EN BETON PRECONTRAINT.

B. GODART
Division Surveillance et Pathologie des Ouvrages d'Art, Laboratoire Central des Ponts et Chaussées, Paris
R. LAFUENTE
Groupe Constructions, Laboratoire Régional des Ponts et Chaussées de Bordeaux

Résumé
Lors de l'assemblage par collage des voussoirs préfabriqués dans les ponts en béton précontraint, la résine époxydique subit pendant sa réticulation des contraintes mécaniques importantes. L'objectif des essais entrepris est de répondre à un certain nombre de questions posées à l'occasion de la rédaction de recommandations sur l'assemblage des voussoirs. L'article décrit les résultats obtenus à l'issue de deux types d'essais, puis en tire quelques conclusions.

Le premier type d'essai cherche à déterminer l'épaisseur résiduelle d'un joint de colle soumis à divers cycles de serrage. Celle-ci se trouve être comprise entre 0,4 et 0,7 mm, et il apparaît que le dosage initial de colle ainsi que les cycles de serrage ont peu d'influence sur l'épaisseur résiduelle du film de colle.

Le second type d'essai vise à étudier la résistance en compression-cisaillement d'assemblages collés qui ont été soumis auparavant à des cycles de serrage pendant la réticulation de la colle. Si toutes les éprouvettes périssent par rupture en compression du béton, la présence d'un joint provoque une chute sensible de la résistance en compression de l'éprouvette.

Mots-clés : adhésion, béton précontraint, résine époxydique, voussoirs préfabriqués, assemblage, essais, compression-cisaillement.

1. Introduction

Lors de la construction par encorbellements successifs de ponts en béton précontraint comportant des voussoirs préfabriqués, le rythme de pose des voussoirs assemblés par collage peut avoir des conséquences importantes sur l'adhésion entre le béton et la résine époxydique. Dans ce type de construction, la colle peut être soumise dès son jeune âge à des sollicitations qui viennent perturber sa réticulation. En effet, les contraintes d'origine mécanique supportées par la résine époxydique lors de l'assemblage des voussoirs peuvent casser les macromolécules tridimensionnelles récemment formées ou faire glisser les chaînes moléculaires les unes par rapport aux autres.

Dans le cadre de l'établissement de recommandations françaises relatives à l'assemblage des voussoirs préfabriqués, il convenait d'éclaircir quelques problèmes liés à l'influence du mode de construction sur la tenue du collage. C'est pourquoi ont été entrepris au Laboratoire des Ponts et Chaussées de Bordeaux deux types d'essais dont les résultats vont maintenant être présentés.

2. Conditions générales des essais

Les résultats obtenus par Benneton au Laboratoire des Ponts et Chaussées de Lyon à l'occasion d'essais de compression réalisés sur des échantillons d'une résine époxydique destinée à l'injection de fissures d'un ouvrage en béton précontraint, montrent que la durée minimale de réticulation nécessaire à une sollicitation sans dommage de la résine sous des contraintes égales à 3,5 MPa ou 7 MPa peut atteindre plusieurs heures, et que cette durée est particulièrement sensible à la température. Ce dernier point est d'ailleurs mis en valeur dans les travaux de Franke et Rehm.

C'est pourquoi nous avons respecté pendant toute la durée de nos essais un maintien de la température à 20° C.

2.1 Résine utilisée.

La résine utilisée au cours de tous nos essais est une résine époxydique chargée couramment utilisée pour la construction des ponts en France. C'est une résine bi-composant (base 50 %, durcisseur 50 %) dont les éléments ont les caractéristiques chimiques suivantes :

- la base est une résine époxyde du type bisphénol épichlorhydrine ayant un indice d'époxyde de 2,6 et contenant une charge siliceuse
- le durcisseur est un mélange d'amidoamine et d'un constituant aromatique ayant une basicité de 2,3 meq/g et contenant aussi une charge siliceuse.

Les caractéristiques mécaniques mesurées lors d'un essai de traction sur haltère de cette résine sont les suivantes :

- à 7 jours d'âge, contrainte de rupture : 18,95 MPa, allongement à rupture : 2,7 %
- à 28 jours d'âge, contrainte de rupture : 20,2 MPa, allongement à rupture : 1,8 %

Enfin, après 24 heures, sa dureté mesurée est de 50 (shore D).

2.2 Cycles de serrage.

En France, les recommandations actuellement en cours de rédaction interdisent la mise en traction du film de colle entre voussoirs pendant la construction du fléau, et préconisent l'utilisation d'un brêlage provisoire assurant une compression minimale dans toutes les zones d'un joint donné. Ces considérations nous ont conduit à appliquer sur nos éprouvettes d'essais, des efforts de compression variables dans le temps pendant la réticulation de la résine époxydique. Afin d'appliquer des efforts se rapprochant le plus possible des contraintes subies par la colle dans un joint de pont en construction, nous avons fait quelques recherches parmi les ponts construits par encorbellements successifs, et nous avons finalement retenu deux cycles de serrage représentatifs des cycles d'assemblages rencontrés. Le premier cycle, appelé cycle "doux", a été observé pendant la construction du pont de Saumur et correspond à l'assemblage de 1 paire de voussoirs par jour ; le second cycle, ou cycle "dur", observé pendant la construction du pont de Ternay correspond à l'assemblage de 3 paires de voussoirs par jour. (voir figure 1)

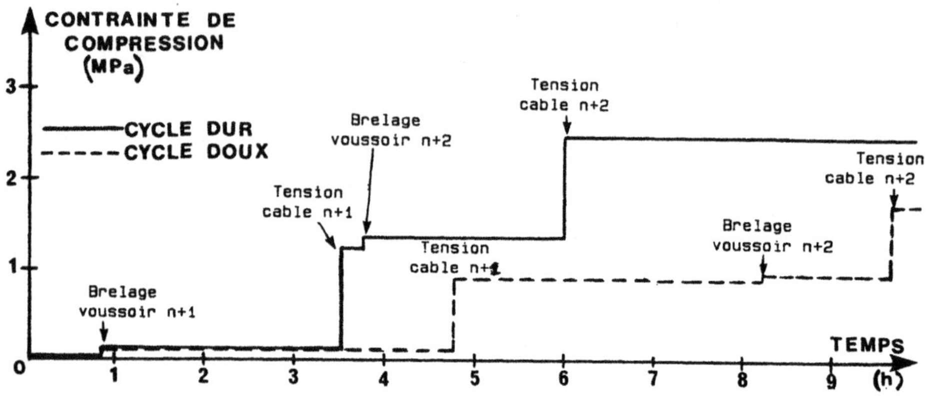

Figure 1 : Cycles de serrage retenus (Effet de l'assemblage des voussoirs sur la contrainte dans le joint n/n+1)

3. Recherche de l'épaisseur résiduelle des films de colle

3.1 Principe de l'essai.

L'épaisseur résiduelle d'un film de colle dans un joint conjugué entre voussoirs étant particulièrement difficile à mesurer in situ, nous avons tenté de la déterminer en laboratoire, en faisant varier les paramètres qui peuvent influer sur cette grandeur :

- quantité de colle utilisée
- cycles de serrage appliqués
- dimensions en plan des éprouvettes (forme du joint)

L'essai consiste à suivre l'évolution de l'épaisseur d'un joint de colle situé entre deux dallettes en béton serrées entre deux plaques métalliques.

3.2 Description de l'essai.

Deux dallettes en béton (A et B), coulées à joints conjugués et non sablées, sont enduites de résine époxydique au moyen d'une spatule, puis sont assemblées et serrées entre deux plaques métalliques épaisses (D) (Voir figure 2). Le serrage des plaques métalliques est assuré par la mise en tension de quatre barres filetées en acier dur de diamètre Ø = 26 mm, équipées de jauges de déformation et étalonnées sous une presse de traction ; cet ensemble permet de contrôler avec précision les cycles de serrage appliqués sur les corps d'épreuves.

Pour obtenir un joint similaire au joint existant dans les constructions, le processus suivant de fabrication des éprouvettes a été respecté :

- coulage de la 1ère dallette, dont la face inférieure est en contact avec un contreplaqué bakélisé et enduit d'un produit démoulant
- retournement de la première dallette
- bétonnage de la seconde dallette

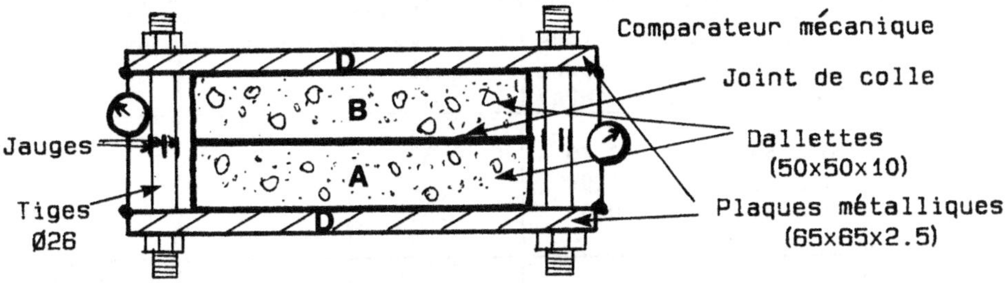

Figure 2 : Schéma de principe de l'essai

La composition du béton retenu pour cet essai est la suivante :

ciment CPA 55	450 kg/m^3
sable roulé (0/2,5)	585 kg/m^3
gravillons roulés de Garonne (2,5/7)	450 kg/m^3
granulats roulés de Garonne (7/22)	770 kg/m^3

3.3 Description des mesures.

En cours d'essai, les variations d'épaisseurs du joint sont mesurées à chaque phase du serrage de 2 façons différentes :

- à l'aide de comparateurs mécaniques mesurant le rapprochement des bords des plaques métalliques (Voir figure 2)
- par récupération de la colle extrudée, puis pesée sur une balance de précision.

La première méthode de mesure ayant posé des problèmes d'interprétation en raison du fléchissement des plaques métalliques et du glissement relatif entre les dallettes (rôle lubrifiant de la colle pendant le serrage), la seconde méthode a été préférée. Celle-ci fournit directement une bonne estimation de la variation d'épaisseur moyenne à partir du volume de colle extrudée, en divisant la masse de colle expulsée par la masse volumique de la colle après durcissement (égale à 1,60 g/cm^3), et par la section de l'éprouvette. Nous retiendrons donc cette méthode en sachant que la valeur de la densité prise en compte ainsi que l'existence éventuelle de bulles d'air dans la colle du joint peuvent être des sources d'erreurs.

Après durcissement du film de colle, une vérification de l'épaisseur résiduelle du film est opérée à l'aide d'une loupe micrométrique. Pour cela, l'éprouvette est sciée le long de ses deux médianes, et, de manière à apprécier les variations locales d'épaisseur, celles-ci sont relevées sur les faces sciées tous les 2 cm. L'examen des résultats détaillés montre des différences significatives entre les bords et le centre de l'éprouvette ; en raison du fléchissement des dalles, les épaisseurs mesurées sont plus grandes vers le centre.

3.4 Résultats des mesures et conclusions.

Le tableau 1 présente les épaisseurs des films de colle mesurées pour trois dosages de colle différents (1,5 - 2 et 2,5 kg de colle par m^2 de joint), et pour 2 cycles donnés.

TYPE	CYCLE DE SERRAGE: Temps écoulé depuis l'origine*	CYCLE DE SERRAGE: Contrainte de compression (MPa)	Dosage initial de colle (kg de colle/m^2 de joint): 1,5	2	2,5
DOUX	0	0	0,94	1,25	1,56
	0 h 50 mn	0,10	0,83	0,97	0,75
	4 h 45 mn	0,90	0,725	0,72	0,60
	8 h 15 mn	0,95	0,665	0,59	0,50
	9 h 40 mn	1,70	0,665	0,59	0,50
DUR	0	0	0,937	1,25	1,56
	0 h 50 mn	0,10	0,77	0,885	0,775
	3 h 30 mn	1,25	0,695	0,63	0,49
	3 h 45 mn	1,35	0,645	0,595	0,46
	6 h 00 mn	2,45	0,62	0,56	0,44

* L'origine des temps est prise à la fin du malaxage des 2 composants de la résine époxydique.

Tableau 1 : Epaisseur du film de colle (en mm)

Tout d'abord, on constate que les épaisseurs résiduelles du film varient entre 0,4 mm et 0,7 mm. Le dosage initial ainsi que le type de cycle ont peu d'influence sur l'épaisseur résiduelle du joint ; néanmoins, on constate que l'épaisseur a tendance à diminuer quand le dosage initial augmente. Cette constatation,ajoutée au fait que la majeure partie de la colle est évacuée dès le début des cycles, nous amène à penser que le rôle lubrifiant de la colle non polymérisée et mise en quantité importante est favorable à son extrusion du joint.

Pour chacun des trois dosages testés, le cycle "dur" donne des épaisseurs résiduelles plus faibles que le cycle "doux". Il semble que la contrainte de compression égale à 1,25 MPa appliquée après 3 h 30 dans le cadre du cycle "dur" ait joué un rôle important.

Des essais complémentaires effectués sur des dallettes de forme différente (dallettes rectangulaires de dimensions 100 x 16) permettent de conclure que l'allongement des éprouvettes ne provoque qu'une très légère augmentation de l'épaisseur résiduelle du film de colle.

4. Essai de compression-cisaillement sur éprouvettes collées

4.1 Objectif de l'essai.

Les joints entre voussoirs préfabriqués sont soumis au cours de leur existence à des contraintes de compression et de cisaillement. L'objectif des essais est d'étudier l'influence des contraintes subies par le film de colle en cours de polymérisation pendant l'assemblage des voussoirs, sur sa résistance en compression-cisaillement quand la résine a achevé sa réticulation. Les deux paramètres de l'étude sont les suivants :

a) le cycle de serrage : cycles dur et doux définis précédemment, ainsi que le simple accostage des deux parties de l'éprouvette ; ce dernier mode de serrage, considéré comme un serrage témoin, résulte de la mise en place d'un serre-joint et correspond à l'application d'un effort maintenu constant et égal à 300 daN (soit une contrainte de compression de 0,15 MPa sur le plan du joint) pendant 24 heures.

b) la préparation de la surface du joint : simple agent démoulant compatible chimiquement avec la résine époxydique, ou agent de démoulage accompagné d'un sablage à sec.

4.2 Principe de l'essai de compression-cisaillement.
Pour pouvoir appliquer des efforts sur le film de colle en cours de polymérisation, nous avons dû concevoir une éprouvette de forme spéciale, inspirée des éprouvettes testées lors de la construction du pont de Rio Niteroi et décrites par Ernani Diaz. La figure 3 représente l'éprouvette dont l'angle α entre le plan de collage et l'effort de compression est égal à 30° ; la valeur de cet angle a été choisie en fonction de l'inclinaison des efforts couramment rencontrée sur les joints des ouvrages d'art. En outre les dimensions de cette éprouvette se rapprochent de celles des éprouvettes destinées à évaluer les produits de réparation des ouvrages en béton, dans le cadre des procédures mises en oeuvre par le Laboratoire Central des Ponts et Chaussées (Voir article de Mme Paillère et M. Rizoulières).

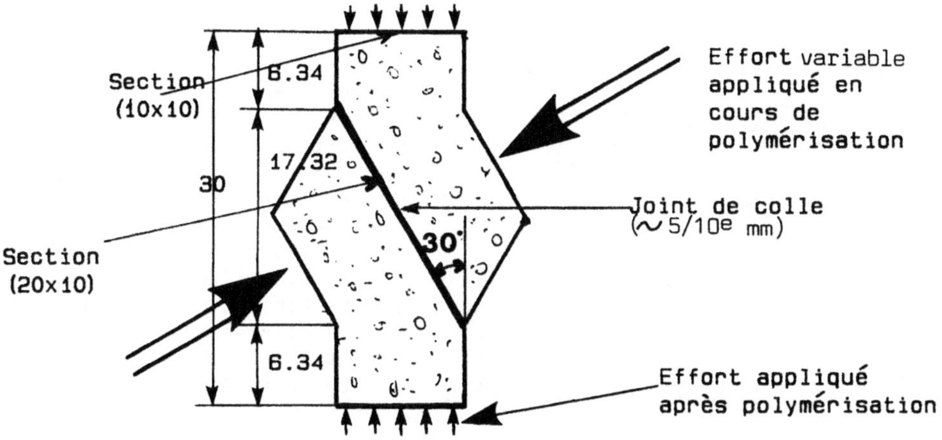

Figure 3 : Schéma de l'éprouvette de compression-cisaillement

4.3 Confection et essai des éprouvettes
Trois éprouvettes étant testées à chaque configuration d'essai, au total 24 éprouvettes ont été confectionnées. Celles-ci étaient fabriquées dans des moules métalliques placés sur une table vibrante, en deux phases espacées de 2 à 3 jours ; la seconde moitié de l'éprouvette était coulée à joint conjugué contre la première moitié. La composi-

tion du béton était similaire à celle utilisée dans l'essai précédent, et pour chaque fabrication successive de corps d'épreuves, trois éprouvettes cylindriques de diamètre 16 cm et de hauteur 32 cm étaient fabriquées afin de déterminer la résistance conventionnelle du béton au moment des essais.

Comme le montre le tableau n° 2, les éprouvettes ont été coulées à deux périodes différentes, puis conservées en salle humide (90 % d'humidité et 20°C) pendant 4 ou 12 mois. Juste avant l'encollage, les éprouvettes ont été ramenées dans une atmosphère à 50 % d'humidité, en gardant la même température. La moitié des éprouvettes ont alors été sablées puis la résine époxydique a été appliquée sur les deux faces à assembler en respectant un dosage de 2 kg de colle/m^2 de joint. Les essais précédents avaient en effet montré qu'il était inutile de faire varier le paramètre dosage. Les éprouvettes ont ensuite été serrées, soit à l'aide d'un bâti permettant de respecter les cycles de serrage définis sur la figure 1, soit à l'aide d'un serre-joint maintenu pendant 24 heures.

Les essais ont eu lieu 14 jours après l'encollage. Au préalable, les faces supérieures et inférieures des éprouvettes ont été surfacées au soufre. Cette technique avait été retenue lors de la phase de mise au point de l'essai, phase qui avait en outre conduit à éviter la réutilisation d'emballages de colle ayant déjà servis.

Pendant l'essai de compression dont le pilotage était réalisé en fonction de la charge, le glissement relatif des deux demi éprouvettes a été mesuré à l'aide d'un comparateur mécanique placé à cheval sur le joint.

4.4 Mode de rupture des éprouvettes.

Photo 1 : Mode de rupture des éprouvettes

D'après Sasse et Fiebrich, trois types de rupture peuvent survenir lorsqu'on examine l'adhésion d'un béton et d'une résine. Dans nos essais, aucune rupture de type cohésif n'a eu lieu à l'interface béton-colle. L'ensemble des éprouvettes ont péri par rupture en compression du béton, des fissures verticales ou légèrement inclinées se propageant à travers le joint de colle (Voir photo 1). Après la rupture en compression du béton, on a pu constater que certaines fissures verticales obliquaient à proximité du joint puis le suivaient (rupture de type cohésif dans le béton au voisinage de la colle). Mais ce type de rupture ne semble être qu'une conséquence de la rupture en compression de l'éprouvette.

Cependant, il faut se demander si la présence du joint constitue une faiblesse pour la résistance en compression-cisaillement de l'éprouvette. En effet, du point de vue mécanique, la différence entre les modules de déformation de la colle et du béton peut entraîner des contraintes transversales de traction dans le béton au voisinage du joint. D'après les résultats d'essais compilés par Frank et Rehm, il semble que la présence du joint ait un effet positif, mais qu'il puisse devenir négatif avec des bétons de qualité supérieure.

4.5 Résultats des essais de compression-cisaillement

Les résultats sont résumés dans le tableau ci-dessous :

DATES			PARAMETRES de l'essai		Force de rupture (KN) moyenne de 3 éprouv.	R*
Coulage	Encollage	Essai	Sablage	Serrage		
4.11.82 au 19.11.82	15.12.83	3.1.84	NON	Serre-joint	454	0,66
				Cycle dur	484	0,71
22.11.82 au 6.12.82	11.01.84	30.1.84	NON	Serre-joint	397	0,63
				Cycle doux	467	0,73
7.11.83 au 21.11.83	12.03.84	28.3.84	OUI	Serre-joint	366	0,69
				Cycle dur	303	0,57
24.11.83 au 9.12.83	14.03.84	30.3.84	OUI	Serre-joint	408	0,80
				Cycle doux	393	0,80

$$*R = \frac{\text{Contrainte moyenne de compression à la rupture dans l'essai}}{\text{Contrainte moyenne de compression à rupture sur cylindres 16x32}}$$

Tableau 2 : Résultats des essais de compression-cisaillement

Si nous examinons les valeurs du rapport R égal à la contrainte moyenne de rupture en compression du prisme collé divisée par la contrainte moyenne de rupture en compression de l'éprouvette cylindrique de référence, on constate que celles-ci sont relativement dispersées (de 0,57 à 0,80) et qu'elles se situent toutes en dessous de la contrainte moyenne de rupture d'un prisme monolithe dont le coefficient R est proche de 0,85. Ainsi, quels que soient les paramètres, nous aboutissons aux mêmes conclusions que Diaz, à savoir que la présence d'un joint provoque une chute de la résistance en compression cisaillement d'une éprouvette.

L'examen des valeurs du rapport R ne permet pas de tirer des conclusions définitives sur les paramètres étudiés. On peut seulement remarquer que les cycles de serrage n'engendrent pas de conséquences fâcheuses comparativement au simple accostage.

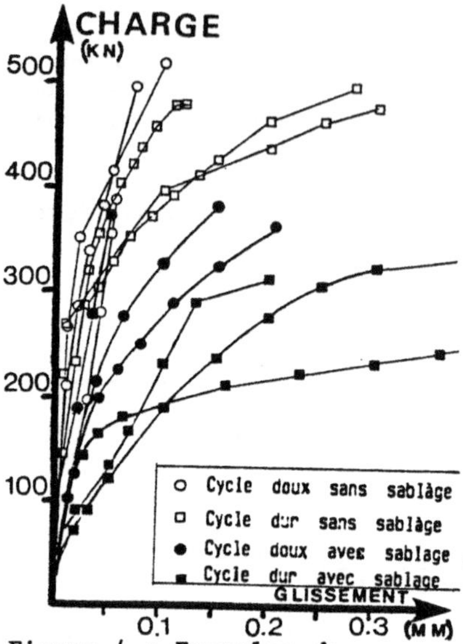

Figure 4 : Exemples de courbes glissement-charge obtenues

Si l'on observe les courbes de glissement obtenus pendant les essais (voir figure 4), on constate une certaine dispersion parmi les trois éprouvettes d'une même série d'essais, ainsi qu'une grande diversité dans le comportement des éprouvettes. Les comportements extrêmes sont constitués d'une part par des éprouvettes à forte charge de rupture et faible glissement au droit du joint de colle, et d'autre part par des éprouvettes à faible charge de rupture et glissement important du joint. Il semble cependant se dégager une tendance, à savoir que les éprouvettes sablées présentent des glissements plus importants que les éprouvettes non sablées, pour un même cycle de serrage. Une explication de ce phénomène peut être recherchée dans la déconjugaison apportée par le sablage qui, non seulement modifie les conditions d'accrochage de la résine sur le béton, mais aussi augmente l'épaisseur du film de colle et en conséquence provoque un glissement relatif des deux parties de l'éprouvette plus important.

Si la présence du joint de colle semble provoquer une chute de la résistance des éprouvettes, les contraintes supportées par la colle au moment de la rupture restent cependant importantes. En effet, un calcul fondé sur l'équilibre des efforts horizontaux et verticaux appliqués sur le joint ainsi que sur l'hypothèse d'une répartition uniforme des contraintes normales (σ_N) et tangentes (τ) sur le plan du joint, permet d'estimer les contraintes selon les formules suivantes :

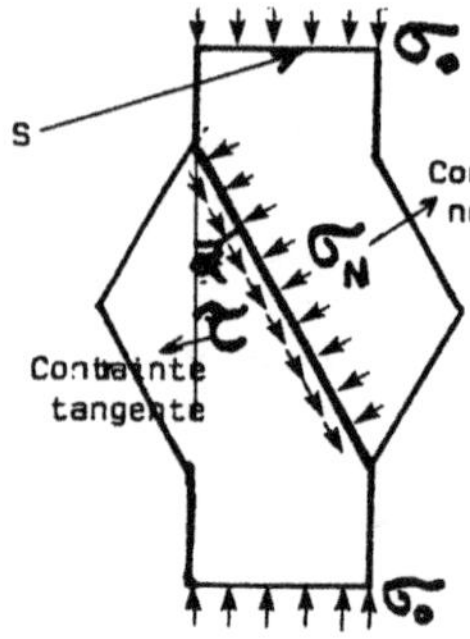

$$\begin{cases} \sigma_N = \sigma_0 \cos^2 \\ \tau = \sigma_0 \sin\alpha \cos\alpha \end{cases}$$

$$\text{et } \sigma_0 = \frac{P}{S} = \frac{\text{Effort vertical appliqué}}{\text{Section droite de l'éprouvette}}$$

Figure 5 : contraintes appliquées sur le plan du joint

Les charges expérimentales de rupture étant comprises entre 303 KN et 484 KN, on obtient des contraintes normales et tangentes à rupture comprises entre les valeurs extrêmes suivantes :

$7{,}6 \text{ MPa} < \sigma_N \text{ (rupture)} < 12{,}1 \text{ MPa}$
$13{,}1 \text{ MPa} < \tau \text{ (rupture)} < 21{,}0 \text{ MPa}$

Eu égard aux contraintes de cisaillement admises dans les joints de ponts en béton précontraint, les contraintes de cisaillement obtenues à l'issue de ces essais semblent confortables si l'on tient compte que la majeure partie du cisaillement est reprise par les clés.

5. Conclusion

Les essais réalisés permettent de conclure que les cycles de serrage envisagés dans cette étude n'apparaissent pas néfastes vis-à-vis de la résistance d'assemblages collés sollicités en compression-cisaillement. S'il ne paraît pas indispensable de procéder à un sablage des faces des joints des voussoirs avant l'encollage et si le dosage de colle retenu n'est pas un critère décisif, en revanche, la qualité de la résine époxydique utilisée et le respect de ses modes de mise en oeuvre restent les conditions nécessaires d'un collage structural réussi, c'est-à-dire un collage permettant de s'assurer que le joint collé n'est pas le point faible de la structure.

Bibliographie

Ernani Diaz, B. (1975) The technique of glueing precast elements of the Rio-Niteroi bridge. Matériaux et construction Vol 8, N°43, pp43-50

Franke, L., Rehm, G. (1982) Kleben im Konstruktiven Betonbau, Teil 1. Betonwerk, Fertigteil-Technik n° 10 pp 632-637

Franke, L., Rehm, G. (1983) Kleben im Konstruktiven Betonbau, Teil 2. Betonwerk, Fertigteil-Technik n° 1 pp 31-55

Mouton, Y. (1979) Réparation des structures en béton fissurées par injection de liants époxydiques. Rapport de Recherche LPC N° 86, Laboratoire Central des Ponts et Chaussées, 103 pages

Paillère, A.M., Rizoulières, J. (1981) Critères d'efficacité des produits de réparation des ouvrages en béton. Colloque international sur la gestion des ouvrages d'art Bruxelles-Paris 13 au 17 avril 1981, Vol. II, pp 479-484 Editions ENPC

Sasse, H.R., Fiebrich, M. (1983) Bonding of polymer materials to concrete. Matériaux et construction Vol 16, N° 94, pp 293-300

INFLUENCE DE LA NATURE ET DE LA GRANULARITE DE LA CHARGE SUR L'ADHERENCE DES RESINES EPOXYDES AU BETON DURCI SEC ET HUMIDE

A.M. PAILLERE Laboratoire Central des Ponts et Chaussées PARIS
J.J . SERRANO Laboratoire Central des Ponts et Chaussées PARIS
J.L. AGUIAR Université de Minho PORTUGAL
R. LAMOURI Laboratoire Central des Ponts et Chaussées PARIS

Résumé
Les recherches menées au LCPC en vue d'étudier l'adhérence des résines époxydes sur des surfaces de béton durci, sèches ou humides ont permis de constater que les polymères non chargés (résines pures) peuvent adhérer de façon satisfaisante au support béton à condition que celui-ci soit sec. En effet lorsque il y a présence d'eau (support humide) les résines pures présentent dans le collage du béton durci sur béton durci une rupture adhésive, alors que les mêmes produits convenablement chargés conduisent à une bonne adhérence puisqu'il y a rupture dans le béton.

Par ailleurs les modes de ruptures adhésive, dans le béton ou cohésive sont fonction de la nature minéralogique de la charge, de sa granularité et de l'épaisseur du joint.

De même l'adhérence sur support humide serait liée à l'absorption, plus ou moins élevée, de l'eau présente sur la surface du béton et mise en contact ave le polymère.
Mots clés : Polymère, époxyde, adhérence, béton durci, support, humidité, charge, granularité, épaisseur, joint, absorption, rupture.

Introduction

Dans les techniques de réparation les polymères sont utilisés purs ou chargés avec des fillers divers. Habituellement on injecte les fines fissures avec une résine pure et l'on procède au collage structural béton durci/béton durci ou béton frais/béton durci à l'aide d'une résine chargée.

Les arguments avancés pour l'utilisation d'une charge sont d'ordre économique (plus il y a de filler moins il y a de résine), d'ordre tixotropique, ou d'ordre modérateur de certaines caractéristiques des polymères, en effet, on cherche presque toujours à modifier, voire même à diminuer, le module d'élasticité, le coefficient de dilatation linéique, le retrait, le fluage, l'exothermie etc. Ainsi les charges jouent un rôle non négligeable dans le comportement des polymères. Nous avons donc voulu, dans la présente recherche connaître si lesdites charges intervenaient aussi dans les mécanismes conditionnant l'adhérence des polymères au support béton.

Dans ce but, nous avons étudié d'une part, l'influence d'une charge donnée sur l'adhérence du système béton durci - polymère - béton durci et d'autre part, l'adhérence en fonction des caractéristiques et dosage de la charge.

Conduite des essais

Toutes les déterminations de l'adhérence des polymères au béton ont été effectuées à l'aide de l'essai de traction directe, conformément à la norme française NFP 18-871, relative aux produits pour collage structural entre deux éléments en béton avec surface d'application horizontale.

Le béton durci utilisé dans le collage structural avait plus de 90 jours d'âge et sa résistance à la compression, sur cylindres, 16 x 32 cm, 28 jours était de l'ordre de 48 MPa.

Afin de minimiser, ou d'annuler, l'effet d'un hétérogénéité ponctuelle dans la couche de polymère mis en oeuvre en vue de collage, nous avons travaillé sur des demi prismes 14 x 9 x 30 cm. L'application des polymères a eu lieu sur des surfaces de 126 cm² sciées ou sablées, sèches ou humides. On a reconstitué le primes 14 x 18 x 30 cm par collage de l'autre demi prisme de béton durci correspondant (fig. 1).

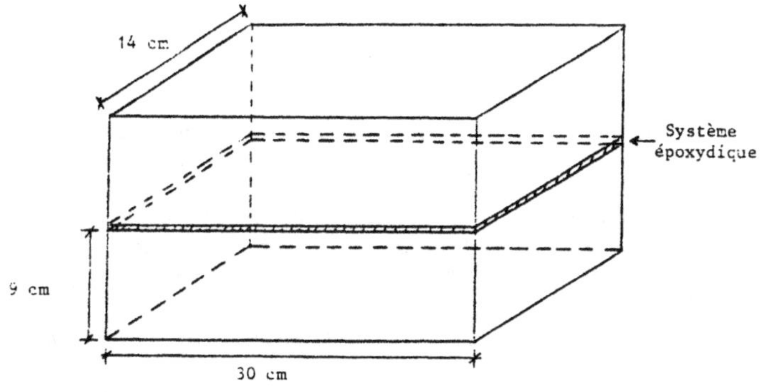

Fig. 1. Eprouvette prismatique reconstituée par collage, à l'aide d'un système époxydique, de deux demi-prismes 14 x 9 x 30 cm de béton durci.

Les éprouvettes destinées à la détermination de l'adhérence ont été obtenues à 28 jours d'âge de l'assemblage, par carottage de cylindres 8 x 16 cm dans le prisme reconstitué (fig. 2).

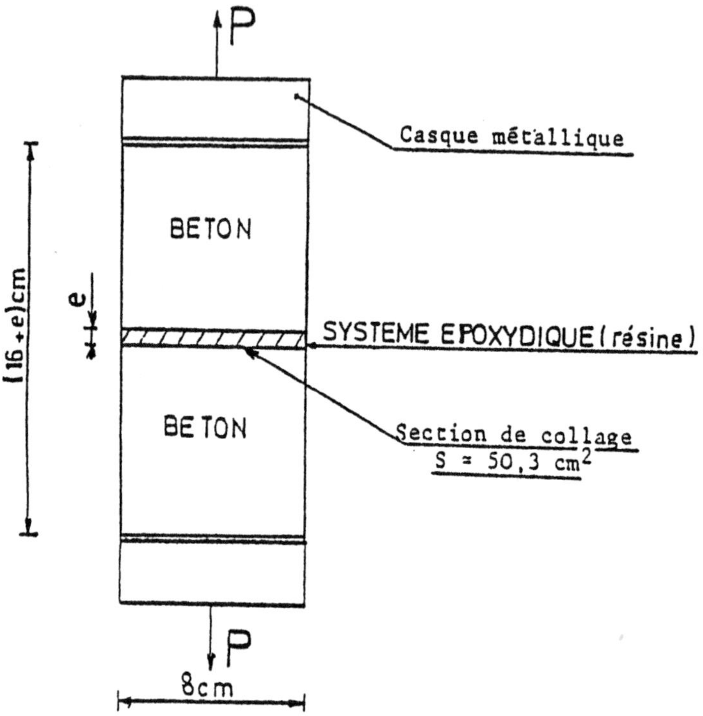

Fig. 2. Eprouvette cylindrique béton - résine - béton

Dans la détermination de l'influence de la charge sur l'adhérence on a utilisé les sytèmes époxydiques suivants :

Système A Résine époxydique A + Durcisseur A_1 à base de polyamine aliphatique

Système B Résine époxyde A + Durcisseur A_1 + 10 % de diluant réactif.

Le diluant réactif a été utilisé dans le but d'augmenter la mouillabilité du système époxydique A ($\theta \sim 40°$ pour A et $\sim 30°$ pour B).

Dans la deuxième partie concernant l'étude de l'adhérence en fonction du dosage et de la nature de la charge nous avons utilisé le même système epoxydiques.

Système A (Résine epoxydique A + Durcisseur A_1)

et le système X Résine époxydique X

Durcisseur Y à base d'un mélange d'amines cycliques et d'alcool benzilique.

Les modes de rupture des éprouvettes reconstituées ont été classés en quatre type (fig. 3).

. Type 1 : Rupture plein béton

Le cylindre reconstitué ne se sépare pas suivant le plan de collage horizontal. Le béton casse sous l'effet des contraintes de traction exercées sur l'éprouvette, dans le cas du collage de béton frais sur béton durci, soit du côté du béton durci servant de support, soit du côté du béton rapporté frais.

. Type 2 : Rupture mixte béton-plan de collage

La cassure intéresse à la fois le béton et le plan de collage.

. Type 3 : Rupture cohésive dans le produit de collage

Le produit reste adhérent sur chaque demi-éprouvette de béton et la rupture a lieu dans l'épaisseur du produit de colla ge suivant un plan parallèle au plan d'assemblage.

. Type 4 : Décollement (Rupture adhésive)

La rupture se produit à l'une des interfaces produit de collage-béton par défaut d'adhérence du produit soumis aux essais.

Fig. 3. Différents types de rupture

Type I - rupture dans le béton, type II rupture mixte béton - plan de collage Type III- rupture cohésive, type IV rupture adhésive (décollement).

A. Influence de la charge sur l'adhérence de l'ensemble Béton durci - polymère -béton durci. Support béton sec ou humide

La charge expérimentée était du filler calcaire à 99,7 % de $CaCO_3$ présentant 10 % d'éléments compris entre 20 et 100 μm et 80 % d'éléments fins < 20 μm.

Chaque système a été chargé avec 116 parties de ce filler calcaire. L'expérimentation a porté sur quatre produits Système A et Système E (Système A + 116 parties de charge)

Système B et Système F (système B + 116 parties de charge)

Conformément à la norme NF P 18-871 les surfaces d'application ont été obtenues : pour les surfaces sèches, en sortant de l'eau les 20 ± 2° C et 50 ± 5 % de H R ; pour les surfaces humides en sortant les 1/2 éprouvettes de l'eau un quart d'heure avant application du polymère et les protégeant de la dessication en les recouvrant d'un linge humide essoré.

L'étude de l'influence de la charge sur l'adhérence a été effectué en fonction de l'épaisseur de film de polymère

A.1 Principaux résultats

A.1.1 Rôle joué par la charge

Les figures 4 et 5 montrent que dans le cas des surfaces d'application sèches les ruptures ont toujours lieu dans le béton et que par conséquent il n'y a aucune influence de la charge.

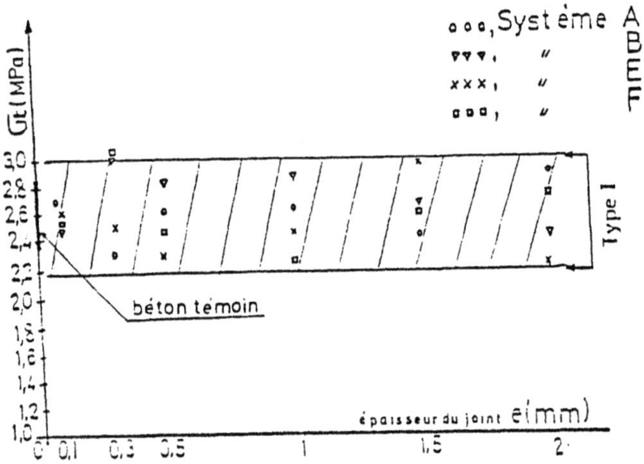

Fig. 4. Contraintes de rupture des éprouvettes béton-résine-béton à surface de collage sciée - sèche

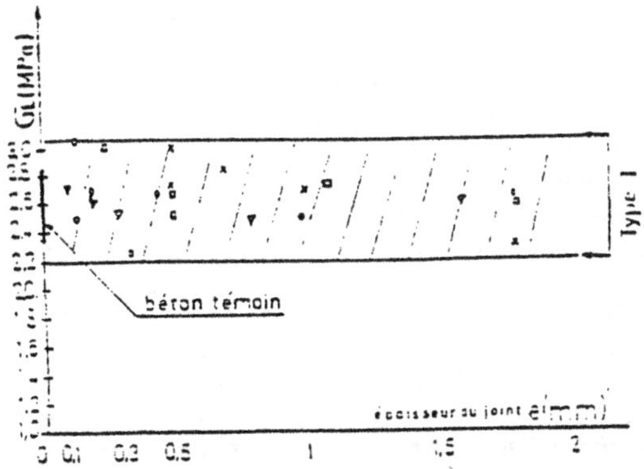

Fig. 5. Contraintes de rupture des éprouvettes béton-résine - béton à surface de collage sablée - sèche

Il en est tout autrement dans le cas des surfaces humides. Ainsi qu'il apparaît dans la figure 6, les systèmes chargés conduisent dans la plupart des cas à des ruptures dans le béton ou à prédominance dans le béton (type I et II). Par contre les systèmes A et B se plaçent dans les types de rupture adhésive. Il semble donc que la charge est indispensable pour obtenir une bonne adhérence sur surface humide

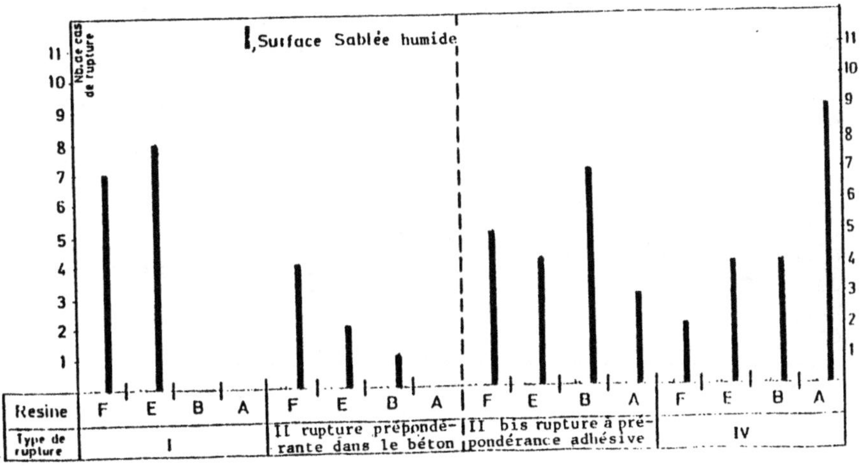

Fig. 6. Nombre de cas de rupture correspondant à chaque système époxydique

A.1.2 influence de l'épaisseur du film de polymère

Le positionnement d'un bloc de béton sur un autre revêtu d'un film, plus ou moins épais, de résine époxydique chargée ou non, ne permet pas de maîtriser l'épaisseur du "joint" de polymère. Celui-ci est extrudé sous l'effet d'un poids du bloc de béton et l'on aboutit pratiquement à une épaisseur de film toujours constante quel que soit le polymère utilisé.

Nous avons donc fait varier l'épaisseur de ce film artificiellement, en maintenant les deux blocs de béton à assembler, à des hauteurs variables, l'un de l'autre et en remplissant l'interstice, ainsi aménagé, avec le système époxydique choisi.

Les figures 7 et 8 confirment les résultats précédents. Les modes de rupture se placent toujours dans le type II (mixte à prépondérance ahdésive) et IV (rupture adhésive) lorsque les polymères ne sont pas chargés. De même la présence de charge entraîne, au fur et à mesure que l'épaisseur du joint de collage augmente, un accroissement de la la rupture qui semble se stabiliser entre 2,4 et 3 MPa pour des épaisseurs comprises entre 1 et 2,3 mm. C'est dans cette zone de valeurs que toutes les ruptures ont lieu dans le béton (type I).

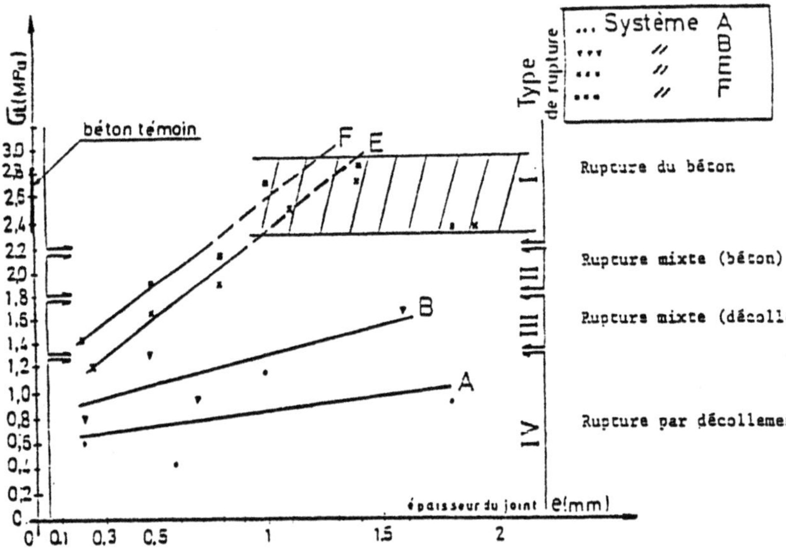

Fig. 7. Contraintes de ruptures en fonction de l'épaisseur du joint - collage sur surface brossée-humide.

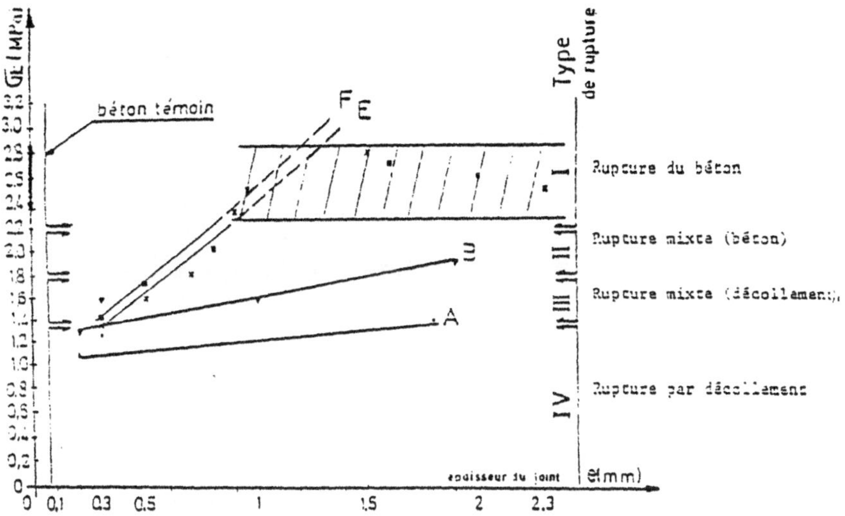

Fig. 8. Contraintes de rupture en fonction de l'épaisseur du joint - collage sur surface sablée-humide

A.1.3 Interaction entre la charge du polymère et l'humidité du support

Compte tenu que la charge était du calcaire et du caractère hydrophyle de celui-ci, il était permis de penser qu'il se produisait une absorption ou succion progressive de l'eau présente sur la surface du béton par ledit calcaire, ce qui entraînait le transfert de celle-ci dans le film de colle et un assèchement superficiel du subjectif. Ceci permettrait donc une meilleure adhérence du polymère au béton.

Les observations effectuées au microscope électronique à balayage semblent confirmer cette hypothèse.

La figure 9 montre l'adhérence parfaite de la résine époxydique A, non chargée, sur un béton à surface d'application sèche. Par contre en présence d'humidité fig. 10 on constate des fissures longitudinales sur le plan de collage et une pertubation du polymère par l'eau avec présence de bulles d'eau emprisonnées par la résine.

Fig. 9. Collage béton durci/béton durci sur surface sèche - bonne adhérence du polymère non chargé A

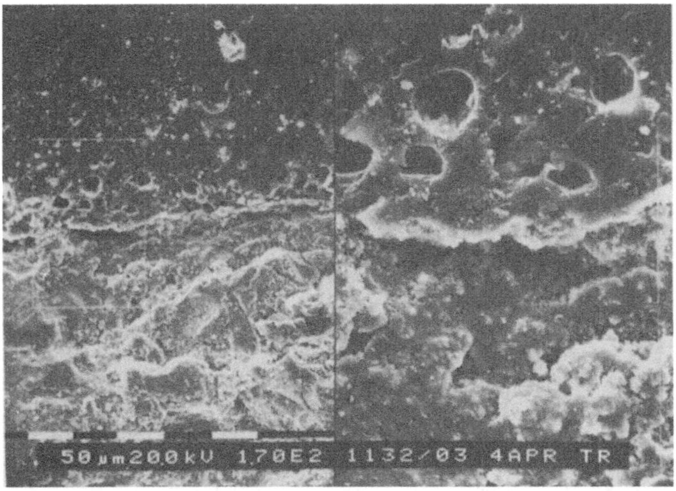

Fig. 10. Collage béton durci/béton durci avec système A non chargé et surface humide. On note des fissures et des bulles à l'interface

La présence de charge dans le système époxydique A modifie l'aspect de la surface de contact ainsi dans les figures 11 et 11 bis on ne constate ni fissures, ni bullages, mais tout au plus (fig. 11) une bande blanche constituée par le système époxydique chargé et l'eau absorbée.

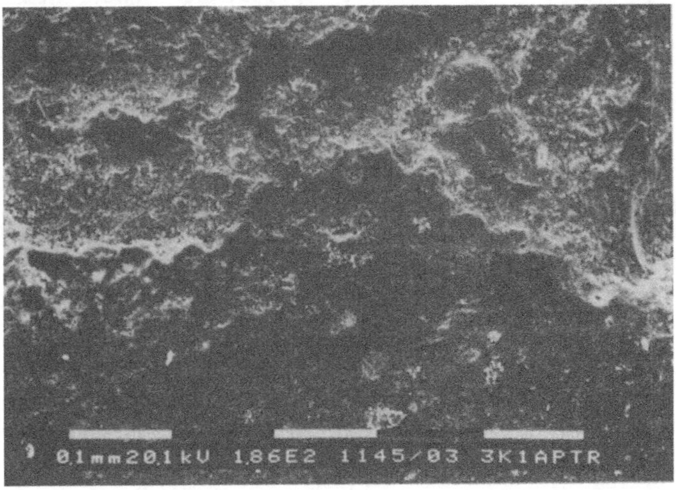

Fig. 11. Bonne adhérence béton -liant par la pénétration de la résine dans les interstices du béton. Apparition d'une bande blande d'eau absorbée par la résine + charge

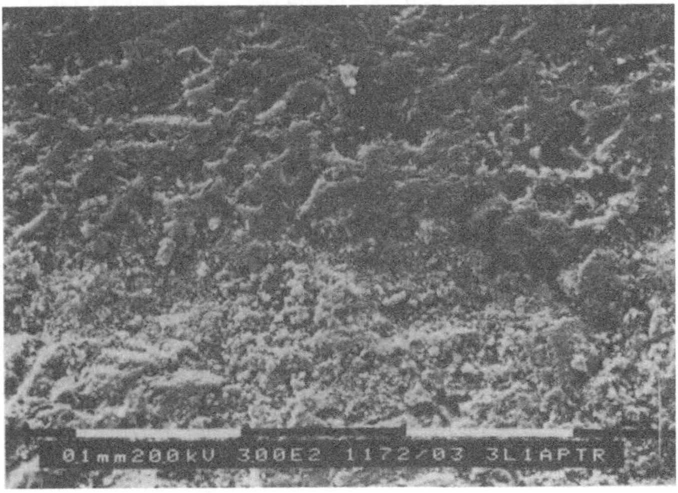

Fig. 11 bis. Très bonne adhérence entre le liant et le béton - l'interface est très peu marquée. Pas de bande blanche visible.

B. Etude de l'adhérence en fonction du dosage et de la nature de la charge

Cette absorption d'eau par la charge semble donc être un des facteurs favorisant l'adhérence des époxydes aux supports de béton humides. Toutefois, il fallait approfondir cette étude afin de connaître, dans quelle mesure cet effet "buvard" était lié à la nature minéralogique du filler, à sa granularité et au pourcentage présent dans le système époxydique. De même il était intéressant de savoir quelle était la quantité d'eau absorbée par la charge qui été compatible avec une bonne adhérence béton durci - polymère - béton durci.

Nous avons donc étudié cette adhérence en expérimentant, avec les deux systèmes epoxydiques A et X définis précédemment, trois fractions granulométriques différentes, d'une charge calcaire et d'une charge 100 % siliceuse.

Ces fractions étaient les suivantes

Nature de la charge	Filler ultrafin	Filler fin	Filler moyen
Calcaire	80 % < 4 m 92 % < 20 m	68 % < 4 m 85 % < 20 m	35 % < 4 m 75 % < 30 m
Siliceux	72 % < 4 m 95 % < 20 m	61 % < 4 m 95 % < 20 m	39 % < 4 m 100 % < 30 m

B.1 Détermination de la quantité de charge incorporable aux polymères

Ainsi qu'il apparaît dans la figure 12, la quantité de filler qu'il est possible d'additionner aux systèmes époxydiques A et X est décroissante au fur et à mesure que la finesse augmente.

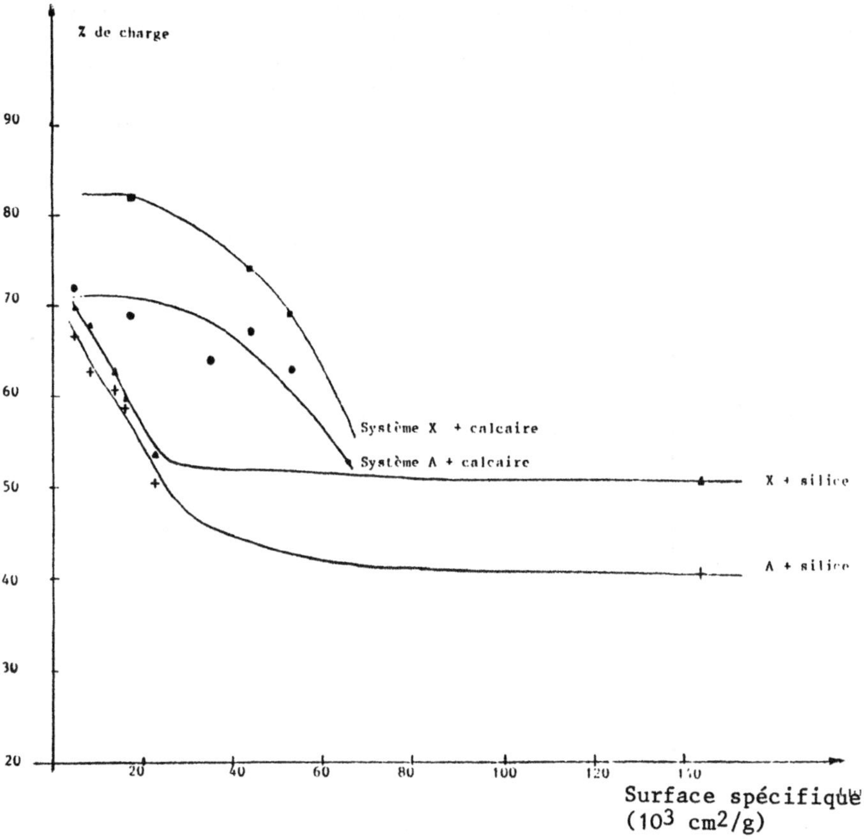

Fig. 12. Quantité maximale de filler incorporable aux deux types de résines époxydiques en fonction de la surface spécifique du filler.

Cette quantité est plus faible dans le cas du filler siliceux. Ainsi pour les fillers moyens (SS < 20 000 cm²/g) on peut atteindre 80 % pour le système X et 70 % pour le A avec du filler calcaire et environ 70 % maximum pour le filler siliceux quelle que soit l'epoxyde utilisée. Pour cette dernière charge on constate qu'à partir d'une surface spécifique de 38 000 cm²/g le teneur maximum de charge est toujours de l'ordre de 52 % pour le polymère X. Dans le cas du polymère A, le palier se situe aux environs de 42 % pour les fillers à SS > 60 000 cm²/g.

B.2 Influence du pourcentage, de la nature et de la granularité de la charge, sur l'adhérence du polymère.

Il est apparu (fig. 13 et 13 bis) qu'il existe un dosage optimum de filler conduisant à un maximum de résistance et d'adhérence. Cet otpimum varie suivant la nature du filler, sa granularité et le système epoxyde expérimenté.

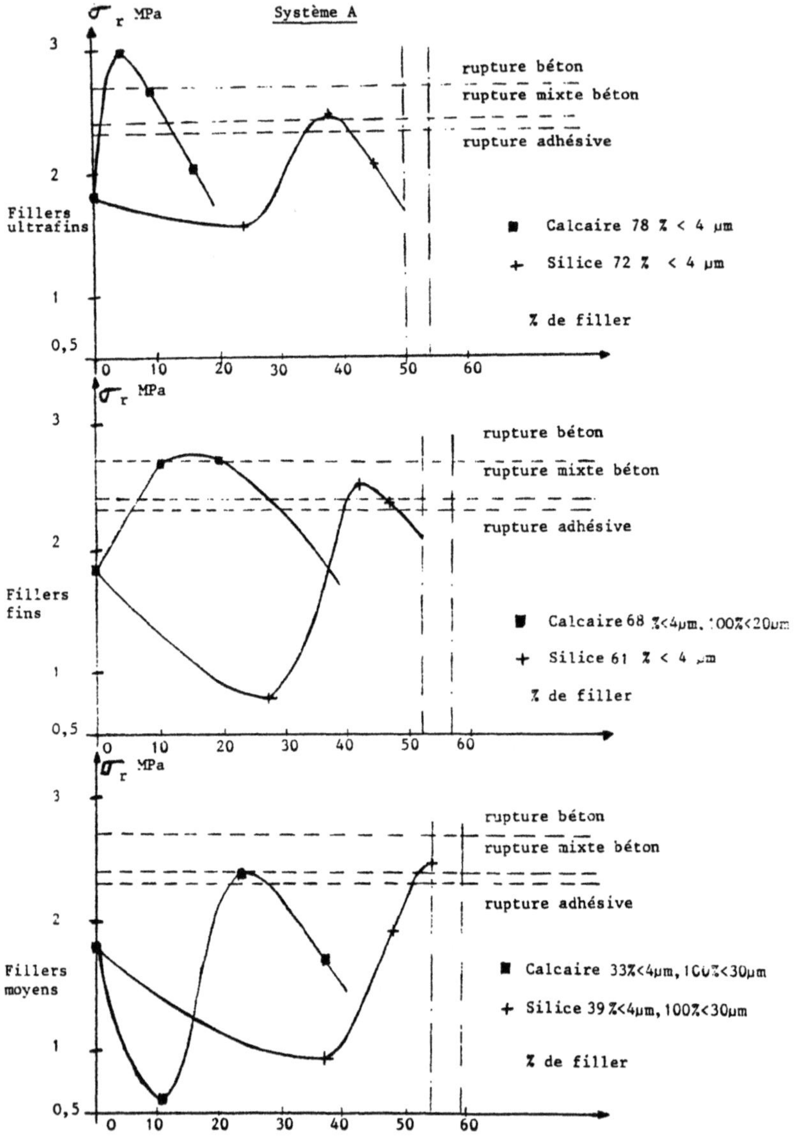

Fig. 13. Adhérence en fonction du dosage en filler (surface humide - épaisseur $\simeq$ 0,6 mm)

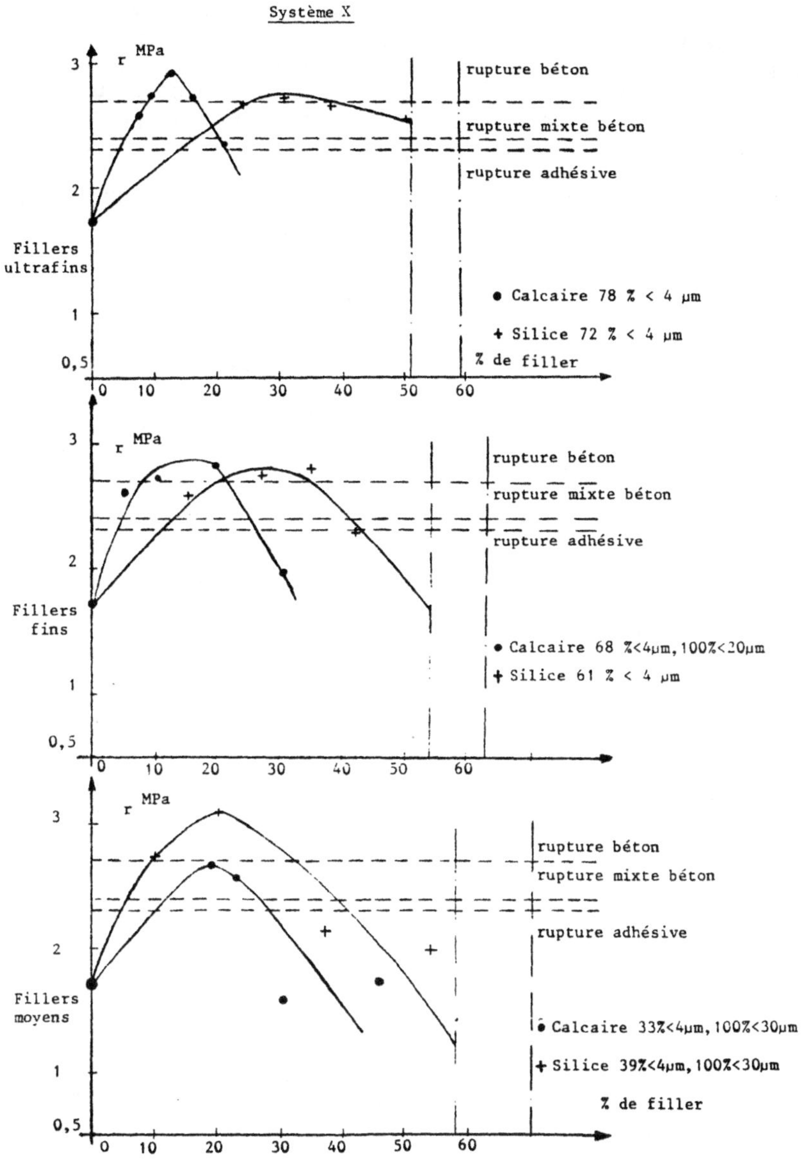

Fig. 13 bis. Adhérence en fonction du dosage en filler (surface humide-épaisseur $\simeq$ 0,6 mm)

Pour la résine A les meilleurs résultats (rupture dans le béton) sont obtenus avec le filler calcaire fin et ultrafin et des dosages compris entre 5 et 16 %. Les fillers calcaire et siliceux semblent conférer le même comportement à la résine X quelle que soit leur granularité, mais les pourcentages optimums, du point de vue de l'adhérence sont compris entre 12 et 18 % pour le calcaire et contre 20 et 30 % pour le siliceux.

Ces résultats sont synthétisés dans la fig. 14 où il apparaît

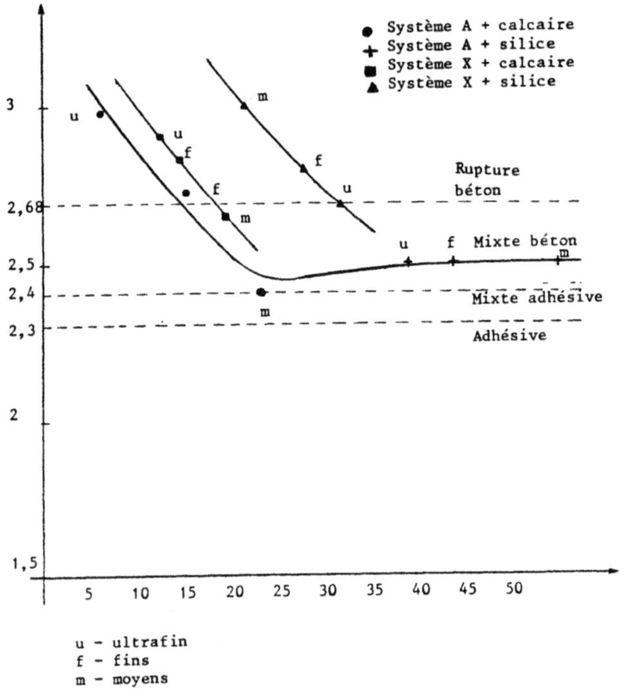

Fig. 14. contraintes maximales à la rupture en fonction du dosage en filler

que les contraintes maximum entrainant une rupture dans le béton, enregistrées avec le système X ont eu lieu indifféremment avec les fillers calcaires ou siliceux, avec une tendance à de meilleures performances dans le cas des éléments moyens. Il en est tout autrement avec le système époxydique A, dont les performances décroissent au fur et à mesure que la dimension des fines calcaires augmente. Les fines siliceuses conduisent, quelle que soit leur granularité, à des ruptures mixtes, pour des résistances de l'ordre 2,5 MPa.

Ces résultats confirment donc l'existence d'un couple optimum polymère charge. L'adhérence d'un système époxydique à un béton humide semble bien être, entre autres, fonction de la nature et granularité de ladite charge.

B.3 Influence de l'aptitude à l'absorption d'eau du système polymère-charge sur l'adhérence

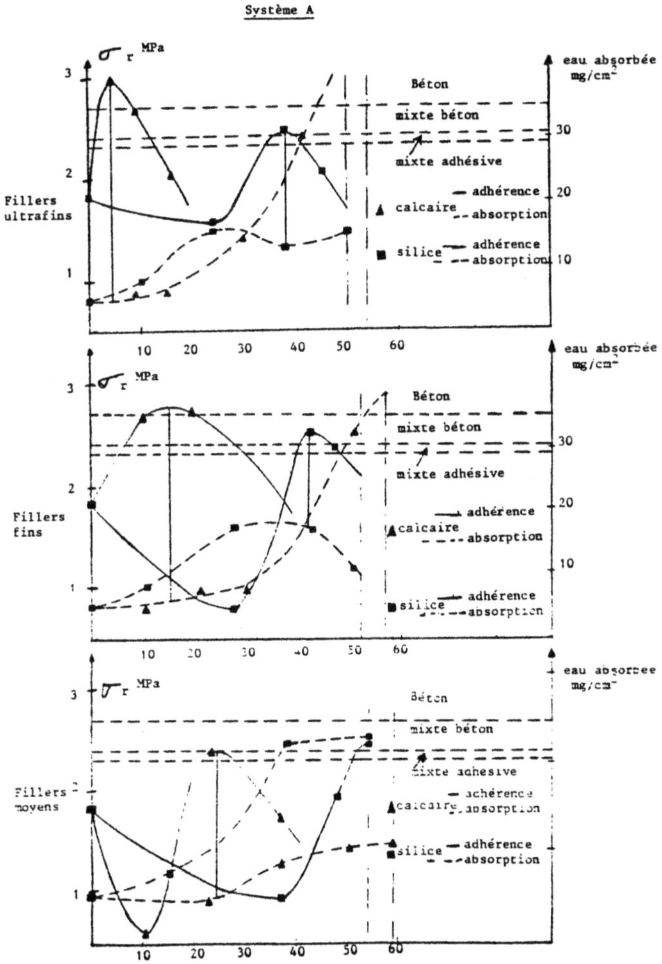

Fig. 15. Adhérence et absorption d'eau en fonction du dosage en filler (surface humide - épaisseur ≃ 0,6 mm)

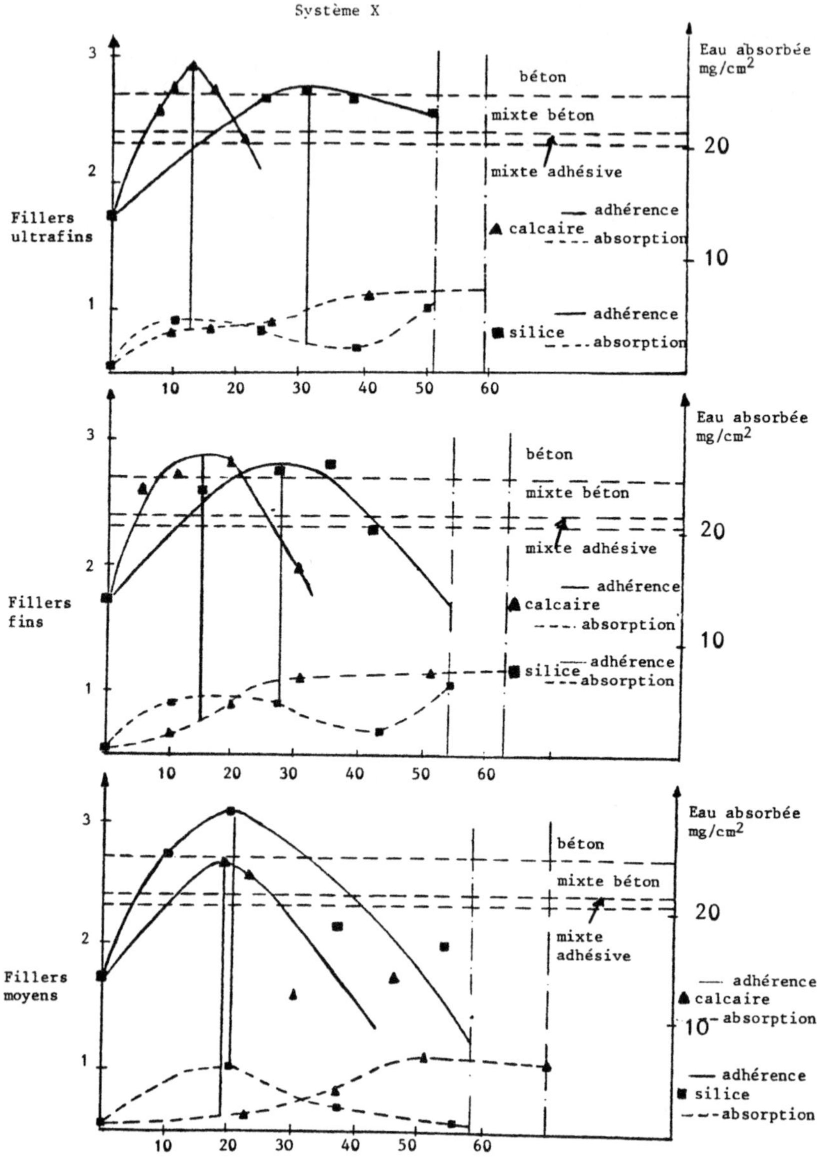

Fig. 15 bis Adhérence et absorption d'eau en fonction du dosage en filler (surface humide - épaisseur $\simeq$ 0,6 mm)

L'étude de l'absorption d'eau des polymères chargés montre (fig. 15 et 15 bis) que la quantité d'eau absorbée croit d'une façon générale, avec le dosage en charge. Mais, toutes choses étant égales par ailleurs, le système A conduit à de plus fortes absorptions d'eau que le système X, dont les quantités mesurées ne dépassent en aucun cas 10 mg/cm² alors qu'avec le polymère A + calcaire elles atteignent jusqu'à 50 mg/cm².

Par ailleurs la figure 16 montre, que les adhérences maximum obtenues avec les divers polymères, correspondent à des pourcentages de charges qui conduisent à des absorptions d'eau très limitées, de 3,5 m à 7,5/cm².

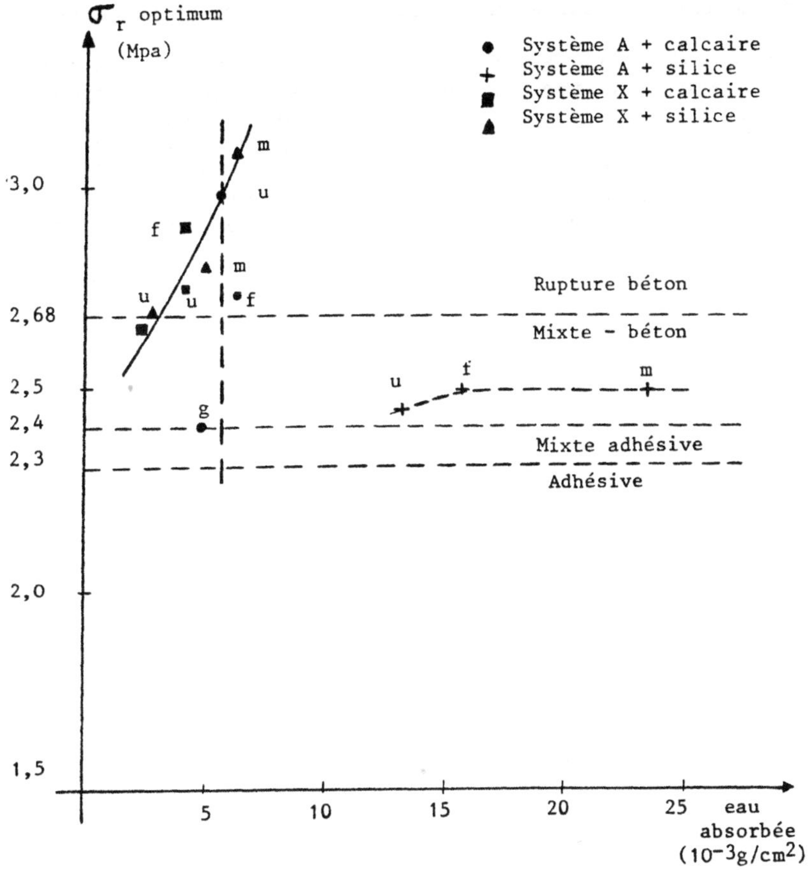

Fig. 16. Contraintes maximales à la rupture en fonction de l'absorption d'eau

Ceci est très marqué pour le système epoxydique A où tous les points maximum se trouvent placés aux environs de 5mg/cm², valeur très faible, si l'on tient compte de l'aptitude élevée à l'absorption d'eau que présente ce polymère chargé. Ces résultats semblent montrer que l'effet bénéfique produit par le pouvoir absorbant de la charge sur l'adhérence est limité et qu'une trop grande quantité d'eau transférée du support au polymère peut être nuisible.

Conclusions

La charge introduite dans un polymère joue un rôle important dans son adhérence sur le béton humide, du fait quelle absorbe l'eau superficielle et permet un contact plus direct du polymère sur le subjectile.

Par ailleurs il importe d'étudier la compatibilité polymère-charge sans la limiter à la nature minéralogique et pourcentage de cette dernière, mais en l'étendant aussi à la granularité.

Il faut aussi signaler que l'absorption de l'eau superficielle par le système, polymère plus charge, n'est bénéfique, du point de vue de l'adhérence, qu'à la condition que la quantité d'eau "déplacée" soit limitée et ne perturbe pas de façon notable, les caractéristiques du polymère durci.

Section 9

Glueing of steel to concrete

Collage de l'acier au béton

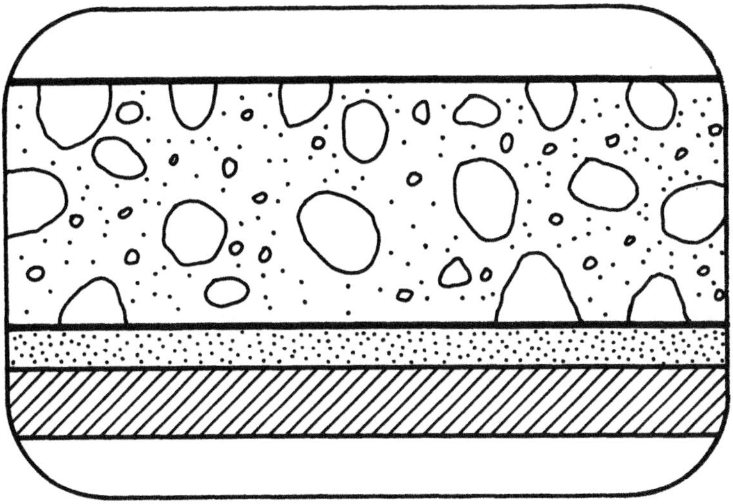

IMPORTANCE DU MODULE DE RIGIDITE DU POLYMERE DANS LES COLLAGES STRUCTURAUX

C. CARON
Laboratoire Géatec - Venelles - France

Résumé
Le paramètre "module de rigidité" est très rarement pris en compte dans le choix de la colle d'assemblage acier-béton ou béton-béton. Plus exactement on estime que plus ce module est élevé meilleur sera l'assemblage. Contrairement à cette opinion courante, nous pensons qu'il est des cas où une minoration raisonnable du module de rigidité est souhaitable pour la tenue de l'assemblage. Ces cas nécessitant des colles "assouplies" sont ceux des ancrages de grande longueur et forte capacité.
Mots clés : Collages structuraux, Colle époxyde, Module de rigidité, Module d'élasticité, Ancrage, Acier lamellaire collé.

1. Introduction

Bien que le module de rigidité de la colle ait une influence incontestable sur la valeur d'un assemblage (cf. les études de l'Hermite, Bresson, Gilibert ...), ce paramètre n'est généralement pas pris en compte dans le choix de la colle d'assemblage et, en conséquence, bien rares sont les formulateurs qui présentent une gamme étendue de colles à module variable.

Or l'interprétation de la formule de base sur la résistance d'un assemblage (§ 2 et 3) permet déjà à elle-même de prouver qu'il faut "moduler ce module" en fonction de la géométrie de l'assemblage.

Dans les exemples d'application que nous décrivons aux § 4 et 5 nous explicitons les raisons qui nous ont fait changer le module G de la colle d'assemblage acier-béton : grande longueur de l'ancrage dans un cas, forte température de service dans l'autre.

2. Rappel des formules de la résistance d'un assemblage collé

En première approximation on admet que la limite de rupture FR d'un assemblage de deux pièces collées est proportionnelle à la surface collée et à la résistance au cisaillement de la colle :

$$FR = \tau S.$$

La relation exacte est bien plus complexe. Elle est donnée par la formule fondamentale de Volkersen :

$$FR = \tau l' \frac{th\left(l\sqrt{\frac{G}{E_p e}}\right)}{\sqrt{\frac{G}{E_p e}}}$$

formule dans laquelle :

FR est la force de rupture au cisaillement de l'assemblage,
τ est la contrainte limite au cisaillement de la colle,
th est la tangente hyperbolique,
l,l' sont les longueur et largeur du collage,
G est le module de rigidité de la colle,
e est l'épaisseur du joint de colle,
E_p est un terme indépendant du collage et intégrant les modules d'élasticité et les épaisseurs des deux pièces à assembler.

3. Conséquence

Si l est petit (cas des scellements courts) $th\left(l\sqrt{\frac{G}{2E_p e}}\right)$ est très voisin de $l\sqrt{\frac{G}{2E_p e}}$ et par conséquent l'expression devient : FR = τll'. On retrouve la formule simplifiée classique FR = τS telle qu'on la mesure dans les essais normalisés en laboratoire. Le seul critère pour la colle d'assemblage est bien sa résistance au cisaillement qu'il faut choisir la plus élevée possible.

Mais plus l augmente, plus G influe sur FR et la relation FR = f(l) cesse d'être linéaire.

De nos expériences de laboratoire et in situ il ressort que, pour des ancrages de plus de 10 cm de longueur, la colle optimale n'est pas celle qui a le τ le plus élevé (d'ailleurs peu variable d'une colle commerciale à l'autre) mais celle qui a le rapport τ/ G le mieux adapté à la longueur de l'ancrage.

Le graphique ci-après explicite ces résultats. Sur ce graphique :

la courbe OA représente la relation normale FR = f(l) conforme à la formule de Volkersen,

la droite OB représente la relation idéal FR = f(l) correspondant à une proportionnalité FR et l quelle que soit la longueur de l'ancrage,

les courbes OC et OD représentent deux types de courbes que l'on obtient en minorant le module de rigidité. Comme cette minoration entraîne une légère diminution de τ ce type de colle est moins performant pour les ancrages courts mais devient plus performant pour les ancrages longs.

De belles réalisations ont prouvé le bien-fondé des résultats précédents. Les deux exemples suivants illustrent des cas où nous avons été amenés à préconiser et concevoir des collages acier-béton avec des époxydes dont G s'écartait notablement du classique 1 200 MPa.

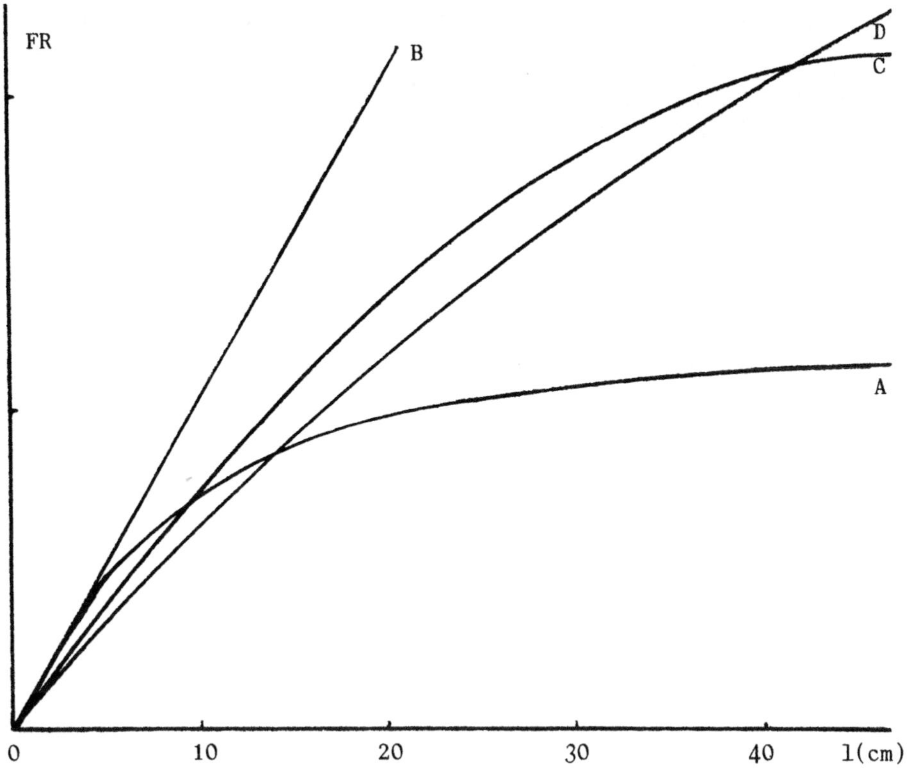

Résistance d'un ancrage en fonction de sa longueur scellée

4. Renforcement de fondations à l'aciérie SOLMER

Il s'agissait de reprendre des efforts de 50 tonnes par ancrage. Chaque ancrage avait un diamètre de 10,5 cm et une longueur de 48 cm. Avec une colle conventionnelle (à haut module élastique) on aurait eu une concentration des contraintes sur les dix premiers centimètres de l'ancrage, signifiant une contrainte de cisaillement de $\frac{50\ 000}{10,5\pi \times 10}$ = 150 bars (15 MPa), ce qui est impensable.

Par contre en assouplissant la résine époxyde on pouvait avoir une répartition presqu'homogène des contraintes sur toute la longueur de l'ancrage, faisant ainsi travailler la colle à $\frac{50\ 000}{10,5\pi \times 48}$ = 32 bars (3,2 mPa), ce qui est réaliste et a été réalisé.

5. Renforcement par acier lamellaire collé à la centrale nucléaire de Tricastin

Il s'agit d'un exemple où, à l'inverse, il était souhaitable d'augmenter le module de rigidité de la colle. Comme ce renforcement ne devait être opérationnel qu'en cas de surchauffe accidentelle du réacteur, la formule de collage devait être étudiée par une température de service de 140°C, température à laquelle on a une nette minoration des modules d'élasticité et de rigidité.

Aussi la formule époxyde a été calculée intentionnellement avec un G plus élevé que le standard, de façon à ce qu'il revienne à 1 200 MPa à 140°C.

THE ADHESION OF RESIN CONCRETES TO REINFORCING STEEL IN REINFORCED CONCRETE ELEMENTS

M. FIERTAK and T. BRONIEWSKI
Instytut Materiałów i Konstrukcji Budowlanych,
Politechnika Krakowska

Abstract
Some experiments have been done in which the adhesion on the resin concrete phase boundary - steel bar was marked. Bars of epoxy resin and polyester rollers were stretched. There were several variants of steel bars of different level of corrosion and surface cleaning as well as different kinds of anticorrosion coating.

None of the variants showed a considerable technique-wise reduction of adhesion. In some cases a significant increase of adhesion was observed.
Key words: Adhesion, Reinforced concrete, Corrosion, Reinforced steel.

1. Introduction

The reinforced concrete elements damaged by corrosion, t.e. bridge beams, floor slab ribs in industrial plants, etc., are often repaired and reinforced with resin concretes. It has been noticed that a new resin concrete cleading of the exposed and corroded reinforcement in the element under repair does not provide an efficient protection. Therefore it was suggested to apply, before the resin concrete, a coat of the rust preventive bottom paint onto the exposed surface of the reinforcement bars. This, however, raised some doubts whether such a coat does not reduce the required adhesion of the resin concrete to steel. Hence the aim of this paper is to evaluate the adhesion of resin concretes to the reinforcement steel bars with various methods of preparation of the steel / resin concrete phase boundary/.

2. Description of investigations

2.1. Tested materials

The adhesion at the resin concrete / bright bar phase boundary was examined for the polyester and epoxy concretes described in Table [1] and for the Ø 8 mm dia. bright bars of R_a = 310 MPa. The bars were prepared in several variants.

Different degrees of corrosion and surface cleaning were included:

I - uncorroded bars
II - corroded bars
III - corroded bars after cleaning with rust remover
IV - corroded bars cleaned with nitric acid, and various types of anticorrosive bottom paints:

- a naked bars
- b bars coated with phthalein paint with ferric pigment
- c bars coated with epoxy paint with amine hardener.

According to the symbols accepted, f.ex. symbol IIIc denotes the examined corroded bar cleaned with a tannin rust remover and coated with an epoxy paint.

Table 1. Characteristic of resin concretes

Type of resin concrete	Binder % by weight	Aggregate % by weight		Apparent density ϱ_o [kg/m^3]		Compression strength R_c [MPa]		Tensile strength R_r [MPa]	
		Type	% by weight	ϱ_o	ν_{ϱ_o} [%]	$\bar{R}_c$	ν_{R_c} [%]	$\bar{R}_r$	ν_{R_r} [%]
epoxy	Epidian 5 amine hardener 15 %	Sand 0-2 mm granite 2-10 mm	-34 -51	2000	0,79	65,6	5,80	8,06	7,60
polyester	Polimal 109 ketonox cobalt naphtenate 15%	Sand 0-2 granite 2-10 mm	32 53	2110	0,75	49,4	5,28	6,07	4,31

2.2 Methods of investigations

From among the different methods of investigations described in literature and concerning the adhesion of reinforcing steel to concrete [1, 2, 3], the following types of tests are most often used:

- "pull-out" test in which the bar is subjected to tension and the concrete to local thrust,
- "pull-out" test in which the bars are embedded in concrete /contact or overlapping type of embedding/ and the concrete is also subjected to a tensile force,
- tests based on observations of the formation of cracks in a sample of concrete enclosing the bar subjected to tension,
- compression test - the steel and concrete are subjected to compression,
- bending test - the test consists in loading of the beam until failure of the adhesion between steel and concrete occurs; the value of the force, which causes cracks, is important here.

The test, which consists in pulling out from the sample of concrete a bar embedded in it, is not the best way to represent the real state of stresses around the bar and in the bar, since the die is compressed which, in turn, causes the formation of the forces increasing the adhesion. The pulled out bar suffers a reduction of area, which decreases the adhesive stresses. Another drawback of this method is also a non-deformable resistance of the machine, on which the sample of concrete is resting during the operation of pulling out the bar. This results in a local concentration of stresses in the extreme sections.

The advantages of this method consist in the simple and easy recording of the bar displacement. During the test the displacement of the free end of the bar is measured, and the force, corresponding to this displacement, is determined. If the pre-determined conditions of the test are strictly observed, it is possible to compare in a simple way the conventional values of adhesion.

The standard specimen, to be used in this method, was elaborated by Paduart [1]. Due to the above mentioned reasons, in the present investigations the method of pulling out the bar from the specimens of concrete was applied. The standard specimens developed by Paduart [1] were used.

According to Paduart, the adhesion was defined as a mean value of the shearing stress, calculated from the three values of the force pulling out the bar with the displacements of the free end of the bar amounting to 0,1 mm; 0,25 mm; 1,00 mm.

2.2.1 Test elements

As test elements, rectangular prismes of dimensions 10x10x9 cm were accepted. They had central, length-wise reinforcements with bars of a total length amounting to 25 cm. The length of specimens amounting to 9 cm resulted from the use of a 1 cm thick washer, which enabled centring of the bar and extending it outside the concrete sample to make possible the measurement of the displacements.

For the accepted specimens, the ratio between the bar diameter and the minimum lateral dimension of concrete sample amounted to 1:12.5.

2.2.2 Description of tests

Applying the method of pulling out from the sample of concrete a bar embedded in it, the adhesion was determined on a tensile testing machine.

The values of the force pulling out the bar were recorded; the force corresponded to a displacement of every 1 μm of the unloaded bar and in respect to concrete. The values of 1 μm, 10 μm, 100 μm were accepted as values deliminating certain stages of the loss in adhesion. And so:

1 μm - corresponds to the moment in which there is a biginning of the slip,

10 μm - it has been assumed that until this moment the adhesion was not disturbed,

100 μm - after acceeding this value, the adhesive forces stop acting.

From the analisis of a relation ship between the slip of the unloaded bar end and pulling out force P it follows that the diagram is of a linear nature until the displacement of about 10 μm is reached. For displacements above 10 μm, an increase in the pulling out force is noted to take place. When the value of 100 μm is reached, it can be assumed that further increase in the force is relatively small.

3. Results of tests

The failure of test pieces occured in 100% due to the destruction of adhesion between the steel and concrete. This enabled an observation of the course of a relationship $\tau = f/u/$, i.e. a relationship between the force pulling out the bar from concrete in function of the displacement of the free bar end, as well as the determination of a maximum adhesion.

The obtained results are compiled in Table 2. The values of adhesive stresses are given during the displacement of the free bar end, amounting to $u = 1\,\mu m$ [τ_1], and maximum stresses τ max.

The relationship $\tau = f/u/$ is shown in Fig.1 for the epoxy concrete and in Fig.2 for the polyester concrete.

Table 2. A comparison of the results of adhesion tests

Type of reinforcement		Type of concrete	Mean value of stresses MPa τ_1	MPa τ_{max}
1		2	3	4
Ia	Uncorroded bar	polyester epoxy	0,55 0,12	2,73 7,55
Ib	Uncorroded bar coated with minimum phthalein paint	polyester epoxy	0,62 0,41	4,41 6,08
IIa	Uncorroded bar coated with phthalein paint with ferric pigment	polyester epoxy	2,74 1,18	7,89 2,77
Id	Uncorroded bar coated with chemoresistant, two-component epoxy paint with amine hardener	polyester epoxy	0,52 0,99	6,16 6,81
IIa	Corroded bar	polyester epoxy	0,40 0,62	13,05 12,39
IIb	Corroded bar coated with minium phthalein paint	polyester epoxy	0,69 0,09	9,85 8,40
IIc	Corroded bar coated with phthalein paint with ferric pigment	polyester epoxy	1,43 3,17	10,69 8,55
IIIa	Corroded bar cleaned with cortanine	polyester epoxy	0,91 8,25	6,75 8,25
IIIb	Corroded bar cleaned with cortanine and painted with chemo-resistant, two-component epoxy paint	polyester epoxy	0,21 7,06	7,83 7,06

Table 2. cntd

1		2	3	4
IVa	Corroded bar cleaned with nitric acid	polyester epoxy	1,16 0,65	11,97 11,19
IVb	Corroded bar cleaned with nitric acid and coated with phthalein with ferric pigment	polyester epoxy	3,46 1,49	10,41 7,27
IVc	Corroded bar cleaned with nitric acid and coated with chemo-resistant, two-component epoxy	polyester epoxy	0,50 0,56	10,03 9,16

An analysis of the results of the investigations enables us to see that in the case of pure steel /variant I/, coating of the steel with phthalein and epoxy paints increased the adhesion to the polyester concrete. Below the relationships are given between maximum stresses τ max and stress τ_0 max / maximum stress for pure steel/

	Polyester Concrete	Epoxy Concrete
Bar coated with phthalein paint /Ib/	τ max = 1,61 τ_0 max	τ max = 0,82 τ_0 max
Bar coated with with epoxy paint /Ic/	τ max = 2,89 τ_0 max	τ max = 0,90 τ_0 max

In the case of epoxy concrete, coating of the bars with anti-corrosive paints resulted in a slight decrease of the adhesive stresses.

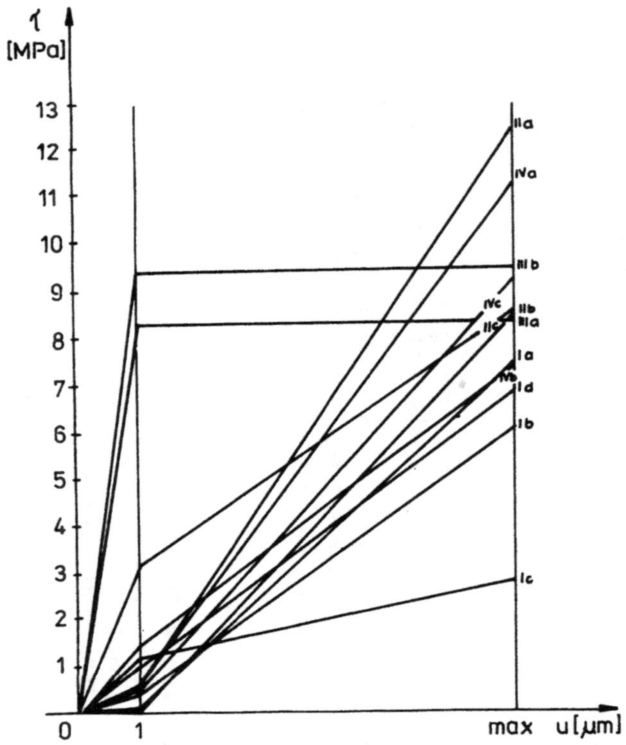

Fig 1. The relationship $\tau = f(u)$ for the eposcy concrete

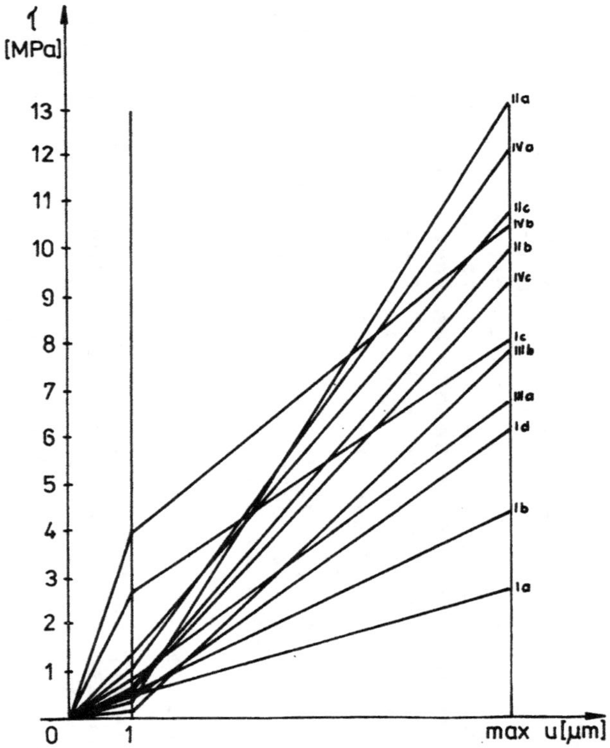

Fig 2. The relationship $\tau = f(u)$ for the polyester concrete

In variants IIb and IIc /corroded bars protected with phthalein and epoxy paints/, coating with paints of both the epoxy and polyester concretes did not reduce the adhesion in respect to IIa /corroded bars without paint/. Hence it follows that no confirmation has been found for the suggestions that coating of the corroded bars with anti-corrosive paints reduce the adhesion to the resin concretes.

The corroded bars, cleaned with a tannin rust remover and then painted /variants IIIa and IIIb/, are characterized by a slightly lower adhesion to the resin concretes than the bars without any anti-corrosive protection. Only in the case of polyester concrete reinforced with bars coated with an epoxy paint, an increase in the adhesion was noted to take place: $\tau_{max} = 1.16 \, \tau_{t\,max}$, where $\tau_{t\,max}$ is adhesive stress at the phase boundary: polyester concrete / corroded steel bar cleaned with a tannin rust remover.

In variants IVa, IVb, IVc, the best adhesion of both types of concrete was obtained in the case when the bar was only cleaned with nitric acid.

4. Conclusions

1. Contrary to the commonly encountered opinions, the adhesion of resin concretes to the reinforcement bars coated with anti-corrosive paints is not, generally speaking, lower from that of the unprotected bars.
2. From the point of view of an anti-corrosive protection, the following solution seems to be advisable:
   - mechanical renoval of the rust from the reinforcement bars
   - additional removal of rust with the tannin rust remover,
   - protection from corrosion with an anti-corrosive paint, with the following re-filling of the losses in a concrete element with the resin concrete.

From the point of view of adhesion, the above described investigations have proved that there are no counterindications for an application of this solution.

References

RILEM Stockholm 1957, Symposium on bond and crack formation in reinforced concrete. vol.4

Takaku, A., Arridge, R.G.C /1973/ The effect of interfacial radial and shear stress on fibre pull-out in

composite materials. J Phys D, Appl Phys, vol. 6, pp 2038-2047
Chanis, C.C. /1974/ Mechanics of load transfer of the interface, Lawrence J., Brautman R., Koch H., Composite Materials, Academic Press, New York pp 31-77

LONG-TERM PERFORMANCE OF EPOXY BONDED STEEL-CONCRETE JOINTS

D. VAN GEMERT
M. VANDEN BOSCH
Department of Civil Engineering,
K.U.Leuven

Abstract
The technique of strengthening reinforced concrete structures by means of epoxy bonded steel plates becomes more and more familiar in renovation industry. It shows to be a reliable and fast repair method, not only under normal laboratory conditions but also under atmospheric and sustained loading conditions. The creep effects in the epoxy joint are very limited, without affecting the mechanical strength of the connection. The preparation of the elements and the execution of the repairing operation turns out to be of great importance. At high temperatures, a weakening of the epoxy-joint appears which calls for insulating measures.
Key words: Epoxy, Reparation, Durability, Strengthening, Concrete, Structures.

1. Introduction

The causes of concrete damage are manifold. Damage as a result of mechanical effects is most striking, for example damage due to accidents, to overloading, to fire, to explosion or to earthquake (ref. 1, 2, 3). Faulty calculations or poor execution usually lead to excessive deflections and severe cracking in reinforced concrete elements. Chemical and galvanic corrosion, however, is the most harmful process influencing durability of reinforced concrete. The effects of carbonation of concrete on the steel reinforcements embedded in the concrete are well known. The chloride ions are able to destroy locally the passivated layer, even in uncarbonated concrete. If the steel reinforcements are corroding by the presence of chlorides, then no superficial remedial treatment is available at the moment.

In a wide variety of structural damage cases as described above, the repairing or strengthening of reinforced concrete structural elements by means of epoxy bonded external reinforcements offers an elegant solution. Research on this subject was started by R. L'Hermite and J. Bresson (ref. 4). The principle of the method is quite simple: steel strips or steel elements with complex geometries are glued to the concrete surface by means of a two-component epoxy adhesive, creating a concrete-glue-steel composite element. A practical example is shown in Fig. 1. It shows the strengthening plates at the bottom of a R/C plate, damaged by explosion and fire.

Fig. 1 : R/C plate, strengthened by means of epoxy bonded, external strips

Although the principle of the method is very simple, the application of the technique in repairing practice does call for care and accuracy. The preparation of the concrete and steel surfaces as well as the bonding operation itself, must be carried out with great care because they are of great influence on the durability of the glued connection. Until now only epoxy adhesives are convenient for gluing steel to concrete. They show an excellent bond both to concrete and to steel, their shrinkage is practically zero, and they are highly chemically inert. The preparation of the steel surface consist in grit-blasting to level SA3 and degreasing of the area in contact with the glue. Also for the concrete surface, a grit-blasting is preferred. In ref. 5 the quality control problem of the appropriate epoxy glues and the conditions of application are discussed. The static short time loading behaviour of the epoxy repaired reinforced concrete elements is excellent. This was confirmed by a great number of real scale tests in the laboratory. Of great importance however is the behaviour of the repaired structure under sustained or cyclic loading, its resistance to high temperatures and its durability in the outside climate. In an experimental program, set up in the Reyntjens Laboratory of the Civil Engineering Department of the K.U.Leuven in 1980, these characteristics are studied. Part of the results of these tests are discussed here.

2. Durability tests

2.1 Resistance to high temperatures

The cold hardening epoxy glue has a poor thermal resistance. The translation of the above statement into practical figures was done by full-scale temperature loading tests in the temperature range from -20°C to +90°C (Fig. 2).

Fig. 2 : Durability test procedure for high and low temperatures

The epoxy adhesive EPICOL U (Resiplast N.V.) was used in the tests. Table 1 summarises the results of these tests.

Beam nr.	Temperature (°C)	Total load at fracture (kN)
4	-20	75
5	20	80
2	50	80
3	80	50
6	90	40

Table 1 : Ultimate loads at different temperatures

It appears that for lower temperatures there is no decrease of the ultimate load. At higher temperatures, however, the situation is different. From a temperature of about 65°C, the glue is becoming weaker and more deformable. This is clearly illustrated by the different crack pattern. At lower temperatures, the crack always starts at the end of the plate and moves on into the concrete (Fig. 3). At higher temperatures, the epoxy joint is not able to transfer the shearing stresses from the steel plate to the concrete, and a crack propagates through the epoxy joint, starting at the plate end (Fig. 4). The safety of the reparation is strongly reduced at high temperatures. In most practical cases, a dowel is placed at the end of the steel plates which is anchored in the concrete. This supplementary safety element will prevent a fast and sudden loosing of the steel plate reinforcement. The application of an insulating cover such as a bituminous felt, or other fire protective measures will be necessary if fire resistance is required.

Fig. 3 : Crack pattern at normal temperatures

Fig. 4 : Crack pattern at higher temperatures

2.2 Cyclic thermal loading tests
A test beam, having the same dimensions as above, was subjected to successive temperature cycles. Firstly, 7 cycles were executed between 20°C and 70°C. Each temperature was holded for 12 hours. Afterwards, 3 cycles were executed with the highest temperature increased to 75°C.

Each time, just before the temperature was changed, the test beam was loaded in a four point bending test - total load of 45 kN - and during the test the stress distribution in the external steel strip as well as the bending stiffness of the beam were measured. The test was aimed to reveal a thermal degradation of the bond. The results of the measurements at 70°C/75°C are indicated in Table 2.

Cycle nr.	Maximum shear stress (N/mm^2)	Maximum tensile stress (N/mm^2)
1	0,75	78,12
2	0,72	78,54
3	0,72	75,29
4	0,74	74,76
5	0,73	75,39
6	0,73	76,23
7	0,76	77,60
8	0,77 (75°C)	79,17 (75°C)
9	0,78 (75°C)	78,33 (75°C)
10	0,79 (75°C)	78,02 (75°C)

Table 2 : Results of the measurements at the high temperature

The results indicate that a limited number of severe thermal loadings do not have a significant effect on the effectiveness of the bond. However, the results obtained are very preliminary, and further research is needed to get them confirmed.

2.3 Tear-off tests on small cylinders

One of the methods applied to check the durability of the glued elements under atmospheric conditions, is the tear-off test on small steel cylinders, glued to the concrete surface. A great number of steel cylinders were glued on a concrete plate in the laboratory, and exposed to the outside climate. Table 3 gives the results of these tear-off tests. The cylinders were glued on 1981-12-09. Fig. 5 shows a cylinder when tearing-off of the concrete plate. These results show no remarkable loss of strength of the glued connection during the first two years.

Testing day	Mean tensile strength (N/mm^2)	Variance (N/mm^2)	Number of cylinders
1981-12-17	2,37 (= reference test)	0,38	6
1982-06-08	2,73	0,83	6
1983-06-08	2,54	0,78	5
1984-01-20	1,78	0,30	5
1984-07-31	1,83	0,75	5
1985-04-02	1,82	0,57	5
1986-01-28	2,73	0,37	5

Table 3 : Mean tensile strength of cylinders glued in the laboratory and exposed to the atmospheric conditions.

The measurement of January 1984 and the two following ones show a small decrease of the mean value. However, this mean value is always calculated from a small series of tests. The result of the last test shows that we may assume that this decrease is not a continuous one

Fig. 5 : Cylinder when tearing-off of the concrete plate

but the result of taking several cylinders with an accidental lower quality.

At the occasion of a repair-work on two bridges a great number of cylinders were glued on the concrete beams. The tear-off tests showed the following phenomenon : after the first six months the tear-off strength did not change. At the tear-off tests the cracks were always in the concrete. After one year, however, the tear-off strength of some cylinders was practically reduced to zero. For these cylinders, cracking always appeared in the contact surface steel-glue. The reason for this phenomenon lies in the preparation of the steel surfaces, which were only sawed and degreased, but not grit-blasted. At the cutting of the steel cylinders continuous cutting lines are carved in the surface. These lines are preferential ways for the water to penetrate between the steel and the glue, as a result of which the steel gets rusty and the bond between steel and glue decreases. With sand-blasted surfaces this phenomenon does not occur. By sand-blasting the surface roughness is composed of a great number of hills and valleys, which constitute an insurmountable barrier for the water.

2.4 Rheological behaviour of a glued connection

The rheological behaviour of an epoxy bonded steel-concrete element is studied on a concrete plate, only reinforced with two epoxy bonded steel strips (Fig. 6). The loaded plate is shown in Fig. 7. The load, corresponding to the normal service load of 9,6 kN, was applied on a 30 days old concrete, whilst the glue had cured for 7 days in laboratory conditions. The creep test started at the 5th of November 1981. The variation of deflection and strains with temperature and relative humidity is surveyed on several places. Two strain measuring bases are placed in the compression-zone, three others in the tension-zone. The remaining eight measure the elongations in the end-zone of the glued plates : four between concrete and concrete, the four others between steel and concrete.

Figure 8 shows the variation of the deflection and the temperature in function of time. The last results date from 10 March 1986.

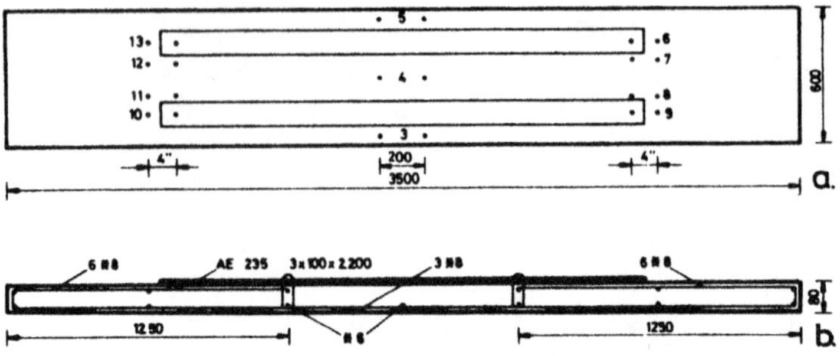

Fig. 6 : Concrete plate with internal and external steel reinforcement. The numbers indicate the different measuring bases

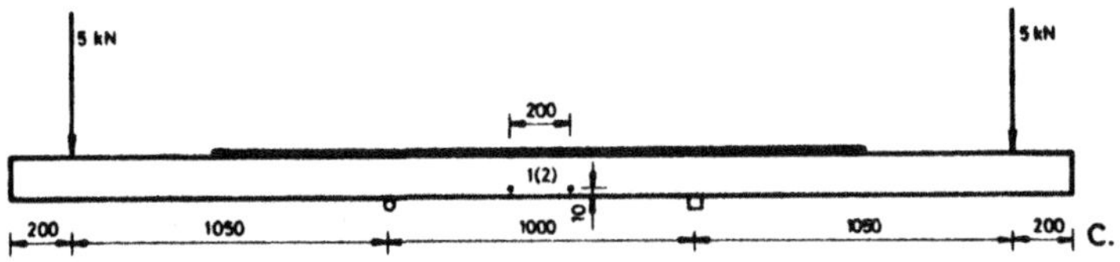

Fig. 7 : Loaded plate. Numbers 1 and 2 indicate the measuring bases in the compression zone

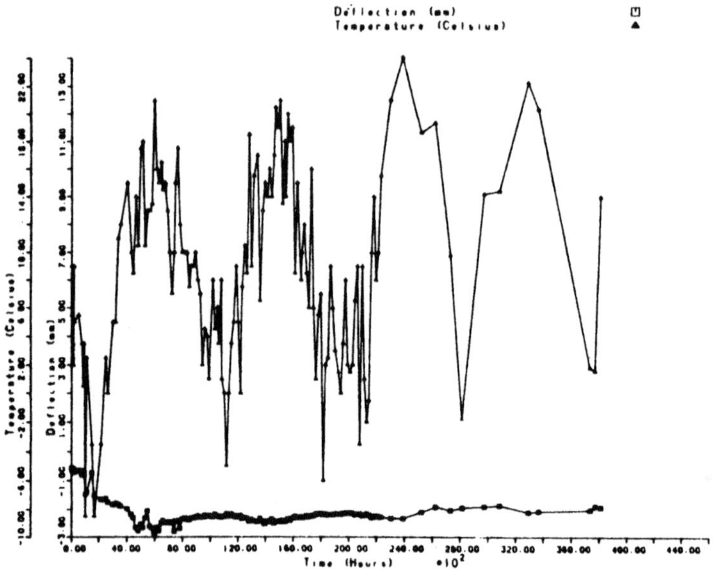

Fig. 8 : Deflection and temperature in function of time

From this figure it can be seen that the plate got an important increase of the deflection simultaneously with the rising temperatures after the winter of 1981. An analogous strong variation was not measured since, but only small variations of the deflection due to a combined influence of temperature and relative humidity.

Figure 9 shows the evolution of the tensile and compressive strains.

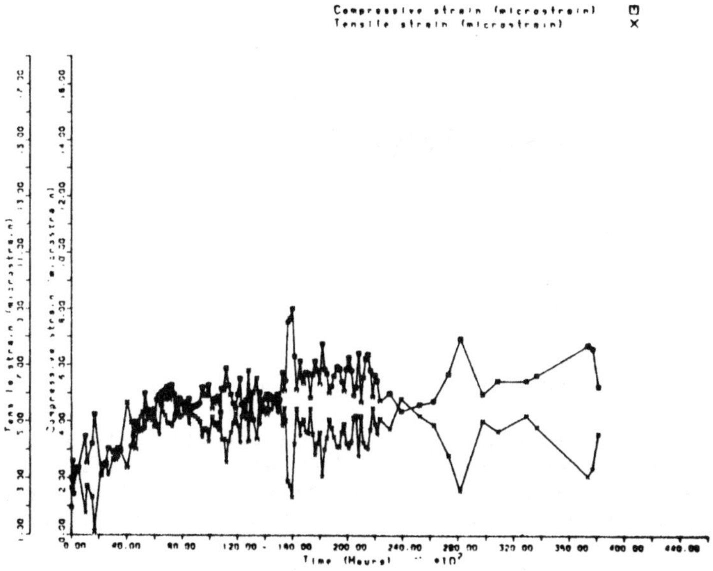

Fig. 9 : Compressive- and tensile strain

During the last years the interval between the measurements was taken greater than in the first years. It can be seen that the plate behaves as a whole under changing atmospheric conditions : with increasing tensile strains, the compressive strains decrease and vice-versa. Figure 10 gives the strains measured with a measuring base between points on the steel plate end and the concrete and between concrete and concrete. Here too an important increase of the strain between steel and concrete can be noticed at an age of about 4500 hours. Later on, only small variations take place. This indicates that the first warm period has an important effect on the behaviour of the glue, without influencing its mechanical characteristics.

3. Conclusions

In all the experiments in the laboratory and in all the practical applications, the epoxy bonded external steel reinforcements have proved to be reliable structural elements. The executed investigations on the behaviour of the glued connection under

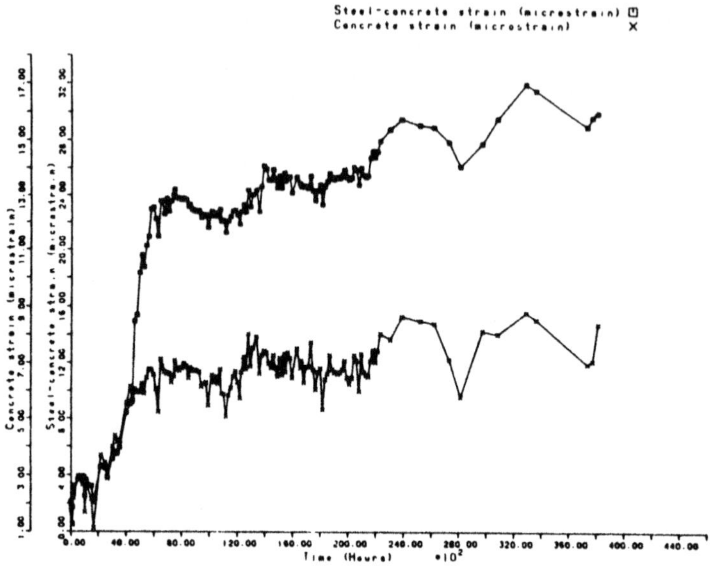

Fig. 10 : Steel-concrete and concrete-concrete strain

atmospheric conditions also show no considerable influence on the mechanical characteristics. This, however, depends for the greater part on the preparation of the concrete and steel plate surface and of the execution of the gluing works. Therefore, to be sure of an optimum and durable result, more specialised personnel and an intense control of the reparation works will be necessary. Until now, a disadvantage is the poor behaviour of the glued connection at higher temperature. More research in improving the fire resistance and on the behaviour in cyclic thermal loading tests will be necessary.

References

1. Van Gemert, D. and Maesschalk, R. (1985) Structural repair of a reinforced concrete plate by epoxy bonded external reinforcements. International Journal of Cement Composites and Lightweigth Concretes, Vol. 5, nr. 4, pp. 247-255.
2. Van Gemert, D. and Maesschalk, R. (1981) Réparation des dommages, occasionnés à des dalles en béton armé par des explosions, au moyen d'armatures collées. Info-Technics, June 1981, pp. 17-19.
3. Van Gemert, D. (1981) Vérification expérimentale des renforcements par plaques collées d'un pont en béton précontraint. Colloque International Bruxelles-Paris: Gestion des ouvrages d'art, Vol. II, pp 515-521.
4. Bresson, J. (1971) Nouvelles recherches et applications concernant l'utilisation des collages dans les structures. Annales de l'I.T.B.T.P., France série BBA/116.

5. Van Gemert, D. (1983) Execution and control of epoxy bonded steel reinforcements (in Dutch). S&E-Publications, nr. 5, Nederlandse Betonvereniging, The Netherlands.

A CONTRIBUTION IN REGARD TO BEHAVIOUR OF REINFORCED TEE-BEAMS WITH BONDED FLATSTEEL COMPONENTS UNDER FATIGUE LOADS

K. HOLTGREVE
Research and Development Group (TuP), Strabag Bau-AG, Head Office Cologne (FRG)

Abstract
Adopting the technique of glueing flatsteel components to existing rc-structures for the purpose of modifying the loadbearing capacities of structural systems in case of changing demands has become a common method. Regulations of such techniques for the FRG are covered under Certificate No Z-26.1-1 of IfBt, Berlin since october, 1979. The method has to date been limited to predominantly static loads according to DIN 1055 P.3, thus excluding special cases of impact loads, frequent repetition of loads or action of unbalanced machinery or other dynamic live loads. It is obvious that a reasonable number of structures are exposed to dynamic load conditions. Adopting the glueing technique as an economic method of preserving existing structures is a future demand.
The forementioned certificate will be reissued in the near future to allow for dynamic loads. The revisions are based on the encouraging results of fatigue tests with subsequent ultimate static bending of tee-beams in the EMPA-laboratories for material testing and research of Switzerland. Some eminent testing results will be described to prove the efficiency of the method.
Key words: Concrete strengthening, Bonded Steel, Fatigue tests.

1. Introduction

Strabag Bau-AG as a user of the technique, in cooperation with the designing civil engineers Tausky-Leu-Müller, CH-Zurich and Ciba-Geigy AG, CH-Basle as the manufacturers of Araldit (R) epoxy resin components have planned and authorized comprehensive tests. The aim was to examine the general behaviour of bonded steel under conditions of frequent repetitive loads. In addition, tests had to verify the subsequent behaviour under static loads in progress to the ultimate state of testing members.

2. Testbeams

Figure 1 shows typical section properties and reinforcing details of the testbeams under consideration. Marks (1) and (2) indicate location of the additional structural steel components.

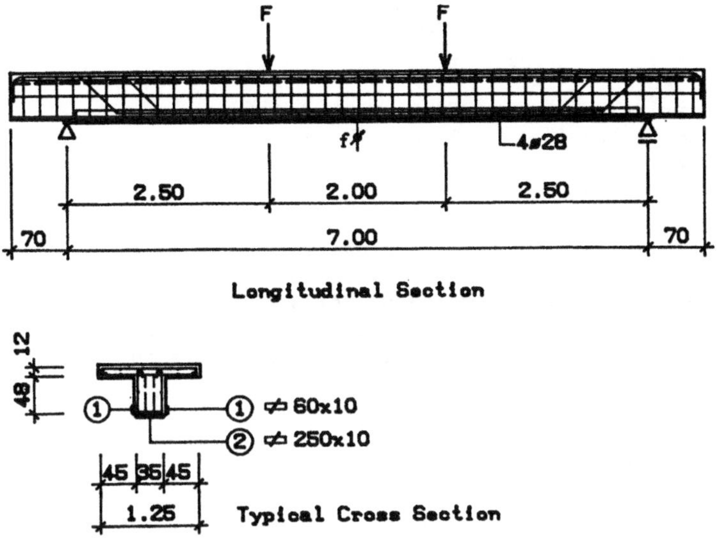

Fig. 1 System and material properties of testbeams

Table 1. Materials.

Description	Designation	fc;fy/fu (N/mm2)	DIN
Concrete	B 25	30	1045
Reinforcing bar	BSt III	420/500	1045/488
Structural steel	St 37-2	235/340	17100
Epoxy resin (Araldit)	XB 3074 (A+B)		

Before glueing works, the test beams were stressed by a pair of concentrated loads of 1oo kN each. Arranged symmetrically about midspan with a 2.00 m spacing, these loads were acting with the aim to initiate cracks and further exclude the cooperation of concrete in the tensile zone during later testing.

The achievement of satisfactory results presumes careful preparation. Roughening of concrete surfaces depending on the results of pull-off tests and cleaning with compressed air; sandblasting, double priming of structural steel and removal of fat immediately before application of resin; disposition of resin in compliance with the ambient temperatures and necessary tools to verify quality control, weighting, mechanical mixing and application of resin to structural steel and concrete within its pot life; aids for correct positioning and the adequate pressure of steel towards concrete during installation as well as early controls of perfect joint filling are further obvious requirements.

Table 2. Results of quality control.

Subject	Test	Strength (N/mm2)
Concrete	pull-off	2.2
Resin	compression	74.8
	bending tensile	28.3
	shear	16.4

Glueing of flatsteel to the testbeams was performed during the dead load case. Testing of beams started approximately one week after the application of steel and lasted for eight days.

The geometrical portion of longitudinal reinforcement increased from originally 0.35 to 0.86 % of the 125x60 cm beam section by lamination.

3. Testing procedure

The testing procedure is shown on figure 2. Readings of deflection were taken for midspan and F-locations at bottom of beam web. Load applications succeded in progress between the upper limit of 254 kN and the lower limit of 117 kN for both static preloading and fatigue test, where readings started under consideration of the dead load condition of beams.

Ultimate stress tests continued upon finalization of fatigue tests. Applied loads were increased by power control in the range of quasi-elastic beam behaviour, whilst upon occurance of considerable creep load increments were governed by deformations.

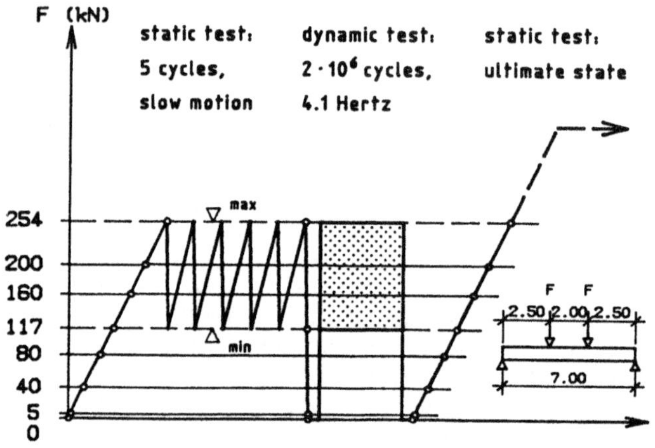

Fig. 2 Schematic graph of testing procedure

4. Results of fatigue tests

The results of fatigue tests may be generally described as follows:

(a) Bond between steel and concrete remained effective during the full testing procedure with insignificant exceptions in small areas.

(b) The increase in deflection due to two million load cycles ranged between 2.3 mm to 2.9 mm.

(c) Insignificant exceptions were detected at side faces of the tee-beam webs.

5. Results of static ultimate stress tests

The force-deflection-diagram (Fig. 3) shows the testing procedure. The early stage of testing is characterized by achieving the reinforcement yield strength in the midsection of span. Load increments reduced in relation. Further deformations of beam were limited, if compared to a ductile behaviour of material bending. Sudden failure of the structural steel laminations, starting with the lower components on bottom of beam web characterize the continuing test. All failures of laminated material occured adjacent to hardened resin components.

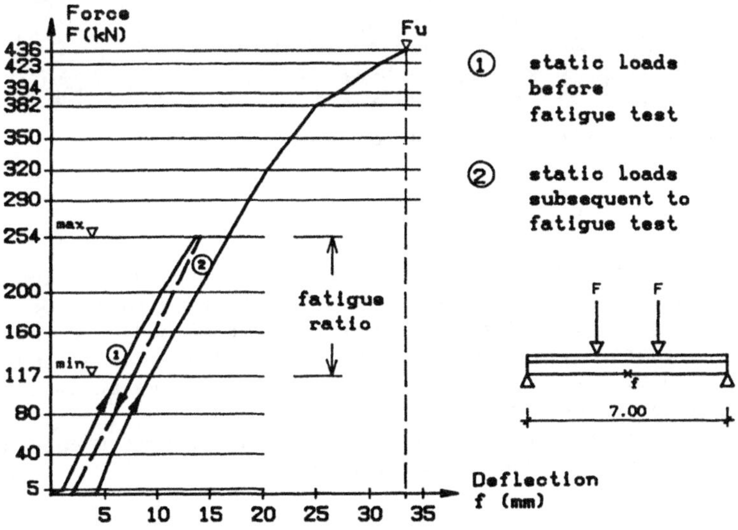

Fig. 3 Force-deflection-diagram

Loosening of the side laminations was activated gradually due to their interlocking into the area of failure.

The maximum structural steel elongations of 0.2 percent were last but one readings. A corresponding yield elongation of 0.14 percent indicates failure at the state of initial plastic deformation.

The following photograph (Fig. 4) was taken after finalization of test. It demonstrates the area of failure in the outer tensile zone of concrete. No failure of resin composite material was detected.

Fig. 4 Beam after static ultimate stress test

6. Conclusion

The application of glueing techniques in constructions under fatigue loads promises to be an alternative method of preserving reinforced concrete in the near future. Some advantages of the method, e.g. neglectible loss of space, "dry" application, minor disturbance during working hours without breaks in production may be decisive arguments in addition to other common methods of repair.

The preceding descriptions of testing results are extracts only of profound investigations in theory and practice.

References

Certificate No Z-26.1-1 (1979) Schubfeste Klebeverbindung zwischen Stahlplatten und Stahlbetonbauteilen, Institut für Bautechnik, Berlin.

Kreitmann, F., Tausky, R., Wolff, R., (1983) Verstärkung von zwei Maschinenbalken für eine Kartonmaschine mit der Methode der "geklebten Bewehrung". Das Papier, 10, 489-492.

Holz, K., (1984) Anschlüsse, Verstärkungen und Konstruktionsänderungen - Flexibel bauen auch mit Stahlbeton. Beton- und Stahlbetonbau, 11, 293-298, 12, 322-327.
Holz, K., (1986) Verstärkung von Betontragwerken des Industrie- und Hochbaus - Bemessung und konstruktive Ausbildung. Österreichischer Betonverein (ÖBV), Fortbildungsveranstaltung Flexibilität im Massivbau.
Strabag Bau-AG, Erhalten, Verstärken, Sanieren mit geklebter Bewehrung. Strabag Report No 5.

COLLABORATION D'UNE ARMATURE COLLEE AVEC LES POUTRES FISSUREES DE BETON ARME

J. JASIEŃKO, J. KARYŚ et A. OLEJNIK
Institut de Génie Civil, École Polytechnique de Wrocław

Résumé
Les polymères et d'abord les époxydes étaient utilisés à joindre l'acier au béton armé pour le renforcer. En renforçant les constructions de béton armé il faut prendre en considération les fissures qui se trouvent dans le béton. Cela peut provoquer une augmentation locale des contraintes de cisaillement dans l'assemblage par collage. Les résultats des essais indiquent que les contraintes de cisaillement formées dans l'assemblage par collage dépassent (dans certains cas) la résistance au cisaillement de cet assemblage.
Les mots clés: Renforcement du béton armé, Armature extérieure, Assemblage par collage, Contraintes de cisaillement, Fissures dans le béton, Établissement d'un projet.

1. Introduction

L'utilisation des matières plastiques dans la construction est liée avec la confiance des ingénieurs en ces matériaux. Un des plus importants domaines d'utilisation est la construction de béton armé, où on utilise les matières plastiques à remplir les fissures dans le béton, à joindre des éléments préfabriqués ainsi qu'à renforcer les constructions qui travaillent en flexion.
On s'occupe depuis 20 ans de la possibilité de renforcer les constructions de béton armé à l'aide du collage de l'armature complémentaire, qui est étirée lors du travail de la construction. Dans certains pays on a même introduit des plans-types des normes pour la realisation du renforcement.

Le profil de l'armature renforçante est calculé d'après les principes de Bernoulli en les limitant avec la réserve de la capacité de charge de la zone comprimée de la section de la construction et avec le pourcentage maximum d'armatures de cette section. La première condition du calcul est la suivante:

la collaboration de l'armature collée au béton, réalisée grâce aux forces d'adhésion de la colle permet de traiter cette armature comme l'armature intérieure.

La deuxième condition du calcul consiste à assurer la longueur nécessaire de l'assemblage par collage. Suivant le modèle du travail de l'assemblage, la plupart des scientifiques considère que la longueur de l'ancrage du couvre-joint renforçant ce sont les tranches extrêmes de dimension de 30 à 100 cm. Dans ces tranches il y a la répartition disproportionnée des contraintes de cisaillement décrite au moyen de la fonction hyperbolique avec la valeur maximum des contraintes à l'extremité chargée de l'assemblage par collage.
La troisième condition de calcul pour le dimensionnement du renforcement, presentée par exemple par Litvinov, concerne la nécessité de contrôler les contraintes de cisaillement dans l'assemblage par collage. Ces contraintes ne doivent pas dépasser la résistance au cisaillement de l'assemblage par collage appliqué entre le béton et l'acier.

L'armature renforçante sous la forme de couvre-joint de tôle ou de bandelette est collée le plus souvent au béton fissuré. L'existence des fissures peut provoquer l'accroissement local des contraintes de cisaillement dans l'assemblage par collage, d'autant plus défavorables quand ils se forment sur les tranches de l'ancrage de l'armature renforçante. Ces accroissements sont très difficils à calculer.

Le problème suivant a attiré l'attention des auteurs - de quelle façon les fissures apparues dans le béton influent sur la répartition des contraintes de cisaillement dans l'assemblage par collage des poutres de béton armé renforcées par le couvre-joint d'acier. C'était notre sujet de recherches.

2. Contraintes et les déformations à l'armature collée aux poutres de béton armé

On a examiné les poutres de béton armé du profilé en T de la longueur de 300 cm. L'armature intérieure des poutres du pourcentage d'armatures (μ) égal à 0,7% était réalisée de barres de 3 ø 8 (15 poutres) et 3 ø 14 - μ=1,8% (9 poutres).
Les poutres étaient chargées statiquement à l'aide de deux forces ponctuelles appliquées dans 1/3 et 2/3 de l'ouverture, qui était égale à 270 cm.
La moitié des poutres étaient chargées jusqu'au niveau 0,5 et 0,9 de la force de destruction, déterminée sur les poutres-témoins. Ensuite après le déchargement,

les poutres étaient renforcées dans la zone de traction à l'aide de tôles d'acier. Les composants de base de l'acier sont les suivants: C=0,20%, Mn=0,58%, Si=0,06%, P=0,04%, S=0,05%, Cr=0,26%, Ni=0,20%, Cu=0,23%.
Le reste des poutres était renforcé sans charge préliminaire (sans fendillement).

Toutes les poutres étaient renforcées (par la colle époxyde) à l'aide de l'armature extérieure sous la forme des couvre-joints de profil de 2x100 mm, 4x100 mm et de longueur 2600 mm. La composition de la colle était la suivante:

- résine époxyde - 100 parties de masse,
- farine fossile siliceuse - 235 parties de masse,
- phtalate de dibutyle - 10 parties de masse,
- durcisseur - 11 parties de masse.

L'épaisseur de l'assemblage par collage était de 2 à 4 mm. Les couvre-joints renforçants ne dépassaient pas l'appui. Cela permettait d'obtenir les conditions réelles du renforcement des poutres.

Lors des essais on a mesuré les déformations de l'armature extérieure sur la longueur entière du renforcement. Dans ce but on a utilisé les extensomètres à fil résistant du type RL 120/10 (résistivité/longueur).
Dans le cas des poutres fendillées et renforcées, les extensomètres étaient introduits dans les sections sous les fissures. L'ouverture des fissures à la hauteur de l'armature intérieure était de 0,05 à 0,20 mm.

La fig. 1 présente les mesures de déformations de tôles renforçantes d'une des poutres. Ces déformations sont caractéristiques pour toutes les poutres étudiées. Cette poutre était chargée préliminairement au niveau 0,9 de la force de destruction.
À cette figure, on présente également l'image du fendillement préliminaire de la poutre, le diagramme des contraintes apparues à l'armature collée (6) pour le niveau différent des charges et le diagramme des contraintes de cisaillement moyennes ($\bar{\tau}$) apparues dans l'assemblage par collage entre les points de mesure pour les mêmes niveaux des charges extérieures. Les contraintes de cisaillement moyennes entre les points de mesure ont été calculées au moyen des déformations mesurées de l'armature extérieure d'après la formule:

$$\bar{\tau} = \frac{E_s \cdot t_s \left(\varepsilon_{i+1} - \varepsilon_i \right)}{l_{i+1,\, i}} \qquad (1)$$

où: E_s - module d'élasticité de l'armature renforçante, N/mm^2,
t_s - épaisseur de l'armature renforçante, mm,
$\varepsilon_{i+1,\, i}$ - déformations de l'armature renforçante aux sections i+1 et i ,

$l_{i+1,\ i}$ - distance entre les sections i+1 et i, mm

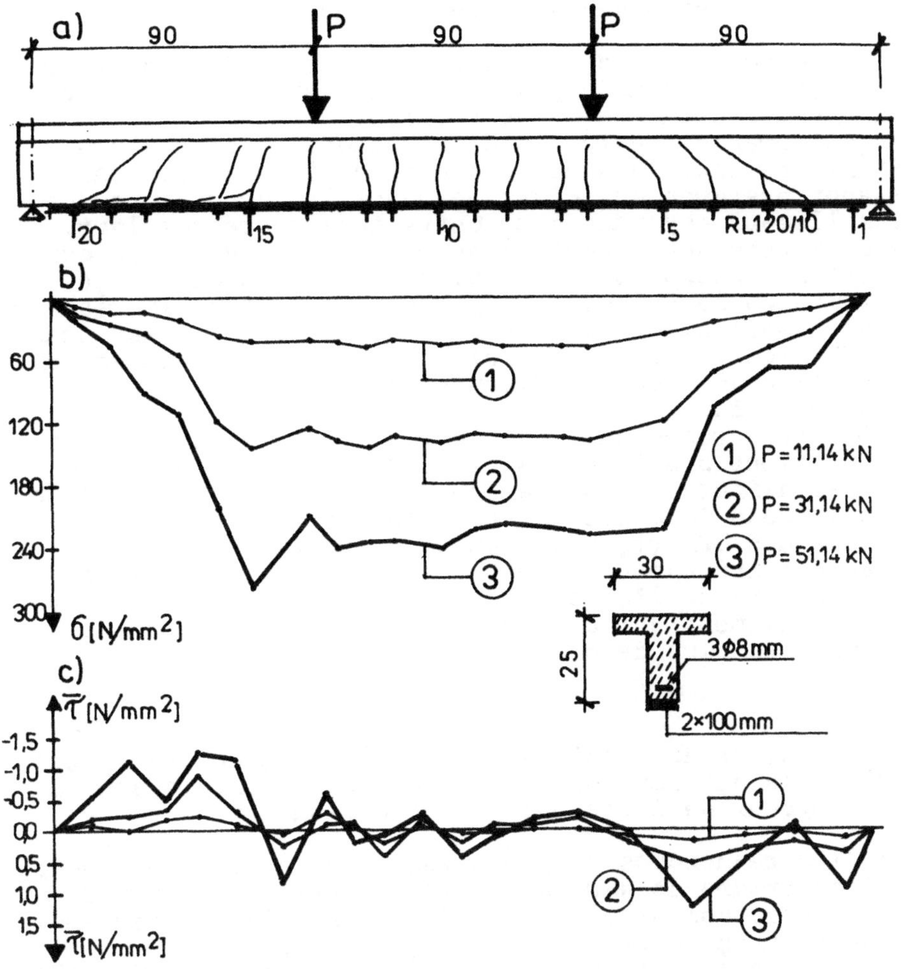

Fig. 1. Déformation et les contraintes dans l'armature renforçante et dans l'assemblage par collage de la poutre fendillée

(a) Image des poutres fendillées,

(b) Contraintes apparues au couvre-joint collé pour les niveaux différents des charges ,

(c) Répartition des contraintes de cisaillement moyennes dans l'assemblage par collage pour les niveaux différents des charges

Le diagramme des contraintes apparues dans l'armature renforçante ressemble, pour le schéma statique utilisé, à l'enveloppe du moment de flexion.
Les différences entre les déformations de l'armature extérieure dans la zone du moment de flexion constant ne sont pas grandes, c'est pourquoi les contraintes de cisaillement moyennes dans cette zone se ressemblent.
Dans la zone des efforts tranchants constants ces contraintes augmentent en atteignant pour la charge de destruction la valeur de 1,5 N/mm^2; cependant la résistance au cisaillement de l'assemblage par collage est égale à 2,87 N/mm^2. Seul l'accroissement des contraintes dans la section 15, sous la fissure détruisante, est très caractéristique.

Les recherches sur les poutres de renforcement sans fendillement préliminaire ont indiqué beaucoup plus de grandes différences de déformations à l'armature collée dans la zone du moment de flexion constant et dans la zone des efforts tranchants constants.
Elles ont indiqué également l'accroissement local des déformations de l'armature extérieure et des contraintes de cisaillement qui atteignaient dans certains zones les valeurs plus de 8 N/mm^2.

3. Influence du fendillement du béton, du collage plein sur la déformation de l'armature extérieure sur la longueur de l'assemblage par collage

Dans les recherches complémentaires on a utilisé 4 poutres de l'ouverture de 120 cm, chargées statiquement par 2 forces ponctuelles placées à 1/3 et 2/3 de l'ouverture.
Le pourcentage d'armatures était de μ=2,93%.
Les poutres étaient chargées préliminairement , fendillées et ensuite déchargées ainsi que renforcées par le collage des couvre-joints de tôles d'acier de la surface du profil de 2x100 mm. L'ouverture des fissures était de 0,05 à 0,15 mm. Les extensomètres ont été placés sous les fissures et entre les fissures sur la tranche où le moment de flexion était constant.
La distance entre les points de mesure était de 20 mm.
Les résultats des recherches sont presentés à la fig. 2.

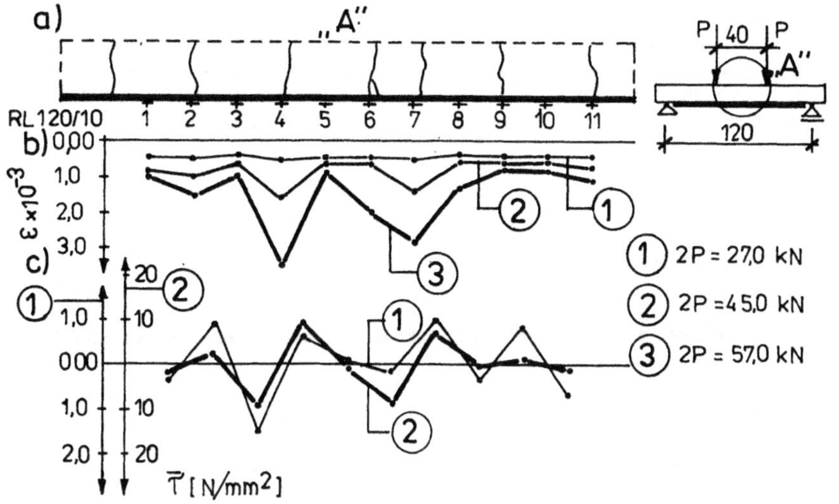

Fig. 2. Déformations de l'armature collée et les contraintes à l'assemblage dans les zones fendillées et dans les zones de collage plein (le détail A)

(a) Répartition des extensomètres à fil résistant sur la poutre fendillée,

(b) Déformations de la tôle collée pour les niveaux différents de la charge,

(c) Diagrammes des contraintes de cisaillement moyennes dans les zones entre les points de mesure

Les diagrammmes de déformations du couvre-joint presentés à la fig. 2 indiquent que dans les zones fendillées ces déformations sont beaucoup plus élevées par rapport aux zones de l'assemblage plein.
L'accroissement de la charge a provoqué l'augmentation locale des déformations de l'armature extérieure.
Cela a provoqué en même temps l'augmentation des contraintes de cisaillement moyennes dans l'assemblage par collage.
Par exemple,pour la charge de 2P=25 kN la valeur maximum des contraintes de cisaillement était égale à 1,5 N/mm², cependant pour 2P=45 kN elle a atteint 10,7 N/mm² en dépassant la résistance au cisaillement de l'assemblage par collage qui était égale à 3,8 N/mm².

Quand la charge augmentait on observait la plastification de l'armature extérieure dans les sections attenantes aux fissures. À cette phase on a observé des fissures horizontales courant dans le béton entre les fissures transversales. En même temps l'armature extérieure travaillait dans la zone centrale comme l'entrait scellé dans le béton, dans les zones finales de l'assemblage.

4. Récapitulation

Les recherches presentées ont indiqué que la formation des fissures dans le béton est liée avec l'augmentation des déformations dans l'armature extérieure.
L'augmentation des déformations locales de cette armature est accompagnée de l'augmentation des contraintes de cisaillement dans l'assemblage par collage qui sont très difficiles à calculer et très défavorables quand les fissures se forment dans les tranches de l'ancrage de l'armature extérieure. C'est pourquoi dans le cas de renforcement par collage de tôles des constructions de béton armé très fissurées il faut appliquer les solutions complémentaires, par exemple l'injection des époxydes dans les fissures ou l'ancrage des extrémités de l'armature extérieure.
Les recherches ont montré également que l'armature renforçante collée au béton collabore avec lui dans la zone fendillée comme l'armature intérieure bien que l'assemblage entre le béton et l'acier soit formé seulement grâce aux forces d'adhésion du liant de colle.

Littérature

Bresson,JJ. (1971) Nouvelles recherches et applications concernant l'utilisation des collages dans les structures. Annales ITBTP, 278, 22-55.

Ciesielski,J. (1976) Łączenie i naprawa konstrukcji betonowych za pomocą żywic epoksydowych. Arkady.

Gamski, K. (1980) La technique de renforcement et de réparation des ouvrages en béton à l'aide des liants époxydes structuraux. Archiwum Inżynierii Lądowej, 26, 67-82.

Van Gemert,D.A. (1980) Force transfer in epoxy Steel--Concrete Joints. J. of Adhesion and Adhesives, 2.

Van Gemert, D.A. (1981) Repairing of Concrete Structures by externally bonded Steel plates, in Plastics in Material and Structural Engineering,RILEM Elsevier, 519-526.

L'Hermite, R. (1967) L'application des colles et résines dans la construction. Annales ITBTP, 239.
Jasieńko, J., Olejnik , A. (1985) Analiza pracy zginanych konstrukcji żelbetowych wzmocnionych zbrojeniem doklejanym, Raport Instytutu Budownictwa Politechniki Wrocławskiej, SPR 29.
Ladner, M., Weder, Ch. (1981) Concrete Structures with bonded external Reinforcement. EMPA Report n° 206.
Litvinov, A.G, Krasulin, N.N. (1978) Issledovanije zhelezobetonnych balok usilennych proklejkoj dopelnitelnoj armatury. Stroitelstvo i Architektura 5, 153-156.

CRACK CONTROL OF REINFORCED CONCRETE BEAMS THROUGH EPOXY BONDED STEEL PLATES

R. JONES, R.N. SWAMY and J. BLOXHAM
Department of Civil and Structural Engineering, University of Sheffield, England

Abstract
Strengthening of concrete structures in situ by bonding steel plates to concrete surfaces using epoxy resins is now recognised to be an effective and convenient method of improving their performance under service loads or to increase their ultimate strength. The main advantages of the technique are that the operation can be carried out relatively quickly and simply even while the structure is in use. Although the technique has been widely used in various parts of the world, there is very little systematic information on the service load behaviour of plated beams, particularly in relation to cracking.

In this paper comprehensive test data are presented on the effect of glued plates on the cracking behaviour of reinforced concrete beams strengthened with such plates. The variables studied in this project include plate thickness, glue thickness, layered plates and lapped plates. The structural behaviour of these beams was closely monitored during testing to destruction, and extensive crack measurements were made to evaluate the effectiveness of the glued plates on crack spacing and crack width.

The paper will present a detailed analysis of the cracking data. It will be shown that maximum crack widths are substantially reduced in the plated beams compared to unplated beams. The test results are compared to existing crack width equations in design codes. New equations are developed to predict crack width in plated reinforced concrete beams.
Keywords Epoxy bonding, Plated beams, Service conditions, Crack spacing, Crack widths, Concrete strain, Tensile strength.

1. Introduction

It is sometimes necessary to modify existing structures in order that they will satisfy the design requirements of any future loading. The modifications may be needed to increase the strength of the structure and/or to improve its performance under service conditions. A technique for carrying out such improvements, which has already been used in a number of countries, is that which involves the bonding of steel plates to the surfaces of the structure. The bonding is usually achieved by the use of an epoxy glue.

The technique was first introduced in the 1960's by research workers such as L'Hermite. In more recent years, reports of practical applications and experimental investigations have been published. It has been shown that full composite action can be achieved between the steel plates and concrete members and that the ultimate strength of a plated reinforced concrete beam can be calculated by normal reinforced concrete methods. It has also been shown that the service conditions of cracking and deflection are both considerably improved by the use of this technique. So far however there is no published data which indicates how crack widths and deflections can be calculated.

An extensive test programme on the use of the bonded plate technique has been in progress in the Department of Civil Engineering at the University of Sheffield since 1976. As part of this investigation, both cracking and deflection have been studied in some detail. The data presented in this paper give typical details of those results relating to cracking, obtained from a recent series of tests.

2. Test Programme

2.1 Beam details

A series of twenty four reinforced concrete beams was used for the tests. All the beams were of the same size; 2500 × 355 × 155mm, and had the same internal reinforcement; three 20mm diameter deformed high tensile steel bars at an effective depth of 220mm. Shear failure of the basic reinforced concrete section was prevented by 6mm diameter mild steel links at 75mm centres in the shear spans.

One beam was left unplated to act as a control and one had a 3mm thick layer of glue only. The other beams were modified using various combinations of glue and plate thicknesses. Of these, some had multiplates or lapped plates, three beams were loaded while the plates were applied, one beam had Vee notches to act as stress raisers and one beam had glue of variable thickness. All the plates used for strengthening and lapping were 125mm wide.

Instrumentation of the beams consisted of ERS gauge on the internal bars and the bonded plates and Demec points on the surfaces of the concrete.

Full details of all the beams are given in Table 1 and Figure 1.

2.2 Materials

The concrete used for these beams was made using ordinary portland cement and washed river gravel and sand. The strength (see Table 1) was chosen to represent that used in bridge beams and is therefore relatively high.

Table 1. Details of test beams

Beam no.	Plate thickness (mm)	Glue thickness (mm)	No. Plate layers	Lap position	Load at plating % Ult.	Cube strength N/mm^2
201	–	–	–	–	–	63.9
202	–	3.0	–	–	–	70.9
203	1.5	1.5	1	–	–	63.3
204	3.0	1.5	1	–	–	66.4
205	6.0	1.5	1	–	–	70.2
206	1.5	1.5	2	–	–	68.8
207	1.5	3.0	1	–	–	70.2
208	3.0	3.0	1	–	–	68.8
209	6.0	3.0	1	–	–	72.2
210	6.0	3.0	1	–	–	70.3
211	1.5	3.0	1	C (2)	–	72.0
212	1.5	3.0	1	ℓp (3)	–	63.9
213	1.5	3.0	2	C	–	67.2
214	1.5	3.0	2	ℓp	–	69.2
215	3.0	3.0	1	C	–	71.5
216	1.5	6.0	1	–	–	68.4
217	3.0	6.0	1	–	–	68.9
218	6.0	6.0	1	–	–	71.5
219	6.0	6.0	1	–	–	73.0
220	1.5	2.0-8.0	1	–	–	68.2
221	1.5	3.0 (1)	1	–	–	65.9
222	1.5	3.0	1	–	50	65.2
223	1.5	3.0	1	–	90	72.0
224	3.0	3.0	1	–	50	71.1

(1) – V notches on tension face of concrete
(2) – Central lap
(3) – Load point lap

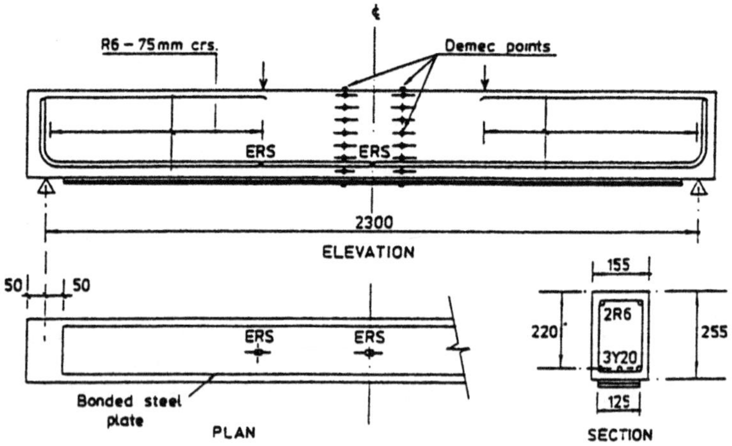

Fig. 1 Beam details

The main tension bars in the beams had a 0.2% proof stress of 470 N/mm^2, an ultimate strength of 507 N/mm^2 and an elastic modulus of 200 kN/mm^2. The mean corresponding values for the plates were 248 N/mm^2, 302 N/mm^2 and 200 kN/mm^2 respectively.

The bonding agent used in the tests was a two part epoxy glue with the following properties :

	Tension	Compression
Elastic Modulus (N/mm^2)	2000	3050
Ultimate Strength (N/mm^2)	16	40
Poissons ratio	0.33	0.36

2.3 Testing

The beams were tested in bending over a span of 2300mm with loading at third points. Load was applied in equal increments and at each increase, readings were taken of the strains on the concrete surface, steel plate and the internal bars, all readings being at midspan. Developing cracks were marked and, for the constant moment region, measurements were made of the heights, spacings and widths, at the level of the internal reinforcement. A considerable quantity of data was obtained from the twenty four beams, but for the purpose of this paper, only typical results will be presented.

3. Results and discussion

3.1 General

The results of crack width, spacing and height were analysed statistically. Typical results for beams 201 and 208 are shown in Table 2. From the full set of results, the following general comments can be made regarding the effects of bonded plates :

(a) Crack development is delayed and mean crack height is reduced. The ratio of the mean crack height to that obtained from normal elastic analysis was 0.66.

(b) The ultimate spacing of the cracks and hence the number of cracks is relatively insensitive to the bonding of plates. At service load, the spacing was greater for the plated beams due to the factors given in (a) above.

(b) The mean crack width and the maximum crack width for the same load level are considerably reduced by plating.

3.2 Mean crack width

It has been shown by other research workers that for normal reinforced concrete, the crack widths are proportional to the strain in the internal reinforcement. For the present investigation the mean crack widths were plotted against the concrete strains, both measured at the level of the internal reinforcement. Roughly linear curves were obtained, suggesting that the same relationship holds for plated beams. Slopes from these curves were obtained from 'best fit" lines computed by linear regression analysis and these were

Table 2 Crack characteristics

Load (kN)	No. of cracks	Mean conc strain ($\times 10^{-6}$)	Mean crack width ($mm \times 10^{-2}$)	Standard deviation ($mm \times 10^{-2}$)	Max* crack width ($mm \times 10^{-2}$)	Mean crack spacing (mm)	Mean crack height (mm)
60	9	550	2.67	1.33	5.0	83	82
	6	300	1.75	0.82	3.0	143	57
100	12	1110	4.29	2.30	9.0	62	95
	10	600	2.85	1.08	4.0	88	92
130	12	1460	6.67	2.51	12.0	62	113
	12	800	3.67	1.23	6.0	74	97
160	13	1750	8.54	3.34	14.0	57	119
	12	1060	4.75	1.54	9.0	64	109
220	14	3400	16.70	6.02	25.0	52	146
	14	1800	7.71	2.97	14.0	54	117
250	–	–	–	–	–	–	–
	15	2600	10.50	4.90	20.0	52	120

Upper figures - Beam 201 (unplated)
Lower figures - Beam 208 (3mm plate, 3mm glue)

then plotted against the plate thickness in order to relate the cracking to the area of plate used. The results were relatively scattered, but bearing in mind the nature of cracking and the limited data available, this was not unexpected. It was found by plotting mean crack width against thickness of glue layer, that for the tests carried out, the glue thickness also had a limited effect on crack width. An approximately mean linear relationship was therefore obtained, by regression analysis, relating the slope of the mean crack width to concrete strain curve, s, to the thicknesses of the plate, t_p, and the glue t_g as follows :

$$S = 19.5 + 4.72\, t_p + 3.43\, t_g \qquad \ldots (1)$$

In order to make this expression more generally applicable, it has been rewritten as :

$$S = 0.78C + 5.83\, \rho_p\, d + 3.43\, t_g \qquad \ldots (2)$$

where C = concrete cover to the internal bars
d = effective depth
and ρ_p = steel area ratio of the plate, based on the dimensions of the original reinforced concrete section

Theoretical values obtained using the expression are shown plotted against experimental values in Fig. 2. Values obtained from the work by Ang are also plotted for comparison. All the results fall within the range +13% to -5% which for the limited data available is reasonable.

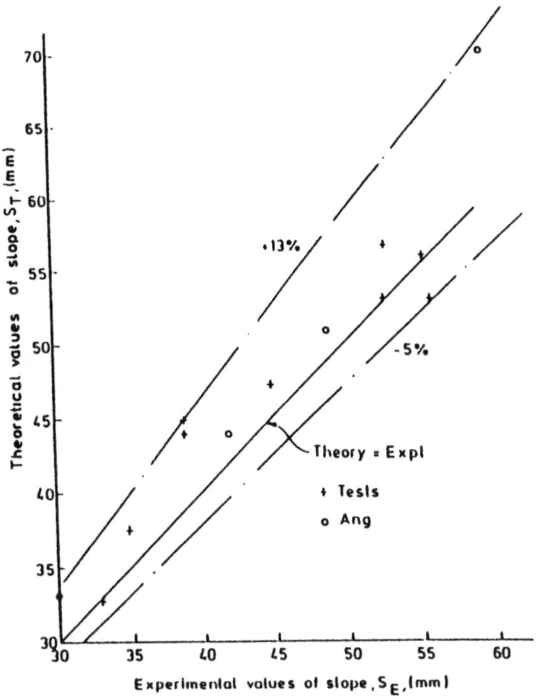

Fig. 2 Theoretical : Experimental slopes

The mean crack width can now be written as :

$$W_{mean} = S \cdot \varepsilon_m \qquad \ldots (3)$$

where S is given by the expression in equation (2)
and ε_m is the concrete surface strain

The value of ε_m depends on the stiffening effect of the concrete between the cracks in the tension zone and is therefore less than the value calculated assuming zero tensile strength for the concrete. Various expressions have been suggested for the determination of ε_m, one of which is that proposed by Beeby, which gives ε_m as:

$$\varepsilon_m = \varepsilon_1 - \frac{4.5.b.(h - x)}{A_s} \cdot 10^{-6} \qquad \ldots (4)$$

where ε_1 is the concrete strain assuming zero tensile strength for the concrete and the second part of the expression allows for the stiffening effect of the concrete between the cracks.

If values obtained using this expression are compared with average test results for each plate thickness, it is found that close agreement is obtained for the 1.5mm plate, but as the plate thickness increases agreement gets progressively worse such that the ratio of experimental to theoretical strain is 0.83 for the six millimeter plate. It therefore seems that a steel plate glued to the tensile face of a flexural member, has a greater effect on the concrete surface strains, than an internal steel bar.

In an attempt to modify the second part of Beeby's formula for ε_m, theoretical steel bar stresses, assuming zero tensile strength for the concrete, were calculated at three load levels and plotted against the measured concrete strains at the same loads. A best fit line was then obtained by linear regression analysis which was compared with the stress:strain curve for the internal bars. The differences in strain between these two lines are equivalent to the second part of Beeby's expression for ε_m. A typical example of these plots, for the 1.5mm plate is given in Fig. 3.

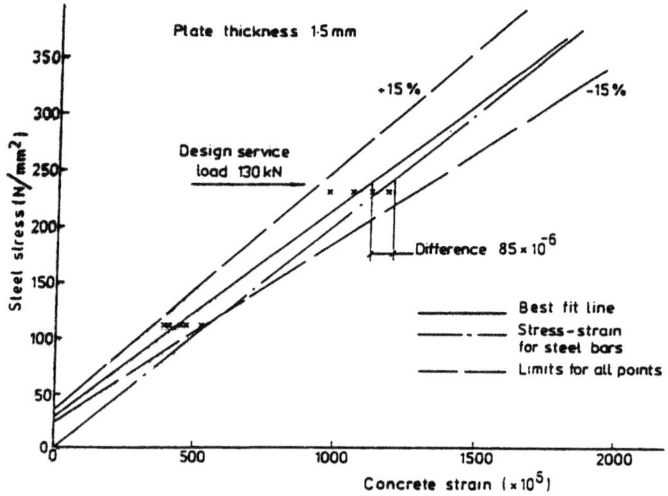

Fig. 3 Calculated steel stress : Measured concrete strain

Using these results, a revised expression for ε_m, which includes for the effect of the plate, was obtained as follows :

$$\varepsilon_m = \varepsilon_1 - b(2,74\ \rho_p.d. + 2.3)\left(\frac{h - x}{A_s}\right) . 10^{-6} \qquad \ldots (5)$$

where b, d and h are respectively the width, effective depth and total depth of the basic reinforced concrete section. A_s, is the total steel area and, x, is the corresponding neutral axis depth obtained from an elastic analysis.

Thus by using equations (2), (3) and (5), the mean crack width for a plated reinforced concrete section can be calculated.

3.3 Maximum crack width

Although the mean crack width is of interest, for design purposes it is normal to specify a maximum crack width at service load which has an acceptably low probability of being exceeded in practice. Typical curves relating the maximum measured crack widths to applied load are shown in Fig. 4. From the full set of results, a number of observations can be made regarding the maximum crack widths at the service load. The unplated beam No. 201 had a maximum crack width of 0.12mm. For the plated beams, without lapped plates or preloading, the 1.5, 3.0 and 6.0mm plates gave reductions in maximum crack widths of 46%, 50% and 58% respectively. Beams with lapped plates had crack widths which on average were 33% greater than similar beams without laps. The preloaded beams with 1.5mm plates both had reductions in crack width of 17% compared with the unplated beam. For the beam with the 3mm plate, the reduction was 42%.

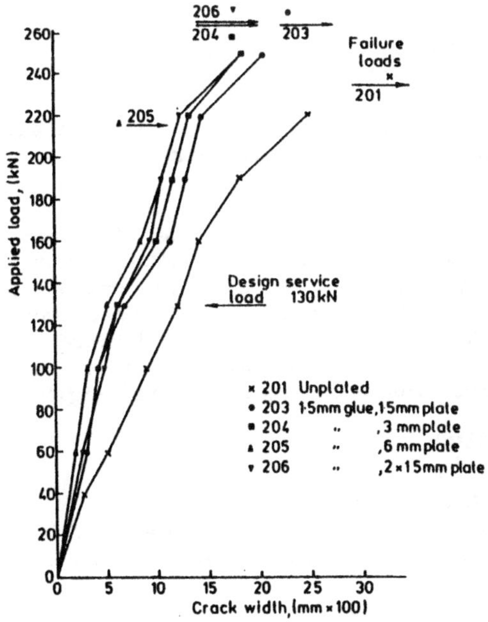

Fig. 4 Maximum crack width : Load

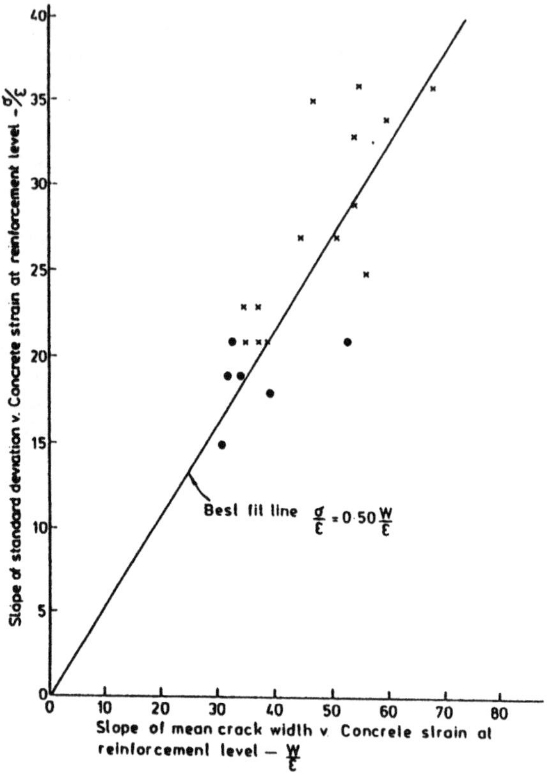

Fig. 5 Slope of standard derivation v. concrete strain to slope of mean crack width v. concrete strain

Various research workers have produced different values for the ratio between maximum and mean crack widths. These range from 1.5 by Hognestadt to 2.0 by Base et al. The corresponding figure for the present investigation was 1.78 which is in good agreement with previous work. In Fig. 5 the slope, S, defined earlier is plotted against the slope of the standard deviation to concrete strain curves. A best fit line forced through the origin in given by :

$$\frac{\sigma}{\varepsilon} = 0.5 \frac{W_{mean}}{\varepsilon}$$

where σ is the standard deviation

If a 5% chance of a certain maximum crack width being exceeded is chosen, then the maximum crack width is given by :

$$W_{max} = W_{mean} + 1.64\,\sigma$$
$$= 1.82.\,W_{mean}$$

Thus the maximum crack width for design purposes can be expressed as :

$$W_{max} = 1.82\ S\ \varepsilon_m \qquad \ldots (6)$$

where S and ε_m are given by equations (2) and (5) respectively.

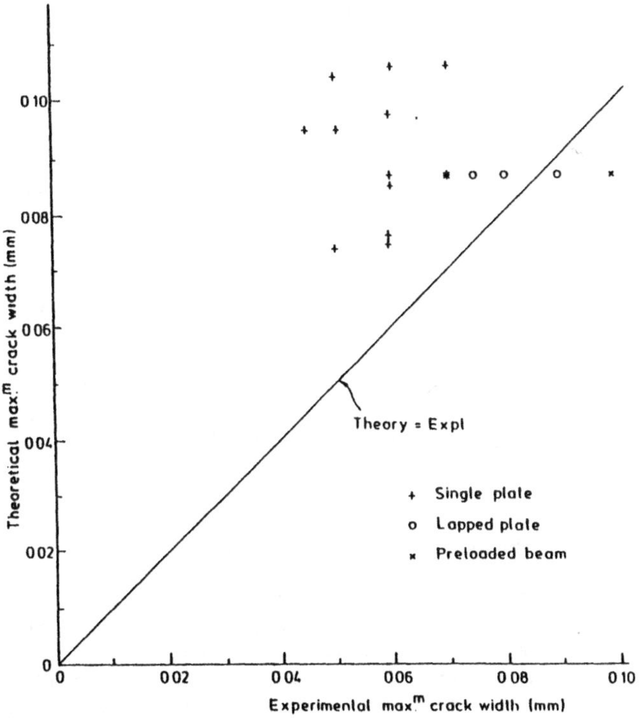

Fig. 6 Theoretical : Experimental crack width

Results obtained using equation (6) are plotted against experimental values in Fig. 6. This shows that within the limitations of the available data the above expression is reasonable for design purposes when considering single plated or multiplated beams, but larger cracks can be expected when the plates are lapped, or the beams are preloaded above 50% of their ultimate strength.

For comparison with the proposed expression, calculations of crack widths were also carried out using the methods given in British Code of Practice CP110 and the ACI Building Code. The results are given in Table 3 together with the values obtained from equation (6) and, for the single plated beams only, the ratios of calculated to experimental widths in each case. It can be seen that for the proposed expression the ratios vary from 0.94 to 1.17 for

CP110 1.12 to 1.47 and for the ACI code 1.08 to 1.55. Within the limitations of the available data therefore, the proposed expression gives a better assessment of maximum crack width, but much more data is needed to verify this.

Table 3 Maximum crack widths at service load

Beam No.	Max crack width (mm × 10^{-2}) Expt. (1)	Theory (2)	CP110 (3)	ACI (4)	$\frac{(2)}{(1)}$	$\frac{(3)}{(1)}$	$\frac{(4)}{(1)}$
203	6.5	7.0	8.8	9.3	1.08	1.35	1.43
204	6.0	6.2	7.4	7.6	1.03	1.23	1.27
205	5.0	4.9	5.6	5.4	0.98	1.12	1.08
207	6.0	7.0	8.8	9.3	1.17	1.47	1.55
208	6.0	6.2	7.4	7.6	1.03	1.23	1.27
209	4.5	4.8	5.6	5.4	1.07	1.24	1.20
210	5.0	4.8	5.6	5.4	0.96	1.12	1.08
216	7.0	6.9	8.8	9.3	0.99	1.26	1.33
217	6.0	6.1	7.4	7.6	1.02	1.23	1.27
218	5.0	4.7	5.6	5.4	0.94	1.12	1.08
219	5.0	4.7	5.6	5.4	0.94	1.12	1.08
220	6.0	7.0	8.8	9.3	1.17	1.47	1.55
221	6.0	7.1	8.8	9.3	1.17	1.47	1.55

3.4 Tensile stresses carried by the concrete

The concrete between cracks in a flexural member will carry tensile stresses and thus increase the stiffness of the member. This gives rise to a reduction in the mean strain in the tension zone as illustrated by the evaluation of ε_m in section 3.2 above. Beeby has produced results from tests on reinforced concrete members subject to pure tension, from which he has calculated the average tensile stress in the concrete which would give rise to the same average strain as measured in the tests. He discovered that for deformed bars, such as used in the tests described here, there was a limiting value of concrete stress of about 1.0 N/mm^2. This value has been adopted in the British Code of Practice CP110 for the evaluation of deflections. In this, the effective resisting moment due to the tensile stress in the concrete is found by taking a linear variation of stress, being zero at the neutral axis and 1.0 N/mm^2 at the level of the tensile steel, and then multiplying the resulting force by its distance from the neutral axis. This moment is then deducted from the applied moment.

The same general approach was used in the present investigation to determine the effective tensile stress in the concrete. The moments due to the total steel(bars and plate) were calculated at three load levels up to the service load. These were then deducted from the applied moments, the resultants being the mean moments due to the tensile concrete forces, from which the tensile stresses were calculated. Mean values of tensile stress were obtained for each combination of steel plate and glue thicknesses. The overall mean results are given in Table 4. For comparison the value of tensile stress for the unplated beam was 1.2 N/mm^2.

Table 4 Mean values of tensile stress in the concrete

Glue thickness (mm)	Plate thickness (mm)		
	1.5	3.0	6.0
1.5	1.4	3.7	5.1
3.0	2.4	5.7	5.9
6.0	4.2	5.5	6.4

From these results, the following observations can be made.

(a) The stiffening effect produced by the application of a steel plate to the tensile face is much greater than would be expected if additional internal bars of the same area were used.

(b) The stiffening effect increases with both the area of plate added and the thickness of the glue layer.

(c) The tensile stresses appear to be tending towards limiting values for each glue thickness.

Before prediction formulae can be suggested, much more data is needed in this area. In the future, it may be possible to relate the effective tensile stress to the mean concrete strain ε_m and then to use this in the calculation of crack widths.

4. Conclusions

From the results presented in this paper, the following conclusions can be drawn regarding the effect of steel plates glued to the tensile faces of flexural members.

(a) The propogation of cracks away from the tensile face was delayed.

(b) Crack heights were reduced and the mean ratio of experimental height to theoretical height was 0.66 with a range of 0.60 to 0.72.

(c) The number of cracks developed and the mean ultimate spacing of these cracks were relatively insensitve to the application of plates. in the tests described the beam size, internal reinforcement and the concrete cover to the bars were kept constant and therefore could not have counteracted the effects of plate and glue on the spacing.

(d) For the same load level, the mean crack widths were considerably reduced by plating.

(e) Tentative proposals have been made for the calculation of the mean crack width, using the relationship

$$W_{mean} = S \cdot \varepsilon_m$$

where $S = 1.3c + 5.19\,\rho_p\,d$

and $\varepsilon_m = \varepsilon_1 - b(2.74\ \rho_p\ d + 2.3)\ (\frac{h - x}{A_s})\ .\ 10^{-6}$

(f) At the same load level, the maximum crack widths were considerably reduced by plating. Beams with single or multiple plates compared with the unplated beam had reductions ranging from 45% to 58%. Where lapped plates were used, the maximum crack widths were on average 33% grater than similar beams without laps. For the preloaded beams, reductions compared with the unplated beam were in the range 17% to 42%.

(g) A statistical analysis was used to obtain a relationship between mean and maximum crack widths. For a 5% probability of being exceeded, this was given by :

$$W_{max} = 1.82\ W_{mean}$$

where W_{mean} is calculated as shown in (e) above.

Theoretical crack widths obtained using this expression were compared with those measured in the tests and also those calculated using design equations from the British and American Codes of Practice. The proposed relationship gave the best agreement with the tests.

(h) The mean effective tensile stresses in the concrete in the tension zone was considerably greater than the 1.0 N/mm^2 suggested for conventional reinforced concrete. The value of this stress increases with both the area of plate and the glue thickness, a value of 6.4 N/mm^2 being obtained with a 6mm plate and a 6mm thick glue layer.

(i) The proposals presented in this paper are based on data obtained from twenty four beams, with a limited number of variables. Much more data is required before they can be applied with confidence. Therefore at this stage they should be used with caution when applied to practical problems.

References

Ang, T. (1979) The strengthening of reinforced concrete beams using glued steel plates. M.Eng. Dissertation, University of Sheffield.

Beeby, A.W. (1972) A study of cracking in reinforced concrete members, subjected to pure tension. Cement and Concrete Association Technical Report No. 42,468, 24 pp.

Bresson, J. (1972) Reinforced par college d'armatures du passage inferieur du CD 126 sous l'autouroute du sud. Annales de l'ITBTP, Serie Beton, Beton Arme, Supplement 297, pp. 3–24.

Davies, B.L. and Powell, J. (1984) Strengthening of Brinsworth Road Bridge, Rotherham. IABSE, 12th Congress, Vancouver, BC, pp. 401–407.

Fleming, C.J. and King, G.E.M. (1967) The development of structural adhesives for three original uses in South Africa. RILEM International Symposium, Synthetic Resin in Building Construction, Paris, pp. 75–92.

Iino, T. and Otokawa, K. (1981) Application of epoxy resins in strengthening of concrete structures. Polymer in Concrete, Third International Congress on Polymer in Concrete, Koriyama, Japan, Vol. II, pp. 997-1011.

Jones, R., Swamy, R.N., Bloxham, J. and Bouderbalah, A. (1980) Composite behaviour of concrete beams with epoxy bonded external reinforcement. The International Journal of Cement Composites, Vol. 2, 2, pp. 91-107.

Jones, R., Swamy, R.N. and Ang, T.H. (1982) Under and over reinforced concrete beams with glued steel plates. The International Journal of Cement Composite and Lightweight Concrete, Vol. 1, 1, pp. 19-32.

Jones, R. and Swamy R.N. (1985) Structural implications of repair using epoxy bonded steel plates. Proc. International Symposium on Structural Faults and Repairs. Inst. of Civil Engineers, London, pp. 75-80.

Jones, R. and Swamy R.N. (1984) In-situ strengthening of concrete structural members using epoxy bonded steel plates. Proc. 4th International Congress on Polymers in Concrete, Darmstadt, pp. 251-255.

L'Hermite, R. and Bresson, J. (1967) Beton arme par collage des armatures. RILEM Symposium, Part 2, Paris, pp. 175-203.

Parkinson, J. (1978) Glue solves a sticky problem for Gestetner. New Civil Engineer, pp. 26-27.

Ryback, M. (1981) Reinforcement of bridges by gluing of reinforcing steel. Materials and Structures, Vol. 16, 91, pp. 13-17 (in French).

Swamy, R.N. and Jones, R. (1980) Behavour of plated reinforced concrete beams subjected to cyclic loading during glue hardening. The International Journal of Cement Composites, Vol. 2, 4, pp. 233-234.

CERTAINS RESULTATS DES RECHERCHES EFFECTUEES SUR LE POUTRELLES EN BETON ARME RENFORCEES PAR UNE ARMATURE EXTERIEURE COLLEE GRACE A UNE RESINE EPOXYDE

N.KNEZEVIC-VUKSANOVIC et B.STANIVUKOVIC
Institut de Máteriaux et Constructions de la Faculté de Gĕnie Civil, Sarajevo

Résumé
En collant une armature extérieure on peut souvent obtenir un renforcement efficace des éléments en béton armé dans les cas de réparation due à différentes causes. On a présenté dans le travail les résultats des recherches expérimentales effectuées sur les poutrelles exposées à la flexion jusqu'à la rupture. On a examiné les cas suivants: poutrelles en béton non armé; poutrelles armées au moyen d'armature extérieure collée, grâce à une sorte de résine époxyde, à la surface inférieure tendue; poutrelles classiquement armées; poutrelles avec armatures classique et extérieure (combinaison d'armatures des deux cas précédents). On a également analysé les données sur l'apparition des fissures, sur l'importance et le caractére de la rupture pour les cas particuliers examinés.
Mots-clés: Armature extérieure, Résine époxyde, Fissure, Rupture.

Introduction

Dans les cas du renforcement des éléments en béton armé apparait comme une possibilité plausible le procédé du renforcement consistant á coller une armature extérieure au béton au moyen des résines d'époxyde. On a effectué, à l'Institut des matériaux et constructions de la Faculté du Génie civil à Sarajevo, des recherches expérimentales portant sur les poutrelles non-armées et armées, renforcées par une armature extérieure, et exposées à la flexion jusqu'à la rupture. Ces recherches ont permis d'aboutir à certains résultats ainsi qu'à certaines recommandations et suggestions en vue des recherches futures.

Caractèristiques des modèles et résultats des essais

Les essais ont été effectués sur les poutrelles en béton présentant des dimensions de 10 x 15 x 70 cm, qui étaient exposées à la flexion jusqu'à la rupture, en appliquant une force concentrique au milieu de la portée de 60 cm. On a examiné en tout quatre cas différents présentant les caractéristiques suivantes:

- modèles en béton non-armé - désignés par A,

- modèles en béton non-armé avec armature extérieure (tôle d'acier épaisse de 0,8 mm) collée au moyen d'une résine d'époxyde à la surface inférieure tendue - désignés par B,

- modèles armés avec 3 ø 6 mm, sans armature extérieure - désignés par C,

- modèles armés avec 3 ø 6 mm, avec une armature extérieure (tôle épaisse de 0,8 mm) collée au moyen d'une résine d'époxyde à la surface inférieure tendue - désignés par D.

Pour chacun des cas cités on a examiné trois modèles identiques,vieux de 28 jours.

L'armature extérieure a été collée à la surface, séchée d'elle - même, des poutrelles de béton, ce dernier étant vieux de 21 jours. Avant le collage, la surface de contact de la tôle d'acier a été polie avec du papier de verre et époussetée, tandis que la surface de contact du béton a été meulée au moyen d'une meuleuse électrique et époussetée grâce à l'air comprimé. L'enduit a été apposé au moyen d'une brosse. La tôle d'acier collée a été exposée à une légère pression durant 24 heures.

Pour la confection des poutrelles on a utilisé le béton de composition suivante:

- agrégat de calcaire broyé, fractionné et lavé (0/4 mm, 4/8 mm, et 8/16 mm - 20 % chacun et 16/31,5 mm - 40 %),
- ciment Portland PC 35(300 kg/m^3 de béton utilisé),
- facteur eau-ciment de 0,55.

La résistance à la compression sur des cubes aux côtés de 20 cm, était en moyenne de 44,3 MPa, tandis que la résistance moyenne à la traction de flexion, des poutrelles présentant les dimensions de 10 x 15 x 70 cm, était de l'ordre de 5,3 MPa (béton vieux de 28 jours).

L'armature ø 6 mm était en acier rond et lisse, avec la contrainte moyenne à la limite des grands allongements de 252 MPa, et avec la résistance moyenne à la traction de 382 MPa.

La tôle d'acier présentait les valeurs des contraintes respectives de 238 MPa, c'est-à-dire de la résistance de 342 MPa.

On a utilisé pour le collage une sorte de résine bicomposite d'époxyde, de production yougoslave.

Les résultats des essais sont représentés sur les tables n° 1, 2 et 3.

Table n° 1

Désignation de la série des poutrelles en béton	Superficie de la coupe de l'armature (mm²)	Coefficient d'armature (%)	Moment de l'apparition des premières fissures (kNm)	Moment de l'apparition de la rupture (kNm)
A	-	-	4,266 4,119 3,824	4,266 4,118 3,824
B	80	0,53	7,355 6,766 6,914	8,237 8,237 7,943
C	85	0,57	5,143 4,854 5,442	9,764 8,679 10,297
D	165	1,10	10,738 8,237 9,561	16,828 16,033 17,505

Table n° 2

Désignation de la série des poutrelles en béton	Nombre de fissures visibles apparues après la rupture totale	Mesure maximale des distances entre les fissures après la rupture totale (cm)	Ouverture maximale de la fissure mesurée lors de la rupture totale (mm)
A	-	-	-
B	1 1 1	- - -	20 28 29
C	1 1 1	- - -	20 25 30
D	2 1 6	45 - 20	3 15 3

Table n° 3

Désignation de la série des poutrelles en béton	Caractéristiques de la rupture
A	Rupture cassante simultanée avec l'appartion de la première fissure
B	Rupture dans le béton et rupture le long de la surface de contact horizontale béton-tôle (rupture mixte)
C	Rupture dans le béton avec "sortie" de l'armature
D	Rupture dans le béton et rupture long de la surface de contact horizontale béton-tôle (principalement rupture cohésive dans le béton, partiellement rupture dans la résine et, dans une mesure très réduite, rupture adhésive dans le lien résine-tôle)

La table n° 4 représente les valeurs moyennes des moments lors de l'apparition des premières fissures et au moment de la rupture, et les valeurs de leurs relations pour les cas examinés.

Table n° 4

Désignation de la série des poutrelles en béton	Valeurs moyennes des moments (kNm)		Relation des valeurs moyennes des moments lors de la rupture et lors de l'apparition des premières fissures
	lors de l'apparition des premières fissures	lors de la rupture	
A	4,069	4,069	1,000
B	7,012	8,139	1,161
C	5,146	9,581	1,862
D	9,512	16,788	1,765

Considérations finales

Les résultats des recherches effectuées nous fournissent les données pour les deux cas suivants:

1. Le cas où l´armature classique est remplacées par une armature extérieure collée, et
2. Le cas où le béton armé classique est renforcé par une armature extérieure collée.

Dans le premier cas il faut comparer mutuellement les résultats de la table n° 4 pour les poutrelles armées avec une armature extérieure (B) et les poutrelles armées avec une armature classique (C). On constate le moment de la rupture du modèle à l´armature extérieure qui est d´environ 15 % inférieur, mais que le moment lors de l´apparition des premières fissures est d´environ 36 % supérieur. En appliquant uniquement l´armature extérieure collée on augmente la résistance à l´apparition des fissures par rapport à l´armature classique. Le défaut essentiel des éléments ainsi armés réside dans le danger considérable de l´apparition d´une rupture cassante, la différence entre le moment de la rupture et celui de l´apparition des premières fissures étant assez petite.

Dans le deuxième cas il faut comparer les résultats des poutrelles classiquement armées (C) et des poutrelles identiques renforcées par une armature extérieure collée (D). Dans ce cas-là les résultats sont considérablement plus favorables. Avec une armature collée, dont la superficie correspond approximativement à la superficie de l´armature intérieure classique, on a constaté l´augmentation du moment lors de l´apparition des premières fissures de l´ordre d´environ 85 % et du moment de la rupture d´environ 75 %, sans danger de l´apparition d´une rupture cassante.

Les résultats énoncés prouvent une vaste possibilité du renforcement des constructions en béton armé exposées a la flexion, grace a une armature extérieure collée au moyen des résines d´époxyde. Il faudrait dans les recherches futures prêter une attention particulière à la protection efficace de l´armature extérieure du point de vue de la corrosion ainsi qu´à la protection de la colle d´époxyde du point de vue des hautes températures, en cas d´incendie éventuel.

Il faudrait également, à une échelle plus élevée, varier les paramètres caractéristiques, tels que la sorte et la puissance de l´armature extérieure et la sorte de la résine permettant son collage, en tâchant de déterminer les comportements de tels éléments constructifs du point de vue de la rupture provoquée par des forces transversales (dans le travail on a examiné les cas de la rupture à partir du moment de la flexion, sous charge de courte durée) de même que les comportements dans les conditions d´une charge prolongée, d´une charge due à la fatigue, les comportements des ces éléments exposés à l´action des facteurs climatiques ambiants, particulièrement du gel etc.

BONDED STEEL PLATES FOR THE REDUCTION OF FATIGUE STRESSES OF COUPLED TENDONS IN MULTISPAN BRIDGES

E.-H. RANISCH and F.S. ROSTASY
Institut für Baustoffe, Massivbau und Brandschutz, Technical University of Braunschweig, Germany

Abstract
The working joints of many span-wise built p/c-bridges are cracked. Thereby, the fatigue stresses in the prestressing steel at the joints, where all tendons were coupled, increase markedly. To prevent premature fatigue failure, strengthening of the joints became necessary.

The injection of cracks has not proved to be a sufficient countermeasure for all bridges. The fatigue stresses can be reliably reduced by additional steel plates bonded to concrete with suitable epoxy resin adhesives. The first successful application of this strengthening method was carried out on an Autobahn-bridge near Dortmund in 1981. Its efficacy was proved by load tests. Other bridges followed. The use of fibre composite plates instead of steel plates is also tested. The rules for the dimensioning and bonding of the plates and for accelerated durability tests of bonded joints are presented.

Key words: Concrete bridges, fatigue stresses, working joints, steel-to-concrete bonding, adhesives, bonded reinforcement, durability testing, field measurements, fibre composites.

1. Introduction

Most of the post-war multispan Autobahn-bridges in the Federal Republic of Germany are posttensioned continuous girders with a hollow box section. Usually, these bridges were built by span-wise construction. Thereby, all tendons were coupled at the working joints, positioned in the points of contraflexure. Many of these coupling joints are cracked.

These cracks were predominatly caused by temperature restraint, leading to tensile stresses in the bottom fiber in the region of the coupling joints. Because of the low concrete tensile strength at the joint and because the percentage of ordinary reinforcement in the bottom slab was often low, wide cracks occured.

These cracks do not endanger the load carrying capacity of the section. However, they transform the section into the cracked state. Thereby, the stress amplitudes in the couplers and the prestressing steel, resp. due to live load increase markedly. The dynamic strength of the coupled prestressing steel may be in danger.

Consequently, repair was urgently warranted. The first countermeasure was the injection of cracks with epoxy resin. This remedy proved to be successful for many bridges, but not for all. In some cases injected cracks reopened or new cracks were formed in the proximity of injected cracks, probably caused by the reoccurance of a high temperature restraint. This problem may be overcome by the provision of additional reinforcement. Epoxy-bonded mild steel plates in combination with crack injection proved to be very efficient to reduce the crack widths and the stress amplitudes as well.

2. Design of bonded steel plate reinforcement

The design of the additional bonded steel plate reinforcement for the reduction of fatigue stresses of the prestressing steel in the couplers of the tendons comprises two independent steps. In the first step, the necessary total cross-section of steel plate has to be determined in such a way, that - under the combination of all relevant actions plus maximum thermal restraint - the calculated stress range of the prestressing steel, anchored in the coupler, does not exceed the admissible fatigue stress range adm $\Delta\sigma_p$ of the post-tensioning system:

$$\text{cal } \Delta\sigma_p \leq \text{adm } \Delta\sigma_p$$

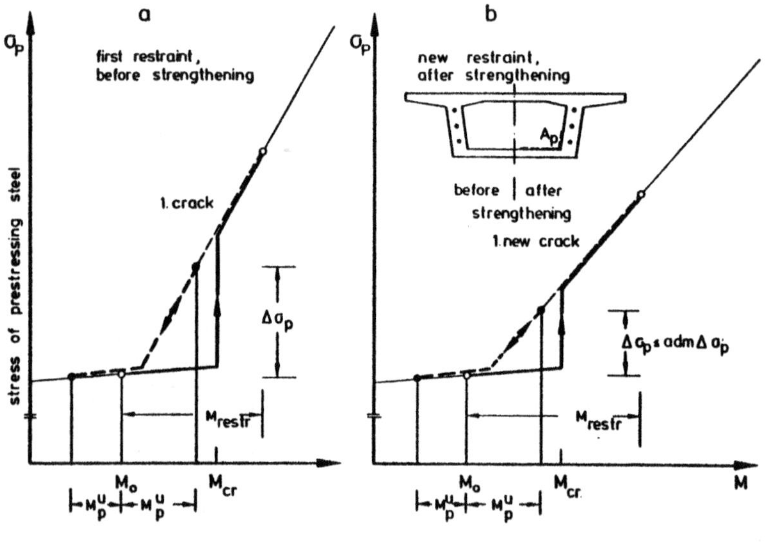

Fig. 1: Effect of strengthening by bonded plates on stress prestressing steel in working joints (schematic)

The design of the cross-section of the steel plate reinforcement is an iterative process. This is due to several reasons. Firstly, the calculation has to be performed for the cracked, posttensioned section. Secondly, neither the real prestressing force nor the exact magnitude of the basic moment M_0 at the joint due to dead load, prestressing force, creep and shrinkage, thermal restraint, etc. but without traffic load are exactly known. Reasonable assumptions within limits must be made. Fig. 1 shows the schematic moment-steel stress-relations before and after strengthening. For the design, the thermal restraint moment may be assumed to be nil at the instant of strengthening, because bonding work must be performed in the early morning hours during which the cross-fall of temperature over the bridge section is negligible. The maximum stress in the steel plate max σ_p is then derived with the sum of all actions including thermal restraint. It renders the force T which has to be anchored by bond.

Many tests and practical applications have shown that bonded external reinforcement is an efficient method for the structural strengthening of prestressed and reinforced concrete structures. The bonded mild steel plates act as additional reinforcement. Tests proved that the crack widths and the deformations of structural elements strengthened in such a way are similar to members conventionally reinforced with ribbed bars of the same cross-section as the steel plates.

Epoxy resin adhesives have been developed which have a higher strength then concrete. Therefore, high quality bondings will never fail in the bond line, but in the concrete underneath. Although the elastic modulus of the adhesive is less than that of the concrete it was found that externally reinforced concrete beams were stiffer than ordinary reinforced beams.

The second step of the design is the determination of the anchorage length. Theory and experiments have shown that the effective anchorage length of metal-to-metal bondings is limited. This means, that the increase of the anchorage length over a certain value does not lead to an increase of the bond failure load.

The principles, however, for the design of the anchorage length of steel-to-concrete bondings were developed by tests on double-lap steel-to-concrete bond specimens and by analysis as well. These tests showed, that the distribution of bond stress along the anchorage length was very nonuniform and dependent on many parameters, with a stress peak in the proximity of the loaded end of plate. This fact forbids the design of anchorage length with a mean bond stress value.

For the analysis a global bond stress-dilation law had to be formulated on the basic of tests. It consists of a lin-

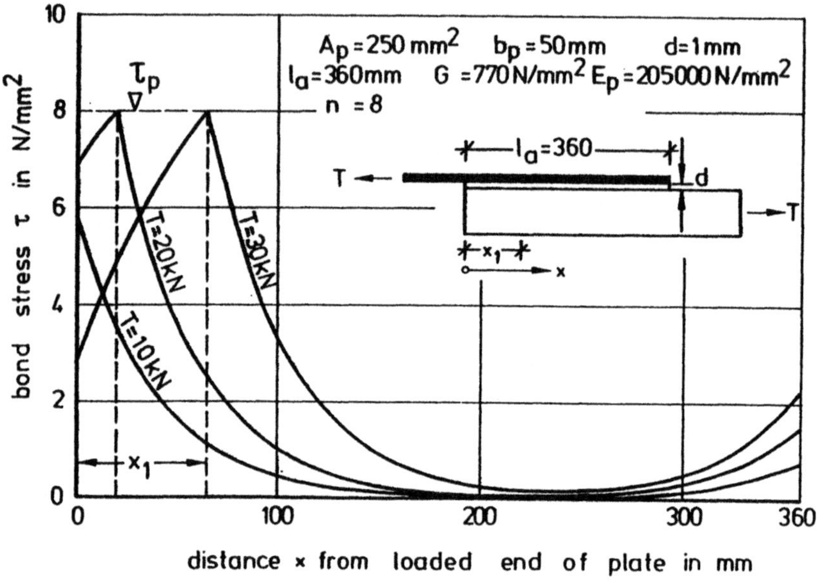

Fig. 2: Calculated bond stress distributions

ear-elastic branch accending to τ_p and a from thereon descending linear branch. The latter representing elasto-plastic behaviour of the bond. The peak value of the bond stress τ_p, representing the global bond strength, depends on the strength of concrete etc. It amounts to 8 MPa for mean cube strength of concrete of 30 MPa. With this material law the destributions of Fig. 2 were calculated. Bond failure occurs if

$$\int_0^{l_a} \tau(x)\ b_p\ dx \geq T$$

and if

$$\tau(o) = 0$$

The figure shows that the peak of bond stresses does not increase with the load, but it travels towards the unloaded end of plate. Even if the bond strength τ_p is locally overcome, the concrete is still able to carry load by a combination of shear and friction. Based on theoretical deliberations, the anchorage length can be expressed by

$$\text{nec } l_a = C \frac{T^2}{E_p \cdot b_p^2 \cdot t_p \cdot \tau_p}$$

with:

$T = \max \sigma_p b_p t_p \le f_{py} b_p t_p / \gamma = T_{py} / \gamma$

b_p, t_p breadth and thickness of plate (mm)

E_p modulus of elasticity of plate material (N/mm²)

τ_p bond peak stress (N/mm²)

C = 18500, constant

In this equation, the global safety factor against bond failure is 1.45. This value is sufficient, because the plates must not increase the bending moment.

It was found that the peak value of bond stress p depends on the tear-off strength of the concrete. A partial load transmission by shear-friction is only guaranteed when the adhesive comes into contact with the peaks of the coarse aggregate (up to 8 mm). This will be achieved by intensive sandblasting of the concrete surface.

3. Durability of bond

The static strength of bond must be maintained in an undiminished way during service life of the structure. Thus, durability of the bond action is a very important property. The ensurance of the durability of the exposed surface of the steel plate renders no problem in practice. The durability of the steel-to-concrete bond depends on several factors: properties of the adhesive, preparation and cleanness of surface of steel plate, properties of primer, and others.

An expertly executed glued joint will always fail, when subject to a short-time load test, in a failure plane within the concrete, located several millimeters below the concrete's surface. The strength of the interfaces steel/concrete, steel/primer, and primer/adhesive will be in excess of the bond strength, as will be the cohesive strength of the adhesive. If, however, the joint will be subject to moisture over a long time, adhesion may be detrimentally affected by the diffusion of water to the interface steel/adhesive and by the resulting corrosion.

Adhesives with equivalent bond strength values in short-time static tests may differ decisively with respect to durability. This is also valid for primers and surface treatments. It is, therefore, necessary to perform durability tests to discern the long-term behaviour of adhesives and primers.

Differences between the durability of adhesives and primers may be disclosed by a time-accelerated test of glued joints subject to a warm and humid climate (40 °C/90 % r.h.). Fig. 3 shows the specimens and the set-up of such a test.

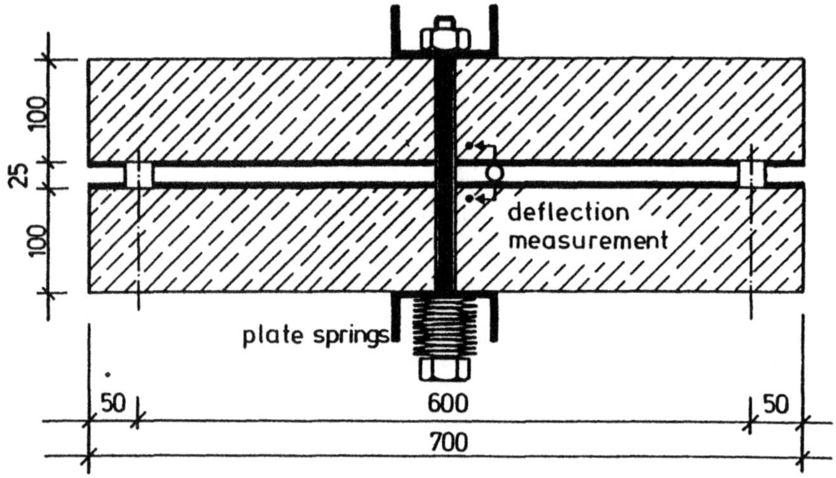

Fig. 3: Specimens and loading system for accelerated durability test

The two prismatic specimens are loaded with a sustained bending moment equivalent to 50 % of the short-time failure load. Then, they are placed in the climate chamber. Adhesives and primers of good quality show after 6 months in 40/90 no loss of static bond strength. In other cases, significant amounts of rust on the steel plates have lead to a loss of adhesion and to premature failure.

We have found that for external reinforcement, plates of fibre composites may used instead of steel plates. These plates need no sandblasting and no corrosion protection after bonding. In the accelerated durability tests these plates showed excellent results.

4. First practical applications

The first bridge to be strengthened by steel plates in Germany was the Sterbecke-bridge in Westphalia-Lippe in 1981. It is a span-wise built, post-tensioned hollow box girder with seven spans of 40 m each. All coupling joints were cracked, exhibiting very wide cracks (up to 1,5 mm) in the bottom plate of the box. The calculated steel stress amplitudes at the couplers were markedly in excess of the permissible value of the applied post-tensioning system. Fig. 3 shows a part of the cross-section of the bridge and the additional external plate reinforcement. The steel plates had sections of 6 x 100 and 8 x 100 mm².

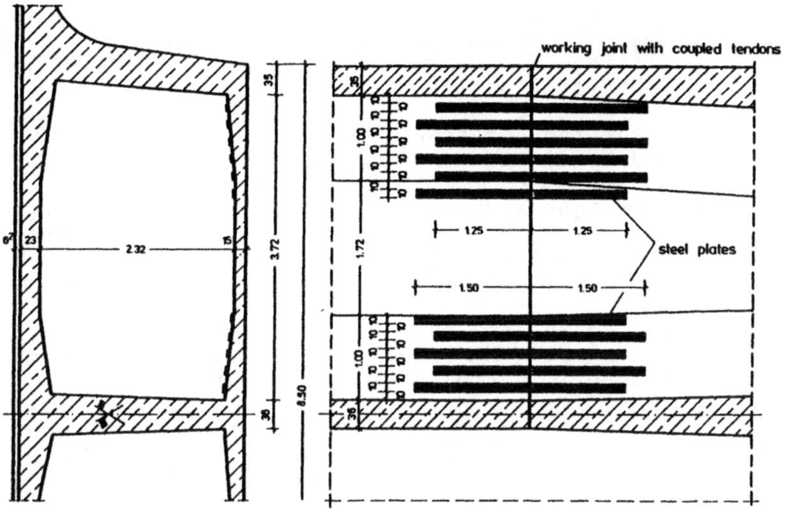

Fig. 4: Cross-section of bridge and lay-out of steel plates

The external reinforcement should be bonded during weather conditions which do not entail thermal restraint. Tests have proved that the crack movements and vibrations due to the traffic load are not detrimental to bond strength of the glued joint. Consequently, the regular traffic may continue during strengthening works. The work was performed in two periods. One half of the joints was reinforced during the summer, the other half in the fall, to study the effect of different weather conditions on the execution of work and strengthening success.

The most difficult part of the work was the preparation of the concrete surface because the bottom slab of the hollow boxes was very uneven. Partially, up to 4 cm of concrete had to be removed carefully without damaging the concrete underneath. Light pneumatically driven hammers and rotating grinders stood the test. Prior to the bonding of the plates, the deeper depressions of the top surface had to be evened-out with epoxy mortar. Then, the bonding area was sandblasted and vacuum-cleaned immediately before bonding. Sandblasting was performed to such an extent that the peaks of the 8 mm grains of the aggregate were visible within the bonding areas.

The concrete surface exhibited a tear-off strength in excess of 1,5 N/mm². Bonding work is suitably performed during the warm season. Relative humidity should not exceed 75 %; the temperature of concrete and ambient air should comply with the corresponding range of the adhesive as specified by its manufacturer. The steel plates have to be grit-blasted to a metallically bright state of surface, cleaned and degreased, resp., immediately before placement of glue on plate. However, there are more reliable alternatives today. Temporary corrosion protection by a shop-primer consisting of suitable dissolved epoxy resins with corrosion inhibiting active pigments is recommended.

After mixing of the components of the epoxy resin, the adhesive should be placed on the plate without delay. After placing of the plate, a uniform pressure must be exerted to ensure the squeezing-out of excess adhesive. This is performed with exactly straight squared timber pieces of the same length as the plate and with suitable props. Uniformity of pressure is more important than its magnitude.

The placement of the plates is shown in Fig. 4. After removal of the props, the steel plates will be protected against atmospheric corrosion in the ordinary way.

Fig. 5: Placement of steel plates in a hollow box girder

Well in advance of the bonding work on site, the identity of the delivered adhesive with the accepted one must be established. The identity tests comprise: infrared spectroscopy, thermal analysis (DTA/TG or DSC), strength development, and other relevant data. On site, quality control comprises the fabrication of specified specimens for each day of bonding work and the ensuing tests: tear-off strength, double lap shear strength, compressive and flexural strength on prisms, and others. Certainly, bonding work requires craftsmen with high skill and experience.

5. Load tests and field measurements

In order to control the efficiency of the steel plate reinforcement to reduce the steel stress amplitudes due to traffic load, several load tests were performed at the Sterbecke-bridge prior and after strengthening. These tests were executed during the early morning hours, because then, no temperature gradient across the section and consequently, no thermal restraint existed.

The load tests were performed with three, gravel loaded trucks, each wheighing 22 tons. These trucks drove along the bridge either side by side or sinqly on different, specified tracks.

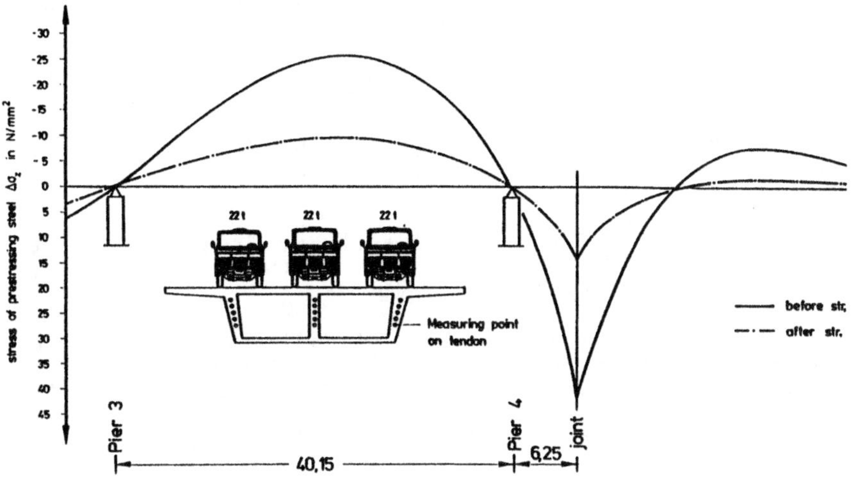

Fig. 6: Influence lines of stress of prestressing steel at coupling joint

During the travel of the trucks, the variation of the crack width at the coupling joint due to live load and of the steel stress at the coupler were measured. Fig. 6 shows the change of steel stresses as a function of the position of the trucks prior and after strengthening. The maximum difference of steel stresses of $\Delta\sigma_V$ = 45 N/mm² before strengthening with steel plates was reduced by the strengthening to a value of $\Delta\sigma_V \approx$ 10 N/mm². This reduction of the steel stress range convincingly proves the success of the described method of repair. Consequently, it led to the strengthening of several other motorway bridges.

6. Summary

The strengthening of continuous post-tensioned concrete bridges which exhibited cracks in the coupling joints with bonded steel plates proved to be efficient and economical as well. An outline of the research work for the dimensioning of the plates and of accelerated durability tests is given. The experience of first practical applications and results of field measurements are presented.

References

L'Hermite, R.: (1977) Le collage structural et le renforcement par résines des structures de la construction. Ann. Inst. Tech. Bat. Trav. Pub. No. 349, 18 - 23

Ladner, M.; Weder, Ch.: (1981) Concrete Structures with Bonded External Reinforcement. EMPA Report No. 206

Lloyd, G.O.; Calder, A.J.J.: (1982) The Microstructure of Epoxy Bonded Steel-to-Concrete Joints. TRRL Sup. Rep. No. 705

Ranisch, E.-H.: (1982) Load Capacity of Steel-to-Concrete Bondings-Bonded Reinforcement-(in German). Doct. Thesis Technical University Braunschweig

Rostásy, F.S.; Ranisch, E.-H.: (1984) Strengthening of Concrete Structures with Bonded Fibre Composites (in German). Technical University Braunschweig, Research Report

Voß, W.: (1984) Building Defects and their Reasons. Betonwerk + Fertigteiltechnik Heft 3/84

BONDING OF STEELBOLTS TO CONCRETE BY UKA 3 - TECHNOLOGY

DR. ING. RUDOLF SELL   Consulting Engineer   Munich / W. Germany

Abstract

Research into the bond and durability of polymer-mortared studs was carried out according to the UKA 3 technology method. The studs were also exposed to extreme weather conditions and chemical influences as well. In the pull-out-tests 18-years old bolts had a shear failure between the mortar and the steel stud. The shear strength after 8 1/2 years of weathering reached an average value of 14,5 N/mm^2 which corresponds to the average value of bond strength reached after one day of curing.

The bond specimens that were sustained under a steady load and cured in alkalic concentrated water for over a period of 8 1/2 years showed a larger slippage in the bond than those cured under dry condition, the strength, however, was the same.

Moreover it was found that cracks of 0,3 mm in concrete caused a strength reduction of 50% for studs of 20 mm in diameter and 80% for studs of 10 mm in diameter.

1. Introduction

In Germany since 1962 the polymer pre-fabricated mortar, enclosed in capsules of glass has been used in order to fasten steel threaded bolts into holes drilled in concrete. This developed technology of the mixing of mortar in drilled holes has become known as the UKA-3-technology method.

When inserting a threaded stud by the UKA-3-technology method the stud acts like a mixing rod turning and compacting the contents of the glass capsule together to form a thick mortar. Also, during the operation of inserting the stud into the drilled hole the threads act to move, distribute and solidify the mortar. Once the mixing rod is at the end of the hole the mixing must be stopped so that the mortar is not forced out of the hole. The pre-fabrication of the mortar quarantees a perfect composition. The sand within the capsule also takes on a large role because

- the sand works to grind the glass of the capsule,
- during the mixing the sand also contributes and distributes the resin within the hole, also any dust that may be on the surface of the hole will be mixed in to create an even mixture,
- it cleans the bonding surfaces by friction,
- it compacts and thickens the mortar,
- a favorable distribution is created by forcing the large particles between the threads,

- it reduces the shrinkage during the hardening of the resin,
- it reduces the temperature created by the reaction of the resin,
- it thickens the fresh mortar,
- makes the bond better and
- reduces the slippage.

The mortar production in restrained areas therefore leads to a different and better distribution of the mortar compared to the method of producing the liquid-resin-mortar outside of the hole, and then to be poured into the restrained areas afterwards. The simple handling of the UKA 3-technology is also the main reason for its widespread use and is of great popularity today.
The load of this fastening-system depends on the bond capacity of the threaded stud with the concrete. Any interference with this bond will weaken the system. Therefore the question of bond is of great importance and will be discussed later in this paper.

2. Composition and properties of the mortar

The components of the mortar are enclosed in a glass capsule so that they will not react until the glass is broken.

The components are
- a high reactant unsaturated polyester resin,
- aggregates of spherical shape and of hard hardness,
- a separate glass tube filled with pressed powder containing a reacting agent.

By producing the mortar in the hole all components will be mixed into the resin. Depending on the diameter and the size of the capsule the ratio of solid material to resin is different. The average ratio by weight of resin to solid material is 1 to 2.5. The mechanical properties of the hard mortar are listed below:

- compressive strength	95 to 115 N/mm^2
- bending strength	20 to 28 N/mm^2
- modulus of elasticity	25×10^3 N/mm^2
- shear strength	9 to 20 N/mm^2

The results here were determined on specimens 4 x 4 x 16 cm relating to the recommendation of DIN 1164. The shear-strength was determined by using threaded studs mortared by UKA 3 technology. They were tested in reinforced concrete specimens mortared over a length of 8 times the width of the diameter of the stud.

3. Bond Strength

3.1 Definition

A threaded stud of normal steel quality fastened by means of the UKA 3-technology is not strong enough to hold the load according to the bond strength of the polymer mortar. In a pull-out-test the stud broke before the bond strength of the polymer mortar was reached.

For that reason it is necessary to use a high tensile steel stud if the bond strength of the polymer mortared threaded stud is to be tested. In order to obtain a definite answer regarding the bond a relative short bond length is necessary for testing purposes. Also a special specimen would have to be available in order to receive reproducable results. This can only be achieved if the stress on the loaded end is at least nearly the same as on the unloaded end. Only a very short bond length can guarantee such conditions. The preparation of specimens with a bond length according to the diameter of the stud is quite a problem. The easier way to obtain a specimen of short bond length is to mortar a stud in the normal length of 8 times of the diameter by means of the UKA 3 technology and then to drill out the stud together with the concrete in a cylindrical shape afterwards. Thin slices have to be cut in right angles to the axis of the stud. In the test the steel stud had to be moved parallel to the stud axis in order to obtain a shearing in the bond surface. The ultimate load in relation to the bond surface is what we call the specific bond strength (see figure 1).

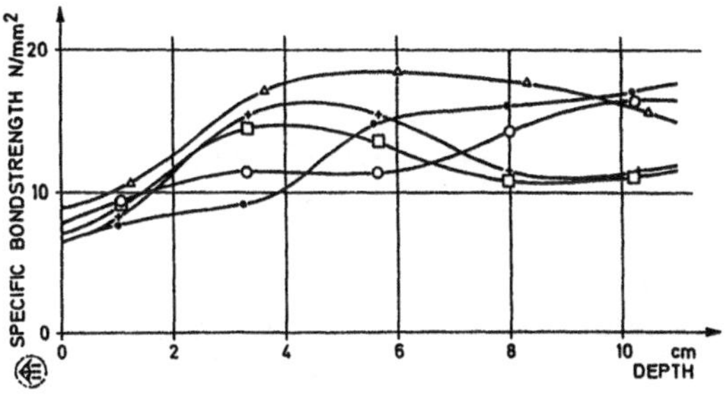

Fig. 1: Distribution of specific bond strength τ_U

Comparing several specimens with one another it is surprising to discover that the distribution of the specific bond strength is rather different. The maximum value of specific bond strength can be found at a different depth, whereas the minimum value of bond strength is always found at the mouth of the hole.

These results can be explained by the technology of the mortar preparation in the hole where the mixing rod is being turned and pushed. The solid parts of the mortar are compacted at the bottom of the hole, whereas the liquid component is being mixed the longest and therefore is transported towards the mouth of the drilled hole.

In a simplified analysis of the results of the pull-out-test the ultimate load is being related to the size of the convex surface of the hole. This ratio shall be regarded as the 'bond-strength'. This simplification can be justified by the fact that the bond length of different studs is always approximately 8 times the size of their diameters.

These values of 'bond-strength' can only be compared with one another if the pull-out-test carried out allows free deformation of the concrete in the stress area.

3.2 Comparison of polymer-bond with the cement-bond

In a series of tests the bond strengths were figured out by taking several types of steel studs and by depositing them 12,5 cm deep into the fresh cement-concrete.

Type A was a high tensile stud with a metric thread, size M16;
Type B was a steel stud like A, but coated with UKA 3-mortar and profiled like the convex surface of a drilled hole in concrete;
Type C had a profile like a stud used for reinforcement, the diameter was 18 mm;
Type D studs were of type A but mortared in holes of hardened concrete according to the UKA 3 technology.

Table 1 below gives a summary of the measured bond-strengths. The load-slippage-diagrams have been recorded in figure 2.

TABLE 1

TYPE OF STUD	BONDLENGTH	BONDSTRENGTH		
		MINIMUM VALUE	MAXIMUM VALUE	AVERAGE
	cm	N/mm^2	N/mm^2	N/mm^2
A	13	5,58	6,86	6,2
B	12 – 13	3,4	4,62	3,9
C	13	6,48	6,73	6,6
D	12,5 – 12,8	9,83	13,45	12,5

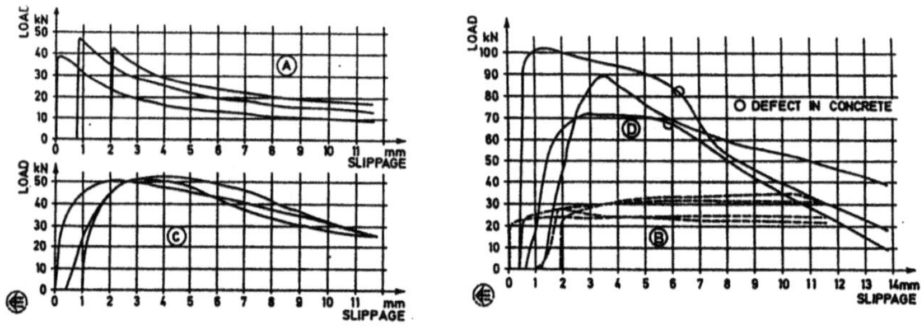

Fig. 2: Load-slippage diagrams of stud types A, B, C and D

The diagram shows distinctly the different behaviours of the stud types A through D under load. The behaviours are remarkable, especially in the range of the ultimate load.

3.3 Critical remarks regarding the bond-strength

Not only the characteristics of the load-slippage curves are different but the ultimate loads as well, even though the bond-lengths were the same, i.e. 8-times their diameters. The reason for that is that the stress distribution on the bond-surface is much higher near the loaded end than at the unloaded end of the stud. At the loaded end of the stud the local bond-strength is consequently being exceeded. In case of type A the result of this local failure is the collapsing of the bond in total; this is evident in the extremely sharp bend in the slippage-curve.

With the first occurrence of a shearing defect the total collapse is being initiated. Once the cement-paste along the first ridge of the threaded stud is damaged the load will be transferred to the next ridge until that ridge is damaged and then onto the next. Ultimately this process will lead to a total failure in the mortar by shearing off between the ridges of the threaded studs. This mechanism of breakage is typical for a brittle-failure.

Just the opposite here, a completely different behaviour can be taken from the slippage-curves of type C. During the increasing load a small increase of slippage can be observed, and will rise more and more until the ultimate load is reached.

While the slippage is increasing a gradual decrease of load is observed. This gradual decrease of load in combination with a remarkable increase in slippage describes a ductile behaviour of breakage.

The most remarkable difference between the slippage curves of type A and type C is that the one for type A does not show a slippage under the ultimate load whereas for type C a considerable amount of slippage is observed under the ultimate load. This comparison makes clear that the bond-strength for type C does include a slippage, in other words the bond-strength given for type C is the value of a "dislocated bond". This shows also that in case of a "dislocated bond" high loads can be endured.

The slippage curve for the specimens of type B are also characterized by a "dislocated bond" after loss of the "stiff bond". The slippage increase under an approximate constant load corresponds to the ideal picture of a ductile failure. - Such a ductile behaviour is a desirable asset for fastenings because the progressing failure is usually being noticed before a total collapse takes place.

The slippage curves of high tensile threaded studs of size M16 mortared according to the UKA 3 technology method are very instructive. The samples of type D show some slippage from the beginning of loading. Under the maximum load the slippage amounts to 1 mm through 4 mm. During the increase of slippage the load decreases until the concrete breaks off. The failure of the mortared studs (type D) is characterized by a dislocated bond and finally by a concrete rupture.

As mentioned before in the pull-out-test the concrete suffers a certain deformation, provided that the pull-out-machine is being situated within a certain distance of the setting-area of the stud. This kind of stress being brought onto the bond corresponds to the stress to be experienced by a mortared threaded stud being used as means of connection in practice.

In case that a failure of concrete should be restrained, a shearing off of the mortar along the threads will occur. - This is a shear-breakage in the polymer mortar which allows to calculate the shear-strength of polymer-mortar.

4. Influences on the bond-strength of polymer-mortar

The polmer-mortared studs, being used as a means of fixing or fastening is not only exposed to mechanical strain, but may also be exposed to extreme weather conditions. Therefore a change in the concrete condition around the stud can take place thus weakening the system; influences caused by varying conditions will create alterations in the material and thus in application.

4.1 Resistence of polymer-mortar regarding age and weather

In October of 1962 crane-tracks were fixed using M16 studs in a reinforced concrete structure. Since 1962 the studs have been exposed to weather and had to endure rigorous conditions of rain, snow, ice and sunshine. They were also exposed to sediments of pollution from the nearby industries. The studs used to secure the crane-tracks were tested over a period of 24 years and the results are given in table 2 below.

TABLE 2

	REF.	DATE OF SETTING	DATE OF TESTING	AGE OF MORTAR DAYS	AGE OF MORTAR YEARS	* SHEARSTRENGTH + BONDSTRENGTH AVERAGE N/mm²	STANDARD DEVIATION N/mm²
IN SITU	/2/	NOV.62	OKT.72	-	10	10,4 *	2,09
	/4/	NOV.62	APR.81	-	18,5	10,6 *	1,26
	/7/	NOV.62	APR.86	-	23,5	11,2 *	1,42
	/2/	OKT.72	OKT.72	1	-	11,8 +)0	2,26
	/3/	MAR.73	MAR.73	1	-	13,4 +)1	1,13
	/7/	MAR.73)2	APR.81	-	8	14,5 *	2,83
LAB	/1/	OKT.66	OKT.66	14	-	9,7 +	0,75

*BREAKAGE IN MORTAR BY SHEARING
+BONDFAILURE ON THE CONCRETE SURFACE
) SHEARSTRESS 0≙ 13,3 ; 1≙ 15 ; 2 PRELOADED UP TO 14 N/mm²

According to these results the following can be stated:

- after the mortared studs were exposed to weather for more than 10 years the breakage occurred in the polymer-mortar;
- there is no significant difference in the shear-strength of threaded studs between the ages of 10, 18 and 24 years;

- 8 year old threaded studs mortared with capsules produced with material from the same resin-batch have the same shear-strength as those threaded studs measured after one day, even though the 8-year old studs have been exposed to all kinds of weather conditions; the failure of the 8 year old studs was observed in the polymer-mortar.

4.2 Chemical resistence of polymer mortar

It is only natural that the question of chemical resistence regarding a UP-resin-polymer-mortar is of great importance when being used as a mortar in concrete of basic components. Therefore a short report is given on creeping-tests with UKA 3 mortared studs of size M10 which had been held under a constant load in $Ca(OH)_2$ saturated water for 8 1/2 years. In order to guarantee a quick and constant soaking of the concrete, a cement conrete with porous aggregates and with a compressive strength of 26 N/mm^2 and density of 1,46 kg/dm^3 was used for the tests.

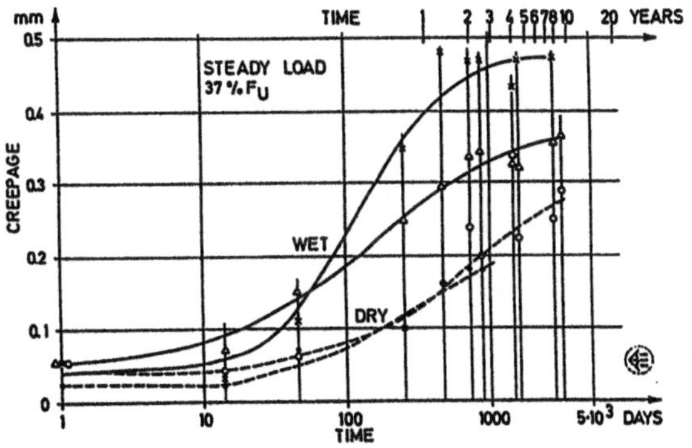

Fig. 3: The creepage-curves of studs in wet and dry cured specimens under comparable load and temperature

The most instructive results shown in figure 3 is the slippage of the stud which had been held in $Ca(OH)_2$ saturated water under a constant load of 37% of the ultimate load. After a period of 8 1/2 years the studs kept in water showed an average slippage of 0,43 mm, whereas the threaded studs kept in air showed only an average slippage of approximately 0,25 mm. - After an interruption of the creep-tests the bond-strengths of the studs were measured; these values will be found in table 3.

TABLE 3

STEADY LOAD DURING THE DEPOSIT IN % OF ONE DAY STRENGTH	AVERAGE BONDSTRENGTH, RESP. SHEARSTRENGTH** AFTER 8,5 YEARS OF SUSTAINED LOADING	
	DRY N/mm^2	WET N/mm^2
18	8,4	11,3 **
28	8,1	10,1 **
37	8,8	9,2 **
ONE DAY BONDSTRENGTH	8,0	-

Surprisingly a bond failure occurred only in the dry specimens whereas in the soaked specimens a failure of the mortar by shearing off between the ridges of the studs was observed.

The remarkable difference between the bond strength and the shear strength can be explained only as a consequence of higher creepage under sustained load in the soaked specimens. The solid compounds of the mortar are compacted more densely under creepage.

4.3 Influence of open cracks in concrete upon the bond-strength

When a stud mortared in reinforced concrete is used as a means of connection one may ask, what influence will a crack in the reinforced concrete structure have on the bond-strength of the connecting element.
In the following figure 4 you will find the slippage-curves of studs without crack-influence (r = 0) and also for studs tested in cracked concrete (r =0,35 and r = 1,1 mm). The fundamental difference of the slippage curves makes evident that the reason of failure was different, too. In concrete without cracks the concrete ruptured after a slippage of 0,8 mm under a load of 40,4 kN. In a test with a crack width of 0,35 mm the ultimate load of 13,7 kN with a slippage of 13 mm finally caused the rupture of concrete. In a comparable test with a crack width of 1,1 mm the stiff bond broke under a load of 17,7 kN and had a slippage of only 0,15 mm. After this event a sudden failure was accompanied by a large decrease in load.

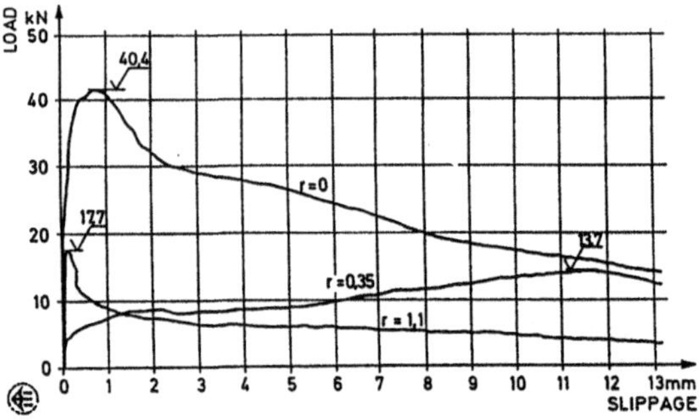

Fig. 4: Slippage diagram of studs of size M10 mounted in crack widths of r = 0, r = 0,35 and r = 1,1 mm

In figure 5 you'll find slippage curves from tests with studs of size M20. A dislocated bond in a crack of 1 mm will withstand a load of 41,8 kN and a slippage of 12 mm before the conrete breaks.

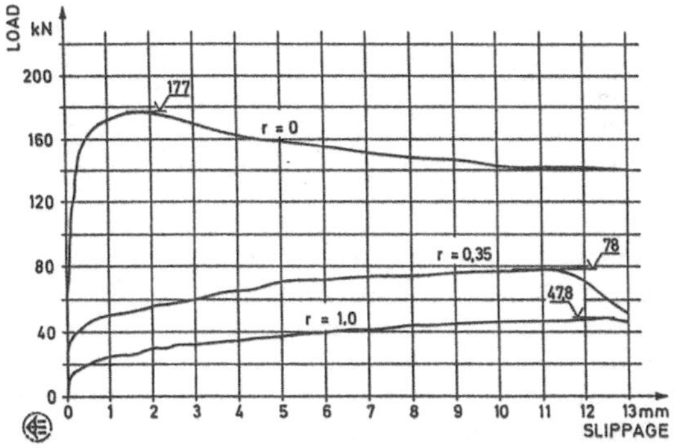

Fig. 5: Slippage-diagram of studs of size M20 mounted in crack widths of r = 0 , r = 0,35 mm and r = 1,0 mm

In order to consider the influences of bond strength in concrete with cracks and without cracks all tests have been related to concrete strength of $ß_W$ = 17,5 kN.

It is obvious that an increase in crack width will decrease the bond strength, so in the case of a wide crack the bond strength is rather weak. In case of an extremely 'toothed' polymer-mortar along the drilled hole surface the mortar is shearing off at one or all sides of the surface of the stud.

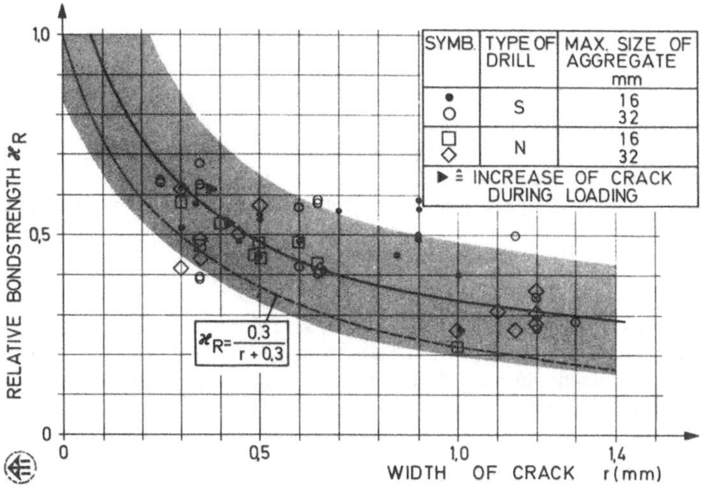

Fig. 6: Comparison of relative bond strength and width of cracks / example of stud size M20

$$\kappa_r = 0,3 / (r + 0,3) \qquad (1)$$

Since the rippled condition of the hole surface shows great variations, the bond strength deviates, too. The scatter-diagram (see figure 6) showsa regression curve and an estimated confidence interval with the lower limit described by the formula (1), valid for crack widths of 0,2 until 1,0 mm. In figure 6, an example of a stud of size M20 is given to show how the relative bond strength decreases upon an increase in crack width.

Concrete structures showing cracks of a width of more than 0,3 mm, are usually regarded as damaged. The damage may have been caused by diverse matters, either by extreme loading or by not being dimensioned correctly in its reinforcement. So, if we take crack widths of 0,3 mm into consideration, which in practice will not be found very often, the relative bond strength within the lower confidence limit for studs of the size M20 is approximately 0,5. The workable bond stress for a mortared stud size M20 will therefore be 0,5 of the workable stress for studs used in concrete without cracks. The corresponding factors for other sizes of studs are given in figure 7 and are to be regarded as valid for a crack width of 0,3 mm, too.

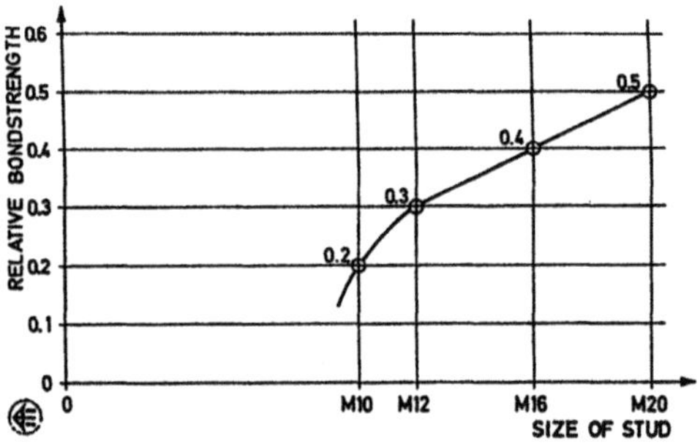

Fig. 7: To derive the relative bond strength for polymer-mortared threaded studs of the sizes M10 to M20, mounted in concrete with cracks of 0,3 mm in width a factor of κ_r is used to determine the workable load.

Conclusions

The short review given on research work in the field of resistance and durability of polymer-mortared studs according to the UKA 3 technology has shown that this method is safe enough also for outdoor use (weather and water proof as well).

Investigations into the influence of cracks in concrete on the bond strength make evident that the slippage-curve is quite different from the one representing the behaviour of studs mortared in concrete without cracks. The 'stiff bond' is no longer intact but is being replaced by a 'dislocated bond'. The steady increase of load in the pull-out-test is accompanied by a steady increase of slippage. The bond turns ductile but is still safe. In the case of a crack-width of 0,3 mm a reduction of the workable load in bond is necessary. In order to consider the influence of cracks in concrete a 'reducing factor' of 0,2 for studs of size M10 and a factor of 0,5 for studs of size M20 is suggested.

With regard to the durability of polymer-mortared studs a perfect mortaring up to the mouth of the hole is of great importance, otherwise - as observed in the tests - a most favorable condition for steel-corrosion on the stud will be produced, which should be avoided.

References

/1/ Prof. Dr. Ing. G. FRANZ (1966) TU Karlsruhe

/2/ Dr. SELL, Report No. 72 019
"Ausziehfestigkeit 10 Jahre in Gebrauch befindlicher, der Witterung und erhöhtem chemischen Angriff ausgesetzter, reaktionsharz-vermörtelter Anker, System Upat UKA 3"

/3/ Dr. SELL, Report No. 73 023
Ergänzungsversuche zum Untersuchungsbericht Nr. 72 019
"Ausziehfestigkeit hochfester Gewindeanker M16, vermörtelt mit UKA 3 Patronen M16, an der Kranbahn der Kohlenbunkerungsanlage des Kraftwerks München-Nord"

/4/ Dr. SELL, Report No. 83 223
"Ausziehfestigkeit 18 Jahre in Gebrauch befindlicher, reaktionsharz-vermörtelter Anker UKA 3 M16"

/5/ Dr. SELL, Report No. 74 048
"Vergleichende Untersuchungen zur Verbundtragfähigkeit im Frischbeton eingelegter Profilstähle und UKA 3 vermörtelter Gewindestäbe"

/6/ Dr. SELL, Report No. 83 231
"Kriechschlupf und Ausziehfestigkeit bei kurz- und langzeitiger Belastung, im trockenen und nassen Normal- und LIAPOR-Beton UKA 3 vermörtelter, trocken und in Wasser gelagerter Anker"

/7/ Dr. SELL, Report No. 85 270 (under preparation)

/8/ Dr. SELL, Report No. 85 257
"Ausziehwiderstand von UKA 3 Ankern M10 bis M20 in Abhängigkeit von der Rißweite"

EFFETS DU FLUAGE SUR LES COLLAGES ACIER-BÉTON

SERGIO TATTONI
Dipartimento di Ingegneria Strutturale, Politecnico di Milano

Resumé
Dans cet étude on a analizé une jonction plaquée tout en considerant que les subjectiles sont élastiques et que l'adhésif est viscoélastique de façon linéaire. Après avoir formulé le probleme sous forme générale, en considerant deux modèles de fluage de l'adhésif, on a developpé des exemples specifiques pour differents épaisseurs. La distribution des contraintes et autres paramètres significatifs ont été calculés au moyen d'une methode discrète d'analyse. Les résultats montrent que la valeur maximale de la contrainte de cisaillement dans la jonction plaquée diminue de façon differente sélon le modèle de fluage consideré et le moment où on applique la force.
Key Words: béton plaqué, fluage, jonction collée, réparation des structures en béton, résine epoxyde.

1. Introduction

Dépuis le congrès Rilem du 1967, pendant lequel l'Hérmite et Bresson présentèrent un rapport relatif a l'application du "béton plaqué", le collage acier-béton est devenu une technique très employée surtout pour ce qui concerne la réparation et le renforcement des structures en béton. Pourtant il arrive très souvent que l'auteur du project concentre son attention sur les proprietés chimiques et mecaniques de l'adhésif tout en negligeant le but de l'application et le comportement global du système plaqué. Un des aspects le plus remarquable d'un collage acier-béton est constitué par la distribution des contraintes d'adhérence à la surface de separation adhésif-metal, et particulierment à celle du béton, qui à l'égard des actions tangentiales montre un comportement du genre fragile. À ce propos on a mis en marche dépuis bien de temps un travail de récherche soit théorique soit expérimental, aboutissant à un modèle de calculation pour les collages acier-béton capable de tenir en compte la variabilite de l'épaisseur de la tôle collée, les effects du confinement passif et actif, les effects sécondaires dûs à la rigidité flexionelle de la tôle meme /1, 2, 3, 4/.

Un autre aspect très interessant est la conduite sous charge constante (fluage) d'un système collé par des résines epoxydes. Tous les études enterpris jusqu'ici dans ce domaine, concernent le problème de la viscosité de prismes en résine epoxyde (plus ou moin chargee avec inertes silicieux) sousmis à une action normale de compression /6, 7/. Ces expériences-ci étaient jusqu'à présent addressées à l'étude de la conduite des formules epoxyde utilisés en façon de couche de pose et jonction de conceaux de poutres de pont post-contraintes et ont montre que la déformation visqueuse dépend de nombreux facteurs tels que le rapport charge siliceuse/résine, témperature ambiente et valeur de la sollicitation appliquée; tout de même on peut conclure que sous sollicitation axiales contenues (10 MPa) la déformation visqueuse d'un prisme réalisé avec résine chargée, s'épuise entre 500 jours à partir de l'application de la sollicitation atteignant une valeur finale environ égale a 10-15 fois la déformation élastique. Couche subtiles d'adhésif sousmises à action normale de compression montrent une déformation visqueuse finale bien inferieure à celle des prismes et tout à fait comparable a celle du béton (2-3 fois la déformation élastique), à cause de l'effet de confinement exercé par les subjectiles (the adherends) /6, 7/. Meme dans le cas de collages acier-béton sousmis à action de cisaillement la deformation visqueuse est bien inferieure à celle des prismes en résine /4/; neanmoins actuellement font défaut résultats expérimentaux complets au sujet.

Le but de cette note est de prévoir sur la base des résultats expérimentaux acquis /6, 7/ la conduite des contraintes et des déformations d'un collage acier-béton à superposition simple employant le schéma de calcul déjà mis à point pour des expériences pareilles dans le domaine élastique. En particulier on évaluera les variations au cours du temps des contraintes d'adhérence, des contraintes normales au plan de collage et de la dilatation normale du collage même. En générale on choisit pour l'adhésif un modèle viscoélatique linéaire /8/, pour lequel soit valable le principe de superposition de Boltzmann /9/. Cependant, en considérant que le procédé de polymerisation ne s'épuise pas pendant les premiers jours après la pose de la formule epoxyde, mais continue avec le temps, de facon analogue à ce qui a lieu pour le durcissement du béton, dans ce travail on a consideré non seulement le modèle héréditaire, mais aussi le modèle de vieillissement, qui tient compte de l'âge du formulé epoxyde à l'instant d'application de la force.

2. Étude du probleme

2.1 Hypothèse et modèle de calcul

On exprime pour le problème proposé les hypothèses ordinaires pour le calcul des collages acier-béton et, en particulier, que:

- la couche de l'adhésif et de la tôle collée soient petites en rapport aux dimensions de l'élément plaqué;
- l'apport dû à la flexion de la tôle collée soit négligeable;
- l'élément en béton soit infinement rigide en rapport à la tôle et à l'adhesif;
- on peut réconduire le problème de la détermination des sollicitations au cas plan.

Ces hypothèses surnommées sont verifiées /1, 2/ dans le cas des collages acier-béton généralment réalisés dans la pratique constructive dans lequels, en plus, la largeur de la tôle collée en rapport à l'épaisseur est telle que, effectivement, sont négligeables les contributions de déformation et efforts d'adhérence perpenduculaires à l'axe de l'élément lui-même (on suppose en conséquence une déformation cylindrique de l'élément infléchi). Dans cet étude on fait l'hypothèse que la déformation visqueuse soit principalement dûe à l'adhésif epoxyde interposé entre acier et béton, hypothèse autant plus valable à mesure que l'operation de renfort par collage est executée sur structure en place dépuis bien d'années (même quelques dizaines d'années).

Supposant valables les hypothèses précédentes, c'est possible schématiser le collage acier-béton à l'aide d'un modèle discret de treillis hyperstatique (fig. 1) dans lequel l'entrait inferieur est répresenté par la tôle, les barres de paroie du treillis par l'adhésif, l'entrait superieur, supposé infiniment rigide, par le béton. Au moyen de l'introduction d'éléments élastique appliqués aux nodes du treillis est en outre possible simuler la présence d'un phénomène de confinement passif (revers, clouages, etc.). Ce modèle-ci a été choisis parce que il se prête bien à l'analyse des sollicitations d'adhérence, ainsi qu'à la détermination des déformations du collage même, même en direction normale à la superficie collée (dilatation du collage) /1, 2/.

2.2 Modèles rhéologique de l'adhésif.

La comportement visqueu de l'adhésif est répresenté par un modèle à trois paramètres (fig. 2) constitué par un élément élastique avec module E_1 en serie avec un élément de Kelvin composé à son tour par un ressort élastique avec module E_2 et un amortisseur visqueux avec constante η_2. Dans le domaine de la viscosité linéaire on étudiera deux differents modèles.

Le premier, ainsi dit héréditaire, suppose que les caractéristique rhéologiques du materiel ne changent pas avec le temps et il est défini par la suivante fonction de fluage /9/:

$$\Phi(t,\tau) = \frac{1}{E_1} + \frac{1}{E_2}\left(1 - e^{-(t-\tau)/\lambda}\right) \qquad \text{(1a)}$$

où $\lambda=\eta_2/E_2$ c'est le temps de rétard, t c'est l'instant lorsque on évalue la déformation, τ c'est l'instant lorsque on applique la force (fig. 3).

Le deuxieme modèle, à difference du précédent tient compte de la variation avec le temps des caractéristiques rhéologiques de l'adhésif et il ne rend pas la déformation visqueuse accumulée (modèle de vieillissement); la fonction fluage est exprimée par la (1b):

$$\Phi(t,\tau) = \frac{1}{E_1} + \frac{1}{E_2}\left(e^{-(\tau-t_0)/\lambda} - e^{-(t-t_0)/\lambda}\right) \qquad \text{(1b)}$$

où avec t_0 est indiqué l'instant de complètement de la polymerisation de l'adhésif epoxyde (temps de maturation).

Au cas général où la sollicitation appliquée à l'élement visqueux soit variable avec le temps, on peut évaluer la déformation au temps t

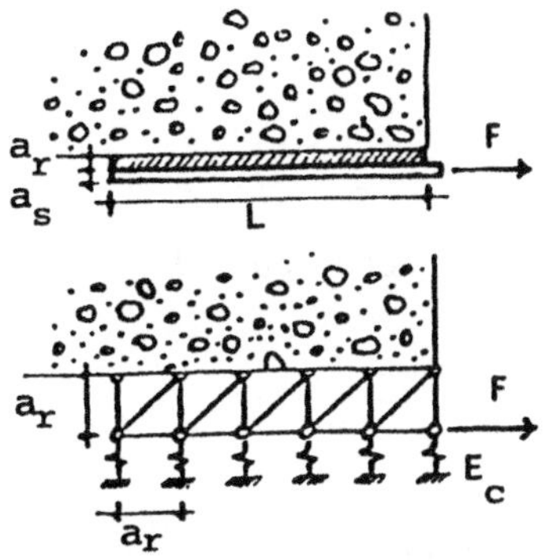

Fig. 1 - Modèle discretisé pour le calcul du plaquage.

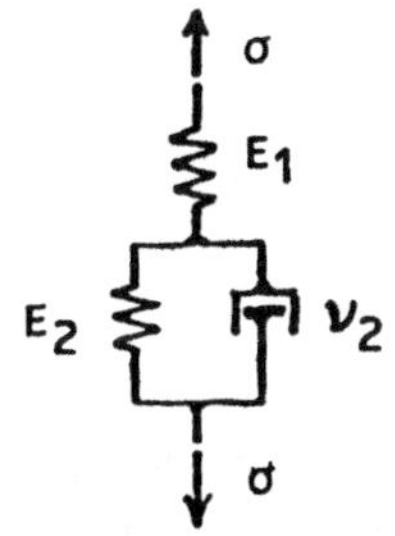

Fig. 2 - Modèle réhologique de l'adhésif.

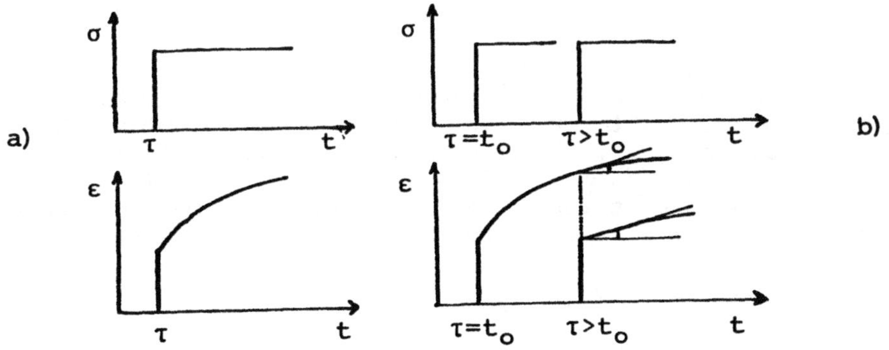

Fig. 3 - Fonctions fluage: a) modèle héréditaire; b) modèle de vieillissement.

à l'aide de l'integral de Stieltjes.

$$\varepsilon(t) = \int_0^t d\sigma(\tau) \cdot \Phi(t,\tau) \tag{2}$$

Auprès d'une réprésentation fondée sur la pseudo-fonction H de Heaviside,

$$\sigma(t) = \sigma_0 \left[H(t-\tau) - H(t-T)\right] \tag{3}$$

pour les deux modeles des figures 4a et 4b, on aura respectivement le suivantes expressions de la deformation

$$\varepsilon(t) = \sigma_0 \left[\Phi(t-\tau) \cdot H(t-\tau) - \Phi(t-T) \cdot H(t-T)\right] \tag{4a}$$

$$\varepsilon(t) = \sigma_0 \{\Phi(t,\tau) \cdot H(t-\tau) - \\ - \left[\Phi(t,\tau) - (\Phi(T,\tau) - \frac{1}{E})\right] \cdot H(t-T)\} \tag{4b}$$

3. Méthode d'analyse

La méthode d'analyse se réduit essentialement à l'application de la méthode des forces au treillis hyperstatique de la fig. 1 en imposant la congruence entre le déplacement relatif de deux éléments adiacents de treillis et le morceau en tôle compris entre eux. Dans ce but on évalue le déplacement horizontal de l'élément i-ème par suite d'une force unitaire (fig. 5) qui sera:

$$\delta_i = a_r (2\varepsilon_{di} + \varepsilon_{mi}) \tag{5}$$

où avec a_r est indiqué l'épaisseur de l'adhésif et avec les indicateurs "d" et "m" les déformations respectivement de la diagonale et du montant, cette dernière comprenente un possible effect de confinement passif /1, 2/. Le déplacement horizontal $\delta_i(t)$ au temps t provoqué par une force unitaire agissante pendent le délai $(t - \tau)<t<(t - T)$, grâce à la (5) prend l'expression suivante respectivement pour le modèle héréditaire et de vieillisement:

$$\delta_i(t) = \left[4\Phi_d(t-\tau) + \Phi_m(t-\tau)\right] \cdot H(t-\tau) - \\ - \left[4\Phi_d(t-T) + \Phi_m(t-T)\right] \cdot H(t-T) \tag{6a}$$

$$\delta_i(t) = [4\Phi_d(t-\tau) + \Phi_m(t-\tau)] \cdot H(t-\tau) -$$
$$-\{[4\Phi_d(t-\tau) + \Phi_m(t-\tau)] - 4(\Phi_d(T) - \frac{1}{E_r}) - \tag{6b}$$
$$- [\Phi_m(T) - \frac{1}{E_m}]\} \cdot H(t-T)$$

car les fonctions fluage pour le montant et pour la diagonale sont exprimées par:

$$\begin{cases} \Phi_m(t-\tau) = \frac{1}{E_m} \cdot [1 + \gamma_m(1 - e^{-(t-\tau)/\lambda})] \\ \Phi_d(t-\tau) = \frac{1}{E_r} \cdot [1 + \gamma_d(1 - e^{-(t-\tau)/\lambda})] \end{cases} \tag{7a}$$

$$\begin{cases} \Phi_m(t,\tau) = \frac{1}{E_m} \cdot [1 + \gamma_m(e^{-(\tau-t_0)/\lambda} - e^{-(t-t_0)/\lambda})] \\ \Phi_d(t,\tau) = \frac{1}{E_r} \cdot [1 + \gamma_d(e^{-(\tau-t_0)/\lambda} - e^{-(t-t_0)/\lambda})] \end{cases} \tag{7b}$$

ou $\gamma_m = E_m/E_2$ et $\gamma_d = E_d/E_2$.

La déformation de l'acier, essentiellement élastique, est éxprimée par la relation suivante /1/:

$$\delta_{si} = \frac{a_r}{E_s \cdot a_s} \tag{8}$$

Comme les déformations de l'adhésif sont fonctions du temps, la congruence des déformations doit être valable pour chaque instant t. Definie avec X l'action axiale inconnue dans le morceau de tôle i-ème, son cours en fonction du temps peut être discretisé (fig. 6) en supposant de diviser l'axe des temps en espaces d'ampleur Δt_m à la fin de chacun desquels on substitue la n-uple $\{X_m\}$ avec la n-uple $\{X_{m+1}\}$, étant n le nombre des inconnues. Le choix du pas d'integration optimal est effectué /10, 11/ par la relation (9)

$$\log t_m - \log t_{m-1} = \log A \tag{9}$$

où A est une constante qui permet d'obtenir avec le moindre effort la meilleure précision. De cette facon c'est possible d'éxprimer les inconnues à travers des "step functions", à l'aide de la pseudo-fonction de Heaviside; pour chaque éspace Δt_m sont pourtant valables les expressions de la déformation visqueuse exprimées par la (6a) et (6b) multipliées par la valeur de l'inconnue X_m. Si on suppose de connêtre le vecteur $\{X_m\}$ qui satisfait la condition de congruence au debut du m-ème éspace de temps Δt_m, c'est alors possible de calculer les déplacements des n treillis d'adhésif $\{\delta_m(t)\}$, dûs à l'application du vecteur $\{X_m\}$ supposé constant dans l'éspace de temps Δt_m. Puisque l'histoire de chargement jusqu'à l'instant t est connue, c'est alors possible de calculer le vecteur des déplacements residuèls $\{\rho_m(t)\}$ des n treillis, dûs à toutes les inconnues

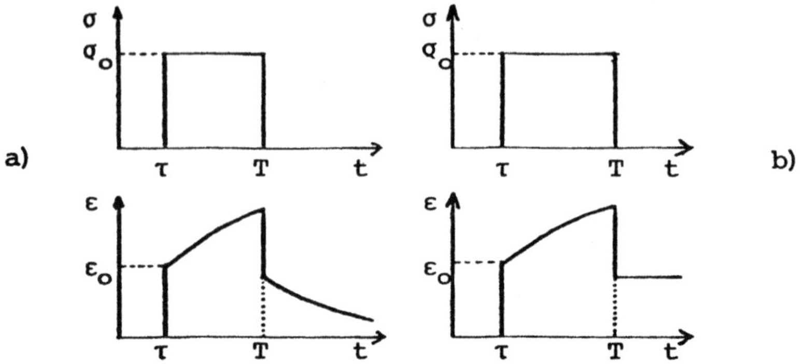

Fig. 4 - Déformation visqueuse provoquée par une "step" force.

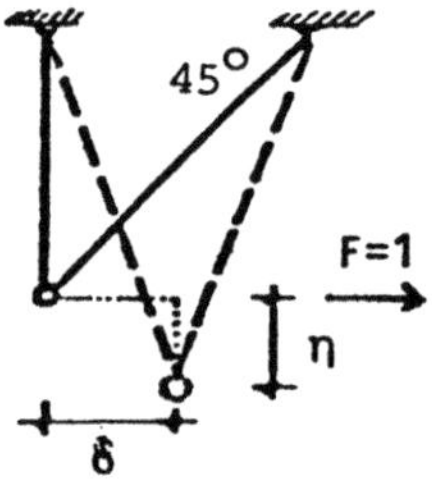

Fig. 5 - Déformation de l'élément i-ème de l'adhésif.

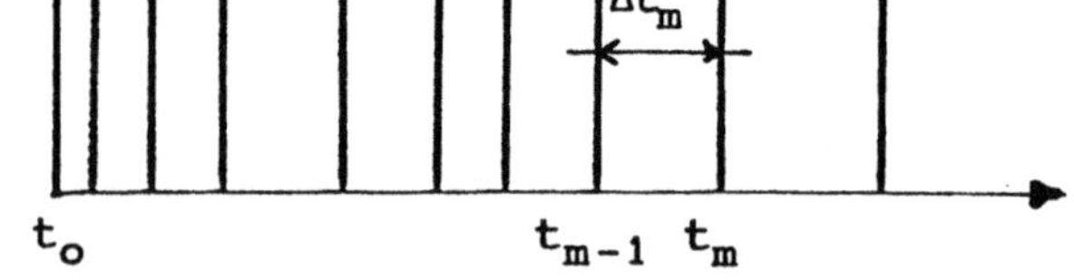

Fig. 6 - Discrétisation de la fonction X(t).

supprimées en précédence à l'instant où on a appliqué le vecteur inconnue $\{X_m\}$

Evidémment les expressions $\delta_{im}(t)$ et $\rho_{im}(t)$ changent en fonction du modèle visqueux choisi sélon ce qui est indiqué par les (6a) et (6b).

La congruence à la fin du m-ème éspace de temps Δt_m est imposé à l'aide d'un système d'equations du genre :

$$\delta_{m,i+1}(t) \cdot X_{m,i+1} - \left[\delta_{m,i+1}(t) + \delta_{m,i}(t) + \delta_{si}\right] \cdot X_{m,i} + \\ + \delta_{m,i}(t) \cdot X_{m,i-1} = \rho_{m,i} - \rho_{m,i+1} \tag{10}$$

par la resolution de laquelle on obtient le vecteur inconnu $\{X\}$ au temps t. De cette façon on peut calculer en correspondence du treillis i-ème la contrainte d'adhérence:

$$\tau_{ad} = \frac{1}{a_r} \cdot (X_i - X_{i-1}) \tag{11}$$

la dilatation normale:

$$\eta_i = \varepsilon_m \cdot a_r \cdot (X_i - X_{i-1}) \tag{12}$$

la contrainte normale au plan du collage:

$$\sigma_c = \frac{1}{a_r} \cdot \left[\eta_i \cdot \frac{E_c}{a_r} - (X_i - 2 \cdot X_{i-1} + X_{i-2})\right] \tag{13}$$

dûe à la présence eventuelle d'un confinement passif /1/.

4. Application numerique

En premier on a contrôlé la convergence de la méthode en verifiant que la solution obtenue avec la méthode exposée en precedence évaluée à un temps t suffissament grand ($t = t_\infty$) coincidait, rélativement au seul modèle héréditatire, avec la solution élastique /1/, obtenue en substituant au module d'élasticité istantané, le module relâché:

$$E_{ril} = \frac{E_1 \cdot E_2}{E_1 + E_2} \tag{14}$$

Ensuite on a evalué la déformation visqueuse d'un collage à simple superposition ayant les caractéristiques suivantes:

- longeur du collage /1/ $L = 50*a_s$;
- épaisseur de l'adhésif a_r = 5 mm;
- épaisseur de l'acier a_s = 5 mm.

L'adhésif epoxyde a les caractéristiques mécaniques et rhéologiques suivantes:

- module élastique instantané E_1 = 3000 MPa;
- coefficent de Poisson m = 0.35;
- déformation à temps infini $\varepsilon_\infty = n^*\varepsilon_{e1}$ avec n = 3 et t_∞ = 300 jours /6/;

Depuis ces données on deduit les valeurs suivantes des paramètres du modèle de la fig. 2:

$$E_2 = \frac{\sigma_0}{\varepsilon_d(n-1)} \tag{15}$$

$$\nu_2 = \frac{t_\infty}{7} \cdot \frac{E_1}{n-1} \tag{16}$$

Les figures suivantes montrent sous forme adimensionale les résultats des élaborations effectuees dans le cas du modèle a) héréditaire sans confinement (fig. 7) et avec confinement passif (fig. 8) et pour le cas du modèle b) de vieillissement (figs. 9 et 10) avec un temps de chargement coïncidant avec le temps de maturation ($\tau = t_o = 0$). Les figs. 11 et 12 montrent le comportement d'un collage réalise avec un adhésif caracterisé par un modèle de vieillisement (modele b) respectivement sans et avec un confinement passif, dans le cas où le collage soit chargé après 300 jours après la prise de la résine (τ = 300 jours).

Les tableaux 1, 2, 3 et 4 montrent pour les deux modèles et pour différents cas d'épaisseur de l'adhésif, de l'acier et pour confinement passif absent ($E_c/E_r = 0$) ou présent ($E_c/E_r=1$), certains paramètres signifiants relatifs aux efforts et aux déformations instantanées et differées dans le temps du collage même, et précisement:

- le coefficient d'utilisation /1/, defini comme le rapport entre la tension maxime de l'acier collé et la maxime contrainte d'adhérence résine-béton;
- la maxime contrainte de confinement (compressions positives) en rapport à la tension σ_s de l'acier;
- la dilatation normale η, en rapport à la longuer du collage;
- le glissement horizontal (fig. 5) du point d'application de la force F à l'acier collé, en rapport à la longueur du collage même.

Cette dernière grandeur, pourvu qui soit convenablement mise en corrélation à l'hauteur de la section et à la déformation du béton, constitue une indication très utile sur l'accroissement local de la courbure de la section plaquée par effect du fluage du collage même.

On fait aussi observer finalment que dans le tableau 1 les grandeurs mesurées à 0 jours à partir du moment d'application de la force, correspondent à celles de la solution élastique.

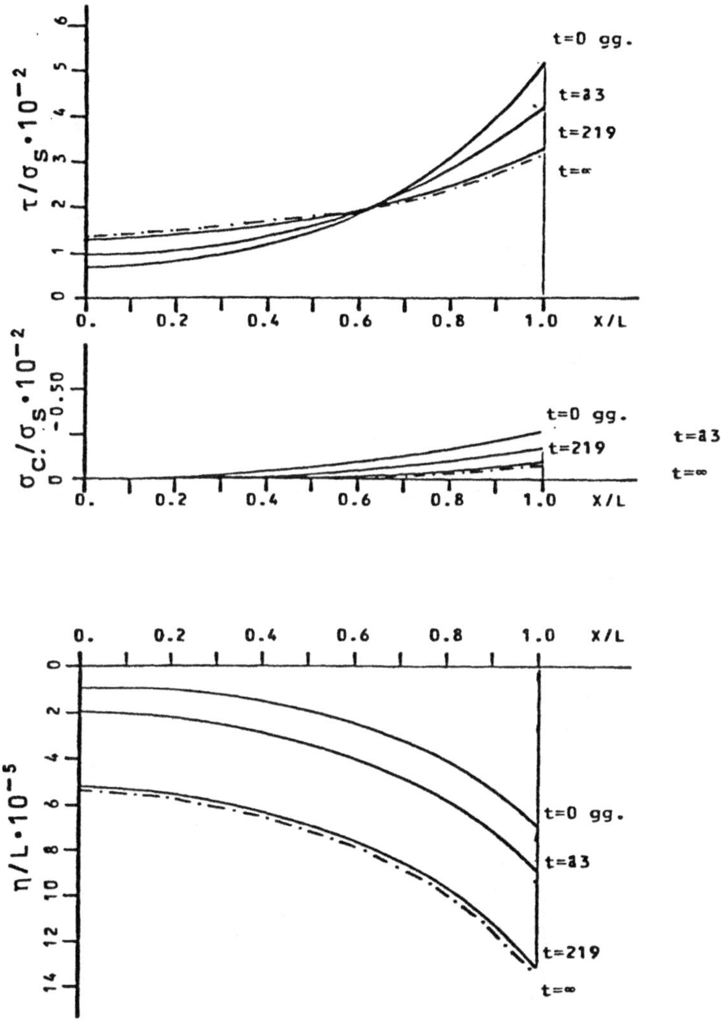

Fig. 7 - Distribution au cours du temps des contraintes d'adhérence τ et de confinement σ_c et de la dilatation normale η. Modèle héréditaire, confinement absent: $E_c/E_r=0$.

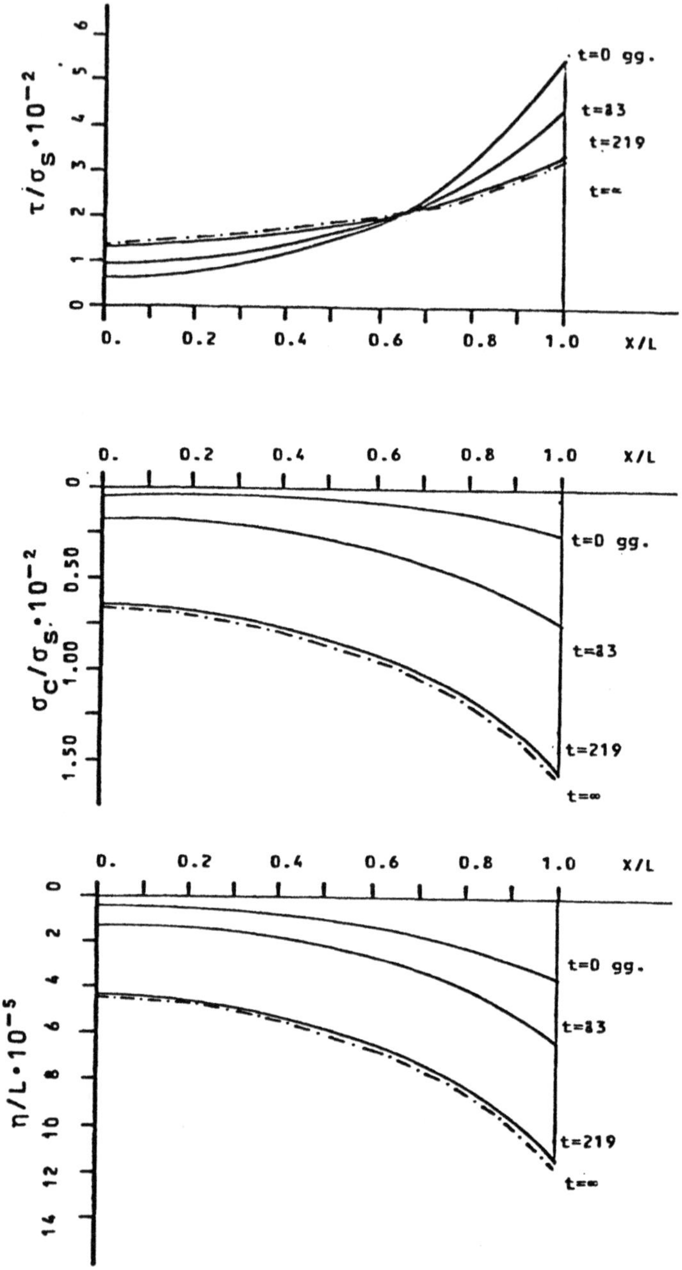

Fig. 8 - Distribution au cours du temps des contraintes d'adhérence τ et de confinement σ_c et de la dilatation normale η. Modèle héréditaire, confinement présent: $E_c/E_r=1$.

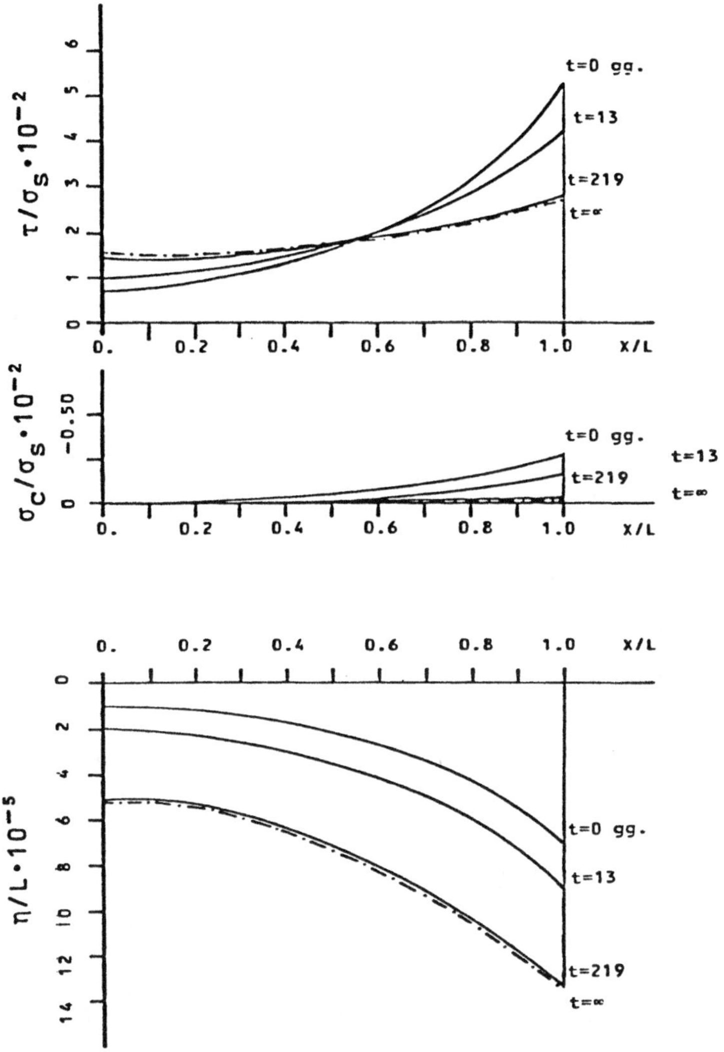

Fig. 9 - Distribution au cours du temps des contraintes d'adhérence τ et de confinement σ_c et de la dilatation normale η. Modèle de vieillissement, confinement absent: $E_c/E_r=0$. Charge appliquée à partir de l'instant de polymérisation.

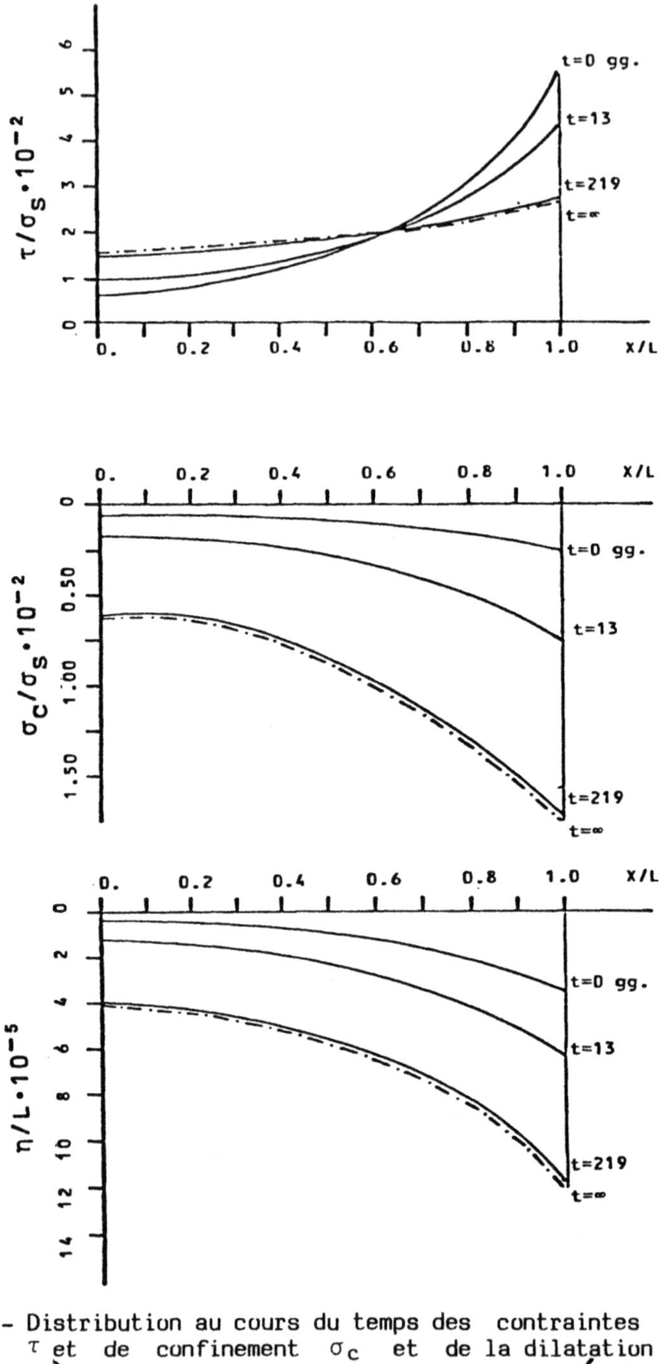

Fig. 10 - Distribution au cours du temps des contraintes d'adhérence τ et de confinement σ_c et de la dilatation normale η. Modèle de vieillissement, confinement présent: $E_c/E_r=1$. Charge appliquée à partir de l'instant de polymérisation.

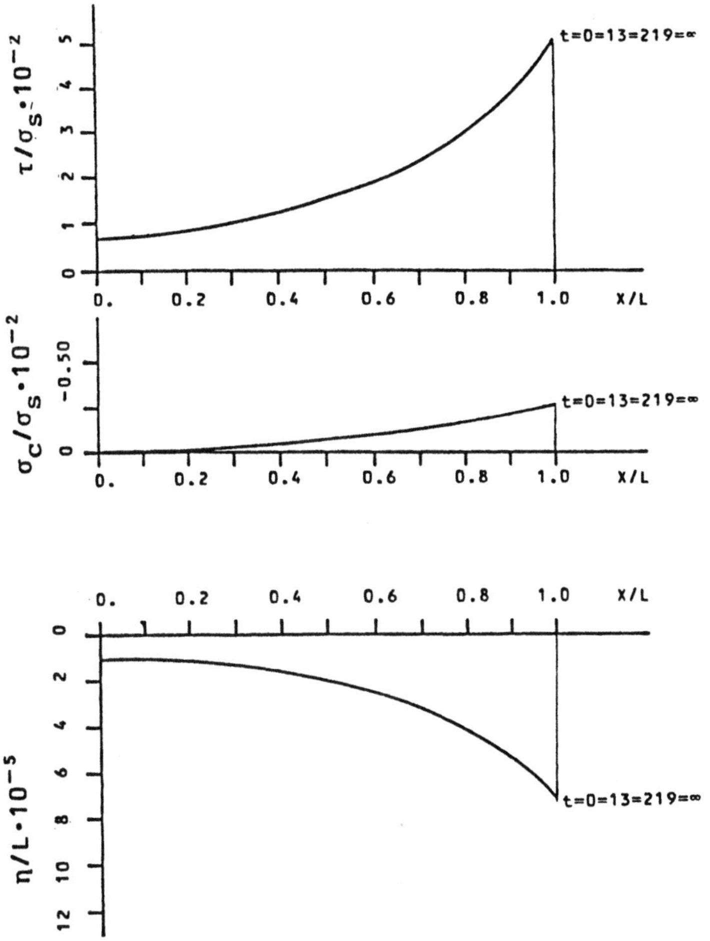

Fig. 11 - Distribution au cours du temps des contraintes d'adherence τ et de confinement σ_c et de la dilatation normale η. Modèle de vieillissement, confinement absent: E_c/E_r=0. Charge appliquée depuis 300 jours à partir de l'instant de polymérisation.

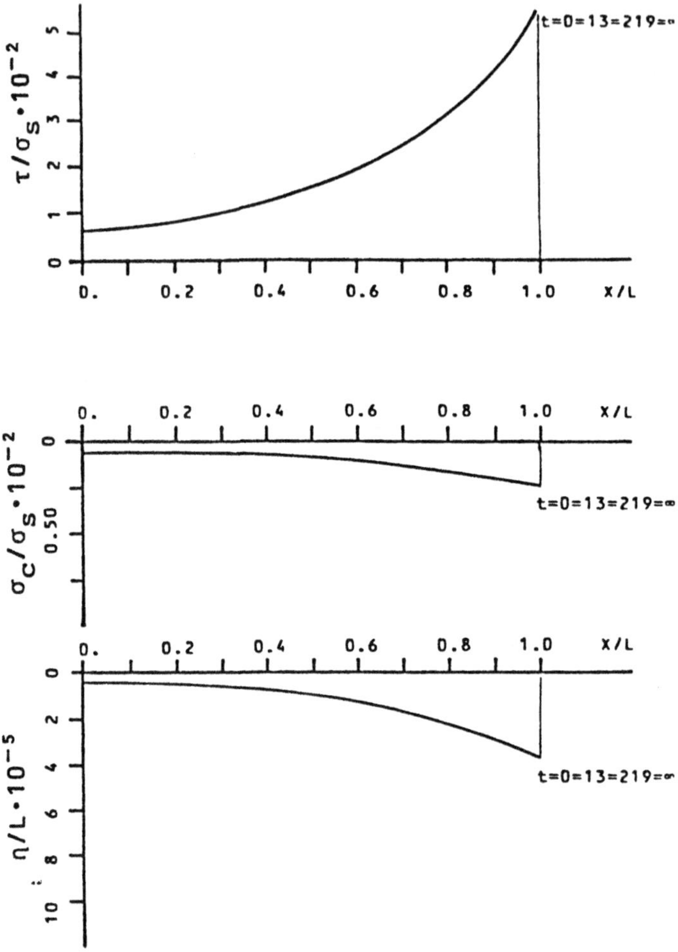

Fig. 12 - Distribution au cours du temps des contraintes d'adhérence τ et de confinement σ_c et de la dilatation normale η. Modèle de vieillissement, confinement présent: $E_c/E_r=1$. Charge appliquée depuis 300 jours à partir de l'instant de polymérisation.

Tableau 1. Modèle héréditaire

a [mm]	E_c/E_r	$\gamma_u = \sigma_s/\tau_{max}$	σ_c/σ_s x10 ^{6}	η/L x10 ^{5}	δ/L x10 ^{5}	t [jours]
a_r = 3.	0	10.41	-2768.	1.9	9.6	0
a_s = 10.	0	17.44	-937.	3.3	16.8	219
a_r = 3.	1	9.88	13.87	1.0	9.1	0
a_s = 10.	1	17.48	46.73	2.8	16.5	219
a_r = 10.	0	32.71	-2560.	22.6	113.4	0
a_s = 3.	0	43.27	-879.	51.0	255.0	219
a_r = 10.	1	31.57	-1241.	11.7	105.8	0
a_s = 3.	1	43.01	4882.	42.8	248.8	219

Tableau 2. Modèle de vieillissement; force appliquée à 0 jours à partir de la polymérisation (τ=0).

a [mm]	E_c/E_r	$\gamma_u = \sigma_s/\tau_{max}$	σ_c/σ_s x10 ^{6}	η/L x10 ^{5}	δ/L x10 ^{5}	t [jours]
a_r = 3.	0	10.41	-2768.	1.9	9.6	0
a_s = 10.	0	21.65	-421.	3.5	17.5	219
a_r = 3.	1	9.88	13.87	1.0	9.1	0
a_s = 10.	1	21.78	51.28	3.1	17.0	219
a_r = 10.	0	32.71	-2560.	22.6	113.4	0
a_s = 3.	0	42.59	-405.	51.4	257.2	219
a_r = 10.	1	31.57	-1241.	11.7	105.8	0
a_s = 3.	1	47.89	5555.	43.8	250.3	219

Tableau 3. Modèle de vieillissement; force appliquée à 100 jours à partir de la polymérisation (τ=100).

a [mm]	E /E	$\gamma_u = \sigma_c/\tau_{max}$	σ_c/σ_s x10 ^{6}	η/L x10 ^{5}	δ/L x10 ^{5}	t [jours]
a = 3.	0	10.39	-2773.	1.9	9.6	0
a = 10.	0	11.41	-2293.	2.1	10.5	219
a = 3.	1	9.87	13.81	1.0	9.1	0
a = 10.	1	10.94	18.95	1.3	10.1	219
a = 10.	0	32.71	-2565.	22.6	113.3	0
a = 3.	0	34.76	-2130.	25.5	127.8	219
a = 10.	1	31.53	-1252.	11.7	105.7	0
a = 3.	1	33.84	-241.	15.3	120.4	219

Tableau 4. Modèle de vieillissement; force appliquée à 300 jours à partir de la polymérisation (τ=300).

a [mm]	E_c/E_r	$\gamma_u = \sigma_s/\tau_{max}$	σ_c/σ_s x10^6	η/L x10^5	δ/L x10^5	t [jours]
a_r= 3.	0	10.39	-2775.	1.9	9.6	0
a_s= 10.	0	10.40	-2771.	1.9	9.6	219
a_r= 3.	1	9.87	13.80	1.0	9.1	0
a_s= 10.	1	9.88	13.86	1.0	9.1	219
a_r= 10.	0	32.67	-2565.	22.6	113.2	0
a_s= 3.	0	32.71	-2561.	22.6	113.4	219
a_r= 10.	1	31.53	-1254.	11.7	105.6	0
a_s= 3.	1	31.56	-1243.	11.7	105.8	219

5. Observations et conclusions

Dans ce travail-ci, on a quantifié l'influence du fluage de la résine epoxyde utilisée comme adhésif dans un collage acier-béton à superposition simple en supposant deux modèles rhéologiques limite pour l'adhésif:

a) un modèle viscoélastique linéaire (héréditaire)
b) un modèle viscoélastique de vieillissement.

Ce choix a été dicté à cause des insuffisantes expériences qu'on avait dans le domaine du fluage des formulés epoxyde, sourtout pour ce qui concerne les petites épaisseur, comme il arrive dans les collages structuraux; on a donc cherché à définir les limites du domaine de variabilite du comportement du materiel réel à travers l'étude de deux comportements extrêmes.

Les résultats qu'on a obtenu montrent qu'il existe une sensible rédistribution des contraintes d'adhérence le long de la superficie collée qui porte à une sensibile réduction de la valeur maximale. On peut considerer que cette rédistribution s'achève entre les premiers 200 jours à partir de l'époque d'application de la force et pour les épaisseurs de l'acier et de l'adhésif de 5 mm ça porte à une réduction de la valeur maximale des efforts d'adhérence du 37% environ pour le modele a), et du 48% pour le modele b), tout en supposant d'appliquer la force pour les deux cas au temps $\tau = t_0$ de polymérisation du formulé epoxyde (figs. 7 et 9). La reduction au cours du temps de la valeur maximale de l'effort d'adherence entre adhésif et béton se répercute favorablement sur le "coefficient d'utilisation" qui, même pour differentes épaisseurs d'adhésif et d'acier (tableaux 1 et 2), a la tendence à augmenter. La valeur de cet accroissement du "coefficient d'utilisation" est variable, dans le cas du modèle de vieillissement, avec l'époque d'application de la force, jusqu'à devenir insignifiante dans le cas où la force soit appliquée bien après la polymérisation de l'adhésif epoxyde (tableaux 3 et 4).

Particulièrment interessant est l'effect du confinement passif sur le phénomène de la dilatation normale et par consequence, sur la distribution et variation des efforts σ_c normaux au plan du collage (figs. 7, 8 pour le modèle a), et figs. 9, 10 pour le modèle b)). Alors qu'en absence de confinement on voit une progressive diminuition au cours du temps des efforts de traction normaux à la surface de separation entre adhésif et subjectile jusqu'à une valeur proche au zero, la présence d'un effect de confinement passif E_c/E_r= 1 jointe à un accroissement de la dilatation normale de la jonction, apporte des efforts de compression normaux à la surface de séparation adhésif-subjectile qui augmente progressivement avec les temps. Ce phénomène-ci est plus remarquable, bien que modeste, dans le modèle de fluage b), si l'instant où on applique la force coïncide avec le temps de polymérisation ($\tau = t_o = 0$). La situation se présente bien differente si la force est appliqué successivement au temps t_o de polymérisation. Les figs. 11 et 12 montrent les distributions des contraintes tangentielles, normales et de la dilatation normale dans le cas du modele b) de vieillissement, en supposant d'appliquer la force à l'instant τ = 300 jours. C'est évident que la variation au cours du temps des distributions est pratiquement nulle car l'adhésif a un comportement essentiellement élastique et les résultats coïncident avec ceux trouvés en précédence /1/ ne considerant pas le fluage.

Des observations ci-dessus formulées on peut donc conclure que:

- le fluage de l'adhésif porte un benefice au collage réduisant la valeur maximale des contraintes d'adhérence et réalisant une redistribution de celles-ci dans le développement du collage même;
- en présence d'un confinement passif (qu'on peut réaliser avec des revers de la tôle ou clouage normal au plan colle) la déformation visqueuse agit de façon d'accroitre au cours du temps les efforts de compression normaux au plan du collage, améliorant ainsi le comportement du béton à l'égard de la rupture par cisaillement;
- les variations dans la distribution des efforts et des déformations le long des superficies colleés dûes au modèle rhéologique (héréditaire ou de vieillissement) sont modestes, et de toute façon il parait que les résultats obtenus pour le modèle héréditaire soient plus en faveur de sécurité.
- la différence entre les deux modèles rhéologiques examinés dévient importante surtout en relation à l'époque d'application de la force au collage même; on pourra pourtant considerer ou non la réduction des contraintes d'adhérence par effet des phénomènes visqueux sélon que la force soit appliquée tout de suite à l'element structural réparé ou rénforcé par du plaquage, et également selon que cette force soit permanente, demi-permanente ou occasionelle.

References

/1/ Tattoni S., "Contributo al calcolo degli incollaggi acciaio - calcestruzzo", Notiziario AICAP, febbraio 1980.

/2/ Tattoni S., "Valutazione del contributo degli effetti secondari nel comportamento di un incollaggio acciaio - calcestruzzo", La Prefabbricazione, marzo 1984.

/3/ Tattoni S., "Riparazione e rinforzo di strutture lesionate in c.a. mediante l'impiego di resine epossidiche. Indagine preliminare", La Prefabbricazione, aprile 1978.

/4/ Tattoni S., "Giunzioni incollate acciaio calcestruzzo presollecitate con bulloni ad alta resistenza", Rassegna Tecnica dei Problemi dell'Energia Elettrica, n. 2, 1980.

/5/ Rostasy F.S., Ranisch H., Alda W., "Verstarkung von Spannbetonbrucken in Koppelfugenbereich durch angeklebte Stahllaschen", Bauingenieur 56, 1981.

/6/ Hugenschmidt F., "Epoxy Adhesives in Precast Prestressed Concrete Construction", Jour. of the Prestressed Concrete Institute, March-April 1974.

/7/ Hugenschmidt F., "Epoxy Adhesives for Concrete and Steel", 1.st Int. Congress on Polymer Concretes, 5-7 may 1975.

/8/ Delale F., Erdogan F., "Viscoelastic Analysis of Adhesively Bonded Joints", Journal of Applied Mechanics, vol. 48, june 1981.

/9/ Nowacki W., "Theorie du fluage", Ed. Eyrolles, Paris, 1965.

/10/ Mola F., "Metodi di analisi di strutture in c.a. e c.a.p. a comportamento elasto-viscoso lineare", Studi e Ricerche, vol. 3, 1981.

/11/ Mola F., "Metodo generale e metodo approssimato per la risoluzione di strutture non omogenee a comportamento elasto-viscoso lineare", Studi e Ricerche, vol. 1, 1979.

REPAIR AND STRENGTHENING OF BRIDGES BY MEANS OF BONDED PLATES

J. N. THEILLOUT
Division Surveillance et Pathologie des Ouvrages d'art, Laboratoire Central des Ponts et Chaussées, PARIS.

ABSTRACT
The design of concrete strengthening by the application of steel plates bonded to the concrete surface requires a very good knowledge of the mechanical behavior of the strengthened structures. The paper deals with following aspects of the behavior of bonded steel plates:

(a) Local behavior of a bonded steel plate in the vicinity of a crack. This study reveals local bending of the plate. The influence of the different parameters has been invistigated.

(b) Study of assmbly failures. It is possible to predict the failure of repaired structures strengthened under normal force, bending moment and with regards to shearing force when using the concepts of brittle fracture mechanics. Some experimental determination of the energy release rate is presented.

(c) Distribution of deformations between reiforced bars and bonded steel plates. Calculations carried out on a numerical model of reinforced concrete tie member show that the hypothesis of plane deformation is not verified. These partials results need to be confirmed by experience.

Key words: Bonded steel plates, Repairing, Strengthening, Fracture Mechanic, reinforcement.

1. DESCRIPTION OF TECHNIQUE

The bonded plate technique allows of concrete structures by the application of steel plates (or plates in another material) to the concrete surface by bonding with an adhesive.

In the particular case of girders, the strengthening arrangement will of course depend on the stresses to which the structure is to be subject.

1.1 Strengthening under normal force (case n°. 1)

This type of strengthening is usually used on reiforced-concrete tie members or columns. In this case, the problems encountered from the mechanical performance standpoint have to do with:

(a) the local behavior of the plate in the vinicity of cracks (in the case of tie members),

(b) determining the limit force causing the "detachement" of the plate.

1.2 Strengthening under bending moment (case n°. 2)

In this type of strengthenig the plates are generally bonded on the most heavily loaded fiber of the girder. Some mechanical problems raised by this type of strengthening are:

(a) Local behavior of plate in the vinicity of cracks.

(b) determination of limit force causing the failure of the assembly by plate detachement.

1.3 strengthening with regard to shearing force (case n°. 3)

In this case the strengthening plates are bonded on the webs of the girders in the vinicity of the bearings.

the main problem in this case is determining the maximum shearing force causing plate detachement

In the following discussion we shall be considering all this points:

* Local behavior of a plate in the vicinity of a crack.
* Study of failure of bonded steel-concrete assemblies.
* Distribution of deformations between bonded plates and steel inside the concrete.

For each of these points we shall look into the methods used, the hypotheses considered and the main result obtained.

2. LOCAL BEHAVIOR OF A PLATE IN THE VINICITY OF A CRACK

2.1 Analytical solution of problem

To solve this problem we shall consider a double-over lap joint (see Fig. 1). This schematic view is a good representation of the functioning of a plate in a vinicity of a crack in the case of tie member or in the case of the strengthening of beams with regard to bending moment.

To study this assembly, the following hypothesis will be made:

(a) The constituent materials of the assembly exhibit a linear elastic behavior.

(b) The edge effects will be neglected, i.e. we shall consider the case of plane elasticity.

(c) The behavior of the plate will be assimilated with that of a beam.

(d) The adhesive layer, considering its natural rigidity, takes up only shearing stresses (noted τ_a) and normal stresses (noted σ_a).

With these hypothesis it is possible to write the equilibrium of a length (dx) of the assembly (see Fig. 2) as follows (sign conventions will be those shown in Figure 3):

(1) ---- $N'_s = \tau_a$

(2) ---- $Q'_s = \sigma_a$

(3) ---- $M'_s + Q_s - \frac{e_s}{2} \tau_a = 0$

(4) ---- $$u'_s = \frac{N_s}{E_s e_s} + 6\frac{M_s}{E_s e_s^2}$$

(5) ---- $$w''_s = 12 \cdot \frac{M_s}{E_s e_s^3}$$

(6) ---- $$\tau_a = \frac{G_a}{e_a}(u_s - u_c)$$

(7) ---- $$\sigma_a = \frac{E_a}{e_a}(W_s - W_c)$$

(8) ---- $$N'_c = -\tau_a$$

(9) ---- $$Q'_c = \sigma_2 - \sigma_a$$

(10) --- $$u'_c = \frac{N_c}{E_c e_c}$$

(11) --- $$W_b = \frac{e_c}{E_c}\sigma_a$$

Where: (u) and (W) designate respectively the displacements along the axis (Ox) and (Oy), the indices (') designate the derivative with respect to (x).
The combination of equations (2), (3), (5), (7), (11) allows to write:

(12) ---- $$\frac{E_s e_s^3}{12} W_s^{(4)} + \frac{1}{\frac{e_a}{E_a} + \frac{e_c}{E_c}} W_s = \frac{e_s}{2}\tau'_a$$

As concerns the combination of the equation (1), (4) (6), (8) and (10) it allows us to write:

(13) ---- $$\frac{G_a}{e_a}\left(\frac{1}{E_s e_s} + \frac{1}{E_c e_c}\right)\tau'_a + \frac{G_a}{e_a}\frac{e_s}{2} W_s^{(4)} = \tau_a^{(3)}$$

Equations (12) and (13) form a coupled system from which it is possible to easily isolate one of the unknown functions. Since we are interested in the local bending of the plate, we shall choose (W_s) as an unknown function.
After calculation we obtain, for this unknown function, a differential equation of the type:

(14) --- $$W_s^{(6)} + \alpha \cdot W_s^{(4)} + \beta \cdot W_s^{(2)} + \gamma \cdot W_s = 0$$

which is easily solved.
The differential equation (12) is the equation of a

girder bearing on a elastic mass subject to a load:

(15) --- $p(x) = \frac{e_s}{2} \cdot \tau'_a$

Thus, any variation in the shearing stress causes a load on the girder along the direction to the normal fiber.

The six integration constants are deduced from the following boundary conditions:

(16) --- $w'_s(0) = 0$

(17) --- $N_c(f) = 0$

(18) --- $N_s(L) = 0$

(19) --- $Q_s(L) = 0$

(20) --- $M_s(L) = 0$

(21) --- $\int_f^L \sigma_a \, dx = 0$

2.2 Influence of geometrical parameters

To discuss the influence of geometrical parameters we shall quickly review the main results obtained relative to the distribution of stresses in the adhesive layer.

2.2.1. Shearing stresses (Fig. 4a and 4b)

An examination of the distribution of shearing stresses shows that they reach their maximum value in the vinicity of cracks. The maximum is independent of the length of the overlap when the bonding length is at the least equal to twice the concrete thickness (e_c). This maximum increases according to the thickness of the concrete and decreases according to the thickness of the steel (e_s) to (e_a) constant. It decreases as the glue film thickness increases, all other things being equal (Fig. 5a and 5b).

2.2.2. Stresses normal to adhesive film (Fig 6a and 6b)

In the vicinity of the crack, the adhesive and hence the concrete undergo compressive stresses over a small distance (10 to 60 mm); then, over a greater distance, they undergo tensile stresses of lesser intensity.

The maximum normal stress is also independent of the bonding length if this length is at least equal to twice or three times the thickness of the concrete (see Fig. 7). It decreases whith the increase in (e_a) (Fig. 8) or (e_c) (Fig.9).

2.2.3. Expression of local bending

Local bending of the plate in the vicinity of the crack results in the appearance of stresses on the extreme fibers

(noted σ_1 and σ_2 with $|\sigma_1| < |\sigma_2|$)

Certains tests have shown that the strengthening could be destroyed by the plasticizing of the plate as soon as the greatest stress (σ_2) reached the elastic limit of the steel making up the plates. The force then taken up by the plates is lower than that caused by their plasticizing if they were subjected to simple tension. Accordingly, we shall look into the local bending below.

We shall call (F_M) the force taken up by the plate (per unit width) if it were subjected to simple tension. It causes an uniform stress (σ_2) along the the thickness of plate:

(22) --- $F_M = \sigma_2 \cdot e_s$

If (F_{av}) is used to designate the tension up by the plate when it subjected to bending:

(23) --- $F_{av} = \dfrac{(\sigma_1 + \sigma_2)}{2} \cdot e_s$

The effect of local bending can be expressed by the relationship :

(24) --- $n = \dfrac{F_{av}}{F_M} = \dfrac{\sigma_1 + \sigma_2}{2 \cdot \sigma_2}$

The ratio (n) can then be translated as the percetage of force caused by the plasticizing of the plate in the case of local bending, compared with the force cause by plasticizing in the case of simple tension.

2.2.4 Influence of steel thickness

Whatever the other parameters which are made to vary, the value of the ratio (n) varies only by about 3% when the thickness of the plate varies from 3 to 10 mm (see Fig. 10 to 13)

2.2.5. Influence of adhesive thickness (see Fig.10)

Local bending decreases as a function of adhesive film thickness. This is explained by the fact that better distribution of shearing stresses is obtained along the gluing as (e_a) increases, hence as the "external load" in equation (15) decreases.

2.2.6. Influence of concrete thickness (see Fig. 11)

The greater the concrete thickness, the greater will be the flexibility of the reaction mass on which the plate bears. This is reflected by an increase in the ratio (n) as (e_c) increases.

2.2.7. Influence of bonding length (see Fig 12)

As we already point out in paragraphe 2.2.1. and 2.2.2. the field of stresses in the adhesive film is independent of the bonding length provided this length is at least equal to twice (or three times) the concrete thickness.

2.2.8. Influence of initial crack opening (see Fig. 13)

The initial opening of the crack has no influence on the local bending of the plate

2.3 Conclusions

Assimiling the operation of a plate bonded on a cracked concrete structure with the functioning of a cover plate bonded in a double-overlap joint, it is possible to reveal local bending of the plate in the vicinity of cracks.

The influence of the different parameters was invistagated on the proposed model. The following results was obtained:

(a) The bonding length has no influence on the value of local bending provided this length is at least equal to twice the concrete thickness.

(b) The plate thickness has no influence on this bending when the thickness involved do not exceed currently used value (3 to 10 mm).

(c) The thickness of adhesive film plays a contradictory role since increasing it tends to decrease the value of local bending but makes the assembly flexible, which can reduce the the effectiveness of the plate in the strengthenig of reinforced-concrete structures.

3. STUDY OF ASSEMBLY FAILURES

In this part we shall be considering the strengthening as structures by bonded plates as well as the strengthening of structures by sprayed concrete (shotconcrete) or by the addition of any material.

The sensitive part of the strengthening of girders by the addition of material is the interface between the material of the girder and the strengthening material. Experience shows that, under the effect of stresses to which the girders are usually subject, the most frequently observed failure mode involves the propagation of one (or more) cracks at the interface.

The description of the method present above is designed to provide engineers responsible for strengthening or repairing girders with the method enabling them to determine the stress limits brining about the impairment of the strengthening area or the repair by interface crack propagation.

The concept to be used are those of brittle fracture mechanics.

3.1. Review of concepts

Let us consider a body (Ω) in which there is a crack of length (ℓ) or area (A). The potentiel energy of this body will be noted (U). Under an external load the length of the crack increases by a value ($\dot{\ell}$). If we assume, as demonstratd by recent investigations, that the crack area (or crack length) is a normal state variable, the power dissipated in the transformation which increases the crack length from (ℓ) to ($\ell + d\ell$)is written:

(25) --- $-\dfrac{\partial U}{\partial \ell}\dot{\ell} \geqslant 0$

by application of the principles of thermodynamics.
The notation ($\dot{}$) indicates the derivative with respect to time.

The preceding formula assumes that the plastic (or damaged) zone at the root of the crack has small dimensions compared with the piece. The only elements intervening in the dissipated power are the asymptotic elastic fields beyond the plastic zone.

Griffith's theory states that the preceding expression is equal to ($2\gamma_s\dot{\ell}$) in which (γ_s) is the surface energy. This is an intrinsic property of the material. The term ($-\partial U/\partial \ell$) is called the elastic energy release rate (G). When the crack propagates we have:

(26) --- $G = 2\cdot\gamma_s$ (Griffith fracture criterion)

In the case of girders the energy release rate (G) is easily expressed as a function of the geometrical and machanical properties of the materials and as a function of loading. The term ($2\cdot\gamma_s$) in a strengthening application will be replaced by a term (R) which will be a property of the interface in the loading mode which produce crack propagation.

thus, if:

(27) --- $G \geqslant R$

there will be crack propagation. It is shown that the equilibrum defined by the equation:

(28) --- $\dfrac{\partial G}{\partial \ell} < 0$

will be unstable, i.e. a small increase in the extrenal load will produce rapid crack extension up to the failure.

The value of (R) depends on the characteristics of the interface in the loading mode consider but also on the crack propagation rate. In the case of bonded plate strengthening, failure occurs by crack propogation in the concrete in the immediat vicinity of the bonding plane. The values of (R) will thus be characteritic of the concrete material in the vicinity of this interface. For the concrete, many investigations hace shown that, in the case of Mode I loading, the variation of (R) as a function of propagation rate was governed by an empirical law of type:

$$(29) \text{ --- } \quad \dot{\ell} = A R^N$$

where (A) and (N) are constants intrinsic to the material; (R) being bounded by a limit value (R_c). As, in our dicussion, what is involved is determining the limit force capable of being taken up by a strengthening girder, we shall consider only the value of (R_c) wheach is easy to determine or at least to approach experimentally. Subcritical propagation (for value of ($R < R_c$)) has not formed the subject of in-depth study by the L.C.P.C. as regards strengthening.

In our case, the problem amounts to comparing the calculated value of the energy release rate (G) (see following paragraph) with the value of (R_c) determined experimentally.

3.2 calculation of energy release rate

In the case of beams and more generally in the case of structures the expression ($-\partial U/\partial \ell$) can be written in the following form:

$$(30) \text{ --- } \quad -\frac{\partial U}{\partial \ell} = \frac{\partial}{\partial \ell}\left(W_{ext} - W_{el}\right)_{Q=cst}$$

where (W_{ext}) designates the work of external forces, (W_{el}) designates the elastic energy and (Q) the external load.

To calculate (G) we shall distinguish tree cases of strengthenig for which we shall assume that, under the external load (Q), an interface crack has propagated over length ($d\ell$).

3.2.1 Case of normal force

What is involved here is the strengthenig of columns or tie members for which we determine the force capable of producing failure by crack propagation from the ends of strengthening. Referring to the notations of Figure 14:

$$(31) \text{ --- } \quad W_{el} = \frac{1}{2b}\left[\int_0^x \frac{P^2}{E_c e_c}\,dx + \int_x^L \frac{P^2}{(Ee)^*}\,dx\right]$$

where (Ee^*) is the homogenized stiffness of the tie member.

$$(32) \text{ --- } \quad W_{ext} = \frac{P}{b}\left[\frac{Px}{E_c e_c} + \frac{P(L-x)}{(Ee)^*}\right]$$

Hence the energy release rate is written:

$$(33) \text{ --- } \quad G = \frac{P^2}{2b}\left[\frac{1}{E_c e_c} - \frac{1}{E_c e_c + E_s e_s}\right]$$

where (b) is the width of the tie member.

3.2.2. Strengthening under bending

Although the reasonig may be extended to girders subjected to any type of loading and exhibiting various boundary conditions, we shall consider here only the simple case (see Fig 15) of a statically determinate girder of width (b) whose central part is subjected to circular bending. According to the Figure we have:

$$(34) \text{ --- } W_{el} = \int_0^x \left(\frac{M^2}{2E_cI_c} + k_1 \frac{T^2}{2G_cA_c} \right) dx + \int_x^e \left(\frac{M^2}{2(EI)^*} + k_2 \frac{T^2}{2(GA)^*} \right) dx + \int_e^L \frac{M^2}{2EI^*} dx$$

$$(35) \text{ --- } W_{ext} = P \left[\int_0^x \left(\frac{M}{E_cI_c} \frac{\partial M}{\partial P} + k_1 \frac{T}{G_cA_c} \frac{\partial T}{\partial P} \right) dx + \int_x^e \left(\frac{M}{(EI)^*} \frac{\partial M}{\partial P} + k_2 \frac{T}{(GA)^*} \frac{\partial T}{\partial P} \right) dx \right.$$

where ($(EI)^*$) and ($(GA)^*$) are the stiffness values under homognized bending and shearing, and where (k_1) and (k_2) refer to the calculation of shearing stresses in the cross section. After all calculation:

$$(36) \text{ --- } G = \frac{P^2}{2b} \left[X^2 \left(\frac{1}{E_cI_c} - \frac{1}{(EI)^*} \right) + \left(\frac{k_1}{G_cA_c} - \frac{k_2}{(GA)^*} \right) \right]$$

3.2.3. Case of strengthening under shearing force

In this case the strengthening is placed on the sides of the girder. It is assumed that thfracture occurs by interface crack propagation whose front is normal to the neutral fiber of the girder. As in the preceding case, we shall deal with the simple case of a rectangular, statically determinate girder whose central part is subjected is to circular bending (see Fig. 16). If we designate by (h) the height of the girder, (n) the ratio of the moduli ($n = E_s/E_c$) and by (B) the thickness of the strengthening, the energy release rate obtained by a similar calculation is expressed as follows:

$$(37) \text{ --- } G = \frac{P^2}{2h^2} \cdot \frac{nB}{b(b+2nB)} \cdot \left(\frac{12X^2}{E_c h^2} + \frac{k_1}{G_c} \right)$$

3.2.4 Discussion

In the last to cases considered the energy release rate can be written in the following form:

$$(38) \text{ --- } G = G_M + G_T$$

(G_M), increasing function, of crack length, coming from bending terms and (G_T), independent of his length, coming from shearing force terms.

The design of strengthening device must consequently be

carried out such that:

(39) --- $G < R_c$

According to Figure 17 this condition can be met for values of (x) between (0) to (X_{MAX}). This has led to the idea of considering this condition as a means of determining over which length of the girder the strengthening should extend. In fact, although the theory does not provide the means of calculating the force required for crack initiation, it may be assumed in the case of concrete and for our purposes that this force is equal to that required for crack propagation. In this case the strengthening may be terminated, if necessary for the stability of the girder, at a distance (x) from the bearing such that:

(40) --- $0 < x < X_{MAX}$

3.3 Experimental determination of (R_c)

Up to now we have determined (R_c) only in two cases: normal force and shearing force.

3.3.1. Case of normal force

A machine was specially developed (see Fig 18) to determine this value. The apparatus is composed of an annular jack which transmits the tensile force to the plate through a rod at a end of which is secured a system of jaws, a bearing plate integral with the fixture and a support frame in direct contact with the concrete slab on which are bonded the plates to be tested.

By recording the applied force and the deplacement of jaws it was possible to follow the evolution of the fracture and to determine experimentally the force level at which the interface crack propagated rapidly. Intoducing this value into the formula (33) we calculated , for different plate thickness values and for different bonding lengths, the value of (R_c). This value falls within the following range:

(41) --- $160 \leqslant R_c \leqslant 240 \ (J.m^{-2})$

and is independent of the parameters, which were made to vary in the study, provided to bending length is at least equal to twice the thickness of the supporting concrete.

3.3.2. Case of shearing force

The value of (R_c) was determined on specimens such as those shown in Figure 19. In this case also we recorded the displacement of two part of the specimen as a function of the applied force. The experiment demonstrated that, in this case, a failure was indeed produced by the propagation of an interface crack in the vicinity of the bonding, the crack front ramaining parallel to the junction between the two specimens. If it is assumed that the two concrete blocks are rigid, the expression of the energy release rate is then:

$$(42) \text{ --- } G = \frac{1}{2} \frac{F^2}{H^2} \cdot \frac{1}{G_s \cdot e_s}$$

see notation Figure 119.

The values of (R_c) found experimentally are independent of (H) and (e_s) and are within the following range:

$$(43) \text{ --- } 50 \leqslant R_c \leqslant 80 \quad (J \cdot m^{-2})$$

3.4 Conclusions

Although considerable experimental investigations are still necessary to determine the value of the critical release rate (R_c), the use of the concepts of fracture mechanics makes it possible to determine with fairely good accuracy the maximium force capable of being supported by a cracked girder. It is also provides information regarding the arrangements to be adopted for these strengthening devices.

4. STUDY OF DISTRIBUTION OF DEFORMATIONS BETWEEN STEELS AND STRENGTHENING PLATES

Deformation measurements carried out on the strengthened girders subjected to circular bending demonstrated that in the vicinity of cracks the sections did not remain plane in their deformation or, more precisely, in certain cases the plate appeared to deform more the steel and, in other cases, the opposite was observed.

The following discussion does not claim to cover the question completely but shows the problems which arise when an attempt is made to combine two steels having fundamentally different concrete adhesion modes: friction for the internal steels of the concrete and adhesion by bonding for the plates.

4.1 Numerical modelling

We modelled a reinforced-concret tie member of 30 cm height and 10 cm width reinforced by HA12 steels (see Fig 20). The HA steels inside the concrete are assimiled by bars which rub on the concrete according to a Coulomb-type friction law. The other constituent parts of the tie rod are simulated by massive elements in which a linear elastic behavior has been introduced. Numerically, three configurations were tested (see Fig.21):

(a) The steel adheres to the concrete before loading over its entire length.

(b) It is assumed that the steel-concrete adhesion is destroyed, from the crack onward, over a length of 61 mm .

(c) Same condition as above but with adhesion destroyed over a length of 125 mmm .

4.2 Results (Figures 22 to 24)

In the case where the hypothesis of plane deformation is true, the results should indicate that the ratio of plate deformation (ε_p) to steel deformations(ε_s) is equal to unit.

Calculation carried out according to the tree cases presented above and for three plate thickness (3, 5 and 8 mm) show that:

(a) In the case where the internal steels adhere over their entire length, the ratio ($\varepsilon_p / \varepsilon_s$) is less than unity and increases as a function of the applied load, tending toward unity.

(b) In the case where the adherence is destroyed over a long length (125 mm), this ratio remains constant, or roughly so, with the applied load. It decreases as a function of plate thickness.

4.2 Conclusions

These partial results, which need to be confirmed by experience, show that the difficulty of estimating the reciprocal participation of the plate and steels in a strengthening device. A parametric study on numerical models, previously calibrated on experimentation, should make it possible to provide relatively simple rules for assessing these phenomena and quantifying them.

5. CONCLUSIONS

The study just presented deals with the mechanical aspects of the behavior of bonded plates. Some practical rules concerning the choice of materials, their application, constructional arrangements and dimensioning can however be deduced from these rules.

(a) Choice of materials:
what is required is an adhesive with a modulus of elasticity which is not too high so as to reduce stress concentrationsin the vicinity of cracks, but sufficiently high to avoid softening the assembly, providing a good distribution of forces between the internal steels of the concrete and the strengthening plates. As concerns the ultimate load, the strengthening plates should be chosen in a metal exhibiting a low modulus of elasticity. However, here too one must avoid excessive softening of the assembly. It is seen that the choice of materials is the result of compromise which must be examined on a case-by case basis.

(b) Application:
the thickness of the adhesive should also be chosen so that stresses in the vicinity of the cracks are not too high, while nevertheless offering sufficiently stiff bonding to

allow the plates to play their role effectively.

(c) Constructional arrangements:
as concerns the often raised problem of the stopping of the plates, one approach to its solution is the application of the brittle fracture mechanics concepts which the fracture behavior of bonded plates appears to comply with.

(d) Dimensioning:
although they do not confirm the hypothesis usually considered for the dimensioning of reinforced concrete sections, sections with bonded strengthening plates can be dimensioned assuming two extreme states of the adhesion between the concrete reinforcing steel and the concrete.

Many studies, on numerical models as well as on physicl models, must still be carried out in order to detail the values to be taken into account for each point considered phere.

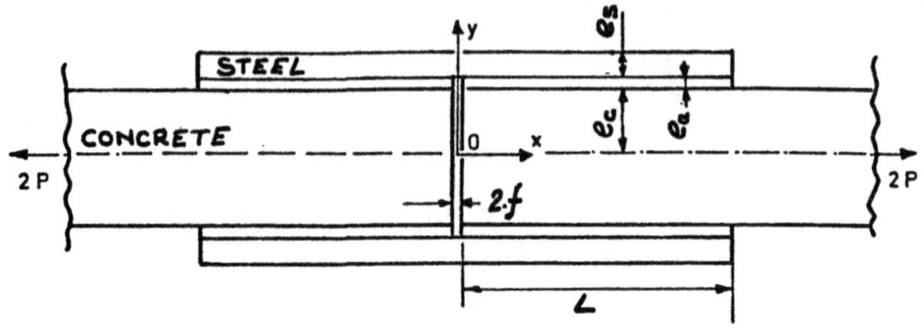

Figure 1 - Double-overlap joint

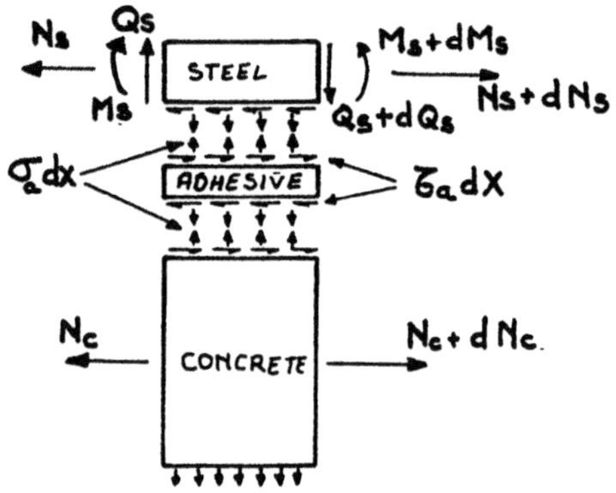

Figure 2 - Equilibrium of a length dx of a double-overlap joint

Figure 3 - Sign conventions

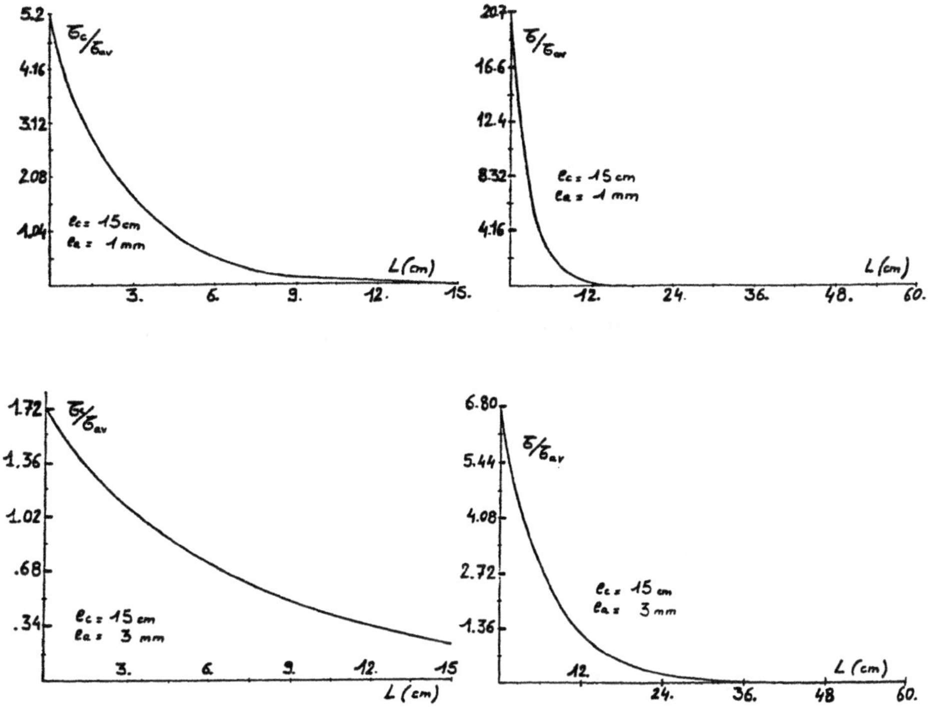

Figure 4 - Distribution of shearing stresses in adhesive film
(E_s = 200 000 MPa, E_c = 38 000 MPa, E_a = 2 000 MPa)

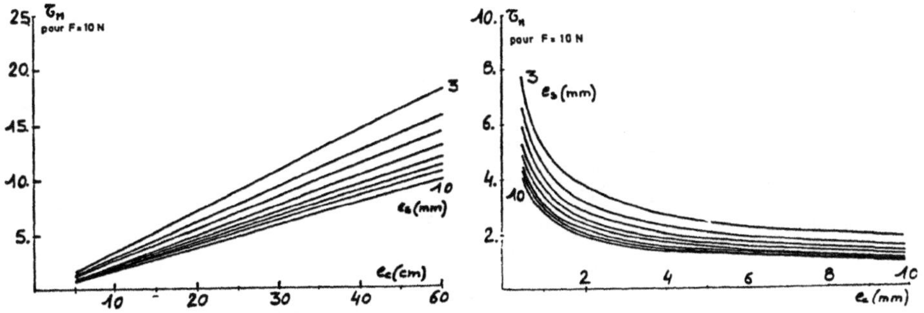

Figure 5 - Maximum shearing stress Influence of parameters
(E_s= 200 000 MPa, E_c = 38 000 MPa, E_a= 2 000 MPa)

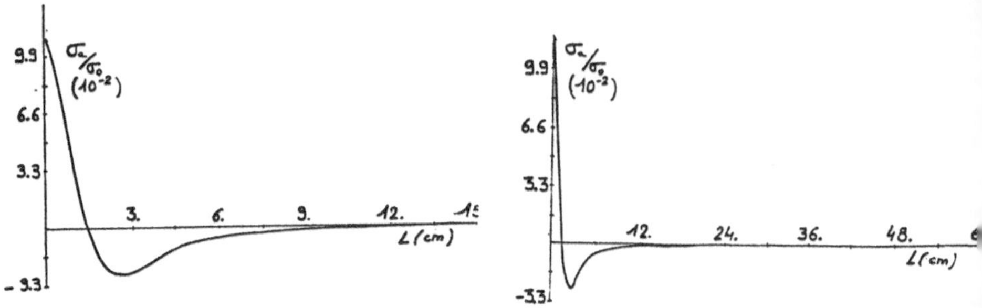

Figure 6 - Distribution of normal stress in adhesive film
(E_s = 200 000 MPa, E_c = 38 000 MPa, E_a = 2 000 MPa)

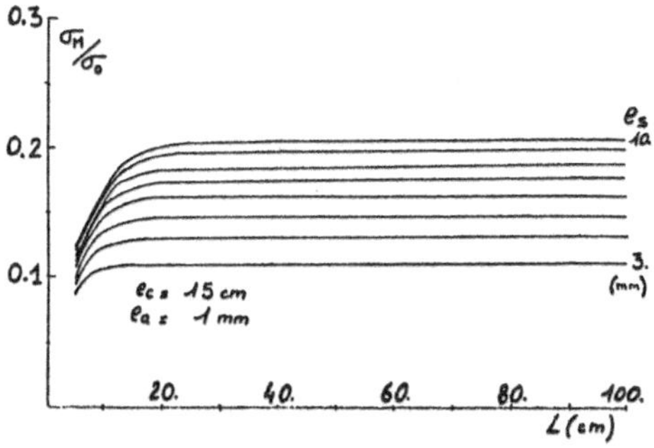

Figure 7 - Influence of bonding length on maximum normal stress

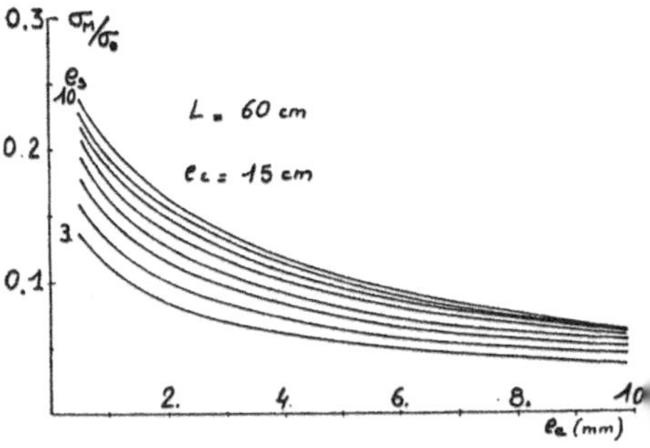

Figure 8 - Influence of adhesive thickness on maximum normal stress

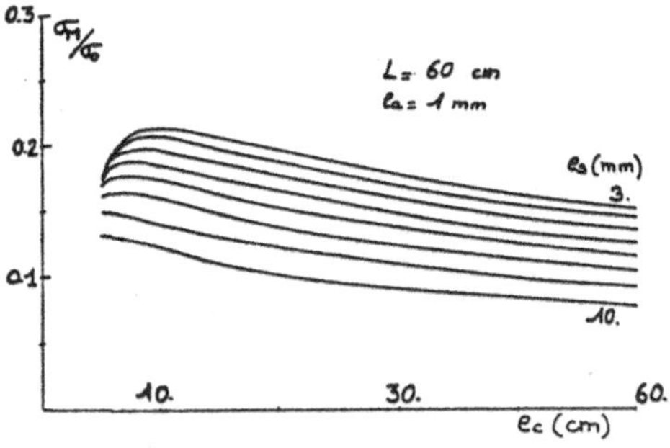

Figure 9 - Influence of concrete thickness on maximum normal stress

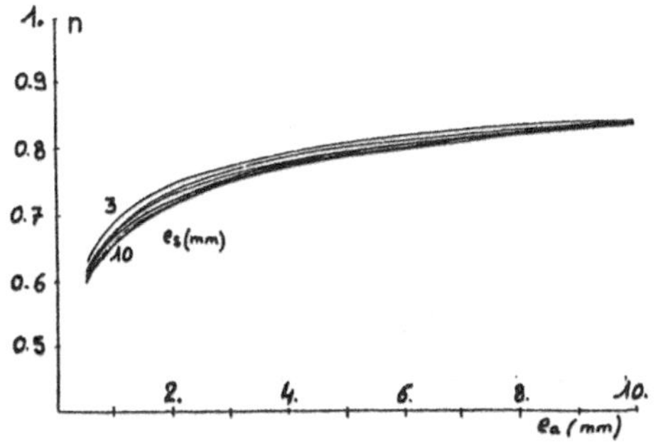

Figure 10 - Influence of adhesive film thickness on local bending

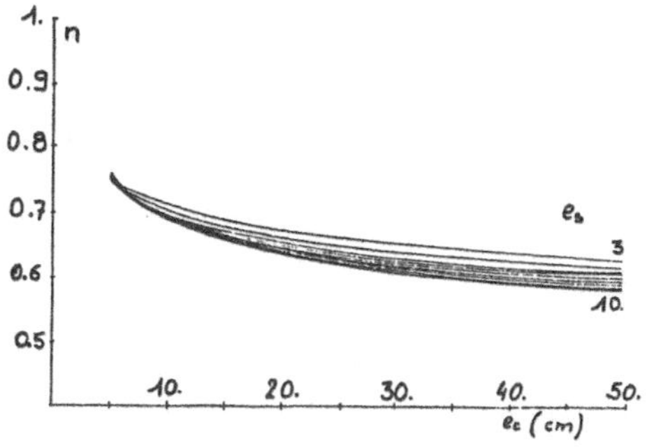

Figure 11 - Influence of concrete thickness on local bending

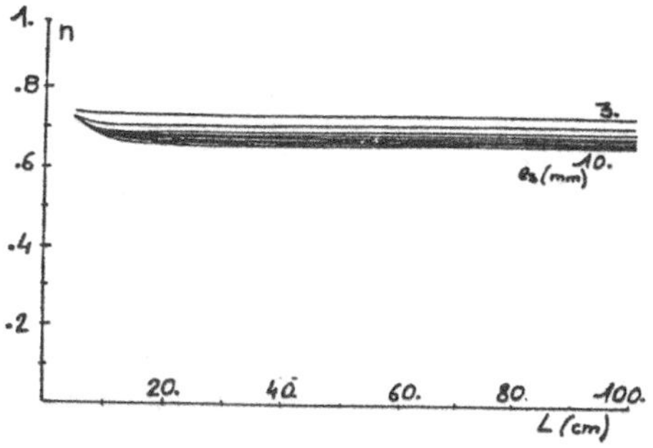

Figure 12 - Influence of bonding length on local bending

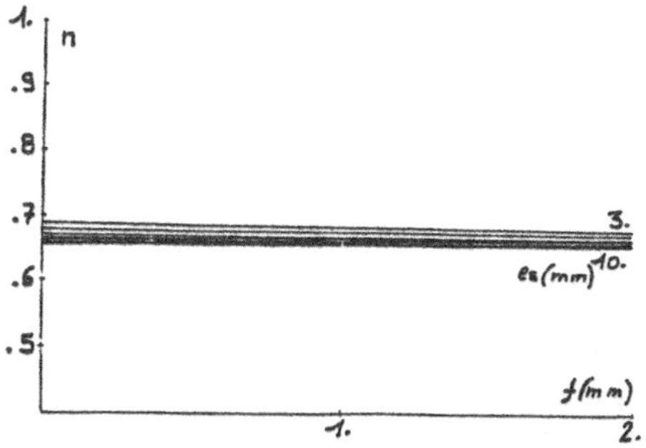

Figure 13 - Influence of initial crack opening on local bending

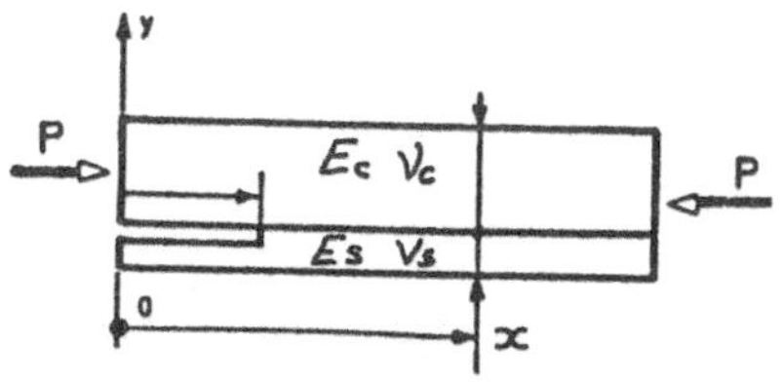

Figure 14 - Strengthening under normal force

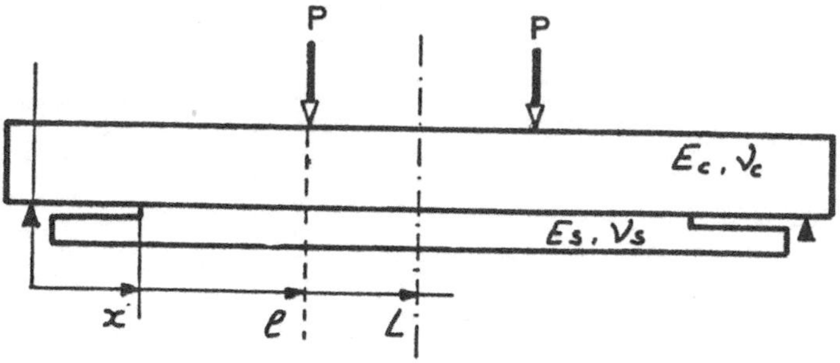

Figure 15 - Strengthening under bending moment

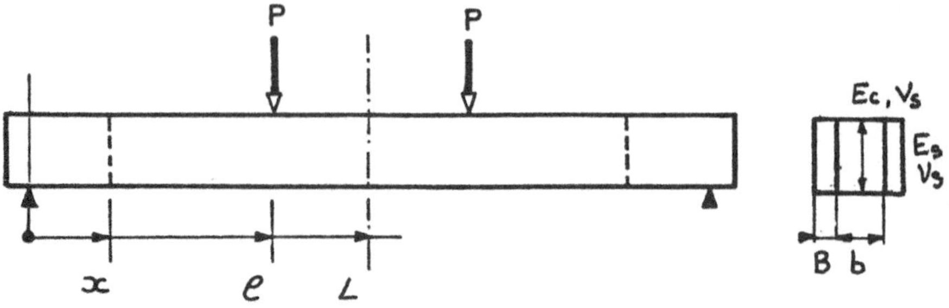

Figure 16 - Strengthening under shearing force

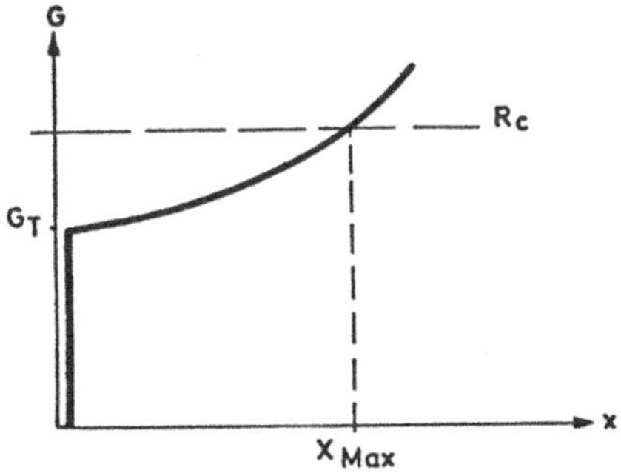

Figure 17 - Diagram showing energy release length rate as a function of crack

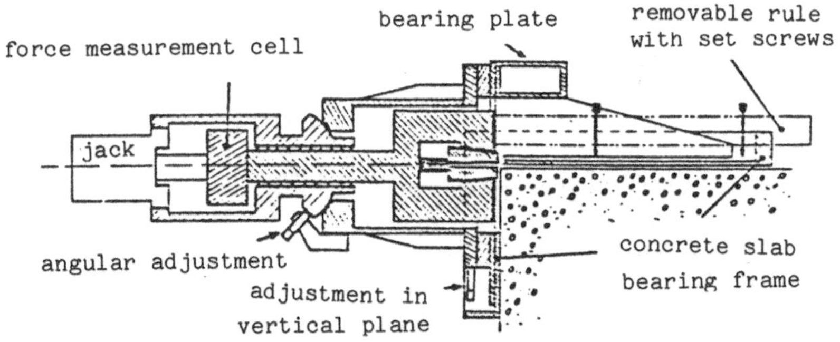

Figure 18 - Diagram of tension machine

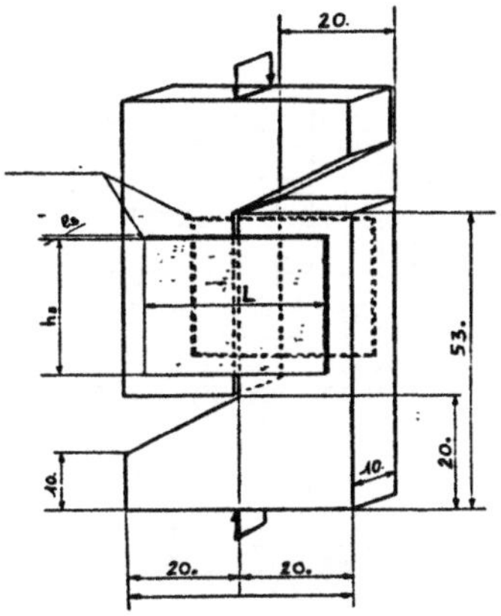

Figure 19 - Test specimen

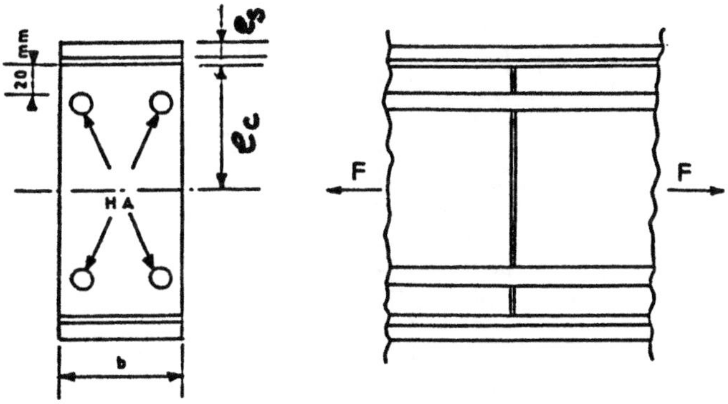

Figure 20 - Section of calculated tie member

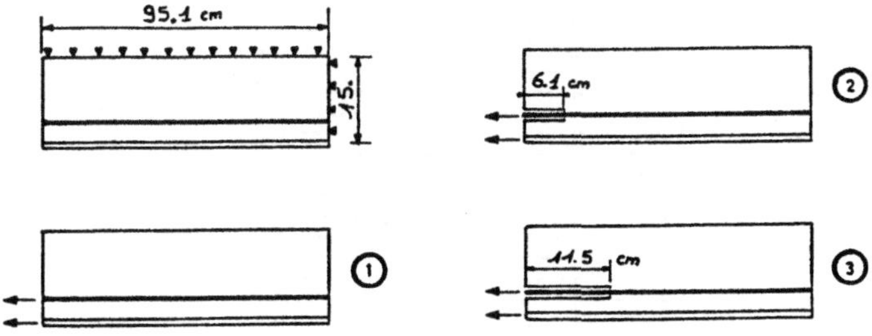

Figure 21 - Calculation configurations

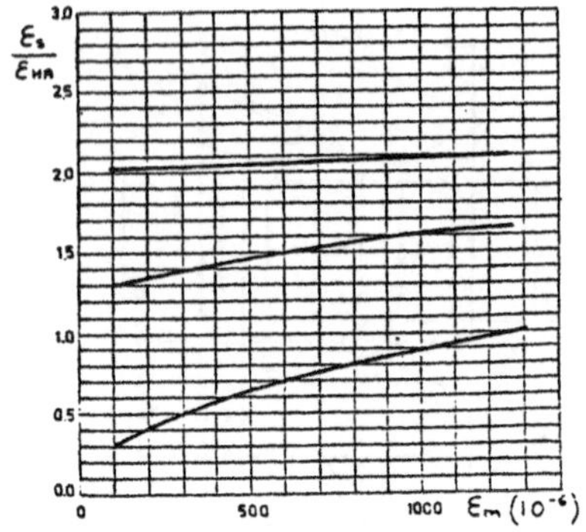

Figure 22 - Ratio of plate-HA steel deformations
(3 mm plate)

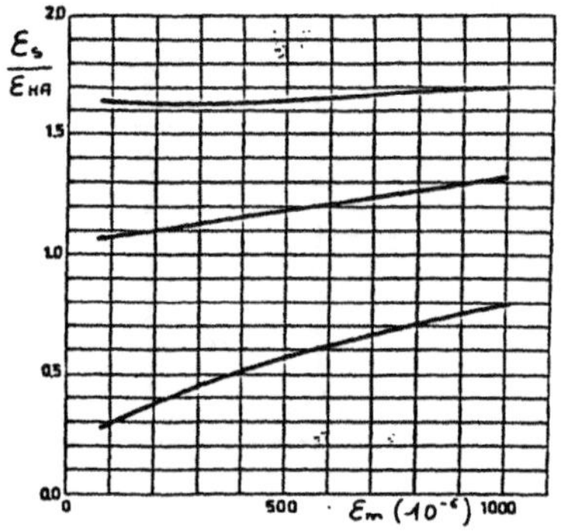

Figure 23 - Ratio of plate-HA steel deformations
(5 mm plate)

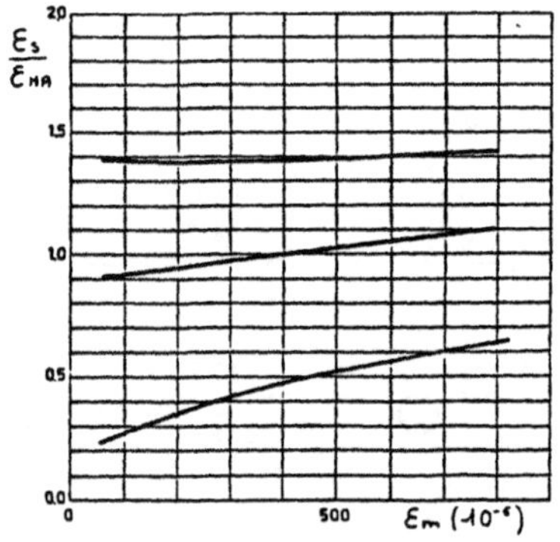

Figure 24 - Ratio of plate-HA steel deformations
(8 mm plate)

References

/1/ L'HERMITE R., L'application des colles et resines dans la construction. Le béton à coffrage portant, Ann. ITBTP série béton-béton armé, n°239, Nov. 1967.

/2/ BRESSON J., Nouvelles recherches et applications concernant l'utilisation des collages dans les structures. Béton plaqué, Ann. ITBTP, n°278, Fév. 1971.

/3/ Le collage structural et le renforcement par résines des structures de la construction, Ann. ITBTP, n°349, Avr. 1977.

/4/ ROSTASY F.S. - RANISCH H., Verstärkung von Stahlbetonbauteilen durch angeklebte Bewehrung, Betonwerk Fertigeil - Technik, n°2, 1981, pp82-86.

/5/ LADNER M. - WEBER C.H., Concrete Structures with Bonded External Reiforcement, EMPA Report n°206, Dübendorf, 1981.

/6/ IRWIN C. A. K., The strengthening of concrete beams with bonded steel plates, TRRL Supplementary report, n°160, 1975.

/7/ MAC DONALD M. D., Strength of Bonded Shear Joints Subjected to Movement during Cure, International Journal of Cement Composite, Vol. 3, n°4, Nov. 1981, pp 267 - 271

/8/ GOLAND M. - REISSNER E., The stress in cements joints, Journal of applied Mechanics, Vol. 11, 1944, pp a-17, a-27

/9/ VOLKERSEN o., Recherche sur la théorie des assemblages collés, Construction Mécanique, n°4, Dec. 1969, pp 3-13

/10/ YUCEOGLU U. - D.P. UPDIKE, Stress Analysis of Bonded Plates and Joints, Proc. of Ame. Soc. of Civ. Eng., Vol 106, n° em1, Feb. 1980, pp 37 - 56.

/11/ BUI M. D., Mecannique de la rupture Fragile, Masson, Paris, 1978

/12/ KENDALL K. Crack Propagation an Lap Shear Joints, Jour. Phys. D.: App. Phys., Vol n°8, 1975, pp 512 - 522.

/13/ MAUGIS, Mécanismes Physiques de l'Adhésion des Solides, Le Vide, n°186, Jan. - Fév. - Mar. 1977, pp 1-19.

/14/ THEILLOUT J. N., Renforcement et Réparation des ouvrages d'art par la technique des tôles collées, Thes. Doct.-ing., ENPC, Dec. 1983.

Section 10

Adhesion test procedures

Méthodes d'essais des mesures de l'adhérence

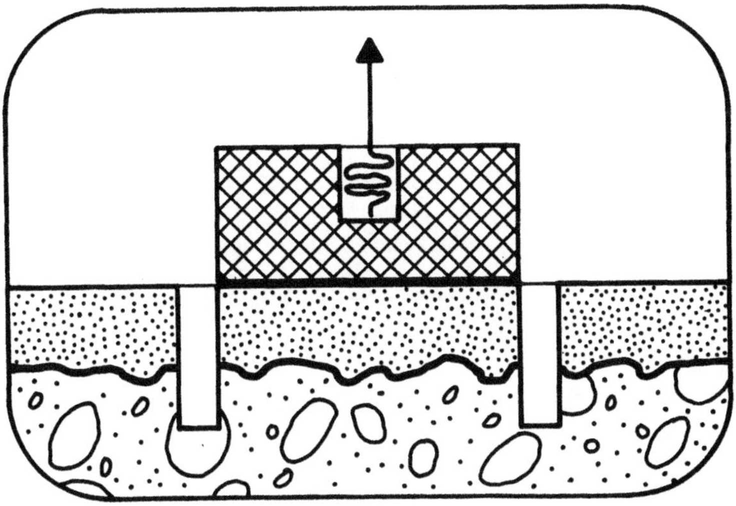

PROCEDURE DE QUALIFICATION DES PRODUITS DE REPARATION DU BETON APPLIQUEE EN FRANCE

A. CORNEILLE E.D.F - C.E.M.E.T.E Quartier Valcros CD 65
13090 AIX EN PROVENCE

Résumé
En france, les produits de réparation du béton sont qualifiés par le LCPC qui édite une liste d'aptitude nationale. La procédure de qualification, élaborée par des représentants des utilisateurs et des fabricants, est présentée dans cette communication sous l'aspect de son fonctionnement, son organisation et son contenu, en particulier en ce qui concerne les essais d'efficacité effectues pour les différents types de produits de réparation existant sur le marché.
Mots-clés : Béton, Produits de Réparation, Qualification

1. Introduction et objet de la procédure de qualification

Devant le développement des travaux de réparation des ouvrages en béton et compte tenu de la multiplication des natures et types de produits rencontrés sur le marché, il est apparu opportun de caractériser et de sélectionner les matériaux susceptibles de répondre aux fonctions demandées.

En 1980, l'ensemble des principaux utilisateurs : PONTS ET CHAUSSEES, SNCF, CEBTP, AEROPORT DE PARIS, COPREC, EDF, etc, a décidé de s'associer aux fabricants représentés par le SNFORES (SYNDICAT NATIONAL DES FORMULATEURS DE RESINES) et SNPMI (SYNDICAT NATIONAL DES PRODUCTEURS DE MORTIERS INDUSTRIELS) pour définir en commun les modalités de mise en place d'une procédure unique d'évaluation, puis de qualification des produits. L'ensemble des travaux étant piloté par le LCPC (LABORATOIRE CENTRAL DES PONTS ET CHAUSSEES).

Cette démarche ne concerne que des produits industriels prêts à l'emploi a base soit de liants hydrauliques, soit de résines synthétiques et non leur mise en oeuvre. Celle-ci fait l'objet de recommandations éditées en Janvier 1985 par le STRESS (SYNDICAT NATIONAL DES ENTREPRENEURS SPECIALISTES DE TRAVAUX DE REPARATION ET DE RENFORCEMENT DES STRUCTURES), associé à :

- L'AFPC (ASSOCIATION FRANCAISE POUR LA CONSTRUCTION)
- LA FNTP (FEDERATION NATIONALE DES TRAVAUX PUBLICS)
- LE SNBATI (SYNDICAT NATIONAL DU BETON ARME ET DES TECHNIQUES INDUSTRIALISEES).

2. Organisation

2.1 Procédure LCPC

Un groupe pleinier composé des organismes cités ci-dessus, a mis en place des groupes de travail chargés chacun de définir les essais de laboratoire destinés à déterminer les caractéristiques et les performances des produits selon les profils d'emploi suivants :

- Collage structural du béton frais sur béton durci et du béton durci sur béton durci
- Injection des fissures
- Réparation du béton dégradé
- Scellement et calage
- Identification des produits.

Ce groupe pleinier, après avoir défini les principes de fonctionnement décrits au chapitre 3, a chargé le LCPC de coordonner les échanges entre laboratoires chargés d'essais et fabricants de produits. La liste des laboratoires chargés d'essais étant entérinée par le groupe pleinier.

Dans un premier temps, à l'issue des essais, les produits testés ont fait l'objet d'une liste d'aptitude publiée par le LCPC.

Par la suite, la mise en place de spécifications reconnues par l'ensemble des membres, a permis de sélectionner les produits pour aboutir à la publication d'une liste de produits qualifiés.

2.2 Normalisation AFNOR (ASSOCIATION FRANCAISE DE NORMALISATION)

Compte tenu de l'état d'avancement, il s'est avéré opportun de normaliser les modes opératoires des essais, eux mêmes associés à des normes de spécifications et à une norme générale de produits.

Ces travaux, pilotés par l'UNICEM et par l'AFNOR sont en cours. Ils sont conduits selon une organisation similaire, a celle décrite au chapitre 2.1.

3. Fonctionnement

La qualification des produits, conformément à la procédure, se déroule selon les étapes suivantes :

3.1 Demande écrite du fabricant au LCPC comportant notamment :

- Le dossier technique du produit (nom, usine, plan assurance de la qualité, etc)
- Un échantillon du produit à essayer
- Le nom du ou des laboratoires choisis pour effectuer les essais.

3.2 Envoi par le LCPC de la demande d'essais au(x) laboratoire(s) accompagnée d'un dossier technique et de l'échantillon de produit nécessaire aux essais

3.3 Exécution des essais conformément aux modes opératoires définis dans la procédure :

- Efficacité à l'état neuf
- Efficacité à l'état vieilli
- Limites d'emploi
- Identification comportant deux volets :
  - Identification chimique, physique et mécanique effectuée

par le LCPC pour établir "la carte d'identité du produit". Identification rapide par le(s) laboratoire(s) en vue des futurs essais de réception sur chantier.

Dans tous les cas, l'application des produits pour les essais d'efficacité est effectués par le fabricant ou son représentant.

3.4 Elaboration par le LCPC du rapport d'essai de synthèse, envoyé au fabricant et présenté au groupe pleinier pour jugement des résultats et inscriptions du nom du produit sur la liste d'aptitude publiée par le LCPC sur la plan national.

3.5 Visite d'usine par des membres du groupe pleinier pour s'assurer du fonctionnement correct de la fabrication ainsi que de l'application du P.A.Q (Plan d'Assurance Qualité) ; en particulier :

- Contrôles matières premières
- Contrôles fabrication
- Contrôles produit fini.

3.6 Prélèvement conservatoire et essais de réception (identification rapide) sur les lots de produits approvisionnés sur les chantiers.

Le groupe pleinier, qui deviendra prochainement une commission ministérielle, a également pour mission de juger les anomalies ou dérives relevées par les utilisateurs sur les chantiers.

4. Description des essais d'efficacité par profil d'emploi des produits

4.1 Collage structural

4.1.1 Définition

Il s'agit de produits qui, appliqués sur une surface d'un élément en béton durci en vue de coller ce dernier à un autre élément de béton durci ou de le lier à du beton frais, assure la continuité mécanique de l'ouvrage.

4.1.2 Essais d'efficacité à l'état neuf et à l'état vieilli

Description de l'essai	Référence	Principe de l'essai	Schéma
Traction directe sur cylindre scié et reconstitué	NFP 18.871	Reconstituer par collage une éprouvette cylindrique de 16 cm de diamètre et 32 cm de hauteur puis exercer une force de traction directe jusqu'à rupture	Annexe 1

Description de l'essai	Référence	Principe de l'essai	Schéma
Compression cisaillement sur prisme scié et reconstitué	NFP 18.872	Reconstituer par collage une éprouvette prismatique de 100 cm^2 de section et d'élancement 3 puis exercer une force de compression suivant l'axe longitudinal jusqu'à rupture	Annexe 1
Tenue verticale	NFP 18.873	Vérifier la tenue d'une colle sur surface verticale	/
Temps limite d'assemblage	NFP 18.874	Déterminer le temps limite d'assemblage des pièces à coller par traction directe de pastilles collées en fonction du temps	/

4.2 Injection des fissures

4.2.1 Définition
Il s'agit de produits qui, injectés dans les fissures du béton dégradé, contribuent à rétablir essentiellement la continuité mécanique de l'ouvrage.

4.2.2 Essais d'efficacité à l'état neuf et à l'état vieilli

Description de l'essai	Référence	Principe de l'essai	Schéma
Injectabilité à la colonne de sable en milieux sec et humide	NFP 18.891	Injecter sous pression constante le produit à tester dans une colonne verticale remplie de sable calibré et mesurer le temps nécessaire à l'injection	Annexe 2
Traction par fendage (Brésilien) d'éprouvettes cylindriques de mortier	NFP 18.892	Déterminer la résistance au fendage d'éprouvettes cylindriques de mortier provenant de l'injection d'une colonne de sable	/

Description de l'essai	Référence	Principe de l'essai	Schéma
Fragilité aux chocs	NFP 18.893	Evaluer la fragilité d'un produit d'injection par mesure de la résistance aux chocs produits par des masses tombant d'une hauteur de 20 cm	Annexe 2
Adhérence avec ou sans cycles thermiques	NFP 18.894	Evaluer la charge et le mode de rupture d'éprouvettes de mortier 4 x 4 x 16 cm^3 cassées puis recollées à l'aide d'un produit d'injection et soumises à un essai de flexion centrée	/
Vieillissement de l'adhérence par fatigue mécanique	NFP 18.895	Evaluer la charge et le mode de rupture par flexion d'éprouvettes de mortier 4 x 4 x 16 cm^3 cassées et recollées à l'aide d'un produit d'injection après avoir subi des cycles de fatigue	/

4.3 Réparation du béton dégradé

4.3.1 Définition

Il s'agit de produits destinés à rétablir l'aspect géométrique et esthétique d'un ouvrage ainsi qu'à restaurer les conditions de protection et de tenue initiale de la partie de l'ouvrage réparé.

4.3.2 Essais d'efficacité à l'état neuf et à l'état vieilli

Description de l'essai	Référence	Principe de l'essai	Schéma
Flexion centrée sur prisme évidé et reconstitué	NFP 18.851	Combler un évidement pratiqué sur une éprouvette prismatique de béton puis soumettre cette éprouvette à un essai de flexion sous moment constant, la reconstitution étant placée en fibre tendue	Annexe 3

Description de l'essai	Référence	Principe de l'essai	Schéma
Adhérence	NFP 18.852	Appliquer le produit sur une dallette de béton présentant un état de surface conventionnel puis arracher par traction directe des éléments cylindriques de produit	Annexe 3
Adhérence après cycles thermiques	NFP 18.853	Même essai que le précédent sur dallettes soumises à 100 variations cycliques de température entre -20 et + 55°C	/
Tenue aux chocs répétés	NFP 18.854	Evaluer le comportement d'un produit soumis à 50 chocs répétés obtenus par la chute d'une sphère métallique tombant d'une hauteur de 1 m	Annexe 3
Perméabilité aux liquides	NFP 18.855	Mesurer la perméabilité à l'eau ou à l'air d'un produit appliqué sur un support en mortier poreux	Annexe 3
Tenue aux rayonnements ultra-violets (liants à base de résines synthétiques)	NFP 18.856	Apprécier l'altération d'un produit à base de liant résineux soumis à un rayonnement ultra-violet	/
Tenue aux cycles gel-dégel	NFP 18.857	Evaluer le comportement d'un produit soumis à 50 chocs répétés (cf. essai chocs répétés) avant avoir subi 100 cycles de gel-dégel entre - 25° et + 20°C	/

4.4 Scellement

4.4.1 Définition
Il s'agit de produits qui, mis en oeuvre à l'état liquide (consistance fluide ou pâteuse) assurent la fixation d'une pièce

dans un élément en béton hydraulique.

4.4.2 Essais d'efficacité à l'état neuf et à l'état vieilli

Description de l'essai	Référence	Principe de l'essai	Schéma
Arrachement selon différentes configurations : plancher (horizontal), mur (vertical), plafond	NFP 18.831	Vérifier l'adhérence d'une tige d'acier scellée dans un bloc de béton puis à exercer sur la tige une force de traction	Annexe 4
Fluage en traction (liants à base de résines synthétiques)	NFP 18.836	Soumettre une tige d'acier scellée dans un bloc en béton à un effort permanent de traction et à mesurer le déplacement de la tige en fonction du temps	/
Tenue à l'eau de mer et/ou à forte teneur en sulfates (liants hydrauliques)	NFP 18.837	Mesurer en fonction du temps les variations de longueur d'éprouvettes prismatiques 2 x 2 x $16cm^3$ immergées dans une solution d'eau agressive	Annexe 4

4.5 Calage

4.5.1 Définition
Il s'agit de produits qui, mis en oeuvre à l'état liquide (consistance fluide) entre des éléments préalablement positionnés, assurent la continuité mécanique entre eux.

4.5.2 Essais d'efficacité à l'état neuf et à l'état vieilli

Description de l'essai	Référence	Principe de l'essai	Schéma
Aptitude à la mise en place . Liants hydrauliques . Résines synthétiques	 NFP 18.832 NFP 18.833	Introduire le produit par simple écoulement gravitaire sur une surface horizontale sous une plaque de verre puis observer son écoulement jusqu'au remplissage complet du dispositif d'essai.	Annexe 5

Description de l'essai	Référence	Principe de l'essai	Schéma
Ressuage en volume confiné . Liants hydrauliques	NFP 18.834	Injecter gravitairement le produit dans un cylindre métallique étanche puis mesurer 48 h après injection le volume de vide résiduel en partie supérieure du cylindre	Annexe 5
Fluage en compression . Liants à base de résines synthétiques	NFP 18.835	Soumettre un échantillon à une pression et une température constantes puis mesurer sa déformation dans le temps	Annexe 5
Tenue à l'eau de mer et/ou à forte teneur en sulfates . Liants hydrauliques	NFP 18.837	Mesurer en fonction du temps les variations de longueur d'éprouvettes prismatiques $2 \times 2 \times 16$ cm^3 immergées dans une solution d'eau agressive	/

5. Essais d'identification

Ces essais ont pour but d'établir la liste des valeurs constituant les références qui permettront d'identifier le produit. Ces valeurs peuvent, par la suite, être utilisées pour vérifier la constance de fabrication. Ils sont effectués sur l'échantillon utilisé lors de l'épreuve de qualification.

Les essais d'identification rapide sont prévus pour :

1/ Noter l'importance des différences éventuelles entre les valeurs des principales caractéristiques des échantillons de produit reçus par les laboratoires chargés des essais,

2/ Effectuer des essais de réception sur chantier pour s'assurer que le produit livré présente les mêmes caractéristiques que l'échantillon qualifié.

La liste et les modes opératoires de ces essais sont précisés dans les fascicules de procédures édités par le LCPC pour chaque type de produit.

6. Spécifications

Pour les essais d'efficacité à l'état neuf et à l'état vieilli, des spécifications ont été élaborées par chaque groupe de travail, puis entérinées par le groupe pleinier.

C'est sur la base de ces specifications qu'est jugée l'aptitude à l'emploi d'un produit qui conditionnera sa publication sur la liste d'aptitude.

L'ensemble des spécifications est récapitulé dans les normes suivantes :

- Collage structural	NFP 18.861
- Injection des fissures	NFP 18.881
	NFP 18.882
- Réparation du béton dégradé	NFP 18.841
	NFP 18.842
- Scellement	NFP 18.821
- Calage	NFP 18.822

7. Norme de produit

Une norme générale de produit est en cours d'élaboration. Elle couvrira l'ensemble des produits ci-dessus et précisera :

- la définition des produits,
- les références aux normes d'essais et normes de spécifications,
- les modalités d'application de la norme puis d'une future conformité à la marque NF,
- les conditions de marquage et de conditionnement,
- les conditions de réception des produits sur chantier.

TRACTION DIRECTE SUR CYLINDRE SCIE ET RECONSTITUE

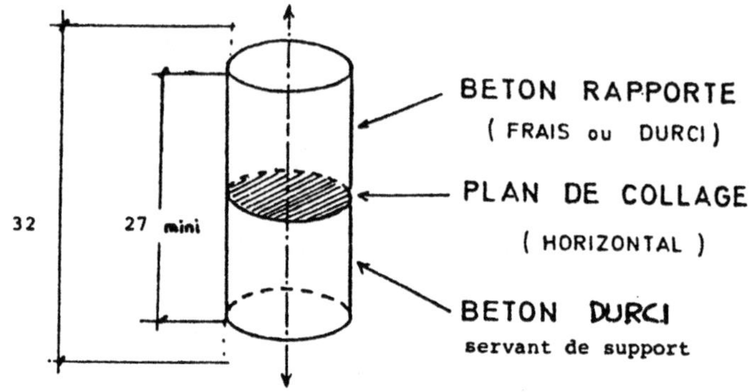

COMPRESSION - CISAILLEMENT SUR PRISME SCIE ET RECONSTITUE

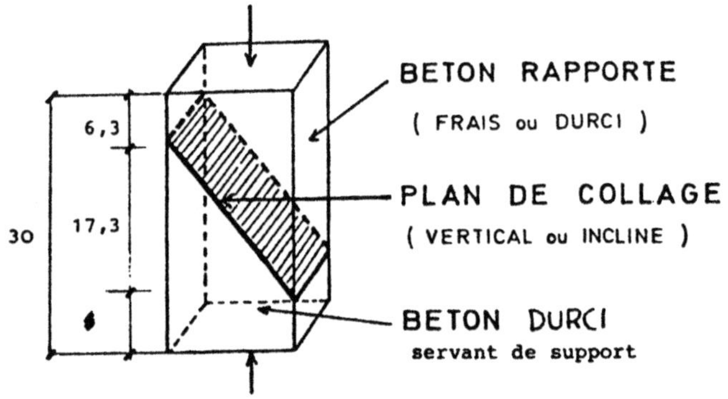

INJECTABILITE A LA COLONNE

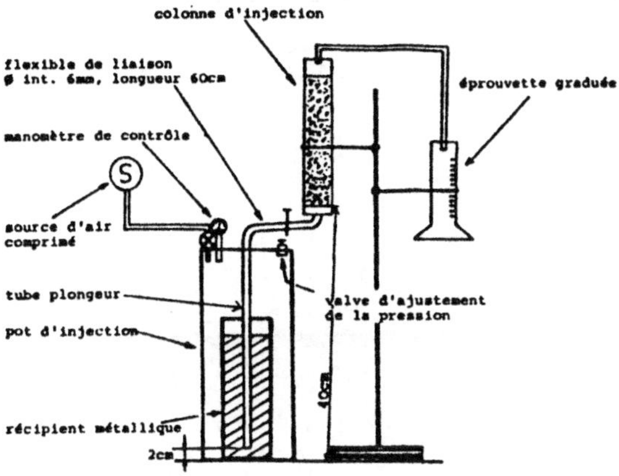

FRAGILITE AUX CHOCS

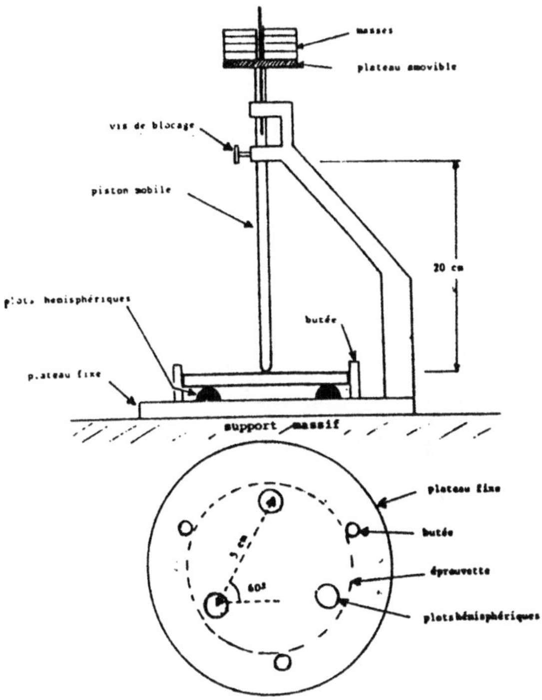

FLEXION CENTREE SUR PRISME EVIDE ET RECONSTITUE

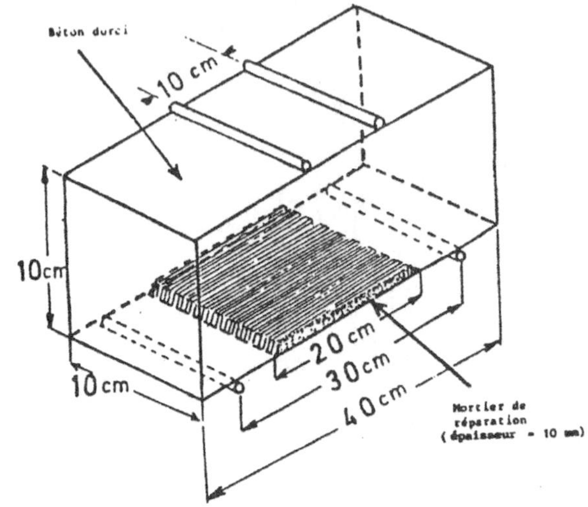

ADHERENCE

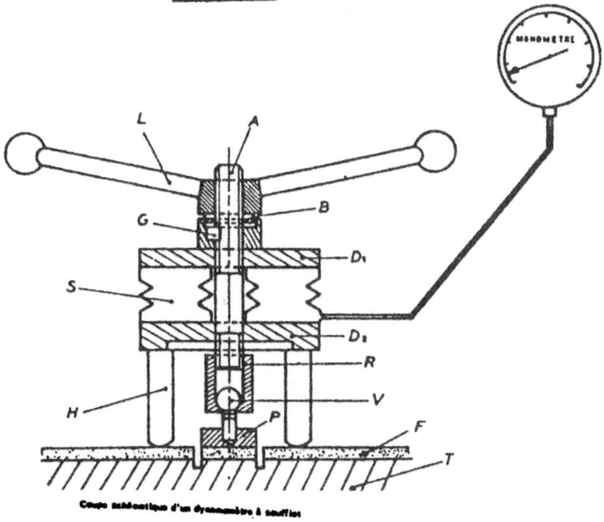

Coupe schématique d'un dynamomètre à soufflet

TENUE AUX CHOCS REPETES

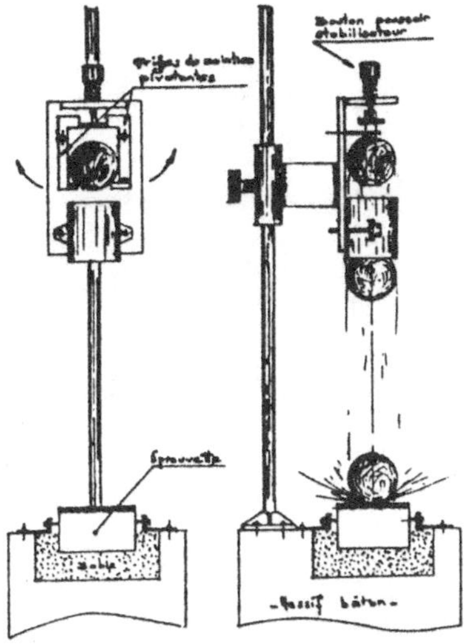

PERMEABILITE AUX LIQUIDES

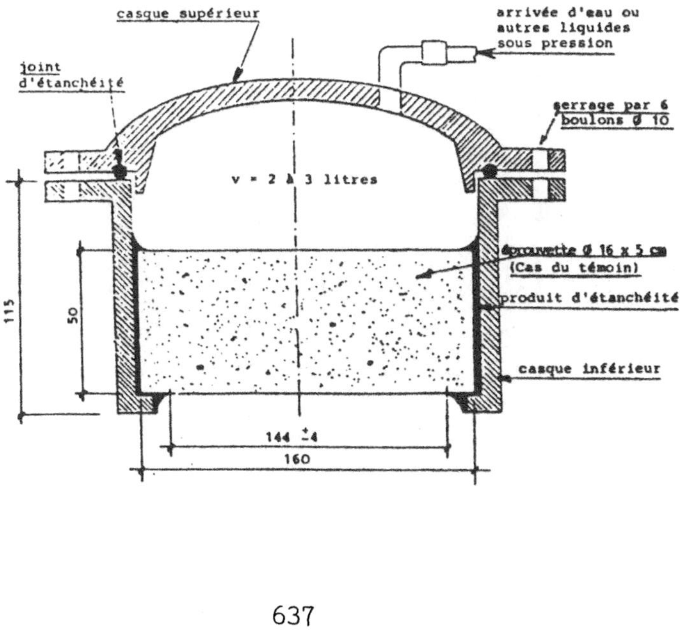

ANNEXE 4

ARRACHEMENT

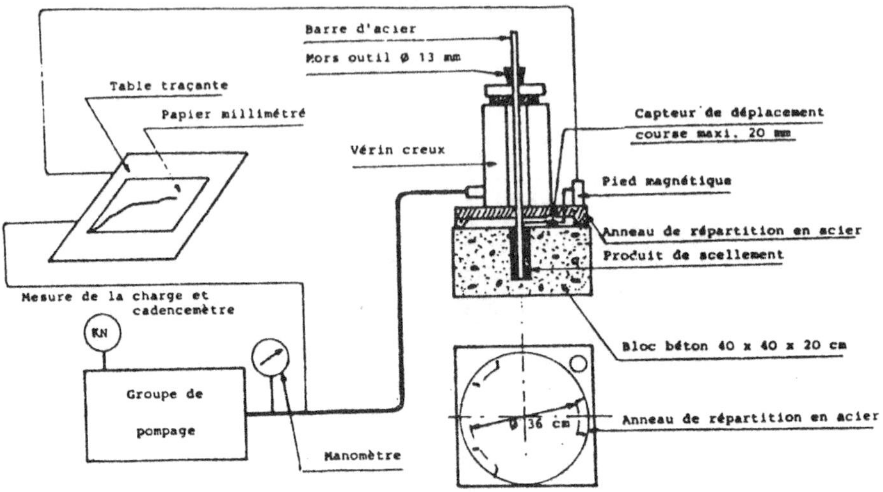

FLUAGE EN TRACTION

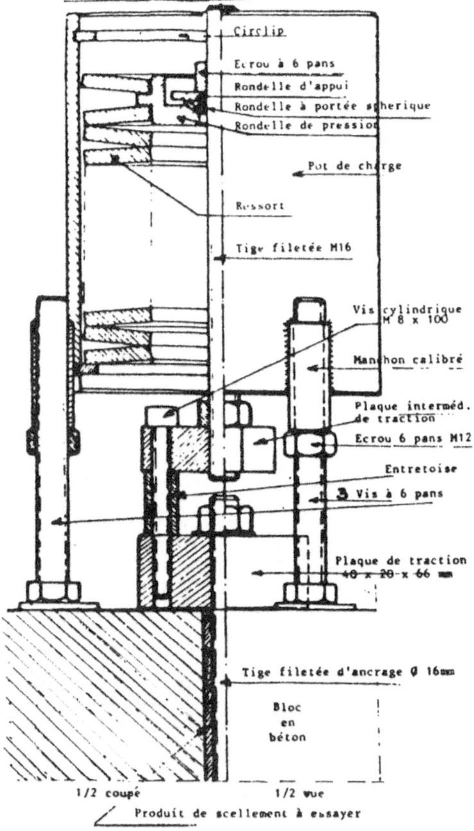

ANNEXE 5.1

APTITUDE A LA MISE EN PLACE (LIANTS HYDRAULIQUES)

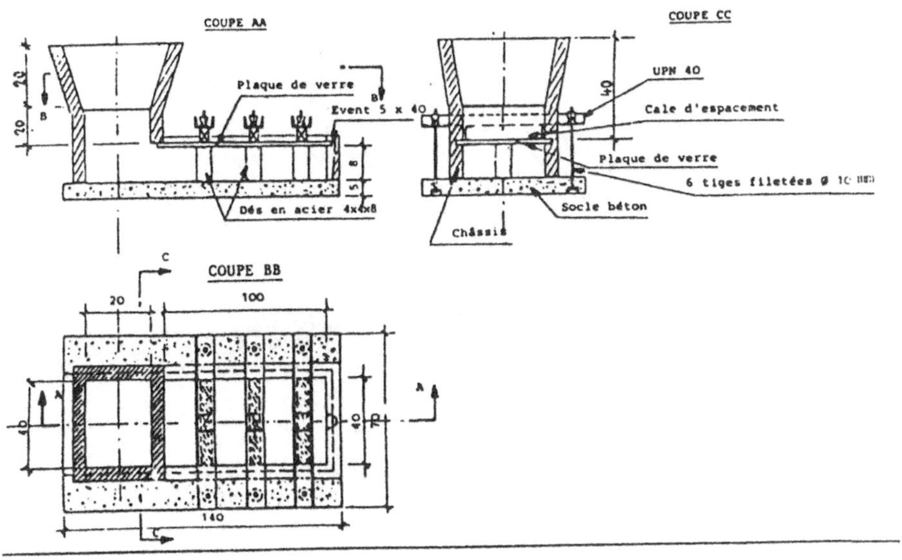

APTITUDE A LA MISE EN PLACE (RESINES SYNTHETIQUES)

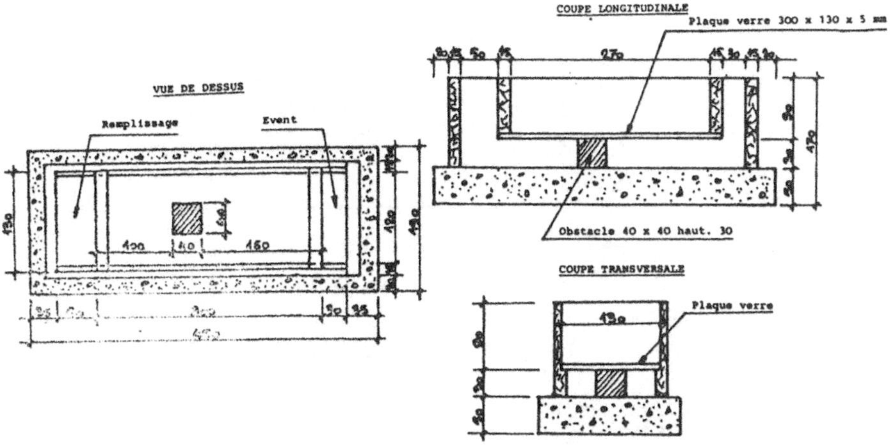

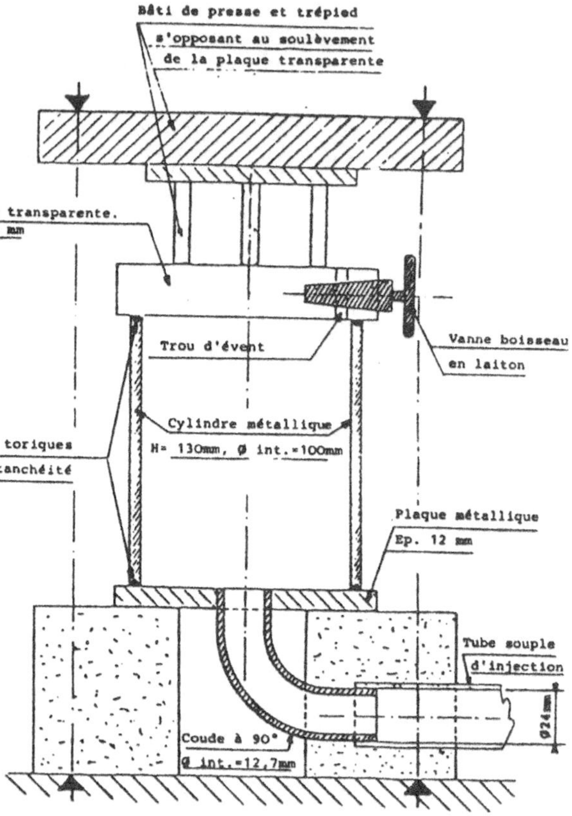

FLUAGE EN COMPRESSION

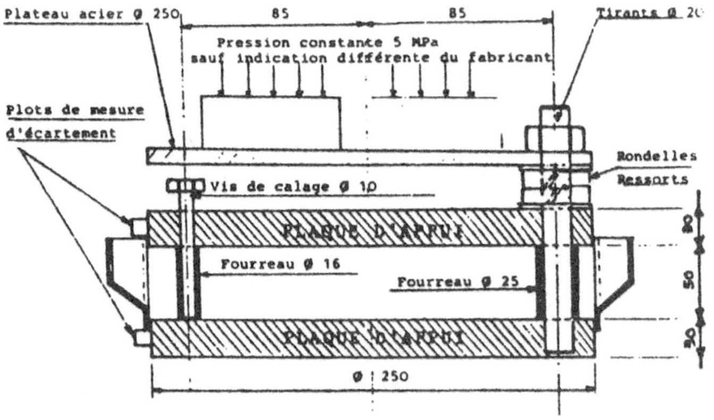

RESISTANCE A ARRACHEMENT DES ANCRAGES FIXES DANS LE BETON

K. GAMSKI
Institut du Génie civil, Université de Liège,
6, quai Banning, B-4000 Liège, Belgique.

Résumé
Pour assurer la continuité entre les éléments d'une structure en béton et ceux à rapporter, l'utilisation des ancrages en acier est souvent en application. Selon le principe généralement admis dans la construction en béton armé, la rupture eventulle et accidentelle devrait être précédée des signes précurseurs de la défaillance : fissures, déformations,. C'est pourquoi il convient que le scellement soit dimensioné de façon à ce que l'arrachement soit précédé de la plastification de la barre (dépassement de sa limite élastique apparente ou conventionnelle) accompagné éventuellement d'un glissement limité. Les recherches du Comité Technique RILEM-CEB-FIP ont conduient, en outre, à l'établissement des Recommandations : RC 5: Essais d'adhérence, 1. Par flexion, 1978 (Bond test for reinforcing steel, 1. Beam test, 1978) et RC 6: Essai d'adhérence, 2. Par traction, 1978 (Bond test for reinforcing steel, 2. Pull - out test, 1978). Le texte qui suit présente une tentative de comparaison d'efficacité d'amarrage des barrés à adhérence améliorée dans le béton frais et des barres lisses fixées par l'adhérence ou mécaniquement dans le béton durci. En particulier sont examinés la répartition des contraintes de cisaillement le long d'ancrage adhèsif d'une part et le mécanisme de rupture de l'ancrage à expansion d'autre part. Quelques remarques soulignent l'importance des soins à apprter lors de préparation de scellemnt des ancrages au mortier de résine ou encore au mortier à liant hydraulique modifié.
<u>Mots clef</u>: Acier, Adhérence, Ancrage, Béton, Mortier de résine, Mortier à liant hydraulique modifié.

1. Introduction

La continuité entre les éléments d'une structure en béton armé et ceux à rapporter peut être matérialisée à l'aide de plusieurs techniques. Si un raccord a été prévu il y a des barres d'attente fixées dans le béton frais pendant la construction. Si cela n'était pas le cas il convient de fixer la barre d'ancrage au béton durci de la construction existante. Le principe généralement admis dans la construction en béton armé veut que la rupture accidentelle soit précédée d'une déformation perceptible annonçant le danger. C'est

pourquoi les barres de continuité ou d'ancrage devraient être calculée et dimensionées de façon que leur arrachement soit précedée d'une plastification voir d'un faible glissement. Pou réaliser l'ancrage les barres à adhérence améliorée torsadées, crénelées, nervurées sont utilisées. Lorsque la barre est fixée dans un béton frais la lonqueur d'ancrage est spécifiée en tenat compte de la nature du béton, de sa résistance, de son armature destinée à prévenir la fissuration transversale, ainsi que de caracteristiques de l'ancrage, limite élastique Re, son diamentre D, relief de sa surface. La verification expérimentale de la lonqueur d'ancrage dans le béton durci demande la préparation d'un forage de diamètre et longueur appropriés afin que l'ancrage fixé puisse satisfaire à l'exigeance évoquée plus haut pour les barres fixées dans le béton frais. La vérification expérimentale de la longueur d'ancrage fixé dans le béton frais est réalisée par pull-out ou beam test élaborés par RILEM-CEB-FIP.
La fixation des ancrages dans le béton durci peut être matérialisée par: A - frottement, B - verouillage, C - expansion, D - adhérence ou par une combinaison de ces dispositifs.
Dans cette étude la résistance à l'arrachement d'ancrage fixé par adhérence dans le béton durci est comparée à la résistance de l'ancragefixé à l'aide d'une cheville à expansion et celui fixé dans le béton frais.

2. Rupture de scellement

Le diametre D de l'ancrage étant imposé par la sollicitation extér. il convient de définir la longueur de scellemnt L = nl (fig 1) compte tenu de son relief simulé ici par des crénelures d'épaisseur r, de longueur l. La rupture de scellement peut se produire:

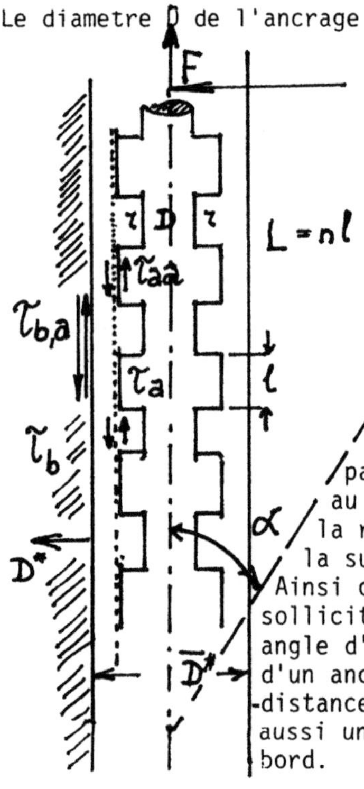

fig.1

F1 - sur le pourtour du forage cylindrique de diametre D#, par dépassement de la résistance au cisaillement tba, à l'interface béton mortier adhésif sur la surface $\pi D\#L$, par F2 - sur le pourtour des crênelures de diametre D+2r, par dépassement de la résistance taa et simultanement de ta, par F3 - sur le pourtour d'un cône de rayon R au sein du béton (fig 1), par dépassement de la résistance du béton au cisaillemnt tb sur la surface $S = \pi R\sqrt{R^2+L^2}$
Ainsi chaque ancrage implanté à une profondeur L, sollicite le béton au cisaillement selon un demi angle d'ouverture α. Il en resulte une interaction d'un ancrage sur son voisin. Pour l'éviter l'inter-distance $I = 2L\ tg\alpha$ est à respecter. Ceci impose aussi une distance min. d'implantation d'ancrage du bord.

Pour obtenir la résistance à l'arrachement $F > Fp$ ou $Fp = ReD^2\pi /4$ est la force de plastification, il convient de choisir un mortier de scellement à ciment hydraulique, très sec, à liant hydraulique modifié voir à liant résineux, offrant une adhésion suffisante à la paroi de forage tba, à la barre d'ancrage taa, une bonne résistance au cisaillemnt ta et à la compression ca, caractérisé aussi par l'absence du retrait et par une stabilité dans le temps à l'humidité et à l'alcalinité du béton.
Pour obtenir la plastification de la barre avant glissement notable dans le mode F1 il faut que $Fp < F1 < F2 < F3$, que la résistance à la plastification Fp soit plus faible que la réssitance au glissement dans le mode F1. Il vient: $\frac{Re\ D^2}{L\ 4D\#} < tba$. Dans le cas d'implantation d'un ancrage dans le béton frais, sans adhésif il vient pour barre lisse $D = D\#$ d'où : $L > DRe / 4\ tba$.
Pour une barre crénelée $L > D^2\ Re / 2\ (tba + tb)(2r + D)$.

L'expérience enseigne que l'ancrage en barre crénelée pour être efficace $Fp < F1 < F2 < F3$ doit être placé avec une armature hélicoïdale préservant l'éclatement transversal du béton. Cependant l'implantation d'une armature hélicoïdale lors de fixation d'ancrage dans le béton durci, oblige de réaliser un forage de gros diamètre D#. C'est pourquoi il semble parfois préférable d'utiliser l'ancrage lisse ou à très faible relief à l'aide d'un mortier à liant résineux caractérisé par une forte adhérence tba, taa et ta.

3. Mécanisme de résistance d'un ancrage adhésif

La contrainte d'adhésion moyenne à la rupture est $tmr = F/(D+T/2)\pi L$ ou T est l'épaisseur du mortier adhésif.
Il vient $L > D\ Re / 4\ tmr$.
On sait que dans le cas d'application d'une force Fa à l'ancrage, la contrainte locale de cisaillement dans l'adhésif est:

$$t\ (z) = Fa\ K\ \frac{ch\ K\ z}{sh\ K\ L}$$

Dans le cas d'ancrage nous adoptons pour l'operateur K:

$$\frac{G}{T}\ \frac{4}{\Pi}\left(\frac{1}{D\ Ea} + \frac{1}{(D^* - D)\ Eb}\right)^{\frac{1}{2}} = K ,$$

D, D*, Ea, Eb, étant respectivement diametre de l'ancrage et diametre de l'épaisseur collaborant de béton ainsi que les modules de Young de l'acier et dubéton.
G, est le module de cisaillemnt et T, l'épaisseur de l'adhésif, L, est la longueur de l'ancrage et z, la distance où agit la contrainte tz.
Fz, est la force locale agissante à l'ancrage et comptée par unité du périmètre de l'ancrage (N/mm).
En présence de l'adhésion les déplacements de l'acier, wa et du béton wb, sont liés: $wa - wb = tT / G$, ainsi

$$wa - wb = K\ \frac{T}{G}\ \frac{chKz}{shKL}\ Fa$$

Le déplacement compté à partir de $L = 0$ à $L = z$ sera

$$\Delta\ (z) = \int_0^z (wa - wb)dz \qquad \Delta\ (L) = \frac{Fa\ K}{sh\ K\ L}\ \frac{T}{G}\int_0^L ch\ K\ z\ dz$$

avec $\Delta\ (L) = F\ T/G$, $tz = dFa/dz = -\ dFb/dz$ et d'autre part

$dFa / dz = (G / T)(wa - wb) = - dFb / dz$, il vient alors :
$d^2F / dz^2 - K^2F = 0$ dont la solution générale est

$$F = A_1 \, shKz + A_2 \, chKz$$

pour $Z = 0$, $Fa = Fb = 0$ d'où $A_2 = 0$
pour $z = L$, $Fa = F = A_1$ sh Kz donc $A_1 = F / sh \, K \, L$, par la:

$$Fa = F \frac{sh \, K \, z}{sh \, K \, L} = - Fb.$$

Fa est la force par unité de la périférie d'ancrage $Fa = Fp / \pi D$, équilibrée par l'éffort repris par l'adhésif. La repartition exponentielle des contraintes de cisaillement le long d'ancrage dépend de l'opérateur K, donc des caractéristiques de l'adhésif (G), de son épaisseur (T), de la qualité du béton (Eb) et de son diamètre collaborant (D*) et du module d'ancrage (Ea) et de son diamètre (D). Pour calculer la répartition Fz et tz, on calcul d'abord la longueur de scellement en adoptant une contrainte moyenne de résistance au cisaillemnt de l'adhésif, puis K, Fz et tz,
pour fixer les idées examinons deux ancrages Ø = 10 et Ø = 30 mm, en acier Re = 300 MPa, $Ea = 2,1 \, 10^5$ MPa, $Eb = 4 \, 10^4$ MPa, $G = 2 \cdot 10^3$, tmr = 10 MPa, T = 2 mm et D* = 5D,
Il vient pour Ø = 10 mm, l'opérateur K = 0.037444/ mm
Fp = 23561,945 N, $Fa = Fp/ \pi D = 750$ N/mm et L = Fa/t = 75 mm
Fz = Fa (sh K L/ sh K L) et t(z) = Fz K $z = \alpha L$, $0 \leq \alpha \leq 1$

α	0.I	0.2	0.3	0.4	0.5	0.6	0.7	0.8	0.9	0.95	I
Fz	25.7	53	85	124	173	235	316	422	562	648	750
tz	1.0	2.2	3.2	4.6	6.5	8.8	11.9	15.8	21	24.3	28.1

Pour Ø = 30 mm, l'opérateur K = 0.0241048 avec T = 3 mm,
Fp = 212057.5 N, Fa = 2250, L = 225 mm,

α	0.1	0.2	0.3	0.4	0.5	0.6	0.7	0.8	0.9	0.95	1
Fz	11.3	26.0	48.6	85.8	149.4	257.0	442.	760.	1308	1715.	2250.
tz	0.3	0.6	1.2	2.1	3.6	6.2	10.5	18.3	31.5	41.4	54.24

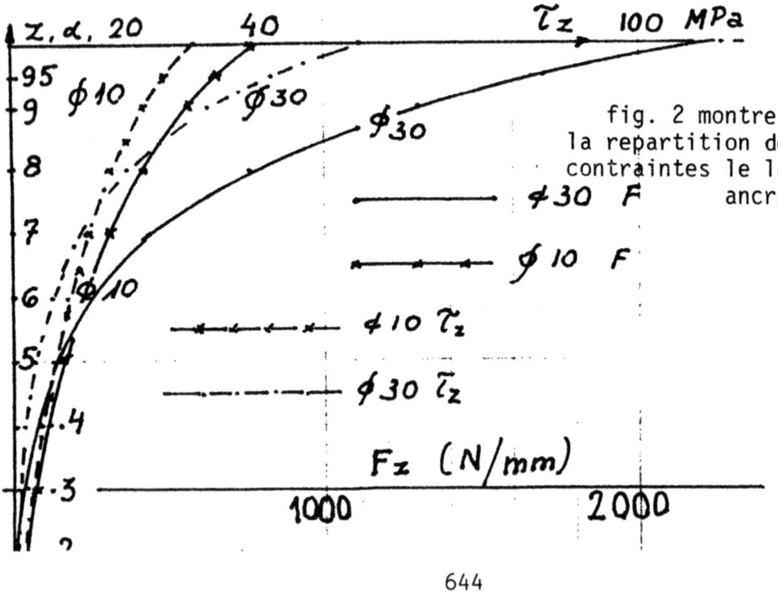

fig. 2 montre la repartition des contraintes le long des ancrages

4. Ancrage mécanique à l'expansion.

La fixation resulte de l'action de cône sur une douille dont l'expansion provoque serrage sur la paroi de forage réalisé dans le béton. La projection sur l'axe de Z (voir fig 3) donne:

$F = \Sigma Ni \sin\alpha + \Sigma Ti \cos\alpha$, avec $Ti = Ni \, tg\,\delta$, ou δ est l'angle de frottement au glissement cône-douille.

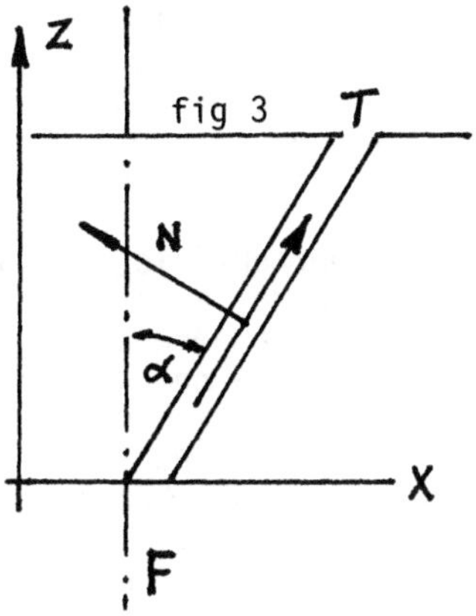

fig 3

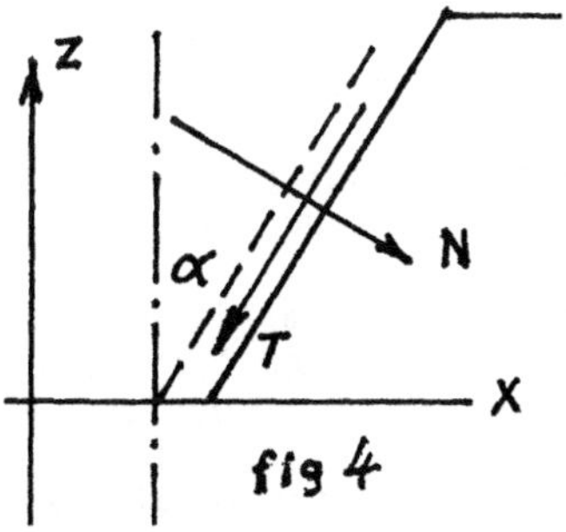

fig 4

Il vient: $F = \Sigma Ni(\sin\alpha + tg\,\delta \cos\alpha)$

La réaction sur la douille du cône d'une part et d'autre part du béton s'équilibrent. Celles du béton sont remplacées par des forces normales et tangentielles Pi, Ci, respectivement dont la projection sur l'axe vertical est: (voir fig 4)

$\Sigma Ci - \Sigma Ni\sin\alpha - \Sigma Ti\cos\alpha = 0$

$\Sigma Ci - \Sigma Ni\sin\alpha - \Sigma Ni\,tg\,\delta\cos\alpha = 0$

$\Sigma Ci = \Sigma Ni(\sin\alpha + tg\,\delta\cos\alpha) = F$

La projection sur l'axe horizontal:

$\Sigma Pi - \Sigma Ni\cos\alpha + \Sigma Ti\sin\alpha = 0$

$\Sigma Pi - \Sigma Ni\cos\alpha + \Sigma Ni\,tg\,\delta\sin\alpha = 0$

$\Sigma Pi = (\cos\alpha - tg\,\delta\sin\alpha)$ et

$\Sigma Pi = F\,(\cos\alpha - tg\,\delta\sin\alpha)/(\sin\alpha + tg\,\delta\cos\alpha)$

ou encore $\Sigma Pi = F\,cotg(\delta + \alpha)$ d'où

$Fc = \pi D\# LR'b/cotg\,(\delta + \alpha)$.

Pour que la rupture se produise par plastification de l'ancrage sa longueur de fixation L, doit resulter de la limite $Fc \geqslant Fp$. Il vient alors:

$L = ReD^2 cotg(\delta + \alpha)\; / \; 4D\#R'b$. Toutefois ici, en realité, le cisaillemnt se supperpose à la compression. On sait que la fissuration dans ce cas aura lieu suivant un angle $tg\,2\beta = 2\tau / \sigma = 2F / P$ et comme $P = Fcotg\,(\delta + \alpha)$ il vient:

$\beta = 0,5\ arctg((2\ /(cotg\,\delta + \alpha))$.

Le frottement de la douille s'oppose à la force de traction $F = \Sigma Ci + \Sigma Pi\ tg\theta$ ou θ est l'angle de frottement de la douille sur béton. Il vient après remaniement: $F = \pi\ D\#\ L^*\ tb + Fcotg\,(\delta + \alpha)tg\theta$, avec $tb = 0,3\ R'b$ il vient la charge de rupture par cisaillemnt:

$$Fs = \pi\ D\#\ L^*\ 0,3\ R'b\ /\ 1 - cotg\,(\delta + \alpha)\ tg\theta$$ où

L^* est la longueur de la douille au contact avec la paroi du béton. En conclusion pour que la rupture se produise par dépassement de la limite élastique de l'ancrage Re, il faut que la longueur de implantation de l'ancrage soit telle pour que:

$$Fp > Fc > Fs$$

5. Ancrage fixé dans le béton frais

Rappelons que l'objet des essais Pull-out test (POT) et Beam test (BT) est la mesure d'adhérence des barres au béton durci, les barres ayant été placées dans le béton frais. Le relief de surface relative de ces barres est exprimé par la surface relative des nervures, calculée selon certaines regles. La résistance au glissement relatif acier-béton est mesurée et exprimée à trois niveaux : t,0.ol, (adhérence au décollage), tm = (1/3)(t,0.ol + t,0.1 + t,1.0)adhérence moyenne, t,r adhérence à la rupture au glissement. Les courbes force glissement des barres telles fig 1, sont représentées à la fig 5, courbes A et B. Pour la barre crênellée (A) le glissement reste imperceptible jusq'au point I. Après le décollage de la barre du béton, sur toute la longueur d'adhérence pour une force correspondante à I, la charge monte pour vaincre la résistance du béton au cisaillemnt et à la compression sous les nervures sur toute la longueur d'adhérence. Dans la région II - III, le glissement a lieu pratiquement à charge constante et à partir du point IV (charge max) l'effort décroit alors que le glissement se poursuit de plus en plus facilement dans la gaine du béton de diametre D# (fig 1).La courbe B de la fig 5 schematise la résistance d'une barre lisse. Au poit I, il se produit le décollage sur toute la longueur d'adhérence pour une charge plus faible par rapport à la barre nervurée (A); la résistance au glissement dimunue alors que le glissement se poursuit.

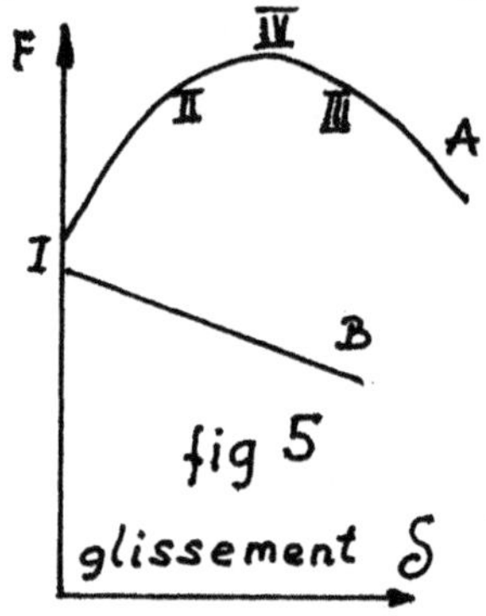

fig 5

L'objet de ces mesures n'est donc pas l'étude de la repartition des contraintes de cisaillement le long de la barre mais bien l'évaluation de l'état limite d'adhérence. Dans le cas d'ancrage qui nous occupe il s'agit d'évaluer la profendeur d'implantation telle pour obtenir la plastification de l'ancrage $F_p = R_e D^2 \pi /4$ avant que le glissement important (fig 5) ne soit atteint.
Prenons pour fixer les idées t,m comme valeur de référence à partir du tableau ci-après, qui rassemble les résultats extrêmes des essais publiés dans Matériaux et Constructions,N°35,1973, sur barres 8-30mm, Re 400-500 MPa et béton 30 MPa. Entre paranthèses diamètres barres:

t,0.ol	t,m	t,r
(14) 1.9 ≤ POT ≤ 6.5 (10)	(10) 2.8 ≤ POT ≤ 11 (14)	(10) 3.1 ≤ POT ≤ 22(16)
(12) 2.8 ≤ BT ≤ 10.7 (10)	(10) 4.7 ≤ BT ≤ 15 (10)	(10) 4.2 ≤ BT ≤ 20(8)

Les valeures extrêmes d'adhérence limite sont données en MPa et correspondent à des surfaces relatives des nervures échellonées de 0,011 à 0,125. Vu la grande disperssion des resultats des mesures de cette recherche aucune relation certaine n'a pu être truvée entre les rapports POT / BT et la surface relative des nervures ou le diamètre des barres. En outre, les resultats de 900 essais d'adhérence publiés dans Matériaux et Constructions N°43,1975, montrent que la

résistance d'adhérence exprimée comme plus haut, t,0.o1, t,m et t,r, augmentent moins vite que la surface relative des nervures. Selon diverses préscriptions la surface relative des nervures augment avec le diamètre de la barre mais t,m diminue lorsque Ø augmente (exemple CEB: t,m = 80 - 1,2 Ø, en kg/cm², Ø en mm). En nous y référant nous avons la longueur d'implantation avec $Fp = Re\ D^2\ \pi / 4 = \pi\ D\ L\ tm$ d'où: $L = Re\ D / 4\ tm$, avec Re = 400 MPa, D = 10 mm, t,m = 6,8 MPa il vient L = 147 mm.
N.P.Roberts (Matériaux et Constructions N° 32,1973 précise que selon le Code de bonne Pratique de British Standard CP 114, un ancrage doit résister avec un glissement qui n'éxcède pas 0,2 mm (de coté libre de la barre) à la charge correspondante à la limite élastiques exercée sur la barre. Pour les barres hélicoïdales L = Re D / 21, pour les barres crénellées L = Re D / 28.
En revenant au cas de la barre Ø lo mm plus haut il vient avec la Re =400 MPa, la longueur d'implantation de L = 400 . 10 / 28 =143 mm

6. Conclusions

L'éffort appliqué à l'ancrage est transmis au scellement et au béton; toutefois le mode de sa tarnsmission dépend du système de fixation. Il est donc souhaitable d'adapter le système de fixation aux conditions techniques et économiques du problème posé. Il apparait que le système de fixation mécanique demande, pour les ancrages de même diamètre, la plus court longueur de scellemnt; par la cet ancrage exerce des fortes contraintes sur la paroi de forage et peut être exploité dans le cas de fixation dans un béton de haute résistance par exemple de 40 MPa et plus. Les ancrages fixés au mortier de résine de type PC (polymer concrete) ou au mortier hydraulique modifié de type ECC p.ex. (epoxy cement concrete) peuvent être utilisés de bétons de résistance moyenne et faible. En effet l'effort appliqué à l'ancrage se reparti selon une courbe exponentielle sur sa longueur; la forme de cette courbe, nous avons vu, dépend notamment du module d'élasticité du béton, de la longueur de l'ancrage ainsi que de son diamètre, mais aussi du module d'élasticité de mortier de scellement, de son épaisseur voir de son pouvoir d'adhèrence au béton et à l'acier. En outre le mortier de scellement, suffisamment fluide est à même d'imprégner la paroi de forage devant recevoir l'ancrage, et même de refermer la porosité éventuelle du béton et le consolidér. De cette façon l'utilisation d'un tel mortier de scellement peut éventuellement améliorer localement la qualité du béton. Malgré l'apparence la fixation des ancrages au mortier adhésif, pour être couronnée de succé doit être réalisé par le personnel spécialisé ou tout au moins entrainé et soigné. Nombreux échecs qui ont été constatés dans ces applications sont imputables à de non respect des regles ci après:

(1) les dimensions, diamètre et longueur de forage doivent être adaptées au diamètre d'ancrage d'une part et d'autre part au type du mortier de scellemnt pour ses caractéristiques mécaniques et rhéologiques, fluidité, suil d'écoulement.

(2) la paroi de forage doit être parfaitement propre, exempte de l'huile et de poussière et être seche de façon à être à même d'absorber le liant par succion capilaire.

(3) le mortier adhésif doit être parfaitement homogénéisé au moment d'emploi de façon à favoriser la réaction de polymèrisation et le développent d'adhèrence.

(4) le mortier adhésif doit être suffisamment liquide au moment d'emploi pour mouiller et imprégner au mieux la paroi (basse tension superficielle) et remplir le forage.

(5) le volume de mortier préparé doit présenter en excé par rapport au volume à remplir.

(6) le mortier adhésif doit être caractérisé par un faible retrait ou nul pour sa partie physique (absence de solvant volatil), pour sa partie chimique (formation des liaisons plus courtes) et pour sa fraction thermique c.e.d. la réaction de solidifica--tion doit se faire à la température relativement basse; l'addition éventuelle d'un produit expansif susceptible de componser le retrait doit être fait avec prudense car souvent l'expansion s'accompagne d'abaissement de résistance au cisaillemnt et d'adhérence.

(7) le passage de l'état liquide à l'état solide doit être bref.

(8) le mortier adhésif devra se caractériser par un fluage faible sous l'effet d'une charge permanente exercée sur l'ancrage.

(9) l'ancrage, au moment de son insertion doit être parfaitement propre, exempte de l'huile et d'humidité.

(10) la force portante de l'ancrage doit être vérifiée sur un échantillon représentatif par démonstration que son limite élastique est atteint avant que un glissement notable se produit.

Références

Roberts, N.P. (1973) A limit state anchorage bond test for reinforcement in concrete. Matériaux et Constructions.32,119-122

Bony, J.C.,Claude, G. et Soretz, S.(1973) Comparaison des essais d'adhérence par flexion (beam test) et par traction (pull-out test) Matériaux et Constructions,35,395-401.

Bony, J.C., Claude, G. et Soretz, S. (1975) Comparaison des prescriptions relatives à la résistance d'adhérence et à la surface spécifiques des nervures des barres à béton. Matériaux et Constructions.43,33-41.

Bresson, J. (1971) Béton plaqué.Annales ITBTP.N°278,23-55

Gamski K. (1986) Evaluation de la qualité du joint collé. Rapport N°1401 de la Soc.Belge des Mécaniciens: Collage, Matériaux, technologie et applications.

PULL OUT RESISTANCE OF STEEL ANCHORAGES EMBEDDED IN CONCRETE

K. GAMSKI
Civil Engineering Institute, University of LIEGE
6 quai Banning, B-4000 LIEGE Belgium

Abstract.
In order to ensure continuity between both old and new elements of a reinforced concrete structure, steel anchorages are very often used. It is generally accepted that possible or accidental failures should present warning signs : cracks, deformations ... For this reason, the sealing should be designed in such a way that plasticization of the bar occurs before the pulling out - eventually together with a limited slip. Moreover, the research carried out by the technical Committee RILEM-CEB-FIP has resulted in the formulation of the following Recommendations : RC5 : Adherence tests, 1.Flexion Tests - 1978 (Bond Test for Reinforcing Steel,1.Beam Test, 1978) and RC6 : Adherence Tests,2.Tensile Tests,1978 (Bond Test for Reinforcing Steel,2.Pull out test, 1978). In the following text, we have tried to draw a comparison between the efficiency of anchorage of bars with a better adherence in unset concrete and of smooth bars set in the hardened concrete either by adherence or mechanically. We have considered more precisely on one hand the spreading of shear stress along adhesive anchorage and on the other hand the failure mechanism of expansion anchorage. A few remarks underline the importance of the care to be taken during the preparation of anchorage sealing with resinous mortar or even with modified hydraulic binder mortar.
Key words : steel, adherence, anchorage, concrete, resinous mortar, modified hydraulic binder mortar.

FAILURE CRITERIA FOR STRUCTURAL JOINTS

MARKO HRANILOVIĆ
Civil Engineering Institute Zagreb, Yugoslavia

Abstract
Structural joints are often subjected to complex multiaxial state of stresses. In design, however, the ultimate load carrying capacity is roughly estimated by limiting the simple tensile, compressive and shear stresses of substrate or adhesive materials.

Adhesive materials have different properties from concrete. Generally their tensile to compressive strengths ratio is more favourable than for concrete. The evaluation of adhesive materials is done in laboratory tests under simplified states of stresses, different from the states in application.

This paper presents an analysis of the state of stresses in certain laboratory tests applying the failure criteria for each part of the joint. The limitation of these tests in application is also pointed out.

Key words: Structural joints, Bonding material, Carrying capacity, Direct shear test, Slant shear test, Biaxial strength envelope, Maximum strain criteria, Failure.

1. Introduction

Structural joints are often subjected to complex multiaxial state of stresses. In design, however, the ultimate load carrying capacity is roughly estimated by limiting the simple tensile, compressive and shear stresses of substrate or adhesive materials.

The characteristics design values are obtained from laboratory tests under simplified states of stresses. Different laboratory tests, are adopted for evaluation of adhesive materials, but slant shear and direct shear tests are widely used as most suitable /1, 2 and 3/.

In both tests the high quality substrate concrete is used to confine the possible area of failure of the joint. If the cause of joint failure is the weak bond on the interface between concrete substrate and adhesive material, then the Coulomb law is the most adequate model.

But in the case of good bonding materials the typical mode of failure is characterized by the rupture in substrate concrete: It is obvious that in these cases simple distributed compressive load causes in joints of slant or shear tests uneven distributed shear, compressive and tensile stresses, both in the adhesive material and in concrete.

This paper will give the results of FEM analysis of both above-mentioned laboratory tests. The analysis has covered the optimisation of the dimension for direct shear test, the influence of rigidity of adhesive materials, i.e. their thickness and E moduli.

The results of FEM analysis are compared with experimental data for biaxial strengths of concrete. These experimental data are obtained for proportional biaxial stresses, and usually are presented with an envelope according to the Fig.1 or by means of maximum strains vs. strengths ratio ($\alpha = \sigma_1/\sigma_2$) /5/. For the sake of better comparisone of results of FEM analysis with experimental data is made a diagram of maximum strains vs. principal stresses. In this way is obtained simple criteria which includes Poisson's ratio for all directions.

The similar diagram is drafted as the criteria for bonding material-epoxy resin.

2. Maximum tensile strain criteria

The failure criteria will be given with the following assumptions:

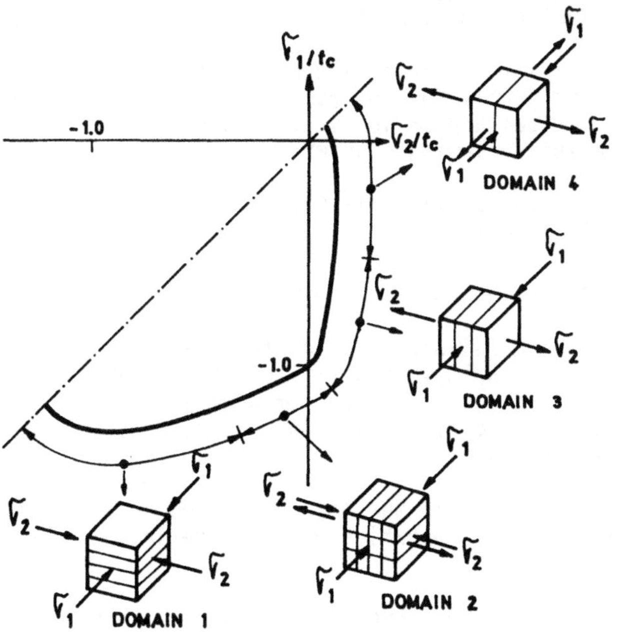

Fig. 1. Failure modes of biaxially loaded concrete (Nelissen,1972) from /4/

- the failure of joint is either in concrete or in adhesive material, and not in the interface;
- the materials are homogeneous and isotropic continuums;
- E moduli and Poisson's ratio are constants equal for tensile and compressive domain;
- the proportionality of stresses is fulfilled ($\sigma_1 = \alpha . \sigma_2$).

For the biaxial strength envelope of concrete for proportional loading the typical modes of failure for concrete are shown in Fig.1 /4/.

Since in all modes of failure the planes of breaks in concrete are perpendicular to tension or parallel to compression, it can be taken that direct tension and/or accompanying tension caused by Poisson's ratio control the failure. Besides the Poisson's ratio this results from the well-known tensile/compression strengths ratio for concrete.

For the sake of simple evaluation of concrete failure criteria by means of maximum tensile strain, the envelope of biaxial strengths (Fig. 1) will be recalculated with the equations:

$$\varepsilon_1 = 1/E\ (\sigma_1 - \nu \sigma_2)$$
$$\varepsilon_2 = 1/E\ (\sigma_2 - \nu \sigma_1)$$
$$\varepsilon_3 = -\nu/E\ (\sigma_1 + \sigma_2)$$

The obtained results, which are based on Kupfer's and Calzona's experimental data /5 and 6/ are shown in Fig.2.

It can be seen that the maximum compressive strain (ε_2) is not greater than fivelfold maximum tensile strain (5x ε_1 or 5x ε_3).

Diagrams for maximum strains are given only for the tension - compression part and partly for compression - compresssion, because these states of stresses appear in slant or direct shear tests, which will be shown in the following chapter.

By the similar calculations is drafted for epoxy resin Fig. 3 on the basis of data from the paper /7/.

3. Stress-Strain Analysis of Shear Tests

In order to avoid failures outside the joint area in a direct shear test, the test L parts were optimized Fig. 6. Stresses in characteristic cross-sections of the speciman were calculated by means of the FEM analysis and by application of the constant and variable values quoted below.

The constant values were the load expressed as τ=P/S N/mm^2 and the uniaxial mechanical properties of the materials as given in Table 1. /3, 5 and 8/.

The variable values were the dimensions of L parts, the joint surface S and the joint thickness t.

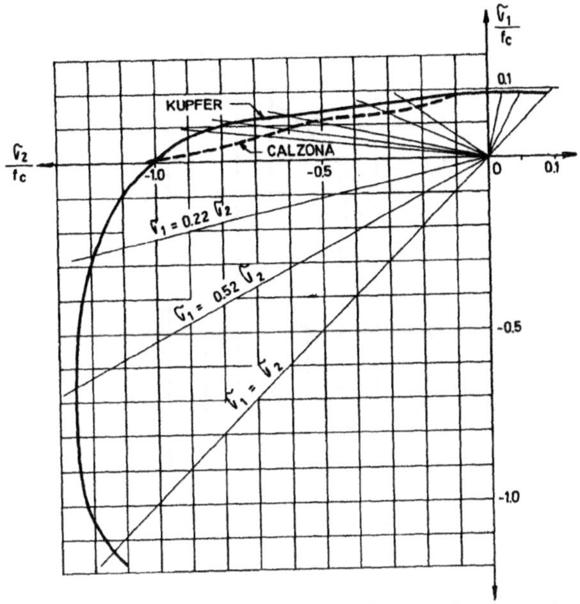

Fig.2 a) Biaxial strength envelope of concrete

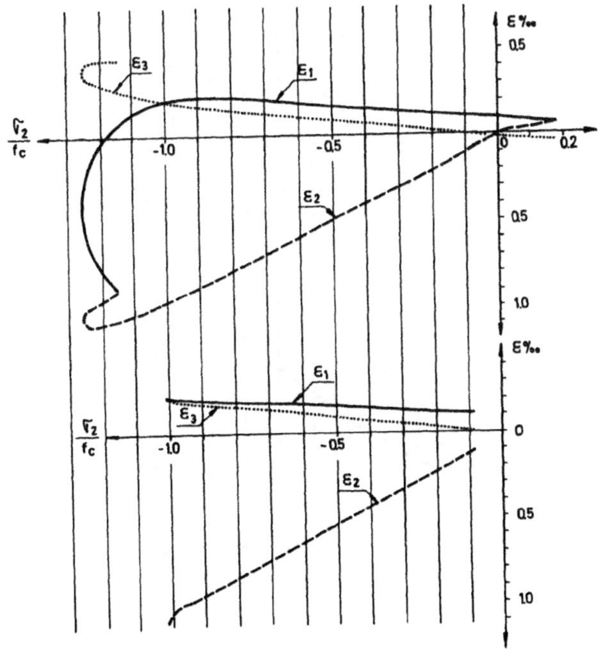

Fig.2 b) Maximum tensile strain criteria

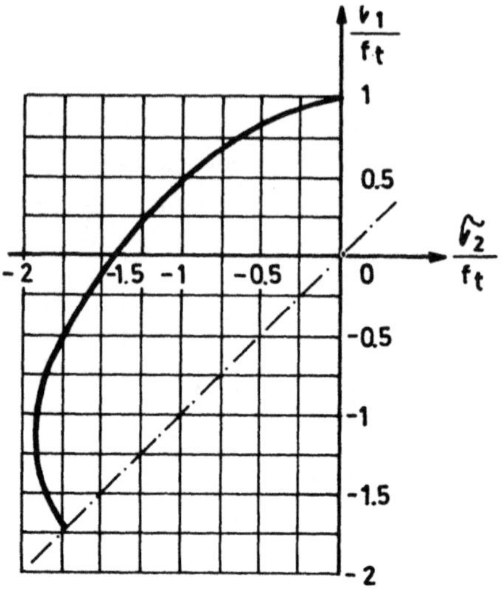

Fig.3 a) Biaxial strength envelope of epoxy resin

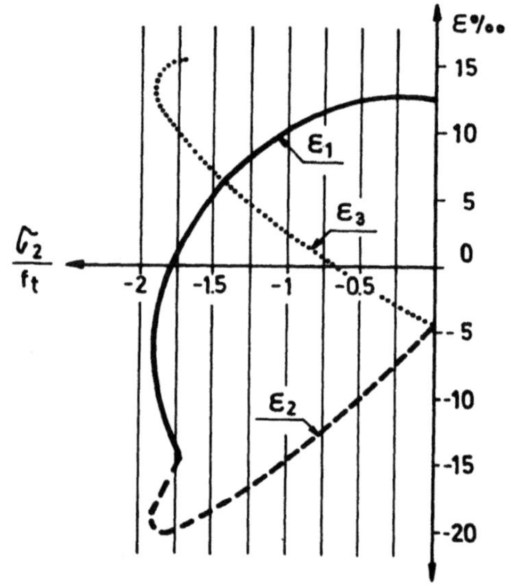

Fig.3 b) Maximum tensile strain criteria

Table 1. Uniaxial mechanical properties of the materials

Material	f'_c $\frac{N}{mm^2}$	f'_t $\frac{N}{mm^2}$	Ult.strain ε'_{fc} (‰)	ε'_{ft} (‰)	Modul E (N/mm^2)	Modul G (N/mm^2)	Poisson ν	Temp. koef. α (/°C)
Concrete	30	3	2	0.2	$3 \cdot 10^3$	$13 \cdot 10^3$	0.2	$7.5 \cdot 10^{-6}$
Polymer--cement mortar PCM	32	3.6	3.3	0.37	$13 \cdot 10^3$	$5.2 \cdot 10^3$	0.2	$8.2 \cdot 10^{-6}$
Epoxy resin ER	72.5	50	14	10	$4 \cdot 10^3$	$1.4 \cdot 10^3$	0.35	$70 \cdot 10^{-6}$

In the first step, dimensions of L part (width, thickness and height) as well as the way of loading by concentrated or continuous loads were varied. By comparing the amounts of main stresses in border points and joint area points, basic dimensions of the L parts were obtained which assures the failure in the joint area.

In the second step, the joint surface S was varied from 4.10^3 to 10.10^3 (mm2). The influence of varying the joint surface is shown in Fig.4 in relation to the maximum shear and the maximum principal tensile stresses in the adhesive material.

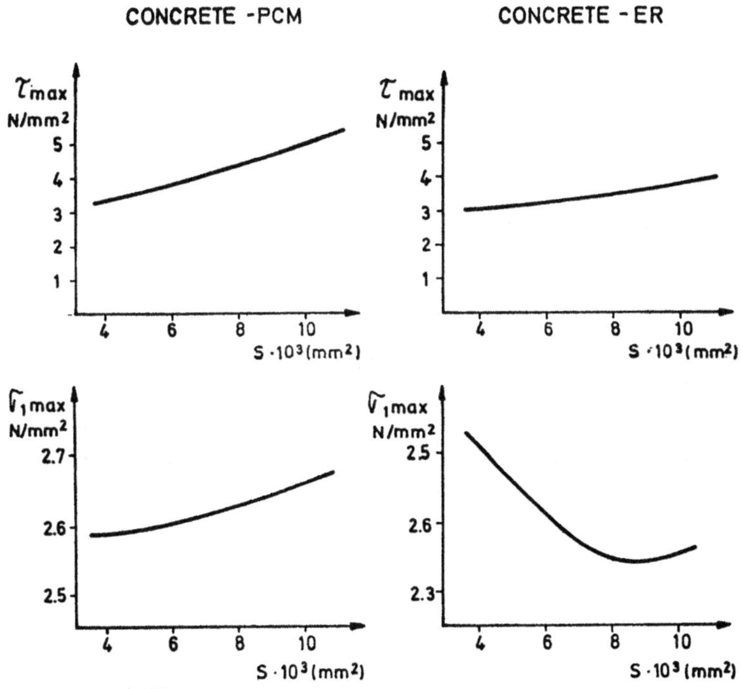

Fig.4 Changes in σ_1 max and τmax stresses depending on "S"

In the third step, the joint thickness was varied from 1 to 5 mm,and its influence to the variation of the same stresses in the adhesive was observed (Fig.5). It can be noticed that the change of adhesive thickness t changes the diagram shape and the place where main tensile stresses appear in the adhesive material. In case of a 5 mm thickness, maximum main tensile stresses appear in the middle of the joint height, and in case of a 1 mm thickness, they appear at the ends.

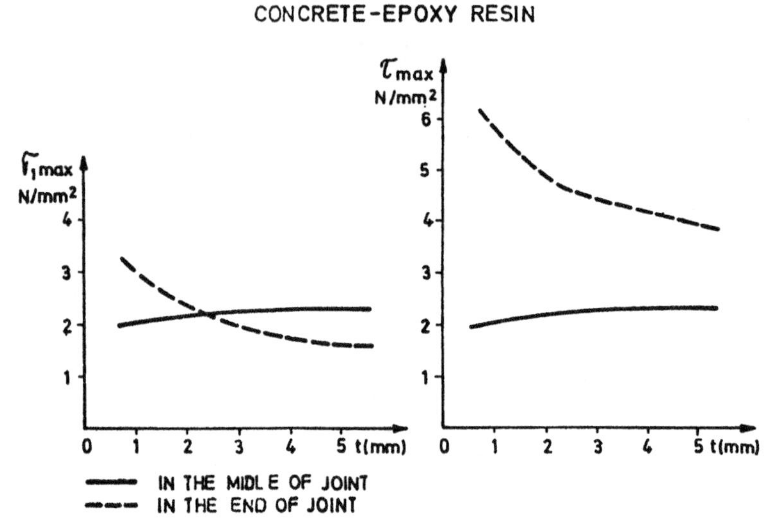

Fig.5 Changes in max τ and max σ_1 stresses depending on "t"

Finnaly, on the basis of the above FEM analysis for the direct shear test and the standard BS 6319 for the slant shear test, the following dimensions of test samples were chosen (Fig.6). Joint slant γ=63.43^{0} in the slant shear test, causing the greatest weakening of the joint, was taken from the paper /1/.

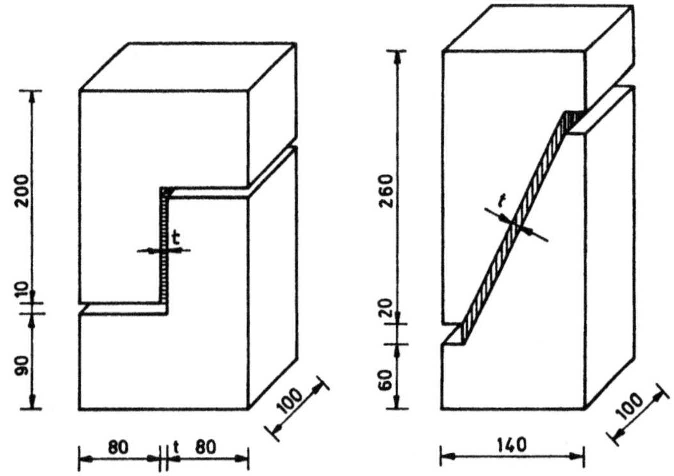

Fig.6 Shapes and dimensions of the direct shear and slant shear samples

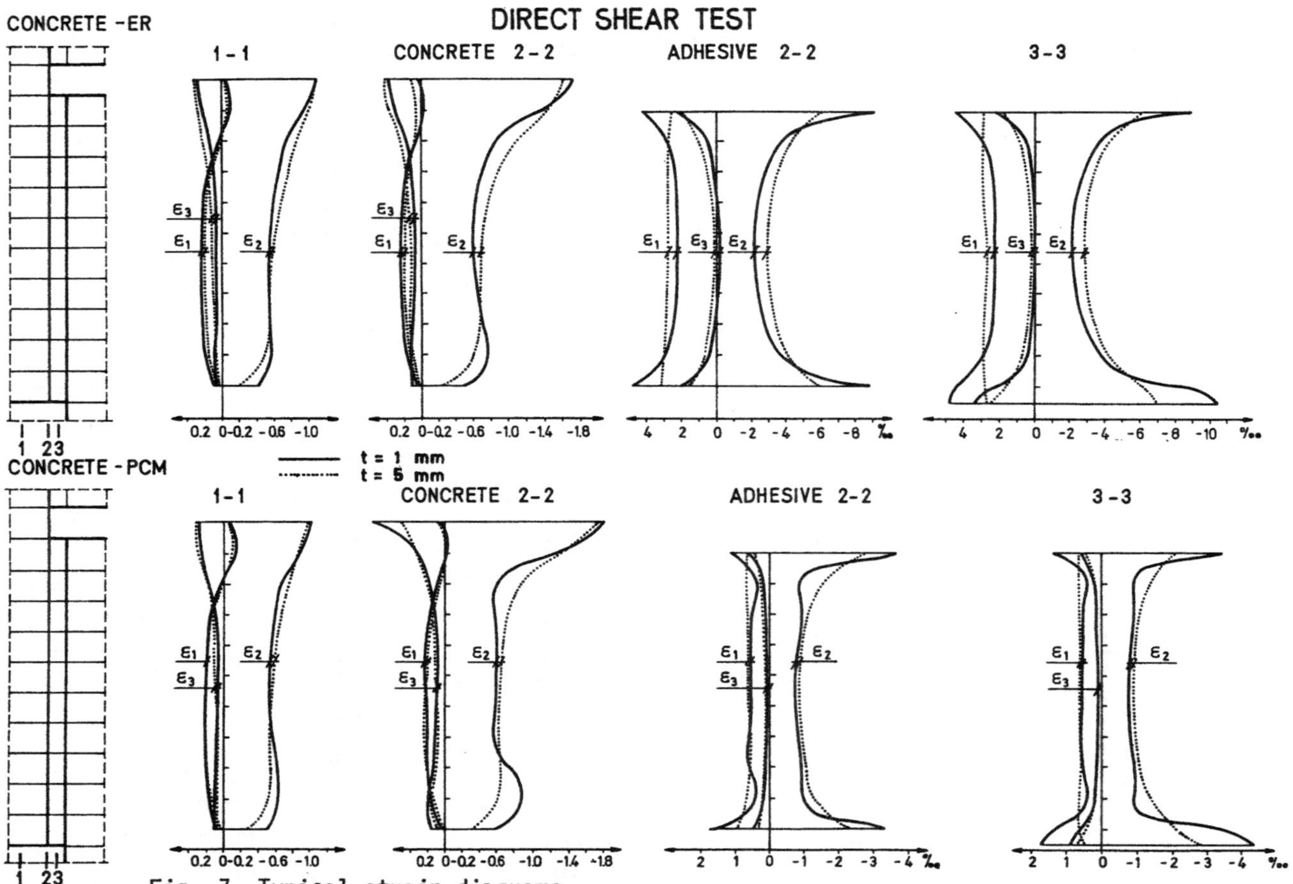

Fig. 7 Typical strain diagrams

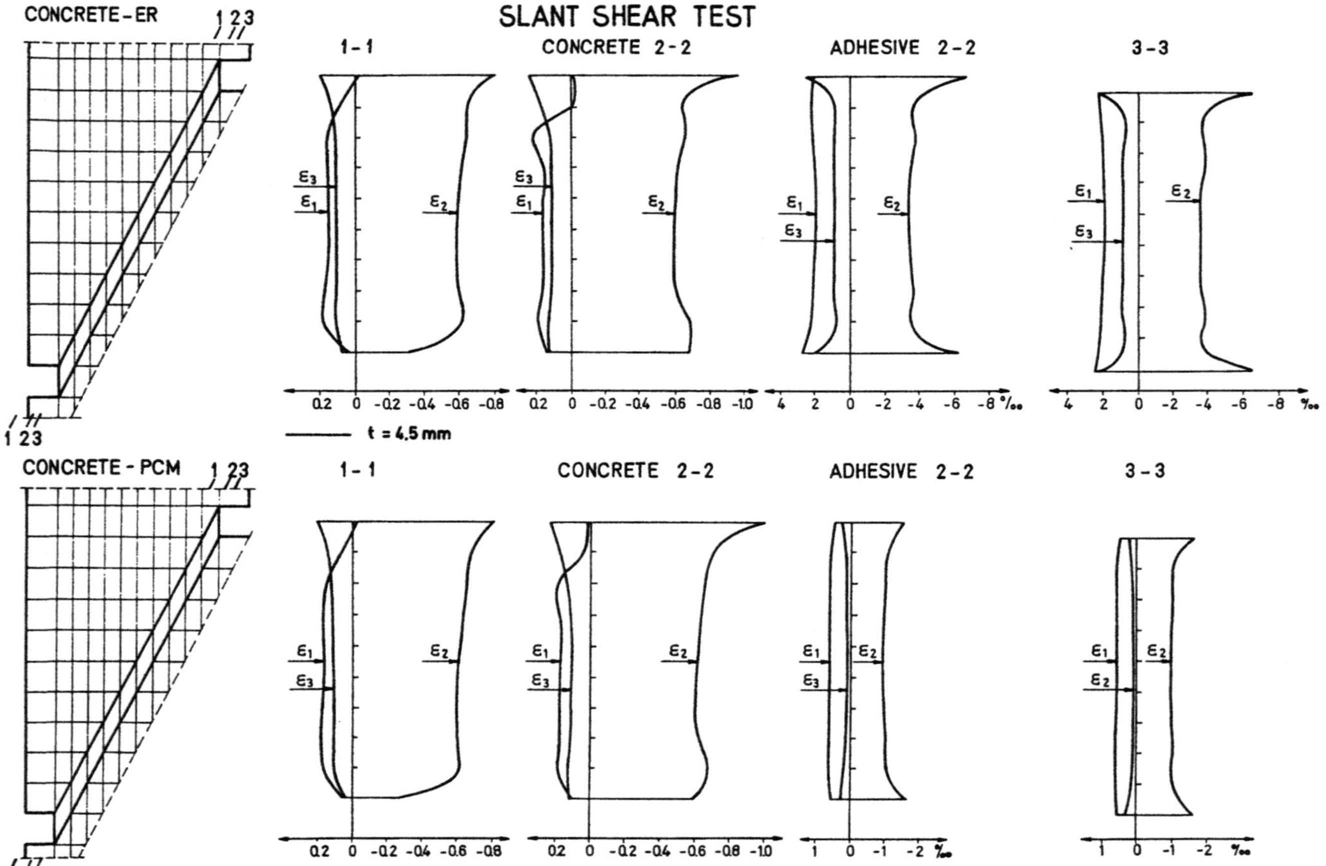

Fig. 8 Typical strain diagrams

The test samples with selected dimensions were used for further stress-strain analyses in three characteristic cross-sections of the joint: 1-1, 2-2 and 3-3, varying the joint stiffness by varying the E moduli and joint thickness. Diagrams of principal strains in both tests (direct and slant shear) for a certain load τ=P/S=10 N/mm^2 in direct shear test, and τ=P_t/S=10 N/mm^2 in slant shear test are shown in Fig. 7 and 8.

As a result of the analysis the following can be stated:

In both shear tests in the larger part of the 1-1 and 2-2 cross-section and in the whole 3-3 cross-section, biaxial tensin-compression stress apears in the following proportions of α.

- in the 1-1 cross-section for C-ER $-\alpha$=0.005 to 0.21
  for C-PCM $-\alpha$=0.001 to 0.22
- in the 2-2 and 3-3 cross-section for C-ER $-\alpha$=0.06 to 1.06
  for C-PCM $-\alpha$=0.02 to 0.54

Biaxial compression-compression stress appears only in a narrow area at the ends of the 1-1 section in the following proportions:

for C-ER α = 0.053 to 0.30
for C-PCM α = 0.036 to 0.33.

This proves the condition of stress as in the joint area are rectricted to the part of the envelope as assumed in chapter 2 (Fig.2).

The maximum strains ε_1 or ε_3 and ε_2 in the concrete and adhesives are:

- in the direct shear test for

C-ER	t = 1 mm	ε_{3C} = 0.37 ‰ $> \varepsilon'_{ft}$,	ε_{2R} = -10.9 ‰ $< \varepsilon'_{fc}$
	t = 5 mm	ε_{3C} = 0.39 ‰ $> \varepsilon'_{ft}$,	ε_{2R} = - 6.82 ‰ $< \varepsilon'_{fc}$
C-PCM	t = 1 mm	ε_{3C} = 0.81 ‰ $> \varepsilon'_{ft}$,	$\varepsilon_{1\ PCM}$ = 1.56 ‰ $> \varepsilon'_{ft}$
	t = 5 mm	ε_{3C} = 0.43 ‰ $> \varepsilon'_{ft}$,	$\varepsilon_{1\ PCM}$ = 0.98 ‰ $> \varepsilon'_{ft}$

- slant shear test for

C-ER	t = 4.5 mm	ε_{3C} = 0.27 ‰ $> \varepsilon'_{ft}$,	$\varepsilon_{2\ R}$ =-8.85 ‰ $< \varepsilon'_{fc}$
C-PCM	t = 4.5 mm	ε_{3C} = 0.25 ‰ $> \varepsilon'_{ft}$,	$\varepsilon_{1\ PCM}$ = 0.61 ‰ $> \varepsilon'_{ft}$

This means, if ER is applied the joint will fail in both tests in the concrete substrate and if PCM is applied as adhesive material the joint will fail in the adhesive material. In both cases it is due to exceeding the maximum tensile strain.

Stress concentrations and strain values at the ends of joints are characteristic, especially greater are in the direct shear test for small t values. Concentration values significantly decrease with increased adhesive thickness, which is of special importance in practical application . At the same condition increases the tensile strain in concrete.

4. Conclusion

For slant and direct shear tests in shown, that the failure in substrate concrete is caused by exeeding the tensile strains. In the case of PCM adhesive the failure could be located in adhesive, if its strength were lower than the strength of concrete. In both cases the good bond adhesive/concrete is assumed.
The drafted diagrams of maximum tensile strains for biaxialy loaded concrete and epoxy resin enable simple evaluation of results of FEM analysis of shear tests.
The biaxial strengths envelopes for other adhesive materials should be experimentaly investigated.
The rigidity, of joint (E-moduli and t-thickness) highly influence the tensile strain and stress distribution.
The stress concentration at the ends of the joint in direct shear test is higher than in the slant shear test, hence the stress distribution in direct shear test is more uneven.

References

1. Rehm,G., Franke,L., Kleben im Konstruktiven Betonbau, Teil 1, 1a, 2. BETONWERK+FERTIGTEIL-TECHNIK, Heft: 10/82, 11/82, 1/83.
2. BS. 6319: Part 4 1984. Testing of resin compositions for use in construction. Part 4. Method for measurement of bond strength (slant shear method)
3. Jitendra, K.Bhargava, Polymer-Modificed Concrete for Overlays: Strength and Deformation Characteristics. ACI-Publication SP 69-14/ 1981 Detroit/Michigan: AC.
4. Chen,W.F., Plasticity in reinforced concrete, McGraw-Hill Book Company 1982 USA
5. Kupfer,H., Hilsdorf,K.H., Behavior of Concrete under Biaxial stresses, ACI Journal/August 1969 No 66-52
6. Calzona,R., Behaviour of dense and lightweight concrete under multi-axial stresses, Instituto di Scienza delle Costruzioni report II-222. Presented at the 8th International Congress of the F.I.P. London, April 30 - May 5, 1978.
7. Schneider,W., Bardenheler,R., Versagenskriterien für Kunststoffe, Teil I und II, Z.f.Werkstofftechnik/J. of Materials Technology 6. Jahr g 1975/Nr.10.
8. Hranilović,M., Comparison of the epoxy laminate elasticity constans. Gradjevinar 37/1985, 6, 273-283.

BONDING SYSTEMS FOR CONCRETE REPAIR - AN ASSESSMENT OF COMMONLY USED MATERIALS

A.I. JUDGE, L.W. CHERITON & R.W. LAMBE
Repairs & Flooring Section, FOSROC Technology.

Abstract
Commercially available aqueous polymer dispersions of various types are compared with respect to their bonding properties when used to adhere repair mortars to cementitious substrates. Polymer dispersions tested include a number of acrylates and styrene-acrylates, a styrene butadiene, a vinyl acetate, an acrylate versatate and a vinyl acetate acrylate versatate. A composite bonding aid, an epoxy bonding system and the use of cement and polymer/cement slurries are also examined. Three different test methods are used to measure bond and a simple comparative test has been developed to monitor 'grab'. The merits of these tests are discussed. The best all round performance including tolerance to site use has been found to be given by an acrylate dispersion.
Keywords: Polymer dispersions, Bond, Concrete, Repair, Mortar, Test methods.

1. Introduction

The development of a sound bond between a repair mortar and mature hardened concrete can readily be achieved in controlled conditions such as in the laboratory environment. Here, certain requirements can easily be met. These include good surface preparation, adequate repair mortar compaction and sufficient curing control. It would be unrealistic to expect these ideals to be met fully in most site situations. Additionally, even a well prepared substrate may not have sufficient mechanical key or may be too porous to allow the repair mortar to 'grab' effectively, particularly on overhead or high-build vertical applications. Therefore, although a good bond may theoretically develop with time it is necessary to have good initial 'grab' in order to adhere the cementitious repair mortar to the substrate concrete in the first instance.

For these reasons a variety of bonding promoters have been developed. These can be simply a painted or trowelled-on slurry coat of the mortar 'knocked-up' with extra water. This concept has been further developed by the inclusion of a water dispersed polymer to promote greater adhesion and sealing of the substrate surface. Simplifying the system to reduce potential problems on site has resulted in the use of paint-on polymer aqueous dispersions in order to achieve the same effect. Thermo-plastic polymers supplied in this way as concrete bonding aids include vinyl acetate homo and co-polymer (PVAc), vinyl acetate acrylate versatate terpolymer (PVAc/VeoVa/Acrylate), acrylate, styrene acrylate and styrene butadiene (SBR). These systems contain approximately 50% solids by weight of polymer dispersed in water and are discussed further in Section 2. Also available are thermo-setting, two-component epoxy resin systems containing around 100% solids usually with a small quantity of solvent to improve the application properties.

Our experience has shown that a good bond is achieved with a slurry based on an SBR dispersion and OPC. However, the system is operator sensitive and, in severe drying conditions is prone to premature film-forming. If this occurs the bonding aid acts as a release agent and it is in fact necessary to remove the material by mechanical means and start again! A preferred system is one based on an acrylate polymer which, for simplicity, is brush (or spray) applied directly from the pot. This removes the need for having part bags of repair mortar or OPC lying around the repair site which could lead to mistakes being made. The advantage of acrylate systems is that even if allowed to film-form a good long term bond can still be achieved when a fresh mortar is applied. Obviously, as with all the systems, the best time to apply the repair is when the bonding aid has become 'tacky' in order to achieve maximum 'grab'. For these reasons this research has concentrated on acrylate systems and the work presented in this paper represents part of that research. Prior to this work preliminary screening was carried out on many systems and only the most promising were selected for this study.

2. Basic chemistry of polymer dispersions commonly used in the construction industry.

The use of water-based polymer dispersions and emulsions has been established for many years in the construction industry. This is illustrated by the British Patent (1) issued in 1923 describing a mixture of natural rubber latex and cement. Current applications involve the use of synthetic polymers dispersed in water. These are used as additives to cement mortars, as bonding or priming agents in, for example, the repair of spalled, reinforced concrete, and as curing membranes for cementitious systems.

Water based polymer dispersions are used in many other industries including paints, paper coatings and carpet backings. They have also been employed as model colloidal dispersions for academic studies into colloid stability and polymer adsorption (2,3,4). In each case the properties of the polymer system are designed to meet the highly specific requirements of the individual application. For example, a dispersion may be required to be stable (unflocculated) in highly-alkaline media and exhibit a minimum film forming temperature (MFFT) of 5°C, as is typically required in the construction industry. Other parameters of importance to the end user are solids content, pH, viscosity, surface tension, particle size, particle size distribution, mechanical stability, residual monomer content and mechanical stability.

To meet specific requirements the polymer manufacturer will modify the polymer dispersion systems chemically and physically. As a result the customer is required to carry out extensive testing and screening of new products to ensure that the polymer dispersion is compatible with the other components of the formulation or application in which it is to be used. One example of this may be the effect of chemical additives in the dispersion on the hydration rate of cement. In some cases severe retardation is observed. Increased levels of air entrainment due to the surfactants present in the polymer formulation have also been observed. It is necessary to ensure that the other components in the system do not modify the properties of the polymer.

The types of polymers in dispersion form used in the construction industry have been introduced in Section 1.

To begin to develop an understanding of the mechanism of bonding in priming systems it is necessary to consider the formulation of water-based polymer dispersions in more detail. These systems are often referred to as emulsions yet consist of polymer particles, typically of 200nm median particle diameter, dispersed in an aqueous phase. Hence the term 'dispersion' is more correct since the term 'emulsion' is usually used to describe a system consisting of droplets of one liquid dispersed in another liquid.

The preparation of polymer dispersions is described in detail elsewhere (5,6). The first stage of preparation involves the polymerisation, under controlled conditions of temperature, pressure and stirring rate, in a mixture typically of monomers, water, electrolyte, dissolved polymer and surfactants. Being relatively water insoluble, the monomer forms droplets in the aqueous phase. The surfactant molecules are adsorbed at the interface formed between the droplets and the aqueous phase ie. at the droplet surfaces.

At this stage the system could be described as an emulsion. The presence of the adsorbed layer imparts stabilization to the emulsion. Initiator is then added and polymerisation of the stirred emulsion commences. The adsorbed polymer/surfactant layer also plays a part in controlling the polymerization process. When polymerisation of the monomer is complete and a dispersion is formed the adsorbed polymer/surfactant also imparts stability to the system. This is achieved by the introduction of repulsive forces occurring between particles. These are steric repulsion (arising from non-ionic species), electrical double layer repulsion (arising from ionic species) or a combination of both. The repulsive forces overcome weak attractive forces which may arise on close approach of two particles and hence prevent flocculation.

In the second stage of preparation, additives such as anti-foaming agents, bacteriocide, anti-oxidant and perhaps further surfactant are introduced. The final product can thus be envisaged as a dispersion of spherical particles, each surrounded by a layer of polymer/surfactant, dispersed in an aqueous phase containing a wide range of dissolved species. Fundamental studies on colloid stability (2,3,4,5) have shown that the type of surfactant/polymer system used and whether the stabiliser molecules are physically adsorbed at the polymer/liquid interface or chemically anchored at the interface during polymerisation can dramatically influence the particle interactions occurring within the system. In terms of colloid stability, the properties of the adsorbed layer can dominate the properties of the particle. Of great relevance to this paper are the implications of these factors on the film-forming properties of polymer dispersions. These will affect the bond to the concrete substrate and to the subsequently applied layer of fresh, wet mortar.

The film-forming process in polymer dispersions is complex. However, it is known that during evaporation or loss of water from the dispersion the interparticle spacing decreases and particle-particle contact eventually occurs. As stated above, the surfactant system can influence the overall forces as particles approach and hence modify film formation and concomitantly affect mobility or adhesion of particles on initial contact with the substrate. Reference 7 describes how, on coalescence of particles to form a continuous film, stabilizer is either incorporated into the film or stripped from the particle surfaces forming surfactant clusters situated between particles. This tends to cause the formation of coherent, strong, water-insoluble films. The degree of coalescence depends, therefore, on the nature and amount of stabilizer present. As the degree of coalescence affects the nature of the film it is also the case that the amount and type of stabilizer will affect the tendency of the film to revert on coming into contact with fresh, wet mortar. Other properties of the system such as particle size, surface tension, MFFT, Tg and viscosity will also influence film formation.

Therefore, components of the polymer dispersion other than the polymer itself can play a major part in determining the behaviour of the system. As the polymer manufacturer cannot reveal details of his formula the customer has no choice but to carry out large scale screening exercises as reported here. Thus, it is necessary to test a range of products from different manufacturers as it is very likely that differences in formulation and hence performance will occur.

3. Test methods

There are various test methods which have been developed to test the ability of a mortar (with or without a bonding agent) to form a monolithic system when used to repair concrete. It is the opinion of the authors that no one test accurately reproduces the conditions experienced in practice on site. However, a repaired composite can be tested in the B.S. 6319:Part 4(8) slant shear bond test or the French L.C.P.C. Annexe No.5 (9) test.

The test methods used during this work include the B.S. 6319:Part 4, direct tensile pull-off strength measured on 20mm diameter cores and a repaired 'dogbone' test. Since this work was initiated to screen a variety of available products it was felt that a comparison of direct tensile bond strengths would be sufficient to identify the promising systems. Most of the testing was therefore carried out using the relatively simple pull-off and repaired 'dogbone' tests.

A simple comparative test was developed for measuring the 'grab' of various primers when used with repair mortar.

3.1 Slant shear bond strength

This test is performed by casting repair mortar onto a primed, angled fracture surface of good quality concrete, saw cutting the cured monolith and loading the resulting 50 x 55 x 150mm test specimen on a compression machine. The specimen geometry is illustrated in Figure 1. For further information on the test method the British Standard (8) should be consulted.

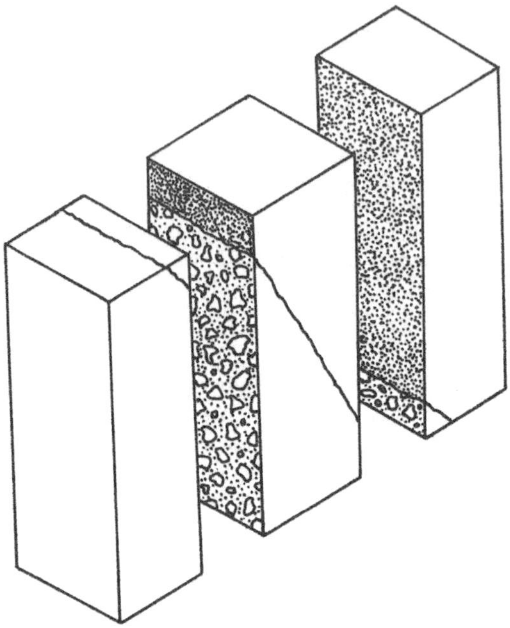

FIGURE 1. Specimen geometry for the BS 6319:Part 4 slant shear bond test.

3.2 Pull-off strength

The Elcometer pull-off tester is used to measure the direct tensile strength of 20mm diameter cores drilled into the test specimen. This is achieved by adhering special test 'dollies' onto the surface of the core to be tested with a suitable adhesive. The full-scale load of the tester is $70kg/cm^2$. The test apparatus is illustrated in Figure 2. The use of the tester in relation to this work is discussed in Section 4.2.

3.3 Tensile strength

B.S. 6319:Part 7(10) describes a method for determination of tensile strength. Test specimens are cast in a dogbone-shaped mould giving a cross-sectional area at the waist of $645mm^2$. The geometry is such that during testing the specimens can be held at each end by the use of specially shaped jaws and under tension will break across this narrowest width. The specimen geometry is illustrated in Figure 3. The adaptation of the test used for this work is described in Section 4.3.

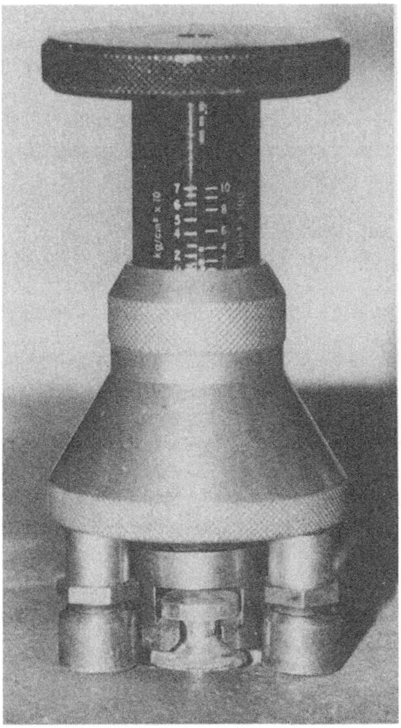

FIGURE 2. Elcometer direct tensile pull-off tester.

4. Experimental

In order to eliminate any contribution to the bond other than from the priming system or from cementitious reaction non-polymer mortars were used preferentially. However, the use of bonding promoters with a polymer-modified mortar was also examined. FOSROC mortars were used in preference to a sand-cement mortar because they enabled very early-age tensile bond strengths to be measured in the repaired dogbone test. The non-polymer materials were normal and rapid setting cementitious repair mortars having compressive strengths of 35MPa at 24 hours and 15MPa at 1 hour, respectively.

Aqueous dispersions of polymers were obtained from a number of manufacturers who had specified these systems as being compatible with cement. Three different acrylates, two styrene acrylates, an acrylate versatate and a PVAc/VeoVa/acrylate terpolymer were included in this work.

Proprietary, concrete repair bonding systems also tested included a two component epoxy resin, an SBR dispersion, a PVAc dispersion and a composite system thought to be based on a mixture of cellulose ether in a blend of PVAc and acrylate dispersions. In addition to screening the various systems, the use of slurry coats and the effect of allowing the bonding agent to dry were examined. This latter being a simple method of assessing one aspect of tolerance to site applications.

4.1 Testing slant shear bond

A modification of the B.S. 6319:Part 4 test was used in that the substrate concrete was prepared with ordinary rounded concreting aggregate rather than the crushed aggregate specified. This provided a more typical substrate having a less angular fracture surface than that obtained with the crushed aggregate. In all other respects the testing was carried out according to the standard using various bond promoting systems. These bonding agents were painted onto the fracture surface of the concrete substrate and allowed to go tacky prior to application of the normal-setting non-polymer repair mortar. The composite plaques were cut and the specimens tested when the repair mortar and primer had been dry cured for 14 days at 20°C and 65% r.h..

4.2 Pull-off testing

In general, test samples were prepared in the following manner for each bonding system.

A 600mm x 600mm x 50mm concrete paving slab was acid etched, washed well with water and, when dry, wire brushed to provide a clean, laitence free surface. The prepared slab was then primed by painting on the bonding agent in three 600mm x 200mm strips: (i) onto the dry substrate and allowing to dry over 24 hours; (ii) onto the dry substrate; and (iii) onto the pre-wetted substrate. Toppings of 20mm deep repair mortar were then cast onto the primed slabs such that the bonding agents were dry, tacky and wet, respectively, when the mortar was applied. The specimens were then cured for 1 day under polythene followed by a further 27 days at 20°C and 65% r.h.. Curing under polythene was adopted to prevent premature drying out and was a substitute to a spray applied curing compound which may have affected the subsequent adhesion of the test dollies.

For some of the available systems the testing was limited to the ideal (recommended) situation of repair mortar applied to the tacky bonding agent only. The same curing regime was used.

Three cores were cut in each section to be tested through the repair mortar to a depth of at least 10mm into the concrete substrate. After drying the repair mortar surface of each core was wire brushed and test dollies were adhered using a fast-setting, proprietary, polyester adhesive to allow for pull-off testing at 28 days.

After testing, the location of failure was noted (see Tables 2,3,4). Since the results quoted were an average of 3 readings, a location of failure written as 'Bond/mortar' indicates that failure was predominantly in the bond but one specimen, for example, failed in the repair mortar.

4.3 Repaired 'dogbone' testing

FIGURE 3. BS 6319:Part 7 (dogbone) mould and tested specimen.

Substrates of 3:1 sand (zone 3)/cement mortar(w/p = 0.12) were prepared by casting into the B.S. 6319:Part 7 moulds. These samples were cured for 28 days under water at 20°C. The complete mortar samples were broken in tension at 28 days to provide two half-dogbone substrates each. These substrates were allowed to dry at 20°C and 65% r.h. for 7 days.

Repaired dogbone specimens were prepared by returning the half-dogbone substrates to the B.S. 6319:Part 7 moulds, priming the fracture surface of each substrate as required and casting repair mortar into the empty cavity of the moulds. The resulting repaired specimens were tested at the required age on a Nene tensometer, at a loading rate of 6mm/min, using a 5kN load cell.

By testing at 1,3,7,14 and 28 days the tensile strength development with time for the bond between the substrate and a repair mortar was monitored. This was examined for one acrylate bonding agent used in conjunction with a polymer-modified or a non-polymer mortar and was compared with the situation where the substrate was simply pre-wetted and no bonding aid was used.

Different systems were screened by testing the bond at 24 hours using a non-polymer-modified mortar applied to the tacky primer. The resistance to on-site abuse was assessed by repeating the test for each system but with the repair mortar cast up to the substrate which had been primed 24 hours previously.

The very early age bond strength was tested at 1 hour by using a rapid-setting repair mortar.

Results generated in a similar test have previously been published (11).

4.4 Initial 'grab'

A simple test to compare the initial 'grab' of bonding systems was developed. The test apparatus consisted of an inverted, acid-etched paving slab as an overhead substrate and a small plastic bucket suspended from a split-pin embedded in a 45mm diameter disc of a cured lightweight mortar. A special cement/fine lightweight filler mortar was used as the interface between the disc and the substrate. The disc was primed with the bonding agent under test and a 5mm layer of the fresh mortar was spread and compacted onto this. The apparatus was then pressed onto the primed substrate and loaded up by adding sand to the bucket. The weight of the load which caused debonding of the mortar from the substrate was used as a relative measure of initial grab.

5. Results

5.1 Data generated in the slant shear bond test

The results of the testing are presented in Table 1. These tabulated data represent the load to failure per unit area over which the load was applied.

Table 1. Slant shear bond results.

Bonding system	Slant shear bond strength (MPa)
Water	38
Slurry (1:3-water:mortar powder)	37
Acrylate A	35
SBR Slurry (1:1:3-SBR:water:mortar powder)	38
Epoxy Resin	39

5.2 Data generated in the pull-off test

The pull-off test results are summarised in Tables 2,3 and 4. The pull-off strengths for each system represent the mean of three results originally measured in kg/cm^2 and recalculated to MPa.

Table 2. Pull-off test results for normal-setting non-polymer mortar.

Bonding promoter	Method of use	Condition of primer at application of topping	Failure stress (MPa)	Location of failure
None	-	Dry	2.3	Mortar
Water	-	Wet	2.0	Mortar
Acrylate A	As supplied	Dry	1.4	Bond
		Tacky	1.6	Bond
		Wet	1.6	Bond
	Slurry(1:1:5 dispersion: water:mortar powder)	Dry	2.2	Slurry/mortar
		Tacky	1.8	Mortar/dollie adhesive
		Wet	2.3	Substrate/ slurry
Acrylate B	As supplied	Dry	1.4	Bond
		Tacky	1.4	Bond
		Wet	1.4	Bond
Acrylate C	As supplied	Dry	2.2	Mortar/ substrate
		Tacky	2.3	Mortar/ substrate
		Wet	1.6	Mortar
Terpolymer	As supplied	Dry	2.7	Bond
		Tacky	2.5	Bond/mortar
		Wet	1.0	Bond
SBR	Slurry(1:1:3 dispersion: water:OPC)	Dry	0.0	Bond
		Tacky	1.4	Slurry
		Wet	1.6	Slurry/ substrate
PVAc/VeoVa	As supplied	Dry	2.5	Bond/ substrate
		Tacky	1.8	Bond/ substrate
		Wet	1.2	Bond
Styrene Acrylate A	As supplied	Dry	2.0	Mortar/bond
		Tacky	2.6	Mortar/bond
		Wet	1.8	Mortar/bond
Composite System	As supplied	Dry	1.7	Bond
		Tacky	0.0	Bond
		Wet	0.0	Bond

Table 3. Pull-off test results for polymer-modified mortar.

Bonding promoter	Method of use	Condition of primer at application of topping	Failure stress (MPa)	Location of failure
None	-	Dry	0.0	Bond
Water	-	Wet	1.9	Bond/mortar
Acrylate A	As supplied	Dry	1.7	Bond/mortar
		Tacky	1.4	Mortar/bond
		Wet	1.5	Mortar/bond
	Slurry(1:2:5 dispersion: mortar powder)	Dry	2.4	Mortar/dollie adhesive
		Tacky	2.2	Mortar/dollie adhesive
		Wet	2.0	Mortar/dollie adhesive
Terpolymer	As supplied	Tacky	1.3	Bond
	Diluted(1:1 with water)	Tacky	2.0	Mortar/ substrate
	Slurry(1:1:5 dispersion: water:mortar powder)	Tacky	0.8	Slurry
Composite System	As supplied	Tacky	1.0	Bond
	Diluted(1:1 with water)	Tacky	0.6	Bond
	Slurry(1:1:5 dispersion: water:mortar powder)	Tacky	1.3	Slurry

Table 4. Pull-off test results for rapid-setting non-polymer mortar.

Bonding promoter	Method of use	Condition of primer at application of topping	Failure stress (MPa)	Location of failure
Water	-	Wet	2.9	Substrate
Acrylate A	As supplied	Dry	1.6	Bond/mortar
		Tacky	2.2	Dollie adhesive/ mortar
		Wet	2.7	Dollie adhesive/ mortar

5.3 Data generated in the repaired dogbone test
The results of the testing are summarised in Tables 5,6 and 7.

Table 5. Bond strength development with time for non-polymer mortar using Acrylate A priming vs. pre-wetted substrate.

Age (days)	Bond strength (MPa)	
	Acrylate A priming	Pre-wetted substrate
1/3	0.3	0.4
2/3	0.8	1.3
1	1.6	1.4
3	2.0	1.5
7	2.4	2.8
14	3.1	3.5
28	3.1	4.3

A bond strength of 2.2 MPa was obtained at 7 days for the mortar which had been applied to 24 hour dried Acrylate A primer.

Table 6. Bond strength development with time for polymer-modified mortar using Acrylate A priming vs. pre-wetted substrate.

Age (days)	Bond strength (MPa)	
	Acrylate A priming	Pre-wetted substrate
1	1.6	2.0
3	2.7	2.1
7	3.2	2.2
14	3.4	2.4
28	3.5	2.7

Using a rapid-setting non-polymer mortar, bond strengths were recorded at 1 hour. On tacky Acrylate A bonding aid and on the pre-wetted substrate the results were 1.1 MPa and 1.8 MPa, respectively.

Table 7. Comparison of bond strength at 24 hours for non-polymer mortar applied to tacky and 24 hour-dried bonding aids.

Bonding aid	Bond strength (MPa)	
	Tacky	24 hours dry
(Water)	(1.4)	(1.1)
Cement Slurry (w/c = 0.5)	1.1	0.8
Acrylate A	1.6	1.7
Acrylate C	0.9	1.3
Terpolymer	1.5	1.1
SBR	1.2	1.4
Styrene Acrylate A	1.5	1.5
Styrene Acrylate B	0.4	0.4
PVAc	0.5	0.7
PVAc/VeoVa	0.6	0.6
Epoxy Resin	2.9	0.3

5.4 Preliminary results obtained in the initial grab test
The results of the preliminary testing are summarised in Table 8. These data have been calculated in units of MPa based on the load applied (weight of sand) onto a bond area of 1590.4mm^2. Each result represents an average of three test readings.

TABLE 8. Initial grab imparted by various bonding systems.

Bonding System	Initial Grab (10^{-5}MPa)
Water	6.1
Styrene acrylate A	6.2
Epoxy resin	7.4
Acrylate A	8.3
Terpolymer	8.9

6. Discussion

6.1 Comparison of test methods

As can be seen from the results in Tables 2,3 and 4 it is difficult to assess the bond strength, imparted by a bonding agent, by pull-off testing due to the limiting factors of the repair mortar and substrate strengths. It is for this reason that B.S. 6319:Part 4 calls for the use of a special high strength concrete for the substrate in the slant shear bond test. Moreover, bonding agents are used to promote early age bond strength/initial grab but these longer term tests merely show whether the adhesive is or is not detrimental to the development of long term bond strength. In view of this the repaired dogbone test seems very attractive. It has a number of advantages over the other test methods. The small sample size means that specimen handling is easier and smaller quantities of materials are required. Owing to the specimen geometry, the bond line is at the narrowest cross-sectional area so that it is generally at this point that the specimen fails thus allowing a direct measurement of bond strength. Results can be generated more quickly once a supply of substrates are available. Also, very early age bond strengths can be measured so that the development of bond strength can be assessed. Additionally, it is intended to monitor the effect on bond of freeze-thaw cycling and this could be readily achieved by using repaired dogbone specimens. None of these methods was suitable for assessing the real advantage of initial grab imparted by bonding agents. A simple initial grab test was therefore developed. The authors fully appreciate the unsophisticated nature of this test. However, the test permits a very important aspect of bonding aid application to be assessed comparatively.

As mentioned previously, no one test takes into account all site conditions and further work is desirable in this direction. It is hoped that Prof. Sasse's committee and the Universities will be able to develop relevant standard test methods for the future. Aspects to be considered are the relationships between size of repair, shrinkage, thermal movement, curling forces, bond and modulus as well as initial grab and longer term bond strength.

At the present time, as has been stated previously (11), the slant shear bond test can be used to assess the efficacy of bonding systems particularly where structural performance of the composite is important.

6.2 Bonding agent performance

The results obtained in the slant shear bond testing indicate that comparable bond strengths are developed between a repair mortar and mature concrete with water, cement slurry, Acrylate (A), SBR slurry and epoxy resin priming. These findings are of the same order as previously reported (11) values for various polymer dispersion slurries and epoxy resins which had been used to bond fresh concrete to mature concrete. All these data indicate that a strong bond can be achieved between mature concrete and a repair mortar, or fresh concrete by using any of the various bonding systems.

The tensile pull-off test is more generally established than the slant shear bond test. In the United States a tensile pull-off strength of 0.7 MPa is deemed to be adequate for a concrete substrate to which an epoxy resin topping is to be applied (12). A pull-off strength in excess of this figure may therefore be considered acceptable. However, it should be noted that a value of 1.4 MPa for a cementitious repair mortar applied to a concrete substrate without the use of a bonding aid (11) has been previously reported. The results presented in Tables 2,3 and 4 show that, with only a few exceptions, the repaired composites gained a tensile pull-off strength after 28 days in excess of 0.7 MPa. Many of the systems also achieved a strength of 1.4 MPa or above.

Very low tensile pull-off strengths were obtained in four cases. These are indicated in Tables 2 and 3 as having no bond (0.0 MPa). In fact, a tensile bond strength could not be measured because it was impossible to drill cores into these composites without breaking the bond.

The poor bond strength found for the polymer-modified mortar applied to the dry, unprimed, concrete substrate (Table 3) is probably due to desiccation at the bond line, the fairly dry mortar having insufficient aqueous phase to adequately 'wet-out' the substrate.

The diluted composite system gave poor all-round performance with both the polymer-modified and the non-polymer mortars except where allowed to dry. This may be due to a severe retarding effect caused by components in the system such as cellulose ether.

As found in the authors' own site experience, allowing an SBR slurry to dry resulted in very poor bonding properties. This has previously been attributed to the film-formed polymer acting as a non-bonding barrier between the mortar and the substrate. However, the repaired dogbone result for 24 hour dried SBR emulsion did not show this effect and is discussed later.

Values falling between 0.7 MPa and 1.4 MPa were obtained for the terpolymer and the acetate versatate dispersions when used wet with the non-polymer mortar. However, the bond strengths imparted by these dispersions when used dry or tacky with the same mortar are excellent. The low results may be due to an excess of water being present initially, at the bond line. These dispersions tend to be fairly viscous so that more of the dispersion, and hence more water, is deposited on the substrate during the primer application than with other water-based systems. This could result in a high water to cement ratio in the repair mortar at the bond line.

The relatively low values of 1.3 MPa and 0.8 MPa for the terpolymer used as supplied and as a slurry, respectively, in conjunction with the polymer-modified mortar may be due to a retardation effect or to the low permeability of the mortar to the viscous primer. Alternatively the result for the dried primer may simply be due to there being insufficient free aqueous phase to 'wet-out' the polymer layer. This layer is thicker than with other systems because of the higher viscosity of the dispersion. When used in a slurry it is possible that the polymer coats the cement grains allowing only partial hydration which results in a weakness at the bond line.

The comparison of bond strengths between Acrylate A priming and simply pre-wetting the substrate for a repair with the non-polymer mortar as generated in the repaired dogbone testing (Table 5) indicates that the bonding aid does not give a significant improvement in bond. This is also shown by the tensile pull-off test results. In fact, excellent bond strengths were obtained in the laboratory with this mortar when applied to a wet or dry unprimed substrate. This is presumably due to the fairly fluid, almost self-levelling mortar achieving good surface contact with, and wetting out of the substrate and therefore developing a sound cementitious bond.

A beneficial trend was found, however, with the polymer-modified mortar which generated a stronger bond more rapidly with Acrylate A priming than with a pre-wetted substrate (Table 6). This may reflect the difficulty in obtaining such good compaction at the bond line than with the more fluid mortar because of the confines of the small moulds. On site good surface compaction is usually achieved with a trowel applied scratch coat of mortar.

The advantages of a bonding aid appear to be the generation of early age bond strength and the initial grab. The repaired dogbone results in Table 7 indicate that, at 24 hours, four of the bonding aids tested imparted an improved bond over simply pre-wetting the substrate. The initial grab results (Table 8) show that the same four bonding systems all have superior grab when compared to a pre-wetted, unprimed substrate.

The action of allowing bonding agents to dry prior to the application of a repair mortar can have a marked effect on the subsequent bond strength. The second column in Table 7 shows that the Acrylate A still imparts a good bond strength whereas the epoxy resin forms a debonding layer. The Styrene acrylate A also gave good results both when tacky and dry and the terpolymer was at least as good as priming with water only. In comparison, bond strengths imparted by the other systems were relatively low. The surprising result is that for the dried SBR emulsion. It was expected this would give a debonding effect in much the same way as the dried SBR slurry as has been reported by other workers (13). However, this dispersion developed a better bond strength when dried than when tacky. This may simply be a mechanical key effect which is not found when the emulsion is incorporated into a slurry because of the thickness of the resulting primer layer.

The three most promising systems are Acrylate A, Styrene acrylate A and the terpolymer. Overall Acrylate A seems to give the best results. It imparts good initial grab, gives good long term bond strengths as the straight emulsion or as a slurry, when wet, tacky and dry. It also gives good early age bond strength used either tacky or dry and promotes good bond strength development.

6.3 The use of bonding aids

Apart from the aspect of initial grab the results obtained during this work may suggest that, in general, bonding promoters impart little benefit. However, the experience within the authors' company has shown that bonding aids are in fact highly desirable on site to ensure a good bond. This is related to applications which are on a larger scale than are practicable in the laboratory. In the former case factors such as shrinkage, thermal movement and curling forces come into play. As has been found by other workers a sound cementitious bond is readily achievable in the laboratory environment where the magnitude of these factors is less significant. Also the use of small laboratory specimens allows thorough surface preparation.

As already stated the reason for the high bond strength results obtained without primers when using the relatively fluid, almost self-levelling, non-polymer mortars is presumably related to the ease with which good surface contact at the bond line is achieved. However, it is often these types of mortars which are used in large scale flooring repairs where bonding agents have been found to be necessary in practice. This is presumably related to shrinkage and thermal movement.

The use of an acrylate bonding agent has been shown to have benefits over the situation where no bonding aid is used in applications of a polymer-modified mortar to a cementitious substrate. It is with the use of these mortars which are usually applied in vertical and overhead repair situations that the aspect of initial grab also becomes particularly important.

Other functions of bonding agents in site applications may include the ability to bond loose dust and laitence to the substrate, the sealing effect to minimise desiccation at the bond line on porous substrates and the possible relief of stresses at the bond line.

7. Conclusions

This paper has described bond strength results using four test methods and eleven basic polymer systems. The slant shear test gives an indication of how the composite material will perform under load. However, for the majority of work described in this paper, other test methods were found to be more convenient. Direct tensile pull-off testing is well established but is limited by the variations in the tensile strengths of the substrate and the repair mortar. The repaired dogbone test allows a direct measure of tensile bond strength and monitoring of bond strength development from early age. The initial grab test allows an important aspect of bonding agent use to be monitored comparatively. It has become apparent that these test methods do not accurately reproduce large scale site conditions. The authors conclude that more representative but still practicable test methods are required.

This research has shown the importance of screening bonding systems. As discussed in Section 2, variations between different manufacturers' dispersions can cause quite marked differences in performance. In comparing the results obtained for the two styrene acrylates it can be seen that dispersions based on polymers of the same generic type can give widely differing performance (Table 7). Variations are also noted between the acrylate dispersions tested (Table 2).

The work has confirmed that a sound bond can be achieved between a repair mortar and mature concrete with or without a polymer bonding aid in the laboratory environment. Provided the bonding system is not detrimental to the development of long term bond strengths, particularly as measured by the slant shear method, then the choice of system should be based on the initial grab imparted and the ability to generate early age bond strength. Additional factors to be considered are the adaptability to different methods of application and the tolerance to site use. For example, the ability to form a sound bond even when allowed to dry prior to mortar application, tolerance to constant water immersion and resistance to freeze/thaw cycling.

The obvious advantage of bonding systems illustrated by this work is the initial grab. This is of great importance in vertical and overhead repair work, particularly where a high build is required.

The ability of the repair to adequately 'wet-out' the substrate has been highlighted as being very important.

Generally the more viscous water-based bonding systems do not perform as well on very wet substrates as they do when dried or tacky.

The most promising bonding aid identified so far in this on-going programme is Acrylate A.

References

1. Cresson, L. (1923) British Patent, 191 747.
2. Vincent, B. (1974) Adv. Coll, Interface Sci., 4, 193.
3. Tadros, Th.F.(Ed.)(1982) The Effect of Polymers on Dispersion Properties, Academic Press, London.
4. Napper, D.H. (1983) Polymeric Stabilisation of Colloidal Dispersions, Academic Press, London.
5. Buscall, R., Corner, T. and Stageman, J.F. (Eds.)(1985) Polymer Colloids, Elsevier, Applied Science Publishers.
6. Blackley, D.C.(1975) Emulsion Polymerisation - Theory and Practice, Elsevier, Applied Science Publishers.
7. Paul, S.(1985) Surface Coatings, John Wiley & Sons.
8. British Standards Institution (1984) Testing of resin compositions for use in construction - Method for measurement of bond strength (slant shear method), BS 6319:Part 4, London.
9. L.C.P.C., Annexe No.5(1982) Laboratoire Central des Ponts et Chaussées, Paris.
10. British Standards Institution (1985) Testing of resin compositions for use in construction - Method for measurement of tensile strength, BS 6319:Part 7, London.

11. Tabor, L.(1985) Twixt old and new: achieving a bond when casting fresh concrete against hardened concrete, in Proceedings of the 2nd International Conference on Structural Faults and Repair, Engineering Technics Press, Edinburgh, pp 57-63.
12. ACI Committee 503(1979) Standard Specification for Repairing Concrete with Epoxy Mortars, ACI 503.4-79.
13. Dixon, J.F. and Sunley, V.K.(1983) The use of bond coats in concrete repair, Concrete, Palladian Publications, London, pp 34-35.

QUALIFICATION TESTS ON PCC SYSTEMS FOR THE REPAIR OF CONCRETE ROAD BRIDGES

R. KWASNY
Institut für Bauforschung, RWTH Aachen (University of Technology)

Abstract
New building materials for repair of concrete structures have to be investigated by qualification tests to estimate their long-term behaviour. This report deals with such tests for polymer cement concrete (PCC) composite repair systems hardening under dynamic loading followed by thermal cycling tests on the cured composite specimens. Test results showed that shrinkage of PCC and/or thermal stressing of the composite system especially in the bonding interface are decisive for durability of PCC layers on concrete, whereas dynamic loading of the fresh composite system seems to be of minor consequence.
<u>Key words:</u> PCC, repair, dynamic loading test, thermal cycling test, adhesion test.

1. Introduction

German highway administration authorities estimate the annual cost of maintenance of bridges to be about 2 % of the cost of building new ones and expect that, for this reason, future budgets will only assign about 20 % of the available money for new constructional works /1/.

Concrete surfaces are coated and repaired, respectively, nowadays in most cases by cement mortar systems containing thermoplastic dispersions or water-compatible epoxi resins to control or influence their properties. But mostly, there is not enough long-term practical experience with these products. There are a lot of examples where repair attempts have proved unsuccessful and demanded expensive repetition of repair. Since there are no generally accepted standardized test methods for these "new materials", qualification tests have to be defined. In Germany (FRG), particularly, administration authorities try to introduce extensive regulations for repair systems /2, 3/. Part of these regulations are test procedures for the components of the repair mortar, its characteristic values in the fresh and hardened state and - last but not least - test procedures to prove durable adhesion of PCC to concrete.

Main point of each qualification test program has to be the adhesion of the repair mortar to the concrete substratum. The adhesion strength tests should be performed after idealized and accelerated ageing treatments.

This contribution reports about first results of a research program /4/ on composite test specimens (PCC and concrete). Application and hardening of the repair mortar take place under dynamic loading and the adhesion tests are performed after thermal cycling test of the cured composite system.

2. Experiments

2.1 Materials

The substratum was a high-strength concrete of correspondingly high density, it consisted of

- aggregate: quartz sand, maximum grain size 8 mm, standard grading A/B (DIN 1045), 1710 kg/m³,
- portland cement: PZ 45 F (DIN 1164), 462 kg/m³,
- water cement ratio: w/c = 0,40,
- plasticizer.

All concrete test specimens have been stored in water up to an age of 7 days, after that at room temperature or at climate 23/50 (prisms).

Characteristic values of the concrete are summarized in table 1, modulus of elasticity has been determined using ultrasonic pulse method (cubes and prisms) and by static testing (cylinder, DIN 1048), respectively.

High strength values, especially surface tensile strength, are not realistic in view of the fact that concrete strengths in most repair cases are considerably lower. On the other hand, this approach does ensure that cohesion failure in the repair mortar and more particularly also adhesion failure at the bonding interface of the composite system will be more probable. The undifferentiated "concrete failure" on a substratum of poor strength will not allow to define a factor of safety against interfacial adhesion failure /5/.

The composition of the two PCC which were used for the composite systems are summarized in table 2. They are characteristic for concrete repair works in Germany. Research is going on with other PCC types. Both repair systems consisted of the PCC mortar and a slurry, made of the same PCC, as coupling agent. The PCC mortars showed good workability whereas the slurry of system II was not stable, it tended to sedimentation and segragation, respectively. Parallel to the coating of concrete beams and slabs, prisms of PCC were produced to control the characteristic material properties in the time of curing. Table 3 gives a summary of these properties. The two PCC differ in the development of strength and modulus of elasticity at early age, cf. fig. 1, as well as in their shrinkage properties, cf. fig. 2. An extensive research program is carried out parallely in the "Bundesanstalt für Materialprüfung, Berlin" to test the properties of these PCC systems in the fresh and hardened state /6/.

Strength - especially bending strength - reached remarkable values compared with "pure" cement mortars. But high strength is not a synonym for a high quality, repair system as will be shown later on.

Table 1. Characteristic values of concrete used for test specimens

Test specimen (mm)	Age at test (d)	Modulus of elasticity (N/mm^2)	Bending strength (N/mm^2)	Compressive strength (N/mm^2)	surface tension strength (N/mm^2)
cube 150x150x150	28	43.000	-	70	-
	84	45.000	-	75	-
cylinder Ø = 70 h = 210	28	40.000	-	66	-
	84	41.000	-	74	-
prism 40x40x160	28	41.000	9,2	89	-
	84	43.000	11	93	-
slab 300x300x100	28	-	-	-	4,1

Table 2. Characteristics of the PCC

PCC	I	II
Maximum grain size (mm)	4	4
Type of cement (DIN 1164)	Z 45 F	Z 45 F
Polymer admixture	acrylat	styrene-acrylat
Mix proportion in weight percent: ready-made cement mortar	87,9	88,4
aqueous polymer dispersion	5,5	11,6
water	6,6	-

Table 3. Properties of the PCC mortars

PCC No	Age at test (d)	Bulk density (kg/dm³)	Dynamic modulus of elasticity (N/mm²)	Bending strength (N/mm²)	Compressive strength (N/mm²)	shrinkage[1)] (mm/m)
I	1	2,20	12.300	1,7	4,7	0,0
	7	2,20	31.400	7,8	38	0,13
	28	2,16	31.300	15	50	0,49
	84	2,15	31.300	15	57	0,69
II	1	2,15	22.000	4,9	17	0,0
	7	2,18	30.100	9,3	41	0,15
	28	2,18	30.700	15	56	0,91
	84	2,18	31.200	16[2)]	65	1,19

1) between 1st and 84th day after mixing, datas from /6/

2) age of test: 110 d

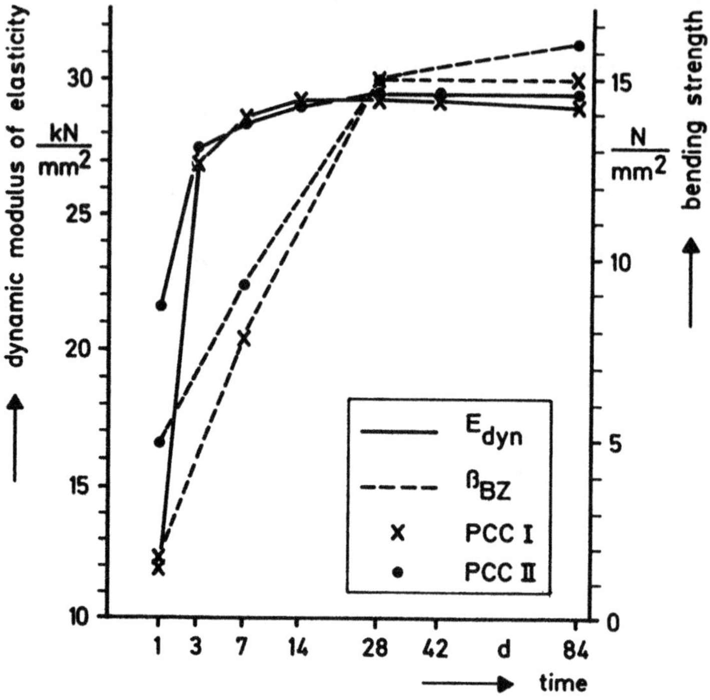

Fig. 1. Dynamic modulus of elasticity (E_{dyn}), development with time

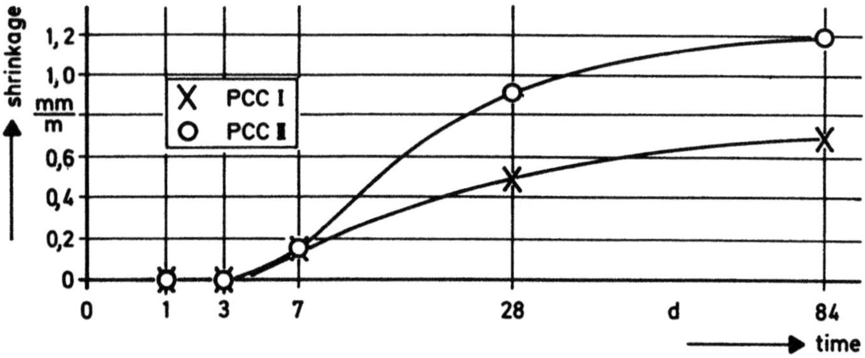

Fig. 2. Shrinkage of the two PCC (first day in mould, second and third day wrapped in PE-foil. (Datas from /6/)

2.2 Composite specimen preparation

The surfaces of the concrete beams and the slabs were prepared by sandblasting when they were 21 days old. Roughness of sandblasted surface was measured indirectly using a defined sand (grain mixture and quantity), a simple test procedure commonly used in road building /7/. Roughness amounted 0,6 mm in average.

Bonding interface of concrete specimens was wetted so that it was faintly moist when slurry was applied at a concrete age of at least 28 days. The amount of slurry was about 1,5 kg/m² for both PCC systems, as recommended by the producers. PCC was applied by hand on the fresh slurry, thickness of PCC layer was 20 mm.

2.3 Test procedure

The test program consisted of

- dynamic loading during the application of the repair mortar and during the following 24 hours (only for the beams)
- storage at room temperature
- thermal shock cycling (normal curing for comparison)
- adhesion strength test.

Essentially, the shape of the beam for the dynamic loading test corresponds to the recommendations of RILEM committee 52-RAC, cf. fig. 3, referring to dimensions and loading conditions.

Different to the 52-RAC proposal parameters of dynamic loading were fixed as:
lower stress 0,3 N/mm² tensile
upper stress 2,5 N/mm² tensile
for the concrete in the bonding interface. The upper value results from an estimation, which has been made in cooperation with German federal authorities of transport. The following assumptions were used to fix the stress limits:

- partial prestressed bridge,
- single-span beam,
- concrete strength class: 45 N/mm²,
- no limitation referring to heavy duty vehicles on the main track,
- half traffic load on secondary track.

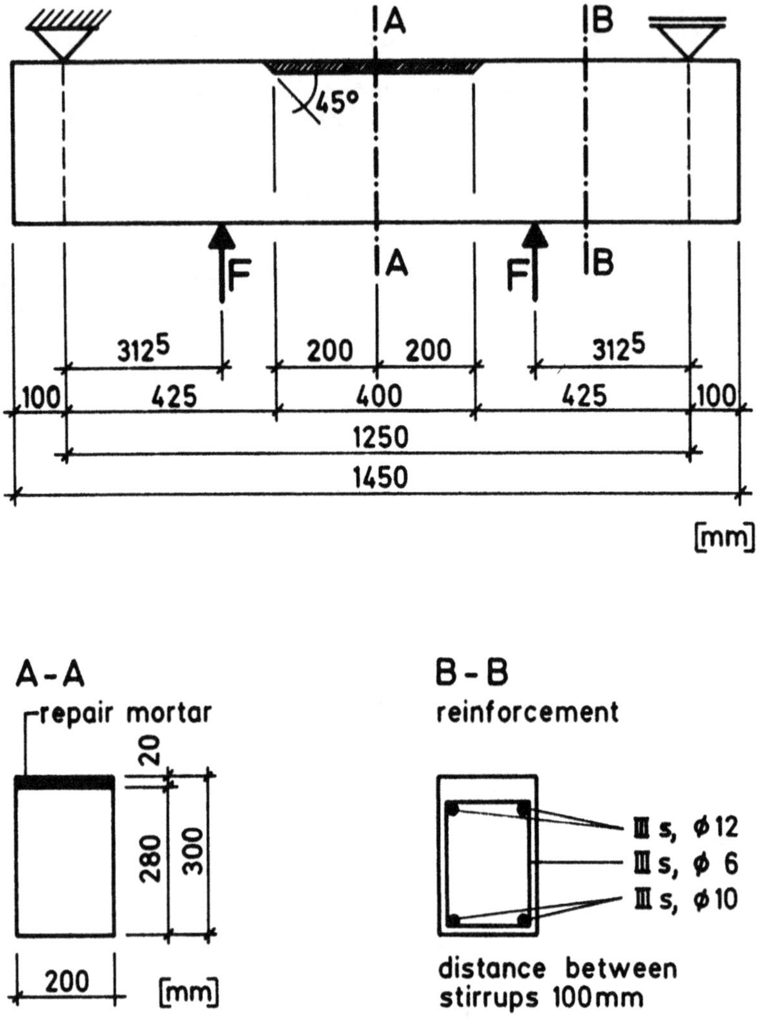

Fig. 3. Beam test specimen for dynamic loading test

This estimation was done because systematic measurements at bridge structures about their dynamic response to traffic vibrations are still missing. Published measurements so far scatter over a wide range, cf. fig. 4. The upper values of frequency and acceleration are not representative for road bridges, they result from extrem stressing of special structures.

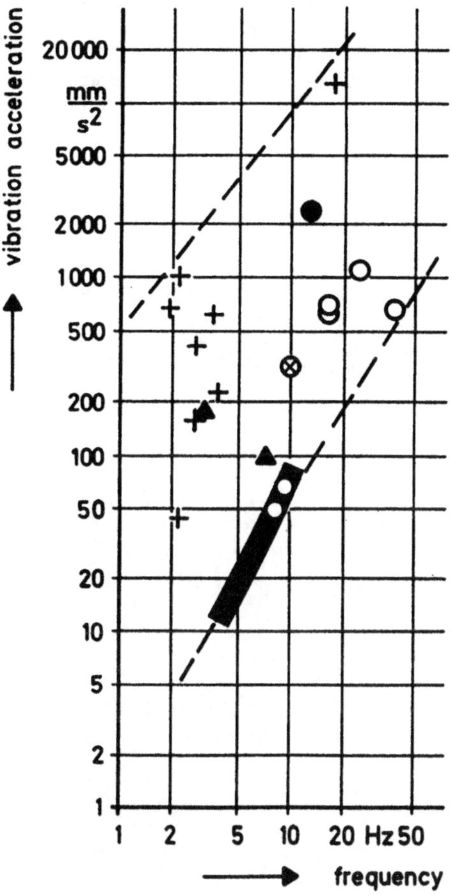

Fig. 4. Vibration acceleration in disconnection of frequency, measurements at road bridges

+ ○ ●... several authors, summary of literature data made in /8/

▲ ... measurements at highvay valley bridges /9/

⊗ ... dynamic loading test for PCC qualification tests /4/

▬... RILEM 52-RAC tentatice proposal for dynamic loading test, possible frequencies 4-10 Hz

Dynamic loading test lasted the first 24 hours of PCC curing with a frequency of 10 Hz and an acceleration of about 300 mm/s^2, fig. 5 shows the testing assembly. After that, the beams were stored at room temperature. Temperature cycling test was carried out 10 times as a thunder-shower simulation:

- heating of composite specimens surface up to about +60 °C for 6 h,
- shocking the surface using tap water, about +10 °C, for 10 min.

Temperatures at the bonding interface came up to 47 °C at the end of heating. At the end of sprinkler irrigation surface temperature felt down to 10 °C, temperature at the bonding interface to 27 °C.

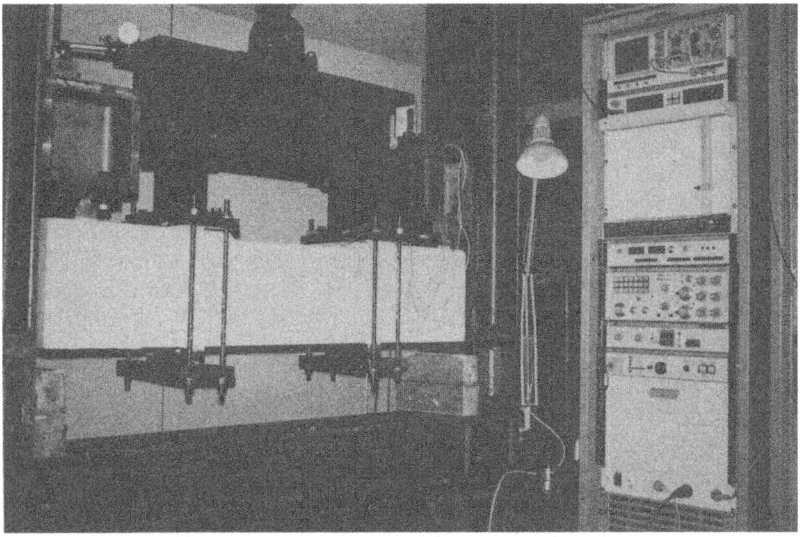

Fig. 5. Testing assembly of dynamic loading test

Adhesion of PCC was tested by knocking with a small hammer, qualitatively, and by pull-off test. Criteria for adhesion assessment were the adhesion strength and the associated fracture pattern. Fig. 6 and 7 show the plans for pull-off tests on the beams and on the slabs.

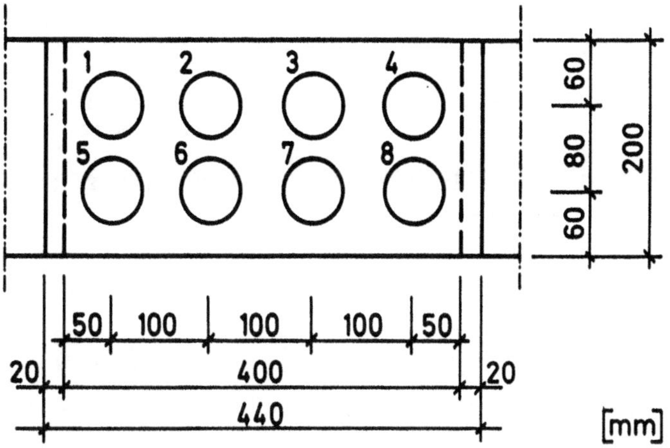

Fig. 6. Plan for pull-off tests on the beams

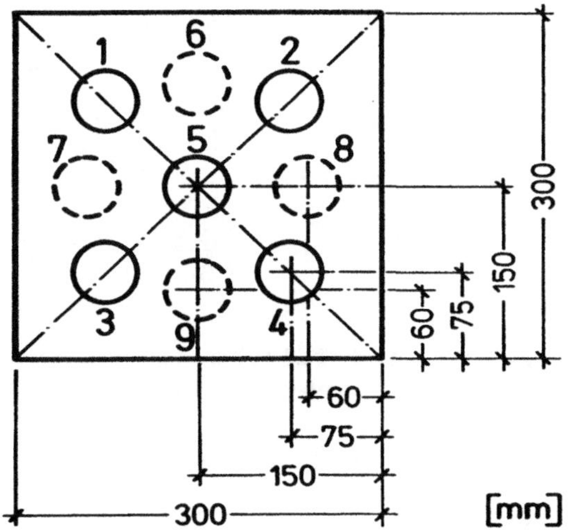

Fig. 7. Plan for pull-of tests on the slabs

3. Results

During storage of coated beams and slabs surface cracks in the PCC mortars could not be detected visually. Observations of the beam test specimens using a light stereo microscope showed a single crack for the PCC system II, it was nearly in the middle of the PCC area transvers to longitudinal axis of the beam. Slab test specimens showed cracks on the side faces between PCC and substratum, especially for PCC II specimens, indicating partial adhesion failure.

After thunder-shower simulation knocking on the PCC surfaces signified hollow areas at the border of nearly all slabs. Extent of hollow areas is roughly shown as darkly marked parts in fig. 8. In the same figure the type of failure is marked.

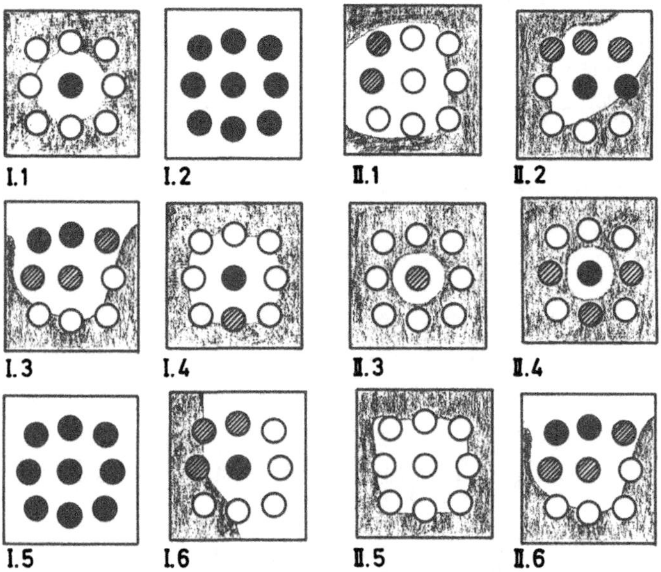

Fig. 8. Extent of bond failure of PCC-systems I and II on concrete slabs and type of failure in adhesion test

- ● failure in pull of test
- ◍ failure during pore-loading
- ○ failure during core-drilling

Some slabs showed surface cracks after the first thermal shock cycle. Fig. 9 shows the crack formation of a PCC II slab after the 10th cycle. There was only one crack in the middle of the slab surface which reached the edge of the specimen. A fine crack net only appeared in the middle part of the slab, that indicates that the edge areas had already been hollow before starting thermal shock.

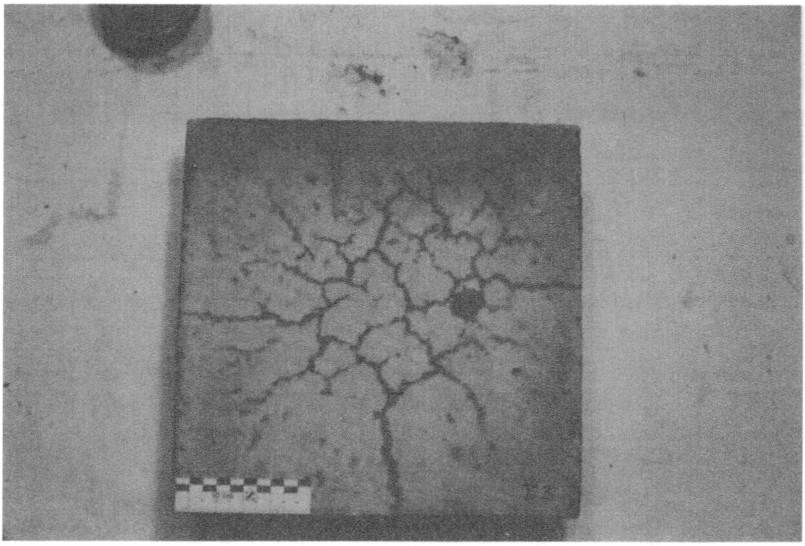

Fig. 9. Crack formation an PCC II surface of a slab after flurnal cacling test

PCC applied to the beams only showed a few cracks with regular distances transvers to the longitudinal axis of the beam.

Whereas bond performance of both PCC systems tested by thermal cycling was bad to the same extent, pull-off tests on composite specimens stored at room temperature showed distinct differences, cf. fig. 8. Adhesion tests results of slab and beam specimens are summarized in table 4. The table contains arithmetical means of test results and informations about the distribution of adhesion and cohesion failure over all test results. Arithmetical mean of adhesion strength is stated, including "zero"-values. Nearly all failures appeared in the bonding interface (adhesion failure). This indicates that both coupling agents were not able to charge stress resulting from shrinkage and thermal shock test. Adhesion strength of PCC I, only stored at room temperature, showed faily good adhesion strengths.

Table 4. Results of pull-off tests

PCC	Specimen shape	Specimen No	Thermal shock	Number of tests	Pull-off strength[1)] (N/mm^2)	failure area: Adhesion interface (%)	failure area: Cohesion PCC (%)	failure area: Cohesion concrete (%)
I	slabs	1-4	+	36	0,36	97	3	0
		5,6	-	18	3,36	77	6	17
	beams	1	+	8	0	100	0	0
		2	-	8	1,94	89	11	0
II	slabs	1-4	+	36	0	100	0	0
		5,6	-	18	0	100	0	0
	beams	1	+	8	0	99	0	1
		2	-	8	1,16	64	0	36

1) arithmetical mean

The influence of the application and hardening during dynamic tensile stressing (beam specimens) is ambiguous: With PCC I a reduction of pull-off strength from 3.4 N/mm² to 1.9 N/mm² (no thermal shocks) was measured. On the other hand, with PCC II the "dynamic" value of the beams was 1,2 N/mm² compared with complete loss of adhesion at the "static" slabs. The zero adhesion probably is the result of the extremely high shrinkage of PCC II (see table 3 and fig. 2). The different behaviour of beams and slabs will also be influenced by the different geometric conditions. Further research is needed to clarify the effects.

The thermal stresses resulting from the idealized thunder-showers exceeded the adhesion strength in nearly all cases, independent from the type of PCC and the shape of the specimen. But it seems to be obvious, that shrinkage plays the major role as a reason for the failures.

4. Closing remarks

Up to this stage of the research program it cannot be stated whether the complex testing of composite systems (dynamic loading and thermal shock) is able to simulate severe conditions in situ, natural weathering in open air, so that a safe quality statement can be derived from the tests.

Results gained with two PCC systems, which are commonly used in Germany, but are not representative for all PCC repair systems, permit the following final observations:

- durability of a repair composite system is mainly influenced by the quality of the coupling agent
- high shrinkage leads to failure of the composite system if the coupling agent doesn't show outstanding properties (PCC systems are known which show the same shrinkage as system II but have prominent adhesion strength obtained with the help of epoxi resin as coupling agent)
- influence of dynamic loading on adhesion strength seems to be of low significance, but a complete assessment is not yet possible
- thermal cycling test realized by thunder-shower simulation seems to be a severe test, which doesn't need much cycles to destinguish qualified from unqualified repair systems.

The test programm is being continued with other PCC systems.

Acknowledgement

Research has been sponsored gratefully by Minister of Transport of the Federal Republic of Germany. Thank also belongs to the institute's laboratory staff, especially to Mr. Albers and Mr. Braun, for their skilful technical assistance.

References

/1/ Rabe, O. (1981) Die Unterhaltung von Stahlbeton- und Spannbetonbrücken. Bauingenieur, 56, 431-437.

/2/ Hansen, U. (1985) Technische Lieferbedingungen für Betonersatzsysteme unter befahrbaren Brückenflächen. Report on the occasion of Fachtagung Umwelteinflüsse auf Ingenieurbauwerke und Erfahrungen bei Bauwerksinstandsetzung und Bautenschutz, Verwaltungsakademie Berlin, Feb 26, 1985.

/3/ Deutscher Ausschuß für Stahlbeton (1985) Aktuelle Arbeiten, Kunststoffe im Betonbau. Annual report 1985.

/4/ Institut für Bauforschung, RWTH Aachen. Untersuchung der Eignung von kunststoffmodifizierten Reparaturmörteln/-betonen als Ersatz von Konstruktionsbeton im Brückenbau - Verbundprüfungen -. Running research project sponsored by Federal Minister of Transport.

/5/ Kwasny, R. and Schulz, R.-R. (1986) What method for testing the bond of coatings to concrete? - An appreciation of commonly used thermal cycling test methods. Concrete Precasting Plant and Technology, 52, 102-106 and 173-179.

/6/ Maultzsch, M., Bundesanstalt für Materialprüfung, Berlin. Untersuchungen an Zementmörtel mit Kunststoffzusatz für den Ersatz von Konstruktionsbeton (Reparaturmörtel). Running research project sponsored by Federal Minister of Transport.

/7/ Kaufmann, N. (1971) Das Sandflächenverfahren. Ein einfaches Verfahren zur Messung und Beurteilung der Textur von Fahrbahnnoberflächen. Straßenbau-Technik, 24, 131-135.

/8/ Weber, J. (1985) Beeinflussung der Betonerhärtung durch Verkehrserschütterungen. Literature exploitation made by Institut für Bauforschung (F 214) sponsored by Deutscher Beton-Verein, Wiesbaden.

/9/ Bundesanstalt für Straßenwesen (1980) Schwingungsmessungen an der Ahrtalbrücke bei Bad Neuenahr. (1985) Kalibrierfahrten auf der Brohltalbrücke. Internal documents, BASt B 3.3.

AN EXAMPLE OF EVALUATION OF BONDING DURABILITY BETWEEN CEMENTED WOOD CHIP BOARDS AND CERAMIC TILES WITH ELASTOMER ADHESIVE

K. MOTOHASHI
Department of Building Materials, Building Research Institute, Ministry of Construction

Abstract
A new type of exterior wall panels have been developed for a prefabricated house system. These panels are produced by bonding ceramic tiles by use of chloroprene adhesive, butyl rubber sheet, and polyurethane adhesive on cemented wood chip boards fixed to light gauge steel frames in a factory. Such panels are less dangerous for falling off of tiles against deformation of panels due to flexibility of an adhesive layer, however, it seems that bonding durability against weathering or creep behavior must be checked. In order to evaluate bonding durability of the panels, various experimental approaches were performed such as an accelerated weathering test, an ozone resistance test, a boiling test, and a creep rupture test. Among these test, the creep rupture test which was conducted at various temperature and loading in both dry and wet condition was successful in obtaining quantitative information regarding with service life time of creep rupture of the bonded tiles. The data obtained were analyzed by means of the Larson-Miller method (F.R. Larson and J. Miller, Trans. ASME, Vol. 74, 765 (1952)), and finally master curves in both dry and wet condition could be obtained for the objected bonding system. It seems that the Larson-Miller method is useful for evaluating bonding durability of such elastomeric adhesive systems.
Key words: Elastomer adhesive, Larson-Miller method, Bonding durability, Creep rupture test.

1. Introduction

A new type of exterior panels for a prefabricated house system have been developed in Japan. These panels are produced by bonding ceramic tiles by use of chloroprene adhesive (solvent type), butyl rubber sheet reinforced by polyester fiber mesh, and polyurethane adhesive (two components type) on cemented wood chip boards specified in JIS A 5417 (Japanese Industrial Standards) as shown Fig.

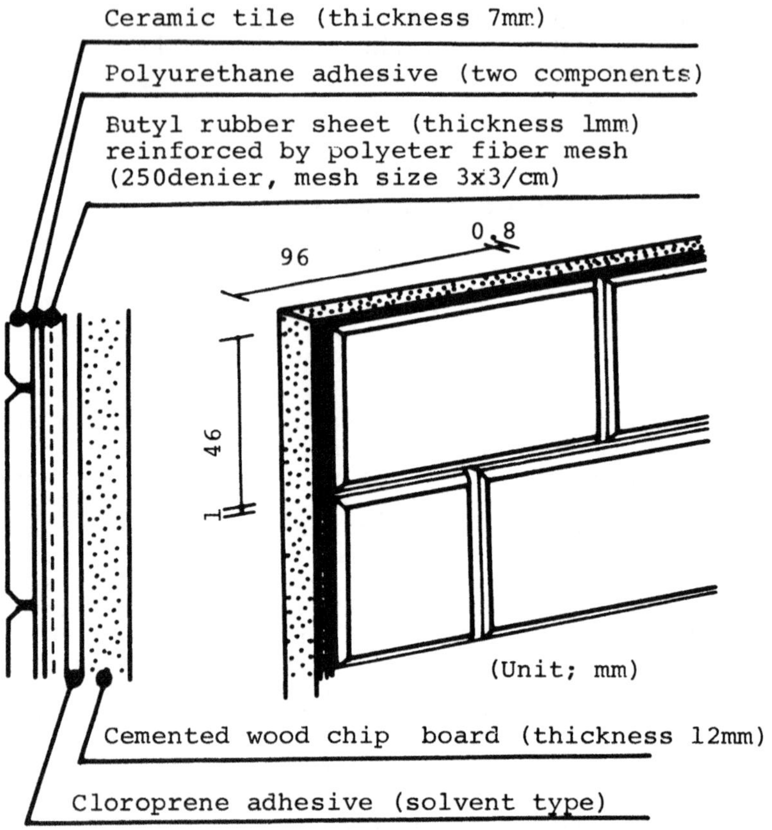

Fig. 1 Objected panel.

1.

In this flexible adhesive system, resistance against falling off of bonding tiles due to movement of panels or due to impact is more than a bonding system by use of cement mortar. However, it seems that bonding durability against weathering or creep must be checked in such a rubber-like adhesive system. Various experimental approaches containing an accelerated weathering test, an ozone resistance test, a boiling test, and creep rupture test were performed.

For example, Fig. 2 shows the result of the boiling test. Two panels were soaked in hot water (95 C) for 28 days. The back of one panel of the two was sealed with epoxy resin for prevention of water immersion through the back. Adhesive tensile strength of the soaked panels decreased with soaking duration. This fact

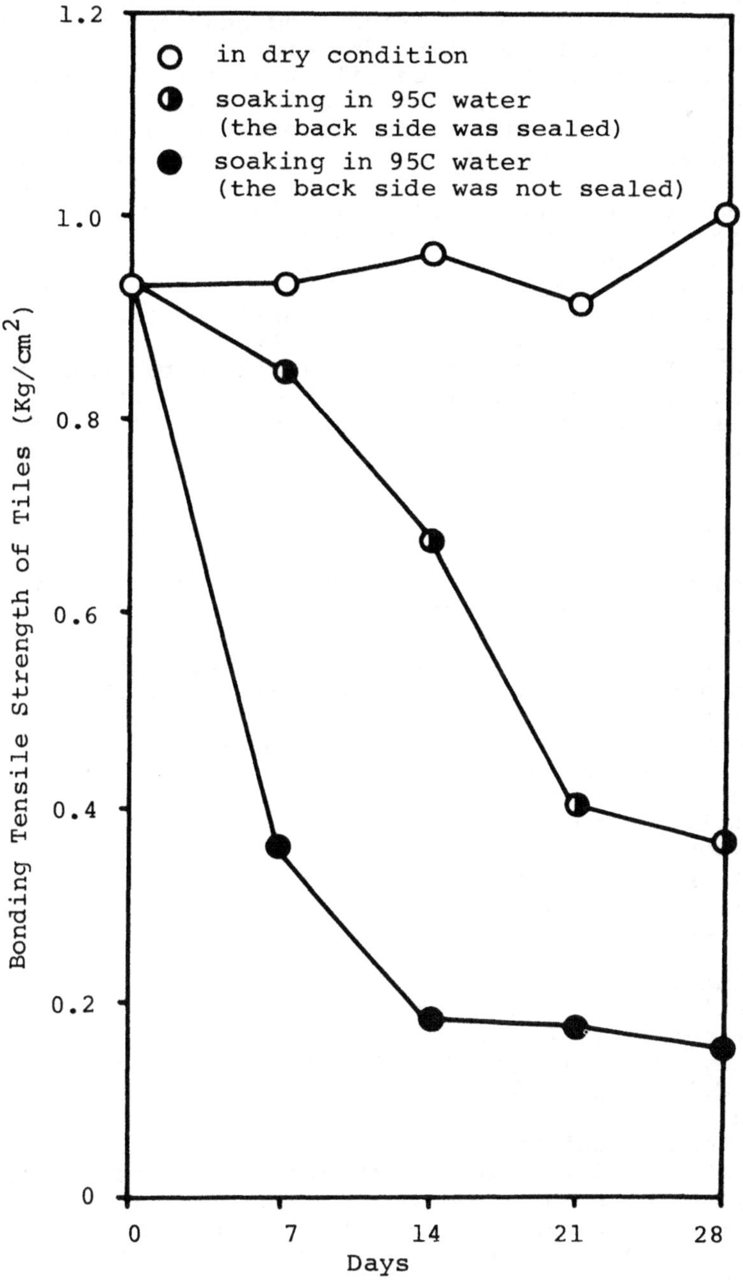

Fig. 2 Decrease of bonding strength in the boiling test.

means that soaking in hot water decreases adhesive strength. Therefore, moisture and heat are considered to be important deterioration factors. However, quantitative prediction of service life time for this adhesive system can not be obtained from the result. In these cases, "relative durability" is often evaluated by comparing with the results after the same acceleration test for the controlled specimens whose durability is rather evident in our experience. Also in this case, it is considered that quantitative data as to service life time can not be obtained.

Among the various experimental approaches, the creep rupture test was successful in obtaining the master curves predicting creep rupture time of the bonding tiles. The data obtained were analyzed by means of the Larson-Miller method (1952). This method can apply in the case that service life is considered to be dominated by creep rupture.

2. Larson-Miller's equation

Larson and Miller (1952) found the following Larson and Miller correlation between creep rupture time and temperature by studying influence of temperature on creep behavior of 8 types steel alloys. This correlation can be obtained assuming that Arrhenius' equation as to chemical reaction is applicable to rate of creep process on condition that creep stress is constant.

$$1/t = A\exp(-E/RT) \qquad (1)$$

where

t: time to rupture

A: constant

E: apparent activation energy

R: gas constant

T: absolute temperature

then

$$-\ln t = \ln A - E/RT$$

$$-RT\log t = RT\log A - E/2.303$$

$$T(\log A + \log t) = E/2.303R$$

finally the following Eq. (2) can be obtained.

$$T(C + \log t) = E/2.303R \qquad (2)$$

where

$C = \log A$ (constant)

Plot of 1/T vs. logt (Arrhenius plot) shows a straight line, and constant C can be obtained as logt when 1/T equals 0 (intercept). Activation energy can be also obtained by multiplying the slope of the straight line by 2.303R.

Gloor (1958) found that this Larson-Miller method was applicable to plastic materials. In Japan, this method was applied to wood-to-concrete bonds with polyvinylacetate adhesive (solvent type) by Inoue (1970).

3. Experimental methods

The creep rupture test was conducted in both the dry condition and the wet condition. The cross-lap type specimens simulating the objected adhesive system were set in zigs which were in a temperature controllable chamber as shown in Photo. 1. The bonding area of the specimen was 65mm by 65mm. The dry condition means the creep test in ambient air of controlled temperature. In the case of the wet condition, each specimen was sealed into a polyethylene bag after 48 hours' immersion in water. The temperature conditions were 20, 40, 60, 80, 100C for the dry condition and 20, 60, 80C for the wet condition.

The other creep rupture test using a real size panel was performed only in the dry condition as shown in Photo. 2. In this case, both tensile and shear stress was applied. This test was performed in both the case that the peripheral adhesive layer of the tile was cut and the case that the peripheral adhesive layer was not cut. The temperature conditions were 20 and 80C.

Photo. 1 The cross-lap specimen set in the zig.

Photo. 2 Creep rupture test for the real size panel

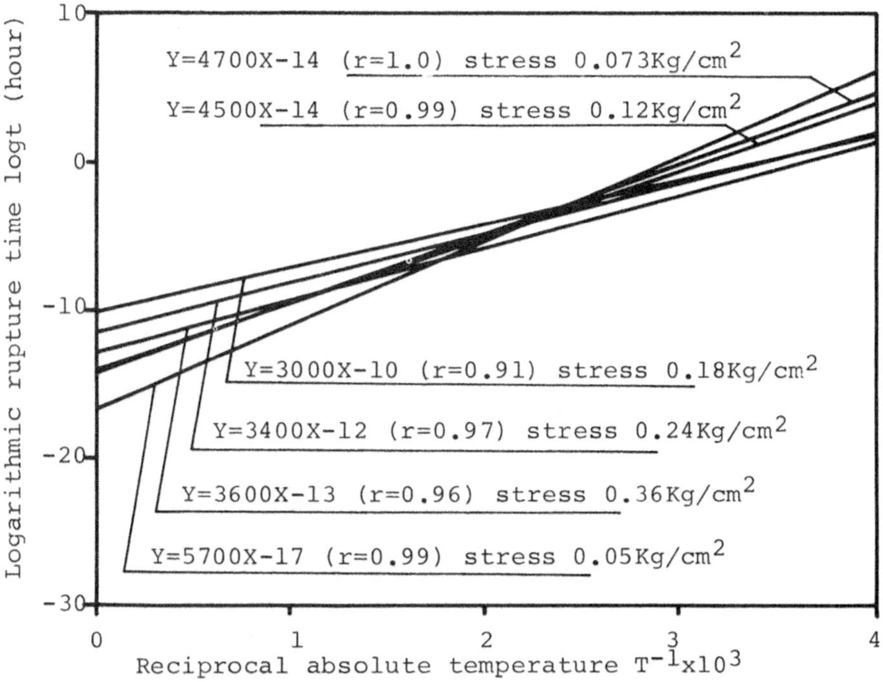

Fig. 3 Creep rupture time vs. reciprocal absolute temperature for the dry condition (Arrhenius plot)

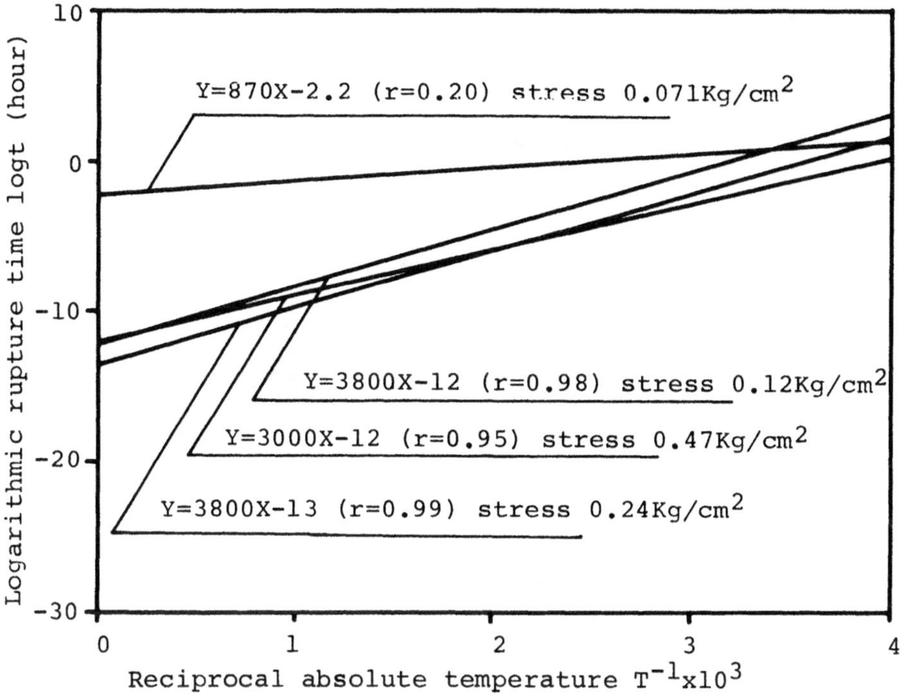

Fig.4 Creep rupture time vs. reciprocal absolute temperature for the wet condition (Arrhenius plot)

4. Result and discussion

Creep rupture time in various conditions could be obtained for the dry condition, the wet condition and the test using the real size panels. The mode of failure was cohesive failure of butyl rubber sheet under all the conditions.

Arrhenius' plots can be obtained for the test using the cross-lap specimens in the dry and wet conditions as shown in Fig. 3 and 4. The number of repetition and conditions were small as to the test using the real sized panels, because the purpose of the this test was to check the creep rupture behavior in more real conditions. Past reports shows that activation energy of creep rupture process decreases with increasing creep stress. This tendency was not so clear in this study probably because of roughness of the experimental conditions. However, it can be said that the activation energy in the wet conditions is smaller than that of in the dry condition. The purpose of this study is to confirm the usefulness of this Larson-Miller method for evaluating service life time of the elastomeric adhesive systems. Although the data have such a little variation, the C values in the

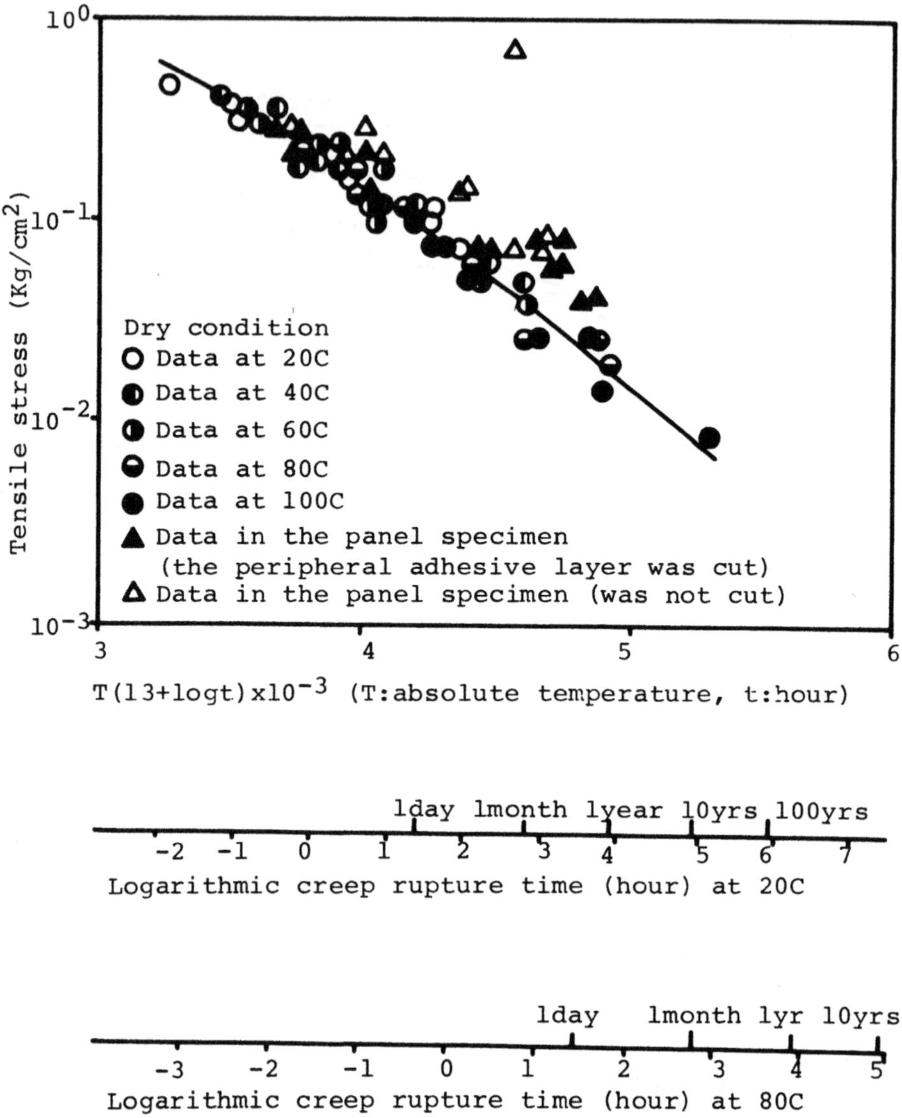

Fig. 5 Master creep rupture curve in the dry condition

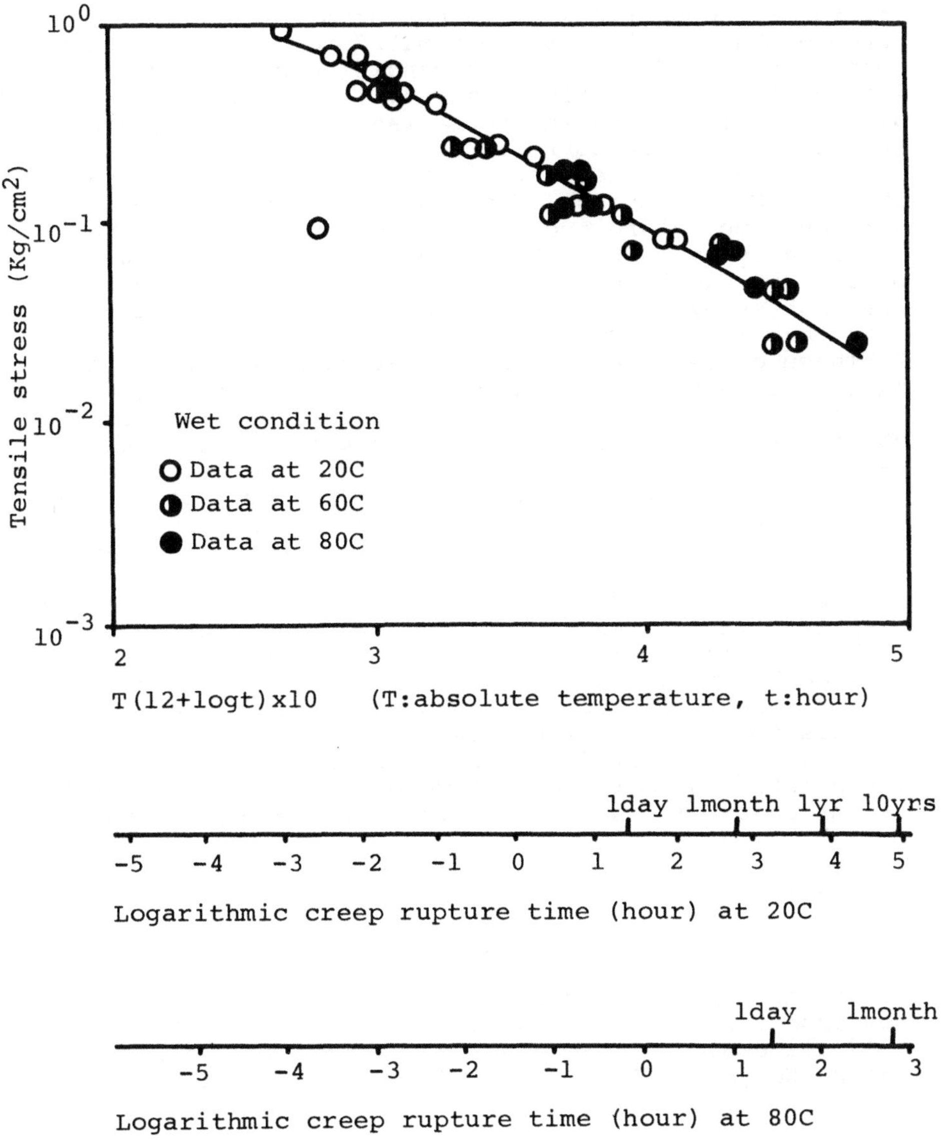

Fig. 6 Master creep rupture curve in the wet condition

dry and wet conditions dared be estimated for this purpose. The C values are estimated to be 13 for the dry condition and 12 for the wet condition by averaging the intercept values in Fig. 3 and 4. The datum at 0.071Kg/sq.cm stress in Fig. 4 was excluded from averaging because of low correlation.

Consequently, the master curves could be obtained as shown in Fig. 5 and 6. The time to creep rupture in various temperature and creep stress can be estimated quantitatively. The data in the real size panel were plotted in Fig.5 for comparison. Time to creep rupture in the test using the real size panels seems to be longer than that in the cross-lap specimen.

These curves seems to be rather satisfactory. The main creep stress is considered to be weight of the tile itself. This value is estimated to be 0.0015Kg/sq.cm in shear stress, and this adhesive system seemed to be rather durable against creep behavior under supposed in-service conditions. It must be checked by further investigation whether these master curves are reliable or not, however, further survey for long time seems to be actually impossible.

The advantage of this method is to be able to obtain quantitative information on time to creep rupture by a rather simple experiment. And, the author considered that this method is applicable for service life time prediction of such a elastomeric adhesive system where creep rupture dominate service life time of the adhesive system. In this study, effects of weathering were not taken into considerations. However, some effects may be estimated by similar tests after accelerated weathering.

Acknowledgment

The author would like to express thanks to Dr. S. Imaizumi, Prof. of Kogakuin University, Mr. R. Naniwa, Associate Prof. of Kogakuin University, and Dr. T. Soshiroda, Prof. of Shibaura Kogyo University for their kind discussion. Thanks are also expressed to Sekisui Kagaku Kogyo Co. Ltd. for kind help in the experiment.

Reference

Larson, F.R. and Miller, J. (1952) A time temperature relationship for rupture and creep stress. Trans. Am. Soc. Mech. Eng., 74, 765.

Gloor, W.E. (1958) Service test data on high density polyethylene. Modern Plastics, 36.

Inoue, S. (1970) Stress-temperature-time relationships for creep in wood-to-concrete bonds. J. of the Adhesion Society of Japan, 6, 5-15.

INSITU TEST METHODS FOR REPAIRED CONCRETE STRUCTURES

M. NADERI) Department of Civil Engineering,
D. CLELAND) The Queen's University of Belfast.
A.E. LONG)

Abstract

The repair and maintenance sector of the construction industry has become a growing sector in recent years. In the market there are many repair systems which has made the choice of the appropriate system very difficult. As far as the repair is concerned, achieving a good bond between the repair and the original concrete is essential. To date a few methods of testing the bond strength between the repair and original concrete have been suggested but mainly these are tests which must be carried out under laboratory conditions. In addition there is the question of the applicability of laboratory test results to actual site conditions. This paper describes insitu testing methods which can be used to determine the tensile and shear bond strengths between the repair and original concrete. Comparisons have been made between these methods and a compression/shear method.

KEYWORDS: Deterioration, Repaired Concrete, Bond Test, Insitu Shear Test, Insitu Tensile Test, Friction Transfer Method, Adhesion, Test Methods.

1. Introduction

The widely held belief that concrete structures remain serviceable for at least their design life, without needing major maintenance and repair has become questionable. During the last 10-15 years there has been increasing reports indicating the deterioration of concrete structures resulting in major expenditure for the owner and concern for the construction industry (5,6). The owner is faced with three choices:

(a) leave the structure to continue its deterioration

(b) Demolish and re-build

(c) Take remedial actions to stop further deterioration and maintain structural safety.

In most cases consideration of safety of the structure and economy dictates that option (c) be chosen. For this reason the repair sector has become a very fast growing part of the construction industry and has provided new challenges not only for Engineers but also for experts from other fields such as chemistry (7).

One of the most important requirements of any repair material is the achievement of a satisfactory bond between the repair material and the original concrete. Despite the production and use of many different materials available as repair systems, a standard bond testing procedure that can be applied to all materials has still not been devised. Consequently many repairs have been carried out without any effective monitoring of their quality. In recent years some attempts have been made to overcome this problem but the results are not completely satisfactory (1,2,3).

In the adhesive industry there are numerous types of test methods for assessing the bond strength between an adhesive and the adherend (4). In most cases the test methods simulate the stress condition and size which is most likely to occur in practice. Further, since the application of most adhesives takes place under laboratory conditions (e.g., airplane industry), results of laboratory tests methods are directly applicable.

However, in the construction industry the situation is different in that:

(a) Loading condition in service may be difficult to simulate in the laboratory

(b) The actual repair must be carried out under site conditions.

Therefore the necessity and usefulness of test methods which can be used on site as well as in the laboratory is self evident.

Recently in-situ methods for testing concrete strength have been developed at The Queen's University of Belfast. These have been adopted to test the bond strength of repairs.

(1) in situ tension test, carried out using "pull-off" method (8).

(2) in situ shear test, carried out using "Friction Transfer" method (9).

The "Friction Transfer" method that has been used to execute the in-situ shear tests reported in this paper can be applied to measure the tensile strength as well.

This paper describes the test techniques and compares the results with the compression/shear method (10).

2. Test Procedure SPECIMENS

To assess the tensile and shear bond strength of the repair, blocks of 600 x 300 x 200mm with compressive strength of about 65 N/mm^2 (at

28 days), were cast with three holes along the mid plane (parallel to 300mm side). After seven days three wedges were driven into the three holes and the block was split into two specimens of 600 x 300 x 100mm dimensions with a fractured/rough surface of 600mm x 300mm. To produce sawn surfaces, some of the split specimens were sawn to produce two specimens of 600 x 300 x 50mm size.

A timber frame was fixed around the specimens to provide a uniform repair depth of 15mm. Various repair systems were used to repair split/rough and sawn surfaces using a trowel. Cementation repairs were covered by wet hessian for seven days to cure.

After curing, a coring drill was used to cut 50mm dia. partial cores from the repair side extending about 5mm into the original concrete, (fig. 1). For the compression/shear test, 100 x 100 x 500mm concrete beams were cast and sawn in such a way that the angle between the sawn surface and the side of the beam (α in fig. 2), was either 30, 45 or 50 degrees. These angled prisms were then placed back into the mould and the remaining half filled with the repair mortar, (fig. 2).

TEST METHODS

In-situ tensile test - with this method, aluminium or steel probes were glued on top of the partial cores by means of epoxy resin adhesives. Then using the limpet apparatus, (8), a tensile force was applied to the probe and the average tensile bond strength was calculated simply by dividing the failure load by the cross sectional area of the core.

Knowing that the adhesive is stronger in tension than either repair or the original concrete, several fracture planes were anticipated, i.e., fracture in the original concrete or in the repair or at the interface between the two and finally combination of these three, depending on their relative strengths, (fig. 1). However, the use of high strength concrete, about 65 N/mm^2 minimised the likelihood of obtaining failures in the concrete. High strength concrete was appropriate for two other reasons.

(a) It is on high strength concrete that greatest difficulty is experienced in achieving a repair which is as strong as the original concrete.

(b) The deterioration of off-shore structures, where concrete of high strength is invariably used was of particular interest.

In-situ shear test - to measure the shear bond strength between the repair and original concrete, the "Friction Transfer" method which has been proposed and developed by M. Naderi was used, (9). In this method a specially devised apparatus fits on to the top of the partial core and is clamped to it. To measure the shear strength of the bond, a torque is applied through the handle and the maximum

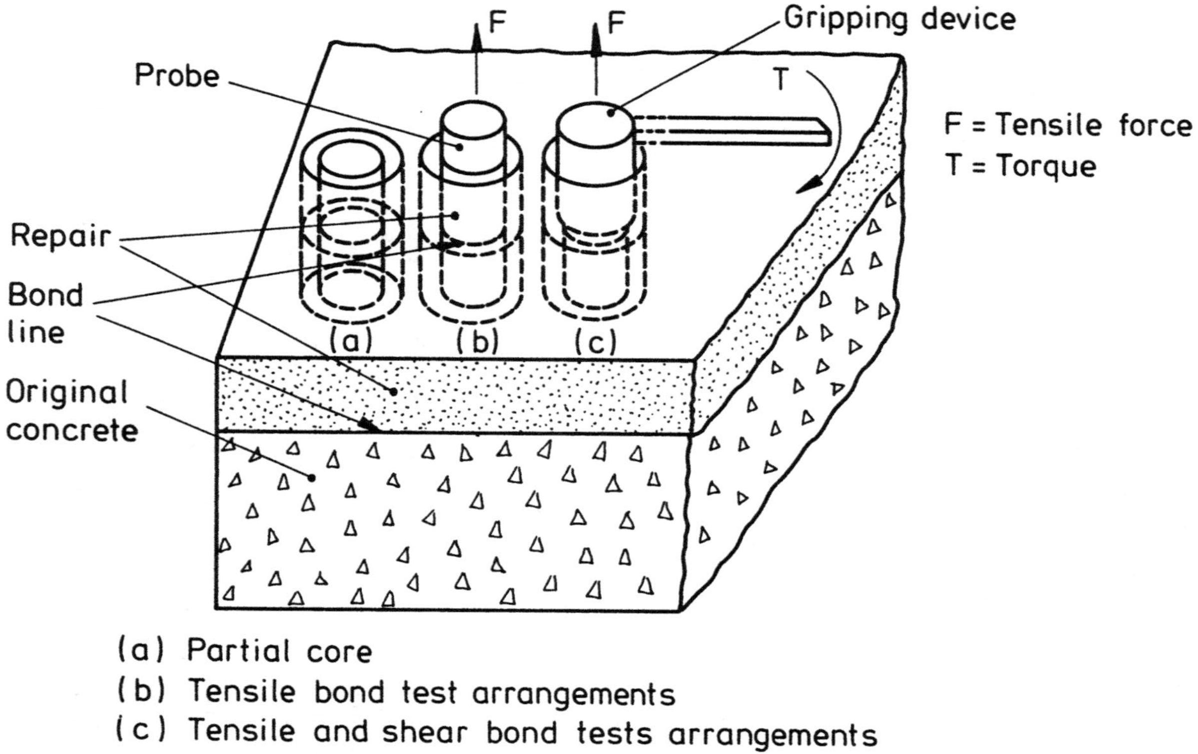

Fig. 1. In situ test arrangement

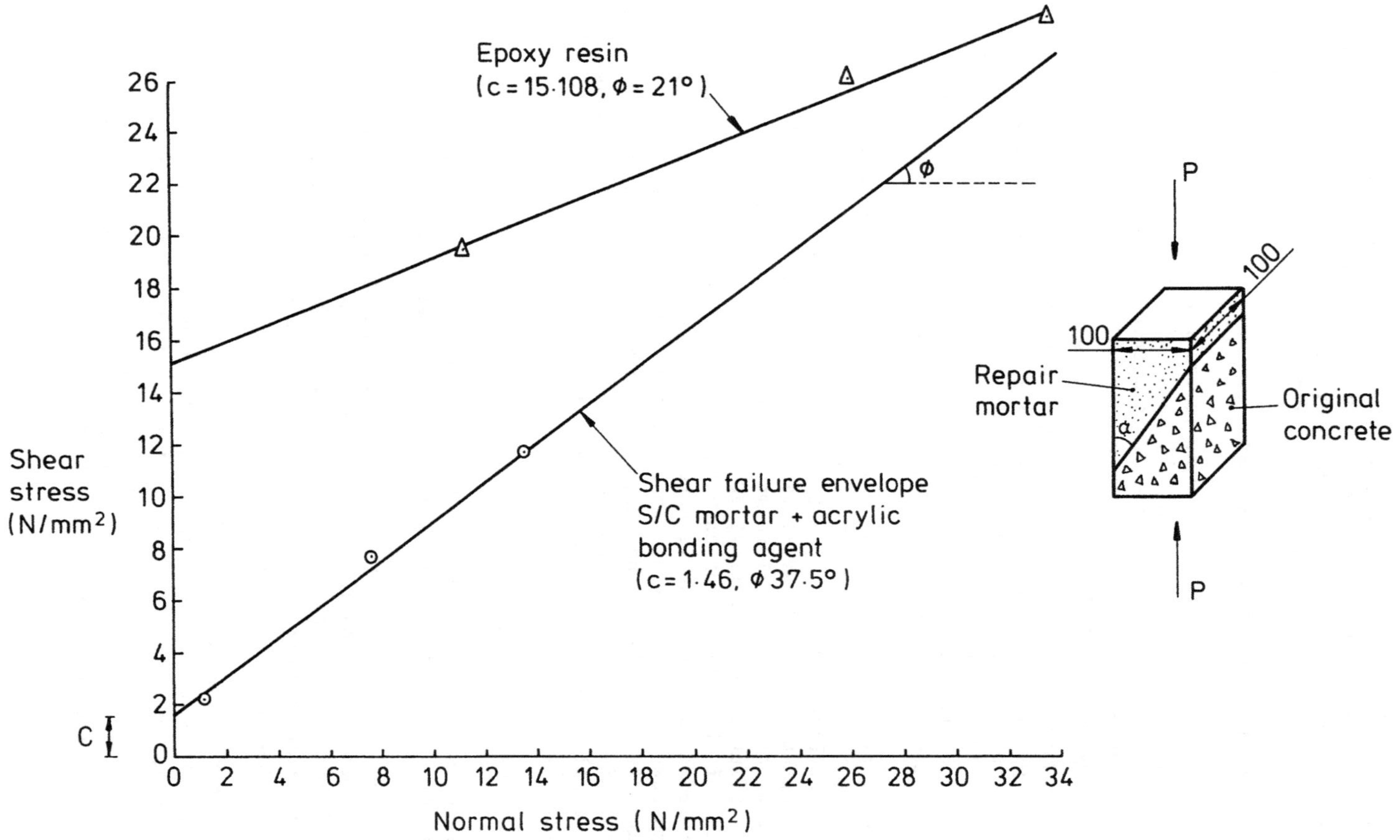

Fig. 2. Compression/shear test arrangements

shear stress at failure is recorded, (fig. 1).

As explained in earlier paragraphs, when testing the bond strength between a repair and the concrete structure failure may take place in the concrete, repair or at the interface or at a combination of these, depending on their relative strengths.

This method has the advantages that:

(a) Because the applied torque is transferred into the partially cored conrete or repair by means of pure friction between the core and the device, this method can be used in more aggressive environments such as under water and chemically polluted environments.

(b) When testing the bond strength between a repair and the original concrete, by carefully positioning the device relative to the bond line failure can be constrained to take place at the interface rather than in the repair or original concrete especially when the shear bond strength on saw cut concrete surfaces is being measured.

Compression/Shear method (10) - in this method composite prisms were subjected to axial compressive forces using a universal compressive testing machine, (fig. 2). When the specimens are loaded axially, both compressive and shear stresses develop along the interface, the magnitude of which at failure may be calculated with this method, three different angles, i.e., 30°, 45° and 50° were chosen and average of three tests for each angle was recorded. The average shear and normal compressive stresses at the interface were ploted to produce the shear failure envelope for each material, (fig. 2). By extending this line, the intercept between this line and the shear stress axis was chosen as the shear bond strength of the material when normal compressive stress on the interface is zero, (shown 'C' in fig. 2).

3. Results and Discussion

The results of the tensile, shear and compression/shear bond strength of different repair systems are shown in Tables 1, 2 and 3.

Shear bond strength on both, saw cut and rough (chisel hammered) surfaces are ploted against their relative tensile bond strengths in figs. 3 and 4. From these figures and table 3 it can be seen that good correlation exists between the shear and tensile bond strengths on both types of surfaces.

Noting that, in addition to the homogeneity problems faced when dealing with concrete, the bond phase introduces new complexity into the system, the coefficient of variations obtained with these in-situ test methods are within reasonable limits which demonstrates the easy repeatability of these methods. To make these methods and their results applicable to practice, no attempt was made to achieve a

TABLE 1 : BOND STRENGTH OF DIFFERENT REPAIR MATERIALS AT 14 DAYS APPLIED TO SAW CUT SURFACES

REPAIR SYSTEM	SHEAR BOND (N/mm^2)	NUMBER OF TESTS *	COEFFICIENT OF VARIATION (%)	TENSILE BOND (N/mm^2)	NUMBER OF TESTS *	COEFFICIENT OF VARIATION (%)	COMPRESSION /SHEAR BOND (N/mm^2)	NUMBER OF TESTS *
SAND/CEMENT MORTAR + SAND/CEMENT BONDING GROUT (S:C = 3:1, W:C = 1:2)	3.13	30	20.8	1.92	17	20.3	2.051	3
SAND/CEMENT MORTAR + ACRYLIC BONDING AGENT	4.0	9	17.23	2.40	2	-	1.46	3
SAND/CEMENT MORTAR + EPOXY BONDING AGENT	4.22	13	13.98	3.06	1	-	5.695	3

* THESE SHOW THE NUMBER OF BOND FAILURES

TABLE 2 : BOND STRNEGTH OF DIFFERENT REPAIR MATERIALS AT 14 DAYS APPLIED TO SPLIT/ROUGH (CHISEL HAMMERED) SURFACES

REPAIR SYSTEM	SHEAR BOND (N/mm^2)	NUMBER OF TESTS *	COEFFICIENT OF VARIATION (%)	TENSILE BOND (N/mm^2)	NUMBER OF TESTS *	COEFFICIENT OF VARIATION (%)
SAND/CEMENT MORTAR + SAND/CEMENT BONDING GROUT (S:C = 3:1, W:C = 1:2)	3.47	10	19.62	1.21	7	46.14
SAND/CEMENT MORTAR + ACRYLIC BONDING AGENT	4.44	5	20.13	1.62	6	13.22
SAND/CEMENT MORTAR + EPOXY BONDING AGENT	4.72	8	14.00	1.88	4	14.07

* THESE SHOW THE NUMBER OF BOND FAILURES

TABLE 3 : RATIO OF SHEAR BOND STRENGTH/TENSILE BOND STRENGTH OF DIFFERENT REPAIR MATERIALS AT 14 DAYS APPLIED TO SAW CUT AND SPLIT/ROUGH (CHISEL HAMMERED) SURFACES

REPAIR SYSTEM	SAW CUT	SPLIT/ROUGH (CHISEL HAMMERED)
	SHEAR BOND STRENGTH/TENSILE BOND STRENGTH	SHEAR BOND STRENGTH/ TENSILE BOND STRENGTH
SAND/CEMENT MORTAR + SAND/CEMENT BONDING GROUT (S:C = 3:1, W:C = 1:2)	1.63	2.87
SAND/CEMENT MORTAR + ACRYLIC BONDING AGENT	1.67	2.74
SAND/CEMENT MORTAR + EPOXY BONDING AGENT	1.38	2.52

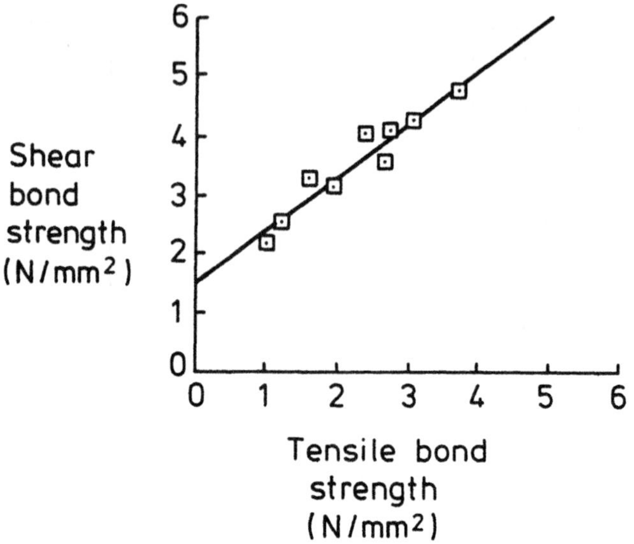

Fig. 3. Shear bond strength v. tensile bond strength on sawn surface at 14 days

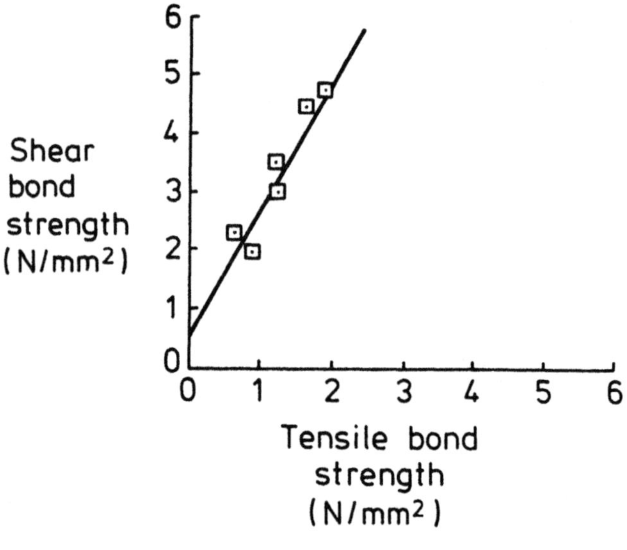

Fig. 4. Shear bond strength v.tensile bond strength on split/chisel hammered surface at 14 days

better statistically appealing results by employing a uniform compaction, etc. To limit the factors which may affect the bond strength such as, surface conditions, compaction, curing, etc., it was decided to chose concrete surfaces of the size 600 x 300mm which represent a patch repair encountered in practice. To be more practical, all the repairs were applied manually using a trowel. Every slab gave 50 no. 50mm partial cores, 25 of which were tested in tension and 25 in shear.

It is worth mentioning here that the mere magnitude of bond strength obtained in, say tension, should not be compared with that of the shear, due to the fact that the stress mechanisms that causes failure are different in both methods and further, the quoted tensile bond strength is the average, while the shear bond strength is the maximum shear stress obtained in torsion, (i.e., that at the outer edge of the core). However, the comparison between the shear and the tensile bond strength can be very useful for prediction of the real performance of a repair system in practice.

From the results, there seems to be some major disagreements between the in-situ methods and the laboratory compression/shear method. The reason may be explained as follows:

(a) Two in-situ methods were conducted on the same repaired slab, while the specimen for the compression/shear method were prepared separately, i.e., different curing, compaction and the size of the specimens may have had an effect.

(b) With the compression/shear method, there is always an element of normal stresses on the bond plane, although we have extrapolated a value for shear bond strength with zero normal stress, however, the best fit line while predicting the failure envelope at the stress levels tested in terms of both intercept and slope may be in error in terms of intercept alone. This reason also casts some doubts on the reliability of the tests conducted according to BS6319 Part 4.

When the results of bond strength on smooth (saw cut) surfaces are compared with that of rough (split, chisel hammered) surface, it can be seen that the shear bond strength is higher on the rough surfaces, the reason may be the fact that interlocking increases the shear resistance of the bonded surface while the tensile bond strength on rough surfaces is lower; this may be due to compaction problems and entraped air in the depressions and further due to the surface cleanness.

4. Conclusion

In-situ tensile and shear tests for assessing the bond strengths between the repair material and the original concrete which can be applied successfully on site and in the laboratory conditions, have been described. From the results of different repair materials, good correlation has been obtained between the tensile and shear bond strengths on both smooth (saw cut) and rough (chisel hammered) surfaces. The repeatability of these methods have been shown to be satisfactory.

The "Friction Transfer" method can be used to measure the bond strength between a repair and the original concrete in more aggressive environments such as underwater and chemically polluted areas (offshore structures, sewage, tanks, etc.). In this respect, a number of tests have been carried out successfully, where the adhesive failure was observed with other method. The compression/shear test method gave less consistent results. Although with the number of results (three for each specimen) it is difficult to judge this method statistically. However, the trend of results casts serious doubts on the credibility of this method, particularly when applied to trowel grade cementations repairs.

References

1. International Union of Testing and Research Laboratories for Materials and Structures, RILEM RECOMMENDATIONS MR 1-21, 1st Edition : December 1982.
2. British Standard Institution, BS5980: 1980, "Adhesives for use with ceramic tiles and mosaics".
3. British Standard Institution, BS6319: Part 4: 1984, "Testing of resin compositions for Use in Construction".
4. British Standard Institution, BS6350: 1982, "ADHESIVES".
5. Rasheedozzafar, Fahd H. Dakhil, Ahmad S. Gathani, "Corrosion of reinforcement in concrete structures in the Middle East", Concrete International/September 1985, P.48.
6. Denis Higgins, "Repairs to reinforced concrete", Concrete, December 1982, P.36.
7. Hewlett, P.C., Hurely, S.A., "Repair Materials - Selection and Testing", The Second International Conference on Structural Faults and Repair, 30th April - 2nd May, 1985, organised by The Institution of Civil Engineers, London.
8. Long, A.E., "Early Detection of Structural Faults in Concrete by using partially destructive tests", The Second International Conference on Structural Faults and Repairs, 30th April - 2nd May, 1985, organised by The Institution of Civil Engineers, London.
9. Naderi, M., "Internal Research Report", Civil Engineering Dept., The Queen's University of Belfast, Oct., 1985.
10. Eyre, J.R., Demone, P.L.J., "The slant shear testing of Bond of Repair Materials for concrete structures", The Second International Conference on Structural Faults and Repairs, 30th April - 2nd May, 1985, organised by The Institution of Civil Engineers, London.

ADHESION OF POLYMER-MODIFIED MORTARS TO ORDINARY CEMENT MORTAR BY DIFFERENT TEST METHODS

Y. OHAMA, K. DEMURA , H. NAGAO and T. OGI
Department of Architecture, College of Engineering, Nihon University

Abstract
This paper deals with the adhesion or bond strength of polymer-modified mortars to ordinary cement mortar, which is tested by different methods, and the effects of polymer types and polymer-cement ratio on the adhesion. Polymer-modified mortars using a commercial styrene-butadiene rubber (SBR) latex, ethylene-vinyl acetate (EVA) emulsion and polyacrylic ester (PAE) emulsion are prepared with various polymer-cement ratios, and tested for adhesions in tension, flexure and compressive shear. Substrate used is a mortar with a ratio of cement to standard sand 1:2. The test results obtained are summarized as follows:

(1) The adhesions in tension, flexure and direct compressive shear of the polymer-modified mortars increase with a raise in the polymer-cement ratio regardless of the types of polymers and specimens.

(2) The adhesion in slant (indirect) compressive shear of the polymer-modified mortars reaches the maximum at a polymer-cement ratio of about 5%.

(3) The failure mode distribution of the polymer-modified mortars remarkably depends on the test methods and polymer-cement ratio.

(4) The adhesion test methods which can successfully reproduce service conditions in the practical applications of the polymer-modified mortars are recommended.

Key words: Polymer-modified mortar, Substrate, Adhesion test methods, Adhesion, Failure mode distribution, polymer-cement ratio

1. Introduction

Polymer-modified mortars using various polymer dispersions, i.e., latexes and emulsions, have been widely used as finishing materials such as floorings pavings, adhesives, integral waterproofings, decorative coatings, repairing materials, anticorrosives and deck coverings in the construction industry because of their superior performance. The adhesion or bond strength of the polymer-modified mortars is one of the important properties required in such practical applications. However, few proper test methods to evaluate the adhesion considering their service conditions in the applications have been established till now. The purpose of this investigation is to examine the adhesion of the polymer-modified mortars to ordinary cement mortar as a popular substrate by different test methods in which their service conditions are considered.

In this paper, polymer-modified mortars using three types of typical polymer dispersions are prepared with various polymer-cement ratios, and tested for adhesions in tension, flexure and compressive shear by using different specimens in which ordinary cement mortar is employed as a substrate. The effects of the polymer types, polymer-cement ratio and test methods on the adhesion and failure mode of the polymer-modified mortars are discussed.

2. Materials

(1) Cement and aggregate

Ordinary portland cement and Toyoura standard sand as specified in JIS (Japanese Industrial Standard) were used in all the mixes.

(2) Polymer dispersions for cement modifiers

Commercial cement modifiers used were a styrene-butadiene rubber (SBR) latex, an ethylene-vinyl acetate (EVA) emulsion and a polyacrylic ester (PAE) emulsion. Their basic properties are listed in Table 1. Before mixing, silicone-emulsion-type antifoamer was added to the polymer dispersions in a ratio of 0.7% of the silicone solids in the antifoamer to the total solids in the polymer dispersions.

Table 1. Properties of Polymer Dispersions.

Type of Polymer Dispersion	Specific Gravity (20°C)	pH (20°C)	Viscocity (20°C, cp)	Total Solids (%)
SBR	1.020	8.6	175	44.9
EVA	1.055	5.2	1600	44.5
PAE	1.074	9.4	40	44.8

3. Testing procedures

3.1 Preparations of mortar substrates

According to JIS R 5201 (Physical Testing Methods for Cement), ordinary cement mortar was mixed with a ratio of cement to standard sand 1 : 2 and a water-cement ratio of 65.0%. Seven types of mortar substrates as shown in Fig. 1 were molded, and subjected to a 1-day-20°C-80% R.H.-moist, 6-day-20°C-water and 7-day-20°C-80% R.H.-moist cure. Then the bonding surfaces of the mortar substrates were rubbed by AA-150 abrasive papers specified in JIS R 6252 (Abrasive Papers), and compressed air was blown them for the purpose of removing dust.

3.2 Preparations of specimens

Polymer-modified mortars were mixed according to JIS A 1171 (Method of Making Test Sample of Polymer-Modified Mortar in the Laboratory) as follows; cement : standard sand = 1 : 3 (by weight), polymer-cement ratios, P/C (calculated on the basis of total solids in polymer dispersions) of 0, 5, 10 and 20%, and their flows were adjusted to be constant at 170±5. The mix proportions of the mortars are given in Table 2. Seven types of mortar substrates were placed in the respective molds to make seven types of specimens as shown in Fig. 1, and patched by the polymer-modified mortars. After molding, the specimens were given a 2-day-20°C-80% R.H.-moist, 5-day-20°C-water and 21-day-20°C-50% R.H.-dry cure.

Table 2. Mix Proportions of Polymer-Modified Mortars.

Type of Mortar	Cement:Sand (By Weight)	Polymer-Cement Ratio (%)	Water-Cement Ratio (%)	Flow
Unmodified	1 : 3	0	77.7	166
SBR-Modified	1 : 3	5	66.7	166
		10	68.5	170
		20	62.8	171
EVA-Modified	1 : 3	5	66.7	166
		10	63.7	169
		20	57.2	169
PAE-Modified	1 : 3	5	59.9	175
		10	47.3	166
		20	45.8	173

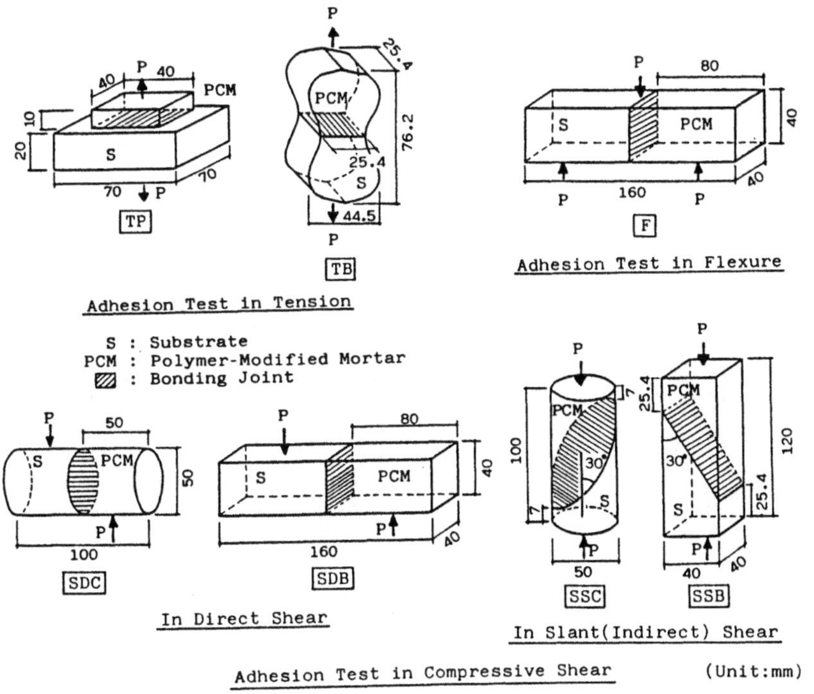

Fig.1. Test Methods for Adhesions in Tension,Flexure and Compressive Shear.

3.3 Adhesion tests
The cured Types TP and TB specimens were tested for adhesion in tension in accordance with JIS A 6915 (Wall Coatings for Thick Textured Finishes) and ASTM C 190 (Tensile Strength of Hydraulic Cement Mortars) respectively. The cured Type F specimens were tested for adhesion in flexure according to JIS A 1172 (Method of Test for Strength of Polymer-Modified Mortars). The cured Types SDC and SDB specimens were tested for adhesion in direct compressive shear at a loading rate of 50 N by use of the Amsler-type universal testing machine. The cured Types SSC and SSB specimens were tested for adhesion in slant (indirect) compressive shear in conformity to ASTM C 882 (Standard Test Method for Bond Strength of Epoxy-Resin Systems Used with Concrete). The adhesion or bond strength of the specimens was calculated by dividing the maximum load (the load carried by the specimens at failure) by the area of the bonded surface as follows:

$$\text{Adhesion in tension, flexure or compressive shear } (N/mm^2) = P/A$$

where P is the maximum load (N), and A is the area (mm^2) of the bonded surface.
After the adhesion tests, the failed crosssections of the specimens were observed for failure modes, which were classified into the following three types:

A: Adhesive failure (failure in the interface).
M: Cohesive failure in polymer-modified mortars.
S: Cohesive failure in substrate (ordinary cement mortar).

The respective approximate rates of A, M and S areas in the total area of 10 on the failed crosssections are expressed as suffixes for A, M and S.

4. Test results and discussion

Figs. 2 and 3 show the relation between the polymer-cement ratio and adhesion in tension of polymer-modified mortars by Types TP and TB specimens. Irrespective of the types of polymers and specimens, the adhesion in tension of the polymer-modified mortars tends to increase with a raise in the polymer-cement ratio. This tendency is more remarkable for PAE- and EVA-modified mortars than SBR-modified mortar, and generally the adhesion in tension by Type TB specimens is larger than that by Type TP specimens. Except for SBR-modified mortars, the adhesion in tension of EVA- and PAE-modified mortars with a polymer-cement ratio of 10% or more reaches about 2 to 4 times that of unmodified mortar. In general, the failure mode of Type TP specimens at the adhesion in tension of $3N/mm^2$ or more is cohesive failure in ordinary cement mortar as a substrate, but that of Type TB specimens is mostly a mixture of adhesive failure and cohesive failure in the polymer-modified mortars.

Fig. 4 exhibits the relation between the polymer-cement ratio and adhesion in flexure of polymer-modified mortars by Type F specimens. Like the adhesion in tension by Types TP and TB specimens, the

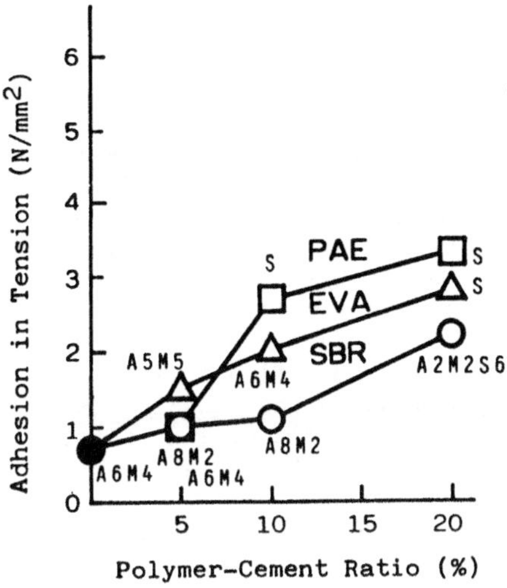

Fig.2. Polymer-Cement Ratio vs. Adhesion in Tension of Polymer-Modified Mortars by Type TP Specimens.

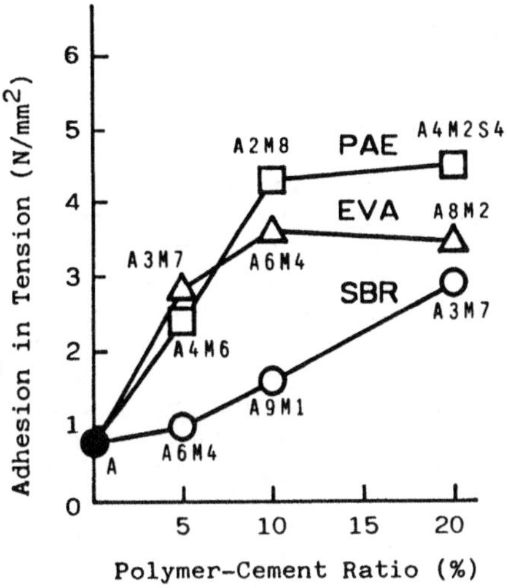

Fig.3. Polymer-Cement Ratio vs. Adhesion in Tension of Polymer-Modified Mortars by Type TB Specimens.

adhesion in flexure of the polymer-modified mortars is inclined to increase with raising polymer-cement ratio regardless of the polymer types. This trend is more marked for PAE- and EVA-modified mortars than SBR-modified mortar. Except for SBR-modified mortar with a polymer-cement ratio of 5%, the adhesion in flexure of the polymer-modified mortars reaches about 3 to 6 times that of unmodified mortar. In particular, improvement in the adhesion of PAE-modified mortar with an increase in the polymer-cement ratio is the most remarkable in these three types of polymer-modified martars. Generally the adhesion in flexure is about 2 to 3 times the adhesion in tension. The failure mode of the specimens in the adhesion test in flexure is generally a mixture of adhesive failure and cohesive failure in the polymer-modified mortars, and the cohesive failure

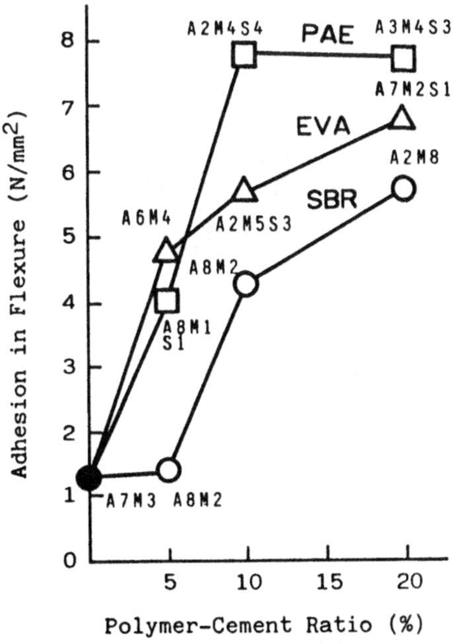

Fig.4. Polymer-Cement Ratio vs. Adhesion in Flexure of Polymer-Modified Mortars by Type F Specimens.

in the substrate is somewhat observed at a polymer-cement ratio of 5% or more.

Figs. 5 and 6 indicate the relation between the polymer-cement ratio and adhesion in direct compressive shear of polymer-modified mortars by Types SDC and SDB specimens. Like the adhesions in tension and flexure, the adhesion in direct compressive shear of the polymer-modified mortars increases with raising polymer-cement ratio regardless of the types of polymers and specimens. This trend is more remarkable for PAE- and EVA-modified mortars than SBR-modified mortar. The adhesion in direct compressive shear by Type SDC specimens is larger than that by Type SDB specimens. Except for PAE-modified mortar with a polymer-cement ratio of 5% by Type SDC specimens, the adhesion shows about 2 to 5 times that of unmodified mortar. The failure mode of Types SDC and SDB specimens in the adhesion tests in direct compressive shear is mostly adhesive failure, and cohesive failure in the polymer-modified mortars or substrate is somewhat observed at a polymer-cement ratio of 5% or more.

Figs. 7 and 8 represent the relation between the polymer-cement ratio and adhesion in slant (indirect) compressive shear of polymer-modified mortars by Types SSC and SSB specimens. As the polymer-cement ratio increases, the adhesion in slant compressive shear of the polymer-modified mortars attains to the maximum at a polymer-cement ratio of about 5% irrespective of the types of polymers and specimens. All the polymer-modified mortars with a polymer-cement ratio of 20% provide a lower adhesion than unmodified mortar. The adhesion in slant compressive shear differs from the adhesion obtained by the

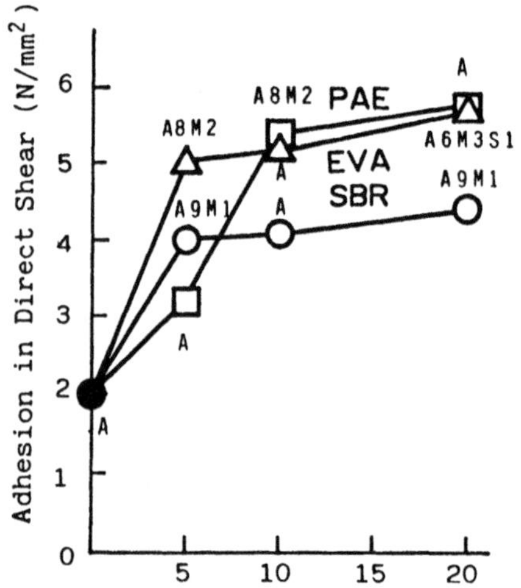

Fig.5. Polymer-Cement Ratio vs. Adhesion in Direct Compressive Shear of Polymer-Modified Mortars by Type SDC Specimens.

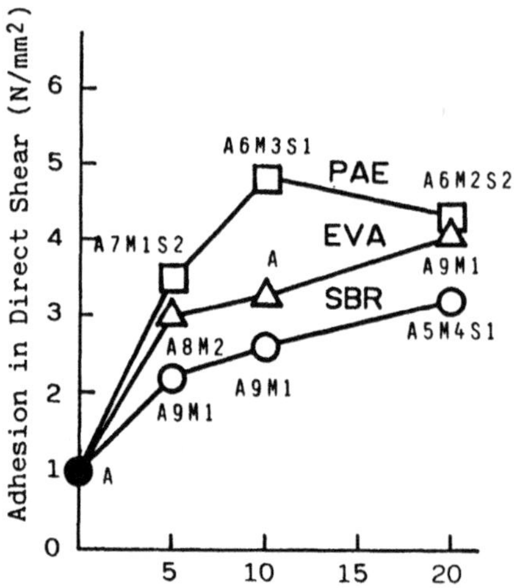

Fig.6. Polymer-Cement Ratio vs. Adhesion in Direct Compressive Shear of Polymer-Modified Mortars by Type SDB Specimens.

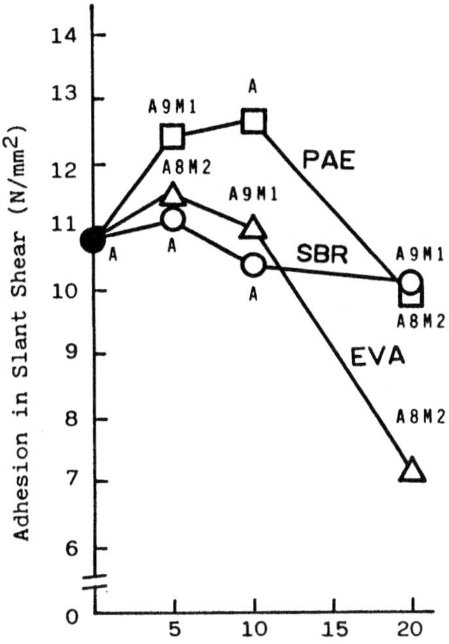

Fig.7. Polymer-Cement Ratio vs. Adhesion in Slant(Indirect) Compressive Shear of Polymer-Modified Mortars by Type SSC Specimens.

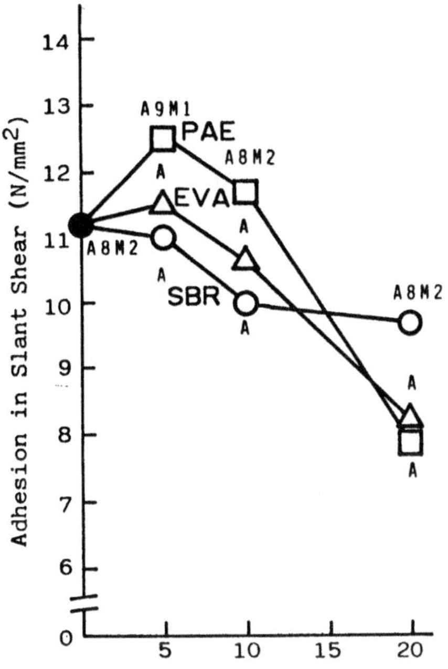

Fig.8. Polymer-Cement Ratio vs. Adhesion in Slant(Indirect) Compressive Shear of Polymer-Modified Mortars by Type SSB Specimens.

above-mentioned other test methods including a direct compressive shear test in this respect. The reasons for this may be explained to be due to the effects of a combination of shear and compressive stresses on the bonding joints of the specimens because of the slant compressive shear tests, and of the relaxation of the stresses by the soft polymer films formed on the bonding surfaces at a polymer-cement ratio of 5% or more. A difference in the adhesion in slant compressive shear between these two types of specimens is hardly recognized. Generally the failure mode of Types SSC and SSB specimens in such adhesion tests is almost adhesive in the interface.

Fig. 9 demonstrates the effects of test methods on adhesions in tension, flexure and compressive shear of polymer-modified mortars with various polymer-cement ratios. The adhesion in slant (indirect) compressive shear of the polymer-modified mortars by Types SSC and SSB specimens is extremely large compared to the adhesion determined by other test methods irrespective of the polymer types and polymer-cement ratio. This tendency is noticeable at smaller polymer-cement ratio regardless of the polymer types. Such an extremely high adhesion is attributed to the effects of the combined shear and compressive stresses and their relaxation as stated above.

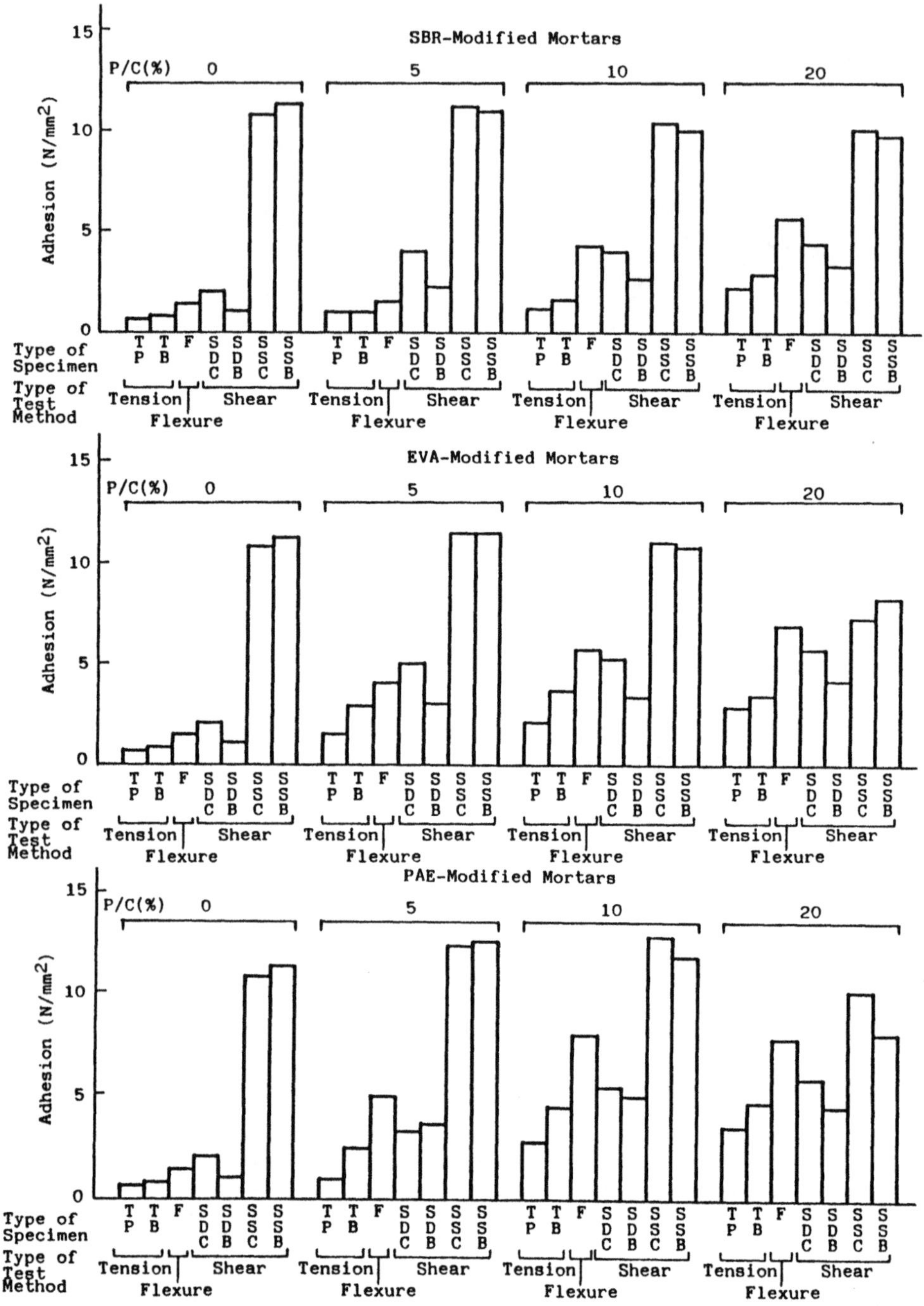

Fig.9. Effects of Test Methods on Adhesions in Tension, Flexure and Compressive Shear of Polymer-Modified Mortars with Various Polymer-Cement Ratios.

Fig. 10 represents the failure mode distribution of polymer-modified mortars in various adhesion tests. The percentage of adhesive or cohesive failure of the polymer-modified mortars in Fig. 10 was caluculated by the following equation:

$$F = (Fa / Ta) \times 100$$

where F is the percentage of adhesive or cohesive failure, Fa is the adhesive or cohesive failure area (cm^2) on the failed crosssections of the specimens, and Ta is the total area (cm^2) of the bonded surfaces of the specimens. As the percentage of adhesive or cohesive

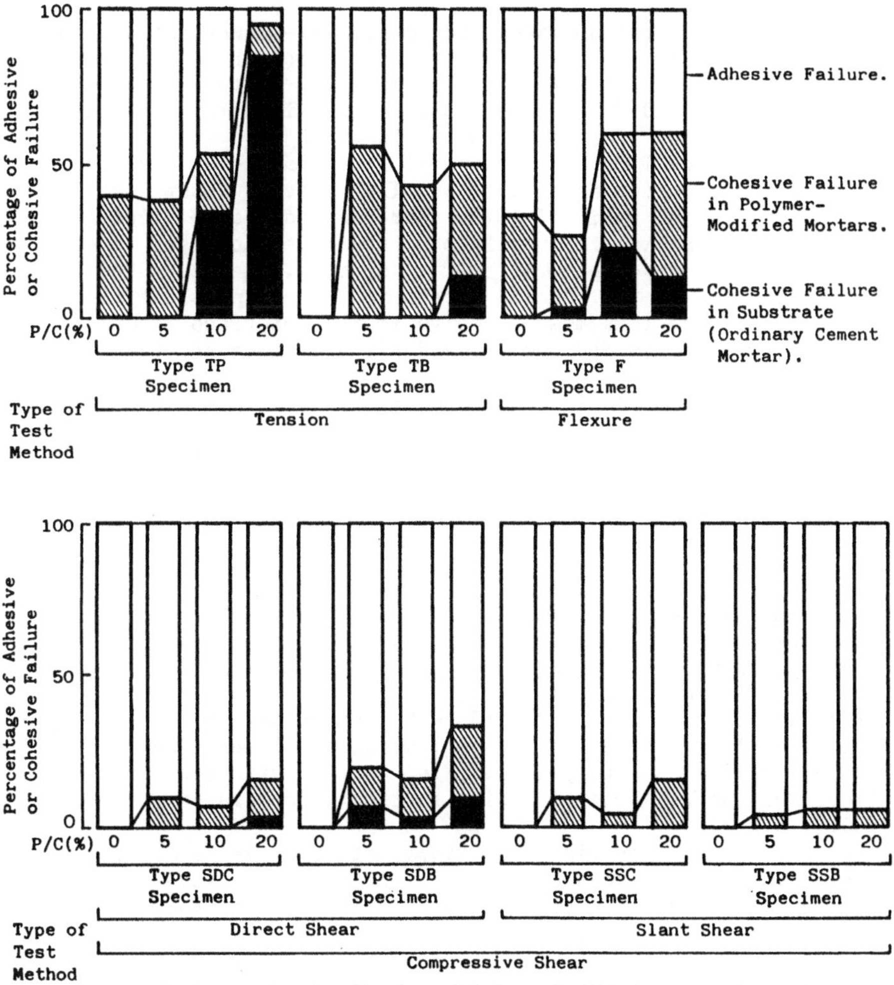

Fig.10. Failure Mode Distribution of Polymer-Modified Mortars in Various Adhesion Tests.

failure was not affected to a lesser extent by polymer types, the averaged percentage at each polymer-cement ratio in the respective test methods is illustrated in this figure. As seen in Fig. 10, the failure mode distribution of the polymer-modified mortars markedly depends on the test methods and polymer-cement ratio. In the adhesion test in tension by Type TP specimens, the percentage of cohesive failure in the substrate increases with a raise in the polymer-cement ratio, and reaches about 90% at a polymer-cement ratio of 20%. At the same time, the percentages of adhesive failure and of cohesive failure in the polymer-modified mortars decrease with raising polymer-cement ratio. On the other hand, in the adhesion test in tension by Type TB specimens, the ratio of the adhesive failure to the cohesive failure in the polymer-modified mortars is approximately 1:1 irrespective of the polymer-cement ratio (except for a polymer-cement ratio of 0%), and the cohesive failure in the substrate is somewhat observed at a polymer-cement ratio of 20%. Such differences in the failure mode between Types TP and TB specimens may be explained to be due to the geometry of the specimens and the tension methods. In the adhesion test in flexure by Type F specimens, the percentage of adhesive failure at a polymer-cement ratio of 5% or less is about 70%, and that at a polymer-cement ratio of 10% or more is reduced to about 40%. The percentage of cohesive failure in the substrate increases to some extent at a polymer-cement ratio of 5% or more. As described above, the failure mode in the adhesion tests in direct and slant (indirect) compressive shears is almost adhesive in the interface regardless of the geometry of Types SDC,SDB,SSC and SSB specimens and the polymer-cement ratio.

Considering the adhesion data and the failure mode distribution in various adhesion tests, it is most important to select the test methods to successfully reproduce service conditions in the applications of polymer-modified mortars.

5. Conclusions

(1) Generally the adhesions in tension, flexure and direct compressive shear of polymer-modified mortars to ordinary cement mortar increase with a raise in the polymer-cement ratio regardless of types of polymers and specimens.

(2) The adhesion in slant (indirect) compressive shear of polymer-modified mortars attains to the maximum at a polymer-cement ratio of about 5%, and all the polymer-modified mortars with a polymer-cement ratio of 20% show a lower adhesion in slant compressive shear than unmodified mortar.

(3) The adhesion in slant (indirect) compressive shear of polymer-modified mortars by Types SSC and SSB specimens is extremely large compared to the adhesion determined by other test methods irrespective of polymer types and polymer-cement ratio. The reasons for this may be considered to be due to the effects of the combined shear and compressive stresses and their relaxation.

(4) The failure mode distribution in the adhesion tests of polymer-modified mortars markedly depends on test methods and polymer cement ratio.

(5) The selection of the test methods to successfully reproduce service conditions in the practical application of polymer-modified mortars is most important for evaluating their adhesion.

References

Ohama, Y. (1973) Study on properties and mix proportioning of polymer-modified mortars for buildings. Report of the Building Research Institute No.65, Building Research Institute, Tokyo.

Sasse, H.R. and Fiebrich, M. (1983) Bonding of polymer materials to concrete. Materials and Structures, Research and Testing, 16, 293-301.

BS 6319 (1983) Testing of resin compositions for use in construction Part 4. Method for measurement of bond strength (slant shear method), British Standard Institution.

ADHESION TESTING OF POLYMER MODIFIED CEMENT MORTARS

W.H. PEIER
Department of Research and Development, SIKA Ltd., Zurich

Abstract
Perfect adhesion of polymer modified cement mortars (PCC-mortars) on concrete is of vital importance in applications such as repair, flooring, waterproofing or corrosion protection of concrete structures. Suitable adhesion test procedures have therefore to be developed. In this paper the complexity of adhesion testing is discussed. A program of adhesion tests, as used in our R + D laboratories for the development of new mortar products, is presented. It consists of three system tests: a preliminary test for qualitative evaluation of the adhesion and cohesion properties of the mortar on concrete; a test to determine the adhesive tensile strength and the durability of adhesion in low and high humidity environment; and an ageing test in which the adhesive tensile strength behaviour is investigated under cyclic temperature changes of the environment including freezing/thawing with deicing salts. The test performance of epoxy and styrene-acrylate modified cement mortars is analysed.
Key words: Adhesion testing, Adhesive tensile strength, Ageing, Durability of adhesion, PCC mortar, Styrene-acrylate.

1. Introduction

Today polymer modified cement mortars (PCC mortars) are successfully used in applications such as repair, flooring, waterproofing or corrosive protection of concrete structures. Compared to ordinary cement mortars they show better workability at a low water-cement ratio, higher tensile and flexural strengths, improved watertightness and freeze-thaw durability, higher chemical and carbonation resistance, as well as better adhesion on concrete.

While for the measurement of most physical properties there have been well-established and straightforward test methods existing for a long time now, adhesion testing due to its complexity was only investigated systematically over the last few years. The aim of this paper is to discuss the problem of adhesion testing and to indicate the requirements to be met by an adhesion test program. A program of tests to measure the adhesive tensile strength and durability of adhesion of PCC mortars subjected to loading and curing conditions encountered in the above-mentioned applications is presented. This program, which is partly based on proposals worked out by German groups (Arbeitskreis "Instandsetzung von Konstruktionsbeton mit Zementmörtel/Beton mit Kunststoffzusatz" (1985), Forschungsgesellschaft für Strassen- und Verkehrswesen (1985)), is used in our R + D laboratories for the development of PCC mortar products. Recently a new assortment of styrene-acrylate modified and epoxy modified cement mortars was formulated with the aid of the adhesion test program described above. The test results obtained are evaluated.

2. Adhesion testing

2.1 System behaviour

A typical repair system which is made of a reinforced concrete, a repair mortar patch, a fine-grained mortar layer and a polymer coating is illustrated in Fig. 1. The different materials stick together by adhesive forces.

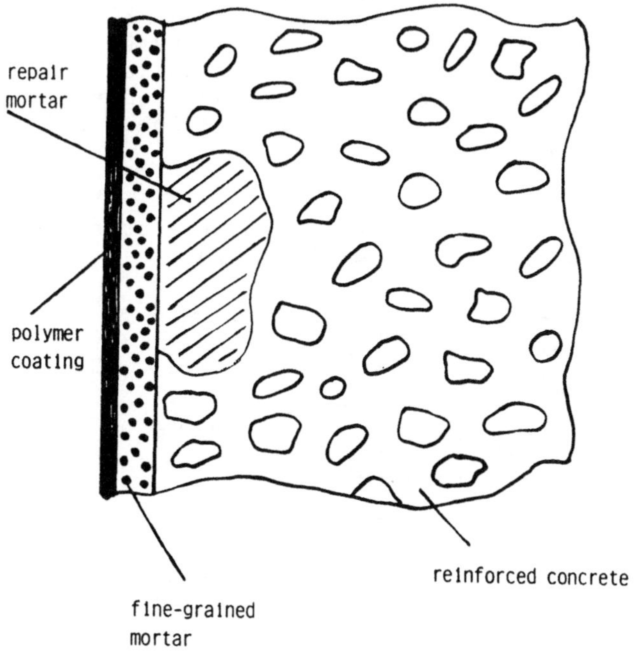

Fig. 1. Configuration of a repair system

If an external force is exerted on the system, a stress state will develop in the system. According to the kind of loading, three groups of stresses can be distinguished:

(a) Swelling and shrinkage stresses, mainly caused by water and water vapour phenomena.
(b) Thermal stresses, caused by temperature changes.
(c) Mechanical stresses, caused by external mechanical forces.

In order to determine a stress state, the mechanical behaviour of the system has to be known. The behaviour is described by a set of characteristic mechanical properties containing the bulk properties of the various materials (e.g. compressive strength, modulus of elasticity, linear thermal expansion coefficient etc.) and the adhesion properties (e.g. adhesive tensile strength) of adjacent materials. If all characteristic properties are known, the response of the system to all possibly applied external loadings can be calculated.

Bulk properties essentially differ from adhesion properties because they only characterize one material, whereas adhesion properties characterize two materials and in addition depend on the application method used in bonding one material to another (wetting procedure, bonding agent etc). The measurement of the bulk properties is therefore in general straightforward following well-established standards. Configuration and size of the system are of no importance. For the measurement of the adhesion properties the situation is different. Configuration and size of the system must be taken into consideration. This will be shown in the next section.

2.2 System testing

Fig. 2 illustrates a simple system made of a concrete slab which is freshly coated with a mortar layer.

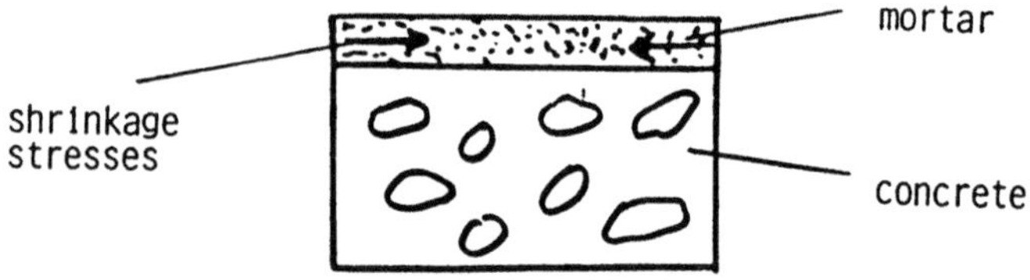

Fig. 2. Mortar-coated concrete slab.

After application shrinkage stresses develop in the mortar layer acting on the mortar-concrete interface. Two types of failure may occur:

(a) Cohesion failure (cracking), if adhesive strength > shrinkage stresses and shrinkage stresses > cohesion

holds.
   (b) Adhesion failure, if adhesive strength $<$ shrinkage stresses holds.

The two types of failure are shown in Fig. 3.

cohesion failure (crack) — adhesion failure

Fig. 3. Failure types.

Only if the two conditions, i.e. adhesive strength $>$ shrinkage stresses and cohesive strength $>$ shrinkage stresses, are fulfilled simultaneously, the system remains intact.

In order to determine the adhesive tensile strength, with a core drill of 50 mm Ø a core cylinder is drilled through the mortar layer ending at a depth of 5 mm within the concrete substrate. A metallic test cylinder is glued on the upper surface of the core cylinder by means of a suitable epoxy adhesive. With a tension test machine the strength necessary to pull off the glued cylinder is measured.

Unfortunately the tensile strength value which is obtained is not independent of the stress situation. It rather represents the difference between the tensile strength in a no stress situation and the stress component acting as tensile force. As the shrinkage stresses increase with increased mortar layer thickness, the measured adhesive tensile strength also depends on the layer thickness. Therefore, two systems which only differ in layer thickness show different adhesive tensile strength. In consequence, configuration and dimensions of the test sample cannot be chosen independently of configuration and dimensions of the represented system.

3. Requirements to be met by an adhesion test program

An adhesion test program should satisfy the following requirements:

(a) The choice of tests has to be such that all possible loading and curing conditions or ageing processes imposed on the system in a practical application can be taken into consideration.

(b) Configuration and dimensions of the test samples have to be representative of reality.

(c) Minimum requirements with respect to adhesive tensile strength provided that no cohesion failure occurs have to be indicated for each mortar type used (classifying the various applications).

4. SIKA adhesion test program

4.1 General

The program used in our R + D labs comprises three tests. A crack-bridging test has been abandoned. The tests will be described here in detail. Three mortar types are distinguished.

4.2 Mortar types

The PCC mortars are classified as follows (the abbreviated names will be used in the description of the program).

(a) Sealing mortars (S) - mortar used to protect concrete structures against the infiltration of water, the action of de-icing salt and the phenomenon of carbonation.

(b) Thin layer mortar (TL) - mortar used in thin layer applications to provide protective finishing coats, or to level surface damage in coats up to 15 mm thickness. No particular strength requirements have to be met.

(c) Concrete replacing mortar (CR) - mortar with a maximum grain size of 8 mm, mainly used to patch concrete chipping and to level or reprofilate damaged concrete surfaces up to about 50 mm thickness. This replacement mortar has to meet particular strength requirements.

4.3 Substrate

Substrate used in all tests is a concrete with a maximum grain size of 32 mm and a mean pull-off strength of at least 3 N/mm2 after 28 days, and every single pull-off strength value higher than 2.5 N/mm2.

4.4 Wedge test

The test is illustrated in Fig. 4.

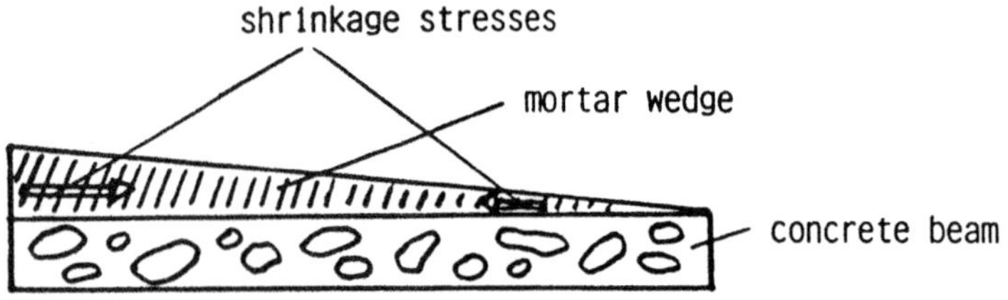

Fig. 4. Wedge test.

This preliminary test provides a quick evaluation of crack sensitivity and adhesion strength of a mortar, thus ensuring quick selection among several candidates.

As substrate we use two beams 100 x 10 x 10 cm stripped from framework after 24 hours, and stored for 7 days at 23°C/95%R.H., and from then up to the 28th day in normal climate of 23°C/50%R.H. After this a rectangular lateral face of each beam is sand-blasted and then coated with a mortar wedge. The mortar layer thickness varies from 0 to 20 times the maximum grain size. One of the beams, after it has been coated, is placed for 24 hours in longitudinal sense into a wind channel, and afterwards stored at 23°C/50%R.H. The other beam is stored at 23°C/50%R.H. right after coating. During the following two weeks the test samples are checked daily whether cracks have formed in the mortar wedge or whether adhesion failure between mortar and concrete has occurred.

No cracks and no layer separation must occur for TL, CR.

4.5 Durability test

This test measures the durability of adhesion of a mortar coat under extreme humidity conditions.

As substrate slabs 30 x 30 x 8 cm, stored as described in section 4.4, are used. After sand-blasting one large lateral face of each slab is coated with mortar. The layer thickness equals the maximum application thickness for S and TL, and amounts to 10 times the maximum grain size, yet at least 2 cm, for CR (for layers exceeding 4 cm in thickness, slabs 30 x 30 x 10 cm are used). 12 slabs receive a coating, 6 of them stripped from framework after 24 hours, are stored at 23°C/50%R.H., while the other 6 ones are stored under water. The adhesive tensile strength is determined on 2 slabs after 7, 28 an 90 days.

No cracks and no blistering must occur. The adhesive tensile strength values after 28 and 90 days must satisfy the minimum requirements: mean value $\geq$ 2 N/mm2 and no single value $<$ 1.5 N/mm2 for CR, mean value $\geq$ 1.5 N/mm2 and no single value $<$ 1 N/mm2 for S and TL.

4.6 Temperature change test with de-icing salt exposure

The test is shown in Fig. 5.

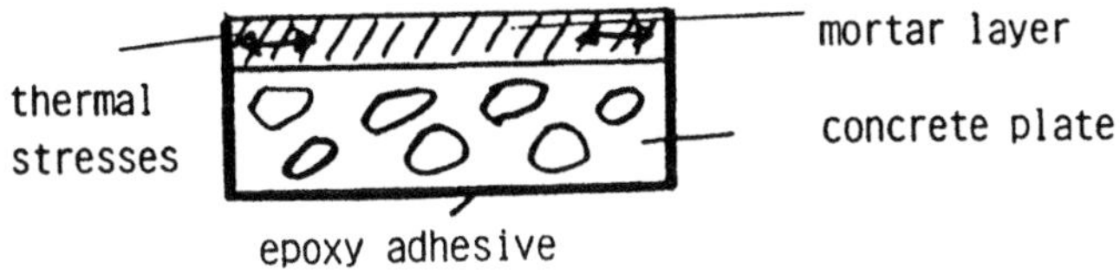

Fig. 5. Temperature change test.

This test measures the durability of adhesion under cyclic thermal loading conditions with additional de-icing salt exposure.

Three concrete slabs with the dimensions and preparation as described in section 4.5 are coated with mortar. The layer thickness is the same as in the durability test. After a 7 days storage at 23°C/50%R.H. the substrate and the lateral faces of the mortar layer are coated with an epoxy adhesive. Prior to the alternating temperature test the samples are stored under water during 3 days. Then they are submitted to 15 cycles with the following steps: 2 hours in a saturated salt solution at -15°C, 2 hours in water at 20°C, 16 hours in air at 60°C, 4 hours in water at 20°C. During the various test steps, the slabs are inspected visually as well as with blows of hammer with respect to chipping, blistering and cracking. After completion of the 15 cycles, the adhesive tensile strength is measured.

The same minimum requirements as in the durabilyty test have to be satisfied.

5. Test results of new mortar products

5.1 General

The adhesion test program described in section 4 was used to formulate a new assortment of styrene-acrylate and epoxy modified cement mortars for repair, flooring and waterproofing.

5.2 Polymer-cement ratio

PCC mortars have good workability at a low water-cement ratio because workability can be kept constant if water is replaced by polymer. The lower water-cement ratio manifests itself, for instance, in lower shrinkage and higher strength. The polymer content however does also influence adhesion properties. In order to find the maximum of the adhesive tensile strength of a styrene-acrylate modified cement mortar as a function of the polymer-cement ratio, the cement mortar mix presented in Table 1 is used as a starting point. The water-cement ratio of 0.48 guarantees good workability.

Table 1. Cement mortar mix.

Composition	Weight (kg)
Filler	0.75
Sand 0-1 mm	1.88
Sand 1-4 mm	4.87
Portland cement	2.50
Water	1.20

In 12 steps the quantity of water in the mortar is gradually reduced and substituted by the same amount of styrene-acrylate. Table 2 shows the polymer-cement ratios and the corresponding water-cement ratios in the substitution.

Table 2. Substitution of water by styrene-acrylate.

PCR	0	0.01	0.02	0.05	0.07	0.10	0.12	0.14	0.17	0.19
WCR	0.48	0.47	0.46	0.43	0.41	0.38	0.36	0.34	0.31	0.29
PCR (cont.)		0.22	0.23	0.24						
WCR (cont.)		0.26	0.25	0.24						

With every mix the durability test and the temperature change test are carried out. The adhesive tensile strength after a storage of 28 days at 23°C/50%R.H. and after ageing in the temperature change test are illustrated in Fig. 6.

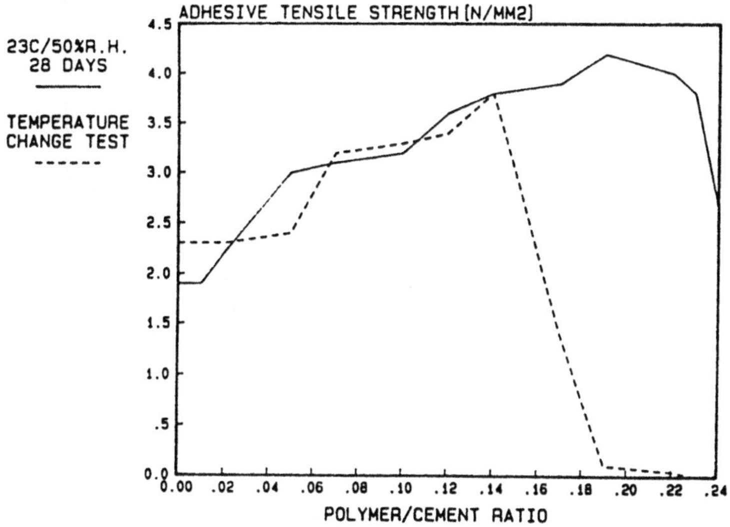

Fig. 6. Adhesive strength behaviour in the substitution.

Both strength curves assume a maximum followed by a steep decline. In the durability test a maximum value of 4.2 N/mm2 at a polymer-cement ratio of 0.19 is obtained, in the temperature change test a maximum value of 3.8 N/mm2 at 0.14. From the strength behaviour in the two tests it can be concluded that, if the polymer-cement ratio exceeds a certain limit value, adhesion drastically deteriorates. Furthermore the temperature change test seems to be more restrictive than the durability test for mixes with a high polymer-cement ratio, probably because of the considerable swelling that takes place under water storage. This interpretation is supported by the results obtained in a swelling test and by the fact that in the durability test a mix with a ratio of 0.15 shows approximately the same adhesive tensile strength behaviour in storage at 23°C/50%R.H. and under water, while a mix with a ratio of 0.22 shows a marked difference, as it is illustrated in Fig. 7.

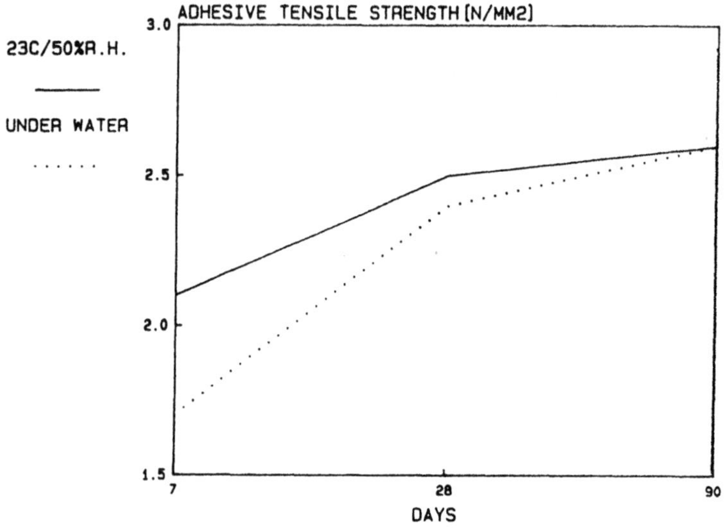

Fig. 7. Adhesion behaviour in the durability test.

Due to swelling stresses the adhesive strength after 7 days is much lower for under water storage.

Best adhesion is achieved at a polymer-cement ratio within the range of 0.13 and 0.15.

5.2 Cement-aggregate ratio

The results from the wedge test indicate that for a sealant or thin layer mortar the optimum cement-aggregate ratio as regards adhesion and shrinkage properties ranges from 0.8 to 1.2. For a concrete replacing mortar with a maximum grain size of 4 to 7 mm and an application thickness of 20 to 50 mm, this ratio has to be reduced, otherwise cracking will occur. The optimum ratio ranges from 0.25 to 0.35.

5.3 Application method

Adhesion of mortar on concrete depends on the application method used for the coating. It is recommended to moisten the substrate with clean water. When the substrate is surface-saturated dry, with no glistening water, coating can be applied. The mortar must be scrubbed into the substrate to be sure to fill all pores and voids.

This recommendation provides a satisfactory bond except for concrete replacing mortars. This type has a large maximum application thickness and a low cement-aggregate ratio. As a lower ratio induces lower adhesive strength, and as in a dry climate large shrinkage stresses develop, adhesion failure may occur. Therefore, for concrete replacing mortars a bonding agent, which is a fine-grained epoxy or styrene-acrylate thin layer mortar, is strongly recommended. Fig. 8 shows the behaviour of a concrete replacing mortar of 7 mm maximum grain size with and without bonding agent in the durability test.

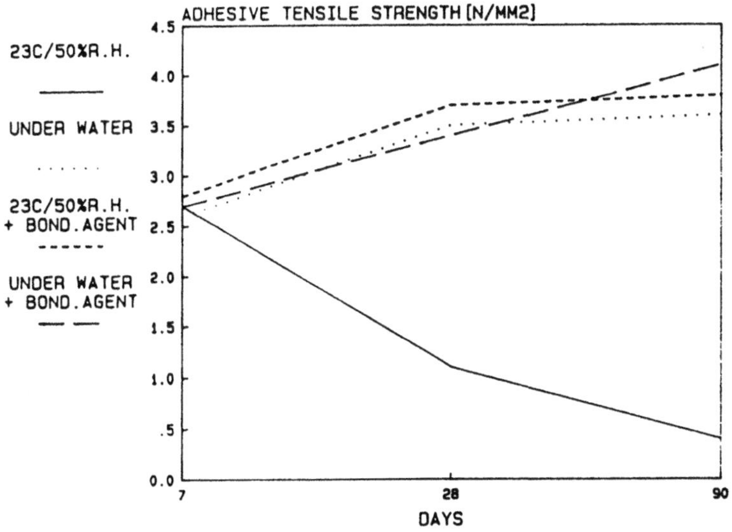

Fig. 8. Effect of bonding agents on adhesion

If a bonding agent is used, the mortar shows perfect adhesion. Without bonding agent it does not satisfy the minimum requirements when stored in a dry climate.

6. Conclusions

A program of tests to measure the adhesion properties of PCC mortars has been presented. From first experience with new formulations of styrene-acrylate and epoxy modified cement mortars it can be concluded that these tests represent a good simulation of the real loading conditions occuring in applications such as repair, waterproofing, flooring etc. Further improvements could possibly be obtained by extending the time interval in the durability test from 3 to 6 or even 12 months.

References

Arbeitskreis "Instandsetzung von Konstruktionsbeton mit Zementmörtel/Beton mit Kunststoffzusatz" (1985) _Technische Lieferbedingungen für Zementmörtel/Beton mit Kunstoffzusatz (PCC) als Betonersatz im Brücken- und Ingenieurbau TL BE - PCC_. Senator für Bau- und Wohnungswesen, Berlin.

Forschungsgesellschaft für Strassen- und Verkehrswesen (1985) der BRD _Merkblatt für die Erhaltung von Betonstrassen_.

SHEAR ADHESION PROPERTIES OF EPOXY RESIN ADHESIVES

R.N. SWAMY Department of Civil and Structural Engineering,
R. JONES University of Sheffield
A. CHARIF Department of Civil Engineering,
Ecole National Polytechnic

Abstract
Tests are reported to investigate the shear bond strength of epoxy resin adhesives to concrete and steel adherents. Two types of tests were carried out on composite steel-epoxy-concrete specimens - pull out tests, in which failure always occurred through the concrete - glue bond, and single and double lap tests to examine the epoxy shear adhesion to steel adherents. It was found that the distribution of the longitudinal forces and shear stresses along the joint was exponential for both types of tests. The shear bond strength of epoxy to concrete varied almost linearly with concrete strength. However, variations in the adhesive layer thickness, in the range 0.5 to 3.0 mm, had practically no effect on the strength of the pull out or double lap specimens. The results obtained show that both the pull out and double lap shear tests can be carried out with a good degree of accuracy and consistency.
Key words: Epoxy resins, Test methods, Bond strength, Pull out tests, Double lap shear tests, Shear stress, Glue thickness.

1. Introduction

The development of synthetic adhesives based as epoxy resin systems has opened up new possibilites for bonding structural materials together. One of the major applications of epoxy resin adhesives is in the strengthening of reinforced and prestressed concrete structural elements through externally bonded steel plates. The engineering properties of epoxy adhesives are of paramount importance in this respect as these depend on several variables including the type of test specimen and the rate of loading.

The mechanical action of the steel-glue-concrete connection in epoxy joints such as plated beams is mainly by shearing. The bond between the resin and the steel/concrete materials is partly chemical (adhesion) and partly mechanical (friction), and depends therefore on the substrates and their surface preparation. The strength of such a joint will always be subject to an upper limit determined by the lowest ultimate shear strength of the three materials or their ultimate tensile (or compressive) strength. Since concrete has the lowest ultimate shear strength, failure of a steel-epoxy-concrete joint should invariably occur through the concrete.

Several investigators have reported the shear adhesiveness of epoxy to concrete by tests on composite steel-epoxy-concrete specimens. Bresson (1971) simulated the behaviour of the joint with a mathematical model from which the shear distribution in the joint can be predicted up to failure. In general, the more highly stressed end of the joint fails first, and then a progressive failure follows. The predicted values by Bresson (1971) and the measured results by Bresson (1971), Solomon (1975), Burkhardt et al (1975), and Van Gemert (1980) show that the distribution of longitudinal forces and shear stresses along the joint in pure shear is exponential. The ultimate shear strength of the joint could be related to either the ultimate shear strength of the concrete as reported by Bresson (1971) and Burkhardt et al (1975), or its ultimate tensile strength as suggested by Solomon (1975) and Van Gemert (1980). The stress concentration factor i.e. the ratio of maximum to mean stress, depends on the geometry of the test specimen, and the elastic moduli of the adherents and adhesive. Mixing glass fibres with epoxy can reduce the stress concentration and thus increase the ultimate load.

The shear adhesiveness of epoxy to concrete and steel is reported in this paper through two series of tests. The first series of tests was conducted on concrete and steel adherents, and pull out tests similar to those of Solomon (1975) were used. As failure in the pull out specimen always occurs through the concrete-glue bond, the epoxy shear adhesion to steel was also studied through tests with steel adherents only. Double lap and single lap tests were adopted for this part of the investigation. The main test variables in these tests were the concrete compressive strength and/or the glue line thickness.

2. Experimental details

2.1 Pull-out tests

The details of the pull out test are shown in Fig.1. Two steel plates 60 x 3mm thick were glued to the opposite sides of a concrete prism 60 x 60 x 150mm. The plates were put into tension through the test arrangement shown in the figure. A spherical seating was used to distribute the load over the concrete prism. The steel plates in tension then transfer their load to the concrete prism through the glue layer.

The test variables studied were glue thickness, 0.5 to 3.0mm, and concrete cube strength, 25 to 75 N/mm^2. For each glue thickness and cube strength, six specimens were tested.

2.2 Double and single lap tests

The test arrangement is shown in Fig.2. When the inner plate (60 x 6mm thick) in the double lap test is put into tension, the load is transferred to the outer plates through the glue layer. The reaction is then taken by the plate P to which the outer plates (60 x 3mm thick) are bolted. In order to ensure that the loads are distributed as uniformly as possible at the ends of the plates, the testing machine was fitted with friction grip jaws capable of accepting the

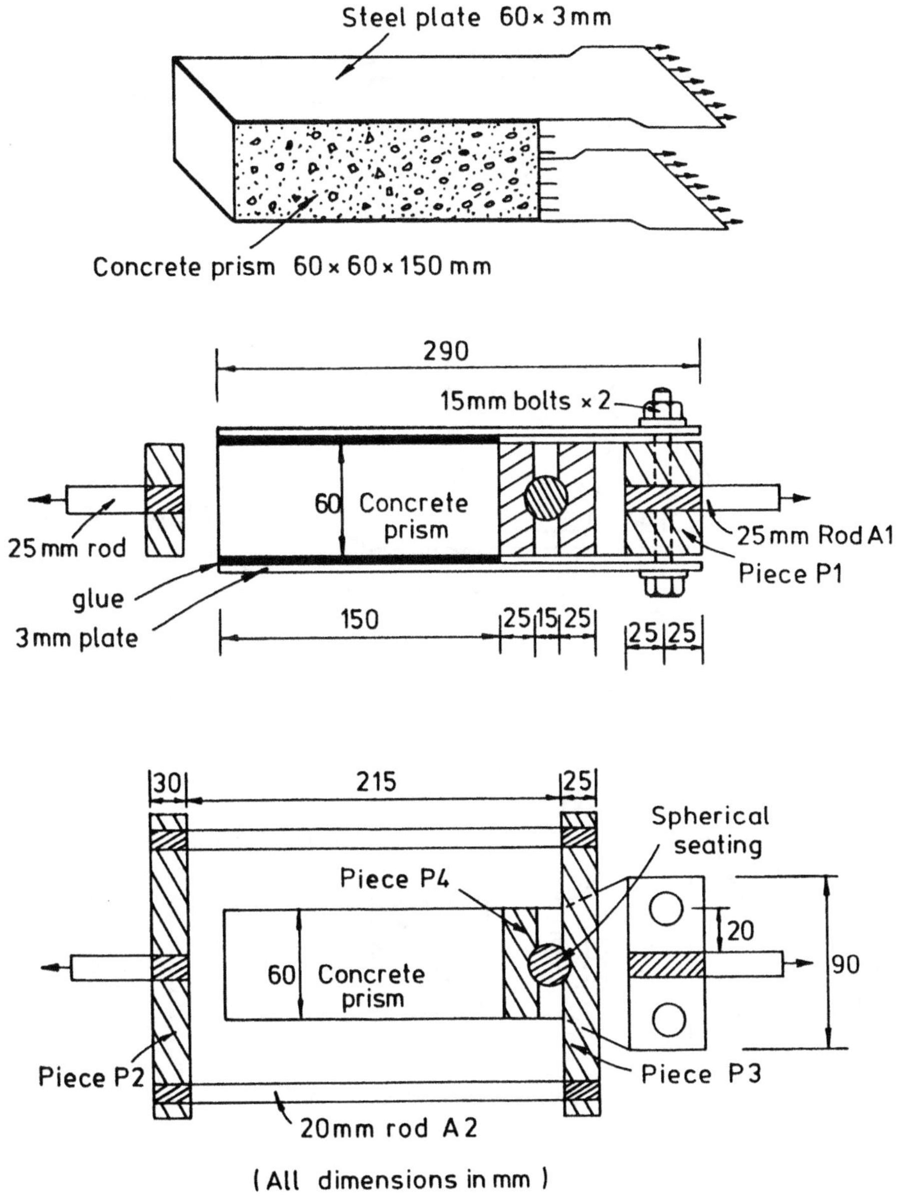

FIG. 1 PULL OUT TEST SPECIMEN

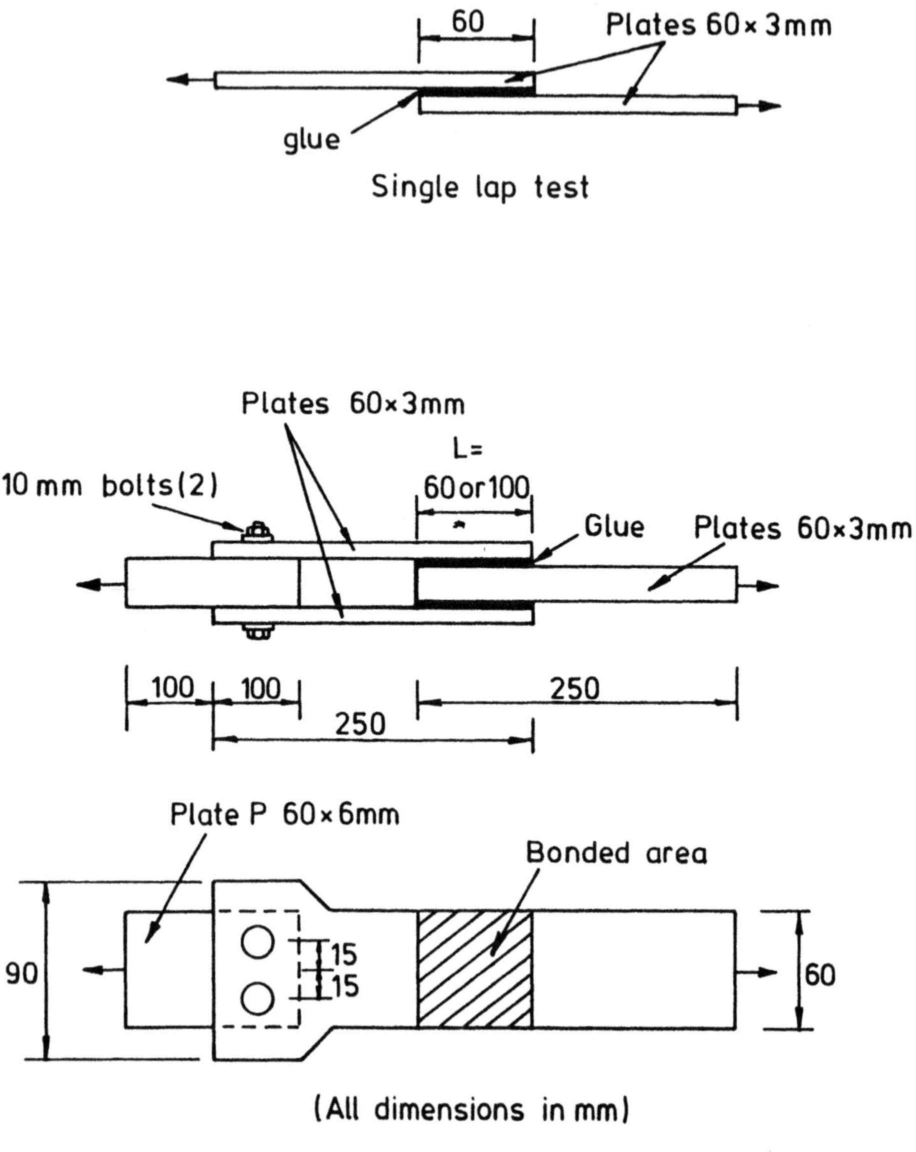

FIG. 2 SINGLE AND DOUBLE LAP TEST SPECIMENS

whole width of the plates.

The main test variable was the glue thickness. For all single lap tests, and for most of the double lap tests, the joints were 60mm long and 60mm wide. Additional double lap tests with 100mm long joints were carried out to examine the behaviour of the bond when the steel yielded.

2.3 Materials

The concrete mixes used in these tests were designed for 28 day strengths of 25 to 70 N/mm^2. Ordinary portland cement, washed natural river sand, and 10mm maximum size crushed gravel were used as concrete constituents. The steel plates used in the tests were all mild steel with an elastic modulus of 200 kN/mm^2, yield stress of 265 to 290 N/mm^2 and an ultimate tensile strength of 365 to 400 N/mm^2. Only one type of epoxy adhesive was used. It was found that both tensile strength and elastic modulus of the adhesive varied with the rate of loading and specimen thickness (i.e. geometry). For a range of loading rates of 0.254 to 25.4mm per min. and specimen thickness of 1.5mm to 5.5mm, the ultimate tensile strength varied from 6.4 to 13.3 N/mm^2, and the initial elastic modulus from about 430 to about 2000 N/mm^2.

2.4 Bonding procedure

The concrete prisms were initially cured in the laboratory for two weeks, prior to glueing the materials. The concrete surfaces were first abraded to remove laitance and expose the aggregates. They were then wire brushed and vacuum cleaned to remove all loose particles. The face of the steel plate to receive the adhesive was shotblasted using steel grit of 340μm mean particle size, and at a pressure of 0.75 N/mm^2. The resin base and hardener was mixed using a low speed drill, at 280 revolutions per minute, and fitted with a paddle. The glue was then applied to both the concrete surface and steel plate. The joint thickness was controlled by hardened epoxy spacers. The plates were then applied and held in position by dead weights. The alignment was ensured by putting bolts through the plate holes during curing.

2.5 Instrumentation

The resins were allowed to cure for four weeks. The instrumentation of the tests consisted of 2mm electrical resistance strain gauges fixed along the centre line of both plates. Typical gauge positions are shown in Figs.3 and 6. The strain gauges were used to record the deformations along the joints from which the longitudinal force and shear stress distribution could be computed. The gauge spacings were made smaller near the most stressed end of the joint to ensure greater accuracy of the computed forces and stresses. A data logger was used for continuous recording of the gauge outputs. Direct tensile tests on steel plates were carried out for calibration checks, and additional gauges fixed outside the joint, were used as references.

2.6 Test procedure and measurements

All the specimens were tested at about 6 weeks, at a stress rate of about 2 N/mm^2 per minute for the pull out tests and 6 N/mm^2 per minute for the lap tests.

3. Test results and discussion

3.1 Evaluation of data

From the recorded voltages, and using the appropriate calibration factors, the following information was evaluated.

(a) Local longitudinal force against applied load for each gauge location.

(b) Force distribution along the joint at different load levels.

(c) Shear stress distribution along the joint at different load levels. The shear stresses were calculated assuming a linear variation of the longitudinal force along the plate between consecutive strain gauge positions:

$$\text{Shear stress} = \frac{\Delta F}{b.\Delta L}$$

where ΔF is the variation of the longitudinal force between two consecutive gauge positions, ΔL apart, and b is the width of the joint.

3.2 Pull out tests

The results of the pull out tests are summarised in Table 1 and typucal data shown in Figs.3, 4 and 5. Fig.3 shows that the local longitudinal forces (for each gauge location) are proportional to the applied load, and that they change rapidly at regular intervals probably due to microcracking at the interfaces. At high loads, the local longitudinal force near the stressed end of the joint increases rapidly until it reaches the value of the applied load. This implies that no force is transferred from the plate to the concrete at that section, due probably to cracking in concrete or the joint or both. Fig.4 depicting the force distribution along the joint at different load levels confirms this.

The computed shear stress distribution along the joint at different load levels is represented by histogrames (Fig.5). At higher loads the shear stresses are reduced in the stressed end region, and near the ultimate load, the bond stress is reduced to zero near the ends of the joint (Fig.5).

Both the longitudinal force and shear stress distributions along the joint may be seen to follow exponential trends. At about 60% of the ultimate load, the joint begins to fail at the most stressed end, but the trend of exponential variation in the remaining part of the joint continues. Near the ultimate load, the most stressed end fails completely reducing the effective length of the joint by about 10 to 20mm, or by about 7 to 14% of its initial length.

The maximum shear stress recorded near the edge of the joint is about 4.1 N/mm², about 60% of the value recorded at about 20mm from the edge (Fig.5). The joint thus seems to offer a weaker bond strength at its edge near failure. Similar phenomenon has also been reported by Van Gemert (1980): his tests showed that the bond strength reduction occurred over about 42% of the initial joint length. This

TABLE 1. Pull out tear results.

EFFECT OF CONCRETE STRENGTH (glue thickness = 1.0mm)				EFFECT OF GLUE THICKNESS (concrete grade = $45 N/mm^2$)			
Concrete cube strength (N/mm^2)	No.of spec.	Mean shear stress (N/mm^2)	Max. shear stress (N/mm^2)	Glue line thick. (mm)	No.of spec.	Mean shear stress (N/mm^2)	Max. shear stress (N/mm^2)
25.1	6	2.17	6.0	0.5	6	2.42	6.9
44.9	6	2.52	6.8	1.0	6	2.52	6.8
60.2	6	2.77	7.4	1.6	6	2.38	6.7
71.3	6	3.33	8.3	3.0	6	2.83	6.9
Maximum standard deviation (N/mm^2)		0.21	0.38	Maximum standard deviation (N/mm^2)		0.17	0.42
Maximum coef.of variation (%)		8.43	5.59	Maximum coef.of variation (%)		7.14	6.27

Joint (epoxy) age = 28 days
Concrete age = 42 days

Mean stress = Ult.load/bond area
Max.stress: From recorded distribution

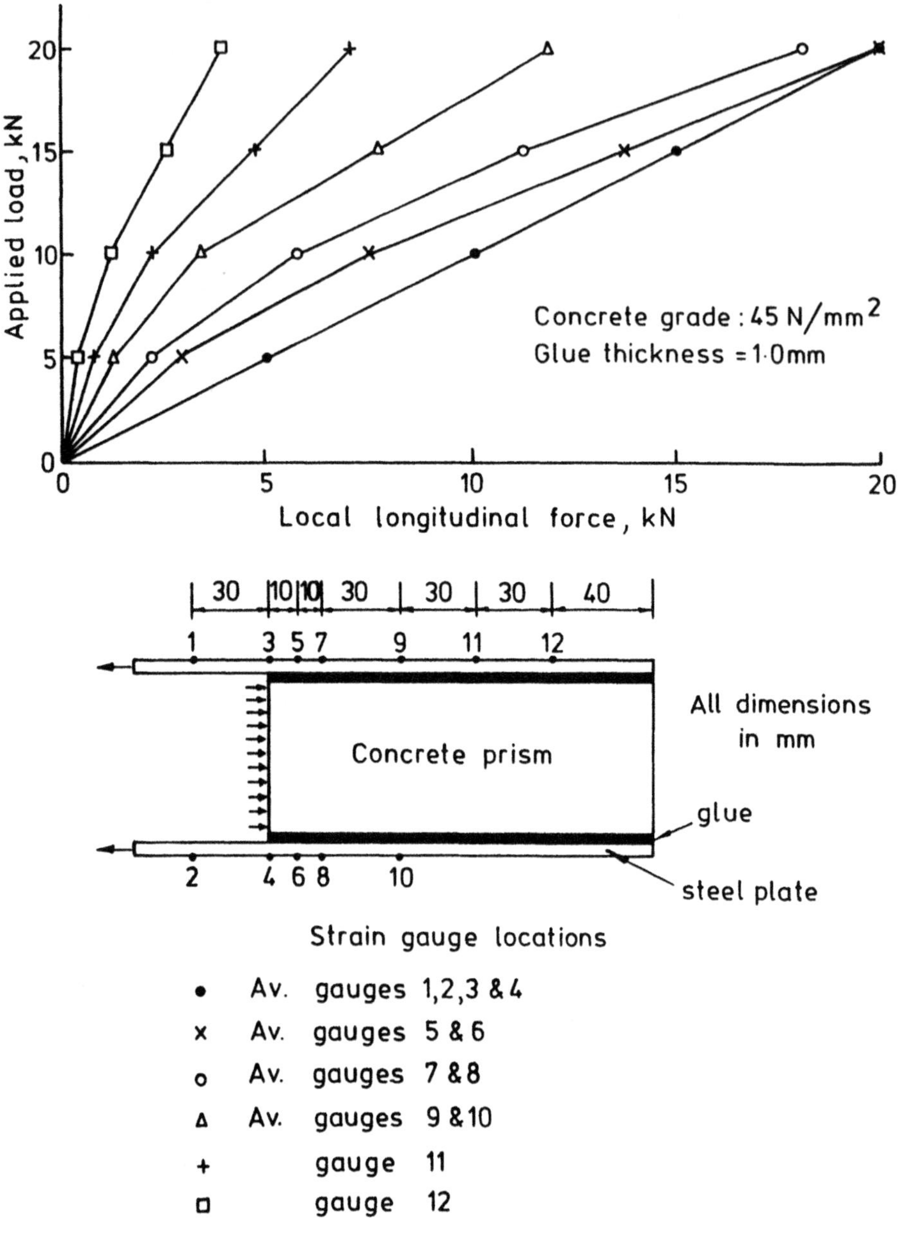

FIG. 3 LOCAL FORCE DISTRIBUTION IN PULL OUT TEST

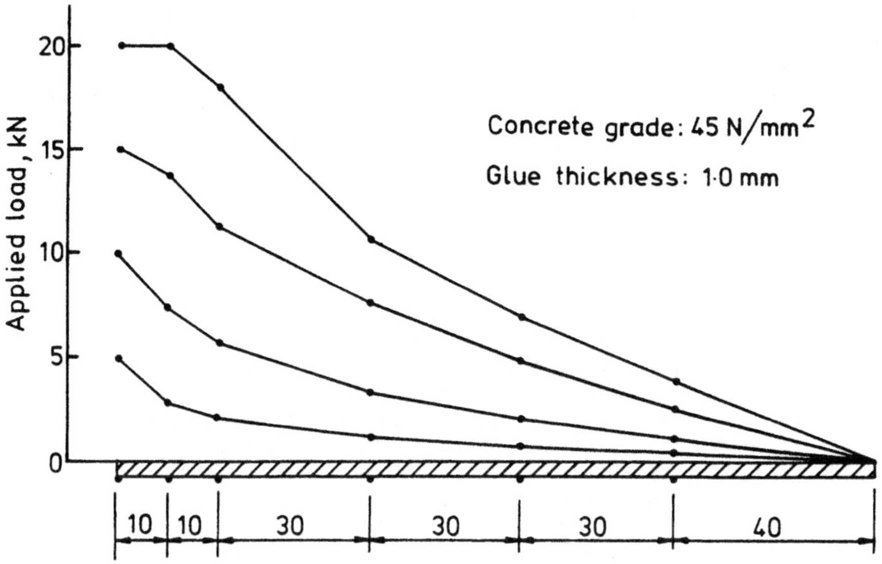

FIG.4 FORCE DISTRIBUTION ALONG PULL OUT JOINT

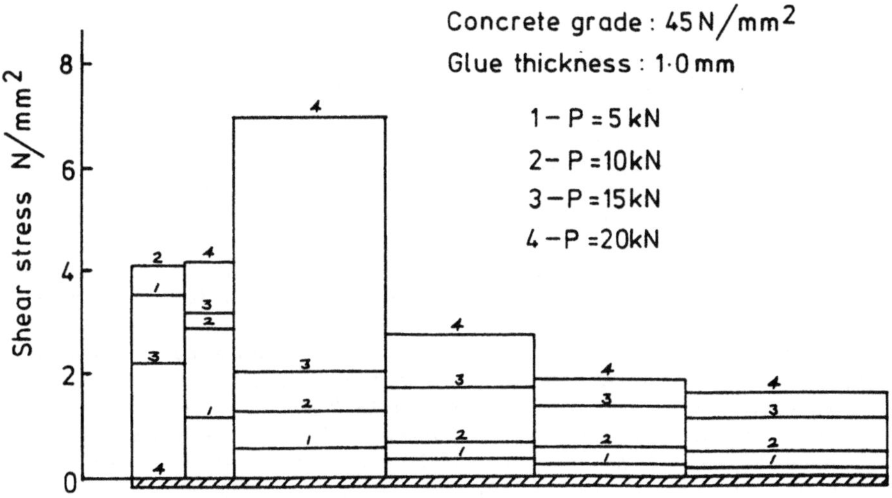

FIG. 5 SHEAR STRESS DISTRIBUTION ALONG PULL OUT JOINT

greater reduction in the effective length of the joint may be due to differences in the elastic properties of the materials used or the greater joint length (300mm as opposed to 150mm).

For the range of concrete strengths used in these tests namely, 25 to 70 N/mm^2, both the mean and the maximum shear stress varied almost linearly with the concrete strength - the mean stress from 2.17 to 3.33 N/mm^2 and the maximum shear stress from 6.0 to 8.3 N/mm^2 (Table 1). These ultimate shear stresses were higher than the tensile splitting strength of the concrete used, which varied from 3.1 to 4.0 N/mm^2, although, other investigators have related the bond strength to tensile strength.

Surprisingly, the variation in glue thickness had little effect on the shear stresses. For a range of glue thicknesses from 0.5 to 3.0mm, the mean shear stress varied from 2.4 to 2.8 N/mm^2 whilst the maximum shear stress varied between 6.7 and 6.9 N/mm^2. Theoretical models would suggest that the joint strength should increase with glue thickness: the apparent contrdiction of the test results may be accounted for by the shrinkage deformations during setting in thick resin layers or the presence of normal stresses in joints with thick adhesive layers as shown by Guess and Allred (1977).

The data in Table 1 also show that pull out tests can be carried out with a good degree of accuracy and consistency. The standard deviations are all low, the maximum value being less than 0.5 N/mm^2. The maximum coefficient of variation is also well below 10%.

All the pull out specimens failed through concrete proving that the epoxy resin can offer a bond stronger than concrete in shear. The ultimate shear strength of the concrete therefore seems to be the upper limit of the joint strength rather than the tensile resistance of concrete.

3.3 Double and single lap test results

The results of the lap shear tests are summarised in Table 2, and typical data are shown in Figs.6, 7 and 8. Unlike the pull out tests, the local longitudinal forces measured in the double lap test remained proportional to the applied load right up to failure. The distribution of the longitudinal forces and shear stresses along the joint, however, showed exponential trends as before. Theoretically, the stress distribution should be symmetrical about the central axis of the joint, with maximum values at both ends of the joint, and minimum at the centre. The unsymmetry of the recorded stresses is probably due to the gauge spacings used in the tests.

The average of the mean and maximum shear stresses remained relatively unaffected for the range of glue thicknesses of 0.5 to 3.0mm used in these tests - at 12.7 N/mm^2 and 18.2 N/mm^2 respectively. Mean and maximum shear stresses reported in literature from double lap tests do not all agree - as would be expected, since these values depend not only on the type of epoxy, but also on the type of double lap test used. Variations in shear stress capability can thus be largely accounted for by differences in epoxies and test specimens, Gilbert et al (1976), Cusens and Smith (1980).

The double lap joints of 100mm length behaved in a similar manner, but sustained higher loads than the 60mm long joints. However, they

TABLE 2. Single and double lap test results.

DOUBLE LAP TESTS					SINGLE LAP TESTS				
Joint length	Glue line thic	No.of spec.	Mean shear stress	Max. shear stress	Joint length	Glue line thic	No.of spec.	Mean shear stress	Max. shear stress
(mm)	(mm)		(N/mm^2)	(N/mm^2)	(mm)	(mm)		(N/mm^2)	(N/mm^2)
60	0.5	4	12.86	18.2	60	1.0	3	10.2	18.1
60	1.0	4	12.50	18.0					
60	1.6	4	12.71	18.2					
60	3.0	4	12.89	18.3					
100	1.0	3	9.68	13.8					
	Max. standard dev. (N/mm^2)		0.85	1.07		Max. standard dev. (N/mm^2)		0.98	1.31
	Max. coef.of var. (%)		6.69	5.88		Max. coef.of var. (%)		9.61	7.24

Joint age = 28 days (epoxy cured for 28 days)
Mean stress = Ult.load/bond area
Max.stress: from recorded distribution

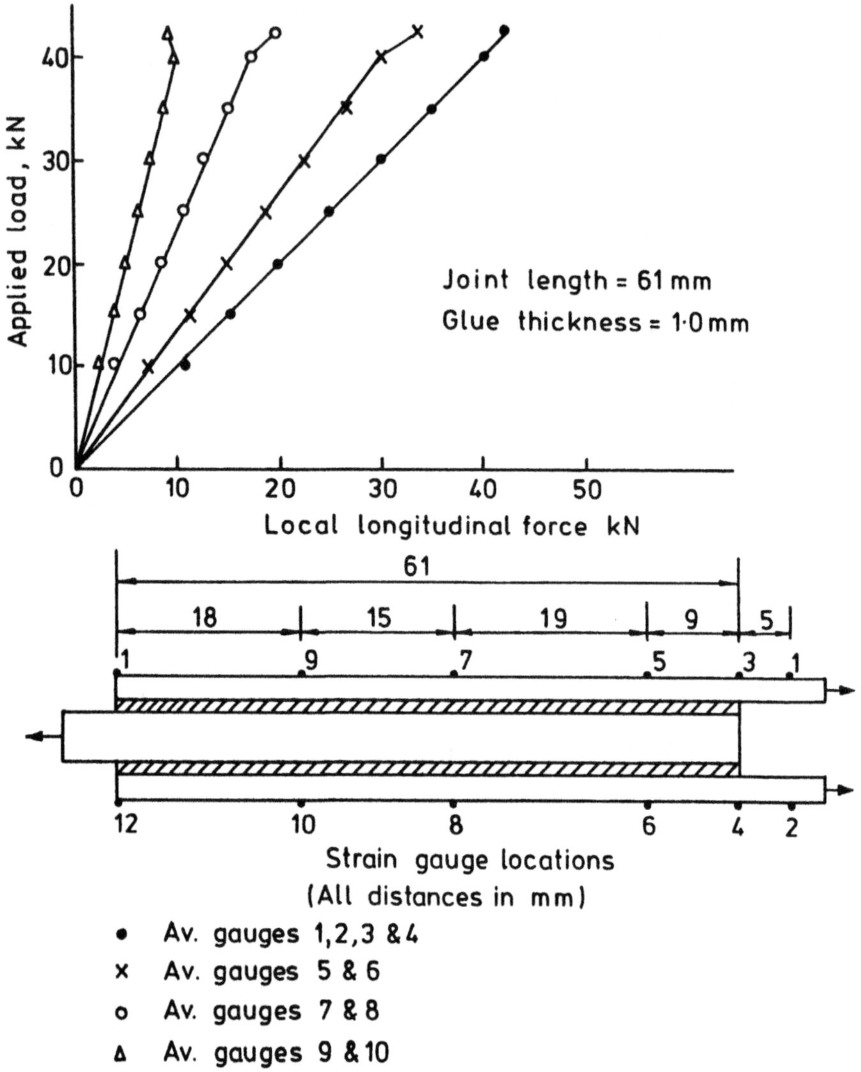

FIG.6 LOCAL FORCE DISTRIBUTION IN THE DOUBLE LAP SPECIMEN

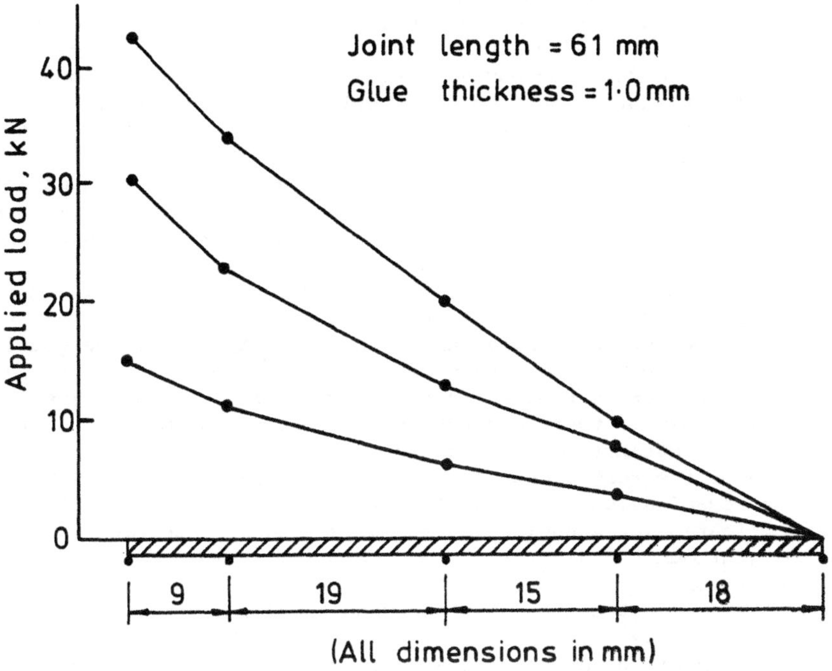

FIG.7 FORCE DISTRIBUTION ALONG DOUBLE LAP JOINT

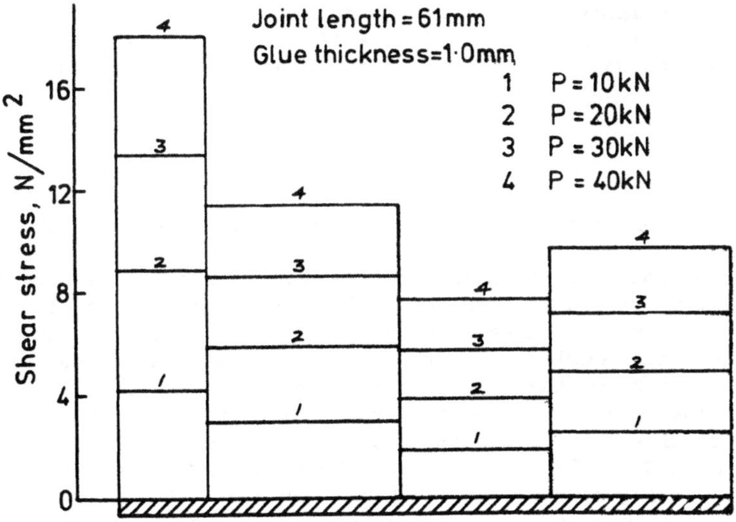

FIG. 8 SHEAR STRESS DISTRIBUTION IN DOUBLE LAP JOINT

failed at lower mean and maximum shear stresses of 9.68 and 13.8 N/mm^2 respectively. This early separation of the joint, compared to the shorter ones, was caused by the yielding of the steel plates at the higher failure loads. Obviously there appears to be limitations for epoxy joints in sustaining high deformations or higher stress concentrations resulting possibly from a stress redistribution following the yielding of the plates.

The single lap specimens, by comparison, recorded the same maximum shear stress but failed at an average shear stress of 10.2 N/mm^2, some 20% below the double lap test value. The manufacturer's single lap tests with the same adhesive gave an average result of 12.25 N/mm^2 - but again, no complete comparison can be made because of the influence of test parameters such as specimen size, loading rate etc. The lower shear resistance of the single lap specimens, compared to the double lap tests, is probably the result of the combination of shear stresses and peeling forces present in the unsymmetrically loaded specimen. A single lap test is therefore not as reliable a test method to study the shear bond strength of epoxies as double lap tests.

Failure of the test specimens occurred by separation of the plates. Subsequent examination of the failure zones suggested that failure was a combination of adhesion (at the steel - glue interface) and cohesion (through the glue) failures. Similar failure patterns have also been reported by others.

Like the pull out tests, the test results showed no apparent effect of the adhesive layer thickness, over a range of 0.5 to 3.0mm, on the joint strength of the double lap specimens (Table 2). As before, this apparent contradiction from theoretical models can only be explained by the presence of high shrinkage stresses during setting, and/or the presence of normal stresses in thick adhesive layers.

The results in Table 2 show that the double lap tests also gave a satisfactory coefficient of variation, well below 10%, although the standard deviation was higher at about 1.0 N/mm^2. Nevertheless, both the tests and the results are very satisfactory.

4. Conclusions

From the results presented in this paper the following conclusions can be drawn.

The distribution of longitudinal forces and shear stresses showed an exponential variation in both the pull out test and the double lap shear test. Near the ultimate load, the most stressed end in the pull out test fails completely, with the bond stress reduced to zero. The effective length of the joint was thus reduced by about 10%. For the range of concrete strengths of 25 to 70 N/mm^2 used in these tests, the maximum shear stress varied from 6.0 to 8.3 N/mm^2. Variation in glue thickness had little effect on the shear stresses, the maximum shear stress varying from 6.7 to 6.9 N/mm^2 for glue thicknesses of 0.5 to 3.0mm. Failure in the pull out test always occurred through the concrete-glue bond.

In the double lap tests, the maximum shear stresses remained relatively unaffected by glue thicknesses of 0.5 to 3.0mm at an average value of 18.2 N/mm^2. When the steel plate yielded, the maximum shear stress was reduced to an average of 13.8 N/mm^2. The single lap test also gave the same value of maximum shear stress as the double lap test, but the mean shear stress was some 20% lower. The presence of unsymmetrical loading conditions in the single lap test appears to be responsible for this. The lap shear test failures showed a combination of adhesion and cohesion failures.

The data presented here show that both the pull out test and the double lap shear test can be used to determine the shear bond strength of epoxy adhesives very satisfactorily. The tests can be carried out with sufficient accuracy and consistency.

Acknowledgements

The authors would like to record their thanks to the Algerian Ministry of Higher Education for their financial support to the third author.

References

Bresson, J. (1971) Nouvelles recherches et applications concernant l'utilisation des collages dans les structures. Beton plaque. Annales de l'ITBTP serie beton, beton arme, Suppl. 278, 23-54.

Burkhardt, P. et al (1975) Experiences sur les poutres mixtes en acier-beton liees à l'aide d'adhesifs epoxydes. Materials and Structures, 261-178.

Cusens, A.R. and Smith, D.W. (1980) A study in epoxy adhesive joints in shear, The Structural Engineer, 58A, 13-18.

Gilbert, Y. et al (1976) Mesure des deformations et des contraintes engendrees lors d'un essai de cisaillement en traction à la surface des plaques d'acier collees. Materials and Structures, 9, 255-260.

Guess, T.R. and Allred, R.E. (1977) Comparison of lap shear test specimens, J. of Testing and Evaluation, 5, 84-93.

Solomon, S.K. (1975) Steel-concrete-steel sandwich beams. Report WBRU/IR2, University of Dundee.

Van GEmert, D. (1980) Force transfer in epoxy bonded steel-concrete joints, Inter. J. of Adhesion and Adhesives, 1, 67-72

Author index
Table des auteurs

Exhibitors
Exposants

AEROPORTS DE PARIS

Orly Sud 103
F-94396 Orly Aérogare Cedex

DURAL INTERNATIONAL CORP.

95 Brook Avenue
Deer Park
USA-11729 New York

ELECTRICITE DE FRANCE

C.E.M.E.T.E.
Quartier Valcros
F-13090 Aix-en-Provence

FOSROC TECHNOLOGY LTD.

Bourne Road, Aston
GB-Birmingham B6 7RB

FREYSSINET INTERNATIONAL ET CIE

Résidence San Remo
Traverse le mee
F-13008 Marseille

GEATEC

5 Rue de la Touloubre
Les Logissons
F-13770 Venelles

LABORATOIRE CENTRAL DES PONTS ET CHAUSSEES

58 Boulevard Lefebvre
F-75732 Paris Cedex 15

RESINES SYNTHETIQUES

27 Avenue Kennedy
F-77793 Nemours

S.N. FORES

Syndicat national des formulateurs de résines synthétiques
3 Rue Alfred Roll
F-75849 Paris Cedex 17

TEROSON S.A.

Tour objectif
2 Rue Louis Armand
F-92607 Asnières Cedex

Acknowledgements
Remerciements

The symposium was financially sponsored by:
Le symposium e été subventioné par:

Electricité de France
C.E.M.E.T.E.
Quartier Valcros
F-13090 Aix-en-Provence

Bakelite Gesellschaft mbH
Postfach 12 02 44
D-4100 Duisburg

Bayer AG
Bayerwerk
D-5090 Leverkusen 1

Ernst Peiniger GmbH
Unternehmen Bautenschutz
Postfach 10 09 80
D-5090 Leverkusen 1

Weserland – v. Höveling GmbH
Postfach 21 05 69
D-3000 Hannover 21

With gratitude
Avec reconnaissance

The Scientific Committee
Le Comité Scientifique